P9-DDE-830

WELDERS GUIDE

by James E. Brumbaugh

THEODORE AUDEL & CO.

a division of

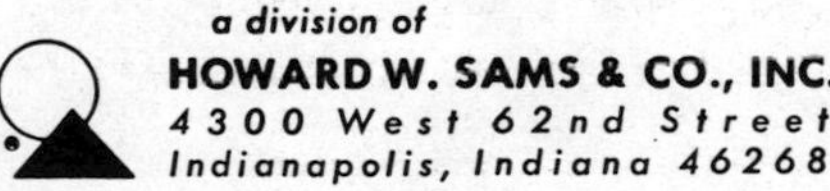

HOWARD W. SAMS & CO., INC.

4300 West 62nd Street

Indianapolis, Indiana 46268

SECOND EDITION

SEVENTH PRINTING — 1979

International Standard Book Number: 0-672-23202-2
Library of Congress Catalog Card Number: 72-97632

Foreword

Operations formerly difficult are now made easy as a result of extensive experiments and testing made in the research laboratories of the leading manufacturers of welding equipment and supplies. A broadening field is open to men who wish not only to master the basic or fundamental practice of welding but also to engage in more specialized technique in the joining and cutting of various metals and thermoplastic materials.

The purpose of this book is to provide an introduction to the fundamentals of welding which includes gas welding, arc welding, resistance and thermit welding, stud and submerged-arc welding, solders and soldering, metal cutting and flame hardening, and welding apparatus and supplies.

A special effort was made to remain consistent with the terminology and definitions standardized by the *American Welding Society*.

The author wishes to acknowledge the cooperation of the many welding organizations and manufacturers for their assistance in supplying valuable data in the preparation of this book.

JAMES E. BRUMBAUGH

Contents

CHAPTER 1

Introduction to Welding

Welding may be defined as the process of joining metals or plastics together through the coalescence of the surface at the point of contact. Coalescence is generally produced by heat, or pressure, or a combination of the two. Filler substances (metal or plastic) may or may not be used, depending on the particular welding process.

There are a number of different methods of classifying the various welding process. One of the more commonly accepted groupings is that offered in Table 1, by the *American Welding Society*. Another classification of welding processes is that proposed, in Table 2, by the *James F. Lincoln Arc Welding Foundation*. The latter is somewhat broader in basic categories than the *A.W.S.* classification. Note that soldering is not included in either of these classifications. It should be pointed out that new or revised *A.W.S.* definitions for the various welding processes appear in the *Welding Journal*, the official publication of the *American Welding Society*. All changes or additions are approved by the *A.W.S.* Committee on definitions and symbols.

Welding classificatory systems should not be regarded as being perfect or all-inclusive. They are simply convenient methods of organizing materials to indicate the origin and interrelatedness of individual items within the system. The best classificatory system, of course, is the one that permits the fewest exceptions. For

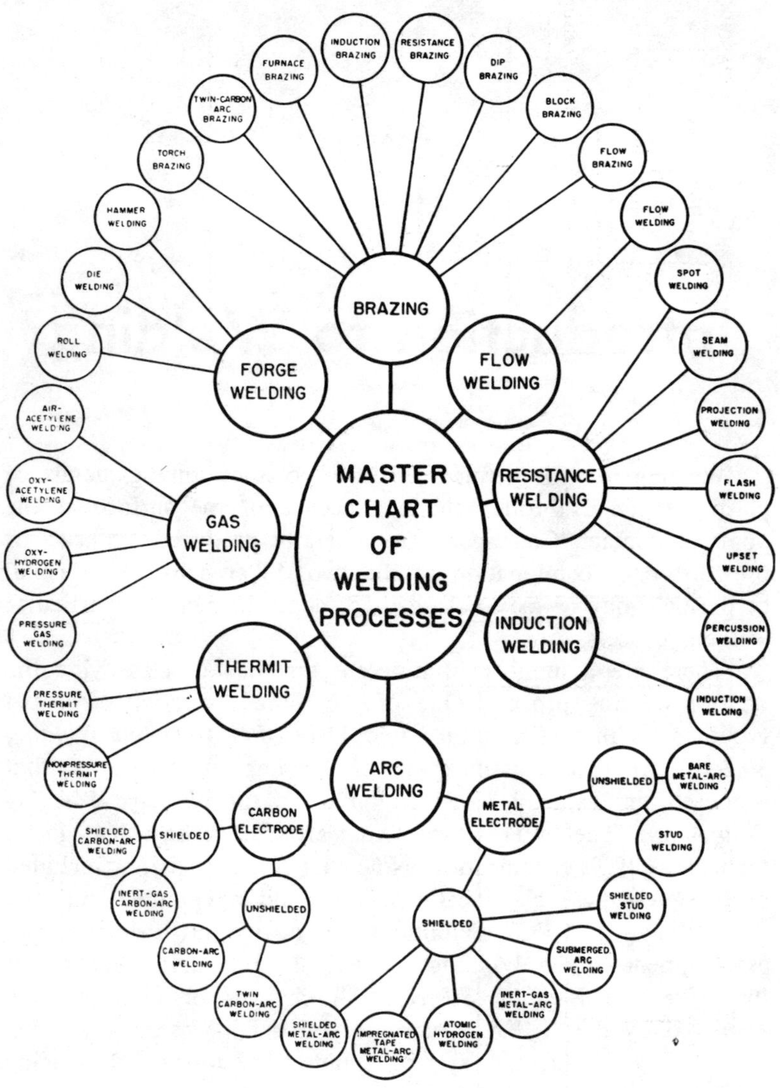

Courtesy American Welding Society

Table 1. Classification of the welding process.

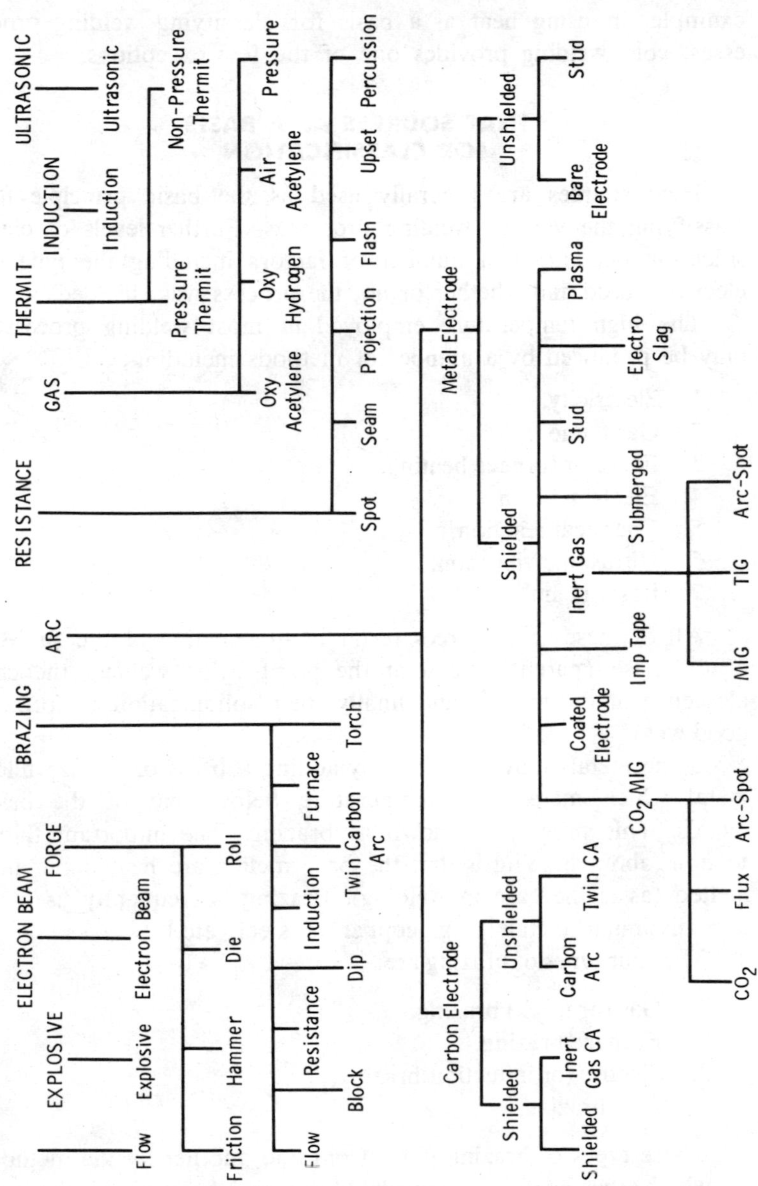

Courtesy James F. Lincoln Arc Welding Foundation

Table 2. Classification of the arc welding process.

example, in using heat as a basis for classifying welding processes, cold welding provides one of the few exceptions.

HEAT SOURCES AS A BASIS FOR CLASSIFICATION

Heat sources are generally used as the basic principle for classifying the various welding processes. Further levels of classification depend on a number of factors including the type of electrode used and whether or not the process was shielded.

The high temperature employed in most welding processes may be produced by a number of methods including:

1. Electricity,
2. Gas flame,
3. Forge or furnace heating,
4. Electron beam,
5. Chemical reaction,
6. Ultrasonic vibration,
7. Laser beam.

All of these heat sources result in the melting of the surface of the base (parent) metal at the point being welded, the coalescence of the metals, and finally their solidification to form a good weld.

Some metals may be joined by adding a brass or bronze filler metal which melts at a temperature below that of the base metals. This process is known as brazing. The important thing to note about brazing is that the base metals are heated but not melted (as is the case in welding). Brazing is frequently used to join dissimilar metals (e.g. copper to steel, etc.).

The four types of brazing are:

1. Gas (or torch) brazing,
2. Furnace brazing,
3. Electric (or induction)brazing,
4. Dip brazing.

These types of brazing differ from one another by the method in which the heat is obtained. However, the end results are basically the same.

Braze welding (or bronze welding, as it is sometimes referred to) is similar to brazing in that it uses a nonferrous filler metal at temperatures above 800°F but below the melting point of the base metal. It differs in that the filler metal is not distributed throughout the joint by capillary attraction (as it is in the case of brazing).

Soldering is similar to brazing in that a filler metal is used to join the metal surfaces without melting the base metals. It differs from brazing in that the filler metal melts at temperatures below 800°F. Brazing filler metals melt at temperatures above 800°F. Soldering is the weakest of the metal joining processes and is used primarily where strength is not important (e.g. joining wires in electrical circuits, etc.).

Metallizing is the process of spraying a thin layer of metallic or metallic oxide (ceramic) particles onto a roughened metal surface with a specially designed gun. It is sometimes variously referred to as metal spraying, ceramics spraying (if a metallic oxide is used), or metal surfacing. Plastics and metal alloys may also be sprayed onto a metal surface using this method. Metallizing is used to provide protection and attractiveness to a metal surface. It is not intended, however, to withstand the physical punishment for which hard surfacing is designed.

Hard surfacing is the process of welding a metal coating, edge, or point onto surfaces that are subject to wear. Hard surfacing is sometimes referred to as hard facing.

Electricity As A Source of Heat

Electricity will produce the high temperatures necessary for welding by means of:

1. An electric arc (arc welding),
2. The resistance of the metal to the flow of an electrical current (resistance welding),
3. An induced electrical current (induction welding).

Electric arc welding produces the high temperatures necessary for melting the metal by means of an electric arc that jumps a gap between a metal or carbon electrode and the metal surface being welded. This is illustrated in Fig. 1. Arc welding is the most widely used process for joining metals.

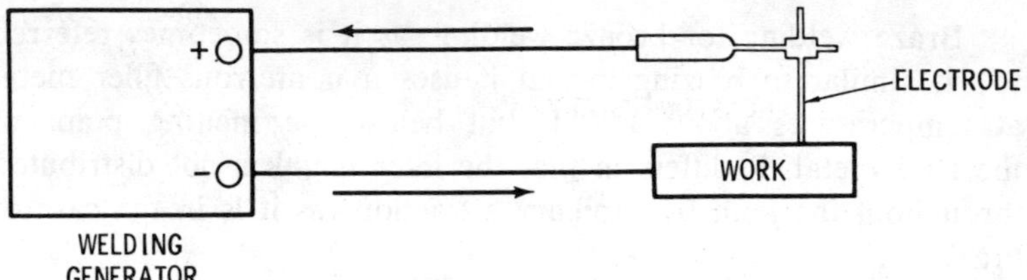

Fig. 1. The electric arc welding process.

Depending on the type of electrode used, electric arc welding can be divided into two main categories:

1. Carbon-electrode arc welding,
2. Metal-electrode arc welding.

Carbon-electrode arc welding is the official designation of the *American Welding Society* for a group of arc welding processes that share the common characteristic of using a carbon electrode. The carbon electrode arc welding classification can be divided into two main subgroups according to whether or not the welding process is shielded or unshielded. These subgroups appear as follows:

A. Unshielded processes;

   1. Carbon arc welding,
   2. Twin carbon arc welding.

B. Shielded processes;

   1. Shielded carbon arc welding,
   2. Inert gas carbon arc welding (CIG Welding).

Carbon arc welding (Fig. 2) is a puddling process in which an electric arc is formed between a single carbon (nonconsumable) electrode held in an electrode holder and the work. The heat produced by the arc creates a small pool of molten metal. Additional metal may be added to the pool by heating a filler rod. Because carbon arc welding is a puddling process, it is not suitable for either overhead or vertical welding. Carbon arc welding is very adaptable for use in automatic welding machines. It also provides an economical means of flame cutting.

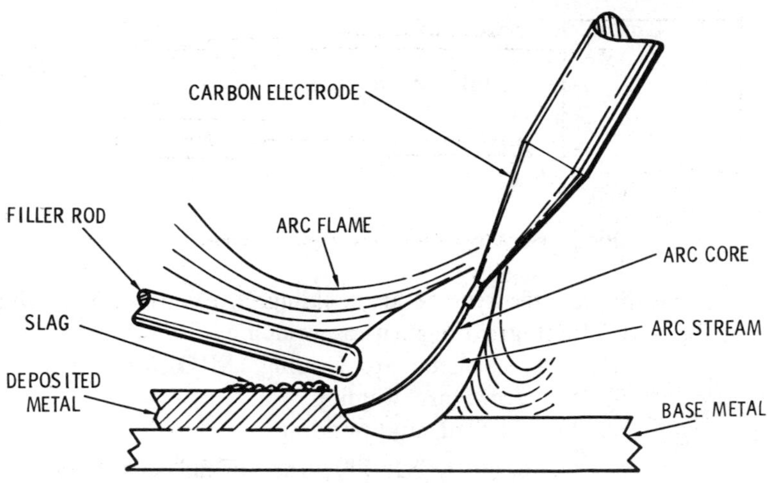

Fig. 2. The carbon arc welding process.

Twin carbon arc welding is an electrical arc welding process in which the arc is formed between twin carbon electrodes. When required, the filler metal is provided with a welding rod. No pressure is used.

Shielded carbon arc welding operates on the same principle as the other carbon arc welding processes; namely, the heat necessary for coalescence is produced by an electric arc formed between a carbon electrode and the work. An additional feature is the use of a shielding agent. The arc is shielded by a flux blanket, a solid material fed into the arc, or both. Pressure or a filler metal may or may not be used.

Inert gas carbon arc welding is a carbon arc welding process in which shielding is provided by an inert gas (usually helium or argon). This process is also referred to as gas carbon arc welding or CIG (carbon-inert-gas) welding.

The second of the two major divisions of electric arc welding is metal electrode arc welding. This is the official designation of the *American Welding Society* for a large group of welding processes all of which use the metal electrode. The metal-electrode arc welding classification can also be divided into two main subgroups depending on whether or not the welding process is shielded. These subgroups appear as follows:

A. Unshielded processes;

1. Bare metal arc welding,
2. Arc seam welding,
3. Arc spot welding,
4. Stud welding.

B. Shielded processes;

1. Shielded metal arc welding,
2. Inert gas tungsten arc welding (TIG Welding),
3. Inert gas metal arc welding (MIG Welding),
4. Submerged arc welding,
5. Gas shielded stud welding,
6. Impregnated tape metal arc welding,
7. Atomic hydrogen welding.

Bare-metal arc welding is an arc welding process in which the heat necessary for coalescence is produced by an electric arc formed between a bare or lightly coated metal electrode and the work. This is an unshielded welding process. Filler metal is obtained from the electrode.

Bare-metal arc welding was first used in the late nineteenth century. The electrode at this time consisted of a bare metal wire which produced an unstable arc and frequently unsatisfactory welds. Arc instability was solved through the development of coated electrodes. The use of these coated electrodes resulted in higher quality welds. Most manual welding today is done with coated electrodes (see Chapter 6, METAL ARC WELDING). Bare-metal arc welding is very limited in use and the electrode is usually light coated.

Seam welding and spot welding may be either arc welding or resistance welding processes depending on how the heat is obtained. Their widest application is found in resistance welding (see Chapter 9). If their welding heat is obtained from an electric arc, they are referred to as arc seam welding or arc spot welding.

Stud welding (Fig. 3) is an arc welding process in which the electric arc is drawn between a metal stud (or a similar piece of

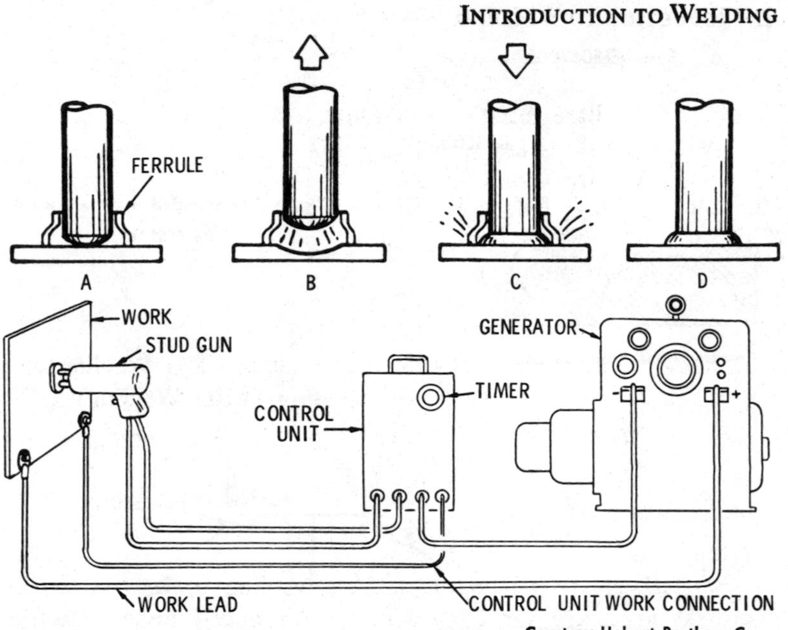

Courtesy Hobart Brothers Co.

Fig. 3. The stud welding process.

metal) and another portion of the work. Pressure is applied after the metal to be joined has reached a suitable temperature.

There are a number of metal electrode arc welding processes in which the arc is shielded from contamination by the surrounding atmosphere. These contaminants will adversely affect the quality of the weld. This is particularly true of oxygen which results in oxidation when it comes into contact with metal. Oxide film must be removed from the surface before a strong weld can be produced. The shielding agent may be an inert gas, a granular flux blanket, or a vapor. In manual arc welding, the shielding action is produced by large quantities of gases given off by the covered electrode during the welding process. The gases envelop the arc and shield it from contaminants. This process is referred to as shielded metal arc welding. The process represents one of the most widely used forms of welding (Fig. 4).

The two principal inert-gas-shielded arc welding processes are:

1. TIG (tungsten-inert-gas) welding,
2. MIG (metal-inert-gas) welding.

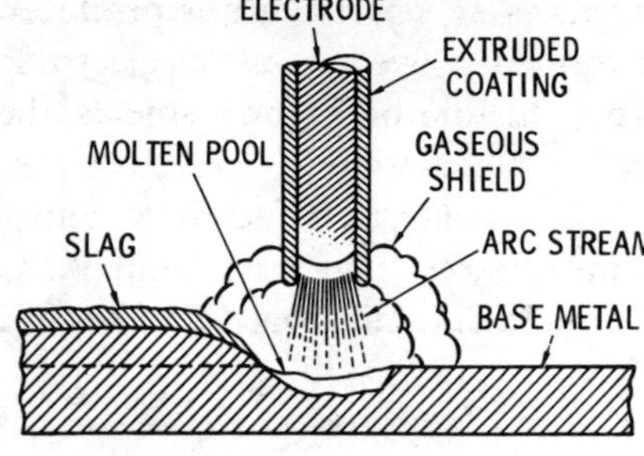

Courtesy Hobart Brothers Co.

Fig. 4. The shielded metal-arc welding process.

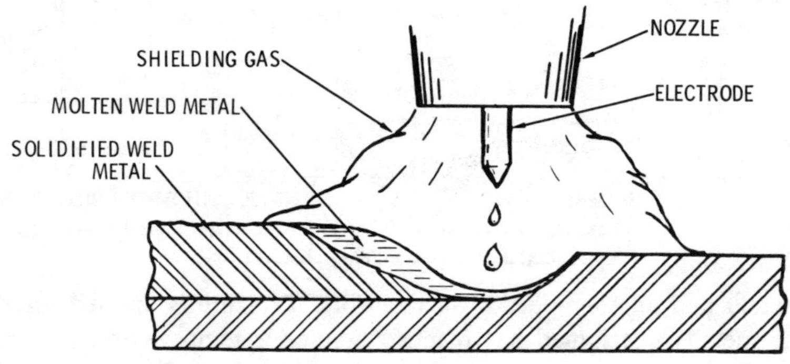

GAS METAL ARC WELDING

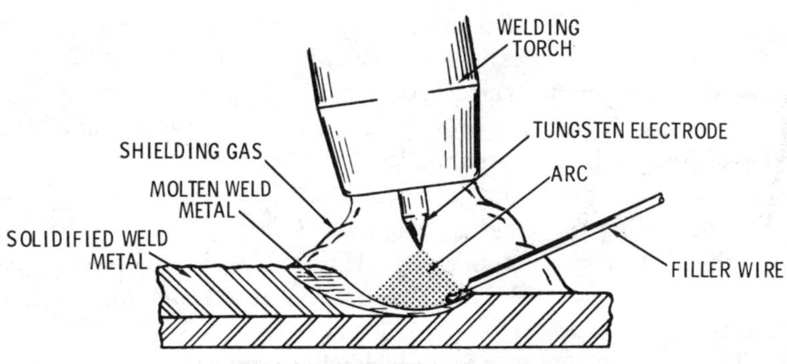

GAS TUNGSTEN -ARC WELDING

Courtesy Hobart Brothers Co.

Fig. 5. MIG and TIG welding process.

As in the other arc welding processes, coalescence is produced by heating with an electric arc formed between a metal electrode and the work. An inert gas (e.g. helium or argon) shields the arc from contamination. Both of these welding processes are used primarily in the joining of such nonferrous metals as aluminum, magnesium, copper, stainless steel, and cast iron. MIG welding is also referred to as inert-gas metal arc welding or gas-metal arc welding. See Fig. 5.

A granular flux blanket is used to shield the arc in submerged arc welding (Fig. 6). A consumable electrode is fed into the joint during the welding process. This electrode also provides the filler metal.

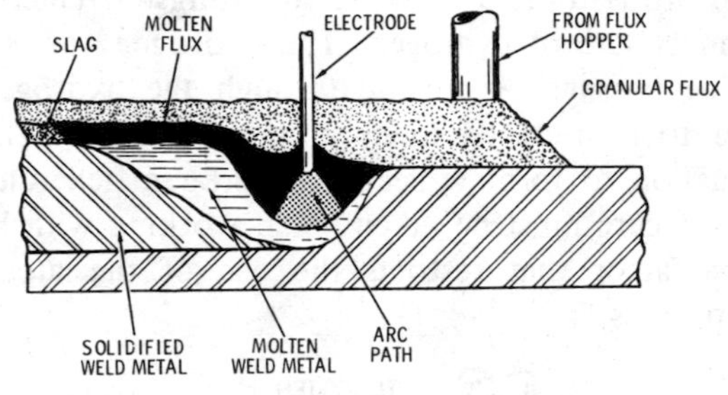

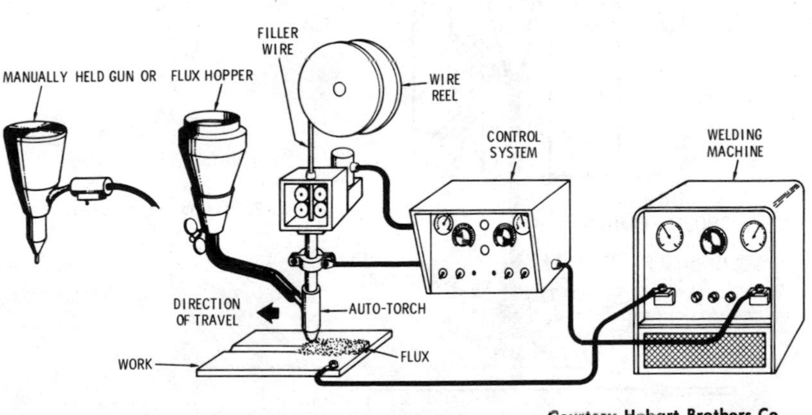

Courtesy Hobart Brothers Co.

Fig. 6. Submerged arc welding.

Gas-shielded stud welding is an arc process in which the electric arc is produced between a metal stud (or a similar piece of metal) and another portion of the work within a shield of inert gas.

Impregnated-tape metal arc welding is another shielded metal arc welding process; that is, the heat is produced by means of an electric arc formed between the metal electrode and the work. The gradual disintegration of an impregnated tape wrapped around the electrode produces the shielding agent. Disintegration occurs as the electrode is fed into the arc. The filler metal is obtained from the electrode. No pressure is used in this welding process.

Atomic hydrogen welding is a shielded welding process in which an arc is formed between two tungsten (metal) electrodes in an atmosphere of hydrogen. The hydrogen gas is the shielding agent. The passage of the arc through the hydrogen gas causes the latter to be reduced to its atomic structure. The subsequent recombination to form molecular hydrogen generates an intense heat. Great care must be taken when working with hydrogen gas. This is a factor that restricts the use of the atomic hydrogen welding process.

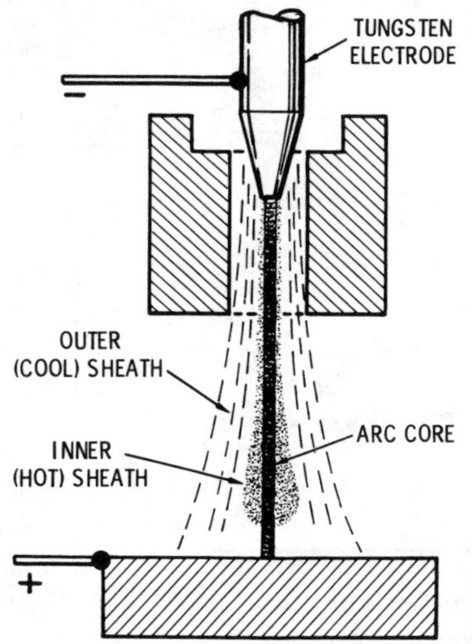

Fig. 7. Plasma arc welding.

Plasma arc welding (Fig. 7) is of relatively late development. This is an arc welding process that draws an arc between a tungsten electrode and a water-cooled torch nozzle. A gas feeds into the torch and through the arc, and, having changed to ionized particles, emerges from the nozzle as a narrow highly concentrated plasma stream. The temperatures produced by the plasma stream are adaptable to welding, cutting, or spraying.

Resistance welding generally combines heat and pressure. For this reason, it is sometimes referred to as a "heat and squeeze" welding process. The heat is produced by the resistance of the metal to the flow of an electrical current. When the parts to be welded are raised to fusion temperature by the passage of a heavy electrical current through the junction, pressure is applied mechanically to accomplish the welding. Examples of resistance welding are spot, seam, flash, upset, butt, percussion, and projection welding.

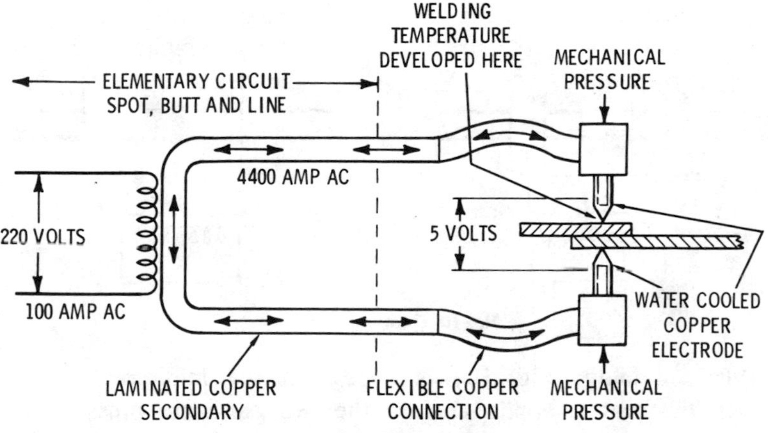

Fig. 8. The spot welding process.

In spot welding (Fig. 8), the weld is made by overlapping the parts and gripping the overlapped sections between two electrode points. An electrical current is passed through these points and pressure applied to make the weld in a single spot (hence the name).

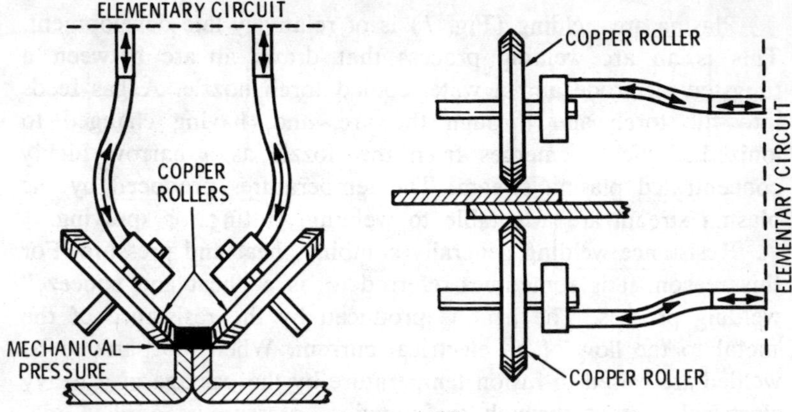

Fig. 9. Seam welding.

Seam welding (Fig. 9) is similar to spot welding, except that a circular electrode is used to produce the effect of a continuous seam weld.

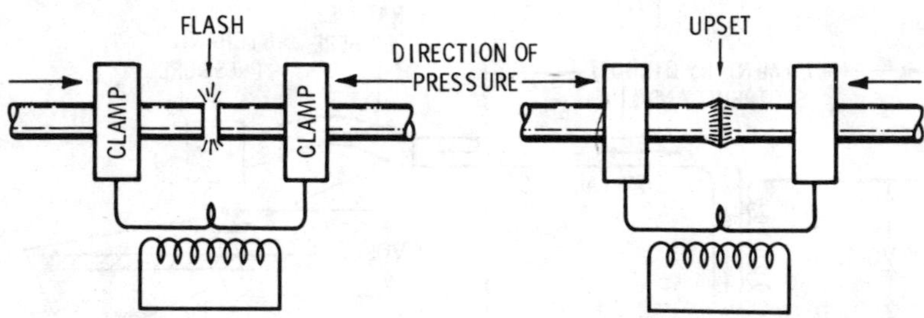

Fig. 10. Flash welding.

Flash welding (Fig. 10) is a resistance welding process in which the heat is applied before the two parts are pressed together. When the pressure is applied, molten metal is squeezed out of the weld.

In upset welding (Fig. 11), the heat is applied after the two metal pieces are clamped in the electrodes (thus distinguishing it from flash welding in which the heat application precedes the pressure for the weld).

Butt welding (Fig. 12) is not a welding term officially recognized by the *American Welding Society*. Actually it is a form of

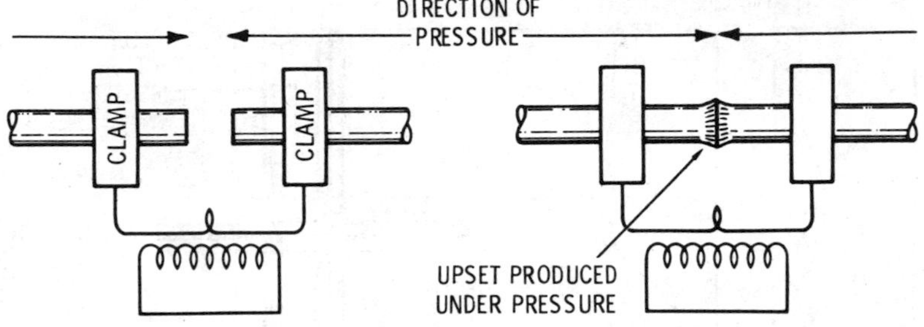

Fig. 11. Upset welding.

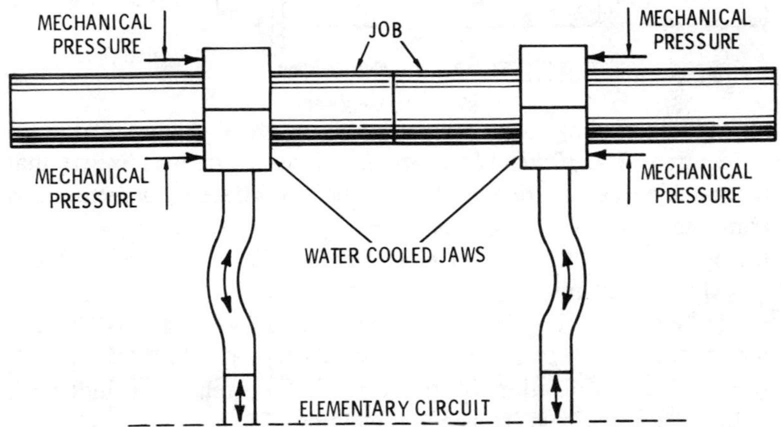

Fig. 12. Butt welding.

either upset welding or flash welding depending on the process used. In butt welding, the parts are placed end to end heated electrically and pressed together to form the weld.

Percussion welding (or percussive welding) is a resistance welding process in which the electrical energy is stored up and released with sudden explosive impact. Fusion occurs simultaneously over the entire contact area.

In projection welding (Fig. 13), the position of the weld is established by the predetermined projection on one of the two metal surfaces to be joined. A flow of electric current causes the projection to melt and fuse the two metal surfaces together.

Resistance welding has a limited application in general manufacturing because special equipment is usually required for each

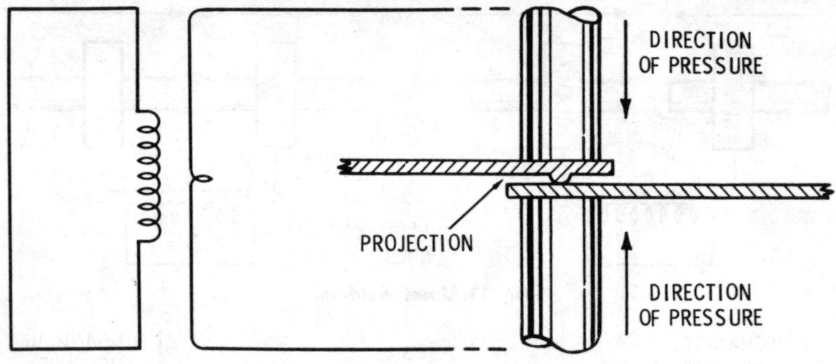

Fig. 13. Projection welding.

individual welding job. It is therefore chiefly used for mass production. The principal advantage of resistance welding is that the machine can be operated by unskilled workers. For the most part, the machine settings determine the quality of the weld and these are set by supervisors or engineers. Welds can be produced quickly and cleanly.

Induction welding is an electrical welding process that utilizes the resistance of metal to the flow of an induced electrical current. It is in the application of the electrical current that induction welding differs from resistance welding. In resistance welding, the work is placed between two electrodes. The electric current passes through the metal on its path from one electrode to the other. Induction welding, on the other hand, involves the placing of the work in a radio frequency field. The radio frequency field is produced within radiating coils designed and constructed to approximate the shape of the work. Induction welding is essentionally a production welding process.

Gas Flame As A Source of Heat

In some forms of welding, the high temperatures required for melting the metal is produced by igniting a mixture of two gases. The two most commonly used gases are oxygen and acetylene (oxyacetylene welding). A mixture of oxygen and hydogen (oxyhydrogen welding) is also used sometimes as is oxygen in combination with such gases as butane, propane, or methane (Fig. 14).

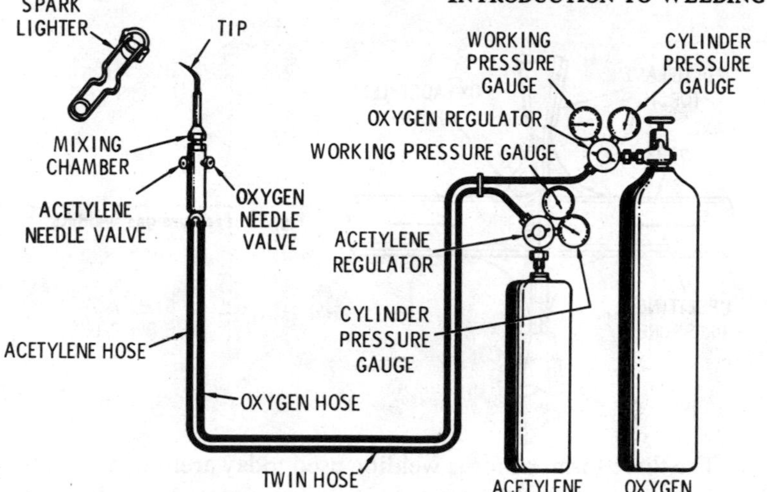

Fig. 14. Oxyacetylene welding process.

Gas welding is a puddling process; that is, the molten metal forming the weld is in a small pool over which the flame is constantly played. Making a weld by this process consists largely in causing this pool to move by melting metal away ahead of the pool and letting the metal cool behind it.

Gas welding has been used more extensively than forge, thermit, or resistance welding, due to the flexibility of the equipment and the small investment it requires. Notwithstanding the comparatively low initial cost of equipment, the high operating cost usually makes it less economical than arc welding for production work.

Pressure gas welding (Fig. 15) is a special gas welding process in which the pieces to be joined are abutted, brought to the plastic stage, and forged together under pressure. The pressure creates an upset action at the point of contact. After the weld is formed, the excess metal caused by the upset action is filed away.

Forge or Furnace Heating

Forge welding operates on the same principle as the method employed by the blacksmith. The metal is heated in a furnace or forge and pressure is applied through impact (hammer blows) or squeezing.

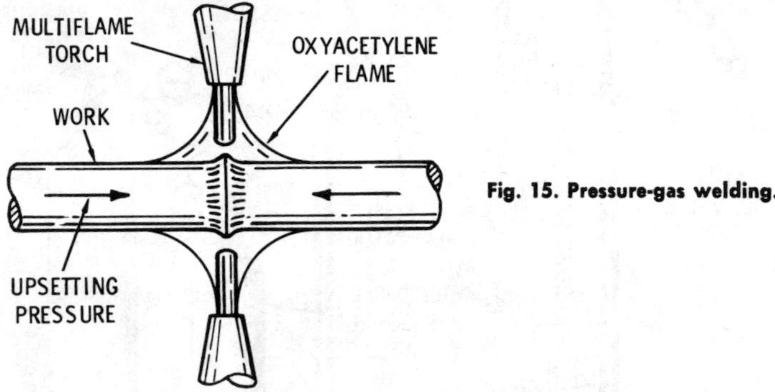

Fig. 15. Pressure-gas welding.

The three forms of forge welding used today are:

1. Hammer welding,
2. Die welding,
3. Roll welding.

As the name suggests, *hammer welding* applies pressure through impact (hammer blows) after the metal has been heated to a fusion temperature in a forge or furnace. Some forms of hardware manufacture still use the hammer welding process.

Die welding is a process wherein the metal is heated to fusion temperatures and pressure applied with a die to produce coalescence. This form of welding is used in the manufacture of pipe or the butt welding of round parts.

Roll welding is so named because pressure is applied by means of rolls. In roll welding, the metal is also heated in a furnace.

The various forge welding processes are done primarily today with machines and are essentially a die forming process.

Electron Beam As A Source of Heat

The electron-beam welding process (Fig. 16) involves the bombardment of a metal surface with a concentrated high-speed stream of electrons. The concentration of the electrons is produced by means of an electron gun. The entire process takes place in a vacuum which contributes to making the weld free of contamination.

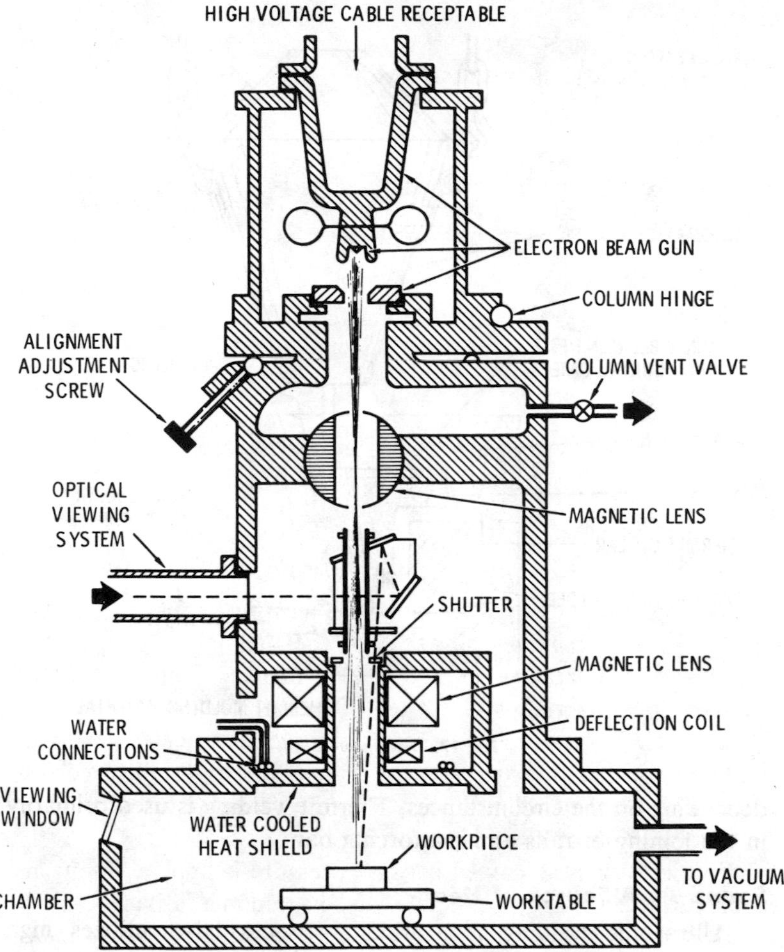

Fig. 16. The electron beam welding process.

Chemical Reaction As A Source of Heat

Thermit welding (Fig. 17) refers to a group of processes in which the high temperatures necessary for fusion are produced by means of a chemical reaction between a metal oxide and aluminum. The chemical reaction produces a quantity of superheated molten steel that not only heats the base metals but also acts as the filler metal in the mold. Pressure may or may not be used

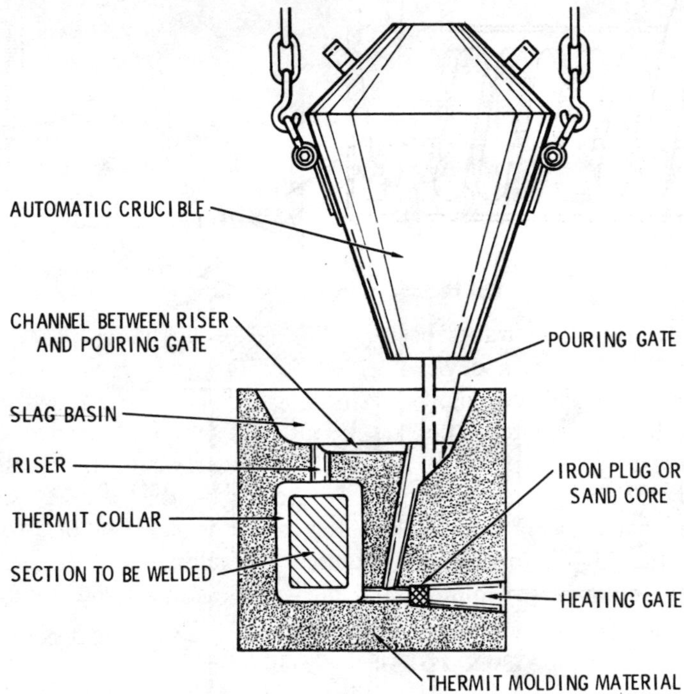

Fig. 17. Thermit welding.

depending on the circumstances. Thermit welding is used primarily in the joining of rails and reinforcing bars.

Sound As A Source of Heat

Ultrasonic welding is a welding process that utilizes high intensity vibrations to produce the temperatures necessary for fusion. This process is used primarily in welding aluminum, magnesium, and their alloys, but is also adaptable to welding plastics.

Light As A Heat Source

Recent technological developments have made it possible for light to be used as a source of heat in welding. This welding process is known as *laser beam welding* (Fig. 18). Laser is the acronym for *light amplification by stimulated emission of radiation.*

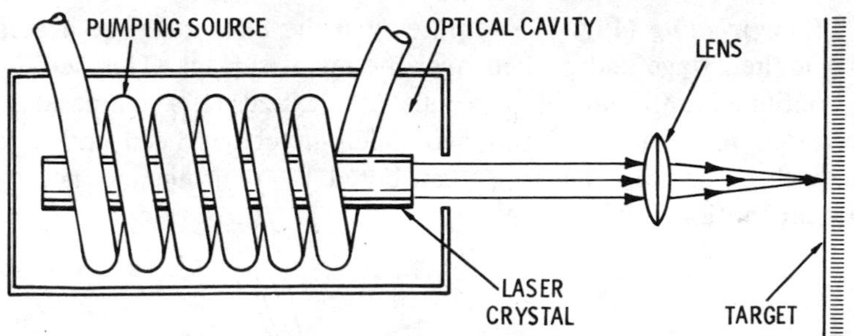

Fig. 18. Laser beam welding.

Ordinarily, light waves become diffused once they leave their source. The laser is a device for concentrating these light waves into narrowly defined highly intense beam. The energy waves emitted by a stimulated atom are organized by the laser device in such a way that they travel in the same direction at the same frequency. The narrowly defined frequency band produces the intense concentrated beam. The laser beam is capable of producing the high temperatures required in welding and is particularly applicable to joining hard high-temperature metal alloys.

OTHER JOINING PROCESSES

Heat was used as a basis for classifying the welding processes just described, because it is essential to the successful coalescence of the metals in almost all welding processes. Neither the use of pressure nor filler metals is as essential to as many forms of welding as is heating.

The classification system just described was not intended to be a complete all-inclusive one. It was simply offered as an example of how the various welding processes could be classified according to a characteristic common to each (i.e. the origin of heat).

Welding processes not included in the previous discussion but of which some mention should be made are:

1. Flow welding,
2. Friction welding,
3. Cold welding,
4. Welding plastics.

Flow welding (Fig. 19) requires that the filler metal be heated to a molten stage and poured over the metal surface. This activity is continued until the base metals are heated to a temperature necessary for fusion and sufficient filler metal has been added to the weld. The filler metal is distributed by pouring and not by capillary action.

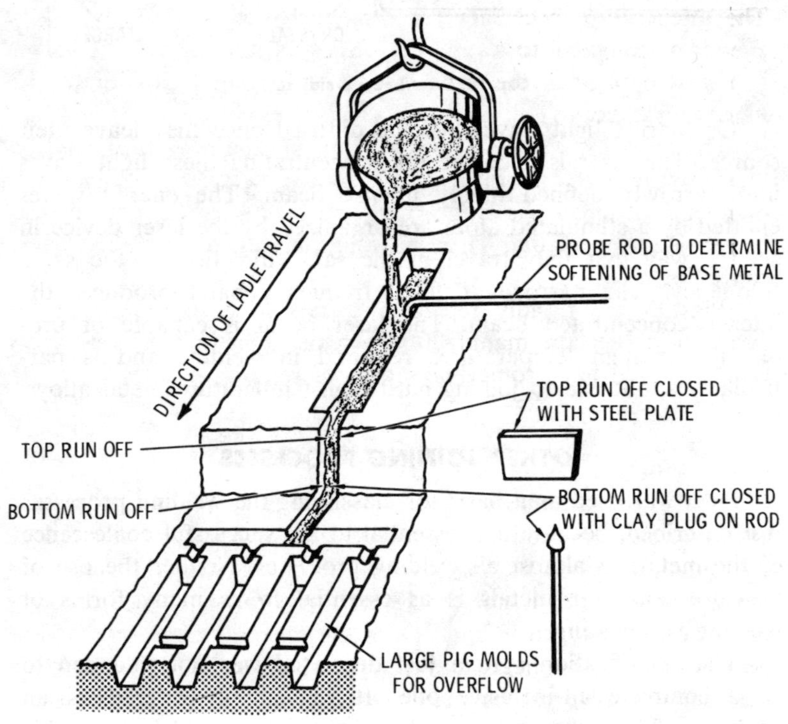

Fig. 19. The flow welding process.

Friction welding is a forge-type welding process in which the temperatures necessary for coalescence are created by rotating one of the surfaces to be joined against the other. Pressure is applied during the friction welding process until coalescence is achieved. Friction welding is particularly suited to automated systems designed for the joining of rods and tubing.

Cold welding (or cold pressure welding) is a welding process that requires no heat. Pressure is used to produce coalescence through upset. It is sometimes used to permanently join aluminum and a number of other metals.

Welding plastic has much in common with welding of metal. Filler rods are used to add additional material (in this case a plastic) and pressure is applied throughout the welding process. The temperature necessary for coalescence is obtained by: (1) a chemical designed to cause the edges of the plastic to soften; (2) heat applied to the surface; or (3) friction (see Chapter 12, FRICTION WELDING).

The two basic types of plastics are: (1) thermoplastic, and (2) thermosetting. The thermoplastic type can be welded in the same manner as a metal. As is the case with metals, thermoplastics can be reheated repeatedly. Thermosetting plastics, on the other hand, cannot be reheated.

Plastics are rapidly replacing metals in portions (or in the entirety) of certain manufactured items. For this reason, every welder should have some knowledge of welding plastics.

CUTTING METALS

There are a number of different methods of cutting metals. One common method is to use oxygen and acetylene gases in combination with a cutting torch. The cutting torch differs from the welding torch in that the former provides a stream of high-pressure pure oxygen in addition to the oxyacetylene mixture. The oxyacetylene flame heats the metal surface to the ignition point; the metal is then cut by the oxygen stream. Hydrogen is sometimes used instead of acetylene (Fig. 20).

The electric arc process (using either the metal or carbon electrode) can also be used for cutting metals. This is a melting process which produces a cut somewhat inferior to the precision and smoothness achieved with oxyacetylene cutting. As a result, it is used primarily in salvage operations.

Plasma arc cutting is faster than oxyacetylene cutting but does not produce a cut of similar quality and economy. A plasma forming gas, such as hydrogen or nitrogen, is heated electrically to such

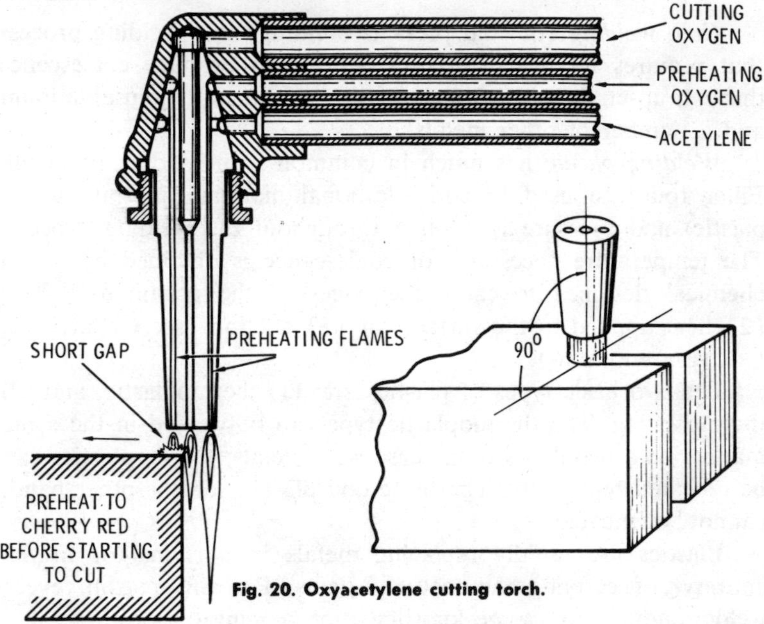

Fig. 20. Oxyacetylene cutting torch.

a high temperature that its molecules are transformed into ionized atoms. These atoms possess extremely high energy in their plasma state. When they are changed back into a gas, the energy is released in the form of heat. The heat is of such intensity that it can cut any known metal.

Compressed air in combination with an electric arc is still another method for cutting metals. The arc is used to melt the metal and the compressed air stream blows the molten metal away from the arc. Both alternating- or direct-current may be used, although the latter is preferred. This is referred to as the *Arcair* cutting method.

There are a number of other cutting methods, particularly the various semiautomatic and automatic systems, that were not included in this brief discussion of metal-cutting processes. These will be described in full detail in the appropriate chapter.

This chapter was concerned primarily with brief descriptions of the various joining (and cutting) processes and their positions in a given classificatory system. Greater in-depth coverage for each of these processes is found in later chapters.

CHAPTER 2

Fundamentals of Welding

The purpose of this chapter is to introduce certain fundamental concepts common to all welding processes. Detailed descriptions of the application of these fundamental concepts will be found in any of the chapters dealing with the specific welding processes. Most of the terminology used in this chapter is based on definitions established by the *American Welding Society.*

SELECTING THE APPROPRIATE WELDING PROCESS

Selecting the appropriate welding process is an important decision to be made. It is particularly necessary in determining cost, ease of accomplishment, and the quality of the weld. Unfortunately, there is no one welding process suitable for all welding situations. For this reason, it is necessary to weigh the advantages and disadvantages of each welding process.

In order to best determine the type of welding process most appropriate for the welding job at hand, the following information is necessary for the welder:

1. A correct and precise identification of the metal,
2. An understanding of those metallurgical aspects that apply to the metals being welded,
3. The total cost of the welding operation.

METALLURGY

Metallurgy is the technology and science of extracting metals from ores, refining them, and preparing them for use. Included in metallurgy is the study of the composition, structure, and properties of metals, and how they behave under a variety of conditions.

MECHANICAL AND PHYSICAL PROPERTIES OF METALS

Metal properties may be either physical or mechanical. A *physical property* is an inherent characteristic of the metal (e.g. magnetism, fusibility, etc.) and is not dependent on external pressure or force to determine its limits. There is no physical change in its structure. A *mechanical property,* on the other hand, is measured by the extent to which it reacts to external pressure (e.g. stress or strain) or to applied forces.

For example, the compressive strength of a metal is determined by the amount of compression or pressure that must be applied to a metal before it will crack or crumble. For purposes of weld metallurgy, the mechanical properties of a weld are generally more important that the physical ones.

The metal properties described in this chapter were chosen on a selective basis and have not been classified according to their physical or mechanical characteristics.

METAL STRENGTH

The strength of a metal is probably one of the most important properties for a welder to know. By knowing the strength of the base (parent) metal, it is possible to determine the strength needed in the weld metal.

Strength is the ability of a metal to withstand a number of different forces or pressures without breaking down. All metals will eventually break down if enough force is applied but most metals have different levels at which this will occur.

The three types of strength that are most important for the welder are:

1. Tensile strength,
2. Compressive strength,
3. Fatigue strength.

Tensile strength indicates the degree to which a metal will resist being pulled or torn apart. It is an indication of the greatest longitudinal stress, and is measured in pounds per square inch, (psi). If a metal has a tensile strength of 70,000 psi, this is the maximum load (applied force) it will bear without failure.

Compressive strength indicates the degree to which a metal will resist being crushed. The load or applied force is gradually applied until the maximum compressive strength of a metal is reached. Beyond this point, the metal will fracture. Both tensile and compressive strength are tested by a gradual application of force.

Fatigue strength indicates the ability of a metal to withstand repeated or alternating stress without breakdown. *Metal fatigue* (or *fatigue failure*) occurs at a comparatively low level of stress. Maximum fatigue strength is commonly reached much more quickly than maximum tensile or compressive strength.

EFFECT OF HEAT ON METAL

The structure or properties of the base (parent) metal adjacent to the joint being welded will be altered during the welding operation. This is referred to as the *heat-affected zone*. The metal in the heat-affected zone does not reach the melting point, but the temperatures may have been high enough to have changed certain characteristics of the metal. These changes are generally harmful and should be taken into consideration when dealing with the ultimate strength of the welded joint. Heating the metal around a welded joint in this manner (without bringing the metal to the melting point) is referred to as *heat treating* or *heat treatment*.

Heating surfaces to high temperatures just short of the melting point frequently causes oxides to form which can be included in (and thereby weaken) the weld. The heat may also cause gases to be dissolved which will likewise be included in the weld. There are a number of welding techniques and developments designed to prevent these contaminant inclusions in the weld. These are discussed in the appropriate sections of this book.

Another important interreaction between applied heat and the surface of the metal is the degree of thermal conductivity. *Thermal conductivity* is an indication of the rate (or speed) at which heat will pass through a metal. Other factors such as the size of the area, time, and temperature differences, also affect the rate of flow.

HEAT-TREATING PROCESSES

As was mentioned in the previous paragraph, heat treating will change certain characteristics of the metal. More specifically, it will change such physical properties as toughness, strength, ductility, and so on. Under controlled conditions, heat treating will result in changes desired by the welder.

There are several types of heat treating processes that contribute to more efficient welding practices, among which are included:

1. Annealing,
2. Stress relieving,
3. Tempering,
4. Hardening.

Annealing

Annealing is a heat treatment process that consists of heating a metal to a predetermined temperature and then causing it to cool at a predetermined rate. Different metals and metal alloys require different rates of cooling (e.g. fast cooling for stainless steel and slow cooling for carbon steel). Annealing is designed to relieve stress in weldments, create a welded joint that is dimensionally stable, and minimize hard spots. Metals that have been cold-worked may require annealing. This is particularly true if the metal has become too hardened to be worked any further. The annealing will soften the metal and for this reason it is frequently referred to as a softening process.

Stress Relieving

In welding complex structures of steels in the higher carbon ranges, the weld metal is stressed either to the rigidity of the members forming the structure or to the fact that the metal, having a high yield point, will not flow plastically under the stresses involved. Therefore, it remains in a state of tension. Unless these stresses are removed, the welded joint may fail in service. That is why structures are *stress-relieved* after welding.

Stress-relieving consists of heating the metal to approximately 1200° F. and holding it at this temperature for a length of time depending upon its size and thickness. This is a compromise between the minimum temperature required to permit a plastic flow to compensate for part of the stresses, and the maximum temperature to which this metal may be heated without severe distortion.

Weld metal has a higher degree of hardness and greater tensile strength than hot rolled steel of identical analysis. This characteristic is partly due to stresses set up in the weld metal. When the stresses are removed, the ultimate strength of the metal is reduced proportionate to the reduction of these stresses.

In mild steel, stress-relieving sometimes reduces ultimate strength as much as 2000 lbs. per square inch from the as-welded condition. However, in some of the new high tensile alloy steels whose strength is derived from elements other than carbon, the ultimate strength of weld metal may be actually increased as much as 10,000 lbs. per square inch by heat treatment. This is due to the precipitation hardening effect of the element used as a ferrite strengthener. The temperature used, however, is below that used for relieving mild steel, or approximately 950° F.

Tempering

Tempering is a heat-treating process designed to both soften and increase the toughness of steel. Once the steel has been hardened (and cooled) it is reheated to a point below the critical point, held at this temperature for a specific length of time, and allowed to cool to room temperature. A particular side effect of tempering is internal stress relief.

Hardening

Hardening is designed to increase the strength of a steel by heating the metal to a point above the critical temperature and cooling it rapidly. Certain heat-treating processes will result in the hardening of the metal. Among these, one may list: (1) quench hardening, (2) induction hardening, and (3) flame hardening. Either the surface layer (case) or both the surface layer and the core may be hardened. The former is referred to as case hardening and the latter as *full hardening*.

In *quench hardening,* the steel is heated to a temperature approximately 100° F. above the critical point and held there for a predetermined period of time. It is then rapidly cooled by quenching it in water, oil, or some other medium.

An alternating magnetic field is used to harden the metal in *induction hardening*. This may be either a case-hardening or full-hardening process. Cooling is accomplished by quenching.

Flame hardening uses a gas flame that is continuously played across the metal surface until the desired temperature is reached. Quenching is accomplished through a variety of methods.

PREHEATING

Preheating the metal surface is a procedure used in several welding processes (the gas welding of thick sections, thermit welding, and welding metals having high thermal conductivity) to reduce the degree of expansion and contraction. Uneven expansion and contraction will result in serious welding problems (breaking, distortion, etc.). Consequently, preheating must be *evenly* applied over as much of the surface area as possible. It would be most desirable if the entire surface can be preheated. This is not difficult to do (using a preheating furnace or gas flame) if the piece is small enough. However, larger pieces are more difficult to preheat.

Preheating will also produce other results that will add to the efficiency of the weld. For example:

1. A reduction of hardness,
2. The cleaning of combustionable elements from the surface,
3. An increased machinability and welding speed.

STRESS AND STRAIN

Stress and strain are two frequently used terms with which the welder should be familiar. *Stress* refers to the amount of force (i.e. load) applied against a metal surface. *Strain* is the resulting physical effect of this force. The amount of stress is expressed in pounds per square inch (psi) and the strain in measured deformation (elongation or distortion).

BRITTLENESS

Brittleness refers to the tendency of a metal to suddenly fracture. It may be viewed as being the opposite of both ductility and toughness.

DUCTILITY

Ductility is the ability of a metal to become permanently stretched without breaking or deforming. Each metal has a limit to its ductility beyond which fracture will occur. Prior to reaching this limit, however, the metal simply remains elongated. In this manner, ductility contrasts with elasticity. The latter being the ability of a metal to return to its original shape.

ELASTICITY

Elasticity refers to the ability of a metal to return to its original shape after having been subjected to a deforming force. All metals possess varying degrees of elasticity. Each metal also possesses an *elastic limit*. This is the point of maximum load beyond which the metal will not return to its original shape.

TOUGHNESS

The tensile and compressive strengths of metals are determined by the gradual application of force until the critical fracture point is reached. *Toughness,* on the other hand, is the ability of a metal to resist a rapid or sudden application of force. Toughness is sometimes classified as a physical property of a metal, but it would be more correct to regard it as a general quality that is often closely associated with tensile strength. Toughness is, in reality, a combination of several properties (elasticity, ductility, etc.) all of which contribute to the resistance of the metal to a force suddenly applied.

HARDNESS

Hardness indicates the ability of a metal to resist denting, scratching, abrasion, or penetration. If the welder can determine the hardness of a metal, the tensile strength can also be determined through a conversion table.

The four methods for testing the hardness of a metal are:

1. Brinell test,
2. Rockwell test,
3. Vickers test,
4. Shore Scleroscope test.

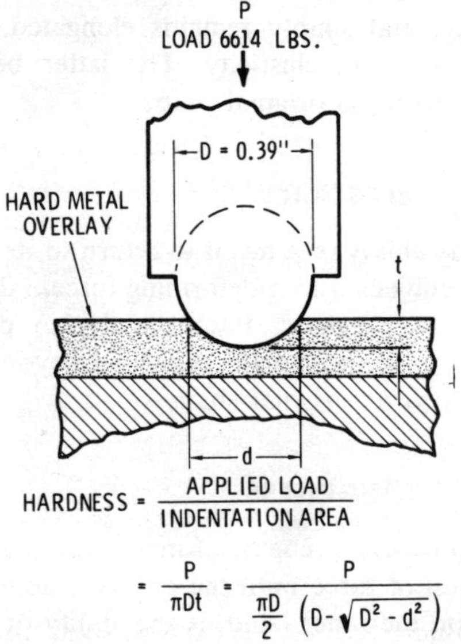

Fig. 1. The Brinell test.

Courtesy James F. Lincoln Arc Welding Foundation

The *Brinell test* (Fig. 1) is one of the oldest testing methods and is used to measure indentation hardness. A steel ball having approximately a ⅜-inch diameter is pressed into a metal surface

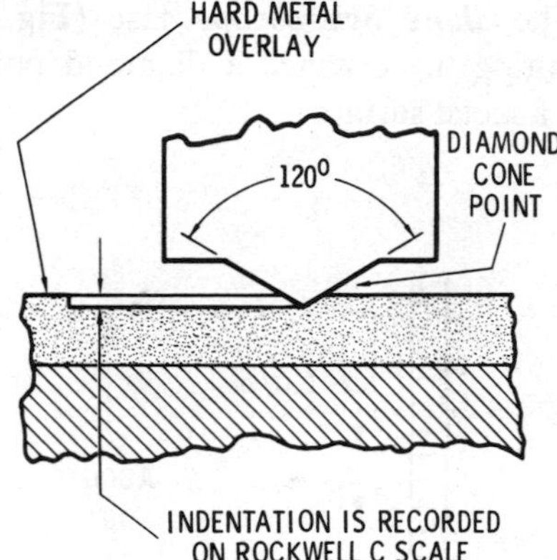

Fig. 2. The Rockwell test.

Courtesy James F. Lincoln Arc Welding Foundation

with a predetermined load (about 3000 kg. for steel and 500 kg. for softer metals). The pressure time varies from 15 seconds for steel to longer periods for the softer metals. The Brinell test is limited to relatively large surface areas.

The *Rockwell test* (Fig. 2) also tests for indentation hardness. A diamond cone or steel ball is pressed into a metal surface with a predetermined load. The load values are considerably lower than those used in the Brinell test (ranging from 60 to 150 kg.). The Rockwell hardness number is read from a dial on the machine. There are nine Rockwell scales each having a different letter designation (A through H and K). These scales differ from one another according to whether or not a diamond cone or a steel ball is used—the diameter of the steel ball (⅛ to 1/16 inch) and the size of the load. The Rockwell B and Rockwell C scales are the most commonly used. The Rockwell test can be used to determine the hardness of much smaller surface areas than the Brinell test.

The *Vickers test* measures indentation hardness by pressing a diamond-tipped point into a metal surface with a predetermined load. The load does not exceed 120 kg. The toughness of the diamond tip enables this test to be used on the hardest steels. The Vickers test is similar to the Rockwell test in that it can be used to determine the hardness of small surface areas.

The *Shore Scleroscope* test (Fig. 3) measures hardness by measuring the distance a diamond pointed hammer will rebound from a metal surface.

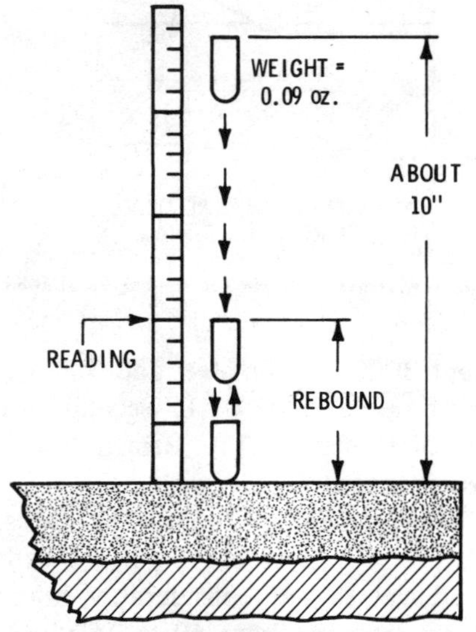

Courtesy James F. Lincoln Arc Welding Foundation

Fig. 3. The Shore Scleroscope test.

Each of the four testing methods described in the previous paragraphs use machines that would be too expensive for the individual welder to purchase. A far less expensive method is the simple scratch test with a hard metal file. The file is drawn across the metal surface and the depth of the scratch determines the hardness. In Table 1, the *James F. Lincoln Arc Welding Foundation* has presented some data indicating the relationship between Brinell hardness values and different types of scratches made with a metal file.

Table 1. Brinell Values Compared with a File Scratch Test

Brinell Hardness	File Action
100	File Bites into Surface Very Easily
200	File Removes Metal with Slightly More Pressure
300	Metal Exhibits its First Real Resistance to the File
400	File Removes Metal with Difficulty
500	File Just Barely Removes Metal
600	File Slides over Surface without Removing Metal
	File Teeth are Dulled

Courtesy James F. Lincoln Arc Welding Foundation

CORROSION RESISTANCE

Corrosion resistance is the ability of a metal to resist structural disintegration. This chemical reaction results from elements within the metal combining with elements and chemical compounds foreign to the metal. A knowledge of which metals are most resistant to corrosion is important to the welder. The stainless steels are high on the list of being most resistant, but other metals and metal alloys also share this characteristic.

ELECTRICAL RESISTANCE

Electrical resistance is the resistance offered by a metal to the flow of an electrical current. Different metals offer a differing degree of resistance to electricity. This is an important factor to consider when using one of the welding processes that obtain their heat from electricity (e.g. electric arc welding, resistance welding, etc.). The greater the resistance offered to the flow of electricity, the greater the amount of heat generated in the metal.

ELECTRICAL CONDUCTIVITY

Electrical conductivity is the degree to which a metal will conduct an electrical current across or through its surface. It is a characteristic essentially opposite that of electrical resistance. Copper exhibits a high degree of electrical conductivity.

WELDABILITY

Weldability is a metal characteristic that represents the combination of several physical and mechanical properties. It indicates the degree of ease or difficulty exhibited by each metal in forming a weld. Some metals are best welded by using special welding processes; others may be welded by the most common methods.

METAL IDENTIFICATION

The first step in any welding process is the correct identification of the metals to be welded. This must not be a matter of guesswork. Metals differ from one another with respect to their mechanical properties, composition, and structure. Consequently, one metal may not be compatible with another in a weld unless special steps are taken to bridge these differences. These special steps will range anywhere from a filler metal composed of elements that will provide a transition between the two metals to a welding process capable of producing the desired results. The inability to correctly identify the metals greatly increases the possibility of a poor weld. Thus, correct metal identification is an absolute necessity to ensure a good weld.

The easiest and most positive method of identifying a metal would be sending it to a laboratory for chemical analysis. This would also be the most expensive and time consuming method; a factor that makes it highly impractical for small welding shops and the individual welder.

Metals may also be identified by their use. Engine blocks, gears, shafts, etc. can be expected to be constructed from certain types of metal (e.g. iron, forged steel, etc.). This may be specified by the manufacturer in the literature accompanying the metal part(s), or such information may be gained by writing to the company.

The appearance of a metal, particularly its color, is frequently used as a means of identification. Table 2 indicates color identifications for a number of commonly used metals and metal alloys. These colors are of unfinished, nonfractured metal surfaces. This

Table 2. Color Identifications of Unfinished, Nonfractured Metal Structures

Metal	Color
Iron	
1. White Cast Iron	Dull Gray
2. Gray Cast Iron	Very Dull Gray
3. Malleable Iron	Dull Gray
4. Wrought Iron	Light Gray
Steel	
1. Low Carbon Steel	Dark Gray
2. High Carbon Steel	Dark Gray
3. Cast Steel	Dark Gray
4. Alloy Steel	Dark Gray
Copper	Reddish Brown to Green
Brass and Bronze	Green, Brown, or Yellow
Aluminum and Aluminum Alloys	Very Light Gray
Monel Metal	Dark Gray
Nickel	Dark Gray
Lead	White to Gray

Table 3. Fracture Colors of Common Metals

Metal	Color
Iron	
1. White Cast Iron	Silvery White
2. Gray Cast Iron	Dark Gray
3. Malleable Iron	Dark Gray
4. Wrought Iron	Bright Gray
Steel	
1. Low-Carbon Steel	Bright Gray
2. High-Carbon Steel	Very Light Gray
3. Cast Steel	Bright Gray
4. Alloy Steel	Medium Gray
Copper	Red
Brass and Bronze	Red to Yellow
Aluminum and Aluminum Alloys	White
Monel Metal	Light Gray
Nickel	Almost White
Lead	White

means of identification leaves much to be desired insofar as the various steels are concerned (or, for that matter, monel metal and nickel). A further differentiation among the various metals can be obtained by comparing the color of their fractured surfaces. For example, the very light gray of high-carbon steels contrasts with the bright gray of both low-carbon and cast steel. Table 3 lists the various fracture colors.

Another method of identifying metals is by means of the spark test. This is neither a difficult nor an expensive method of testing, but it does require some knowledge of different spark stream characteristics.

The spark test is administered by lightly touching the surface of the metal to a spinning grinding (emery) wheel. The individual spark stream is then observed against a black background (e.g. a piece of cardboard painted black). The length of the spark stream depends more on the amount of pressure exerted against the grinding wheel than on the properties of the metal itself and should not be regarded as one of the distinguishing factors. The different spark streams may be distinguished from one another by color, by volume, and by the nature of the spark (forked or fine).

Fig. 4 illustrates the application of the spark testing method to the various types of steel. The results are further explained in the following paragraphs:

1. *Wrought iron* (with no carbon content). This metal creates a spark stream that consists of small particles flowing away from the point of contact in a straight line. The stream becomes broader and more luminous some distance from the source of heat and then the particles disappear as they started. This is probably due to the action of the oxygen of the air on heated particles requiring some time to act.
2. *Mild steel* (with a small percentage of carbon). The spark stream is characterized by a division or forking of the luminous streak. This is due to the presence of carbon which is acted upon by the maximum heat of the iron spark, which then burns explosively causing a break in the original heavy lines.

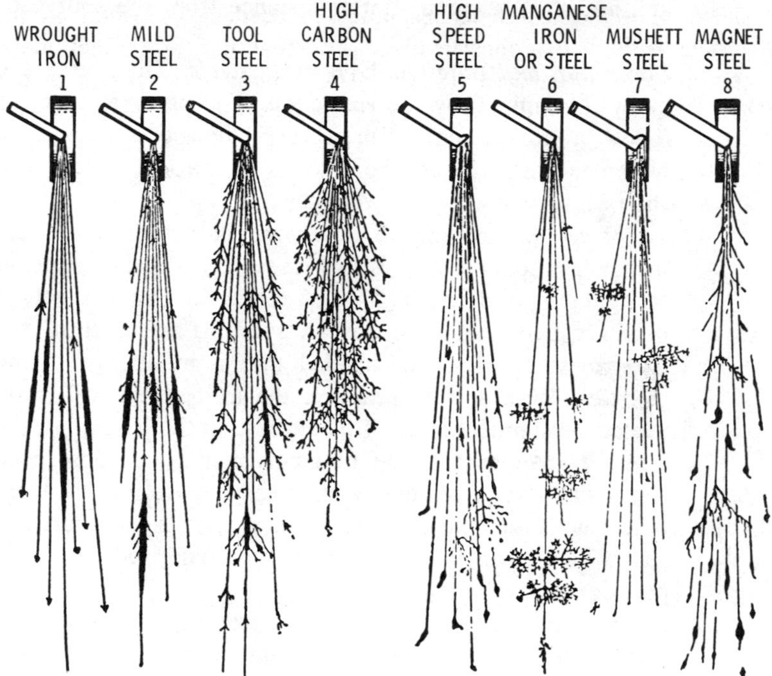

Fig. 4. The spark test for various types of steel.

3. *Lower grades of tool steel* (50-100 point carbon content). The spark stream is characterized by iron lines that become less and less conspicuous, the forking of the luminous streak occurring very much more frequently (and often subdividing). The lower the carbon content, the fewer the sparks and the further these sparks occur from the source of heat.
4. *High grade carbon steel.* The iron lines of the spark stream are practically eliminated, with an increase of the star-like explosions which often divide and subdivide, causing a beautiful display of sparks. This is due to the iron and carbon becoming so united that they are most easily attacked by the oxygen. Hence the great danger of burning steel in the fire. It would be well to state that the higher the percentage of carbon is, the more profuse the explosions

are, and they occur a shorter distance from the source of heat.

5. *Chromium and tungsten high-speed steels.* These are very easily determined by the spark test. The particles seem to follow a broken line with a very slight explosion. Just before they disappear, the color is of chrome yellow and shows no trace of a carbon spark.
6. *Tool-steel manganese.* The basic characteristic of the manganese spark stream is that it seems to shoot or explode at right angles from its line of force. In this manner, it differs widely from the carbon spark. Each dart of the manganese spark is subdivided into a number of white globules. With a little practice, the trained eye will soon detect the slightest trace of manganese in the iron or steels.
7. *Mushett steel* (the grade of air-hardening or high-speed steel). This type of steel is very easy to distinguish from others with the spark testing method because the particles follow a broken line and are a very dark red (with an occasional manganese spark).
8. *Magnet steel.* The illustration represents the spark as thrown from a special steel manufactured to be used especially for magnets.

TYPES OF JOINTS

There are five basic types of joints to be considered in welding. These are:

1. Butt joint,
2. Corner joint,
3. Edge joint,
4. Lap joint,
5. T-joint.

These basic joints may be used in combination to produce a number of different variations.

The *butt joint* (Fig. 5) is a joint formed between two members that lie approximately on the same plane. This type of joint is formed when two plates or pipes are brought together, edge

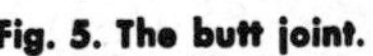
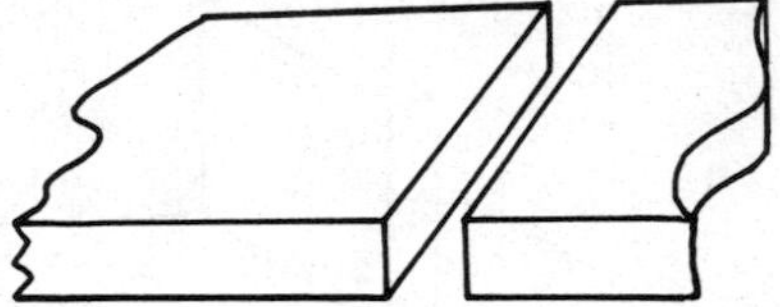

Fig. 5. The butt joint.

to edge, and welded along the seam thus formed. The *corner joint* (Fig. 6) is a joint formed between two members that lie at

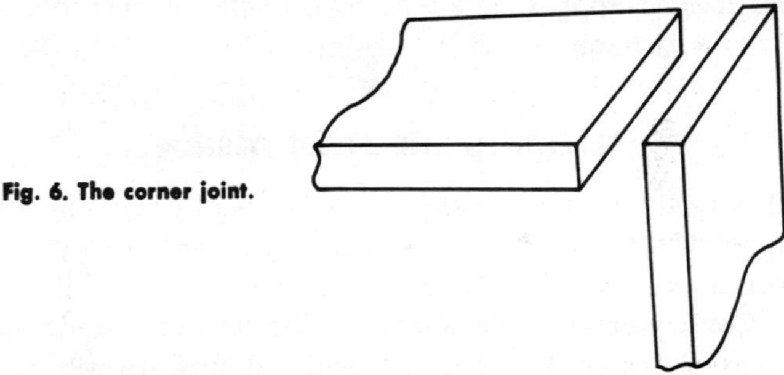

Fig. 6. The corner joint.

approximately a right angle to each other. The *edge joint* (Fig. 7) is a joint formed between edges of two or more members that lie parallel or nearly parallel.

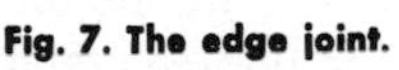
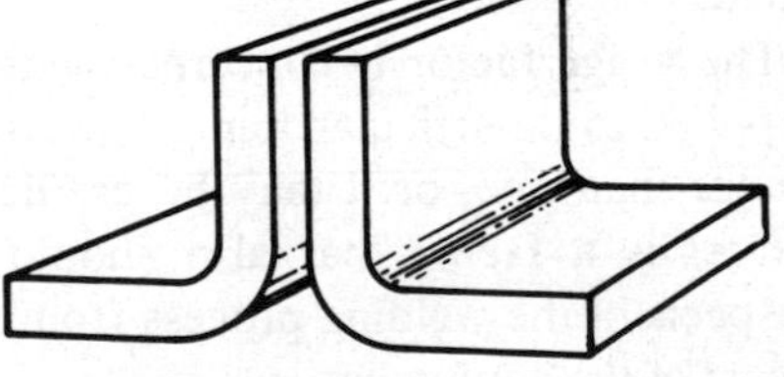

Fig. 7. The edge joint.

The *lap joint* (Fig. 8) is a joint formed by two overlapping members. The welding material is so applied as to bind the edge

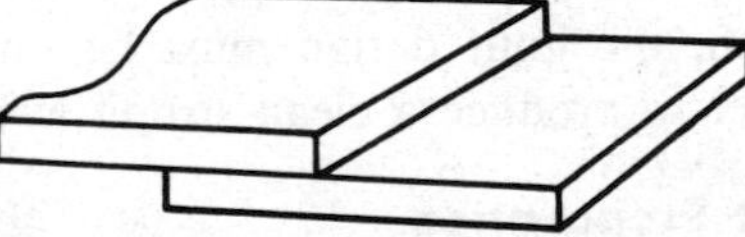

Fig. 8. The lap joint.

of one plate to the face of the other. This is to allow for the contraction stresses set up as the weld progresses.

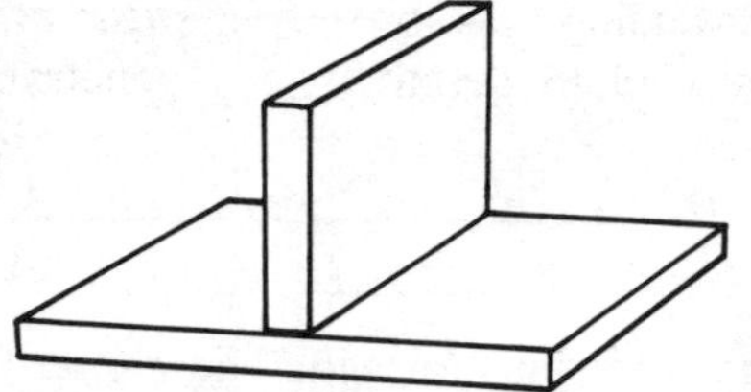

Fig. 9. The T-joint.

The *T-joint* (Fig. 9) is a joint formed between two members that are located approximately at right angles to each other. The resulting connection forms a T-joint.

SELECTING THE JOINT DESIGN

Selecting the most suitable joint design is of crucial importance in welding. Unfortunately, it can be a very complicated procedure. It depends primarily on the use to which the joint will be subjected. Consequently, some joints will be adequate under certain conditions of usage, but will fail under others. However, usage is not the only factor to be considered in selecting a suitable joint design. The welder should also take into consideration such factors as: (1) the cost (both for joint preparation and the welding itself), and (2) the accessibility of the position in which the joint is located.

The usage factor is concerned with the type of stress that will be applied to a particular joint. The stress may be in one direction, or more than one, or it may be predictable or unpredictable.

Cost is a factor that also should be considered. It includes all aspects of the welding process from the amount of filler metal to the cost of the equipment.

The welder must also consider the accessibility of the position to be welded. The weld should not have irregularities that would permit corrosion, cracking, or other welding problems. For this reason, the joint design must be one that would best allow the welder to produce a clean strong weld.

Joint Preparation

Once a suitable joint design has been selected, it is necessary to prepare the joint for the weld. The weld should be thoroughly cleaned before beginning the welding operation. The edges of the joint may also need to be beveled to permit deeper penetration.

Dirt, grease, and other contaminants on the surface of the metal must be removed before welding or they may become included in the weld causing gas or dirt inclusions that will weaken its strength.

If the metal is too thick to permit sufficient penetration by the weld, it may be necessary to bevel the edge (or edges) of the joint. The bevel angle should be great enough to provide the required weld strength but not so great that too much weld metal will have to be used.

TYPES OF WELDS

The two basic types of welds are:

1. The groove weld,
2. The fillet weld.

Both of these basic welds have a number of variations depending upon joint design.

The *groove weld* is a weld made in the groove between two members to be joined. These are also referred to as *butt welds*. Groove welds require edge preparation of the surfaces to be joined.

The standard types of groove welds are:

1. Square groove weld,
2. Single-V groove weld,
3. Single-bevel groove weld,
4. Single-U groove weld,
5. Single-J groove weld,
6. Double-V groove weld,
7. Double-bevel groove weld,
8. Double-U groove weld,
9. Double-J groove weld (Fig. 10).

The *fillet weld* is a weld joining two surfaces approximately at right angles to each other in a lap joint, a T-joint, or a corner joint. The cross section of a fillet weld is approximately triangular (Fig. 11).

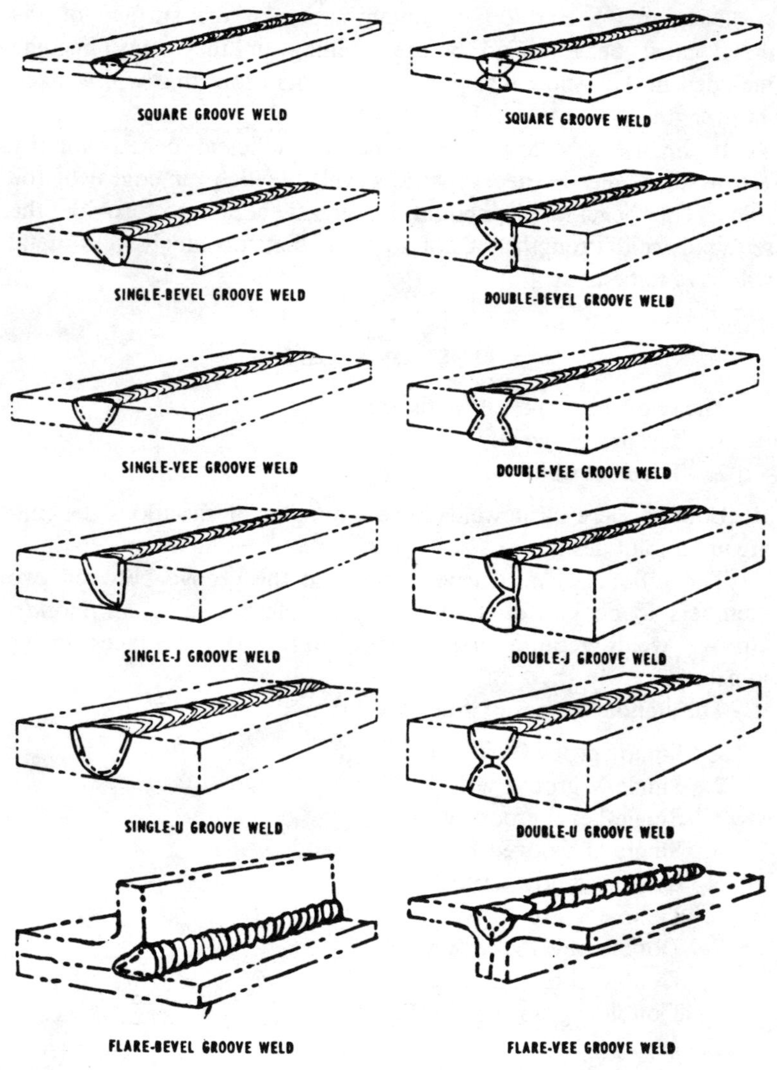

Courtesy American Welding Society

Fig. 10. Types of groove welds.

Fillet welds can also be described according to the concavity or convexity of the weld face. A *concave fillet weld* is one having

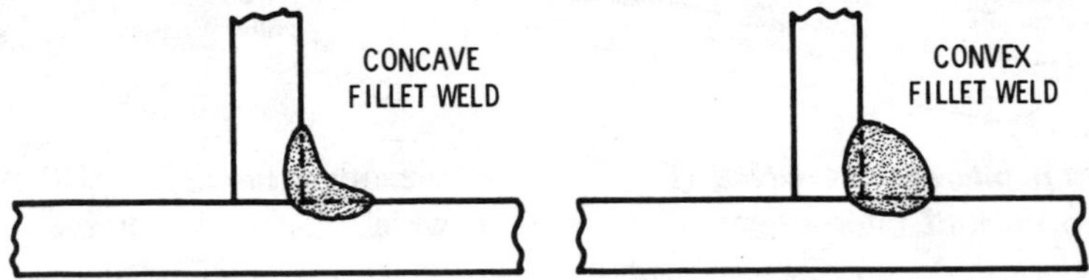

Fig. 11. The various types of fillet welds.

a concave face; whereas, a *convex fillet weld* is one having a convex face. See Fig. 12 and Fig. 13.

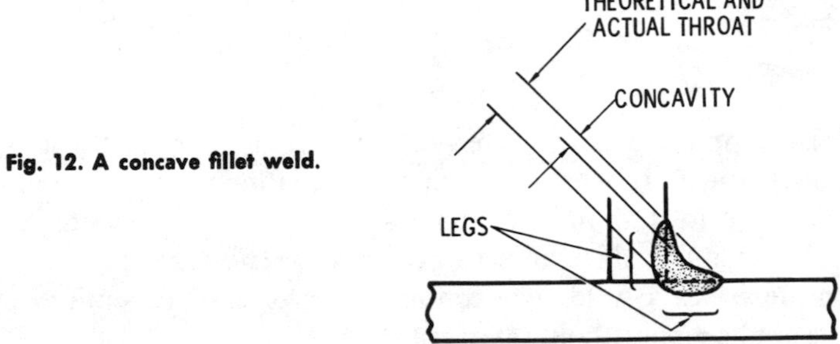

Fig. 12. A concave fillet weld.

A *full fillet weld* is one whose size is equal to the thickness of the thinner of the two members being joined.

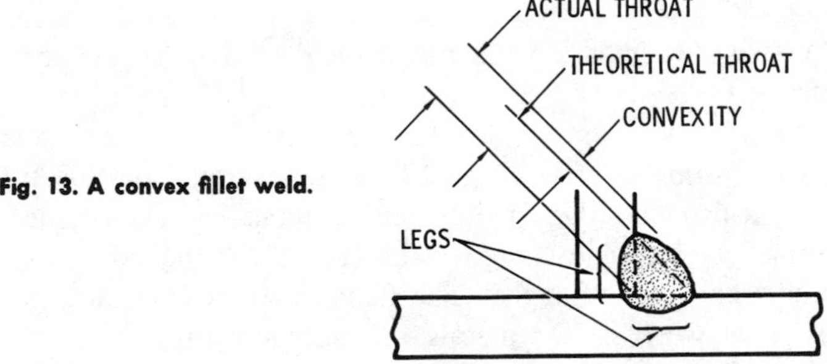

Fig. 13. A convex fillet weld.

Staggered intermittent fillet welds (Fig. 14) consist of two lines of intermittent fillet welds on a joint. The weld increments in one line are staggered with respect to those in the other. The

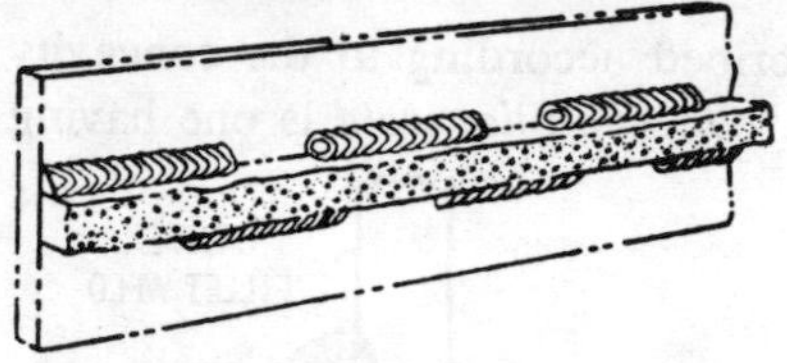

Fig. 14. A staggered intermittent fillet weld.

chain intermittent welds (Fig. 15), on the other hand, consist of two lines of intermittent fillet welds in which the weld increments are located opposite one another.

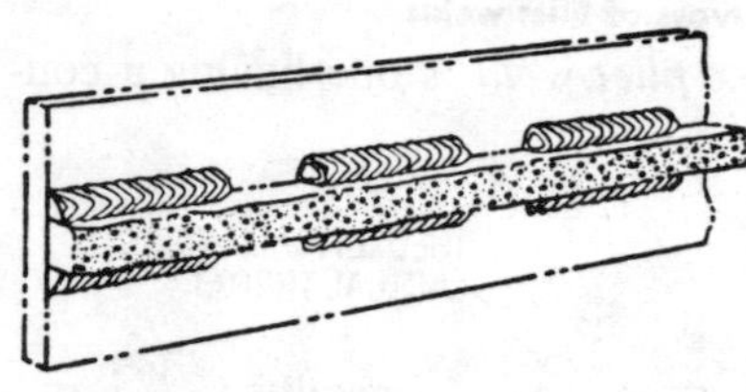

Fig. 15. A chain intermittent fillet weld.

Both of these terms refer to the length of the weld along a seam (particularly whether or not it is continuous or intermittent) rather than to the cross sectional structure as did previous terms. Another term in this same category is the *continuous weld*. This term describes a weld that continues or extends the entire length of its application without interruption.

A *tack weld* (Fig. 16) is a weld made to hold parts or members together in proper alignment until the final welds are made. It is temporarily welded and is not designed for use on a permanent basis.

A *flange weld* is a weld made on the edges of two or more members to be joined at least one of which has been flanged (turned back at right angles to the surface for reinforcement). The *corner flange weld* (Fig. 17) is an example of this type of weld in which the edge of only one member has been flanged. In the case of the edge flange weld (Fig. 18), the edges of both members have been flanged. The flange weld is frequently used in sheet-metal work. Filler metals are not necessary.

A *surfacing weld* (Fig. 19) is a type of weld deposited on an unbroken surface to obtain certain properties or dimensions. These welds consist of one or more string or weave beads. A surface weld is sometimes referred to as a bead weld.

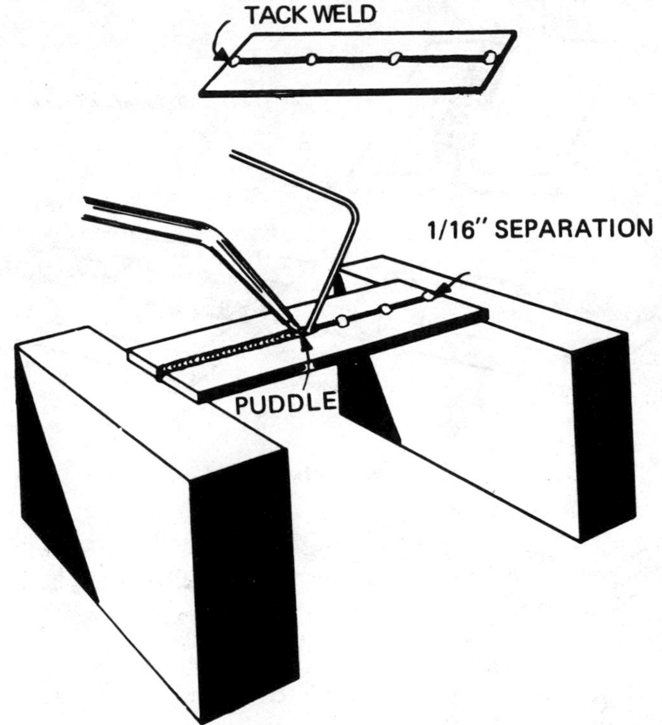

Courtesy Airco Welding Products

Fig. 16. A tack weld.

A *plug weld* (Fig. 20) is a circular weld made through a hole in one member of a lap or T-joint joining that member to the other. A *slot weld* (Fig. 21) is used with the same type of joints (lap or T) but is distinguished from the plug weld by its longer more elongated hole.

Fig. 17. A corner flange weld.

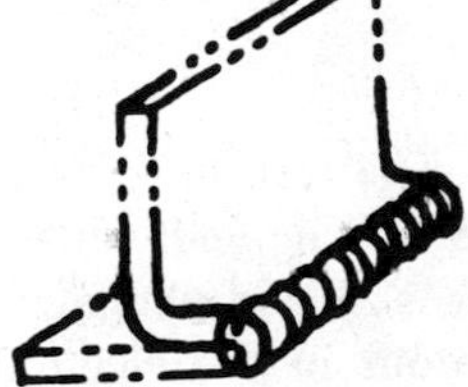

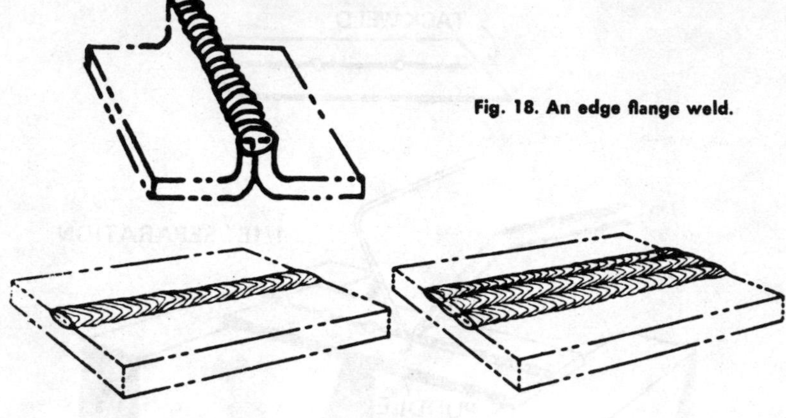

Fig. 18. An edge flange weld.

Fig. 19. A surfacing weld.

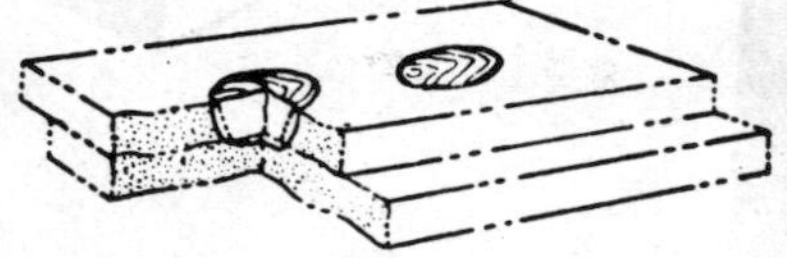

Fig. 20. A plug weld.

There are four types of welds characteristic of the resistance welding process. These four welds are:

1. The resistance spot weld,
2. The resistance seam weld,
3. The flash weld,
4. The upset weld.

Fig. 21. Slot weld.

The *resistance spot weld* (Fig. 22) consists of a series of non-overlapping individually formed spot welds made along a joint. The size and shape of each weld is determined by the size and contour of the electrode. In the *resistance seam weld* (Fig. 23),

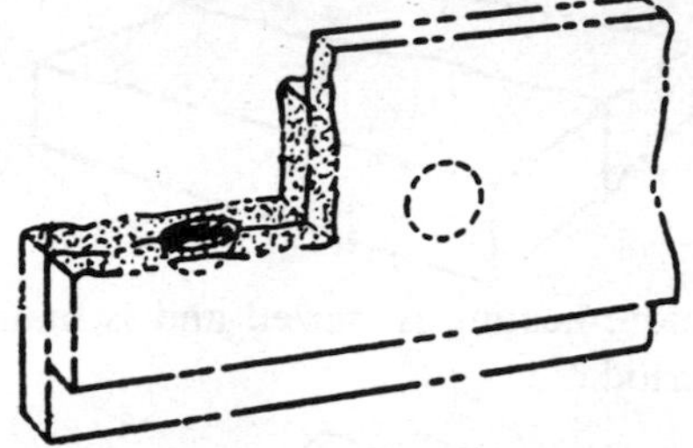

Fig. 22. Resistance spot weld.

the spot welds overlap to form a continuous weld which is made progressively along a seam by rotating the electrode.

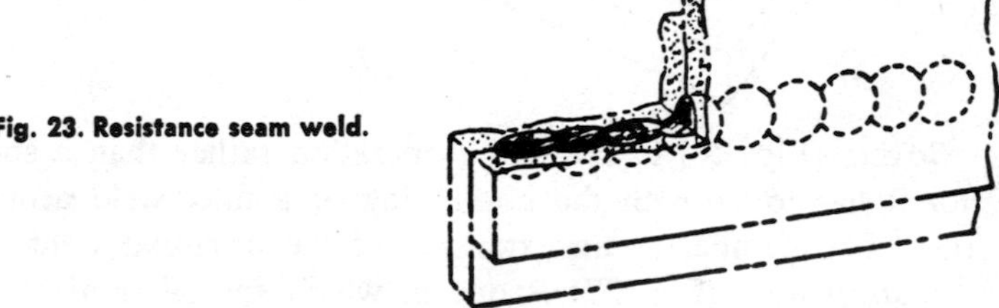

Fig. 23. Resistance seam weld.

A *flash weld* (Fig. 24) is a weld produced in the resistance welding process through simultaneous coalescence over the entire

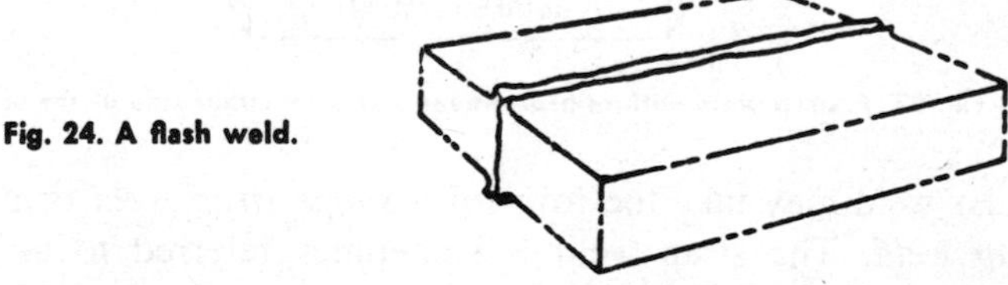

Fig. 24. A flash weld.

area of the abutting surfaces. Pressure is applied after the surfaces to be joined have reached fusion temperatures. An *upset weld* (Fig. 25) is also a product of the resistance welding process. It, too, is produced through simultaneous coalescence over the entire area of the abutting surfaces. However, pressure is applied

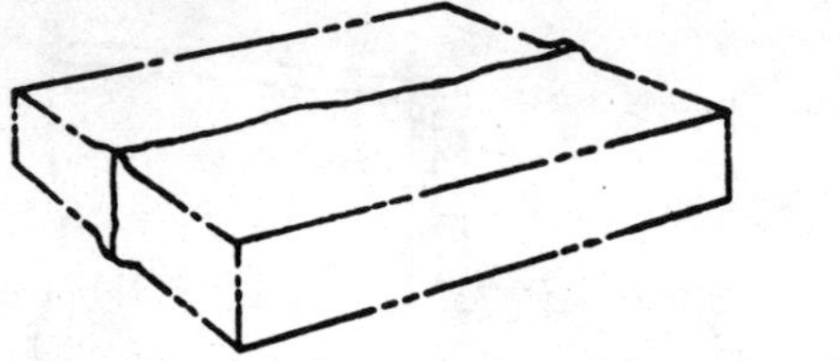

Fig. 25. An upset weld.

before heating is started and is maintained throughout the heating period.

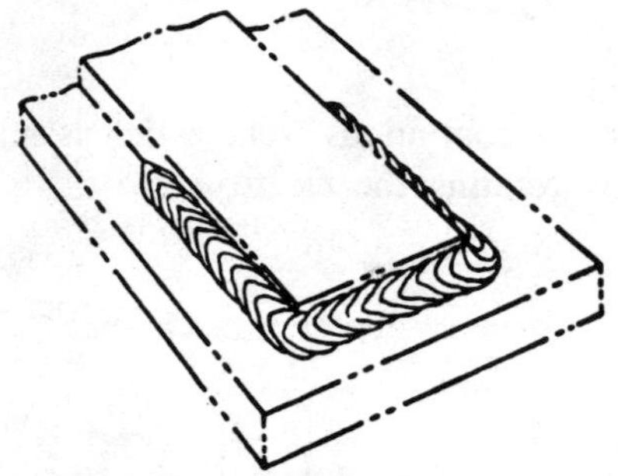

Fig. 26. Boxing weld.

Boxing (Fig. 26) is a welding operation rather than a specific weld. It has to do with the continuing of a fillet weld around a corner of a member as an extension of the principal weld.

A *strap weld* (Fig. 27) is one in which special reinforcement (usually a backing strip or plate) is given to the joint. This par-

Fig. 27. A strap weld with reinforcement plates on either side of the seam.

ticular weld may take the form of a *single strap weld* or a *double strap weld.* The strap weld is sometimes referred to as a *strap joint.*

BASIC WELD TERMINOLOGY

There are certain basic terms used to describe the structure of a weld. These can be divided into two categories depending on whether a groove or a fillet weld is being described. Fig. 28 illustrates the application of these terms. The text that follows explains each term in greater detail.

The *root of the weld* (Fig. 29) refers to the points at which the bottom of the weld intersects the surfaces of the base metal. In a fillet weld, it is specifically the point of deepest penetration of the weld material.

Root face, groove face, and *root edge* (Fig. 30) are terms describing the surfaces either in the groove cavity or along those edges that will touch when the two members are joined. The root face in a groove weld represents that portion of the groove face adjacent to the root of the joint. In the butt and T-joint, the groove face and root face are represented by the same surface. The groove face, then, is that surface portion of a member included in the groove. The root edge is a root face with zero width.

Root opening, bevel angle, and *groove angle* (fig. 31) are all terms referring to space dimensions between the two members to be joined. Root opening refers to the amount of separation between the two members at the root of the joint. The bevel angle indicates the angle formed between a plane perpendicular to the surface of the member and its prepared edge. The groove angle, on the other hand, includes the total angle of the groove between members to be joined by a groove weld.

The term *size of weld* depends on whether a groove weld or a fillet weld is being described. In a groove weld (Fig. 32), the term indicates joint penetration (depth of chamfering plus root penetration when specified). The size of weld in a fillet weld is indicated by the leg length of the fillet (Fig. 33).

Face of weld (Fig. 34) is a term used to describe the exposed surface of a weld on the side from which the welding is done.

The *throat* of a fillet weld can be both theoretical and actual (Fig. 35). *Actual throat* is the shortest distance from the root of the fillet weld to its face. The *theoretical throat* is determined by inscribing the largest possible right-triangle within the fillet-weld cross section. The theoretical throat is the distance from the beginning of the root to the hypotenuse of such a triangle.

The *leg* of a fillet weld (Fig. 36) represents the distance from the root of the joint to the toe of the fillet weld. The *toe* of a weld

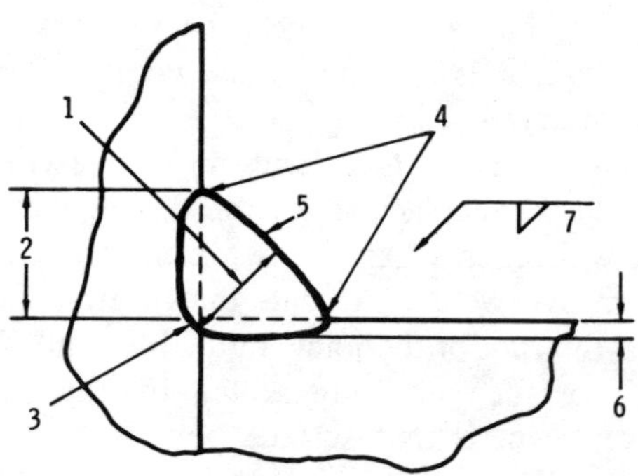

1. THROAT OF A FILLET WELD: The shortest distance from the root of the fillet weld to its face.

2. LEG OF A FILLET WELD: The distance from the root of the joint to the toe of the fillet weld.

3. ROOT OF WELD: Deepest point of useful penetration in a fillet weld.

4. TOE OF A WELD: The junction between the face of a weld and the base metal.

5. FACE OF WELD: The exposed surface of a weld on the side from which the welding was done.

6. DEPTH OF FUSION: The distance that fusion extends into the base metal.

7. SIZE OF WELD(S): Leg length of the fillet.

Fig. 28. Weld terminology of

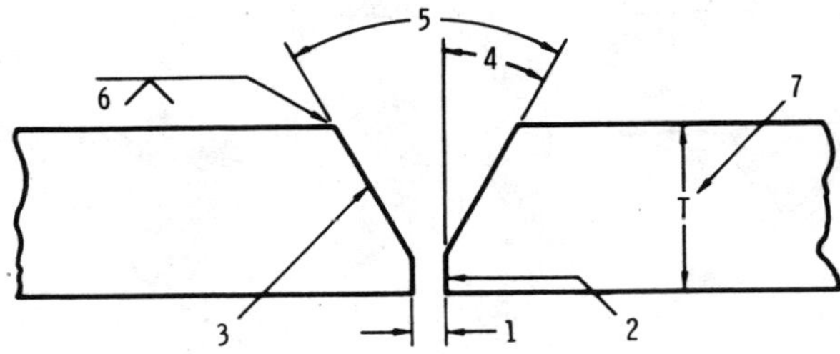

1. ROOT OPENING (RO): The separation between the members to be joined at the root of the joint.

2. ROOT FACE (RF): Groove face adjacent to the root of the joint.

3. GROOVE FACE: The surface of a member included in the groove.

4. BEVEL ANGLE (A): The angle formed between the prepared edge of a member and a plane perpendicular to the surface of the member.

5. GROOVE ANGLE (A): The total included angle of the groove between parts to be joined by a groove weld.

6. SIZE OF WELD(S): The joint penetration (depth of chamfering plus root penetration when specified).

7. PLATE THICKNESS (T) - Thickness of plate welded.

Courtesy Hobart Brothers Co.

a groove and fillet weld.

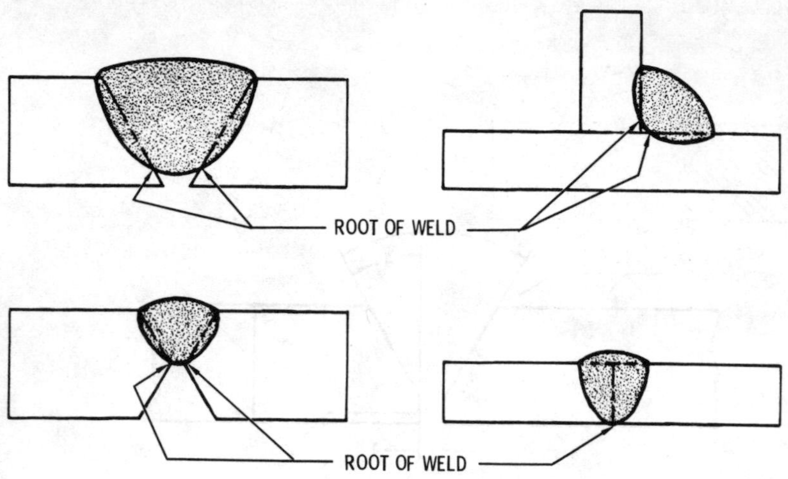

Fig. 29. The root of a weld.

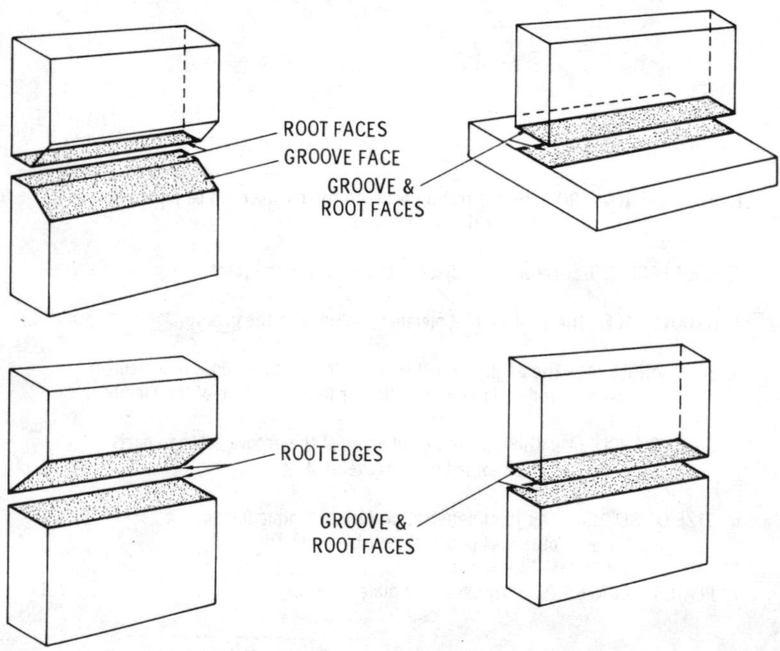

Fig. 30. Root face, groove face, and root edge.

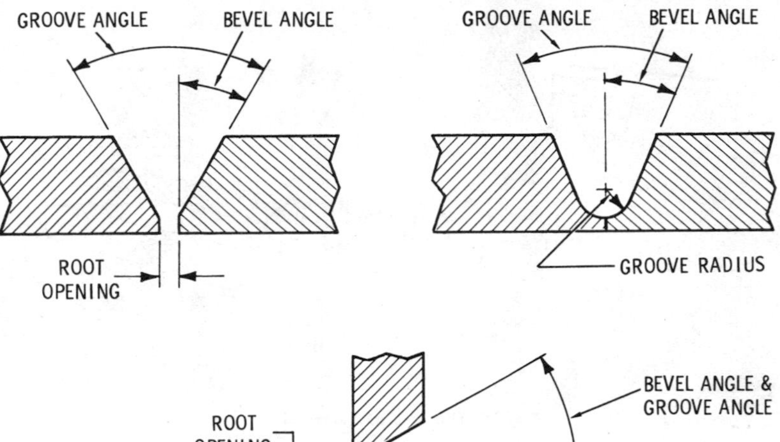

Fig. 31. Groove angle, bevel angle, and root opening.

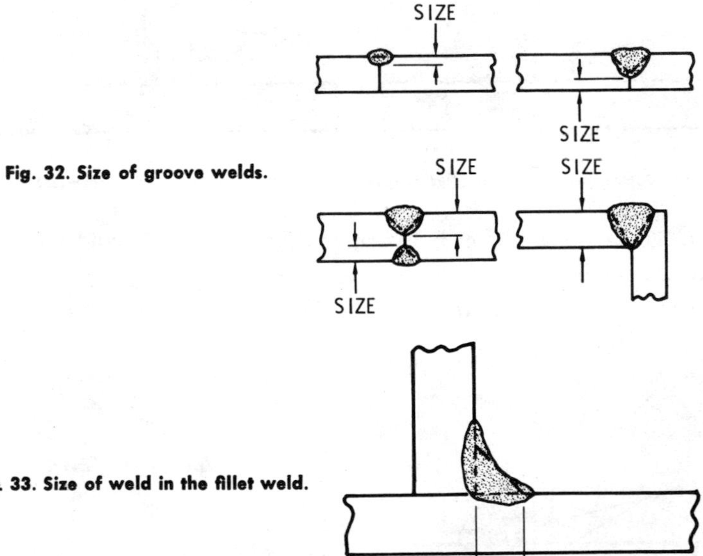

Fig. 32. Size of groove welds.

Fig. 33. Size of weld in the fillet weld.

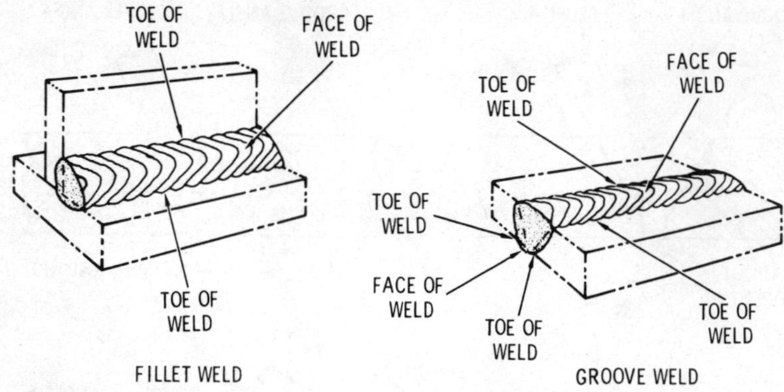

Fig. 34. The face of the weld.

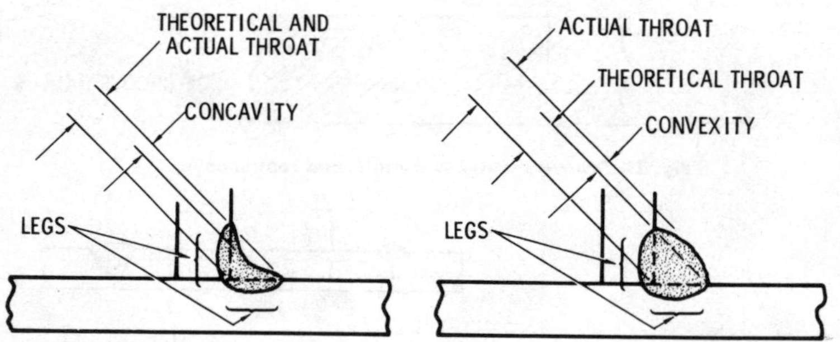

Fig. 35. Convex and concave fillet welds showing both actual and theoretical throat.

(Fig. 37) is the junction between the face of a weld and the base (parent) metal.

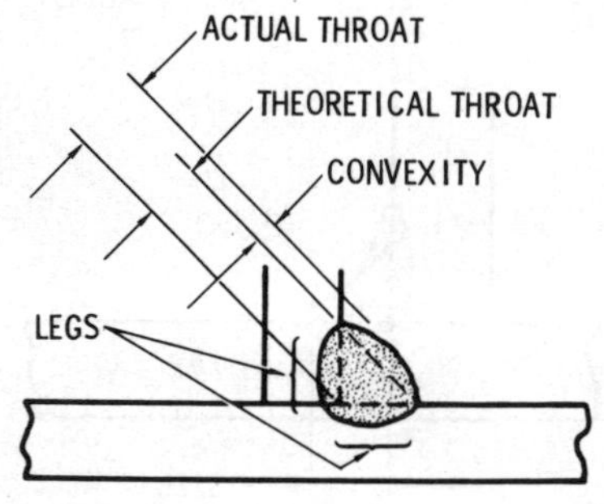

Fig. 36. The leg of a fillet weld.

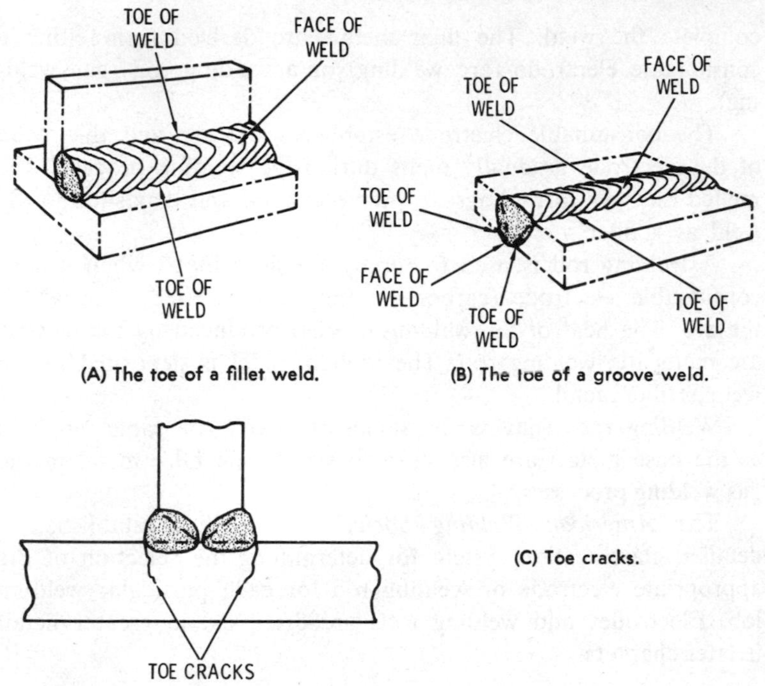

(A) The toe of a fillet weld.

(B) The toe of a groove weld.

(C) Toe cracks.

Fig. 37. The toe of a weld.

The quality of a weld depends largely on the characteristics of the weld metal. The weld metal is derived from two sources:

1. The base (parent) metal,
2. The electrode or filler metal.

If little or no electrode or filler metal is used, the proper selection of the base metal is the important factor. If the weld metal comes mostly from the electrode or filler metal, then the selection of the proper electrode or filler metal becomes of prime importance.

FILLER METALS

Filler metals are used in both the gas and arc welding processes. These are metals added to the base (parent) metals to complete the weld. The filler metals are derived from either a consumable electrode (arc welding) or a welding rod (gas welding).

The consumable electrode establishes the arc and the metal of the electrode gradually melts during the welding process. The melted metal is carried across the electric arc and deposited in the weld as a filler metal.

A welding rod is used to supply the filler metal when a nonconsumable electrode (carbon or tungsten) is used to establish the arc. The heat of the welding process produced by the electric arc melts the welding rod. The molten metal is deposited in the weld as filler metal.

Welding rods (having the same, or nearly the same, analysis as the base metal) are also used to supply the filler metal in the gas welding processes.

The *American Welding Society (AWS)* has established a detailed classification system for determining the selection of the appropriate electrode or welding rod for each particular welding job. Electrodes and welding rods are discussed in greater detail in later chapters.

WELDING POSITIONS

The four positions used in manual welding are:

1. The flat position,
2. The horizontal position,
3. The vertical position,
4. The overhead position.

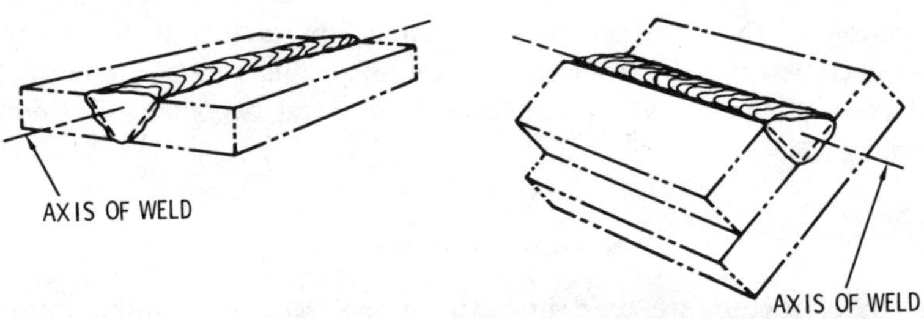

Fig. 38. Flat position weld.

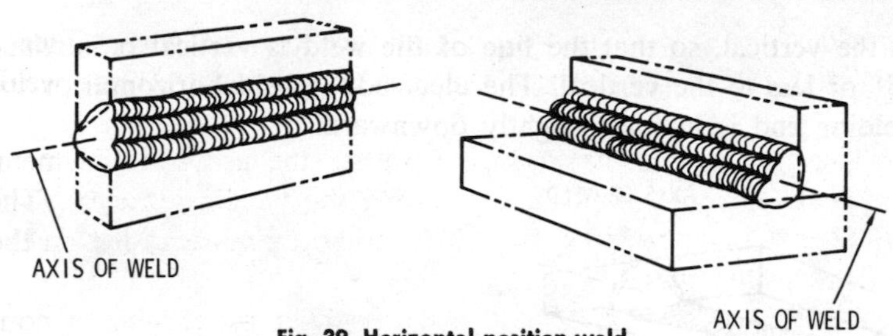

Fig. 39. Horizontal position weld.

The *flat position* (Fig. 38) is one which the welding is performed from the upper side of the joint and the face of the weld is approximately horizontal. The welding material is applied in a generally downward position. For this reason, the flat position is sometimes referred to as the *downward position.*

The *horizontal position* (Fig. 39) has two basic forms depending upon whether it is used for a fillet weld or a groove weld (see TYPES OF WELDS). In a fillet weld, the horizontal position is one in which welding is performed on the upper side of an approximately horizontal surface and against an approximately vertical surface. In a groove weld, the axis of the weld for the horizontal plane and the face of the weld lies in an approximately vertical plane.

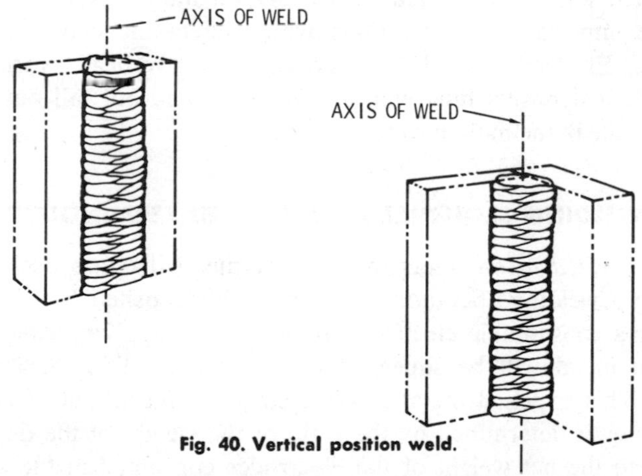

Fig. 40. Vertical position weld.

The *vertical position* (Fig. 40) is one in which the welding material is applied to a vertical surface, or one inclined 45° or less to the vertical, so that the line of the weld is vertical or inclined 45° or less to the vertical. The electrode is held horizontal or the welding end is inclined slightly downward.

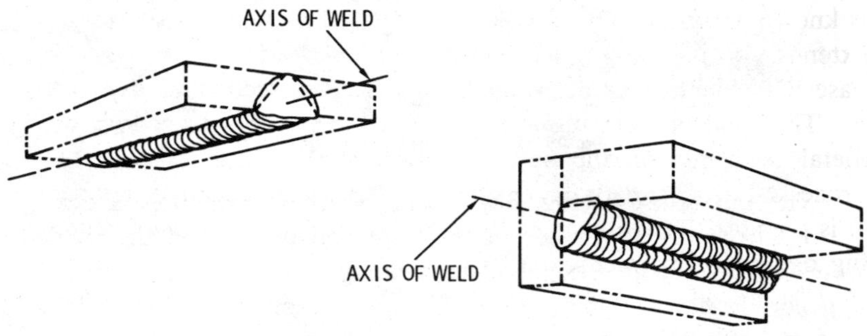

Fig. 41. Overhead position weld.

The *overhead position* (Fig. 41) is one in which welding is performed from the underside of the joint. The electrode is held with its welding end upward.

The best possible fusion is obtained when welding in a downward horizontal plane position on a flat steel plate. The position in which joints are welded, especially on multipass work, is one of the important considerations which affect the choice of electrodes. Electrodes used for vertical and overhead work must provide a deposit which will stay in place and not fall out of the joint while in the molten state.

WELDING SEQUENCE AND RELATED TERMINOLOGY

Deposited metal refers to the filler metal that has been added during a welding operation. This metal is deposited by a welding rod or a consumable electrode. If pressure is applied, some of the molten metal will be squeezed out and will solidify around the weld. This expelled metal is referred to as *flash*. The *deposition efficiency* is determined by the ratio of the weight of the deposited metal to the net weight of the electrodes consumed in the process (exclusive of stubs). Another frequently encountered term is *deposition rate*. This term refers to the weight of metal deposited in a unit of time. Finally, *deposition sequence* indicates the order in which the increments of weld metals are deposited.

The coalescence that results from the melting together of the two base (parent) metals, or of the filler metal and the base metal, is known as *fusion.* The *depth of fusion* is the distance that fusion extends into the base metal. The *fusion zone* refers to the area of base metal melted as determined on the cross section of a weld.

The *bead* or *weld bead* refers to the narrow band or layer of metal deposited on the base (parent) metal during the welding process. Each band or layer of deposited metal is called a *pass.* It is possible to have a single bead pass or a multiple one, depending on the requirements of the weld.

WELDING PROBLEMS

There are a number of different welding problems that may confront the welder. Most of these result in poor weld appearance, but some can actually weaken the strength of the weld. Some of the more serious welding problems are:

1. Weld spatter,
2. Undercutting,
3. Porous weld,
4. Slag inclusion,
5. Incomplete (poor) penetration,
6. Cracking.

Weld spatter ruins the appearance of a weld but has no effect on its strength. This activity is found in both gas and arc welding. The spatter consists of metal particles thrown from the weld during the welding operation. One cause of weld spatter is excessive heat (high current in arc welding).

Undercutting refers to the cutting of a groove into the base metal parallel to the toe of the weld. It primarily serves to distract from the appearance of the weld although it can lead to structural defects. Causes of undercutting are:

1. Excessive weaving,
2. Allowing the puddle to become too large,
3. Excessive heat.

A *porous weld* is one in which gas pockets or voids are included in the weld. This results from gases being dissolved and included in the weld as it solidifies. Porous welds are not necessarily structurally weak welds. It is generally a question of poor appearance. Its cause is often traced to a base metal that produces gaseous combinations which tend to result in gas pockets and blow holes.

Slag inclusion refers to the development of non-metallic solid materials in the weld metal. In arc welding, these inclusions are derived from the coating on the electrode. It particularly tends to occur when filler metal of excessive size is used. These nonmetallic inclusions may be eliminated by thoroughly cleaning the area both before and during welding.

Incomplete (poor) penetration is a direct cause of structural failure in a weld. It is a condition in which the weld metal does not completely fill the weld cavity. It can lead to cracking or the inability of the weld to withstand joint stresses.

Cracking of the weld results from excessive localized stresses. This may be reduced by preheating the metal before welding. The result will be a slower cooling rate of the metal and a reduction of the tendency to crack. The cooling rate may also be reduced by using greater heat, larger electrodes (in arc welding), and a slower welding speed.

PRODUCING GOOD WELDS

A good weld will have superior strength to the base (parent) metal surrounding it. If an overload force is applied to the joint, the base metal will likely give before the weld does. Even the newly trained welder can consistently produce welds of high quality if care is taken in both the preparation of the weld and the welding procedure.

The steps to be taken in obtaining a good weld are as follows:

1. Correctly identify the base (parent) metal of the joint to be welded.
2. Taking into consideration the type of base metal as well as the design and position of the joint, select a welding process capable of meeting these requirements.

3. Use welding equipment and supplies of the highest quality. To save expense at this point is to run the risk of producing poor quality welds.
4. Choose the most suitable electrode of filler metal for the weld. This will depend on the identification of the base metal and the requirements of the welding job.
5. If the metal is too thick for adequate penetration, bevel the edges to be joined.
6. Thoroughly clean the surface of all contaminants (dirt, grease, oil, etc.). If these remain on the surface, they will likely be included in the weld and weaken it.
7. Begin welding by laying the initial bead. Clean the bead, remove any weld slag, and begin the next bead (or pass) until the weld is completed.
8. Examine the weld. If it is defective, remove it immediately and begin again.
9. The finished weld should be clean and have a good appearance. All undercutting and overlapping should be repaired.

CHAPTER 3

Gas Welding Equipment and Supplies

The manual gas welding process requires the use of a number of different types of materials, supplies, and equipment with which the welder must be familiar. These will be thoroughly discussed in this chapter.

Materials or supplies can be distinguished from equipment by the fact that the former are expendable items. Such things as the oxygen or fuel gas supply (acetylene, etc.), welding rods, welding fluxes, welding tip cleaners and even torch lighters and spark-lighters are considered expendable items. Those items with a somewhat longer life expectancy (e.g. cylinders, welding torches, welding hoses, etc.) are regarded as equipment and should be cared for in such a way as to prolong their period of usefulness. The types of equipment found in the typical oxyacetylene welding arrangement are illustrated in Fig. 1.

GASES

There are several gases used in the gas welding process. These can be divided into two basic groups:

1. Oxygen,
2. Fuel gases.

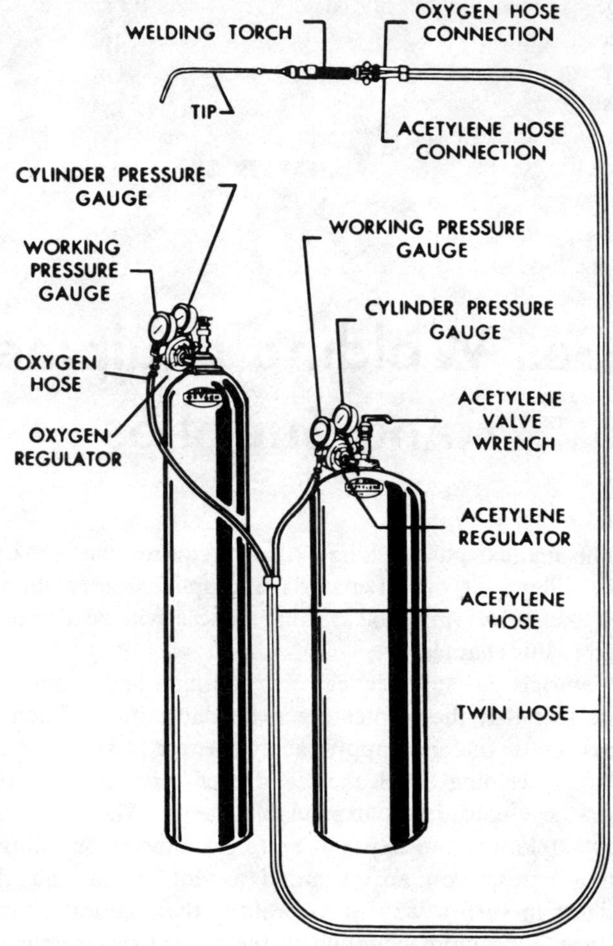

Courtesy Airco Welding Products

Fig. 1. Typical oxyacetylene welding arrangement.

Oxygen

Oxygen is a colorless and tasteless gas with no detectable odor. It should not be referred to as "air", since it does not contain the other types of gases present in the latter. Oxygen is available commercially with a guaranteed minimum purity of 99.5% and may be purchased in either liquid or gaseous form. Commercial oxygen is most commonly derived from air through a liquefaction process. This method is generally preferred because it produces the purest oxygen (containing only about .5% nitrogen). Under pressure, oxygen is very active and can react violently in combination with contaminants (particularly combustionable ones) and other substances.

Oxygen is a nonflammable gas that enters into chemical combinations with the combustionable elements of other substances, particularly those of the various fuel gases (e.g. acetylene, propane, LPG, etc.)

In the following paragraphs, these fuel gases will be described in detail. Acetylene is the most widely used fuel gas. The others (LPG, propane, etc.) are used much less frequently than acetylene.

Acetylene

Acetylene is a colorless combustionable gas with a characteristic odor. It is a product of a chemical reaction between water and calcium carbide.

Acetylene is available for purchase in cylinders as a pressure-regulated gas or as a chemical raw material (calcium carbide). The calcium carbide may be purchased in drums of 100 lbs. or more for acetylene generators. Acetylene cylinders contain a solvent that can absorb many times its own volume of acetylene under pressure. Because of the danger of explosion, acetylene should not be used at pressures above 15 psi.

Acetylene can be combined with oxygen (oxyacetylene welding) or air (air-acetylene welding). The former is by far the most commonly used manual gas welding process. The flame used in the air-acetylene welding process has a temperature much lower than that produced by other fuel gases. Consequently, air-acetylene welding is limited to light brazing and soldering.

Other Fuel Gases

As was mentioned previously, acetylene (in combination with oxygen) is the most commonly used fuel gas in manual gas welding. There are some other fuel gases in use but these are used much less extensively than in acetylene. These other fuel gases include:

1. Hydrogen,
2. *MAPP®*,
3. LPG,
4. Natural gas,
5. Propane,
6. Butane.

Hydrogen is an odorless colorless gas used in combination with oxygen in the *oxyhydrogen gas welding process.* The temperature of the oxyhydrogen flame is about 5500°F., considerably below that of the oxyacetylene flame. As a result, the oxyhydrogen flame is used to weld metals with low melting points (e.g. aluminum, magnesium, lead, etc.). It is commonly used in lead welding (lead burning).

Hydrogen is available commercially in cylinders of approximately 100 and 200 cubic foot capacity. The oxyhydrogen flame is very difficult to see which makes impossible the type of flame adjustment found in oxyacetylene welding. The ratio of the hydrogen to the oxygen is controlled through the hydrogen regulator.

MAPP® is the trademark for a liquified acetylene compound (stabilized methylacetylene and propadiene) used as a fuel gas in welding. It has heating characteristic somewhat similar to acetylene, but is much safer to handle. *MAPP®* gas is available in 12- by 44-inch cylinders or smaller. These cylinders hold about 70 pounds of liquid and give 620 cubic feet of gas at 60°F. *MAPP®* is both cheaper and easier to use than acetylene but is limited to the extent that it cannot weld steel plate over 5⁄16 inch thick.

LPG or *liquified petroleum* (LP) gas is used as a fuel gas for brazing, soldering, and cutting metals. It is available commercially in a liquid form and is purchased by the pound in containers ranging in size up to 100 pounds. Pressure regulators designed for use with liquified petroleum gas are used to control the pressure.

Natural gas is a low pressure fuel gas and must be used with specially designed low pressure injector-type torches. Natural gas is frequently used in soldering, brazing, preheating and cutting.

Propane and *butane* (two fuel gases derived from petroleum) are used primarily in brazing. It is difficult to obtain a nonoxidizing flame with either of these two fuel gases. As a result, they are limited in application.

GAS WELDING RODS

Gas welding rods are thin strips of metal used to add filler metal to the weld in most oxyacetylene welding and certain other welding processes. Since it is absolutely necessary to match the composition and properties of the welding rod as closely as possible to those of the base metal, the selection of the appropriate welding rod for the job is extremely essential. During the welding process, the welding rod melts and deposits its metal into the puddle where it joins with the molten metal to form a strong weld. Choosing the wrong metal rod will result in a weak and ineffective weld.

Welding rods are available in a variety of sizes and compositions. The sizes range in diameter from 1/16 inch to 3/8 inch. The cast-iron rods are sold in 24-inch lengths. All others are available in 36-inch lengths.

The variety in the composition of gas welding rods is such that there is an appropriate rod available for almost every kind of metal or metal alloy.

GAS WELDING FLUXES

A clean metal surface is essential to the formation of a good weld. All contaminents such as dirt, oil, and grease must be removed from the surface prior to welding. In some metals, the formation of oxides present a serious problem. This is particularly true of metals whose oxides have a higher melting point than the base metal itself (e.g. aluminum). The oxides will not flow from the area being welded. Instead, they remain in the welding zone, obstruct the addition of a filler metal and become a part of the solidifying metal. These factors all contribute to the formation of unacceptable welds. Oxides and other contaminants must be removed from the surface being welded and the welding zone protected from further contamination. Fluxes have been developed for this purpose.

A *flux,* according to the definition offered by the *American Welding Society,* is a *material used to prevent, dissolve, or facilitate removal of oxides and other undesirable substances.* This material is fusible and nonmetallic. As a result of the chemical reaction between the flux and the oxide, a slag is formed which floats to the top of the molten puddle of metal during the welding process.

The chemical composition of a flux depends on the metal or metals on which it is to be applied (Fig. 1). Fluxes may be divided into three main categories:

1. Welding fluxes,
2. Brazing fluxes,
3. Soldering fluxes.

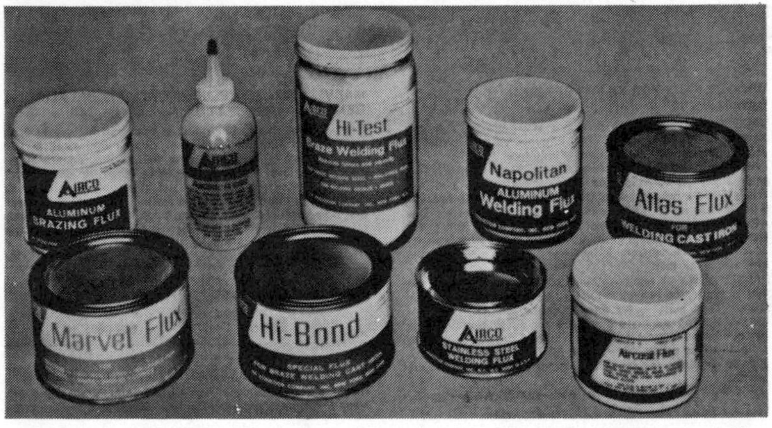

Courtesy Airco Welding Products

Fig. 2. Various types of fluxes.

The welding flux category is again subdivided according to use into fluxes specifically designed for use in stainless steel welding, braze welding, cast-iron welding and aluminum welding. Brazing fluxes and soldering fluxes are discussed in the appropriate chapters.

Fluxes are sold as powders, pastes, or liquids (frequently in plastic squeeze bottles). Quantities are available from as small as ½ to 5 lb. cans, or jars to 25 lbs. or more sold in drums.

Some powdered fluxes may be applied by dipping the heated welding rod into the can. The flux will stick to the rod. The flux powder may also be applied directly to the surface of the base metal. Some powdered fluxes are mixed with alcohol or water and applied to the surface as a paste. Paste or liquid fluxes are applied in the form they are purchased.

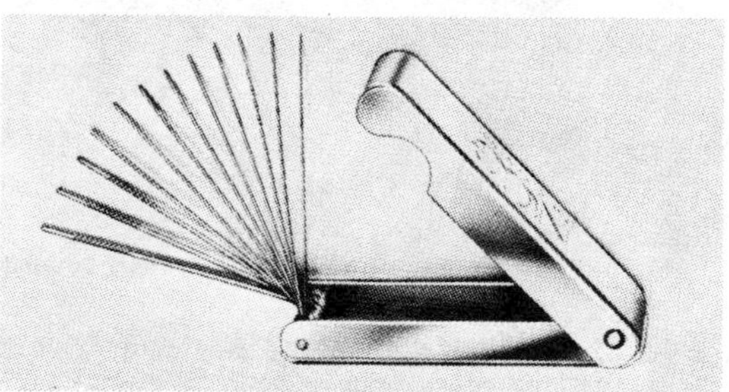

Courtesy National Cylinder Gas, Div. of Chemetron Corp.
Fig. 3. Welding tip cleaners.

WELDING TIP CLEANERS

Welding tip cleaners (Fig. 3) are stainless steel wires that are produced in a number of different sizes to accommodate the different tip orifices. Welding tips must be cleaned regularly in order to prolong the life of the tip and to provide consistent high performance. All deposits on the inside of the tip orifice should be removed. This must be done without enlarging the size of the orifice. Welding tip cleaners have been developed to satisfy this need and are shown in Fig. 3.

TORCH LIGHTERS AND SPARKLIGHTERS

The oxygen and fuel mixture should not be ignited with a match. A sudden flare-up on ignition could cause the hands or other parts of the body to be burned. To insure against such danger, the gas mixture should be ignited with a device that provides the required degree of safety for the welder. One such device is a commercially produced torch lighter (Fig. 4). A specially designed torch holder is attached to the top of this

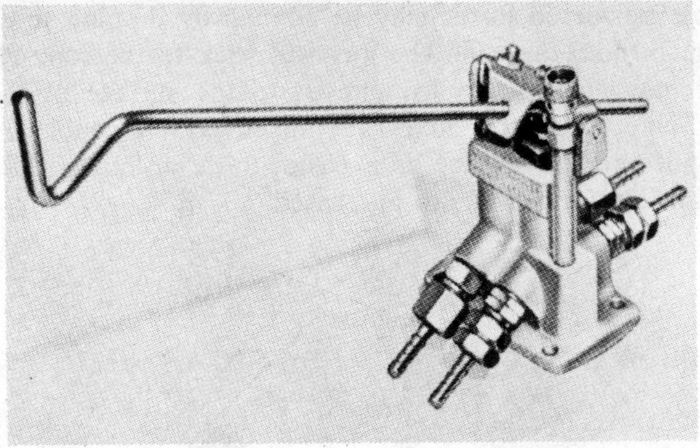

Courtesy Dockson Corp.

Fig. 4. A torch lighter.

device. This provides a convenient and safe place to put the torch when not in use. The torch holder also functions as a gas shutoff lever. The weight of the torch shuts off all gas flow except for a small flow of acetylene to the pilot light (which continues burning). Removing the torch from the holder releases the oxygen and fuel gas. The torch may then be ignited by passing it across the pilot light.

There are twin input-output valves at the base of the torch lighter for the attachment of the oxygen and fuel gas hoses. Most torch lighters are designed to be used with acetylene but can be adapted for use with propane.

The torch lighter reduces the consumption of gases by as much as 30%. This economizing factor coupled with the reduction of fire and accident danger make it an important addition to most welding shops. The torch lighter can be installed at any point in the lines but will perform best at points closer to the torch.

Sparklighters (Fig. 5) are also used to ignite welding torches. These are simple devices constructed from flint and steel. Because sparklighters are both reliable and inexpensive, they are more commonly used than torch lighters. They are available in a variety of types ranging from the simple to the complex. Some sparklighters are equipped with pistol-grips and shoot a shower of sparks at the gas flowing from the torch. Others have a rotating flint holder that permits longer use before having to insert a new flint.

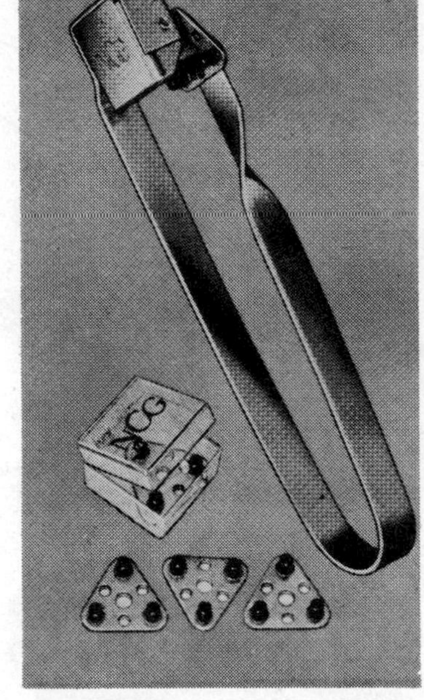

Fig. 5. A torch lighter with a supply of renewal flints.

Courtesy National Cylinder Gas, Div. of Chemetron Corp.

OXYGEN CYLINDERS

The oxygen cylinder (Fig. 6) is approximately 9 inches in outside diameter, 54 inches in height, and weighs (empty) between 104 and 139 pounds. The difference in weight depends upon the type of steel used to construct the cylinder. A full cylinder will increase the weight by about 20 pounds.

Oxygen cylinders are seamless steel containers that hold about 244 cubic feet of oxygen at a pressure of 2200 psi at 70°F. Smaller cylinders holding about 122 cubic feet are also available.

The oxygen supplied through these cylinders is about 99.5% pure. The oxygen pressure will vary according to temperature changes but the weight and percentage of oxygen remain the same.

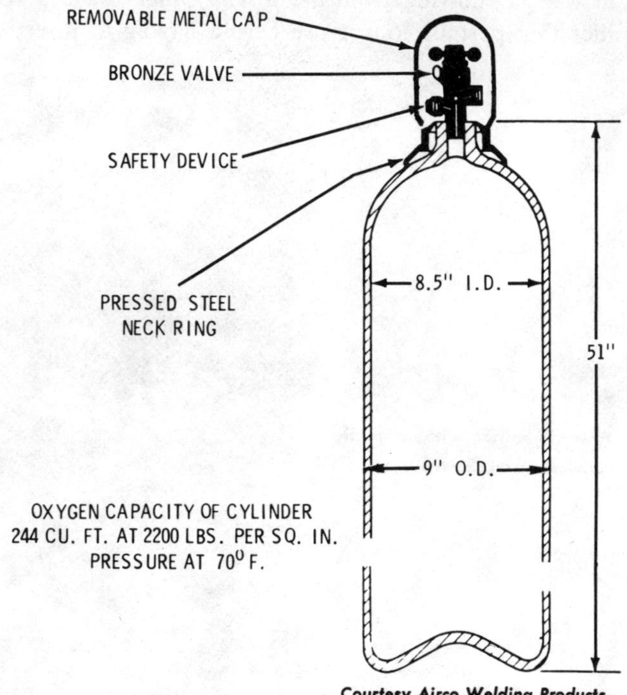

Courtesy Airco Welding Products

Fig. 6. Oxygen cylinder.

A special bronze valve (Fig. 7) is inserted into the top of the oxygen cylinder. A removable cap is placed over it for protection.

The oxygen cylinder valve is designed to withstand great pressures. If the pressure becomes too great, a disc in the safety plug located on the oxygen valve will rupture allowing the excess oxygen to escape before the high pressures rupture the cylinder. Pressure regulators can be attached to the oxygen cylinder valve at an outlet having a standard male thread.

Variations in oxygen cylinder pressures with respect to temperature changes are given in Table 2. Oxygen cylinder content as indicated by gauge pressure at 70°F. (for a 244 cu. ft. cylinder) is listed in Table 3.

Never allow oil, grease, or dirt to come into contact with oxygen cylinders or valves. These contaminants (especially the

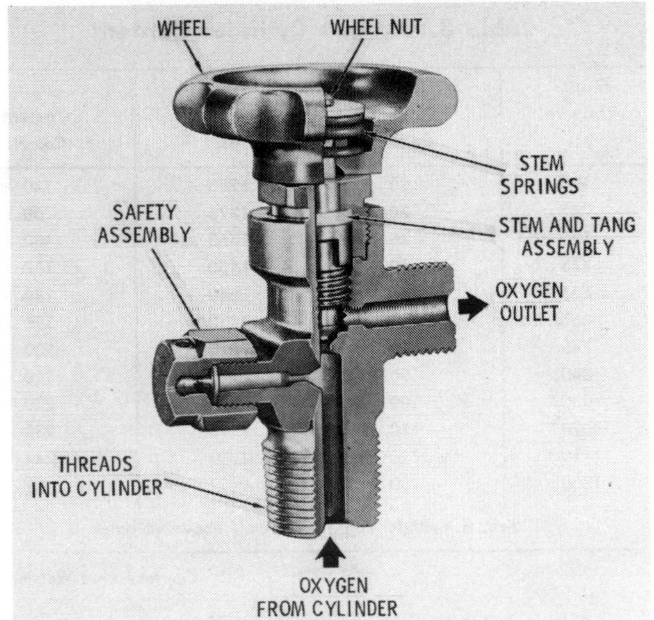

Courtesy Bastian-Blessing Co.

Fig. 7. Oxygen cylinder valve.

Table 2. Variations in Oxygen Cylinder Pressures With Temperature Changes

Gauge pressures indicated for varying temperature conditions on a full cylinder initially charged to 2200 psi at 70 °F. Values identical for 244 cu. ft. and 122 cu. ft. cylinder.

Temperature Degrees F.	Pressure psi approx.	Temperature Degrees F.	Pressure psi approx.
120	2500	30	1960
100	2380	20	1900
80	2260	10	1840
70	2200	0	1780
60	2140	−10	1720
50	2080	−20	1660
40	2020		

Courtesy Airco Welding Products

Table 3. Oxygen Cylinder Content

Gauge Pressure psi	Content Cu. Ft.	Gauge Pressure psi	Content Cu. Ft.
190	20	1285	140
285	30	1375	150
380	40	1465	160
475	50	1550	170
565	60	1640	180
655	70	1730	190
745	80	1820	200
840	90	1910	210
930	100	2000	220
1020	110	2090	230
1110	120	2200	244
1200	130		
122-cu. ft. cylinder content one-half above volumes.			

Courtesy Airco Welding Products

combustible ones) can cause explosions. Always store the cylinders with the valves up.

ACETYLENE CYLINDERS

Acetylene cylinders are available commercially in sizes ranging from a capacity of 10 cubic feet up to 180-360 cubic feet. Cylinders are available on lease or are sold on an exchange basis. Fig. 8 illustrates the construction of a typical acetylene cylinder. Note the removable cap cover over the valve at the top.

Acetylene cylinders must always be stored in an upright position. Damaged cylinders must be immediately removed to the outside of the buildings. No chances should be taken with a leaking acetylene gas cylinder.

The discharge rate of an acetylene cylinder should not be excessive. An excessive rate of discharge results in the drawing off of acetone which results in strong weakening effects on the weld. The discharge rate should be controlled as a ratio to the total cubic-foot capacity of the cylinder (e.g. a 50 cubic feet per hour discharge rate for a cylinder having a 275 cubic-foot capacity).

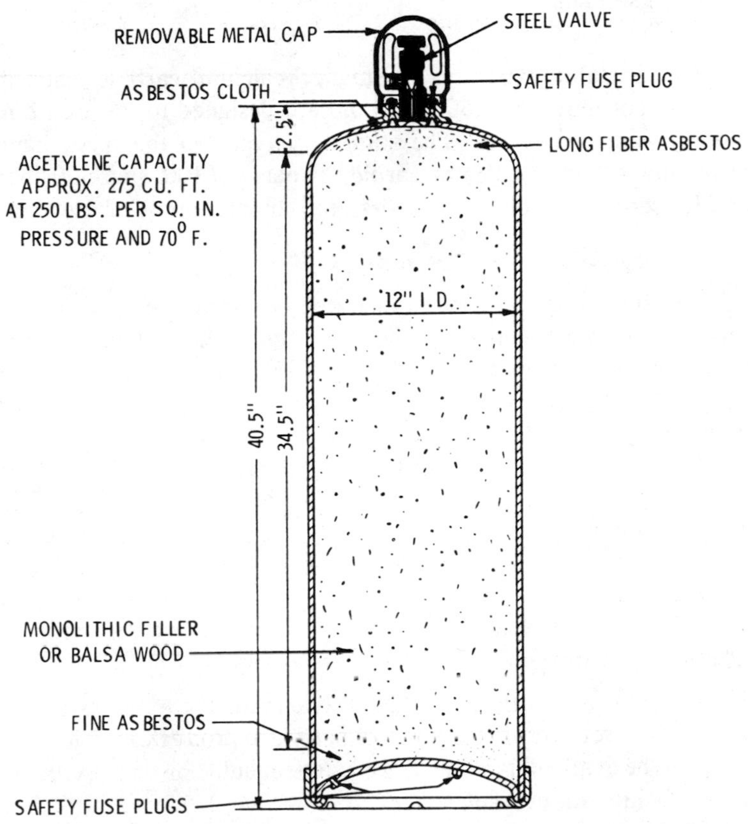

Courtesy Airco Welding Products

Fig. 8. Acetylene cylinder.

Portable Acetylene Generators

An acetylene generator is a device designed to produce large quantities of acetylene gas for welding use. The acetylene gas once produced is piped to the various welding stations.

Acetylene is produced by combining water with calcium carbide. This can be accomplished by adding water to large amounts of calcium carbide or by adding calcium carbide to large volumes of water. The latter is the most extensively used.

Acetylene generators are of two basic types:

1. Stationary,
2. Portable.

The portable type has a smaller calcium carbide capacity (usually not more than 50 lbs.) and are designed to be used both indoors and in the field. The stationary type, on the other hand, has a much higher calcium carbide capacity (300 and 500 lbs.) and functions as a source (indoors) for large amounts of acetylene.

Stationary Acetylene Generators

The stationary acetylene generator is constructed so that the carbide and water are kept separate until the generator is in operation. This is an added safety feature, since it prevents water from splashing into or near the carbide storage area or connections through which it flows.

The larger stationary acetylene generators are designed for welding shops that require large supplies of acetylene. These generators are shown to provide a supply of purer and hotter acetylene than cylinders can at savings of from 50% to 75% of the cost.

Using an acetylene generator provides the welder with the following advantages:

1. Control over the source of acetylene supply,
2. Control over the quantity of acetylene production,
3. The availability of acetylene on demand,
4. Minimum maintenance,
5. Increased space in the production areas (by eliminating the need for several acetylene cylinders around the shop).

HOSES AND HOSE CONNECTIONS

The hoses used in oxyacetylene and other gas welding processes should be strong, nonporous, flexible, and not subject to kinking. The best hoses are constructed from nonblooming neoprene tubing reinforced with braided rayon. The outer coating should be resistant to oil and grease and tough enough to survive most shop conditions. These hoses may be purchased as single hoses (with ⅛ to ½ inch inside diameters), or as twin double-barreled hoses (one line for the oxygen the other for the fuel gas) with metal binders at the base. Gas hoses can be purchased in continuous length to 300 ft. Standard reel lots are approximately 100 ft. In addition, cut lengths (with connections) are available in boxed lengths of 12½ ft., 25 ft., and 100 ft. A short section of twin hose is shown in Fig. 9.

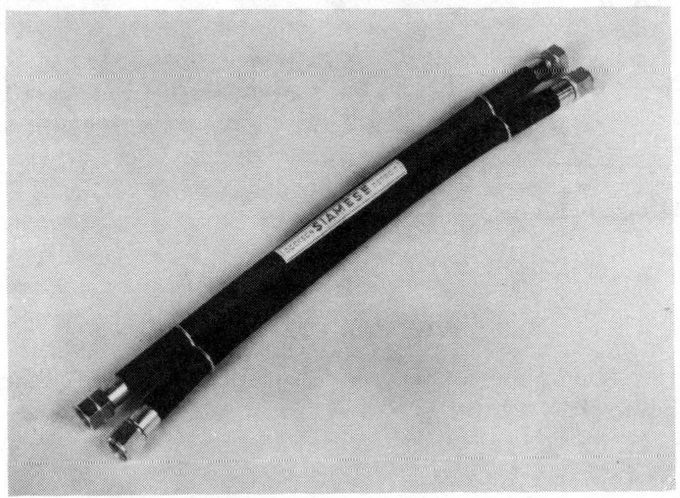

Courtesy Dockson Corp.

Fig. 9. A short section of twin welding hose.

The oxygen hose is usually black or green. The fuel gas hose is generally red. The color distinctions are for safety reasons. For example, using an oxygen hose to carry acetylene could cause a serious accident. No hose should be used for any purpose other than its initial use. A further safety precaution is the design of the threading on the hose connections. The fuel gas hoses have left-handed threaded connections (and a groove around the outside); whereas the oxygen hoses are fitted with right-handed thread connections, and swivel nuts at both ends.

A metal clamp is used to attach the welding hose to a nipple. A nut on the other end of the nipple is connected to the regulator or torch. Sometimes the shape of the nipple will distinguish its use. If such is the case, a bullet-shaped nipple is generally used for oxygen hoses and a nipple with a straight taper for gas fuel hoses. Another method is a groove running around the center of the acetylene nut. This is an indication of a left-hand thread which will only screw into the acetylene outlet (Fig. 10).

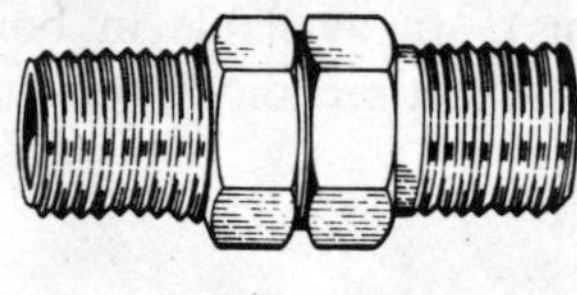

Fig. 10. Acetylene hose coupling and hose connection.

WELDING TORCHES

The welding torch (Fig. 11) is a tool designed to premix a combustible gas with oxygen in certain specified proportions. The

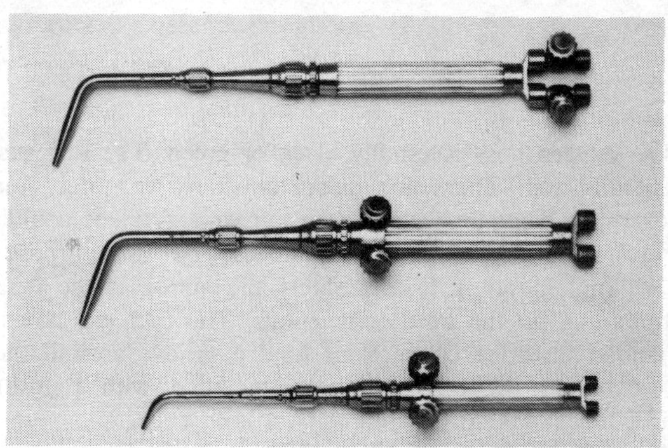

Courtesy Veriflo Corp.

Fig. 11. Typical welding torches.

mixture is ignited and burned at the end of a welding tip attached to the torch. The flame is directed at the area to be welded until the metal reaches the desired temperature.

Torches differ in design from one manufacturer to another but it is possible to list certain features that almost all of them have in common. Torches used in oxyacetylene welding share the following basic components:

1. A handle or body,
2. A mixer or mixing chamber,
3. Extensions,
4. The torch tip.

At the back of the handle are found the valve assemblies. These can be detachable or an integral part of the handle. Fig. 12

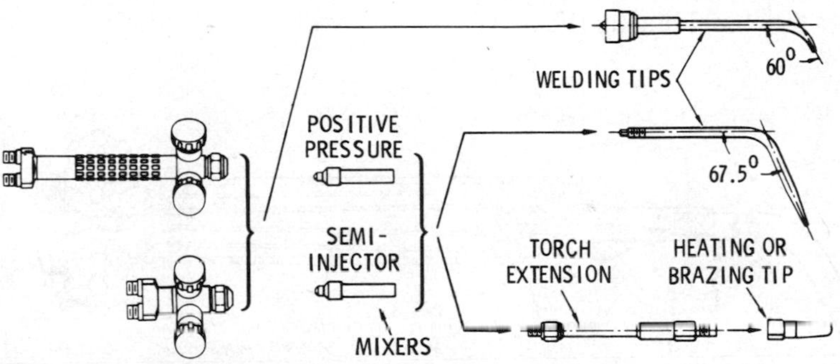

Fig. 12. Two lightweight welding torches illustrating the variety of different attachments that can be fitted to the handle.

illustrates the various attachments and component parts of a typical welding torch.

Some welding torches greatly reduce the use of threaded joints by being completely assembled into a single one-piece unit by silver-soldered construction. This eliminates the need for tightening or repairing threaded joints.

The basic types of welding torches are:

1. The equal-pressure type,
2. The injector type.

The equal-pressure type welding torch is also referred to as the medium-pressure type torch, the balanced-pressure type torch, or the positive-pressure type torch. This torch is used with cylinder gases. Both the oxygen and fuel gas must be delivered with sufficient pressure to force them into the mixing chamber. The equal-pressure torch is not as subject to flashbacks as the injector type. In addition, its flame is more readily adjustable.

The equal-pressure type welding torch in oxyacetylene welding is a relatively simple device consisting of the following principal parts:

1. A handle,
2. A rear end,
3. Two needle valves,
4. Two tubes for the oxygen and acetylene (fuel gas),
5. A mixer,
6. A welding tip (Fig. 13).

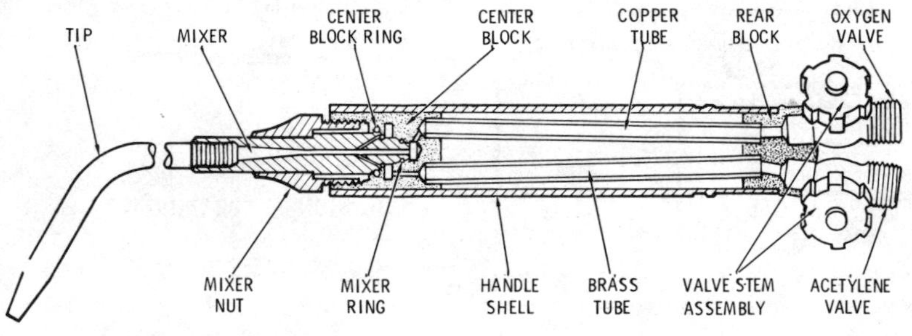

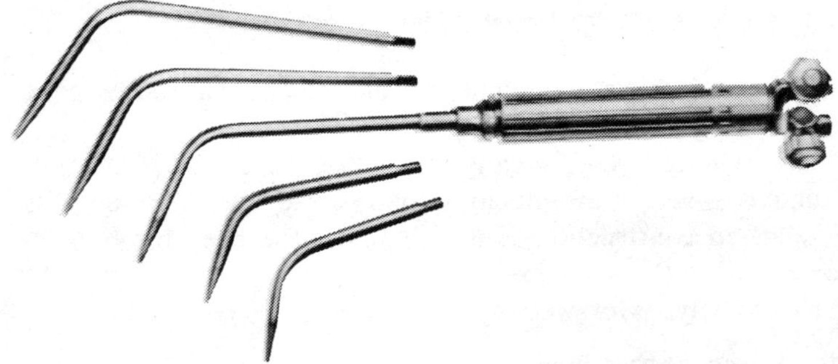

Courtesy Dockson Corp.

Fig. 13. An equal-pressure welding torch.

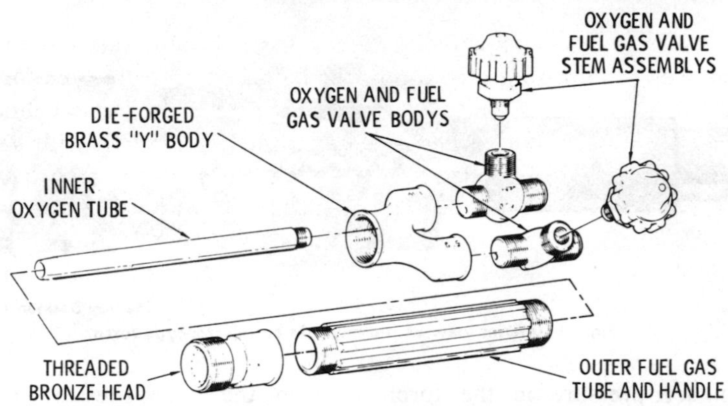

Courtesy Victor Equipment Co.

Fig. 14. A welding torch with an oxygen tube located within a larger fuel gas tube.

The oxygen tube is located within a larger tube carrying the fuel gases in some welding torches (Fig. 14).

The injector-type welding torch (Fig. 15) is also referred to as a low-pressure type torch. Because the injector-type welding torch can be operated at low acetylene pressures, it is frequently used with acetylene generators. These generators are capable of producing large quantities of acetylene, but only low pressure acetylene.

The most distinctive feature in the design of an injector-type welding torch is the construction of the mixing chamber. The oxygen enters the mixing chamber through a passage located in the center of the torch. The oxygen passage is surrounded by the one carrying the acetylene. The oxygen, under considerably

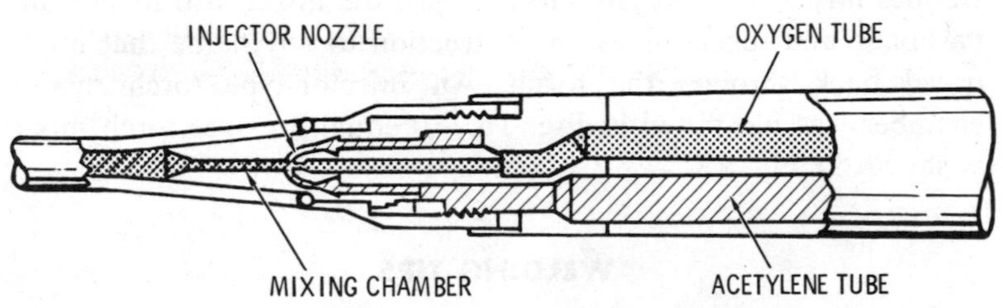

Courtesy Victor Equipment Co.

Fig. 15. Injector-type welding torch.

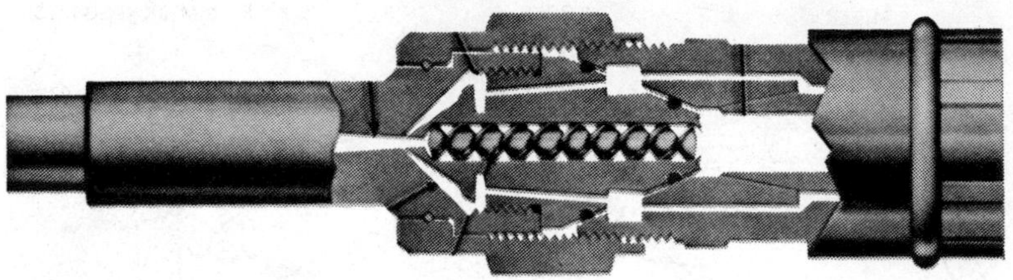

Courtesy Dockson Corp.

Fig. 16. Interior view of the mixer in an injector-type torch.

higher pressure in the torch than in the equal-pressure type, enters the mixing chamber and pulls (or draws) the acetylene in after it (Fig. 16).

Torches may also be classified by the type of welding they are expected to do. On this basis, welding torches can be divided into: (1) heavy duty types, (2) light and medium types, and (3) lightweight types. They can further be modified to meet the specific requirements of the job by a number of interchangeable attachments such as tips, extensions, etc. See Fig. 17.

MIXER

The function of the mixer is to mix, or combine, the oxygen and fuel gas into the form that will feed the torch flame. The mixer may be either a part of the torch body or represent an independent unit, depending on the construction. If the latter is the case, the mixer may be changed each time the tip is replaced. Besides mixing the oxygen and fuel gas, the mixer also blocks any flashback and functions as an obstruction to any flame that might travel back through the torch. An injector-type torch mixing chamber was illustrated in Fig. 16. An equal-pressure torch mixer is shown in Fig. 18.

WELDING TIPS

The welding tip (Fig. 19) is located at the end of the torch and contains the opening through which the oxygen and fuel gas mixture

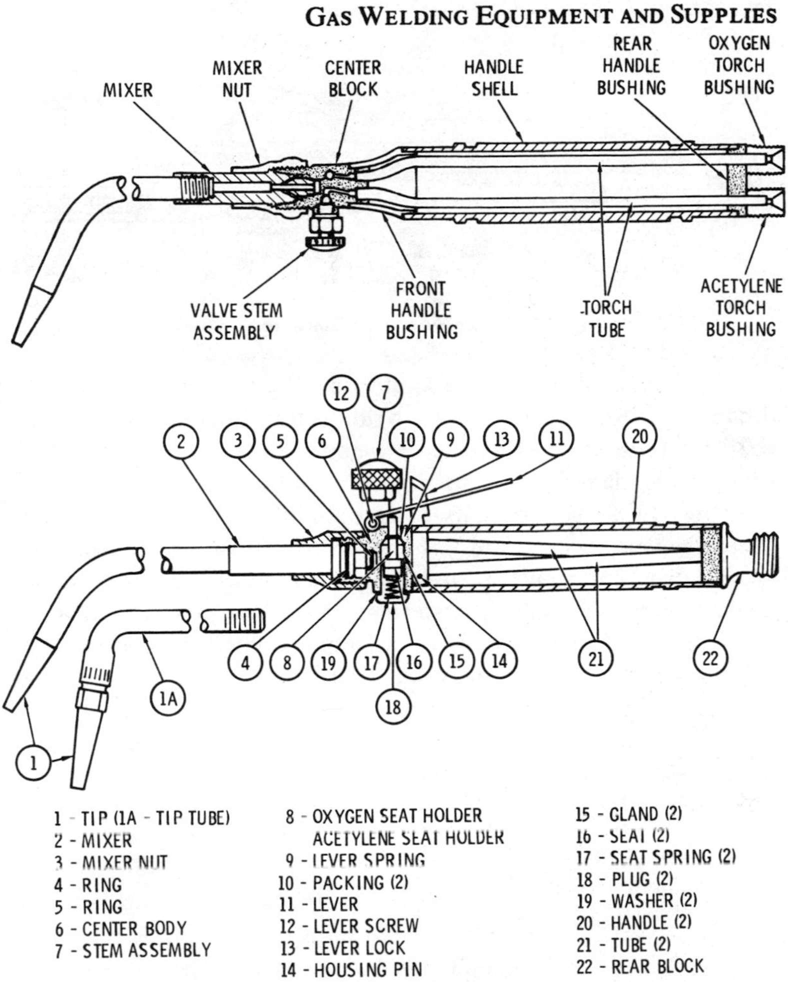

Fig. 17. Exploded view of light-weight torches.

feeds the flame. Between the welding tip and the main body of the torch, there may be an extended length of tubing called the welding torch tube or tip tube. Welding tips may be purchased in a wide variety of sizes and shapes. The suitability of a particular welding tip design depends upon a number of factors including the accessibility of the area being welded, the rate of welding speed desired, and the size of the welding flame required for the job.

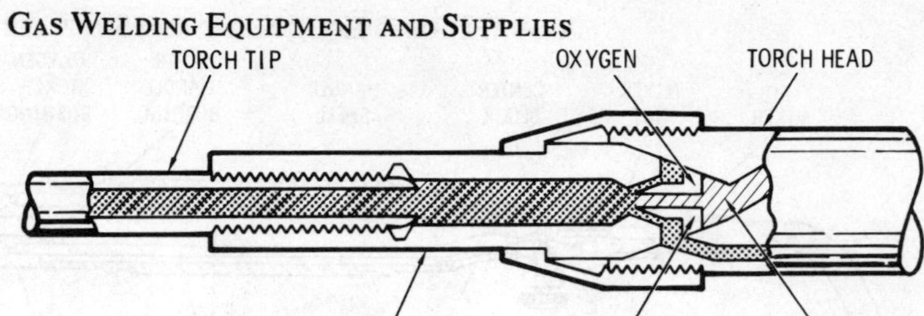

Courtesy Victor Equipment Co.

Fig. 18. Equal-pressure type welding torch mixer.

The majority of torch manufacturers give various numerical designations to their respective welding tips. Such tip size identifications have no bearing on minimum or maximum gas consumption or flame characteristics obtainable.

Drill size alone also fails to give an adequate comparison between the various makes of welding tips with identical tip drill size, since internal torch and tip construction may vary gas exit velocities as well as gas pressure adjustments.

The selection of the type of welding tip is essentially a matter of the purchaser's preference except where special operations definitely recommend either the straight or bent tips.

Factors important to the selection of the appropriate welding tip include:

1. The position of the weld,
2. The material being welded,
3. The thickness of the material,
4. The type of joint.

The size of a welding tip is determined by the diameter of the opening.

Because tips are subject to wear, they must be replaced from time to time. Use an appropriate wrench for this purpose (never pliers). Such malfunctions as "backfire," "blow," and "popping out" will be greatly (if not completely) reduced by using the appropriate tip at the recommended pressure.

Table 4 contains data for the selection of welding tips. These are recommended sizes, and all variables should be taken into consideration before making the final selection.

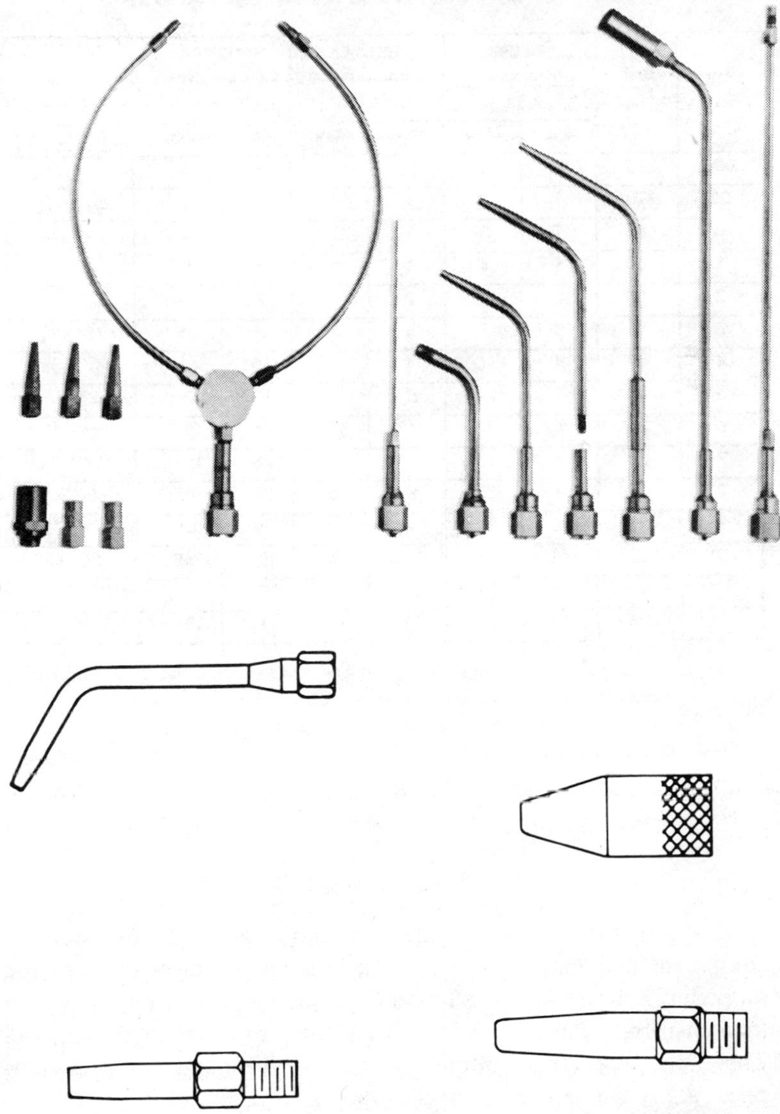

Courtesy National Cylinder Gas, Div. of Chemetron Corp.
Fig. 19. Various welding tip designs.

Table 4. Oxyacetylene Welding Tip Data

Tip Size	Drill Size	Oxygen Pressure P.S.I.		Acetylene Pressure P.S.I.		Acetylene Consumption CFH*		Metal Thickness
		Min.	Max.	Min.	Max.	Min.	Max.	
000	75	1/4	2	1/2	2	1/2	3	up to 1/32"
00	70	1	2	1	2	1	4	1/64" - 3/64"
0	65	1	3	1	3	2	6	1/32" - 5/64"
1	60	1	4	1	4	4	8	3/64" - 3/32"
2	56	2	5	2	5	7	13	1/16" - 1/8"
3	53	3	7	3	7	8	36	1/8" - 3/16"
4	49	4	10	4	10	10	41	3/16" - 1/4"
5	43	5	12	5	15	15	59	1/4" - 1/2"
6	36	6	14	6	15	55	127	1/2" - 3/4"
7	30	7	16	7	15	78	152	3/4" - 1 1/4"
8	29	9	19	8	15	81	160	1 1/4" - 2"
9	28	10	20	9	15	90	166	2" - 2 1/2"
10	27	11	22	10	15	100	169	2 1/2 - 3"
11	26	13	24	11	15	106	175	3" - 3 1/2"
12	25	14	28	12	15	111	211	3 1/2" - 4"

*Oxygen consumption is 1.1 times the acetylene under neutral flame conditions. Gas consumption data is merely for rough estimating purposes. It will vary greatly on the material being welded and the particular skill of the operator. Pressures are approximate for hose length up to 25 ft. Increase for longer hose lengths about 1 psi per 25 feet.

Courtesy Victor Equipment Company

MANIFOLDS

Oxygen and fuel gases are sometimes piped to the welding equipment and machines rather than being fed directly from the individual cylinders. The cylinders are attached to manifolds which then distribute the oxygen and fuel gases at a reduced pressure. There are separate manifolds for the oxygen and fuel gas supplies. As is the case with welding hoses, a manifold must only be used with the gas for which it is designed.

One purpose for distributing the oxygen and fuel gases through manifolds is to centralize the supply. This eliminates the problem of cylinders being scattered throughout the work area and thereby reduces the possibility of accidents. It also increases the available work space. Manifolds are also necessary in those situations where the oxygen or fuel gas consumption, by machines and equipment, is too great to be supplied by individual cylinders.

Oxygen manifolds should never be located near fuel-gas manifolds or areas used for the storage of calcium carbide. It would be best to keep these supplies stored in separate rooms. These rooms should also be of fire-resistant construction. If it is not possible to obtain separate storage rooms, then a fire-resistant partition should be erected between them. In any case, the manifolds should be located at least 50 ft. apart.

Oxygen manifolds having a total connected capacity of more than 6000 cu. ft., and fuel-gas manifolds having a similar capacity in excess of 2000 cu. ft., should be located outside and preferably in separate storage sheds or rooms. Use only approved manufactured manifolds. Manifolds made in the shop can prove to be highly dangerous.

Manifolds can be classified as either portable manifolds or stationary manifolds. A portable manifold (Fig. 20) is simply

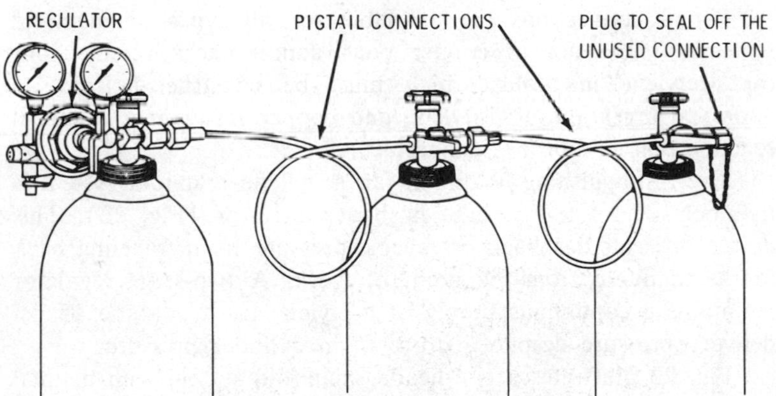

Fig. 20. Portable extendable manifold.

two or more cylinders connected by a single pressure-reducing regulator and pigtail connections. There is no fixed pipeline, as is the case with the stationary manifold. A stationary manifold (Fig.

21) consists of a number of cylinders attached to a stationary pipe by means of pigtail connections.

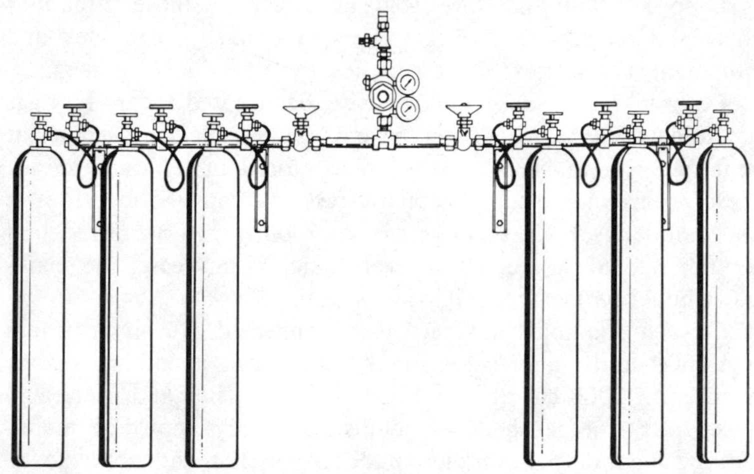

Courtesy National Cylinder Gas, Div. of Chemetron Corp.

Fig. 21. A manifold system with a manual control for oxygen, hydrogen, nitrogen, helium, and argon gas service.

Pigtail connections are available for all types of manifold systems. These are of copper construction except connections for acetylene manifolds which must be constructed from an approved brass alloy. Acetylene and copper result in a chemical reaction that could prove dangerous.

The distinguishing feature of an acetylene manifold system is the inclusion of a hydraulic flashback arrestor (Fig. 22). This device prevents flashback or reverse pressure from traveling back to the manifold from the point of work. A two-stage regulator maintains a continuous supply of acetylene up to a static 15 psi delivery pressure despite a drop off in cylinder pressure.

Fig. 23 illustrates a two-header manifold system with manual control for acetylene service. Note the location of the hydraulic flashback arrestor.

Special elbow assemblies can be purchased for both oxygen and fuel-gas manifold systems to enable corner installation (Fig. 24).

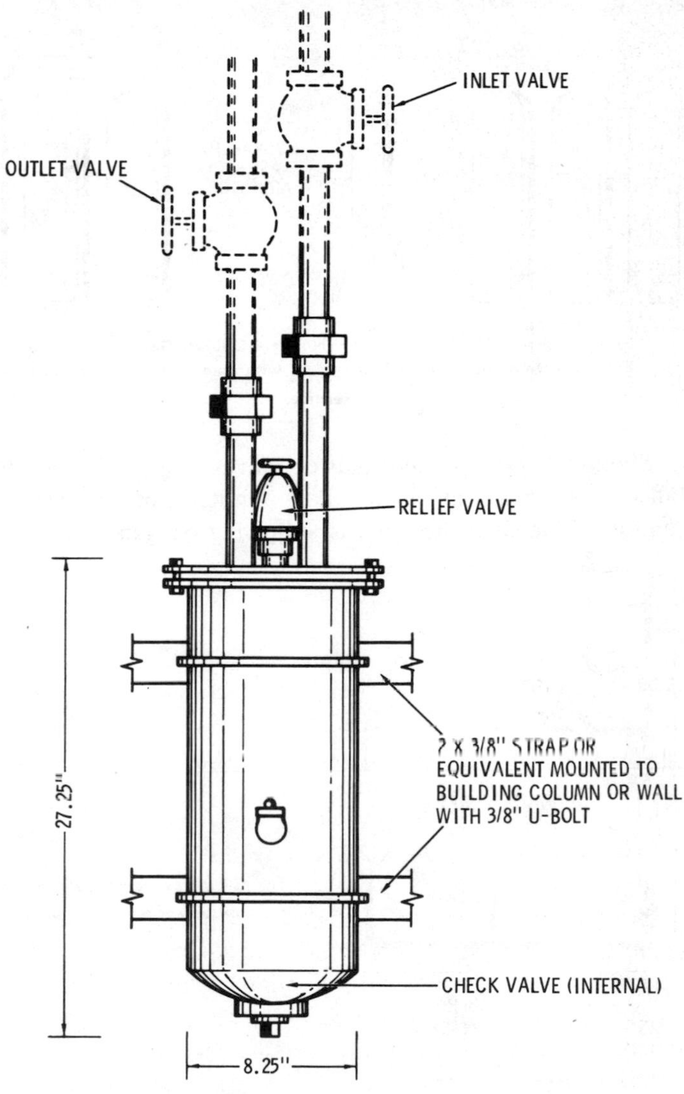

Courtesy National Cylinder Gas, Div. of Chemetron Corp.
Fig. 22. Hydraulic flashback arrestor.

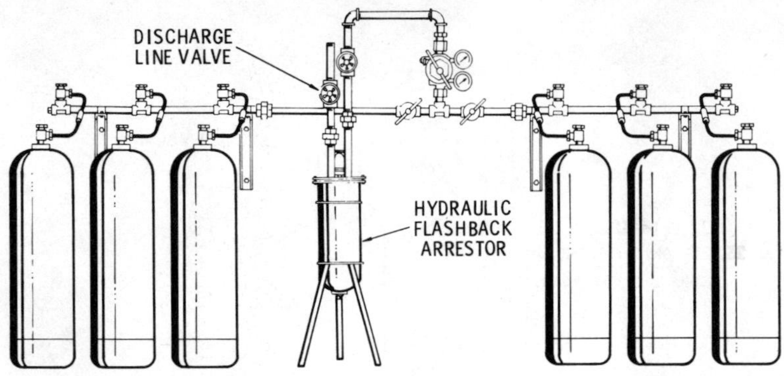

Courtesy National Cylinder Gas, Div. of Chemetron Corp.

Fig. 23. A two header manifold system with manual controls for acetylene service.

Manifold systems have master valves (Figs. 25 and 26) installed in such a way as to allow servicing of one side of the manifold while the other side is supplying oxygen or fuel gas.

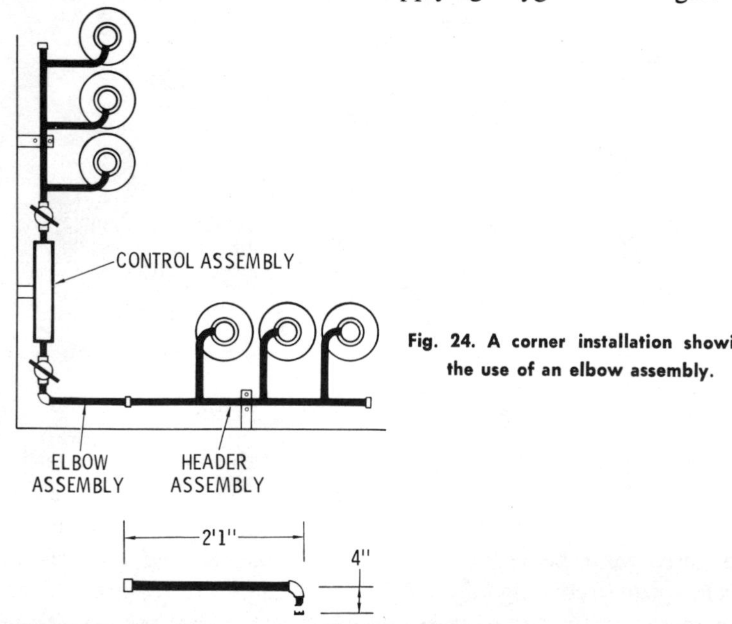

Fig. 24. A corner installation showing the use of an elbow assembly.

Courtesy National Cylinder Gas, Div. of Chemetron Corp.

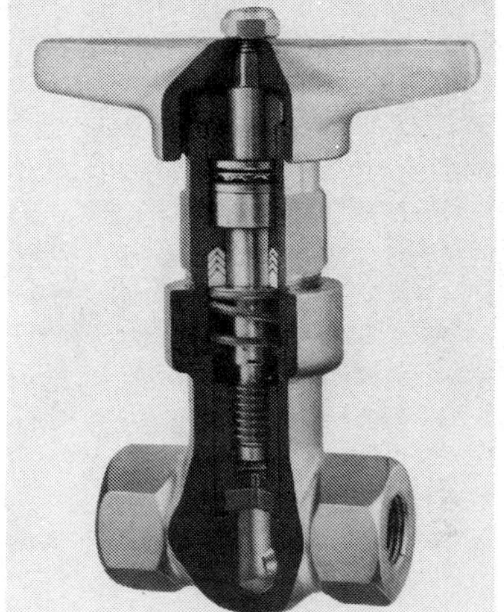

Fig. 25. Shutoff valve used for high pressure oxygen.

Courtesy Bastian-Blessing Co.

Fig. 26. Low pressure oxygen valve.

Courtesy Bastian-Blessing Co.

GAS PRESSURE REGULATORS

The pressure of the gases obtained from cylinders, generators, manifolds, and pipelines is considerably higher than the gas pressure used to operate the welding torch. Since the pressure of the gases must be reduced between the source and the welding torch, special reducing valves called regulators are required. There is a regulator for the oxygen and another for the fuel gas. These

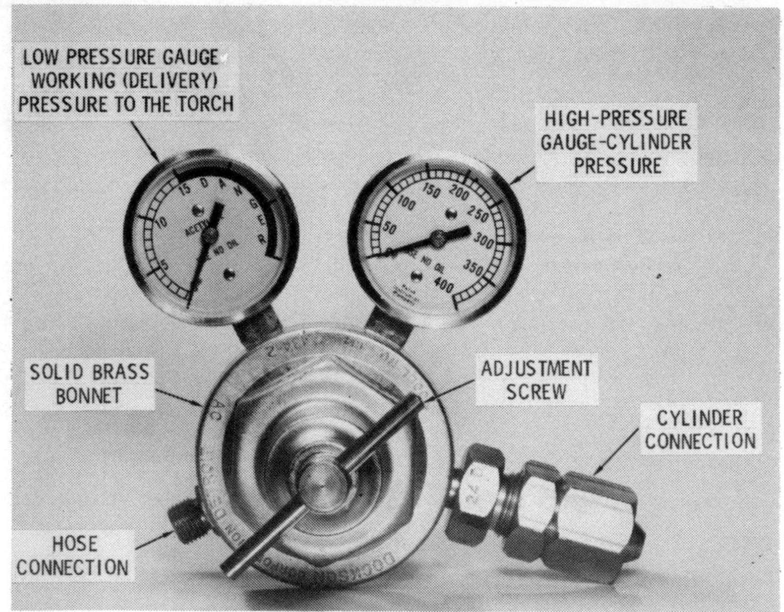

Courtesy Dockson Corp.

Fig. 27. Single-stage regulator (acetylene).

gas pressure regulators are connected between the gas source (cylinder, generator, etc.) and the hose leading to the torch.

Many regulators are constructed with two gauges. One gauge indicates the pressure of the gas in the cylinder (the high-pressure gauge) and the other indicates the working pressure of the gas being delivered to the torch (the low-pressure gauge). Figs. 27 and 28 show examples of fuel gas regulators (acetylene) constructed with two gauges. Note that the maximum pressure of the high-pressure gauge of the single stage regulator (Fig. 27) is considerably below that shown on the high-pressure gauge of the two-stage regulator (Fig. 28). This will be explained further on in this section. Note also that the maximum delivery pressure for the low pressure gauge is 15 psi. Acetylene at pressures in excess of 15 psi becomes very unstable.

Compare the fuel gas regulators shown in Figs. 27 and 28 with the two-stage oxygen regulator in Fig. 29. The low-pressure gauge

Courtesy Airco Welding Products

Fig. 28. Two-stage regulator (acetylene).

is a distinguishing feature on each of these two types (oxygen and fuel gas) of gauges. Differences in connection threads is another distinguishing feature.

There are also gaugeless regulators, and regulators with a single gauge. An example of the latter, a pipe-line station regulator, is shown in Fig. 30.

Fundamentally all gas pressure regulators reduce higher inlet pressures to lower working pressures in the following simple way: an internal seat closes or opens a nozzle orifice through which passes the high-pressure gas into a low-pressure regulator chamber. The seat is fastened onto a yoke device or is actuated in some other manner by a diaphragm (the downward pressure of which is spring-adjustable and the upward or seat-closing movement occasioned by the gas accumulating below its surface) in the low pressure chamber, equalizing or slightly exceeding the

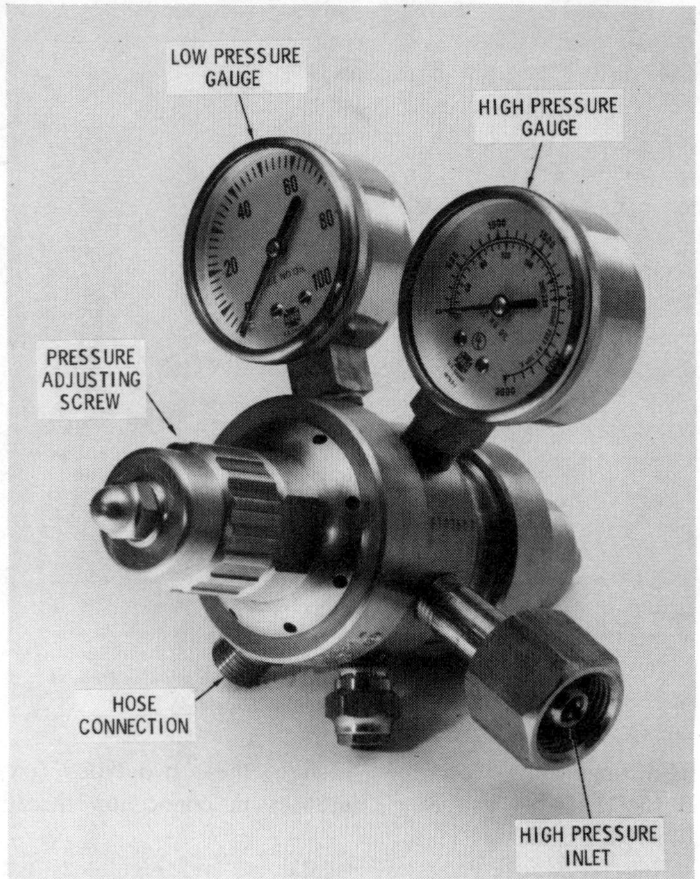

Courtesy Veriflo Corp.

Fig. 29. Two-stage oxygen regulator.

upper spring tension. In other words, the gas is admitted to the low-pressure side through a valve attached to a diaphragm whose movements are automatically controlled by a spring and unbalanced pressures on the diaphragm.

Fig. 31 illustrates the operation of a very simple type of gas pressure regulator. The high-pressure gas from the source (cylinder, generator, etc.) enters the high-pressure gas inlet swivel and inlet passage (A) terminating in the regulator nozzle (C). This nozzle

Courtesy Airco Welding Products

Fig. 30. A single-gauge regulator.

opening is closed by the regulator seat (D) which is held firm by a yoke (E) and securely fastened into a suitable center piece of the regulator diaphragm (G). Above this diaphragm rests a

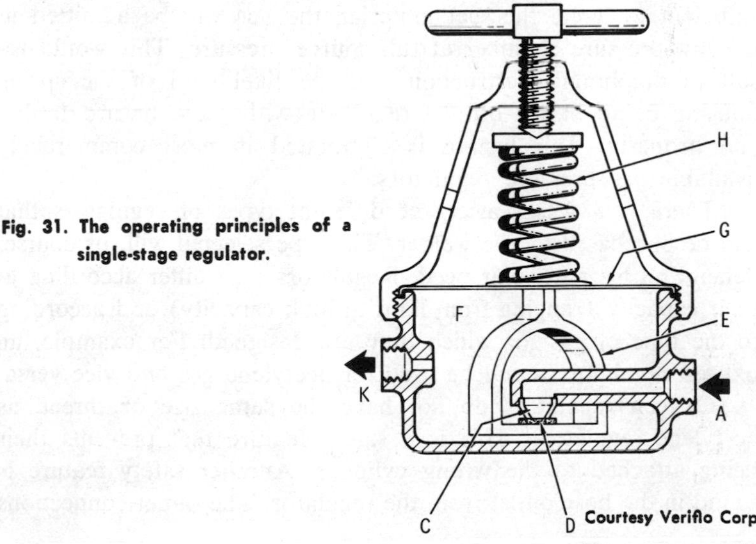

Fig. 31. The operating principles of a single-stage regulator.

Courtesy Veriflo Corp.

spring (H) which is compressible by the regulator tension screw (I). By screwing this tension screw downward (into the regulator housing), the spring tension is increased and in turn forces both diaphragm and seat yoke downward. This causes the opening of the nozzle and permits high-pressure gas to flow into the regulator low-pressure chamber (F). When the gases accumulating in the low-pressure chamber (F) equal in pressure the downward pressure exerted by the spring on the diaphragm, the diaphragm moves upward and pulls with it the seat carrying yoke and the seat; the latter now again closing the nozzle opening.

If the torch needle valve (or any other needle valve) is opened allowing the gas in the regulator low-pressure chamber to escape through the outlet passage (K), the upward pressure against the diaphragm is diminished and the spring tension will force it downward to allow this initial cycle to repeat itself.

It should be pointed out that the pressure regulator described in Fig. 31 is the simplest and most basic type. While it lends itself well to a description of the basic operating principles of a gas pressure regulator, it would not be considered an acceptable regulator in actual service. The major objection is the possibility of the sudden heating of residual air in the regulator by compression. If, as a result of this heating, the air reaches a temperature sufficient to ignite the seat material, the gas will be admitted to the low-pressure chamber at full source pressure. This would result in diaphragm destruction and the likelihood of the spring housing being blown off the regulator with great hazard to life and property. This hazard is eliminated in most commercially available gas pressure regulators.

There is a wide variety of different types of regulators that can be purchased by the welder. The type selected will, of course, depend on his particular need. Regulators may differ according to their capacity (ranging from light to high capacity) and according to the type of gas for which they are designed. For example, an oxygen regulator *cannot* be used for acetylene gas and vice versa.

Oxygen regulators do not have the same size or thread as acetylene regulators. This is a safety feature that prevents their being attached to the wrong cylinder. Another safety feature is found in the hose outlet from the regulator. The outlet connections for oxygen hoses have right-hand threads. For acetylene and other fuel gases the outlet connections have left-handed threads. Caution should still be exercised, however, because older equipment did not offer these safety precautions.

Fig. 32. A two-stage regulator.

SINGLE-STAGE VERSUS TWO-STAGE REGULATORS

The two basic types of gas pressure regulators are:

1. The single-stage regulator,
2. The two-stage regulator (Fig. 32).

A basic single-stage regulator was described in Fig. 31. The two-stage regulator is virtually two regulators in one which operate to reduce the pressure progressively in two stages instead of one. A major objection to the single-stage regulator is the need for torch adjustments. As the cylinder pressure falls, the regulator pressure likewise falls necessitating torch adjustment. In the two-stage regulator, there is automatic compensation for any drop in cylinder pressure. Another objection to the use of single-stage regulators is their tendency to freeze in cold weather. Any regulator is essentially an expansion valve. A sudden expansion and resulting drop in initial pressure causes rapid cooling of the gas involved. Some gases are more refrigerant than others. Oxygen, for instance, carries a substantial percentage of moisture which, upon reaching freezing temperature, deposits itself on or near the regulator nozzle in the form of ice. This ice causes irregular seating of the seat on the nozzle and results in substantial pressure fluctuations. The same phenomenon occurs in the first reduction stage of a two-stage regulator but its second stage is not sufficiently affected to produce a noticeable pressure disturbance.

Single-stage regulators may be used with pipelines and cylinders. Two-stage regulators are used with cylinders and manifolds.

Single-stage gas pressure regulators may be divided into two groups depending on the method used to close the seat mechanism

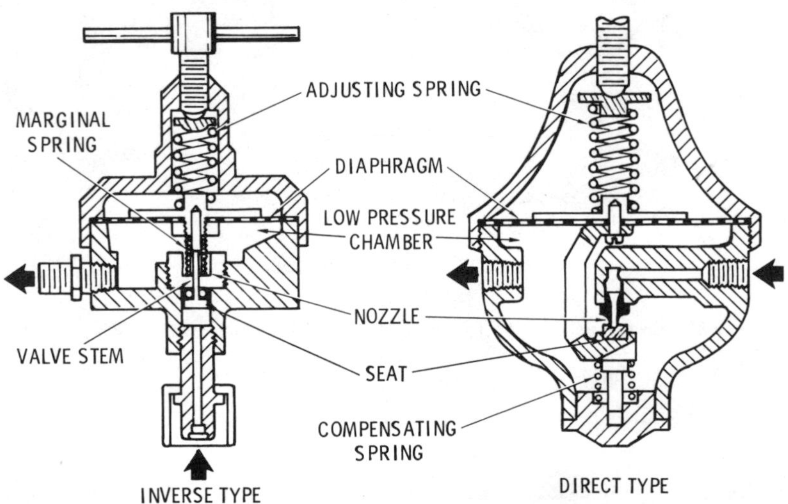

Courtesy Airco Welding Products

Fig. 33. The inverse and direct-type gas pressure regulator.

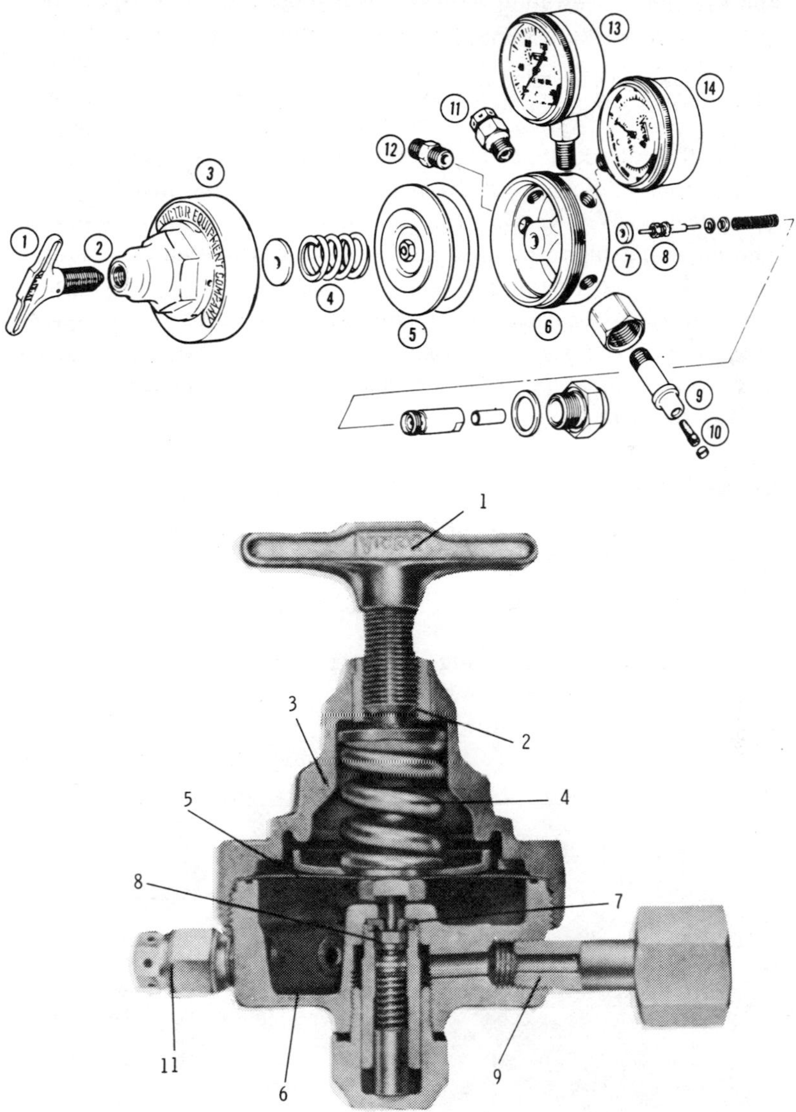

Courtesy Dockson Corp.

Fig. 34. Cut-away view of a single-stage regulator.

and thereby control the gas flow. On the basis of the method used, single-stage pressure regulators may be classified as:

1. Direct or nozzle type,
2. Inverse or stem type (Fig. 33).

In the direct type regulator, the seat is forced to close against the gas flow (away from the nozzle) by the incoming high pressure gas. With the inverse type regulator, the opposite is true. The incoming high pressure gas forces the seat toward the nozzle in the direction of the gas flow.

Two-stage regulators may consist of two direct type stages, two inverse type stages, or one of each. Two-stage regulators of the mixed stage type are the most common. The first stage is usually the direct type.

Single-Stage

The single-stage regulator (Fig. 34) consist basically of a diaphragm, a floating valve, balancing springs and the regulator housing. The diaphragm functions as a flexible dividing wall between two opposing forces—the pre-set pressure above the diaphragm (preset by spring or gas pressure) and the regulated gas pressure below the diaphragm. Single-stage regulators may be either nozzle type or inverse type.

The single-stage gaugeless regulator (Fig. 35) consists basically of a heavy-duty solid forged brass body, a metal diaphragm and

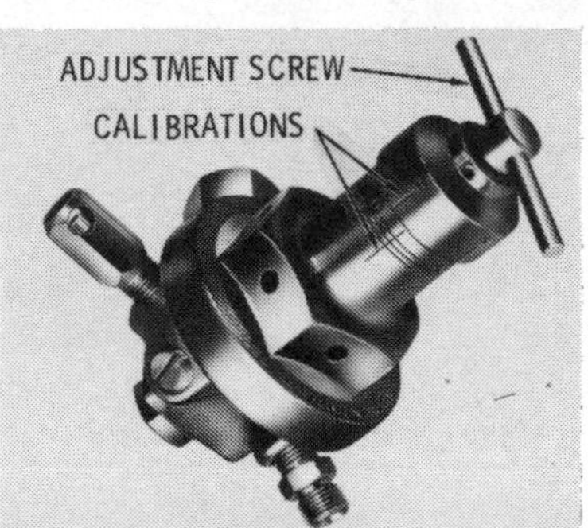

Fig. 35. A single-stage gaugeless pressure regulator.

easily replaceable seats. The working pressure is maintained by adjusting a "T" handle on the pressure regulator cap. In some models, a special viewing hole in the outlet leading to the gas supply enables the welder to determine when the cylinder is empty. This type of regulator is inexpensive to purchase, costs little to maintain, and is not affected by rough handling.

Single-stage gaugeless regulators are designed for use under conditions that require rough handling. The delivery pressure is indicated with a micrometer-type calibration. A high-pressure indicator is recessed in the back of the regulator body and is calibrated to indicate cylinder contents at ¼, ½, ¾ and full.

Fig. 36. A series-line regulator.

Courtesy National Cylinder Gas, Div. of Chemetron Corp.

A series line regulator (Fig. 36) consists of a body constructed from aluminum forgings with an aluminum die cast bonnet. An adjusting screw is provided at the top of the regulator with a nut to lock the setting. These regulators are designed for installation in gas pipelines where large volumes of gas have been reduced from full cylinder pressure by a primary regulator.

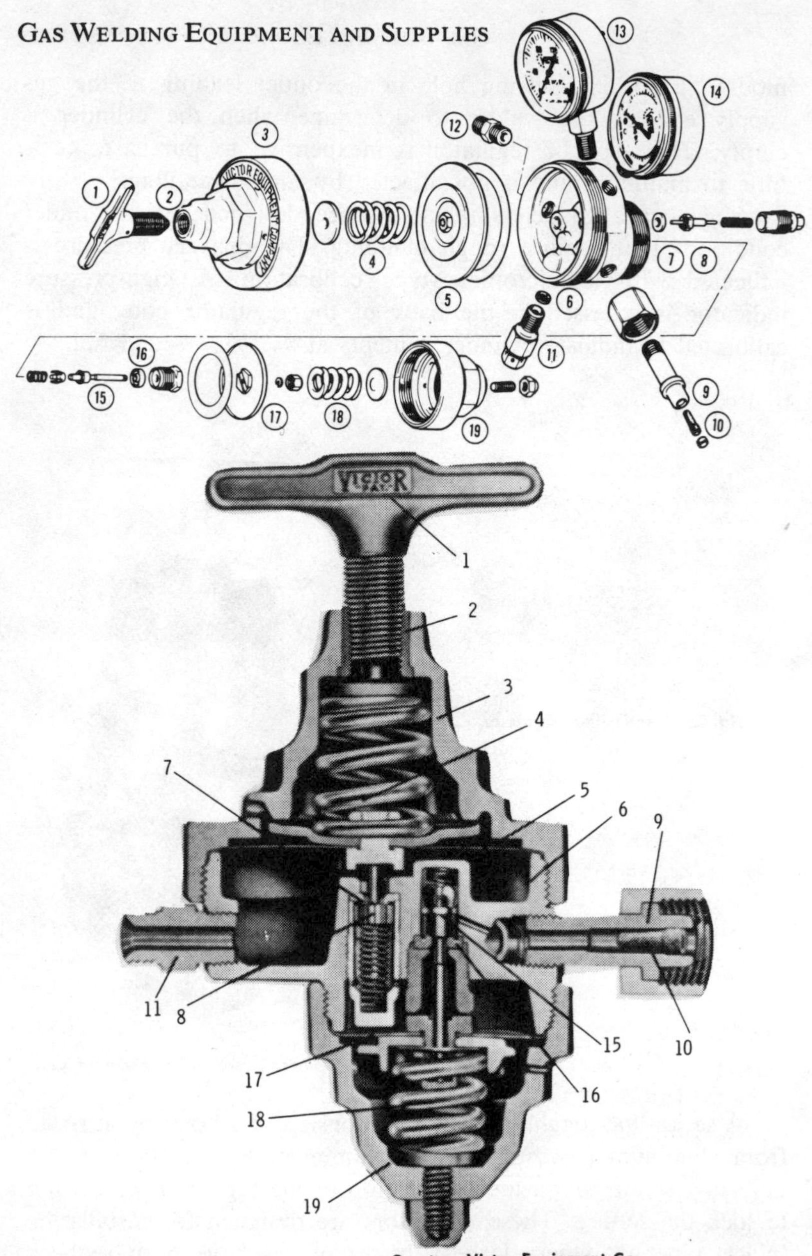

Courtesy Victor Equipment Co.

Fig. 37. Two-stage gas pressure regulator.

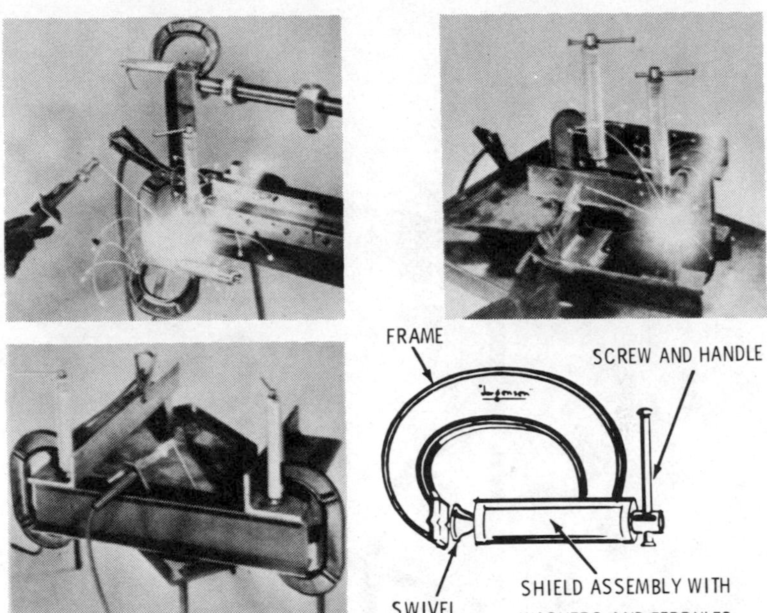

Courtesy Adjustable Clamp Co.

Fig. 38. Welding C-clamps with screw shield assembly.

Two-Stage

Two-stage regulators (Fig. 37) use two separate steps to reduce the pressure between the source and the torch. Two diaphragms and valve assemblies operating independently of one another are responsible for this two-stage reduction. The initial force of the pressure from the source is absorbed by the first of the two regulators. In view of this, the diaphragm and springs in this regulator are of heavier construction than the single-stage regulator. The latter is concerned with reducing a lesser, intermediate pressure.

The first stage in the two-stage regulator serves as a high-pressure reduction chamber. A predetermined pressure is set and maintained by the spring and diaphragm. The gas (at a reduced pressure) then flows into the second of the two stages which serves as a low-pressure reduction chamber. In this second chamber, pressure control is controlled by an adjustment screw.

Courtesy Dockson Corp.

Fig. 39. A cylinder hand truck.

Regulator Safety Features

It is extremely important that regulators never be interchanged in use. For example, an oxygen regulator should never be placed on an acetylene line. The reasons for this are the same as for avoiding the interchanging of welding hoses.

One safety feature designed to prevent the interchange of regulators is a difference in thread size. Oxygen regulators do not have the same size thread as do acetylene regulators. Another safety feature is found in the hose outlet from the regulator. The outlet connections for oxygen hoses have right-hand threads. For acetylene and other fuel gases, the outlet connections have left-hand threads. Caution should still be exercised, however, because older equipment did not offer these safety precautions.

Regulator Maintenance

Proper regulator maintenance is important for both safe and efficient operation. The following points are particularly important to remember:

1. The adjusting screw on the regulator must *always* be released before opening the cylinder valve. Failure to do

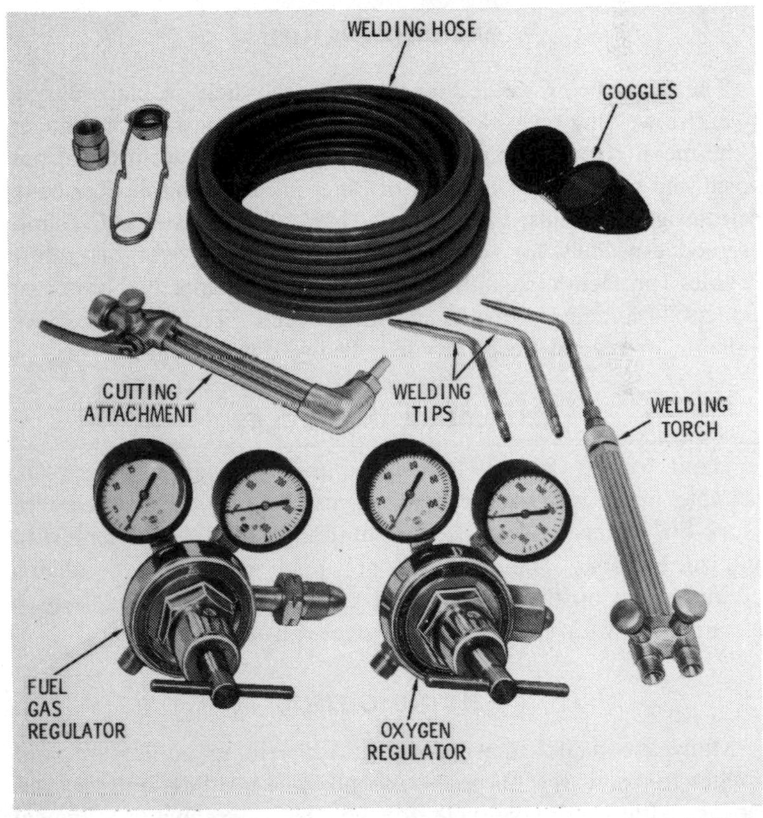

Courtesy Turner Industrial Products

Fig. 40. A typical welding and cutting outfit.

this results in extreme pressure against the gauge which measures the line pressure and may cause damage to the regulator.

2. Gauge-equipped regulators should never be dropped, improperly stored, or, in any way, subjected to careless handling. The gauges are extremely sensitive instruments and can be easily rendered inoperable.
3. *Never* oil a regulator. Most regulators will have the instruction "use no oil" printed on the face of both gauges.
4. All regulator connections should be tight and free from leaks.

WELDING CLAMPS

The metal being welded must be securely held in place during the entire welding operation. Otherwise, there is always the danger of the metal pieces slipping out of place before the weld metal has cooled and solidified. A number of different clamping devices have been designed for just this purpose. Fig. 38 illustrates a C-clamp designed especially for welding. The steel shield over the screw provides protection against weld spatter, hammer blows, or accidental contact with the welding torch. These clamps are available in several deep-throated, heavy duty designs.

CYLINDER HAND TRUCKS

Hand trucks (Fig. 39) designed to carry gas cylinders are available in several sizes ranging from the two-wheeled type to a fork-lift variety constructed to fit under liquid cylinders with rounded bottoms. The cylinders are held in place with chains and supported on the bottom with a steel platform. Models may be purchased with rubber tires or steel-rim wheels.

WELDING OUTFITS

Many companies that manufacture welding equipment and supplies for industrial purposes also produce welding, cutting, and brazing outfits more suitable for smaller operations. A typical welding and cutting outfit is shown in Fig. 40.

Chapter 4

Gas Welding Process

Gas welding is the process of employing a gas fueled torch to raise two similar pieces of metal to their fusion point and allowing them to flow together. A welding (filler) rod is frequently used to deposit additional metal. Oxyacetylene welding (oxygen in combination with acetylene) is the most widely used manually controlled gas welding process. Other less frequently used processes are oxygen-hydrogen welding, oxygen-natural gas welding, and oxygen-LPG welding.

Pressure gas welding is a modification of the gas welding process that shows certain similarities to forge welding. The surfaces to be joined are raised to fusion temperatures by a gas flame. These surfaces remain slightly separated until they reach the fusion point. At that stage, they are forced together under pressure until the weld forms and solidifies.

The equipment and supplies most commonly used by the welder when using a gas welding process are as follows:

1. An oxygen cylinder, regulator, and hoses,
2. A fuel gas cylinder (acetylene, hydrogen, etc.), regulator, and hoses,
3. A welding torch,
4. Proper clothing (safety goggles, asbestos apron, gloves, etc.).

This list will vary both according to the welder's preference as well as specific job requirements. More about gas welding equipment and supplies is found in Chapter 3. A typical gas welding outfit is illustrated in Fig. 1.

SELECTING THE WELDING ROD

Thin metals can be welded by the fused metal on the adjacent parts. For metals over 1/8 inch in thickness, new metal should

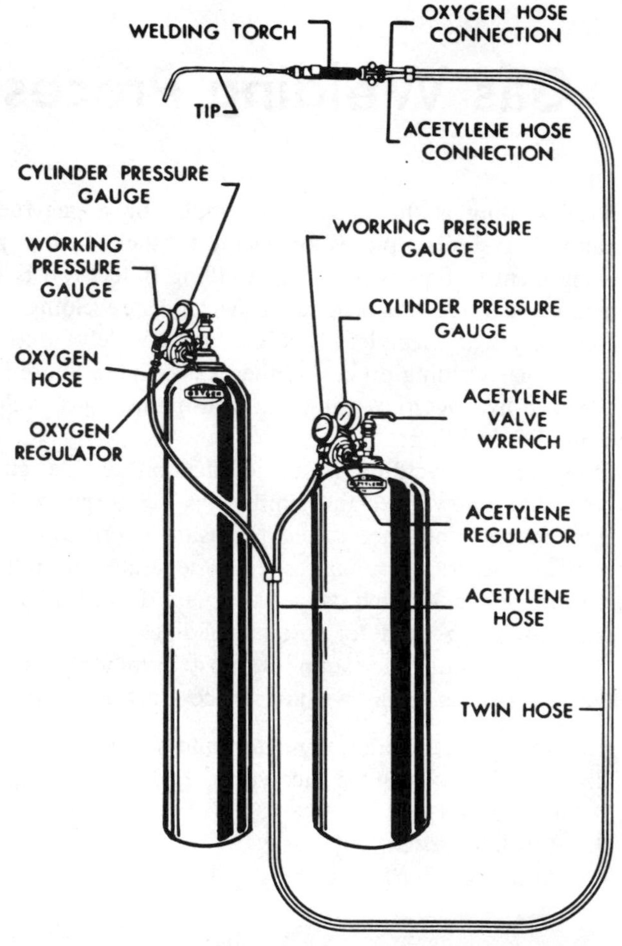

Fig. 1. A typical

be added by melting from the end of a wire or rod known as a welding rod.

A test for the weldability of rods is easily made with the oxyacetylene torch. The rod sample is laid horizontally and partly fused. Its behavior under the flame and its appearance after fusion determine its weldability.

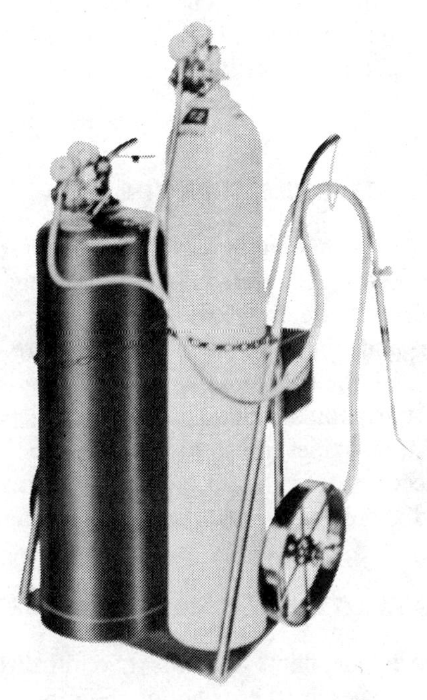

Courtesy Airco Welding Products

gas welding outfit.

A neutral flame (see OXYACETYLENE FLAME CHARACTERISTICS) is used to produce fusion that penetrates to about one half the diameter of the rod. A rod of good weldability will have a surface, after this test is made, somewhat like that of a narrow ripple weld. The metal will have been displaced into smooth blended drops, one merging into another, and none extending much beyond the original rod diameter. There will be little sparking during fusion and no pinholes in the fused metal after cooling.

An unsuitable welding rod will spark excessively during fusion, and the fused metal will spread irregularly, giving evidence of gas inclusions. The fused surface after cooling will be rough and irregular with pinholes. The metal may have a spongy or "bread crust" characteristic in spots. The selection of the correct welding rod depends on the type of metal being welded and the nature of the job.

SELECTING THE FLUX

Fluxes are used in welding cast iron, brass, bronze, aluminum, nickel, monel, and the nonferrous alloys in general, but are rarely or never required in welding low-carbon steel.

Cast iron and the nonferrous metals are welded at lower temperatures and their oxides, as a rule, remain solid and tend to mix with the molten metal. The flux acts essentially as a deoxidizer. These oxides must be removed before a strong weld can be produced. Fluxes should be kept in closed containers when not in use and should, as a rule, be used sparingly for best results.

OXYACETYLENE FLAME CHARACTERISTICS

The oxyacetylene flame exhibits two or three stages of combustion. These stages in the order of their proximity to the torch nozzle are:

1. The inner tip,
2. The beard or brush,
3. The outer nonluminous envelope (Fig. 2).

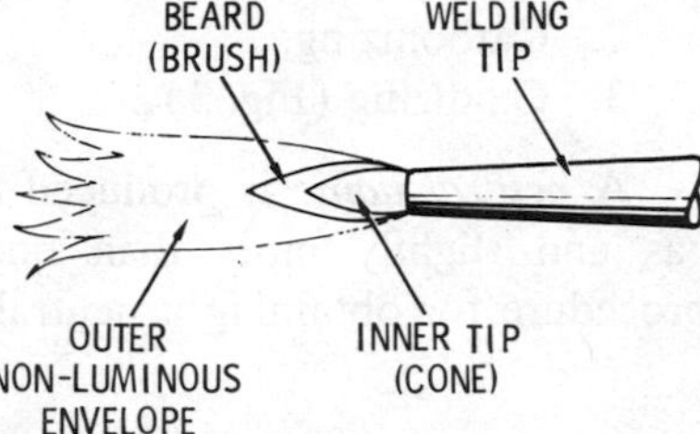

Fig. 2. The three stages of combustion of the oxyacetylene flame.

The *inner tip* is sometimes referred to as the *cone.* The term inner tip is probably more appropriate because it is always located at the bore of the torch nozzle (the innermost position of the flame); however, the inner tip does not always (or even usually) assume the shape of a cone.

The shape of the inner tip of the flame is dependent on several interrelated factors among which one might list:

1. The shape of the bore or nozzle,
2. The cleanliness of the passage,
3. The correct proportion and pressure of the gases.

If all these factors are present, then the inner tip will assume a somewhat conical shape. Under most conditions, one or more of these factors are absent. The result will be a distortion of the conical shape.

The inner tip should never be allowed to touch the work or the results could be disastrous. For example, if the inner tip of the flame touches a steel surface during a welding operation, it will cut the metal instead of welding it and turn it to slag wherever it touches.

The *beard* or *brush* sometimes occurs around the inner tip (cone) and inside the outer envelope. As is the case with the inner tip, the beard should never be allowed to touch the work.

The *outer nonluminous envelope* extends around and beyond both the inner tip or beard. It is usually of considerable volume. This is largely due to the burning of additional oxygen obtained from the surrounding atmosphere which produces both carbon dioxide and water vapor.

The three types of oxyacetylene flames used in welding are:

1. Neutral,
2. Carbonizing,
3. Oxidizing (Fig. 3).

A *neutral flame* is produced by burning one part of acetylene gas and slightly more than one part of oxygen. A suggested procedure for obtaining a neutral flame is as follows:

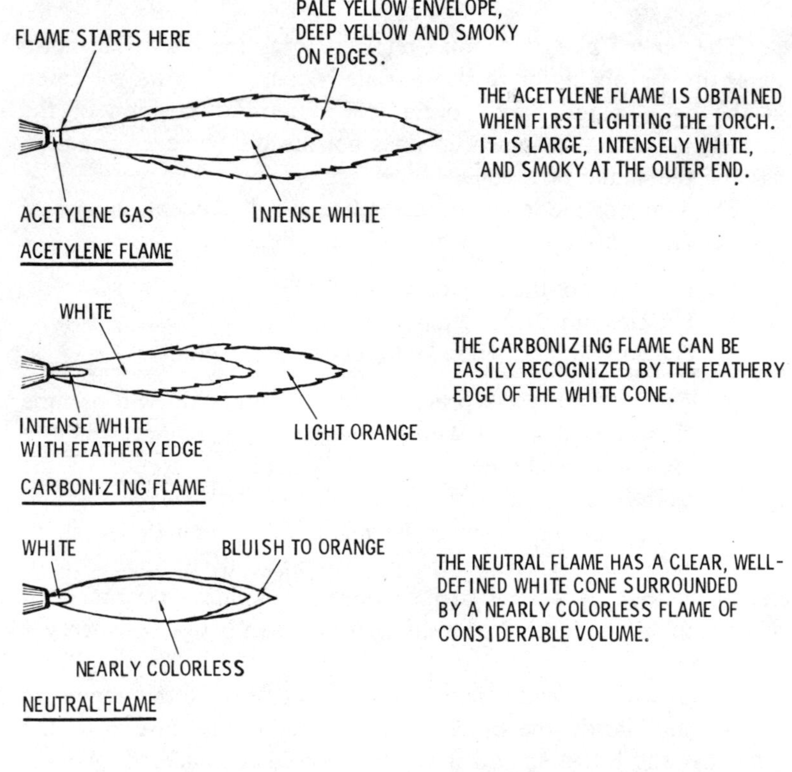

Fig. 3. Types of oxyacetylene flames.

1. Turn the valves on the two tanks of gas so that the minimum amount of oxygen and acetylene enter the torch.
2. Light the torch.
3. Turn the valves on full.
4. Adjust the valves to obtain a flame with a small beard or brush on the inner tip (cone). This is a carbonizing flame and must *never* be applied to ferrous metal (i.e. a metal containing iron elements).
5. Increase the oxygen flow *very* slowly. At a certain point, you will hear a distinct change in the sound of the flame. This will indicate that you have achieved the necessary portion of oxygen to produce a neutral flame. A neutral flame has an approximate temperature of 5850° F.

Courtesy Airco Welding Products

Fig. 4. Cracking the cylinder valve.

A *carbonizing flame* (or *reducing flame*) is produced by burning an excess of acetylene. The flame very often has no beard or brush on its inner tip. When adjusted with a small beard, it may be used on most nonferrous metals (i.e. those not containing iron elements). The outer envelope is usually the portion of the flame used on these metals. A carbonizing flame has an approximate temperature of 5550° F.

An *oxidizing flame* is produced by burning an excess of oxygen. The flame has no beard and both the inner tip and envelope are shorter. Unfortunately, the oxidizing flame is of limited use, being harmful to many metals. An oxidizing flame has an approximate temperature of 6000° F.

Courtesy Airco Welding Products

Fig. 5. Attaching the pressure regulators.

SETTING UP THE GAS WELDING APPARATUS

The beginner should learn a specific procedure for setting up the gas welding apparatus and practice this procedure until it becomes automatic. Mistakes occur when a welder is not consistent in following the various steps for setting up the apparatus in the proper sequence. Figs. 4 through 13 show a typical procedure for setting up gas welding apparatus.

PRACTICE WELDING

The best way to learn to weld is by welding. The beginner should start by practicing with 1/16-inch steel without a welding rod. Two sheets of steel about 6 inches long, 1¼ inches wide,

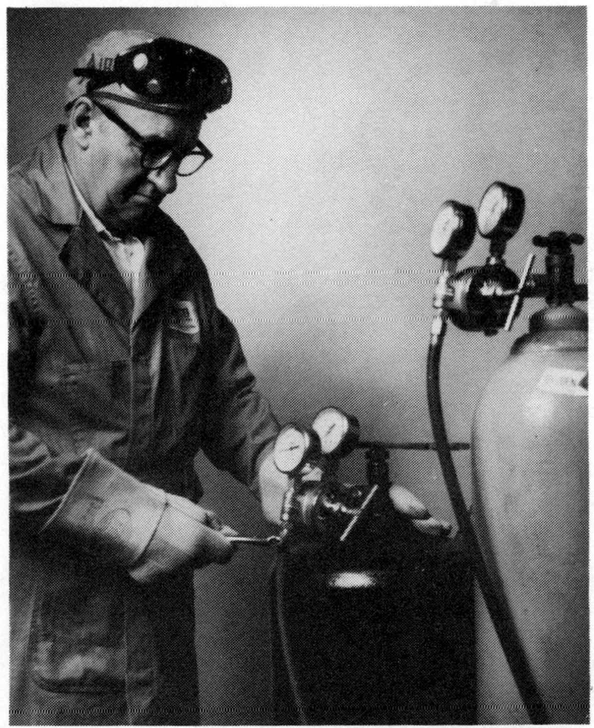

Courtesy Airco Welding Products

Fig. 6. Connecting the hose to the regulators.

and 1/16 inch thick should be adequate. Although it is true that light welding can be done without goggles or gloves, it is advisable that the student wear them from the very beginning. It is necessary that the beginner become accustomed to looking at the weld puddle through colored lenses and to hold the torch and welding rod with gloved hands. Therefore, even when practicing, it is recommended that the beginner dress for welding with suitable goggles, gloves, and anything else deemed necessary by the instructor.

Set the welding outfit alongside the welding table. At this point the beginner should be completely familiar with:

1. The gas welding equipment he is using,
2. The procedure for setting up the gas welding apparatus.

Courtesy Airco Welding Products

Fig. 7. Opening the cylinder valves.

The beginner should start by making a tack weld. As was explained in Chapter 2. A tack weld is a weld designed to hold parts or members together in proper alignment until the final welds are made. It is a temporary weld and the simplest type of weld for the beginner to make.

MAKING A TACK WELD

The tack weld consists of a single spot of melted base metal. One or more tack welds may be used to temporarily join two pieces of metal. The procedure for making a tack weld is as follows:

Courtesy Airco Welding Products

Fig. 8. Blowing out the hose.

Gas Welding Process

1. Lay the two 6-inch pieces of sheet metal side by side on the welding table, bridging across so that about ¼ inch rests on the bricks at each end.
2. Open the acetylene needle valve and light the torch with a spark lighter.
3. Open the oxygen needle valve and adjust for the neutral flame.
4. Pull the goggles down over the eyes.
5. Hold the torch loosely in the right hand, grasping the handle at a point where the hose hanging free nearly balances the overhanging tip.
6. Tack the left end of the plates together as shown in Fig. 14. This is done by holding the flame close to the joint

Courtesy Airco Welding Products

Fig. 9. Connecting the hose to the torch.

until the steel melts and runs together. Remove the flame instantly when the metal flows, as otherwise a hole will be burned through. Warm the two plates by passing the flame over them lengthwise two or three times.

MAKING WELD BEADS

A fundamental exercise for the beginner to master is the making of weld beads. Weld beads are the weld deposits that are produced by a pass of the torch. As with everything else, there is a correct way and an incorrect way to make weld beads.

The beginner must take particular care in maintaining the correct welding speed. This is the correct speed at which the torch

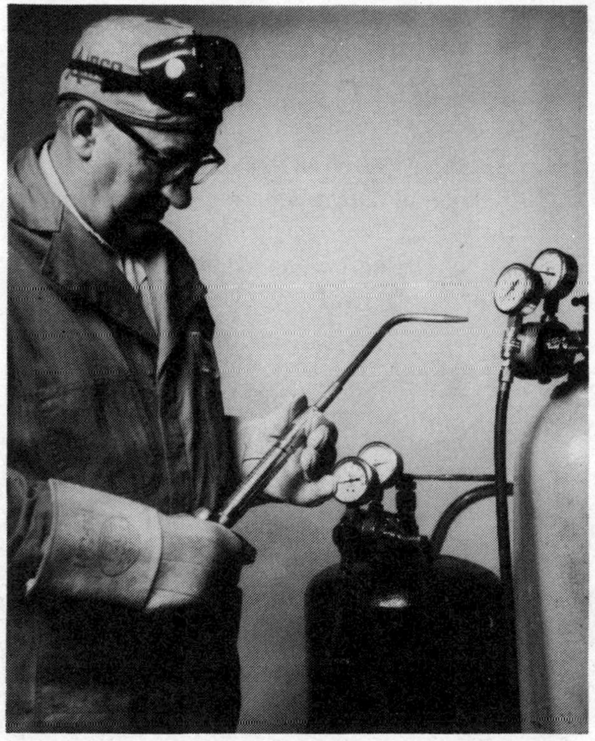

Courtesy Airco Welding Products

Fig. 10. Adjusting the tip.

moves across the metal surface. The speed of the torch must not be too quick nor too slow. By moving too slowly, large puddles of molten metal are produced and there is the danger of overheating. A small puddle is created when the torch is moved too quickly. The result is poor fusion and inadequate penetration. The speed of the torch must also be constant. A varied speed causes the irregular bead shown in Fig. 15.

The beginner should practice making weld beads on the 1/16-inch or 1/8-inch thick steel sheets used in the other welding exercises. The procedure for making weld beads is as follows:

1. Place a steel sheet so that each edge overlaps a firebrick by approximately 1/4 inch.

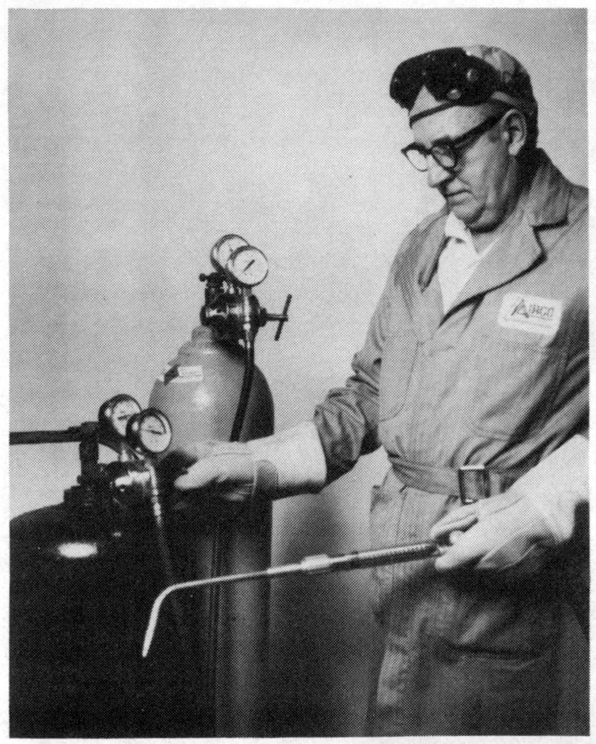

Courtesy Airco Welding Products

Fig. 11. Adjusting the acetylene working pressure.

2. Hold the torch at an angle of 45° from the surface of the steel plate and point the tip in the direction of the welding. Hold the torch so that the inner cone is 1/16 inch to 1/8 inch above the puddle (Fig. 16).
3. The torch tip should be moved slightly back and forth across the weld path during the welding operation (Fig. 17).
4. Fig. 18 shows the results of following the correct technique of producing a weld bead.

TORCH MANIPULATION AND MOVEMENTS

The beginner should be careful to hold the torch at the correct height and angle from the metal surface. In this chapter,

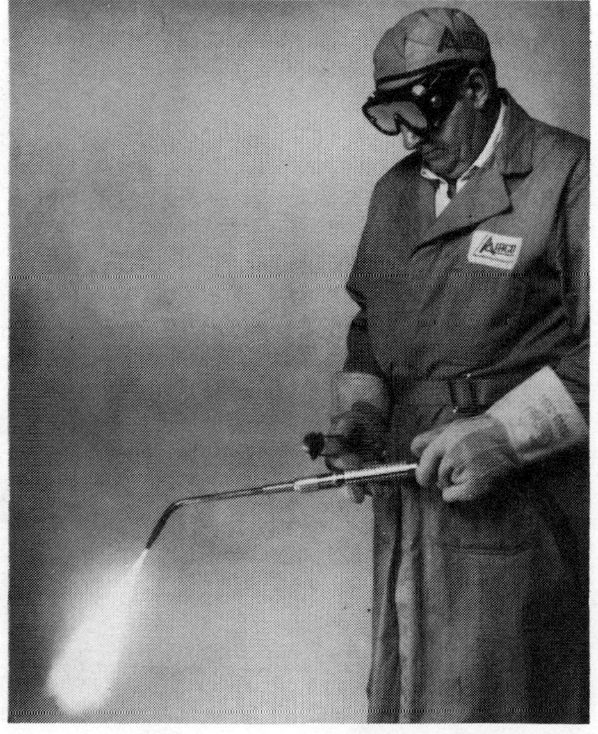

Courtesy Airco Welding Products

Fig. 12. Lighting the acetylene.

the torch height and angle are given in the sections dealing with specific welds. The flame should be directed so as to fuse the base metal evenly, and the torch should be moved to the next spot as soon as fusion has been achieved. *Do not attempt to retrace your steps.* Rewelding should not be attempted except as a last resort.

The torch tip is manipulated to create a circular or oscillating movement. The type of movement usually depends on the preference of the welder (Fig. 19).

The welding rod and the welding torch may be used in one of two different ways:

1. The forehand welding method,
2. The backhand welding method.

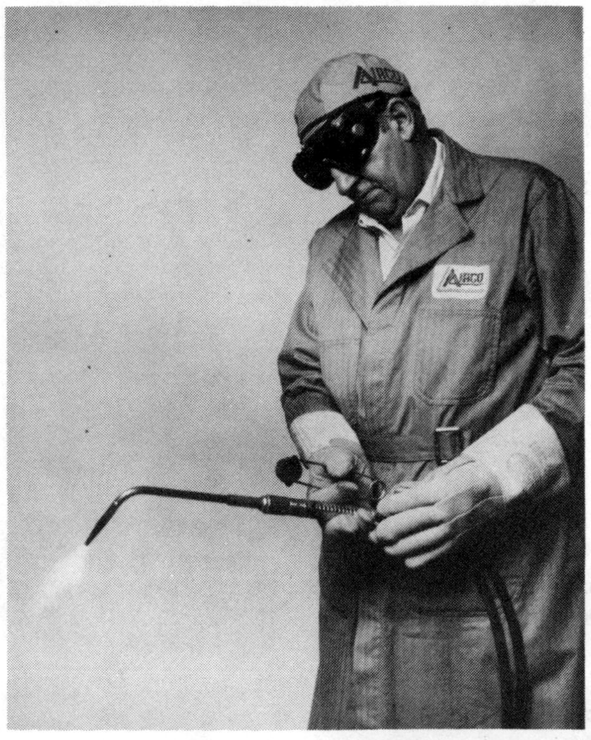

Courtesy Airco Welding Products

Fig. 13. Adjusting for neutral flame.

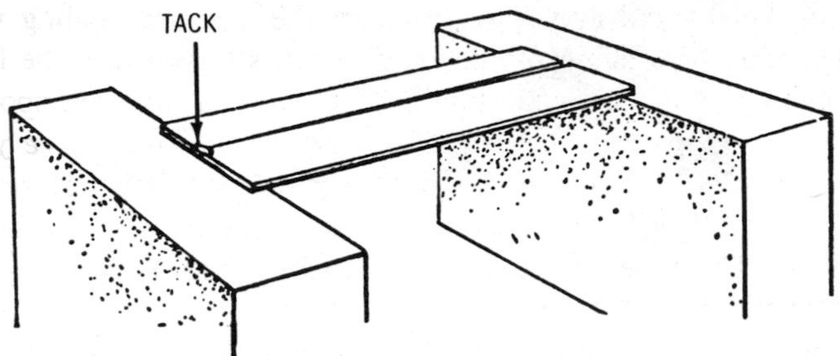

Fig. 14. Illustrating the formation of a tack weld.

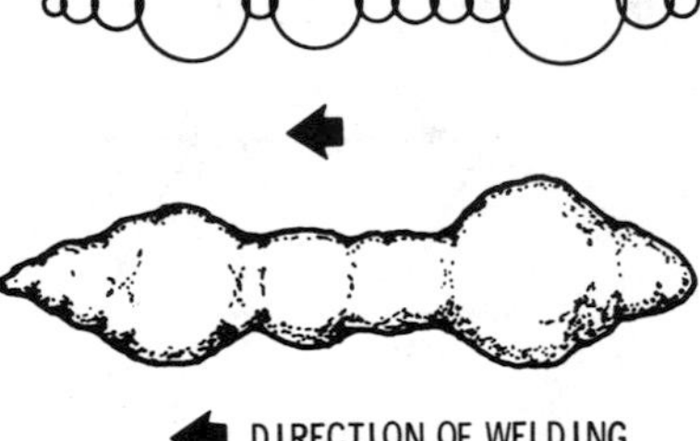

Fig. 15. A bead produced by a varied welding torch speed. The narrow sections are produced by increasing the speed. At slower speeds, a larger puddle forms.

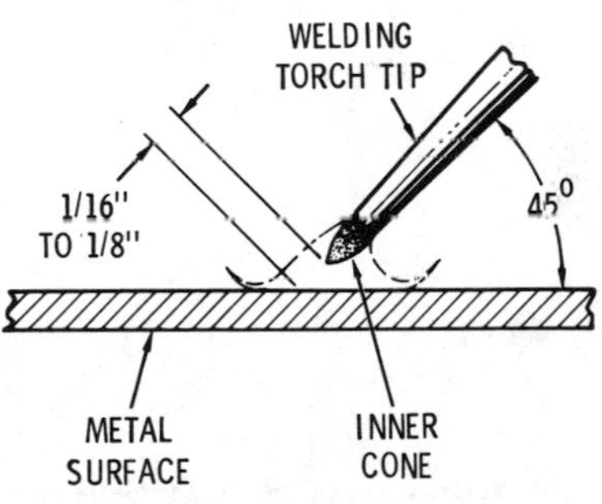

Fig. 16. Torch angle and inner cone distance from the metal surface.

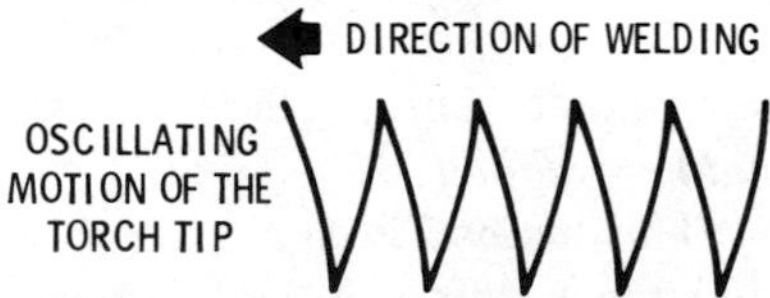

Fig. 17. The oscillating motion of the torch tip. The forward speed of the torch should be constant.

Forehand Welding Method

In the *forehand welding method* (Fig. 20), the tip of the welding torch follows the welding rod in the direction the weld is

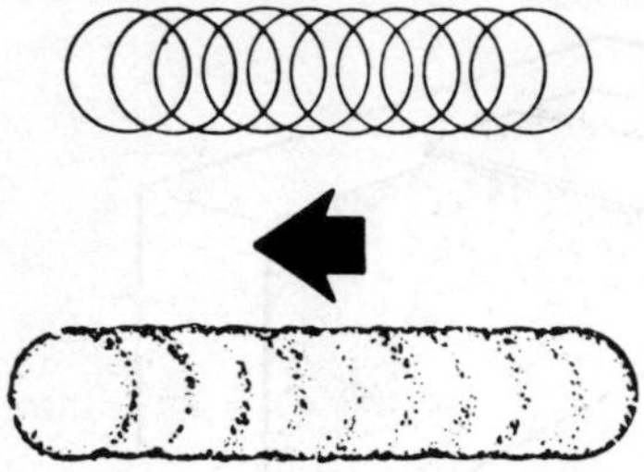

Fig. 18. A correctly formed weld bead.

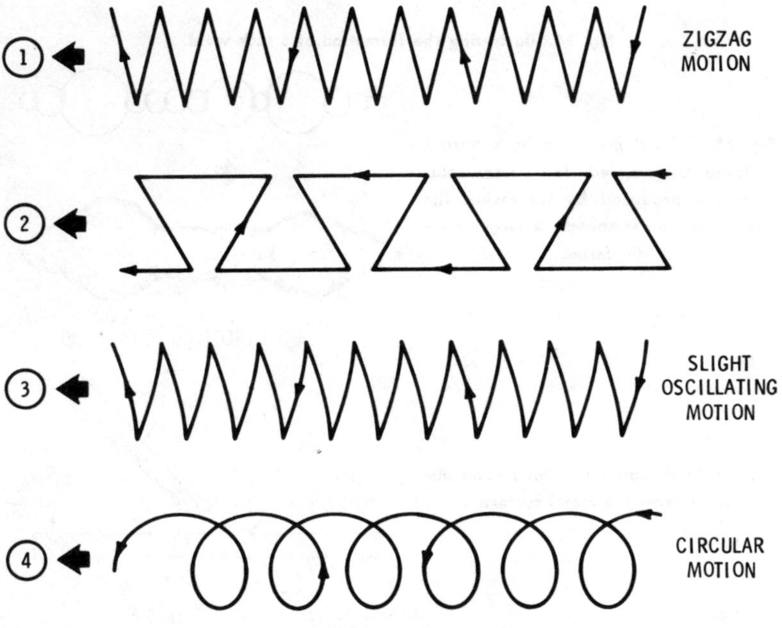

ZIGZAG MOTION

SLIGHT OSCILLATING MOTION

CIRCULAR MOTION

Fig. 19. Various types of torch movements. The welding rod should move in a direction opposite to that of the welding tip.

being made. This method is also referred to as *ripple welding* or *puddle welding.*

This method is charactrized by wide semicircular movements of both the welding tip and the welding rod which are manipulated so as to produce *opposite* oscillating movements. The flame is pointed in the direction of the welding but slightly downward so as to preheat the edges of the joint.

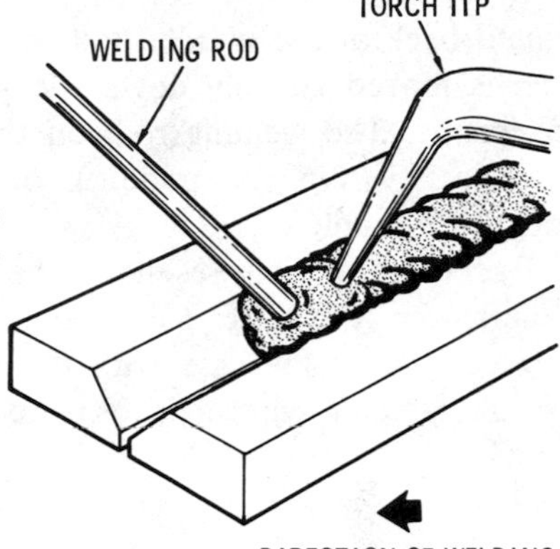

Fig. 20. Forehand welding method. In this method, the welding rod moves ahead of the torch tip.

The major difficulty with the forehand welding method is encountered when welding metals of greater thicknesses. In order to obtain adequate penetration, proper fusion of the groove surfaces, and to permit the movements of the tip and rod, a wide V-groove (90° included angle) must be created at the joint. This results in a rather large puddle which can prove difficult to control, particularly in the overhead position.

Backhand Welding Method

In the *backhand welding method* (Fig. 21), the tip of the torch precedes the welding rod in the direction the weld is being made.

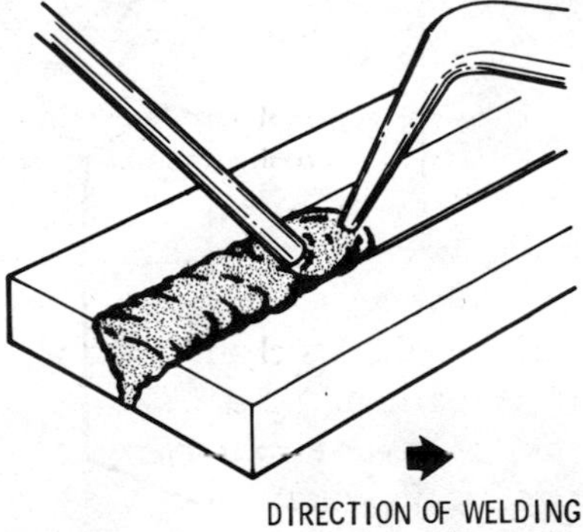

Fig. 21. Backhand welding method. In this method the torch moves ahead of the welding rod.

In contrast to the forehand welding method, the flame is pointed back at the puddle and the welding rod. In addition, the torch is moved steadily down the groove without any oscillating movements. The welding rod, on the other hand, may be moved in circles (within the puddle) or semicircles (back and forth around the puddle).

Backhand welding results in the formation of smaller puddles. A narrower V-groove (30° bevel, or 60° included angle) is required than is the case with the forehand welding method. As a result, greater control is provided as well as reduced welding costs.

MAKING A WELD WITHOUT A FILLER ROD

The tack weld described in preceding paragraphs consisted of a single spot of melted metal. No welding rod was used. The beginner should now practice making a weld without the use of a welding rod. The weld itself will extend the length of the joint formed by placing the previously described steel sheets together. Obviously this will be more difficult than making a simple tack weld. The procedure is as follows:

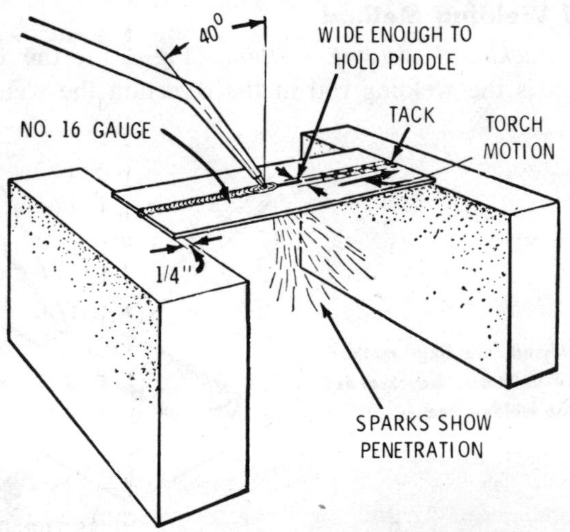

Fig. 22. Making a weld without a filler rod.

1. Start welding at the right (Fig. 22).
2. Hold the torch loosely in the hand and nearly parallel with the top of the welding table.
3. Incline the tip sideways so that it points to the left at an angle of about 45° to 50° to the welding surface.
4. Move the flame across the joint, back and forth, giving it a movement of about ⅛ to ¼ inch. The advance of the torch has to be held down until the metal is flowing freely.
5. As the steel melts and flows together, move the torch to the left slowly following a zigzag path.
6. Keep the torch in motion but do not move to the left along the joint faster than the advance of the puddle.
7. Be sure that each edge receives its due share of the heat.
8. Do *not* hold the handle tightly. Gripping the handle too firmly tires the muscles and makes the arm tremble. Learn to work easily and surely. Do not hurry.
9. Continue a uniform zigzag motion keeping the white hot inner tip of the flame just above the surface of the metal. The rate of welding progress is determined by the puddle. Welding can be no faster than the advance of the puddle; neither can it be slower, because if the advance of the torch and the puddle is stopped the hot metal will drop through.
10. When nearing the end of the joint, lift the torch higher to avoid making a hole in the weld at the end. This is an important point in welding thin metal sheets.
11. Pick up the welded sheets with the pliers and examine them first for uniformity of torch manipulation on each side of the joint. This will be indicated by the heat zone. The heat zone should be of uniform width on both sides of the weld.
12. Turn the piece over and examine for penetration (that is, if the weld penetrated to the bottom of the plates).

An examination of the completed weld will reveal a furrow or depression in the middle. This resulted from the melted metal sinking down and filling the narrow groove between the two metal sheets. It is to fill this depression that a welding rod is used to add more metal to the weld. This is why the welding rod is sometimes called the *filler rod.*

The beginner should continue to practice welding on $\frac{1}{16}$-inch sheets of steel without a welding rod until he is able to make a weld fairly uniform in width, of good appearance, and without holes or protrusions of metal hanging beneath.

Testing First Welds

If the first weld seems satisfactory, the beginner should test it. Cool the welded piece in water, clamp it in an iron vise with the weld parallel with the top of the jaws and just above them.

The under side should be toward the operator. Strike the top with a hammer and bend over until the piece breaks through the weld (Fig. 23).

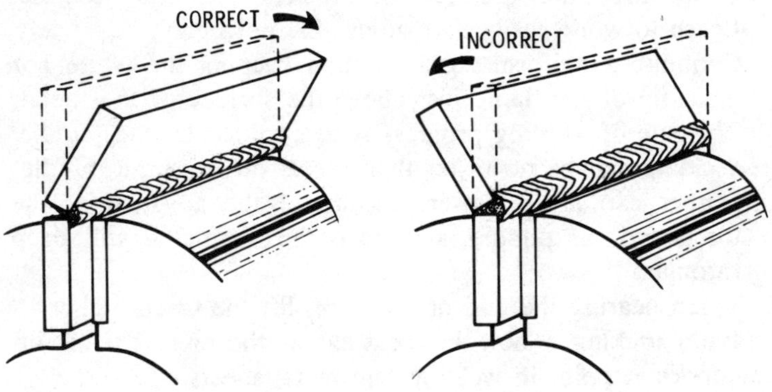

Fig. 23. Testing welded sheets for strength in a vise.

Examination of the first broken weld will generally reveal defects caused by improper manipulation of the torch, incorrect flame adjustment, overheating, underheating, and exposure of hot metal to the air before the puddle is completed.

Common defects of the first welds usually encountered by the beginner are the following:

1. Uneven welds,
2. Fused portion not in the joint, but at one side,

3. Holes in the joint,
4. Oxide inclusions,
5. Adhesions,
6. Brittle welds,
7. Overheating.

Uneven welds are caused by moving the torch along the joint too quickly or slowly, zigzag movements not uniform nor in step with the puddle and zigzag movements overlapping. The result is good penetration in spots and partial penetration between; in other words, incomplete penetration (Fig. 24).

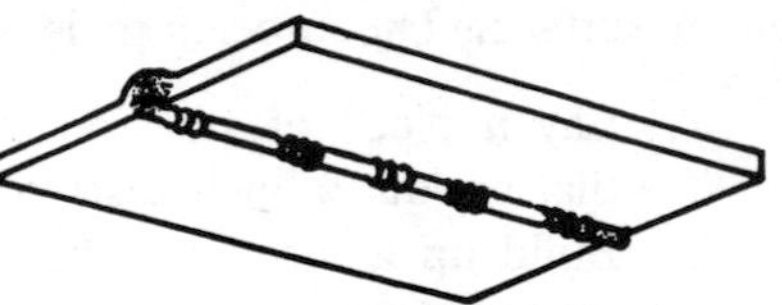

Fig. 24. Underside of weld showing incomplete penetration.

Fused portions at one side of the joint are caused by not playing the torch over the joint equally at each side. *Holes in the joint* are caused by holding the flame too long in one place and overheating the metal. Holes in the joint at the end of the weld are due to not lifting the torch and reducing the heat when the end of the weld is reached.

Oxide inclusions are indicated by black specks on the broken surfaces of the weld. These oxide inclusions are caused by not adequately cleaning the base metal surface (i.e. the surface of the two metal sheets) prior to welding.

Adhesions are an indication that insufficient heat was directed to one side of the joint, or that the welding speed was too high. As a result, the weld metal sometimes will break cleanly from the metal surface on one side.

Brittle welds are caused by using a carburizing flame. It is important for the welder to use the correct flame when welding.

Overheating results from moving the flame so slowly that excess heat is directed into the weld. This results in excess metal forming "icicle" deposits on the bottom of the weld (Fig. 25).

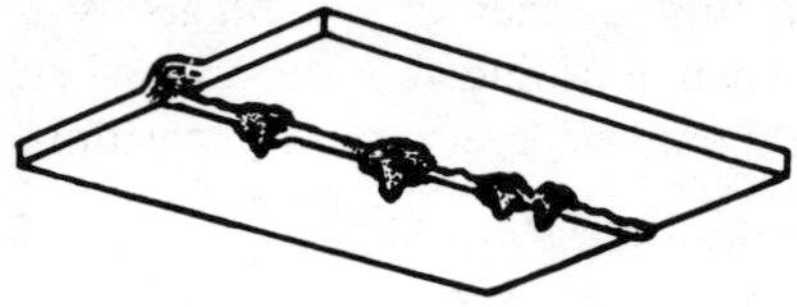

Fig. 25. Underside of weld showing icicle deposits.

PRACTICING WITH THE WELDING ROD

The beginner is now ready to start practicing with the welding rod. However, before attempting the more difficult procedure of filling a groove cavity or a separation between plates, the beginner should practice making mounds of filler metal on a metal surface. The procedure is as follows:

1. Lay a piece of steel on the firebricks bridging across so that about ¼ inch rests on the bricks at each end.
2. Build up a mound of fused welding rod in the middle of the metal surface (Fig. 26).
3. Alternately fuse the top of the mound and drop the rod until the timing of the two operations is satisfactorily acquired.

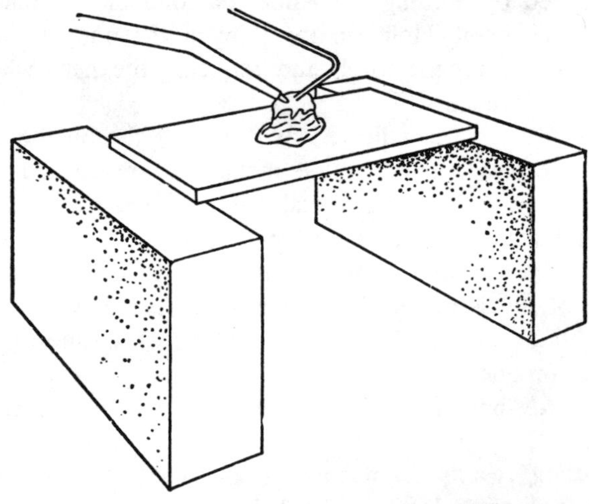

Fig. 26. Building up a mound with filler metal.

4. Practice in building up in this manner is advantageous—also in that it is useful when finishing a weld and when necessary to provide extra reinforcement at any position.

When welding, the end of the welding rod should be melted by keeping it beneath the surface of the puddle. *Never* allow it to come into contact with the inner cone of the torch flame. *Do not hold* the welding rod above the puddle so that the filler metal drips into the puddle.

WELDING SHEET STEEL WITH THE WELDING ROD (NO EDGE PREPARATION)

The beginner is now ready to use the welding rod in joining two pieces of sheet steel. As shown in Fig. 27, these metal sheets have had no edge preparation (beveling); this being a procedure that will be discussed in the next section. The procedure is as follows:

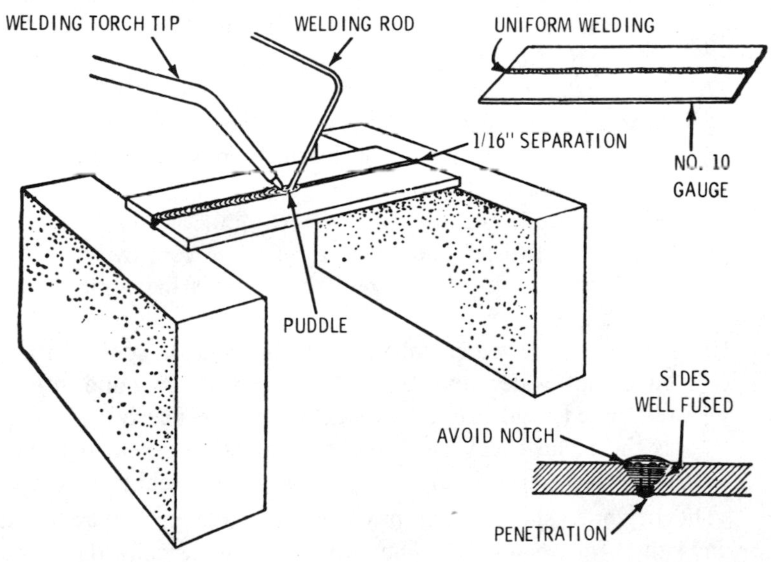

Fig. 27. Welding two steel plates together without edge preparation.

Gas Welding Process

1. Lay two 1/16-inch thick steel sheets on firebricks side by side about 1/16 inch apart (Fig. 27).
2. Select a suitable pair of goggles. If the goggle lenses are too dark, exchange them for a lighter color. This is of utmost importance. The welder must see clearly what is taking place in order to do good work.
3. Light the torch and adjust the flame. Set the goggles in place. Pick up a 1/16-inch welding rod, heat it 4 or 5 inches from the end and bend it to a right angle.
4. Tack the steel sheets together at the left end. Start welding at the right end. Take time to assure penetration to the bottom.
5. Add welding rod to the puddle and build up above the surface about 1/32 inch. Manipulation of the rod consists of holding the end over the joint close to the flame and moving it so that it is *immersed* in the puddle ahead of the flame. The heat of the puddle and flame melts off filler metal which blends into the weld.
6. Hold the torch in the hand loosely; stand relaxed and comfortable. Avoid tension. Acquire the ability to see the welding puddle clearly.
7. The appearance of the puddle is a guide to flame adjustment. If a white scum is seen floating on the puddle, it indicates an oxidizing flame. Turn the oxygen needle valve slightly to reduce the flow of oxygen.
8. When approaching the end of the joint, the torch should be advanced faster and lifted a bit to prevent overheating.
9. Build up the end of the weld so that it is the same thickness as elsewhere.
10. Remove the welded piece with the pliers, cool it in a bucket of water and break it in a vise. A build up of reinforced weld properly made should be nearly as strong as the metal sheet itself and care must be taken to clamp the welded piece firmly and as close to the weld as possible.

The beginner should next practice the same procedure on a 1/8-inch thick piece of steel. The procedure is basically the same as the one outlined for the 1/16-inch thick steel, but with the following modifications:

1. The same zigzag motion is used with the torch, but with a somewhat greater movement across the joint on the thicker metal.
2. The extent of the crosswise movement will be governed by the size of the puddle. The crosswise movement is always governed by the puddle width (the wider the puddle, the greater the swing of the torch).
3. The rate of advance on thicker metals is slower than on thinner metal. It will be slower because more metal has to be heated and fused. The advance of the puddle governs the speed of welding. It cannot be held stationary long nor can it be hurried beyond a certain rate for a given size welding job. By moving too fast, penetration will occur only part way through. The result will be an open seam at the bottom. This can be corrected by working more slowly for full penetration.

WELDING STEEL PLATES WITH THE WELDING ROD (BEVELED EDGES)

Up to this point the beginner has been welding steel sheets without bevel preparation. Thicker metal will require preparation (beveling) of the edges to provide deeper pentration. Fig. 28 shows the type of edge preparation recommended for various metal thicknesses. Note that thicknesses ⅛ inch and under require no preparation at all. The upturned ends of the flanged edge (Fig. 28B) provide enough weld metal to make the use of filler rods unnecessary. These flanged edges are used on metals not exceeding $\frac{1}{32}$ inch in thickness.

Steel sections in thicknesses of ⅛ inch or less are referred to as *sheets*. The term steel *plates* is used to designate thicknesses $\frac{3}{16}$ inch or more.

The beginner should practice bevel preparation and welding with two $\frac{3}{16}$-inch thick steel plates. If possible, these plates should have the same width and length dimensions (6 inches long by 1¼ inches wide) provided for the previously described practicc exercizes.

The procedure for welding two beveled plates is as follows:

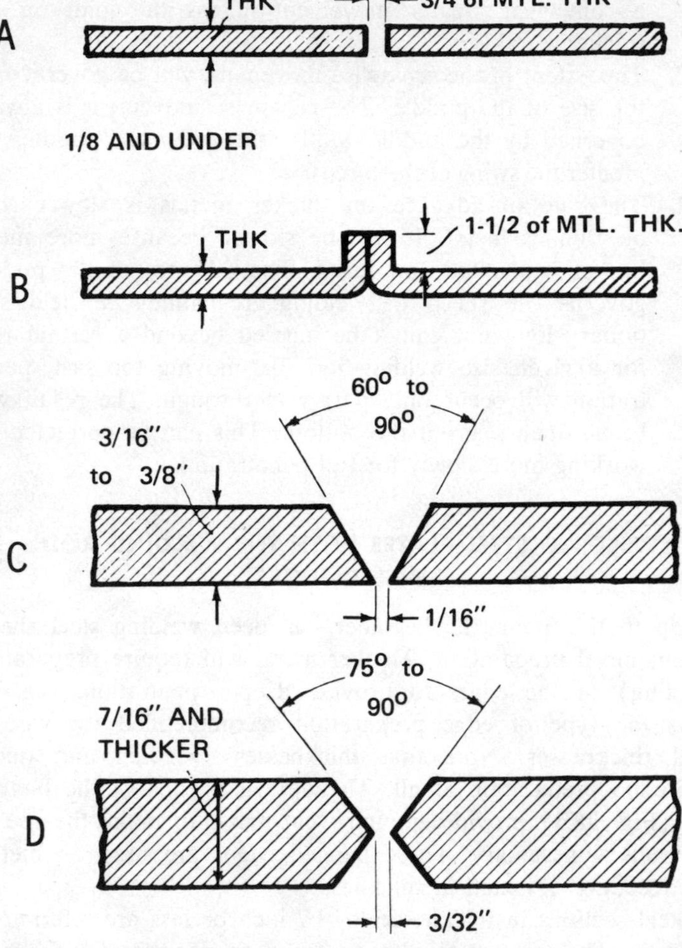

Fig. 28. Various types of edge preparation.

1. One edge of each plate should be beveled to an angle of about 45°. The beveled edges are set together, thus, making a groove of about 90° included angle.
2. Lay the 6-inch by 1¼-inch by 3⁄16-inch beveled plates on the firebricks with the edges parallel and 1⁄16-inch apart (Fig. 29).

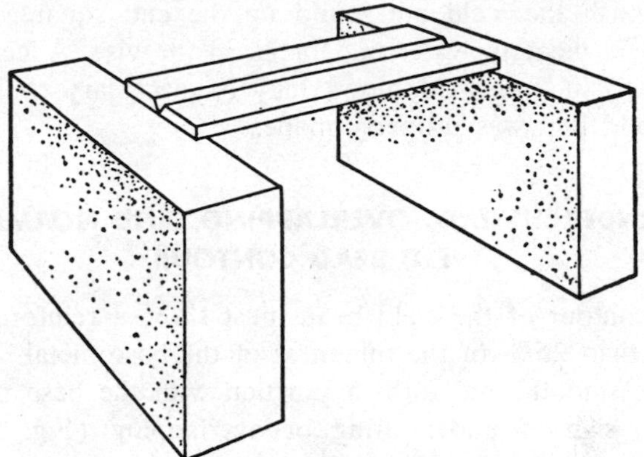

Fig. 29. Steel plate beveled and in place for welding.

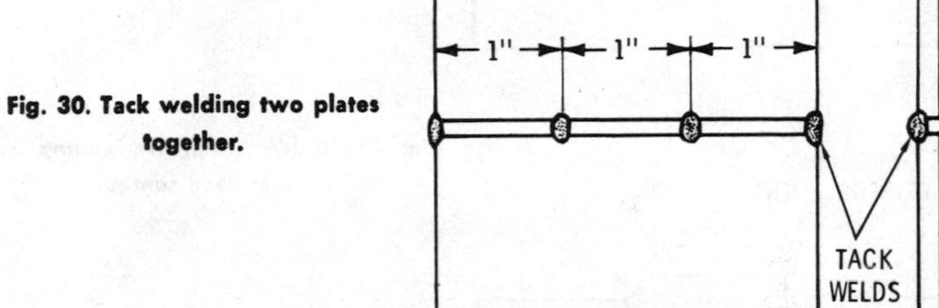

Fig. 30. Tack welding two plates together.

3. Tack at the right end firmly, fusing in a few drops of welding rod (Fig. 30). If the edges are thrown out of parallel by tacking, pry them open to approximately parallel position.
4. Start welding at the left, using a $\frac{3}{32}$-inch welding rod. Be careful to get thorough penetration and to manipulate the torch so as to cover a larger puddle than the beginner has heretofore produced.
5. Build the weld up square at the beginning.
6. Fill the bevel groove with weld metal, making certain that the base metal on both sides is well fused before covering with weld metal.
7. Build up the weld smoothly about $\frac{1}{16}$ inch above the surface of the plates.

8. Finish the weld and build up the end square. Cool the weld piece in water and break in the vise. A heavy hammer and repeated blows may be necessary to break the weld if it was properly made.

UNDERCUTTING, OVERLAPPING, AND NORMAL WELD BEAD CONTOUR

The contour of the weld bead must show a reinforcement of no more than 25% of the thickness of the base metal. The bead should be smooth, and form a junction with the base metal that shows no sign of undercutting or overlapping (Fig. 31). The

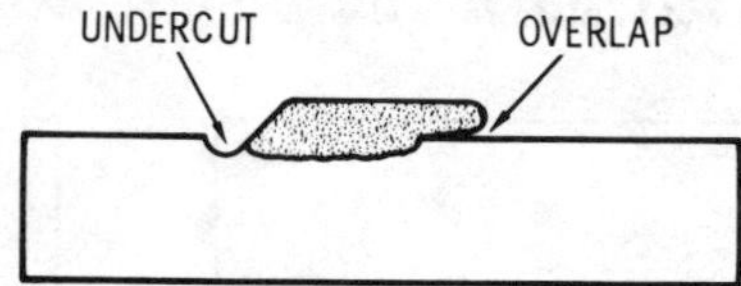

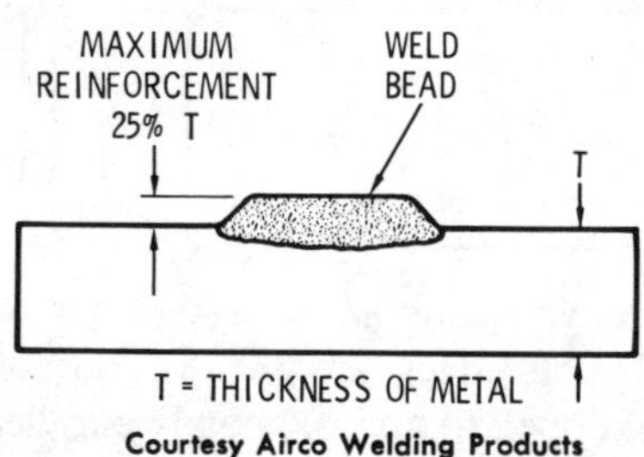

Courtesy Airco Welding Products

Fig. 31. Undercutting, overlapping, and normal head contour.

width of the bead should be uniform, no icicle formations of excess metal should protrude from the bottom of the weld and the bead should be covered (on both the top and bottom) with a thin oxide film.

VERTICAL WELDING POSITION

The vertical and overhead welding positions are the most difficult ones in which to produce satisfactory welds. The obvious reason for this difficulty is the effect that the force of gravity has on the molten metal. The welder must constantly control the metal so that it does not run or drop from the weld. There are several ways in which the weld bead can be controlled in vertical welding. Most important among these are the following:

1. Follow the same instructions (constant speed, tip movement, holding the torch at a 45° angle, etc.) described in the section on making weld beads (Fig. 32).

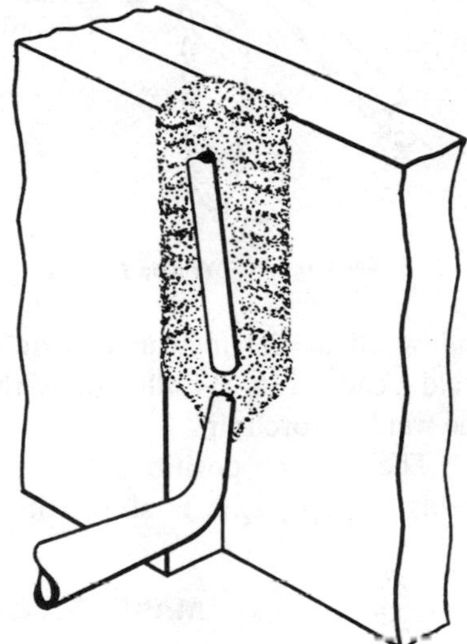

Fig. 32. Vertical weld using the back-hand method.

Courtesy Airco Welding Products

2. Begin the weld bead at the bottom and proceed upward. Allow the lower sections of the weld bead time to cool as you proceed up the weld. This will result in a firmer base for subsequent layers of molten metal (Fig. 33).

OVERHEAD WELDING POSITION

The overhead welding position is probably even more difficult than the vertical position. Here, the pull of gravity against the molten metal is much greater. The force of the flame against the weld serves to counteract the pull of gravity. The welder should

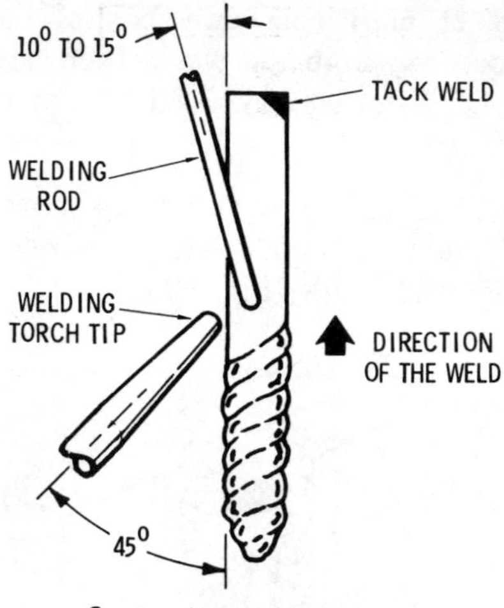

Courtesy Airco Welding Products

Fig. 33. Making a vertical weld.

follow all of the instructions described in the section on making weld beads. Fig. 34 illustrates the angle of the welding rod and the welding torch tip.

The correct positions for the welding tip and rod for fillet welds on lap joints is shown in Fig. 35.

MAKING CORNER WELDS

A corner weld can be either on the inside or the outside of the corner. If it is made on the inside of the corner, it will be in the form of a fillet weld (Fig. 36). A welding rod will be used

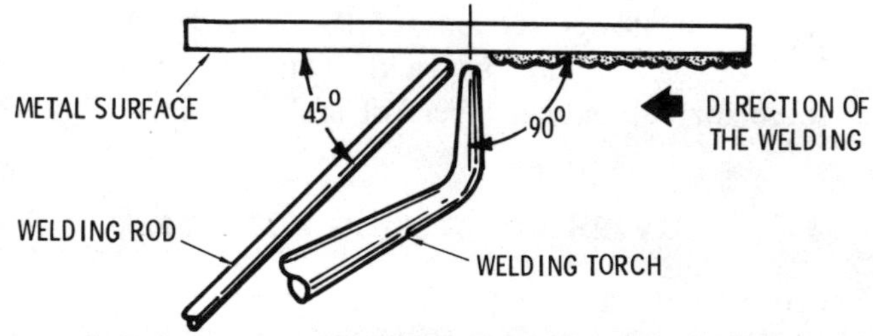

Courtesy Airco Welding Products

Fig. 34. Overhead position welding. The direction of the welding should be toward the operator.

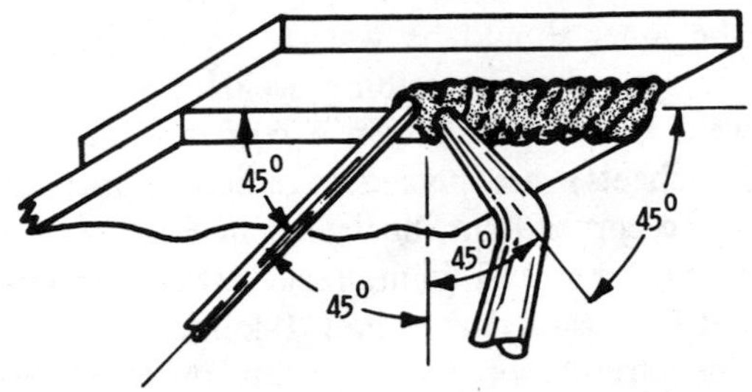

Courtesy Airco Welding Products

Fig. 35. Making a fillet weld on a lap joint in the overhead position.

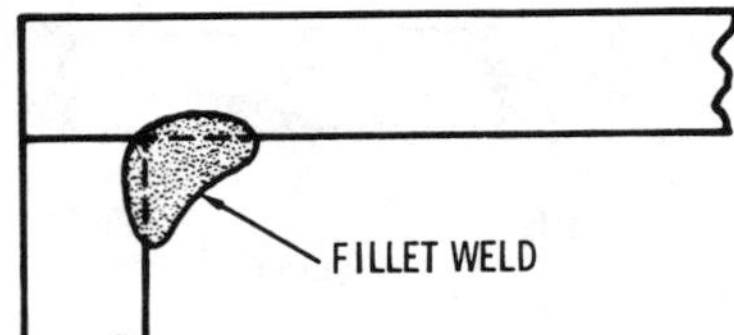

Fig. 36. Inside corner weld.

to supply the additional weld metal. A welding rod will also be used on an outside corner weld if the joined metal plates form a single V-groove (Fig. 37). The use of a welding rod on an out-

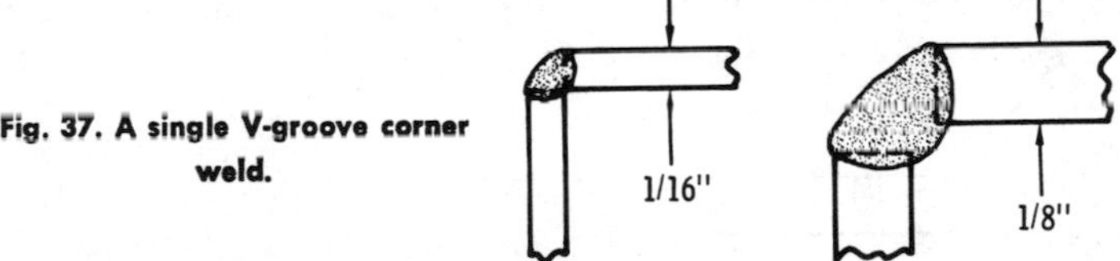

Fig. 37. A single V-groove corner weld.

side corner weld can be avoided by overlapping the plates as shown in Fig. 38. Melting the overlap provides the necessary filler metal. This method should be avoided, however, except on the thinner metals.

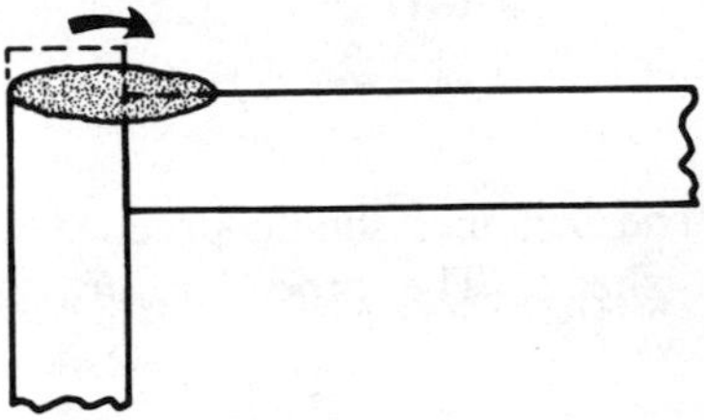

Fig. 38. Overlap corner weld.

The beginner should be warned about the possible confusion of terminology when describing welds and joints. Essentially, a corner weld is a weld made in a corner joint. That is, the metal plates (or sheets) are placed together to form a corner before welding. The corner joint illustrated in Fig. 36 is welded in much the same manner as a T-joint. (see MAKING WELDS IN T-JOINTS). This could be regarded as a half T-joint.

The penetration of corner welds must be complete. Fusion must occur completely through to the bottom of the weld. However, excess metal will cause protrusions on the bottom of the

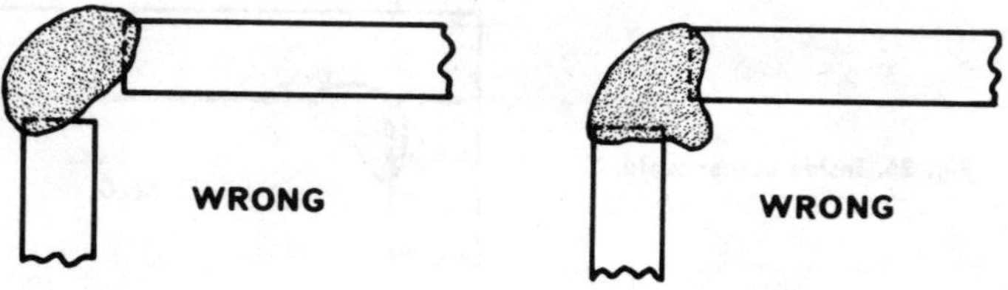

Courtesy Airco Welding Products

Fig. 39. Incomplete fusion and excess metal protrusions in corner welds.

weld which may lead to cracking. Fig. 39 shows examples of both incomplete fusion and protrusions of excess metal. A correct corner weld is shown in Fig. 40.

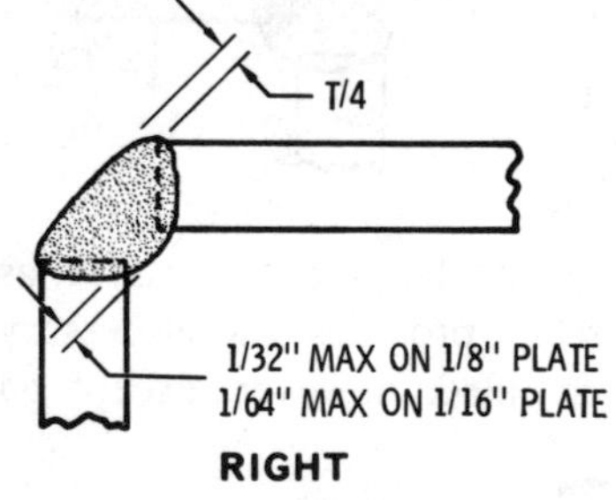

Courtesy Airco Welding Products

Fig. 40. An example of a correct corner weld.

The beginner should practice with $\frac{1}{16}$-inch and $\frac{1}{8}$-inch thick steel sheets. The procedure for making the corner weld is as follows:

1. Place the two steel sheets so that they form a right angle. If the thinner sheets (1/16 inch) are being used, their edges should touch. The 1/8-inch sheets should have approximately 1/16-inch between them (Fig. 41).

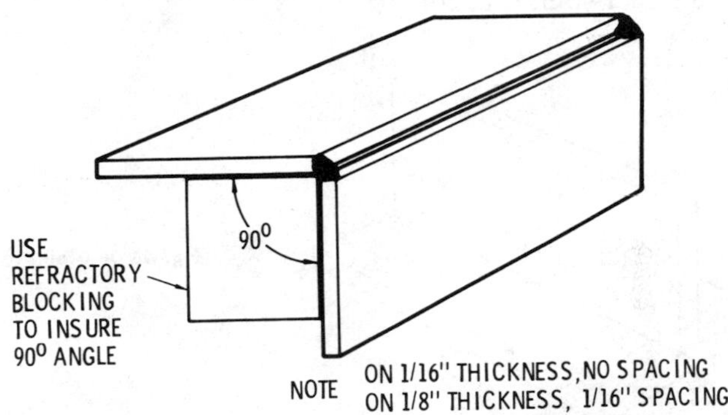

Courtesy Airco Welding Products

Fig. 41. Positioning the surfaces for a corner weld.

2. Tack weld the two steel sheets together, and allow them to cool.
3. Place the joined sheets on the welding table so that the corner faces upward. The weld is made with the forehand welding method. Note the position of the welding rod to the torch nozzle and the angle of the latter (Fig. 42).

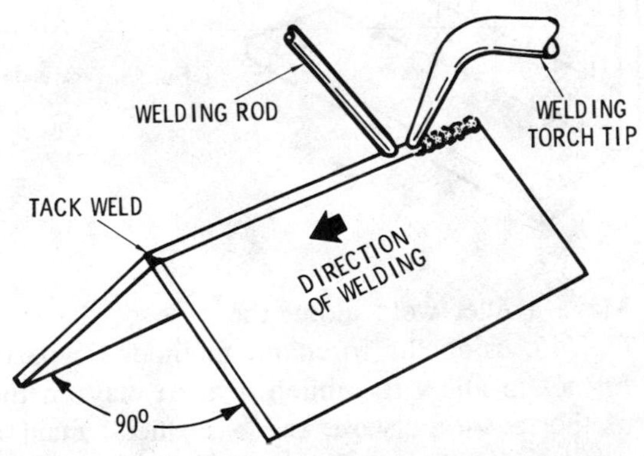

Courtesy Airco Welding Products

Fig. 42. Making a corner weld.

MAKING WELDS IN T-JOINTS

T-joints are formed by placing two metal sheets or plates in such a way that they form a tee (Fig. 43). T-joints (as well as inside corner joints) are joined with fillet welds. The procedure for welding a T-joint is as follows:

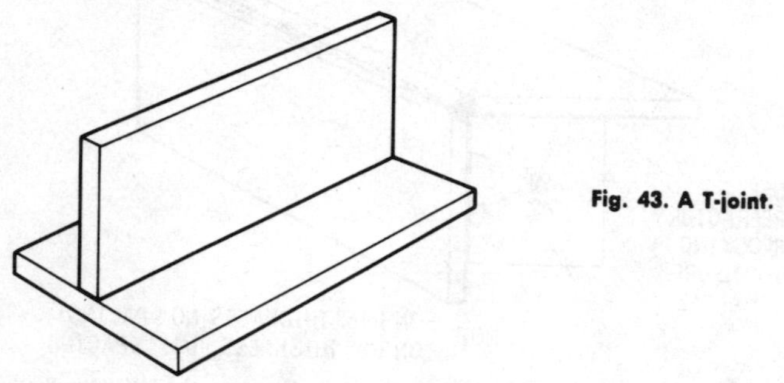

Fig. 43. A T-joint.

1. Situate the two metal plates or sheets to form a tee and tack weld each end. Allow the tack weld to cool before continuing on to the next step (Fig. 44).

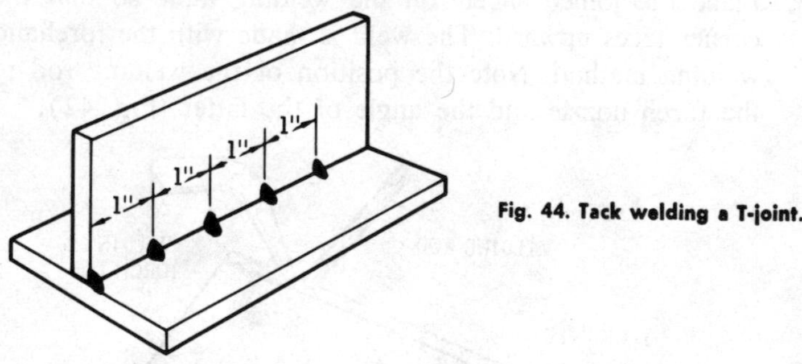

Fig. 44. Tack welding a T-joint.

2. Make a fillet weld along the base of the tee as shown in Fig. 45, using the forehand method. Be particularly careful not to allow too much heat to play on the vertical leg of the tee—otherwise, the base metal may burn through. The backhand method for making a fillet weld in a T-joint is shown in Fig. 46. Note the position of the welding rod.

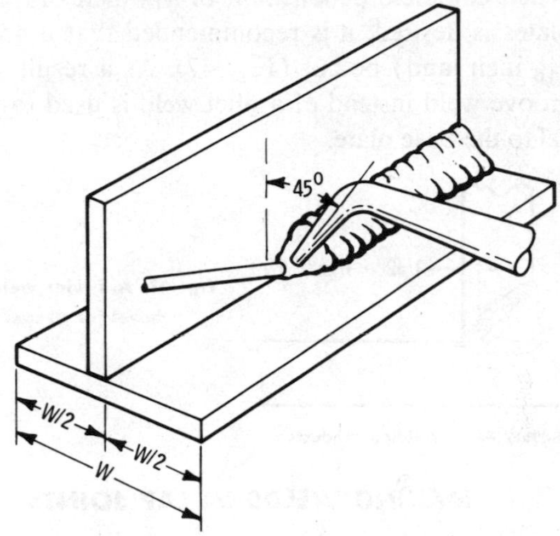

Courtesy Airco Welding Products

Fig. 45. Making a fillet weld in a T-joint.

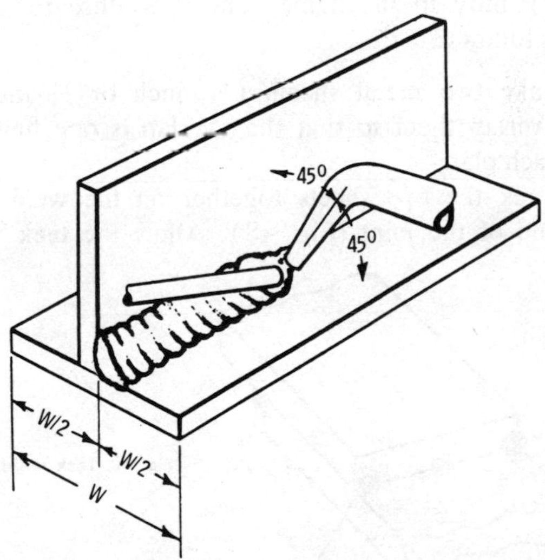

Courtesy Airco Welding Products

Fig. 46. Backhand method of making a T-joint weld.

3. When complete penetration of $\frac{3}{16}$-inch or ¼-inch vertical plates is desired, it is recommended that a 45° bevel ($\frac{1}{32}$-$\frac{1}{16}$ inch land) be cut (Fig. 47). As a result, a single bevel groove weld instead of a fillet weld is used to join the vertical to the base plate.

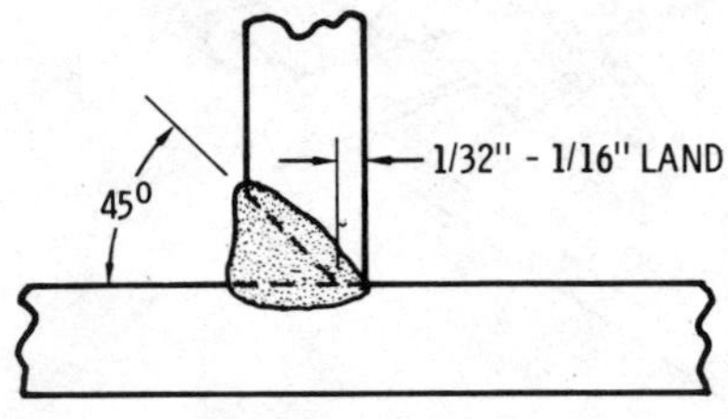

Courtesy Airco Welding Products

Fig. 47. A T-joint weld showing a 45° bevel for deeper penetration.

MAKING WELDS IN LAP JOINTS

Lap joints are formed by allowing the edge of one metal plate, or sheet, to overlap another. The torch flame should be concentrated on the corner (or root) of the joint. This will reduce the possibility of the top plate or sheet melting more rapidly because of its proximity to the flame. The procedure for welding a lap joint is as follows:

1. Take two metal sheets ($\frac{1}{16}$-inch or ⅛-inch thick) and overlap them so that the overlap is one half the width of each plate.
2. Tack the two sheets together on the weld face near the end of the joint (Fig. 48). Allow the tack welds to cool.

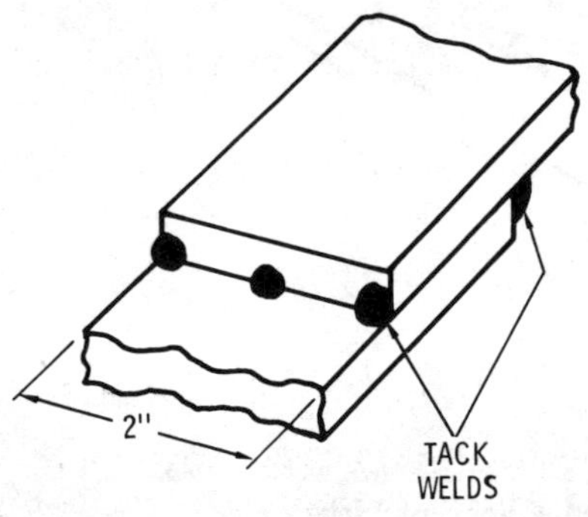

Fig. 48. Tack welding a lap joint.

3. Hold the torch so that the flame is directed at the joint at a 45° angle with the surface of the metal (Fig. 49).

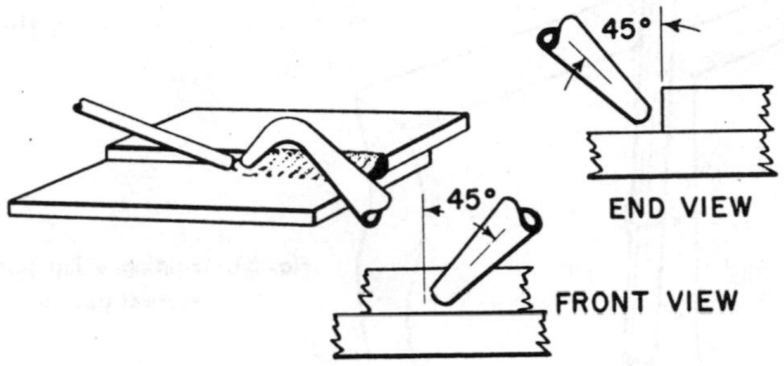

Fig. 49. Torch positions and angles for a weld in a lap joint.

4. A cross section indicating the weld dimensions is shown in Fig. 50. Note the penetration depth.

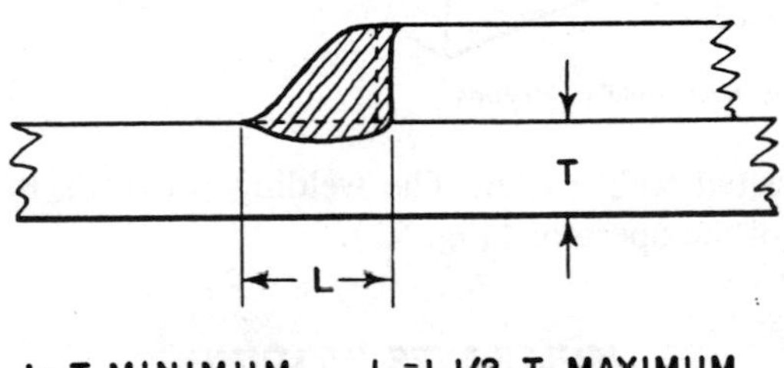

Courtesy Airco Welding Products

Fig. 50. Dimensions and penetration depth of a weld in a lap joint.

5. Making a lap-joint weld in the vertical position is somewhat more difficult than welding on sheets or plates in the horizontal plane. Fig. 51 illustrates just such a weld. Note the angle of the welding rod. The torch tip is held at an angle 45° from each edge, and inclined 45° upward. The weld is built up from the bottom so that a partially solidified base supports the new metal.
6. The overhead position is the most difficult one in which to make a weld in a lap joint. The principal reason for this, of course, is the pulling force of gravity on the newly

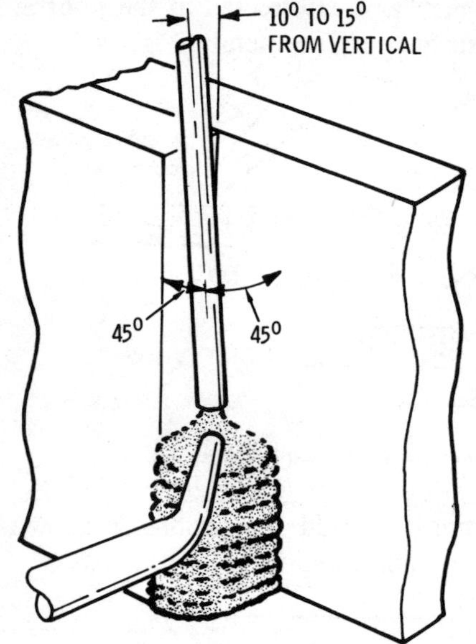

Courtesy Airco Welding Products

Fig. 51. Welding a lap joint in the vertical position.

deposited weld metal. The welding proceeds in the direction of the operator (Fig. 52).

MULTILAYER WELDING

Thicker metals (e.g. ½-inch or ¾-inch thick cast iron) require heavier welds. The size of the puddle necessary to make

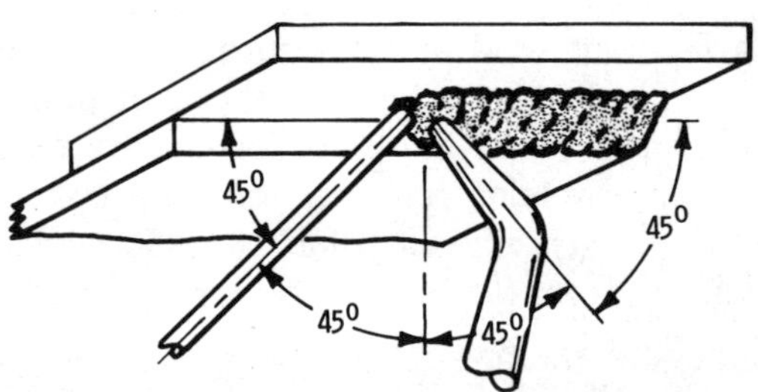

Courtesy Airco Welding Products

Fig. 52. Welding a lap joint in the overhead position.

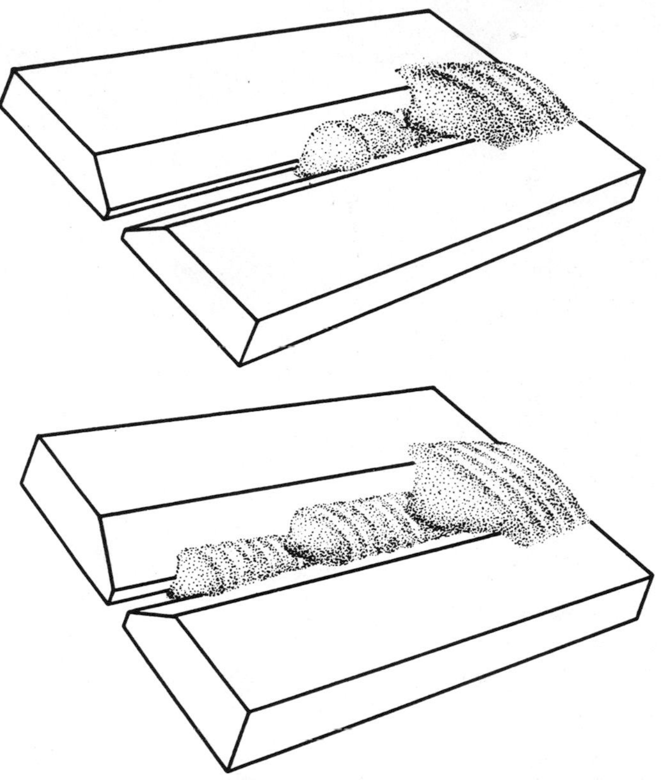

Courtesy Airco Welding Products

Fig. 53. Multiple-pass welding of iron plate.

the heavier welds *in a single layer* is really too large to handle without considerable difficulty. It is much easier to deposit two or more layers of weld metal. This results in smaller puddles and greater control. Multilayer welding can be accomplished by depositing the first layer and then the next, or by building the multilayer weld in steps (Fig. 53).

The first layer in multilayer welding should provide good penetration at the root of the V-groove. Subsequent layers must fuse with the surface of the preceding layer and the walls of the groove. The final layer must provide a good appearance.

CHAPTER 5

Arc Welding Equipment and Supplies

There are certain basic terms that commonly occur in any discussion of electricity or electrical theory. These terms must be defined and explained before proceeding with a description of the equipment used in the arc welding process.

An *electric current* may be defined as the flow of electricity over (or along) a conductor. The two types of electrical current are:

1. Alternating current (AC),
2. Direct current (DC).

An *alternating current* (AC) reverses (alternates) its direction of flow a certain number of times per second. This phenomenon is defined as *frequency*. Most alternating currents are supplied at a frequency of 60 hertz. This means that the current reverses its direction of flow 120 times per second and completes a cycle in 1/60th of a second. Consequently, sixty cycles are completed every second. Such a current is referred to as a 60-hertz alternating current (or a current having a frequency of 60 cycles per second) (Fig. 1). Alternating currents of 25, 30, 40, and 50-cycles per second (hertz) are also found but not to the same degree as the 60-hertz current.

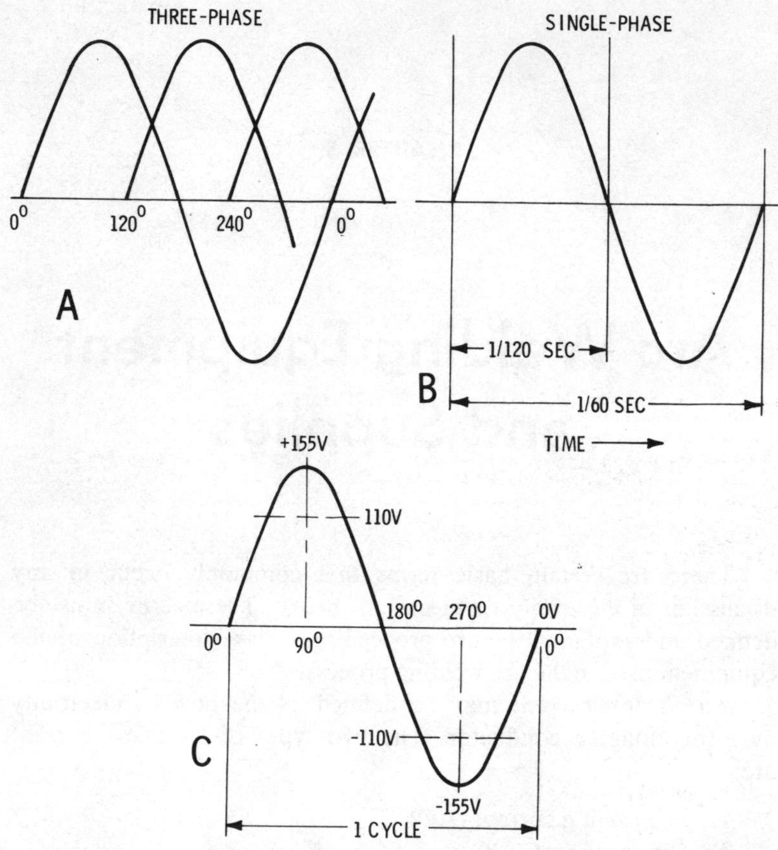

Fig. 1. Line drawing of alternating current.

A *direct current* (DC) is distinguished from an alternating current in that it flows continuously in one direction without reversing itself. To reverse the flow of current in a DC circuit, it is necessary to change the poles to which the welding cables are attached. This can be done manually, or with a switch on the outside of the welding machine. (Fig. 2).

Each generator has a positive and a negative pole. *Reversed polarity* (Fig. 3) is established by connecting the electrode cable to the positive pole and the work cable to the negative pole. *Straight polarity* (Fig. 4), on the other hand, is established by

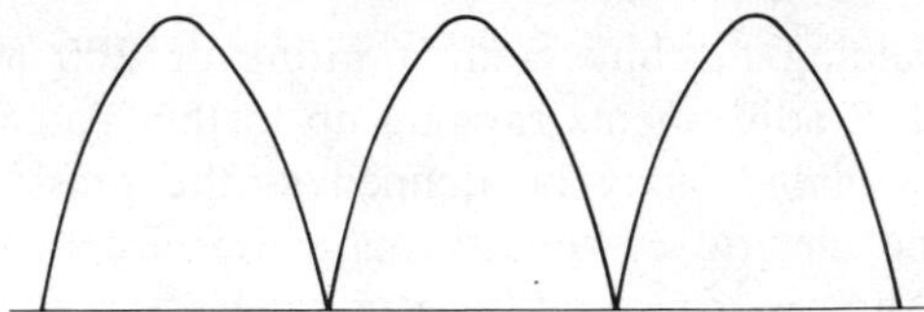

Fig. 2. Line drawing of direct current.

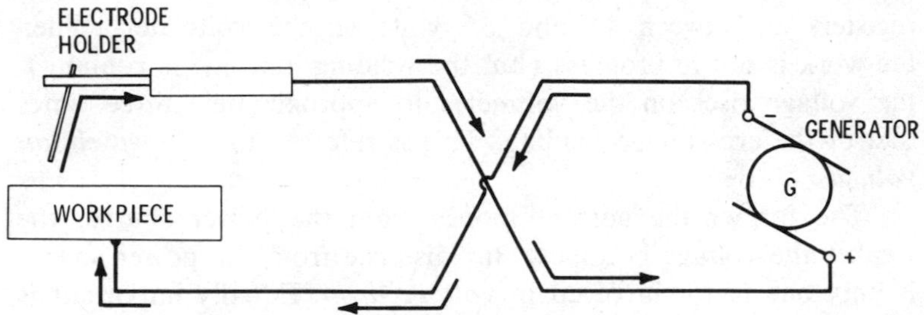

Fig. 3. Showing reversed polarity.

connecting the electrode cable to the negative pole and the work cable to the positive pole.

An electrical circuit is the completed flow of an electrical current from the negative terminal of a power source to the positive terminal of the power source. This phenomenon is illustrated in Figs. 3 and 4.

Amperes (or *amperage*) are units of measure used to indicate the intensity of the current (or the rate of flow in the circuit). The ampere capacity of a welding machine indicates the maximum current that can be supplied from the negative terminal. For

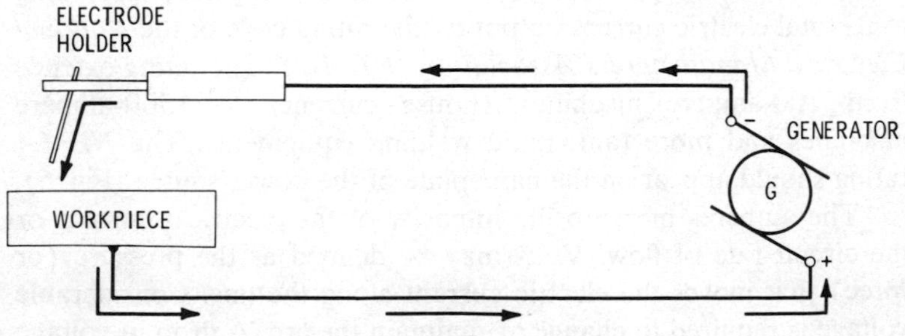

Fig. 4. Showing straight polarity.

example, a welding machine with a rating of 200 amperes can supply current at adjustments ranging up to that amperage rating.

Volts (or *voltage*) may be defined as the pressure or force that moves the electric current through the circuit. Voltage, or electrical pressure, is measured by a voltmeter.

Arc voltage (or *working voltage*) refers to the amount of voltage being used during the arc welding process. It usually registers at between 18 and 36 volts on the voltmeter. When the work is not in progress (but the welding machine is running), the voltage rises on the voltmeter to approximately three times that of the arc-voltage reading. This is referred to as *open-circuit voltage.*

The farther the current moves from the power source, the weaker the voltage becomes. But distance from the power source is only one factor involved in *voltage drop.* Equally important is the resistance found in the conductor (welding cable, wire, etc.). The degree of resistance in an electrical circuit is expressed in terms of another unit of measure—the *ohm.*

A *watt* is the rate at which power is transmitted or used. The number of watts can be determined by multiplying the amperes times the volts. ($P = E \times I$).

P is the power expressed in watts,
E is the voltage in volts,
I is the current in amperes.

WELDING POWER SOURCES

Each welding power source is rated (in amperes) according to its total electric current output by the rating code of the *National Electrical Manufacturers Association* (*NEMA*). The rating extends from 100-ampere machines (house current) to 1500-ampere machines and more (automatic welding equipment). The *NEMA* rating should appear on the nameplate of the power source (Fig. 5).

The amperes measure the intensity of the electrical current or the circuit rate of flow. Volts may be defined as the pressure (or force) that moves the electric current along the line. Considerable voltage is required to change or maintain the arc. A drop in voltage may impair the efficiency of the arc or may even cause it to be

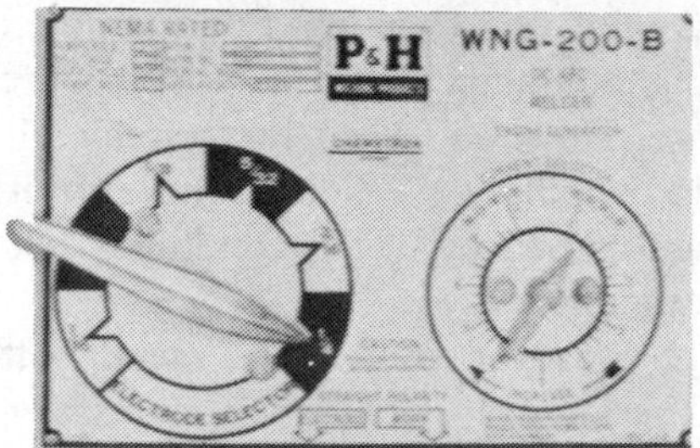

Fig. 5. Showing the NEMA power rating on welding equipment.

Courtesy Chemetron Corp.

extinguished altogether. The duty cycle is the percentage of a ten minute period during which the power source can safely operate at its rated power output.

Assuming that a welding power source is rated at 300 amps., 40 volts, and a 60% duty cycle, the rated power can be supplied six minutes out of every ten, or 60% of the time. Operating at a maximum duty cycle (100% duty cycle) would require a 77½% reduction of the rating of the power source for a current output of 230 amperes. On the other hand, operating at maximum current would require a reduction of the duty cycle to half its rated value, or 30% (three minutes out of every ten).

A 60% duty cycle is the basis for amperage ratings of nearly all drooping characteristic sources. Amperage ratings of constant voltage characteristic sources are based on 100% duty cycle. No power source for welding should be chosen without first considering its average anticipated duty cycle.

The electric arc is basically unstable. This instability is due primarily to variations in the arc resistance. In metallic arc welding, this instability becomes very pronounced. This is caused by the abrupt changes in the arc resistance which occur when metal from the electrode is transferred across the arc.

The electrode is melting constantly. Accordingly beads of metal drop off and as this occurs, and the arc length is continually changing. Since the arc voltage depends on the arc length, the arc voltage is also continually changing. The voltage decreases whenever a drop of molten metal passes across the arc. When there is no actual transfer of material across the arc, its inherent resistance characteristic is the only cause for instability and the requirements are not very exacting. For example, very little more than a drooping volt-ampere characteristic is required to maintain a carbon arc. However, when there is metal transferred across the arc, as in metallic arc welding, and abrupt changes in the arc resistance are caused by it, the conditions become much more complex and exacting.

Steady current melts metal better than a fluctuating current. Therefore, it is extremely necessary that the current be as steady, smooth, and unvarying as possible.

SELECTING THE POWER SOURCE

The welding power source must provide a welding current with characteristics suitable for a wide variety of tasks. The welder should select a welding machine that will produce the best possible results at the most economical cost. In order to do this, consideration must be given to the type of work (e.g. heavy vs. lightweight welding), the type of current (power) supply available (AC or DC, single-phase, etc.), the size of the welding area, the welding position, the type of metal or metals, and the cost factor.

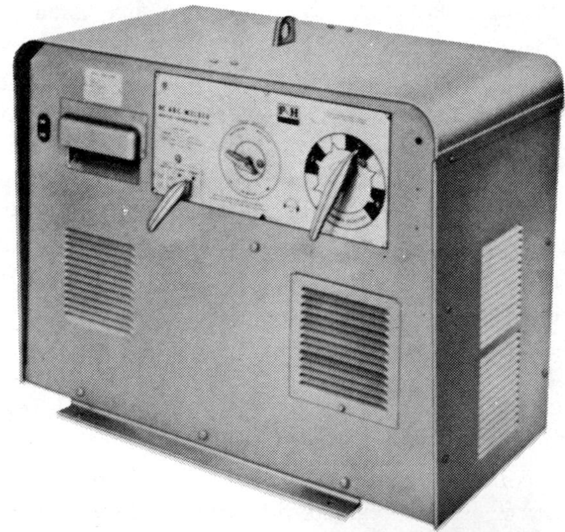

Courtesy Chemetron Corp.

Fig. 6. DC motor generator with a welding current range of 70-550 amps.

The power sources used to supply the electric current in arc welding can be divided into three main categories:

1. Those that supply direct (DC) current,
2. Those that supply alternating (AC) current,
3. Those that supply either alternating (AC) current or direct (DC) current.

DC POWER SOURCES

Power sources that deliver a direct welding current are of two basic types:

1. The DC generator,
2. The DC transformer-rectifier.

The function of either type is to change the input current to a direct current appropriate for welding purposes.

DC welding generators (Fig. 6) are designed to supply direct current (DC) to the electrode and the work. The generator itself may be powered by a diesel engine, an internal combustion engine, or an electric motor. Engine-powered generators are independent of external power supplies; a characteristic that makes them ideally suited for use in the out-of-doors or other areas where power lines are not available. Electric motor-powered generators, on the other hand, are restricted in usage to the availability of electric power lines.

The current supplied by a DC generator is created by an armature rotating in an electrical field. The armature is rotated by a motor or engine. The current is drawn off for welding use by a commutator. A polarity switch on most machines provides straight or reversed polarity.

Special control methods are used with DC generators to obtain current of proper strength and voltage. These control methods may be classified as either constant-voltage types or variable-voltage types.

In the constant-voltage method, a variable resistance in series with the arc is used to regulate the voltage to the proper value for welding. Since this method is wasteful, due to the power lost in the resistance, the variable method is more desirable. In this method, the generator is arranged with such characteristics that the voltage automatically adjusts to the varying voltages demanded by the arc. The variable-voltage machine is the most widely used and the most economical type of welder.

There are several types of external adjustment mechanisms for varying the current in variable-voltage machines. With a wheel or crank, it is possible to adjust the current over a wide range of possible selections (Fig. 7). Levers are also used for this type

Courtesy Chemetron Corp.

Fig. 7. Heavy duty AC welding machine. This movable coil design permits a continuous stepless control of the current.

of adjustment. Many welding machines provide two control mechanisms for varying the current. One is used for coarse current adjustment the other for fine adjustment. One method of coarse current adjustment is through a series of taps each providing a different current value. A voltage rheostat is used to obtain fine current adjustment (Fig. 8).

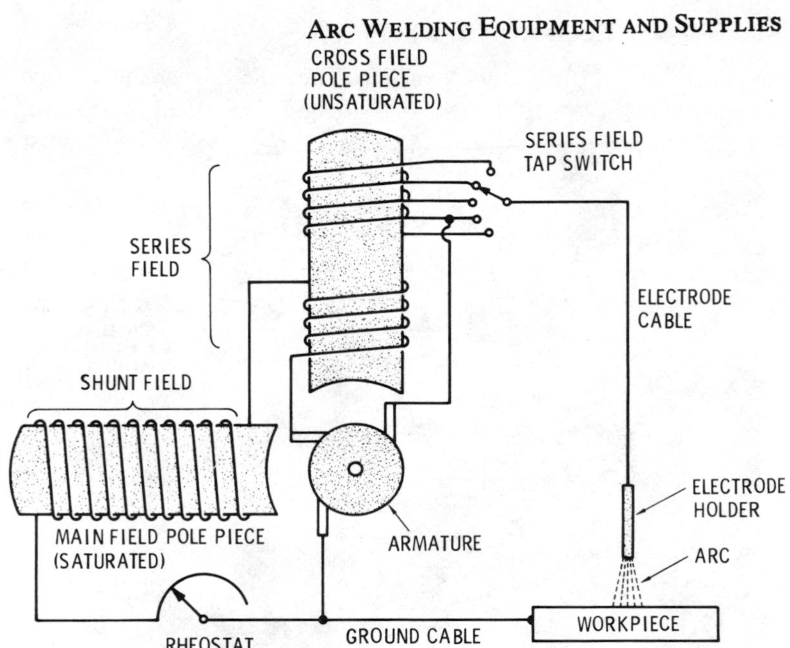

Fig. 8. Wiring diagram of a generator showing location of the rheostat.

Another type of dual control uses shunt field rheostats to provide continuous adjustments of the open-circuit voltage. The current is controlled by another dial which provides continuous amperage adjustments.

DC transformer-rectifier power sources (Fig. 9) are designed to convert the alternating current obtained from the power supply to direct current for welding purposes. These welding machines are essentially single- or three-phase transformers. Selenium rectifiers are employed to convert the alternating current to direct current.

Rectifier-type welding machines are designed to eliminate current drop-off during warm up. Due to their lack of rotating parts (as found in generators), they are characterized by quiet operation as well as low maintenance and operating costs. Furthermore, they provide good performance with low welding currents, and low power input at no load while idling.

Rectifier-type welding machines are adaptable to fine, medium, or heavy welding on all types and thicknesses of metal.

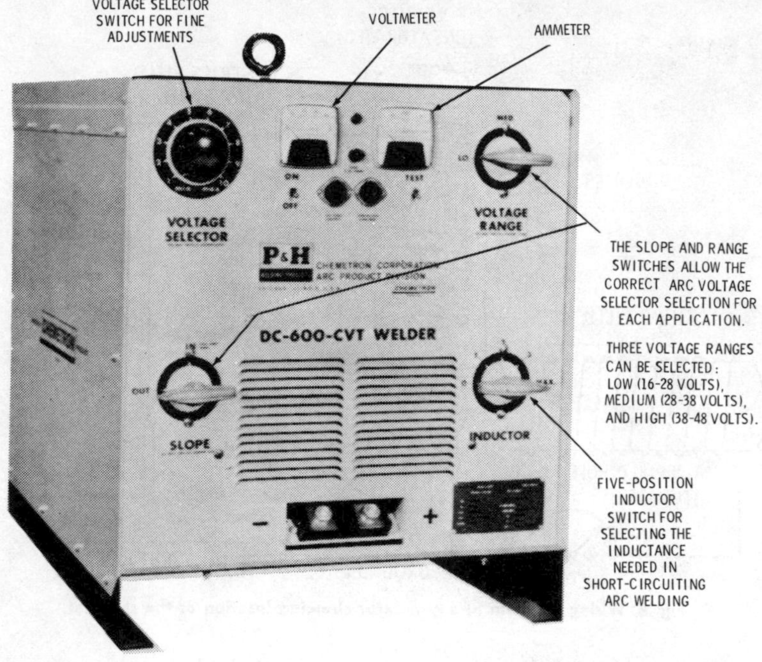

Courtesy Chemetron Corp.

Fig. 9. DC transformer-rectifier type welder.

AC POWER SOURCES

Alternating-current arc welding machines may be divided into two types:

1. AC transformers,
2. AC generators.

AC transformers are characterized as being current conditioning welding machines. AC generators, on the other hand, are regarded as being current generating.

An AC transformer (Fig. 10) consists of an adjustable transforming unit especially designed to deliver alternating current at the proper voltage and amperage for welding with a range of adjustments enabling the machine to meet the varied requirements of the welding circuit.

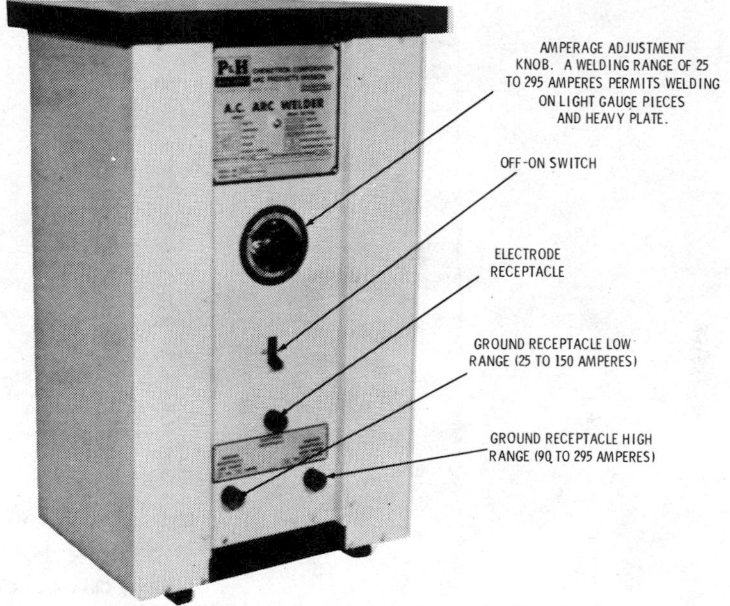

Courtesy Chemetron Corp.

Fig. 10. AC transformer welder.

An AC generator consists of an alternator which furnishes the welding current, driven by an AC motor direct connected. These may be divided into normal-frequency and high-frequency types.

The normal-frequency or transformer-type welding machine delivers 60 hertz. In the high-frequency machine, the alternator is designed to step up the frequency several times (usually three times). Under these circumstances, a 60-hertz welding machine will have an output of 180 hertz.

High frequency is used to reduce the intervals during the cycle when there is very little voltage to keep the arc going. The welding arc stream is conducting as long as energy is supplied to it. It takes about 15 volts to keep it going and if this energy is not supplied it will go out in a short interval of time. (Fig. 11).

Since alternating current passes through zero twice every cycle, it follows that there are two periods every cycle during which the

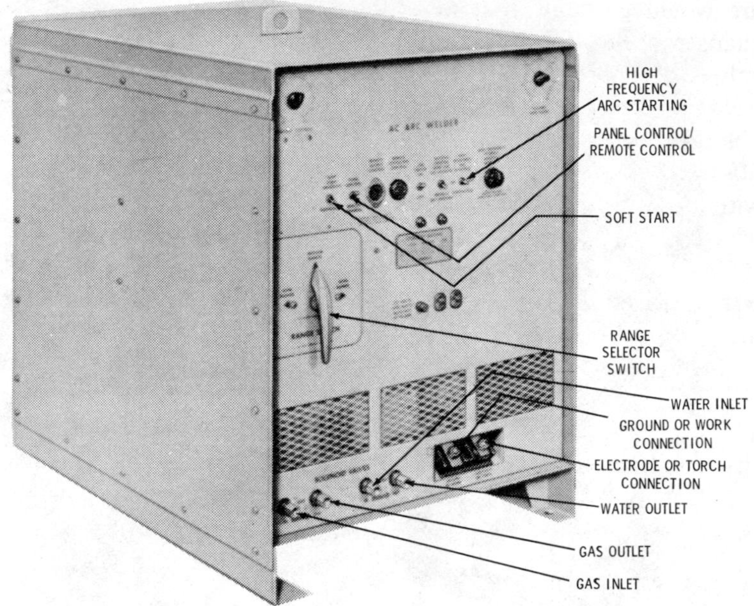

Courtesy Chemetron Corp.

Fig. 11. High frequency welder suitable for TIG welding.

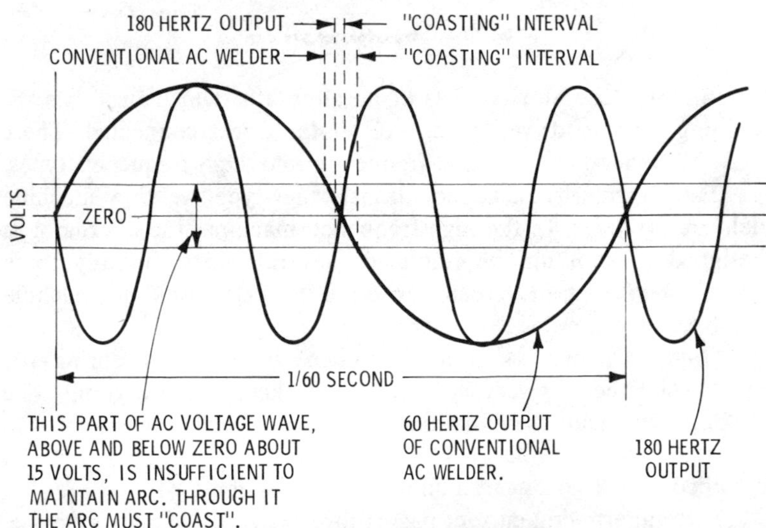

Fig. 12. The coasting interval of an AC cycle.

arc would go "out" if it were not for the speed with which the changes occur. These periods come between the plus and minus values of approximately 15 volts. Since it is necessary for the welding current to "coast" during these periods, it is highly desirable to shorten their duration as much as possible (Fig. 12). On 60-hertz transformer-type welding machines, the arc is improved with built in automatic arc stabilization embodied in the welder windings.

AC generator welding machines in both motor-driven and engine driven models are available, with the latter type being the more common of the two.

An example of an AC generator welding machine is the welder (Fig. 13) produced by *P & H Welding Products Co.* This unit has six main current settings on the control panel. These are changed by a six-position selector switch. Located above and to the right of the selector switch (as one faces the control panel)

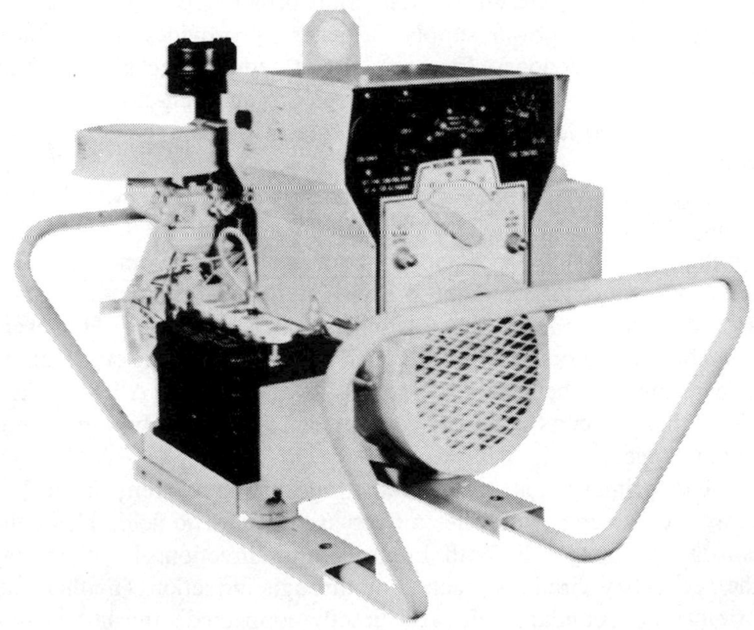

Courtesy Chemetron Corp.

Fig. 13. AC generator-type welder.

is a fine current control. The main current settings are 55, 80, 105, 130, 135, and 180 amperes. This welder uses all 1/16-inch to 5/32-inch welding electrodes and most 3/16-inch electrodes up to the capacity of the unit.

An AC transformer consists of the following principal components:

1. The transformer core,
2. The primary circuit,
3. The secondary circuit,
4. The adjustment mechanism,
5. The ventilating system,
6. The structural frame.

The core consists of magnetic sheets of laminated steel. The core and the two circuits must be insulated from one another according to *NEMA* standards.

The primary circuit of the transformer receives the input current from the power supply. It generally provides single-phase current. The secondary (or arc) circuit reduces the high-current voltage received from the primary circuit to an open-circuit voltage of 80 volts and a closed-arc circuit voltage of between 20 and 40 volts.

Each circuit (primary and secondary) consists of wire wound tightly around a core. The primary and secondary circuits (or windings) must *never* come in contact with one another. AC machines constructed under *NEMA* standards are almost guaranteed to be safe from the possibility of this happening. However, homebuilt transformers run a high risk of just such an occurrence. The results can be highly dangerous to the welder. All too often when these circuits come in contact there is a combining of the total voltage.

The primary coil receives the alternating current from the power supply and establishes a fluctuating magnetic field. This field constantly changes in both intensity and direction. Its effect on the secondary circuit is such that through induction (neither the primary or secondary coils are directly connected) the latter supplies a higher-value output current to the arc. Another result of the fluctuating magnetic field is the creation of an electrical flow in the opposite direction. This is referred to as *reactance* or *counter electromotive force*. The stronger this phenomenon becomes, the weaker is the welding current.

The output current of an AC transformer may be controlled by either: (1) reactor adjustment, or (2) reactance adjustment. In some AC transformers, reactor coils are situated in the secondary circuit between the electrode holder and the secondary coil. Control adjustment may be made by:

1. Tapping the reactor coil,

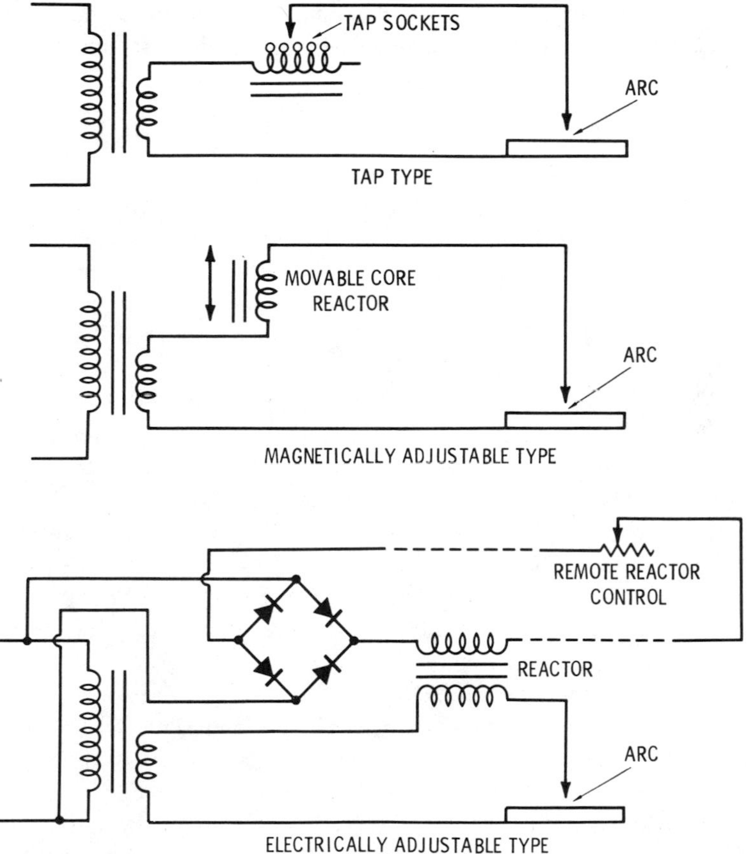

Fig. 14. Types of reactor adjustments.

2. Moving an iron bar within the reactor coil (magnetic adjustment),
3. Electrically saturating the reactor coil with direct current (electric adjustment) (Fig. 14).

The reactance may be adjusted by:

1. Changing the position of the primary and secondary coils relative to one another (coil adjustment),
2. Moving an iron bar between the primary and secondary coils (movable-core adjustment).

AC transformers in which reactance adjustment is the means of controlling output do not have reactors in their secondary circuit.

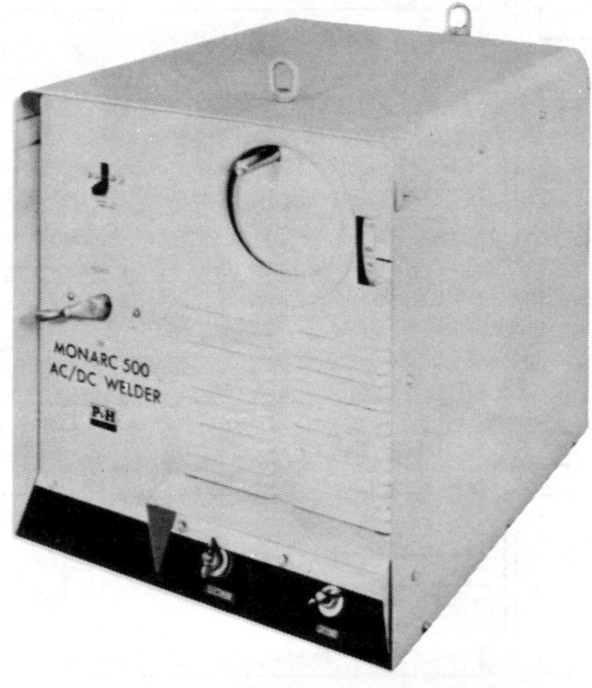

Courtesy Chemetron Corp.

Fig. 15. AC/DC combination welding power source.

COMBINATION AC/DC POWER SOURCES

AC/DC combination welding machines (Fig. 15) use selenium and silicon rectifiers to make the alternating- or direct-current selection. Their construction is similar to the DC transformer-rectifier welding machines. Adjustments for coarse and fine range current can be made. There is also a selection of DC straight polarity (DCSP) or DC reversed polarity (DCRP) current.

WELDING CABLES

In arc welding, two cables are used to create the required electric circuit between the welding machine and the work. One of these cables is referred to as the *electrode lead* and extends from the welding machine to an electrode holder. This cable forms one half of the electric circuit. The other cable is referred to as the *work lead* (or *ground lead*) and extends from the welding machine to the metal bench or the work itself. This cable forms the other half of the electrical circuit. A complete circuit is created by turning on the welding machine and bringing the electrode (in the electrode holder) in contact with the work (Fig. 16).

Welding cables (or leads) are available in a number of different sizes. The size selected for use is dependent on the capacity of the welding machine (stated in amperes) and the maximum

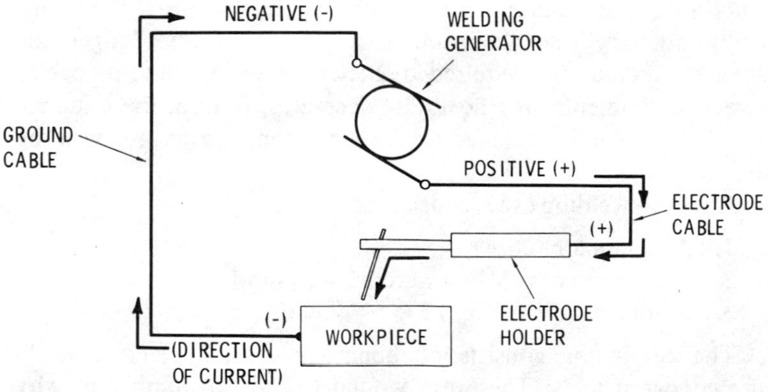

Fig. 16. Location of the electrode and ground cables creating the complete circuit.

length of the cable. Each welding cable size represents a specific conductor (core) diameter. The conductor is that portion of the welding cable through which the electric current flows. Each diameter has a recommended *maximum* cable length. Exceeding this length without increasing the diameter of the cable, results in a serious voltage drop. This in turn will produce a poor weld. Cable diameters and their maximum recommended lengths are given in Table 1.

Another important factor to be considered when using welding cables is their flexibility. A flexible electrode cable is particularly

Table 1. Cable Diameters and Their Recommended Maximum Length

Machine Size	Cable Sizes for Lengths		
in Amperes	Up to 50 ft.	50-100 ft.	100-250 ft.
75 to 100	4	2	2
200	2	1	2/0
300	0	2/0	4/0
400	2/0	3/0	4/0*
600	2/0	4/0	4/0*

* Recommended longest length of 4/0 cable for 400 amp. welder, 150 ft.; for 600 amp. welder, 100 ft. For greater distances cable size should be increased; however this may be a question of cost—consider ease of handling versus moving of welder closer to work.

important to ensure ease of movement for the welder. Flexibility is not that important for the ground cable, because it remains fairly stationary once connected to the work. The longer the cable, the greater the required diameter. Unfortunately, as cables increase in diameter their flexibility decreases. Both of these factors must be taken into consideration when setting up an arc welding site.

A typical welding cable consists of:

1. A conductor (core),
2. A covering of rubber and reinforced fabric,
3. An outer jacket (Fig. 17).

The conductor consists of hundreds of fine electrolytically pure copper strands. These are wound to form a number of wire bunches that are rope stranded to create the specific diameter

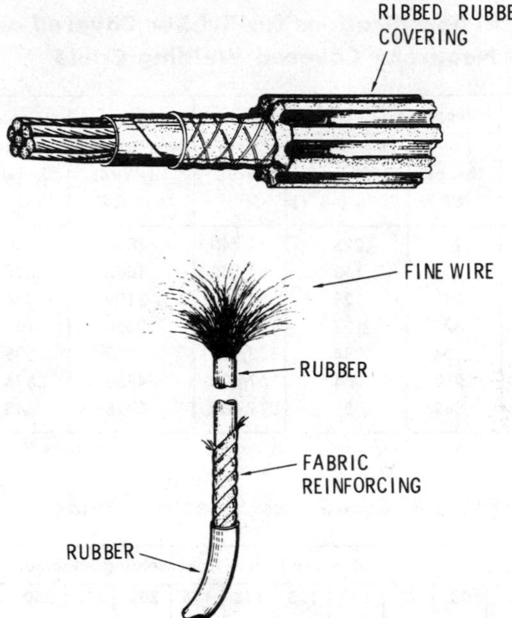

Fig. 17. Composition of a typical welding cable.

size of the welding cable (Fig. 18). Annealing the copper makes thc cablc soft and flexible. Welding cables are also being made with aluminum conductors, but these are not as common as the copper-type.

Fig. 18. Line drawing illustrating the bunching and rope-stranding of a typical cable.

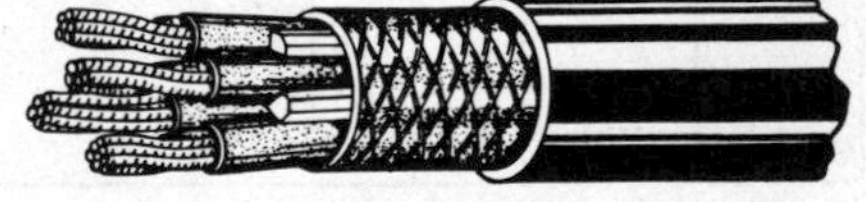

The conductor is covered with a layer of rubber and reinforced fabric which protects the copper from corrosion. The outer jacket consists of a durable layer of rubber or neoprene. Specifications for rubber-covered and neoprene-covered cables are listed in Table 2.

Table 3 is a cable-size selection guide for determining cable requirements in terms of amperage used and the distance of the work from the welding machine.

Table 2. Specifications for Rubber Covered and Neoprene Covered Welding Cable

Rubber Covered		Neoprene Covered		Nominal Area (Circular Mils)	Stranding Nominal No. of Wires #34 AWG	Nominal Conductor Diameter (Inches)	Nominal Cable O.D. (Inches)
Cable Size	Lbs. per 1,000 ft.	Lbs. per 1,000 ft.	Ohms per 1,000 ft. 20°C (68°F)				
4	186	201	.225	41,740	1064	.269	.45
2	288	308	.163	66,360	1672	.337	.55
1	357	383	.129	83,690	2109	.376	.60
1/0	442	472	.102	105,600	2660	.423	.66
2/0	548	584	.084	133,100	3325	.508	.73
3/0	679	718	.064	167,800	4256	.576	.80
4/0	837	882	.051	211,600	5328	.645	.87

Courtesy Airco Welding Products

Table 3. Cable Size Selection Guide

Cable Size	Amps	Distance in feet from welding machine										
		50	75	100	125	150	175	200	225	250	300	350
4	100	4	4	2	2	1	1/0	1/0	2/0	2/0	3/0	4/0
2	150	4	2	1	1/0	2/0	3/0	3/0	4/0	4/0		
	200	2	1	1/0	2/0	3/0	4/0	4/0				
1	250	2	1/0	2/0	3/0	4/0						
	300	1	2/0	3/0	4/0		Based on 4 volt drop					
1/0	350	1/0	3/0	4/0								
2/0	400	1/0	3/0	4/0								
	450	2/0	4/0									
3/0	500	2/0	4/0									
	550	3/0										
4/0	600	3/0										

Courtesy Airco Welding Products

Connections and Accessories

The ground cable may be connected to the work (or metal work bench) by a variety of methods including special clamps (Fig. 19), rotary devices (Fig. 20), a magnet, a hook, or a heavy weight. The last three mentioned are the least secure methods since they can be disconnected merely by jarring the welding cable or the work.

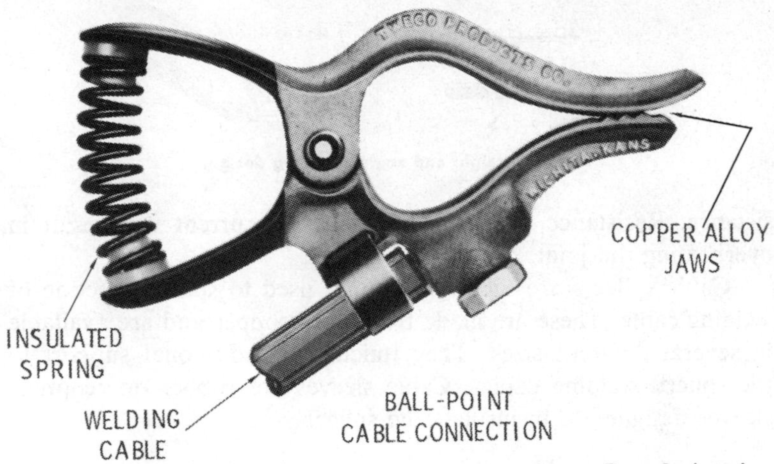

Courtesy Tweco Products, Inc.

Fig. 19. Portable ground clamp (200 amperes).

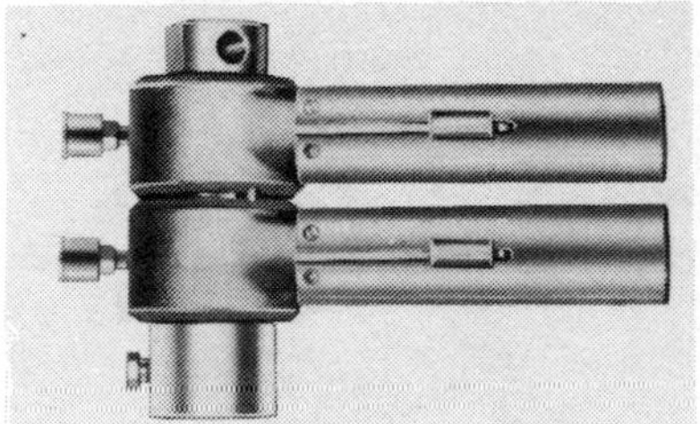

Fig. 20. Heavy-duty rotary grounding device (1500 amperes).

Courtesy Tweco Products, Inc.

The welding cable must be connected to the welding machine with suitable lugs. These are available in a wide variety of sizes depending upon:

1. Welding cable size,
2. The UL ampere rating,
3. Size of the stud hole.

These cable lugs are made of a cast copper alloy and may also be covered with a fiber insulation layer. They are also available in a straight or angle design (Fig. 21).

The lugs are attached to the welding cables by soldering or mechanical methods such as crimping. Whatever the means employed, it must result in a heavy-duty connection offering low re-

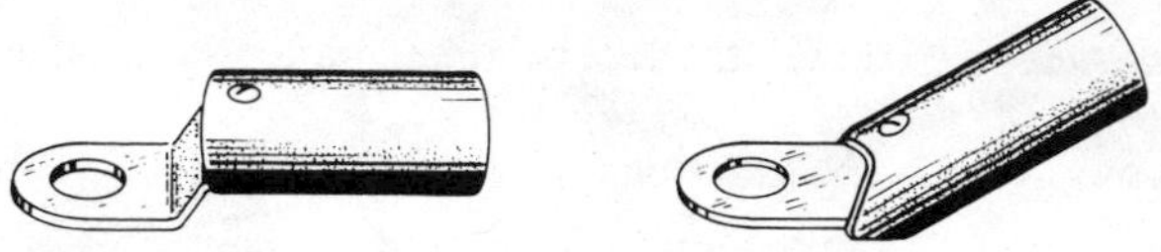

Fig. 21. Straight and angle cable lug designs.

sistance. Resistance to the flow of electric current will result in overheating the joint.

Cable splicers are tube-type fittings used to splice a section of welding cable. These are made from pure copper and are available in several different sizes. They function as additional support to the spliced welding cables. Cable sleeves are rubber or neoprene sleeves designed to in insulate the splice.

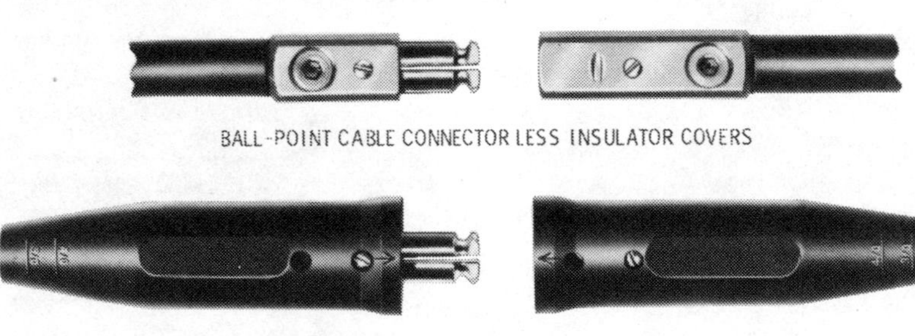

Courtesy Tweco Products, Inc.

Fig. 22. Cable connectors.

Cable connectors (Fig. 22) are fittings especially designed for joining sections of welding cable. The copper conductor is insulated and protected by a rubber or neoprene jacket. Cable connectors are designed to be easily connected or pulled apart.

Whip cable connectors (Fig. 23) have both a whip-end and a plug-end. The plug-end fits the standard welding cable. The

Fig. 23. Whip cable connectors.

Courtesy Tweco Products, Inc.

whip-end is scaled down to accept the smaller diameter cables used for whip leads.

Cable hitches are useful devices for reducing excessive cable weight and drag. The reduction in drag reduces wear on cable connections. The welding cable can be pulled through the hitch, but will not slip back.

Electrode Holders

Electrode holders (Fig. 24) are clamping devices designed to hold the electrode during the arc (manual) welding process. The electrode lead is fastened to the end of the handle with a lug or to a mechanical connection within the handle. There are many different electrode holder designs offered by manufacturers. The welder should select the one most comfortable for use during the welding process.

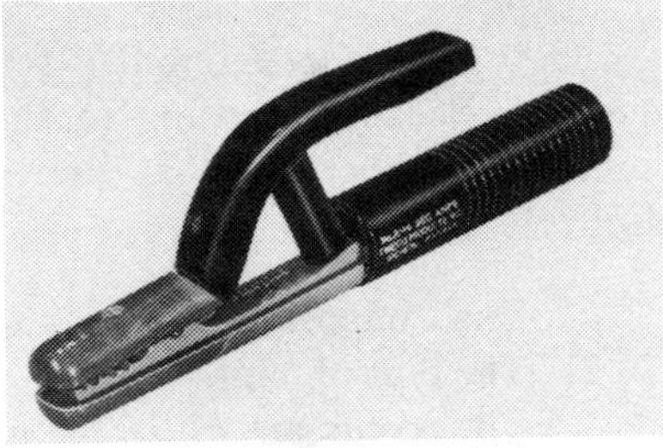

Fig. 24. Electrode holder (350 amperes).

Courtesy Tweco Products, Inc.

The point at which the lead is connected to the electrode holder must be completely insulated or the welder may be subjected to electrical shocks.

Sometimes an electrode holder may become uncomfortably hot during welding. This is frequently an indication of a loose connection between the lead cable and the electrode holder. It can also mean that the electrode holder is the wrong size for the work.

ARC WELDING ELECTRODES

An arc welding electrode is a metal wire or rod used as a terminal (or terminals) in an electrical current for the purpose of producing an electric arc. In the arc welding process, the electrode also provides the filler metal to the weld. The electrodes used are either metallic or carbon, depending on the work to be welded and other requirements.

Electrodes used for welding the commercial grades of wrought iron, plate, structural steel, cast steel, and (to a considerable extent) cast iron, should be a high grade of low carbon steel wire having a carbon content of .20% or less. Practically all commercial welding wire on the market meets this requirement, although there are a number of special electrodes containing greater amounts of carbon.

Selecting the Electrode

There are many different types of electrodes. The one you use must be selected according to the requirements of the job. Five points of comparison that are helpful in selecting the proper electrode are:

1. The composition of the base metal,
2. The type of joint,
3. The position in which the weld must be made,
4. The physical properties required (Xray quality, etc.),
5. The condition of the fit up.

Table 4. N.E.M.A. Color Code for Covered Electrodes

CLASS	END COLOR	SPOT COLOR
Group Color-Blue		
ECu	GREEN	NONE
ECu-Si	RED	NONE
ECu-SnA	YELLOW	NONE
ECu-Ni	NONE	BLUE
ECu-SnC	YELLOW	BLUE
ECu-A 1-A2	SILVER	BLUE
ECu-A 1-B	SILVER	BROWN
ECu-A 1-C	SILVER	GREEN
ECu-A 1-D	SILVER	RED
ECu-A 1-E	SILVER	YELLOW

CLASS	END COLOR	SPOT COLOR
SURFACING ELECTRODES Group Color—Red		
EFeMn-A	BLUE	NONE
EFeMn-B	BLUE	BLUE
EFeCr-A1	WHITE	NONE
EFeCr-A2	WHITE	BLUE
EFe5-A	BROWN	NONE
EFe5-B	BROWN	BLUE
EFe5-C	BROWN	WHITE
ECoCR-A	GREEN	RED
ECoCr-B	GREEN	GREEN
ECoCr-C	GREEN	BLACK
ENiCr-A	RED	NONE
ENiCr-B	RED	BLUE
ENiCr-C	RED	WHITE

Table 4. N.E.M.A. Color Code for Covered Electrodes (Contd.)

CLASS	END COLOR	SPOT COLOR
LOW ALLOY STEEL Group Color—Silver		
E-7020-G	BLUE	GREEN
E-7020-A1	BLUE	YELLOW
E-8013-G	WHITE	BROWN
E-8013-B1	WHITE	WHITE
E-8013-B2	BROWN	WHITE
E-9013-G	BROWN	BROWN
E-10013-G	GREEN	BROWN
MILD STEEL AND LOW ALLOY STEEL XX10, XX11 and all 60XX Group Color—None		
E-6010	NONE	NONE
E-6011	NONE	BLUE
E-6012	NONE	WHITE
E-6013	RED	BROWN
E-6014	NONE	BROWN
E-6015	NONE	RED
E-6016	NONE	ORANGE
E-6018	RED	ORANGE
E-6020	NONE	GREEN
E-6024	NONE	YELLOW
E-6027	NONE	SILVER
E-6028	RED	BLACK
E-7010-G	BLUE	NONE
E-7010-A1	BLUE	WHITE
E-7011-G	BLUE	BLUE
E-7011-A1	BLUE	YELLOW
E-7014	BLACK	BROWN
E-7024	BLACK	YELLOW
E-7028	BLACK	BLACK
E-8010-G	WHITE	NONE
E-8010-B1	WHITE	BROWN
E-8010-B2	WHITE	GREEN
E-8011-G	WHITE	BLUE
E-8011-B1	WHITE	BLACK
E-8011-B2	BROWN	BLACK
E-9010-G	BROWN	NONE
E-9011-G	BROWN	BLUE
E-10010-G	GREEN	NONE
E-10011-G	GREEN	BLUE
E-12015-G	ORANGE	RED
E-12016-G	ORANGE	ORANGE
E-12018-G	ORANGE	BLUE

CLASS	COLOR END	COLOR SPOT
LOW HYDROGEN LOW ALLOY STEEL Group Color—Green		
E-7015	BLUE	RED
E-7015-G	NONE	RED
E-7015-A1	BLUE	WHITE
E-7016	BLUE	ORANGE
E-7016-G	NONE	ORANGE
E-7016-A1	BLUE	YELLOW
E-7018	BLACK	ORANGE
E-7018-G	NONE	BLUE
E-7018-A1	BLACK	YELLOW
E-8015-G	GRAY	RED
E-8015-B1	WHITE	BROWN
E-8015-B2	WHITE	GREEN
E-8015-B4	BROWN	BRONZE
E-8015-B2L	BLACK	GREEN
E-8015-B4L	BLACK	BRONZE
E-8015-C1	WHITE	BRONZE
E-8015-C2	WHITE	WHITE
E-8015-C3	WHITE	RED
E-8016-G	WHITE	YELLOW
E-8016-B1	WHITE	BLACK
E-8016-B2	WHITE	GRAY
E-8016-B4	BROWN	VIOLET
E-8016-C1	WHITE	BLUE
E-8016-C2	WHITE	VIOLET
E-8016-C3	WHITE	ORANGE
E-8018-G	BLACK	BLUE
E-8018-B1	GRAY	BLACK
E-8018-B2	GRAY	GRAY
E-8018-B4	BLACK	GRAY
E-8018-C1	GRAY	BLUE
E-8018-C2	GRAY	VIOLET
E-8018-C3	BLACK	BLACK
E-9015-G	BROWN	RED
E-9015-B3	BROWN	GREEN
E-9015-B3L	BLACK	WHITE
E-9015-D1	BROWN	WHITE
E-9016-G	BROWN	ORANGE
E-9016-B3	BROWN	BLUE
E-9016-D1	BROWN	YELLOW
E-9018-G	VIOLET	BLUE
E-9018-B3	VIOLET	BLACK
E-9018-D1	VIOLET	VIOLET
E-10015-G	GREEN	RED
E-10015-D2	GREEN	YELLOW
E-10016-G	GREEN	ORANGE
E-10016-D2	GREEN	GRAY
E-10018-G	GREEN	BLUE
E-10018-D2	GREEN	VIOLET
E-11015-G	RED	RED
E-11016-G	RED	YELLOW
E-11018-G	RED	BLUE

Table 4. N.E.M.A. Color Code for Covered Electrodes (Contd.)

NICKEL, NICKEL-ALLOY AND HIGH TEMPERATURE ELECTRODES
Group Color—White

CLASS	COLOR END	COLOR SPOT
E3N10	BLUE	WHITE
E4N10	BLUE	BROWN
E3N14	BLUE	RED
E3N1B	WHITE	GREEN
E3N1C	WHITE	VIOLET
E4N12	GREEN	BROWN
E3N11	YELLOW	WHITE
E4N11	YELLOW	BROWN
E3N12	VIOLET	WHITE
E3N19	VIOLET	RED
ENi	ORANGE	BLUE
ENi-Cu	ORANGE	WHITE
ENi-Fe	BRONZE	ORANGE
ENi-CuB	ORANGE	BROWN
Ni-Cr60-13	ORANGE	GREEN
Ni-Cr80-15	GREEN	BLUE
Mil 4N1W	GREEN	WHITE
Mil 3N1L	WHITE	YELLOW
Mil 3N1N	BRONZE	WHITE

Electrode Identification

Color codes were originally used to identify electrodes. These were established by the *NEMA (National Electrical Manufacturers Association)* in accordance with specifications designed to distinguish various electrodes. The color code was painted on the electrode and consisted of:

1. An end color,
2. A spot color,
3. A group color (Fig. 25).

Typical colors for mild-steel arc welding electrodes are listed in Table 4.

Color codes are being phased out as an electrode identification system and replaced by identification numbers marked on the surface of the coated electrode (Fig. 26).

An electrode classification system for most welding electrodes has been devised by the joint committee of the *American Welding Society (AWS)* and the *American Society for Testing Materials (ASTM)*. This classification system is continually under revision and enlargement.

The arc welding electrodes in the *AWS* classification system are classified according to definite filler metal specifications and can be grouped into a number of different categories. These categories are distinguished from one another by their basic metal composition, and include among others:

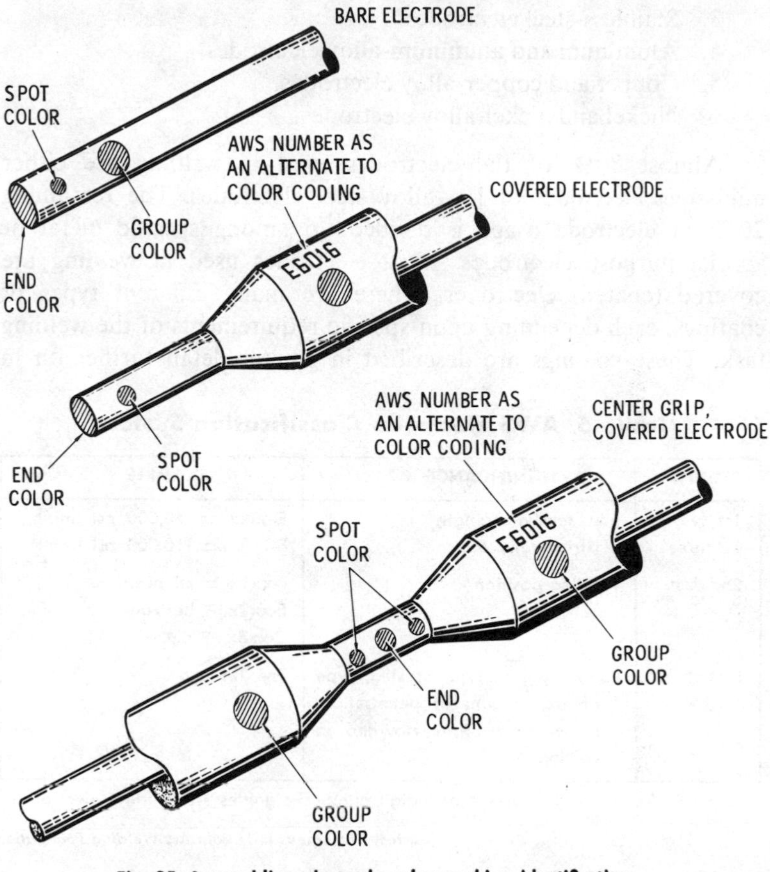

Fig. 25. Arc welding electrode color marking identification.

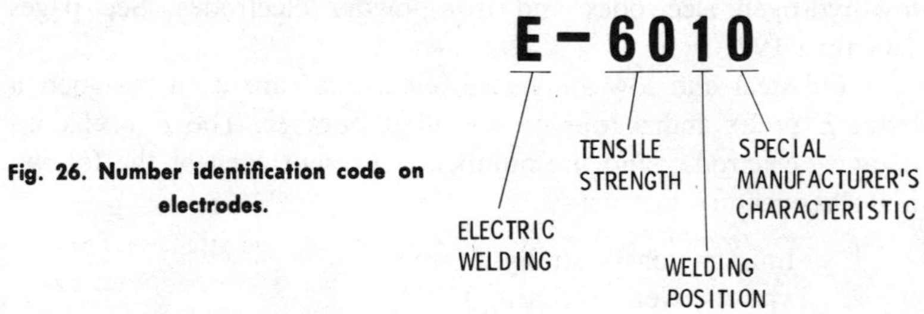

Fig. 26. Number identification code on electrodes.

1. Mild-steel electrodes,
2. Low-alloy steel electrodes,

3. Stainless-steel electrodes,
4. Aluminum and aluminum-alloy electrodes,
5. Copper and copper-alloy electrodes,
6. Nickel and nickel-alloy electrodes.

Almost 80% of the electrodes used in welding are either mild-steel electrodes or low-alloy steel electrodes. The remaining 20% of electrode usage is divided up among specific metal or special purpose electrodes. Most electrodes used in welding are covered (coated) electrodes. There are many different types of coatings, each depending upon specific requirements of the welding task. These coatings are described in greater detail further on in

Table 5. AWS Electrode Classification System

DIGIT	SIGNIFICANCE	EXAMPLE
1st two or 1st three	Min. tensile strength (stress relieved)	E-60xx = 60,000 psi (min) E-110xx=110,000 psi (min)
2nd last	Welding position	E-xx1x =all positions E-xx2x = horizontal and flat E-xx3x = flat
Last	Power supply, type of slag, type of arc, amount of penetration, presence of iron powder in coating	See Table 6

Note: Prefix "E" (to left of a 4 or 5-digit number) signifies arc welding electrode

Courtesy The James F. Lincoln Arc Welding Foundation

this chapter. Specific details are given for such common types as low-hydrogen electrodes and iron-powder electrodes. See pages 188 thru 197.

Mild-steel and low-alloy steel electrodes are each assigned a letter *E* prefix and a four or five digit number. The *E* prefix indicates "electrode" and the numbers represent each of the following classificatory factors:

1. Minimum tensile strength (psi),
2. Type of covering (coating),
3. Electrode welding positions,
4. Type of welding current (Table 5).

The minimum tensile strength is represented by the first two digits. This is the minimum tensile strength of the as-welded deposited weld metal in thousands of pounds per square inch (1000 psi). Expressed in these terms, the electrode classification E60XX designates a 60,000 psi minimum tensile strength; E70XX designates a minimum tensile strength of 70,000 psi; and so forth.

The third digit of the electrode designator indicates the welding position. The fact that *all* welding positions can be used with a particular electrode is indicated by the number 1 (e.g. EXX1X). Number 2 indicates that the electrode is only suitable for use in the flat or horizontal position (e.g. EXX2X). Electrodes suitable for use *only* in the flat position are indicated by the number 3 (e.g. EXX3X).

Table 6. AWS Electrode Classification (Final Digit Interpretation)

LAST DIGIT	0	1	2	3	4	5	6	7	8
Power supply	(a)	AC or DC rev polarity	AC or DC	AC or DC	AC or DC	DC rev polarity	AC or DC rev.	AC or DC polarity	AC or DC rev. polarity
Type of slag	(b)	Organic	Rutile	Rutile	Rutile	Low Hydrogen	Low Hydrogen	Mineral	Low Hydrogen
Type of arc	Digging	Digging	Medium	Soft	Soft	Medium	Medium	Soft	Medium
Penetration	(c)	Deep	Medium	Light	Light	Medium	Medium	Medium	Medium
Iron Powder in Coating	0-10%	None	0-10%	0-10%	30-50%	None	None	50%	30-50%

NOTES: (a) E-6010 is DC reverse polarity; E-6020 is AC or DC
(b) E-6010 is organic; E-6020 is mineral
(c) E-6010 is deep penetration; E-6020 is medium penetration

Courtesy The James F. Lincoln Arc Welding Foundation

The fourth digit indicates the type of coating found on the particular electrode as well as the welding current with which it can be used. If the fourth digit is a zero, then the third digit is used to designate the coating and type of welding current. Table 7 lists the various fourth digit values along with the types of coating and welding current they are used to designate.

Reading the third and fourth digits together (when the fourth digit is a zero) is sometimes referred to as the *usability identification.* Table 7 illustrates how these two digits are read together.

Coated (Covered) Electrodes

The electrode covering (coating) is designed to provide greater stability to the electric arc. It also protects the electric arc from contact with harmful elements (oxygen, nitrogen, etc.) in the surrounding atmosphere. It performs this function by creating a shielding gas as it dissolves during the arc welding process. The shielding gas acts as a neutralizing or reducing agent.

Table 7. Fourth Digit Values

FOURTH DIGIT	TYPE OF COATING	WELDING CURRENT
0	*	*
1	cellulose potassium	AC or DC Reverse or Straight
2	titania sodium	AC or DC Straight
3	titania potassium	AC or DC Straight or Reverse
4	iron powder titania	AC or DC Straight or Reverse
5	low hydrogen sodium	DC Reverse
6	low hydrogen potassium	AC or DC Reverse
7	iron powder iron oxide	AC or DC
8	iron powder low hydrogen	AC or DC Reverse or Straight

* When the fourth digit is 0, the type of coating and current to use are determined by the third digit. For example, E-6010 indicates a cellulose sodium coating and operates on DC Reverse, while E-6070 has iron oxide coating and operates on AC or DC.

Courtesy Hobart Brothers Company

Electrodes and metal filler rods were originally lengths of bare metal wire. However, the instability of the arc produced with bare electrodes led to the development of special coatings to eliminate this problem. As a result, the vast majority of electrodes in use today have some sort of coating (Fig. 27). These coatings are not only designed to produce a stable arc but also provide specific chemical and physical characteristics required by the type of metal being welded or the nature of the job itself.

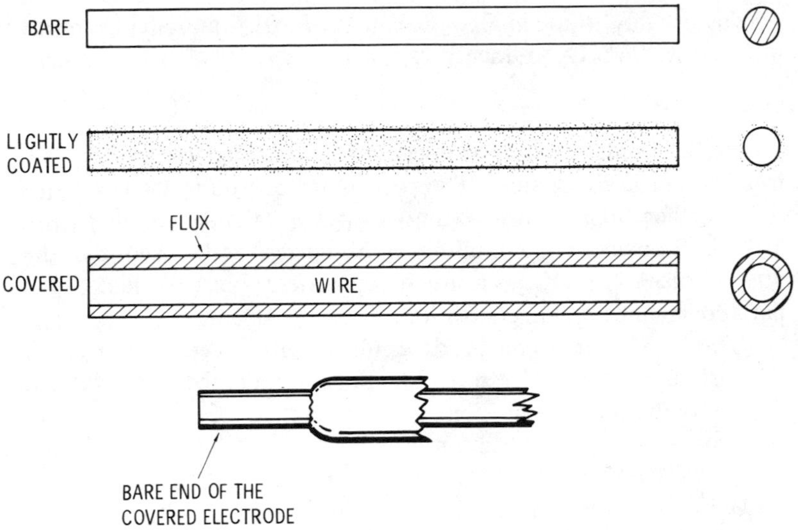

BARE

LIGHTLY COATED

FLUX

COVERED

WIRE

BARE END OF THE COVERED ELECTRODE

Fig. 27. Various electrode coatings.

Low-Hydrogen Coatings (Electrodes)

The presence of hydrogen during the welding of the various alloy steels seriously affects the quality of the weld. Low-hydrogen coatings (electrodes) were developed to eliminate this problem.

AWS E6015 was the first low-hydrogen electrode to be developed for this purpose. This electrode is used with DC reversed polarity current and can be used in all welding positions. The coating designed for the electrode has a very low moisture content, a factor which prevents the introduction of hydrogen into the weld during the welding process. The *AWS* E6015 electrode was specifically designed to be used with high-sulphur and high-carbon alloy steels.

The *AWS* E6016 and *AWS* E6018 electrodes represent further developments of the low-hydrogen type. Both of these electrodes can be used with either AC or DC reversed polarity current. The E6016 has a coating containing potassium salts (an arc stabilizing element). The E6018 has a coating of 30% powdered iron.

The E6015, E6016, and E6018 electrodes are frequently referred to as low-hydrogen electrodes, but the list has been considerably expanded since the introduction of these original three.

To this list can be added E6028, E7016, E7018, and all electrodes with the *AWS* EXXX8 identification.

Iron Powder Coatings

Faster welds can be produced using shielded electrodes with iron powder in the coating. The percentage of iron in the coating is set at either 30% (lime coating) or 50% (titania coating). The 50% iron powder content allows more current to be used than the 30% content. It produces a much faster weld, but is limited to the flat (downhand) welding position.

The iron content can be determined within certain ranges by the last digit of the electrode identification number. EXX4 and EXX8 both indicate an iron powder content of 30% to 50%. Electrodes having identification numbers of EXX0 and EXX2 contain 0 to 10% iron powder, and an EXX7 number indicates a 50% iron-powder content.

Coatings containing significant amounts of iron powder are thicker than other types. As a result, they require a higher current. Iron powder in the coating also contributes to a faster deposition rate.

Sometimes iron-powder coatings are also low-hydrogen types, and posses the special characteristics associated with each element (see LOW-HYDROGEN ELECTRODES).

Stainless-Steel Electrodes

A number of different electrodes have been developed to be used when welding stainless steels. Selecting the proper electrode depends primarily on the composition of the base metal to be welded.

Stainless-steel electrodes are available with either lime or titania coatings. The first is used only with DC reverse polarity current, the second with both AC and DC reverse polarity current.

The lime-coated electrodes produce flat or slightly convex fillet welds. The slag covers the entire weld, spatter is at a minimum, and the impurities are fluxed from the weld metal.

The titania-coated electrodes produce slightly concave type welds with a smoother and more stable arc than that found with the lime-coated type.

The numbering identification system for stainless steel electrodes differs somewhat from that used for mild steel and low-alloy steel electrodes. The prefixed *E* is used to indicate an arc welding electrode. This is followed by a three-digit number that indicates the type of stainless steel (e.g. E308, E410, E502, etc.). Two more digits follow and are separated from the first three by a hyphen. These last two digits are one of the following four number combinations:

1. -15 lime coating, DCRP in all welding positions,
2. -16 titania coating, AC or DCRP in all welding positions,
3. -25 titania coating, DCRP in horizontal and flat positions,
4. -26 titania coating, AC or DCRP in horizontal and flat positions.

Mild-Steel Electrodes

Mild-steel electrodes (Table 8) are general purpose types used in industrial welding on mild-steel fabrications of plate, castings, and piping. These electrodes are frequently used on boiler and pressure vessel work as well as welding the commercial grades of wrought iron and cast iron. Their wide usage is an indication of their adaptability to a broad range of welding tasks.

Both *AWS* E45XX and *AWS* E60XX series electrodes can be classified as mild-steel electrodes. *AWS* E4510 and *AWS* E4520 are bare-metal electrodes and do not receive as wide usage as electrodes in the *AWS* E60XX series.

AWS E6010 is a mild-steel electrode suitable for all welding positions, but limited to DC reverse polarity current. *AWS* E6011 and *AWS* E6012 electrodes function with either AC or DC current. Other mild-steel electrodes are listed and described in Table 8.

Table 9 illustrates the various factors involved in selecting the most appropriate mild-steel electrode for the work.

Low-Alloy Steel Electrodes

Low-alloy steel electrodes (Table 10) arc designed for welding carbon-molybdenum and low-alloy steels. A common application is the welding of high temperature, high pressure piping, carbon moly

Table 8. Mild Steel Arc Welding Electrodes (A233-48T)

Electrode Classification Number	Type of Coating or Covering	Capable of Producing Satisfactory Welds in Positions Shown*	Type of Current
E45 Series.—Minimum Tensile Strength of Deposited Metal in Non-Stress-Relieved Condition 45,000 Psi			
E4510............ E4520............	Sulcoated or light coated	F, V, OH, H H-Fillets, F	Not specified, but generally DC, straight polarity (electrode negative).
E60 Series.—Minimum Tensile Strength of Deposited Metal in Non-Stress-Relieved Condition 62,000 Psi (or higher—See Table 10).			
E6010............	High cellulose sodium	F, V, OH, H	For use with DC, reversed polarity (electrode positive) only.
E6011............	High cellulose potassium	F, V, OH, H	For use with AC or DC, reversed polarity (electrode positive).
E6012............	High titania sodium	F, V, OH, H	For use with DC, straight polarity (electrode negative), or AC.
E6013............	High titania potassium	F, V, OH, H	For use with AC or DC, straight polarity (electrode negative).
E6015............	Low hydrogen sodium	F, V, OH, H	For use with DC, reversed polarity (electrode positive) only.
E6016............	Low hydrogen potassium	F, V, OH, H	For use with AC or DC, reversed polarity (electrode positive).
E6020............	High iron oxide	H-Fillets, F	For use with DC, straight polarity (electrode negative), or AC for horizontal fillet welds; and DC, either polarity, or AC, for flat-position welding.
E6030............	High iron oxide	F	For use with DC, either polarity, or AC.

* The abbreviations F, H, V, OH, and H-Fillets indicate welding positions as follows:

F = Flat
H = Horizontal

H-Fillets = Horizontal Fillets

V = Vertical
OH = Overhead } For electrodes 3/16 in. and under except in classifications E6015 and E6016, 5/32 in. and under.

Table 9. Mild Steel Electrode Selection Guide

	Electrode Class										
	E6010	E6011	E6012	E6013	E7014	E7016	E7018	E6020	E7024	E6027	E7028
Groove butt Welds, flat (> ¼ in.)	4	5	3	8	9	7	9	10	9	10	10
Groove butt welds, all positions (> ¼ in.)	10	9	5	8	6	7	6	(b)	(b)	(b)	(b)
Fillet welds, flat or horizontal	2	3	8	7	9	5	9	10	10	9	9
Fillet welds, all positions	10	9	6	7	7	8	6	(b)	(b)	(b)	(b)
Current(c)	DCR	AC DCR	DCS AC	AC DC	DC AC	DCR AC	DCR AC	DC AC	DC AC	DC AC	AC DCR
Thin material (< ¼ in.)	5	7	8	9	8	2	2	(b)	7	(b)	(b)
Heavy plate or highly restrained joint	8	8	6	8	8	10	9	8	7	8	9
High-sulfur or off-analysis steel	(b)	(b)	5	3	3	10	9	(b)	5	(b)	9
Deposition rate	4	4	5	5	6	4	6	6	10	10	8
Depth of penetration	10	9	6	5	6	7	7	8	4	8	7
Appearance, undercutting	6	6	8	9	9	7	10	9	10	10	10
Soundness	6	6	3	5	7	10	9	9	8	9	9
Ductility	6	7	4	5	6	10	10	10	5	10	10
Low-temperature impact strength	8	8	4	5	8	10	10	8	9	9	10
Low spatter loss	1	2	6	7	9	6	8	9	10	10	9
Poor fit-up	6	7	10	8	9	4	4	(b)	8	(b)	4
Welder appeal	7	6	8	9	10	6	8	9	10	10	9
Slag removal	9	8	6	8	8	4	7	9	9	9	8

(a) Rating is on a comparative basis of same size electrodes with10 as the highest value. Ratings may change with size.

(b) Not recommended.

(c) DCR—direct current reverse, electrode positive; DCS—direct current straight; electrode negative; AC—alternating current; DC—direct current, either polarity.

Courtesy Airco Welding Products

Table 10. Low-Alloy Steel Arc Welding Electrodes (A316-48T)

Classification Number	Electrode: Diameter, in.	Electrode: Current and Polartiy	All-Weld Metal Tension Test[b]	Guided-Bend Test[c]	Fillet-Weld Test[d]	Hydrogen Test[e]
E7010	1/16 to 1/8 incl.	d.c., reversed polarity (electrode positive)	not required	not required	not required	not required
	5/32 and 3/16		F	V and OH	V and OH	not required
	7/32		not required	not required	not required	not required
	1/4		F	F	H	not required
	5/16		F	F	not required	not required
E7011	1/16 to 1/8 incl.	a.c. and d.c. reversed polarity (electrode positive)	not required	not required	not required	not required
	5/32 and 3/16		F	V and OH	V and OH	not required
	7/32		not required	not required	not required	not required
	1/4		F	F	H	not required
	5/16		F	F	not required	not required
E7013	1/16 to 1/8 incl.	a.c. and d.c., straight polarity (electrode negative)	not required	not required	not required	not required
	5/32 and 3/16		F	V and OH	V and OH	not required
	7/32		not required	not required	not required	not required
	1/4		F	F	H	not required
	5/16		F	F	not required	not required
E7015	1/16 to 1/8 incl.	d.c., reversed polarity (electrode positive)	not required	not required	not required	not required
	5/32 and 3/16		F	V and OH	V and OH	not required
	3/16		F	F	H	F
	7/32		not required	not required	not required	not required
	1/4		F	F	H	not required
	5/16		F	F	not required	not required
E7016	1/16 to 1/8 incl.	a.c. and d.c. reversed polarity (electrode positive)	not required	not required	not required	not required
	5/32 and 3/16		F	V and OH	V and OH	not required
	3/16		F	F	H	F
	7/32		not required	not required	not required	not required
	1/4		F	F	H	not required
	5/16		F	F	not required	not required

Table 10. Low-Alloy Steel Arc Welding Electrodes (A316-48T) (Contd.)

Classification Number	Electrode		All-Weld Metal Tension Test[b]	Guided-Bend Test[c]	Fillet-Weld Test[d]	Hydrogen Test[e]
	Diameter, in.	Current and Polartiy				
E7020	1/16 to 1/8 incl.	d.c., straight polarity (electrode negative), and a.c. for horizontal fillet welds; d.c., both polarities, and a.c. for flat-position welding	not required	not required	not required	not required
	5/32 and 3/16		F	F	H	not required
	7/32		not required	not required	not required	not required
	1/4		F	F	H	not required
	5/16		F	F	not required	not required
E7025	1/16 to 1/8 incl.	d.c., reversed polarity (electrode positive)	not required	not required	not required	not required
	5/32 and 3/16		F	F	H	F(3/16 in. only)
	7/32		not required	not required	not required	not required
	1/4		F	F	H	not required
	5/16		F	F	not required	not required
E7026	1/16 to 1/8 incl.	a.c. and d.c. reversed polarity (electrode positive)	not required	not required	not required	not required
	5/32 and 3/16		F	F	H	F(3/16 in. only)
	7/32		not required	not required	not required	not required
	1/4		F	F	H	not required
	5/16		F	F	not required	not required
E7030	1/16 to 1/8 incl.	d.c., both polarities, and a.c.	not required	not required	not required	not required
	5/32 and 3/16		F	F	not required	not required
	7/32		not required	not required	not required	not required
	1/4		F	F	not required	not required
	5/16		F	F	not required	not required
E8010, E9010, E10010	1/16 to 1/8 incl.	d.c., reversed polarity (electrode positive)	not required	not required	not required	not required
	5/32 and 3/16		F	not required	V and OH	not required
	7/32		not required	not required	not required	not required
	1/4		F	not required	H	not required
	5/16		F	not required	not required	not required

Table 10. Low-Alloy Steel Arc Welding Electrodes (A316-48T) (Contd.)

Classification Number	Electrode		All-Weld Metal Tension Test[b]	Guided-Bend Test[c]	Fillet-Weld Test[d]	Hydrogen Test[e]
	Diameter, in.	Current and Polarity				
E8011, E9011, E10011	1/16 to 1/8 incl.	a.c. and d.c., reversed polarity (electrode positive)	not required	not required	not required	not required
	5/32 and 3/16		F	not required	V and OH	not required
	7/32		not required	not required	not required	not required
	1/4		F	not required	H	not required
	5/16		F	not required	not required	not required
E8013, E9013, E10013	1/16 to 1/8 incl.	a.c. and d.c., straight polarity (electrode negative)	not required	not required	not required	not required
	5/32 and 3/16		F	not required	V and OH	not required
	7/32		not required	not required	not required	not required
	1/4		F	not required	H	not required
	5/16		F	not required	not required	not required
E8015, E9015, E10015	1/16 to 1/8 incl.	d.c., reversed polarity (electrode positive)	not required	not required	not required	not required
	5/32 and 3/16		F	not required	V and OH	not required
	3/16		F	not required	H	F
	7/32		not required	not required	not required	not required
	1/4		F	not required	H	not required
	5/16		F	not required	not required	not required

piping used in high pressure, high temperature steam service, and plates or castings with a molybdenum content of approximately .50%.

Electrodes in the *AWS* E70XX series (E7010, E7011, E7013, E7015, E7016, E7020, E7025, E7026 and E7030) are commonly referred to as "low-alloy steel electrodes." *AWS* E8010, -11, -13; *AWS* E9010, -11, -13; and *AWS* E10010, -11, -13 electrodes also belong to this group.

Carbon Electrodes

The electrodes used in carbon-arc welding are made either of carbon or graphite. The carbon electrode possesses the greater strength of the two, and has the added advantage of being the cheaper. The graphite electrode, on the other hand, can carry greater current and will last longer. The carbon electrode should

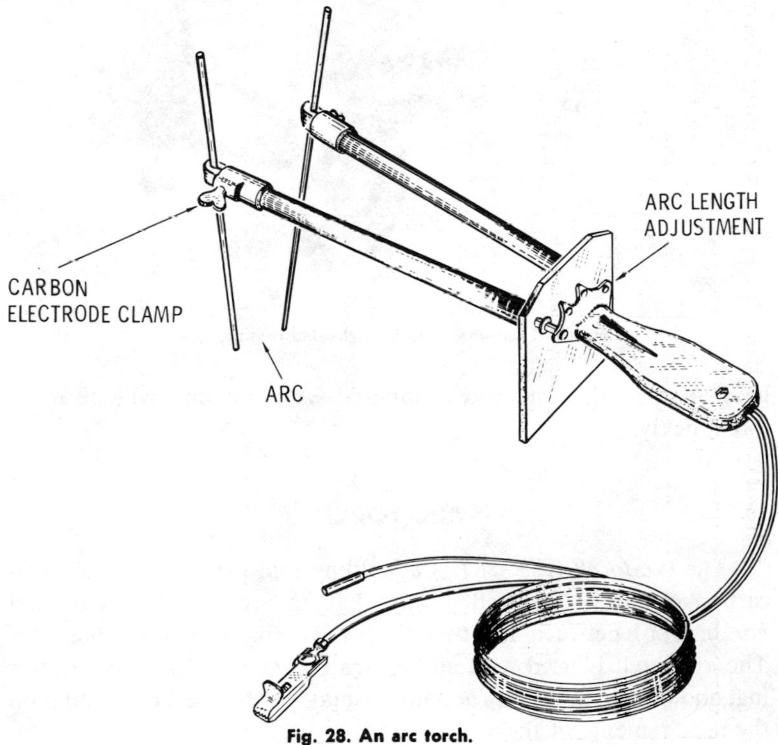

Fig. 28. An arc torch.

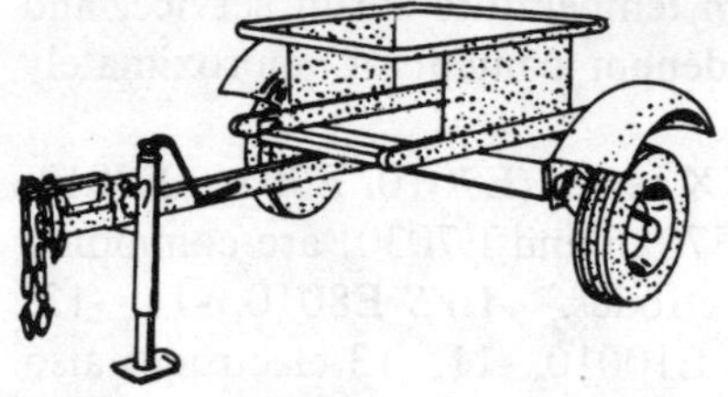

Fig. 29. Two-wheel trailer for transporting welding equipment.

Courtesy Airco Welding Products

Courtesy Chemetron Corp.

Fig. 30. Four-wheel, rubber wheeled running gear.

be shaped so that the end is tapered. A blunt end will burn off too quickly.

ARC TORCH

The *arc torch* (Fig. 28) is a welding tool designed to hold two carbon electrodes in such a way that an arc can be established (or broken) between the two simply by changing their position. The arc torch is used with an AC transformer for heating, soldering, and brazing. A filler rod may or may not be used depending on the requirements of the work.

RUNNING GEAR AND TRAILERS

The *running gear* and *trailers* used with arc welding power sources are available in a number of different models, each designed to satisfy specific job requirements. They are available with steel wheels, pneumatic rubber tires, or solid rubber tired wheels. Body styles for trailers include two and four wheel (tandem) models (Fig. 29). The running gear include two, three, and four wheel models (Fig. 30).

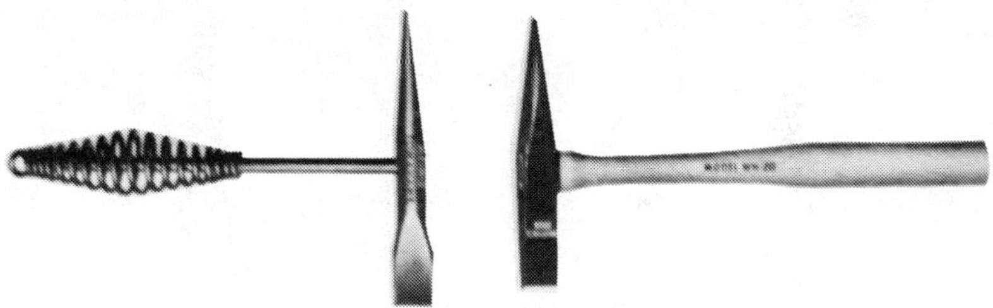

Courtesy Airco Welding Products

Fig. 31. Chipping hammers.

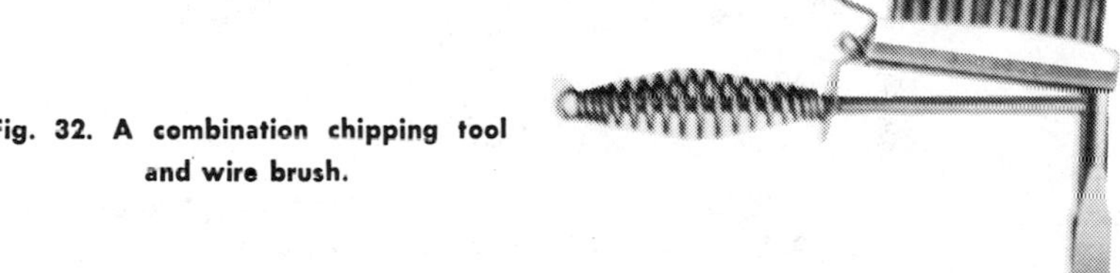

Fig. 32. A combination chipping tool and wire brush.

Courtesy Airco Welding Products

WELD CLEANING TOOLS

Once a weld bead has been laid (prior to the laying of subsequent beads) or a weld has been finished, slag must be removed from the surface of the bead. This is necessary for: (1) appearance, (2) meeting inspection requirements, and (3) preventing impurities from being included in subsequent beads. Two types of tools used for this purpose are chipping hammers (Fig. 31) and wire brushes. The latter are often combined with chipping hammers to form a single dual-purpose tool (Fig. 32).

CHAPTER 6

Arc Welding Process

Arc welding is a term applied to a group of welding processes in which an electric arc is formed and maintained between the work surface and an electrode held in a special electrode holder. The heat resulting from the arc created between the work and the electrode may be effectively concentrated on the area to be welded. Once the arc is established, the welding temperature rises instantly to about 6500° F.

The heat of the arc creates a small pool of molten metal on the surface of the work. Additional metal may be added from the electrode (metal-arc welding) or a filler rod (carbon-arc welding). Both metal-arc and carbon-arc welding are described in greater detail in this chapter.

SELECTING A POWER SOURCE

In welding, since the arc voltage depends on the arc length, it is constantly changing. If the arc demands a high voltage, the generator must respond. The same is the case if it is a low voltage.

The voltage must not be too low or the arc will go out. It is easy to maintain a ⅛-inch arc at relatively low voltage due to the conducting gases in the arc. These gases become nonconducting very quickly and the arc goes out (in a few thousandths of a second) if the voltage does not rise.

If the current rises to a very high value, the results are splatter and explosive action on the bead. If the current does not reach a value sufficient to melt the electrode and metal, there is no fusion.

In both voltage and current, the generator must respond to the arc condition very quickly—in fact, practically instantaneously; there must be no instability. The response must meet the arc demands as they occur, to the degree required, no more and no less. Special attention has been given to the design of modern generating outfits to meet these conditions.

Either AC or DC welding machines may be used for arc welding. The type of machine selected will depend on a number of different factors, including the gauge of the metal to be welded, the efficiency-time factor, and the nature of the welding job (production versus nonproduction welding). The reader should take particular note of the section on ARC BLOW in this chapter.

Most AC welding machines used for manual arc welding in factories and similar areas are static-transformer, single-operator types. They generally have ratings in the 200-400 ampere range and require electrode sizes of 1/16 to 5/16 inch. Welding machines with lower ampere ratings (e.g. a 150-ampere machine) are used for light welding jobs in industry, body shops, and similar small operations. Automatic welding in industry requires machines with 500-ampere ratings and more.

The DC welding machines used for manual metal-arc welding in industrial plants generally have ratings in the 200-400 ampere range, and are of the single-operator type. They are mostly self-regulating units with variable-voltage characteristics.

The selection of a suitable arc welding power source is strongly influenced by the following factors:

1. *Location of the welding.* If the welding is being done in remote areas and no electric power is available, an engine-driven (gasoline, diesel, etc.) welding machine will have to be used. If electricity is available, the type of machine selected is determined by the voltage input (most welding machines are designed to operate on either 230 or 460 volts input).
2. *Type of welding current required.* The welding job may call for AC, DCSP, or DCRP current. The welding ma-

chine selected must be able to deliver the type of welding current required by the job. In addition, the welding machine selected should be able to deliver more amperage than the job requires.

3. *The size of the electrode to be used.* The nature of the welding job may require that a particular size of electrode be used. This will have ramifications in the amount of current used and ultimately in the type of power source selected for the job. See the section DETERMINING THE REQUIRED CURRENT in this chapter.
4. *The amount of welding to be done.* An automatic welding operation usually will have machine requirements different from those required by a manual or semiautomatic operation.
5. *The thickness of the metal to be welded.* Thicker metals require deeper penetration. This necessitates welding machines with the capacity to deliver a sufficiently high current to the electrode.

THE ELECTRIC ARC

Arc welding uses an electric arc to produce the welding heat. An electric arc is a stream of incandescent vapor connecting the terminals of an electric circuit when they have been drawn apart; the current source having sufficient voltage to overcome the extra resistance thus thrown into the circuit.

In the formation of the arc, a small amount of the material forming the terminals of the arc gap is heated to an incandescent vapor. This vapor provides the conducting medium of the arc stream by which the current is carried from one terminal to the other.

The temperature of the vapors in the arc and, consequently, the intensity of the light given out are so great that colored glass must be used in order to protect the eyes. When suitable glass is used, the different portions of the arc can be plainly distinguished. The center is usually referred to as the *arc core* and some observers are able to see that this is divided into two portions designated as *arc core* and *arc stream.* In general, this portion

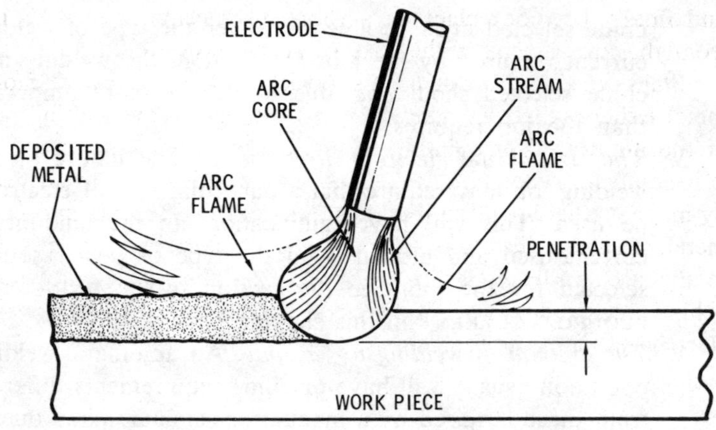

Fig. 1. The various areas of electric arc welding.

of the arc will usually be seen as greenish in color, of comparatively small diameter, and forming a direct line between the two terminals (Fig. 1).

The point where the arc core strikes on either terminal is seen as a light red or yellowish spot considerably brighter and, therefore, hotter than the metal surrounding it. The metal around this spot is molten and is usually seen as a bright red area. This color gradually shades off into a darker red with lower temperatures,

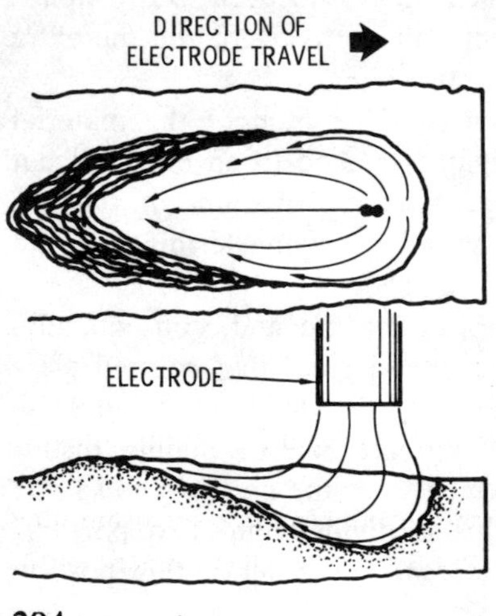

Fig. 2. Showing the flow of metal when welding.

and finally becomes black at a short distance, not over ½ inch from the arc, except in the case of very heavy welding.

Slag, oxides, etc. can be distinguished floating on the molten metal either as light or dark spots, depending on the melting point of the impurity. Surrounding the arc core is the arc flame which is irregular in shape and in constant motion, being easily deflected by magnetic fields caused by the current in the electrode and in the metal, and also by drafts which may arise by reason of the heat in the arc, or by exposure to wind, etc. The weld metal forms behind the arc flowing in a direction opposite to that in which the electrode is moving (Fig. 2).

HEAT DISTRIBUTION OF THE ARC

Electrical arcs may be either direct current (DC) arcs or alternating current (AC) arcs. In the heat distribution of the DC arc, it is generally accepted that approximately ⅔ of the heat is liberated at the positive terminal and ⅓ at the negative terminal. In the heat distribution of the AC arc, approximately the same amount of heat is liberated at each terminal. This is due to the fact that the AC terminals are alternatively positive and negative.

Arc Crater

The depression caused by the penetration of the arc is referred to as the *arc crater* (Fig. 3). Its depth provides a means of observing the penetration during welding and, to a certain extent, of predicting the soundness of the weld, since one requirement of a weld is to obtain good penetration. Crater depth depends on the thickness of the metal welded.

Arc Blow

Arc blow (Fig. 4) is the wavering of the arc from its intended path. It is characterized by undercutting (caused by the deflection of the arc), excessive spatter, and poor fusion. In arc welding, the current which flows through the electrode, arc stream, and the base metal sets up magnetic fields around each of these. The action of these magnetic fields or fluxes on the arc stream may sometimes pull the arc stream out of its intended path resulting in arc blow.

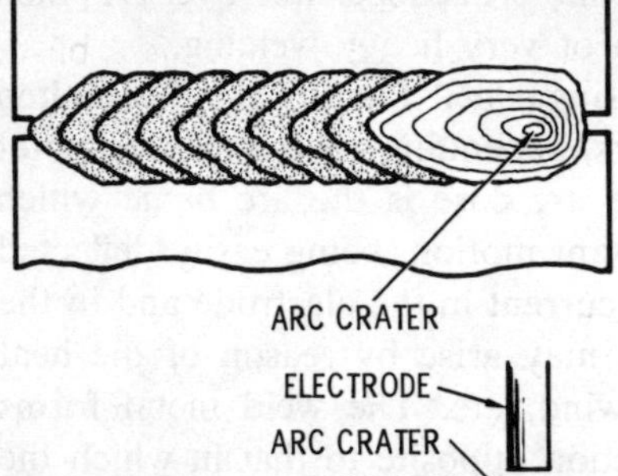

Fig. 3. Metal deposits showing penetration and arc crater.

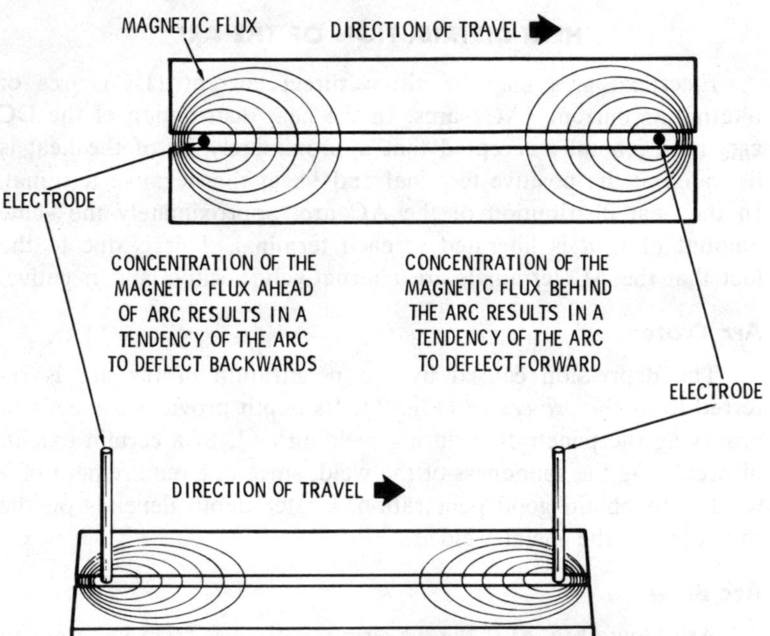

Fig. 4 Showing arc blow.

These magnetic fields can be prevented from building up arc-deflecting strength by using AC rather than DC. The constant reversal of current in the former is the factor that keeps the strength of the magnetic field and the related arc blow to a minimum. Higher currents and larger electrodes are used with AC welding. This results in the aforementioned reduction of arc blow to a minimum, faster production welding of metal plate, and a faster travel speed on the downhand welding of production jobs. A major disadvantage in AC is that it is not as suitable as DC for welding thin-gauge sheet metal. The lower current required for this type of welding is more stable and more easily handled with DC.

The welder must exercise caution in selecting the proper electrode for AC welding. It must be an electrode designed to handle the higher currents required for AC welding. Iron-powder coated electrodes are especially suited for AC welding. Other electrodes (e.g. nonferrous electrodes, stainless steel electrodes, etc.) are more suited for use with DC.

DC may be used, and arc blow considerably reduced, by making a number of adjustments. Because these adjustments are time consuming, they lower the efficiency of production welding (for which reason AC is preferred). However, for nonproduction jobs where speed and time are not essential factors, these adjustments become of minor importance.

Some suggested adjustments for reducing arc blow to an acceptable minimum are as follows:

1. Weld toward a heavy tack or toward a weld already made.
2. Use back stepping on long welds.
3. Place the ground connection as far from the joint to be welded as possible. On small pieces, place the ground connection at the starting end and weld toward a heavy tack if possible.
4. Hold a short arc so that the electrode coating touches the metal surface and direct the tip of the electrode in the direction opposite to that of the arc blow. This enables the arc force to counteract the arc blow.
5. Reduce the current.

Arc Length

The length of the arc is the distance between the end of the electrode and the surface on which the molten globules are deposited. While the correct arc length alone will not insure a good weld, it is agreed that a long arc is almost certain to result in a poor weld (Fig. 5).

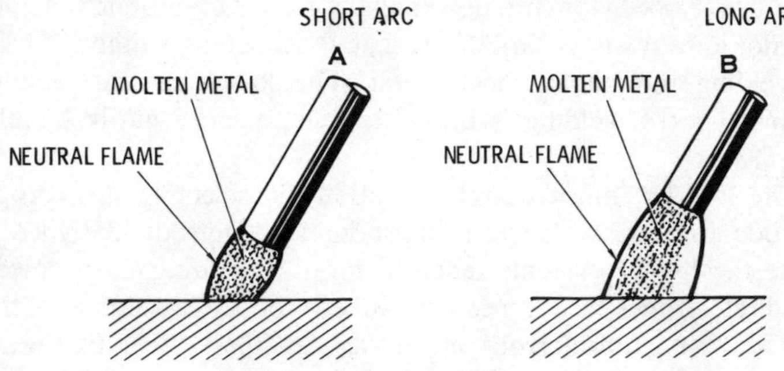

Fig. 5. A short and a long arc length.

Among other factors, the correct length of the arc depends upon:

1. The type and size of the electrode used,
2. The material to be welded,
3. The amount of welding heat necessary.

The ordinary unshielded arc should be short. For the shielded arc, the length is longer.

The main characteristic of a long arc is that it is not as stable as a short arc. The characteristic of a short arc is that it will deposit more metal in the weld at the point needed than a long arc.

The advantages of a short arc are:

1. Maximum penetration,
2. Slight overlap,
3. Maximum strength,
4. Maximum ductility,
5. Minimum porosity,
6. Maximum amount of metal deposited at the point needed,
7. It makes possible the use of alloy electrodes.

The disadvantages of a long arc are:

1. Minimum penetration,

2. Excessive overlap,
3. Minimum strength,
4. Minimum ductility,
5. Maximum porosity,
6. Uncontrolled deposit,
7. Excessive waste of electrode material,
8. It burns out all the alloys in a high grade electrode.

Effect of Arc Length

When a long arc is held, heat is dissipated into the air and the stream of molten metal from the electrode to the work is scattered in the form of spatter; moreover, the arc force is spread over a large area resulting in a wide shallow bead instead of a narrow one with deep penetration. The two best and easiest methods of controlling and directing the arc force are:

1. Moving fast enough to keep ahead of the molten pool,
2. Keeping the tip of the electrode coating lightly in contact with the metal surface and dragging the electrode along the joint.

The effect of arc length is best judged by the appearance of the weld. Good penetration of the welding metal into the base metal is not obtained with a long arc and there will be a bad overlap (Fig. 6). If the arc is short, there will be good penetration

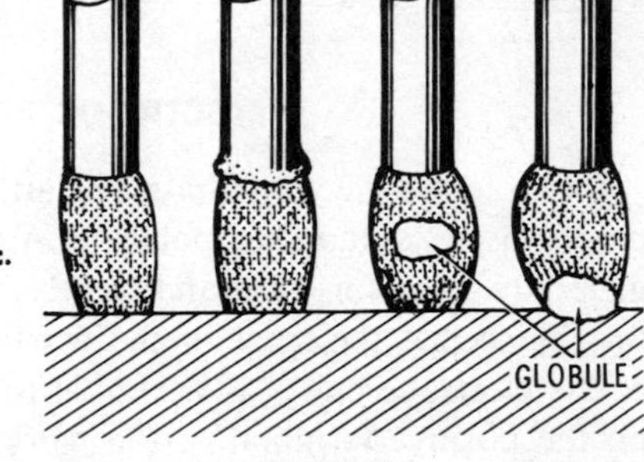

Fig. 6. Characteristics of a long arc.

and a slight overlap (Fig. 7). Another way to determine if the arc is too long is to examine the crater or depression in the base metal on breaking the arc and to see what the penetration looks like.

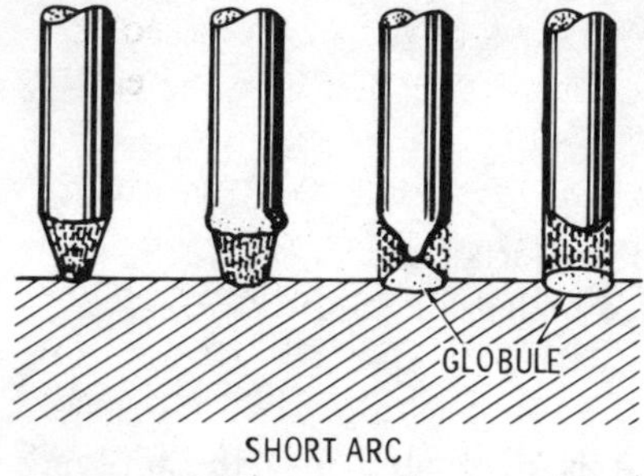

Fig. 7. Characteristics of short arc.

If there is no penetration in the base metal, then the arc is too long (providing, of course, the proper electrode, the required current, and the correct polarity have been used).

Arc Force and Travel Speed

The digging quality of the arc is called the *arc force*. The proper use of the arc force results in faster welding. Just as water forced through a nozzle can be used to dig away dirt, so can the force of the arc stream be used to dig into the base metal.

The arc travel speed should be fast enough to properly utilize the penetrating power of the arc. A slow moving arc results in a puddle of molten metal under the arc, thus, reducing penetration. A fast moving arc results in good penetration. The *limiting speed* is usually the highest speed at which the surface appearance remains satisfactory.

ELECTRODE POLARITY

The terms *electrode positive* and *electrode negative* are commonly used to indicate polarity. A negative electrode polarity is generally used for bare and lightly coated electrodes. The reason for this is that the mass of the work is generally greater than that of the electrode and it is desirable to have the most heat generated on the positive terminal of the work rather than on the electrode. A positive electrode polarity, on the other hand, is desirable for welding certain alloy steels and a few other metals.

The coating on an electrode has an important effect during the arc-welding process. The action of the arc on the coating results in a slag formation which floats on top of the molten weld metal and protects it from the ambient atmosphere while cooling. After the weld metal is sufficiently cooled, the slag may be easily removed.

Electrode negative is used in welding with bare or lightly coated electrodes. The reason for this is that more heat is generated on the positive side of the circuit. The coating on the electrode forms gases in the arc. These gases may alter the heat conditions so that less heat is generated on the positive side. The heat conditions are affected differently by different types of coatings. The polarity to use with a particular electrode is established by the electrode designer.

The burn-off rate is a good guide in determining the correct polarity. Each polarity is used to burn off 6 or 8 inches of electrode. The polarity which results in the longer burn-off time is generally

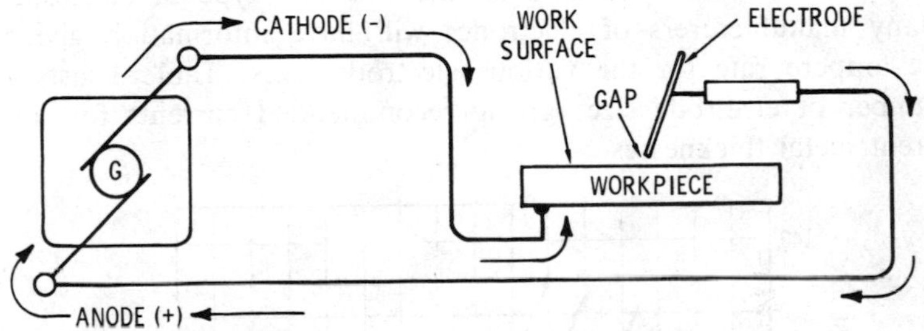

DIRECT CURRENT REVERSE POLARITY (DCRP) WELDING CIRCUIT

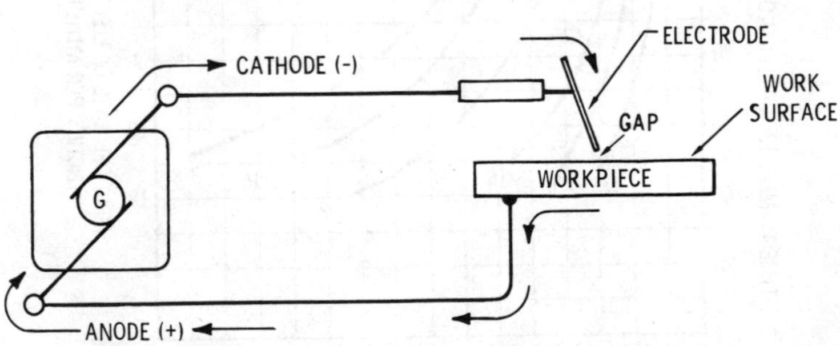

DIRECT CURRENT STRAIGHT POLARITY (DCSP) WELDING CIRCUIT

Courtesy James F. Lincoln Arc Welding Foundation

Fig. 8. Wiring diagram illustrating the DCRP and DCSP welding circuits.

the one which produces greater heat in the metal surface or joint. Whether this or the other polarity is to be used depends on the kind of work to be welded.

Positive and negative electrodes are distinguished by the terminals and their relationship to the arc. The terminal from which the current passes to the arc is termed the positive electrode or anode and the terminal to which the current passes from the arc is called the negative electrode or cathode (Fig. 8). In straight polarity, the electrode is negative and the work positive; in reversed polarity the electrode is positive and the work negative.

DETERMINING THE REQURIED CURRENT

The amount of current required for arc welding will vary within wide limits depending on the size and type of electrode. Many manufacturers of electrodes will have information giving the ampere rate for the various electrode sizes. Table 1 lists a number of electrode sizes giving recommended currents for different metal thicknesses.

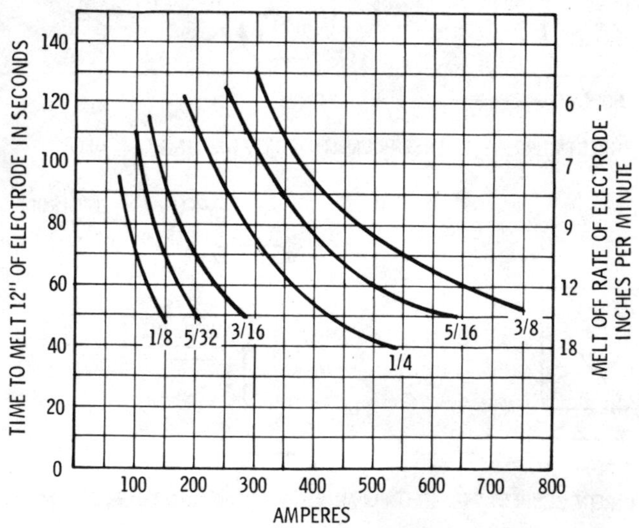

Fig. 9. Current determination.

A very satisfactory method for determining the current is by measuring the number of inches of electrode melted off for a given interval of welding. This is generally the method used when accurate meters are not available (Fig. 9).

An increase in the welding current increases the arc force and penetration. A larger electrode will be required to accommodate the increase in current. As a general rule, the largest possible electrode size (within the limitations of the work) should *always*

Table 1. Electrode Sizes, Currents, and Metal Thicknesses

Electrode Diameter in Inches	Amperes Hand Welding	Corresponding Plate Thickness in Inches
1/16	50-100	Up to 3/16
3/32	100-150	Up to 1/4
1/8	125-175	Above 1/8
5/32	150-200	Above 1/4
3/16	175-350	Above 3/8
1/4	225-400	Above 3/8

be used. Large electrodes increase welding speeds and reduce cost. Table 2 illustrates welding speeds for various electrode sizes on a 3/8-inch thick metal surface. Note the faster speed for the 5/16-inch diameter size.

DETERMINING THE CIRCUIT POLARITY

The simplest method of determining the polarity of a circuit is by means of a voltmeter. Another method is to draw an arc between either a bare metal electrode or a carbon electrode and a steel plate or surface. If the metal surface is positive and the electrode is negative, the arc will be fairly stable. If, however, the circuit is reversed and the electrode is positive, the electrode will heat up very rapidly and the arc will be unstable and difficult to maintain.

SELECTING THE PROPER ELECTRODE

It is most important to select the proper electrode for each welding job. The quality, appearance, and economy of the weld

Table 2. Electrode Sizes and Welding Speeds

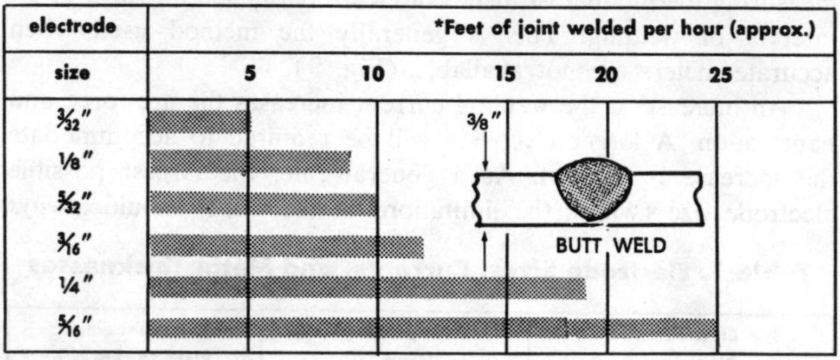
electrode
size
*Feet of joint welded per hour (approx.)
5
10
15
20
25
3/32"
1/8"
5/32"
3/16"
1/4"
5/16"
3/8"
BUTT WELD

*Using 3/8" thick steel plate

Courtesy Hobart Brothers Co.

will depend upon correctly selecting the most suitable electrode. There are a number of factors that must be considered in the choice of electrode. These are as follows:

1. *Metal identification.* It is essential to know not only the kind of metal being welded (e.g. mild steel, cast iron, etc.) but also its mechanical properties. The properties of each electrode are indicated by the identification numbers of the *AWS* electrode classification system. Select the electrode that most closely matches that of the base metal. These electrodes should also be those which can impart the highest ductility and impact resistance to the weld so that it can bear up under most conditions of use.
2. *Welding current.* The electrode selected should be the one that most closely matches the type of power source being used. The type of welding current to be used with a particular electrode is also indicated by the *AWS* electrode identification numbers. As a general rule, the welder should select the maximum current (and the maximum electrode diameter) that can be used with the thickness of the metal being welded. Here, consideration should be given to whether or not the metal has been preheated. Preheated metals require less current than those that have not been preheated.

3. *Welding position.* The *AWS* electrode identification numbers also indicate the welding position for which the electrode is designed. Not all electrodes are designed for use in every welding position. The welder must match the electrode with the welding position being used.
4. *Thickness of the metal.* The thicker the metal, the greater the current that is required to produce a suitable weld. An increase in the amount of current requires a corresponding increase in electrode diameter size. The welder should match, as closely as possible, the welding current being used to the diameter size recommended by the manufacturer.

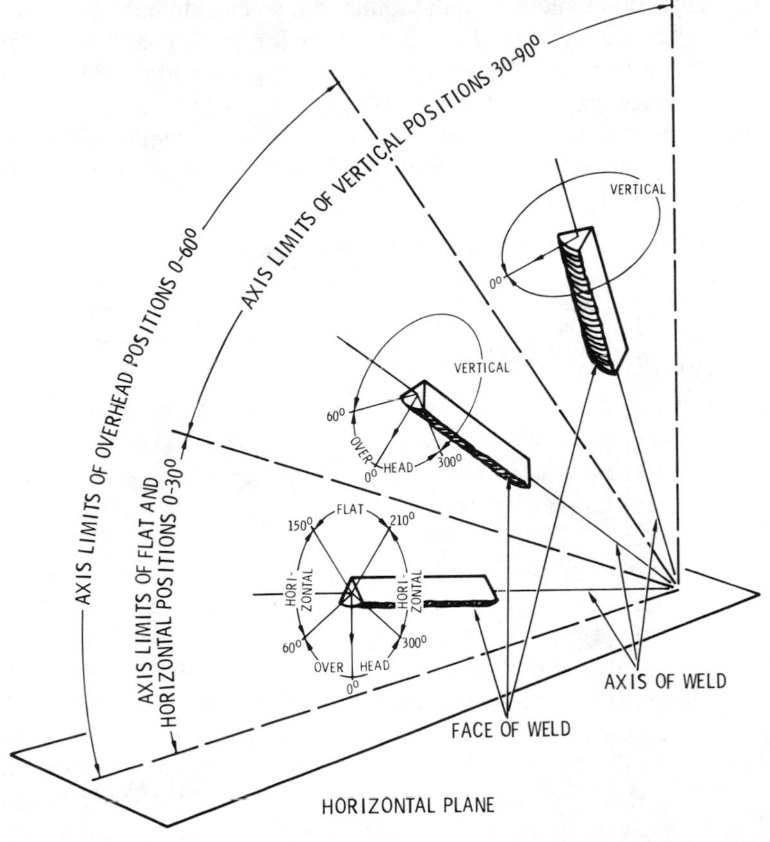

Fig. 10. The joint position. The position in which joints are welded.

5. *Joint design.* The design of the joint (and fitup) determines the degree of arc penetration (deep, medium, light, etc.). This, too, is specified by the *AWS* electrode identification numbers. The welder should select an electrode that gives the required arc penetration.
6. *Welding passes.* The number of passes is also determined by the type of electrode selected. Multiple passes require more current than a single pass.
7. *Joint position.* The position in which joints are welded, especially on multipass work, is one of the important considerations which affect the choice of electrode (Fig. 10). For flat and horizontal joints, the so-called "hot" electrodes should be used. Electrodes used for vertical and overhead work, of course, must provide a deposit which will stay in place and not fall out of the joint while in the molten condition. A deposit of this type usually requires that the electrode not be over 3⁄16 inch in diameter.

WELDING POSITIONS

The welding positions used in arc welding are the same as those used in the gas welding process, namely: flat, horizontal, vertical, and overhead (Fig. 11).

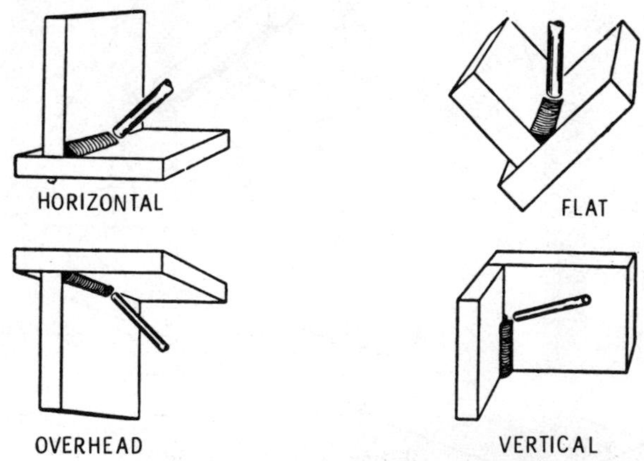

Fig. 11. Welding positions showing electrode angle.

By way of summary, a *flat* position is one in which the welding material is applied in a generally downward position. A *horizontal* welding position is one in which the welding material is applied to a seam or joint on a metal surface, the plane of which is vertical or inclined 45° or less to the vertical and the line of weld is horizontal. The welding material in the *vertical* welding position is applied to a vertical surface, or one inclined 45° or less to the vertical so that the line of weld is vertical or inclined 45° or less to the vertical. The *overhead* welding position is the most difficult. It is one in which the welding material is applied from the under side of any members whose plane is such that it necessitates the electrode being held with its welding end upward.

TYPES OF JOINTS

The metal-arc welding process is used to weld the same types of joints (butt, lap, T-edge, etc.), joined by the gas welding process. These joints are described in considerable detail in Chapter 2.

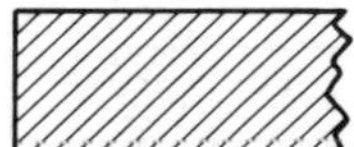

Fig. 12. A simple type of butt joint which is not recommended for sections greater than 1/4 inch in thickness.

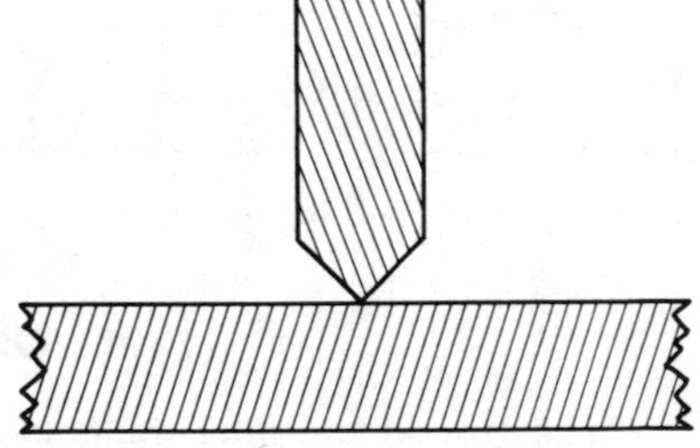

Fig. 13. This type of joint is not recommended except where a square corner is essential after welding.

Figs. 12-19 illustrate the preparation of the most common butt type joints. The same principles of preparation apply to other types of joints.

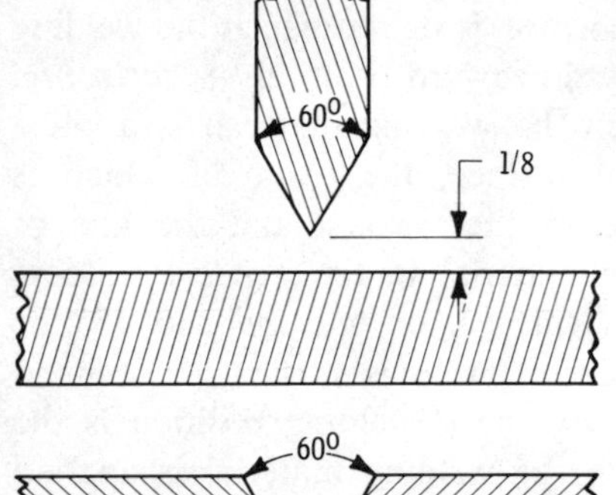

Fig. 14. In this type of joint, great care must be exercised to obtain fusion at the apex of the angle and the vertical plate.

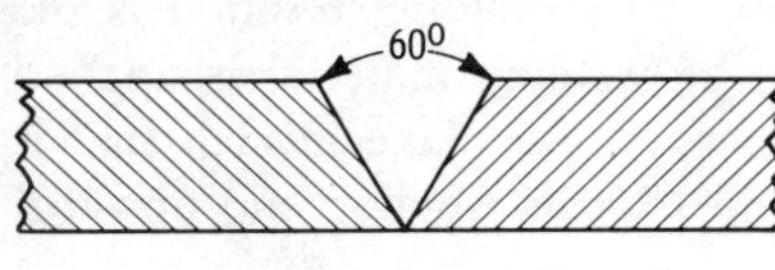

Fig. 15. A joint prepared for welding on one side only.

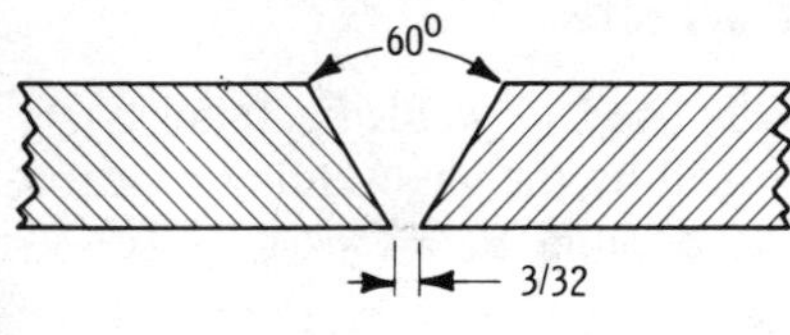

Fig. 16. A gap is provided to allow penetration to the bottom of the plate.

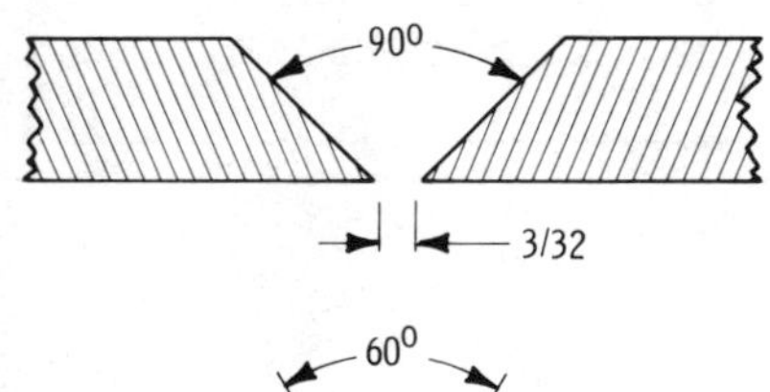

Fig. 17. A V-joint prepared for welding on one side only.

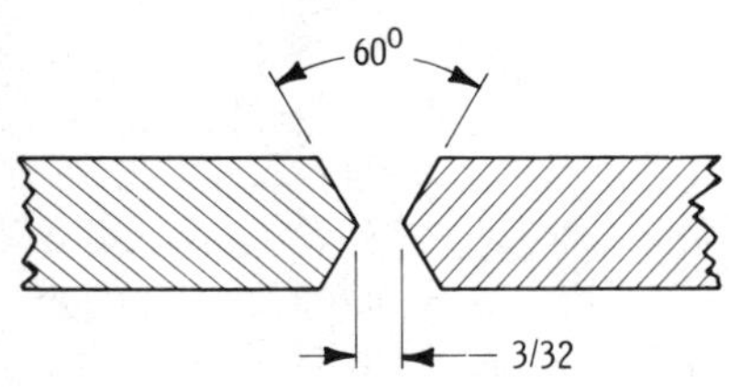

Fig. 18. This joint design is applied on sections which are thick enough to warrant beveling on both sides.

TYPES OF WELDS

The groove and fillet welds are both used in metal-arc welding. The double V-groove and the double-fillet weld both are used when welding heavier and thicker plate for the added strength they impart. However, the single V-groove weld and the smaller fillet

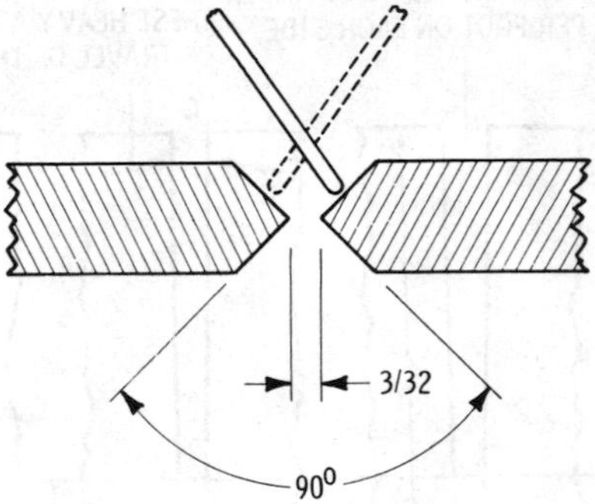

Fig. 19. Warpage is reduced in this type of joint due to the fact that the force on either side is counteracted by the force of the opposite side.

weld are more common for most types of welding. Read carefully the section on welds in Chapter 2.

WELD PREPARATION

There are several factors which must be considered when preparing work for welding in order to get the best results. They may be listed as:

1. Cleaning,
2. Preparing for expansion and contraction,
3. Positioning.

Cleaning is important because good welds can be obtained only when the joints to be welded are kept clean. All rust, scale, paint, dirt, or foreign matter must be cleaned from the surface. This is done to exclude the foreign matter from the weld and to help make the operation of welding as easy for the operator as possible. Foreign matter is usually a poor conductor of electricity and interferes with the control and manipulation of the welding arc.

A provision for expansion and contraction is necessary because the strength of the weld will depend on the correct beveling and spacing of the parts to be welded. Uniform fusion is directly dependent on the proper beveling and spacing. A lack of provision for expansion and contraction may result in stresses that can impair the quality of the weld and even cause fractures on cooling.

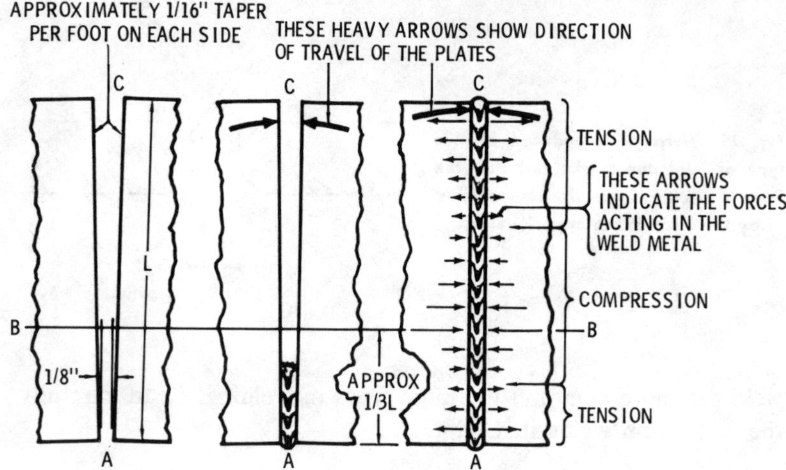

Fig. 20. Locked-in stresses in a weld due to contraction. When the joint is completed, the transverse contraction stresses along the joints will be greatly concentrated at the ends of point A and point C. The stresses impair the quality of the weld and, in many cases, develop a fracture on cooling. When a fracture develops where this method is used, it usually occurs at the end where the joint is finished.

When the joint is completed, the transverse contraction stresses along the joint will be greatly concentrated at the ends. The stresses impair the quality of the weld and, in many cases, develop a fracture on cooling. When a fracture develops, it usually occurs at the end where the joint is finished (Fig. 20).

For ductile materials, where the parts welded are free to come and go by reason of their ductility, extensive precautions to prevent contraction stresses are probably not advisable.

In the case of nonductile materials and castings on large structures, where the contraction effects are liable to be cumulative and to distort seriously the finished product, considerable attention must be paid to eradicating these harmful effects.

Two methods used for reducing contraction stresses are:

1. Locking in the stresses produced,
2. The step-back method of welding to distribute contraction stresses.

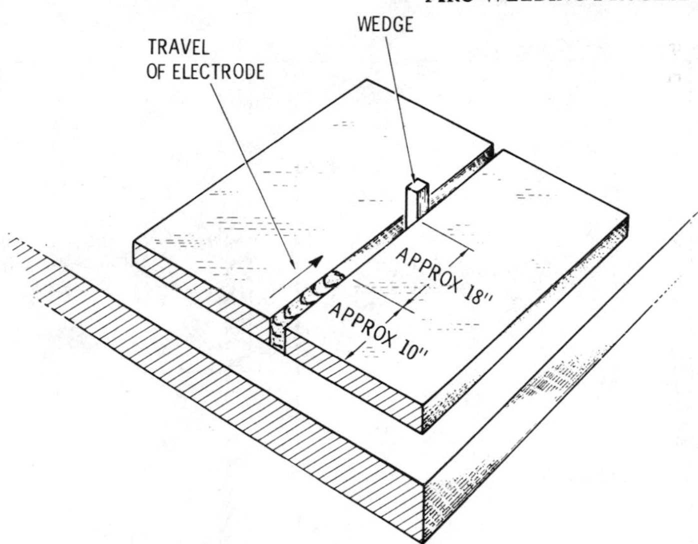

Fig. 21. Method used to reduce contraction. When welding long seams, the drawing may be reduced to almost nothing by the use of spacing blocks or wedges placed in the opening approximately 18 inches from the section being welded.

Spacing wedges or blocks are used in the lock-in method (Fig. 21). When welding long seams, the drawing may be reduced to almost nothing by the use of spacing wedges or blocks placed in the opening approximately 18 inches from the section being welded and toward the end of the seam to which the weld is progressing. This method reduces contraction by locking up the stresses being produced.

In the step-back method (Fig. 22), the deposited metal is finished in sections. Each section is begun approximately 10 inches ahead of the previous one. The metal is deposited in the direction of the previously completed section of weld (that is, away from the end of the seam to which the complete weld is progressing). The step-back method distributes contraction stress along the weld (Fig. 22).

WELDING PROBLEMS

The welder will be confronted with a number of different conditions that can present problems in obtaining a satisfactory

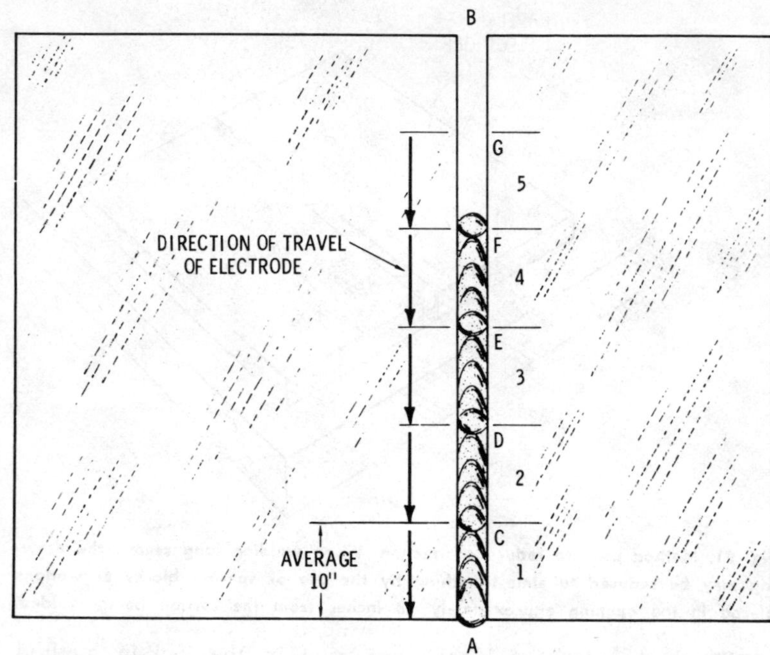

Fig. 22. Step-back method of welding to distribute contraction stresses. The deposited metal is applied in sections. Sections 1, 2, 3, 4, and 5, are welded in numerical order and in the direction shown by arrows. Starting at point C, section 1 progress toward point A. Section 2 would best be welded starting at point D and welded toward point C. Each section should be finished at least flush before starting another.

weld. Some of these are illustrated by their characteristic weld beads in Fig. 23. This illustration also includes an example of a good weld bead for comparison. Arc blow is considered a major problem in DC arc welding and was described in detail in a previous section in this chapter. Some of the other welding conditions over which the welder can exercise control include the following:

1. Welding speed that is too slow.
2. Welding speed that is too fast.
3. Welding current that is too low.
4. Welding current that is too high,

WELDING CURRENT
TOO LOW

Excessive piling up of weld metal.

Overlapping bead has poor penetration.

Slow up progress.

Wasted electrodes and productive time.

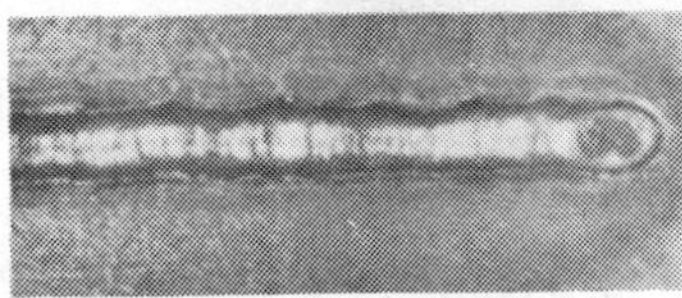

WELDING CURRENT
TOO HIGH

Excessive spatter to be cleaned off.

Undercutting along edges.weakens joint.

Irregular deposit.

Wasted electrodes and productive time.

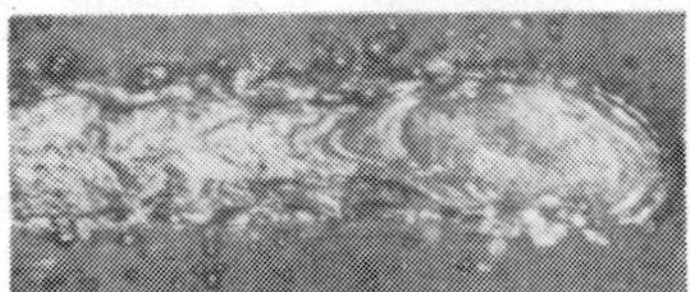

ARC TOO LONG
(VOLTAGE TOO HIGH)

Bead very irregular with poor penetration.

Weld metal not properly shielded.

An inefficient weld.

Waste electrodes and productive time.

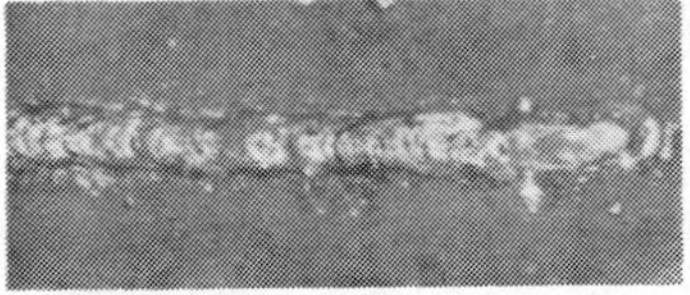

WELDING SPEED
TOO FAST

Bead too small, with contour irregular.

Not enough weld metal in the cross section.

Weld not strong enough.

Wasted electrodes and productive time.

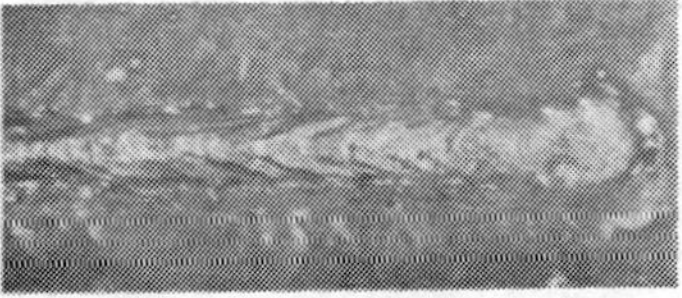

WELDING SPEED
TOO SLOW

Excessive piling up of weld metal.

Overlapped without penetration at edges.

Too much time consumed.

Wasted electrodes and productive time.

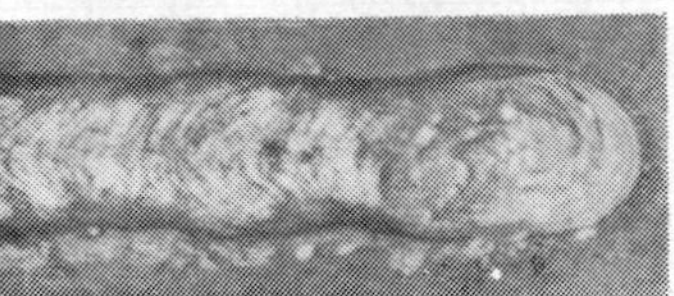

PROPER CURRENT
VOLTAGE & SPEED

A smooth, regular, well formed bead.

No undercutting, over-lapping or piling up.

Uniform in cross section.

Excellent weld at minimum material and labor cost.

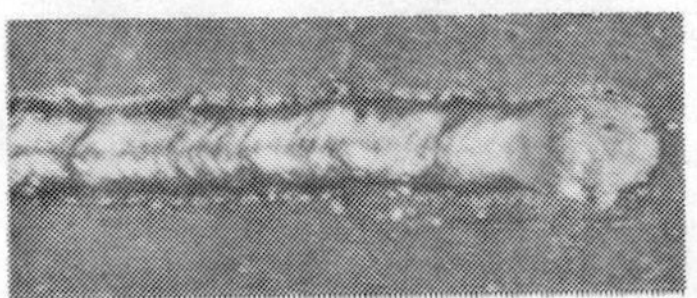

Courtesy Hobart Brothers Co.

Fig. 23. Characteristics of good and bad weld beads and their probable causes.

5. An arc that is too long,
6. Faulty or unsuitable electrodes.

Slow welding speed—A welding speed that is too slow will result in the excessive piling up of weld metal causing it to overlap the surface of the base without adequate penetration of the joint edges. This is a particularly costly problem in welding because it wastes both time and electrodes. The welder should increase the welding (travel) speed until a bead with the correct appearance and penetration has been obtained.

Fast welding speed—A welding speed that is too fast leaves a thin weld bead with irregular contours. This condition results in

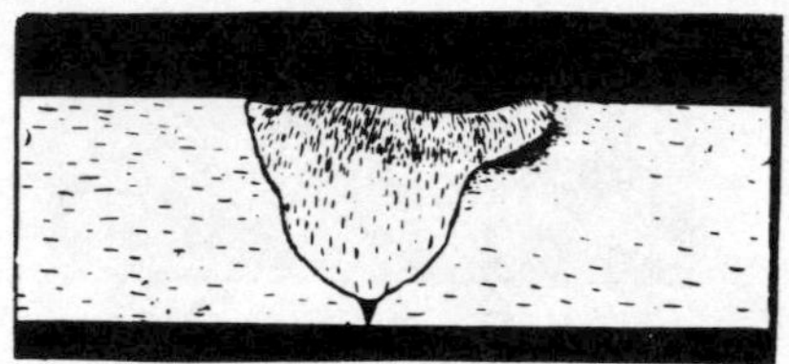

Fig. 24. A weld with poor penetration.

Courtesy Hobart Brothers Co.

poor penetration (Fig. 24), wasted electrodes, and a costly misuse of welding time (the joint will have to be rewelded). The welding speed should be reduced until a correct weld bead is obtained.

Low welding current—A welding current that is too low will also result in poor penetration. The current should be increased until the proper penetration has been obtained.

High welding current—Welding currents that are too high cause undercutting (Fig. 25) and spatter (also present with arc blow) (Fig. 26). Lower the welding current until both conditions are corrected.

Fig. 25. Undercutting a weld.

Courtesy Hobart Brothers Co.

Fig. 26. Weld splatter.

Courtesy Hobart Brothers Co.

An arc that is too long—Excessive arc length also causes spatter in the weld area. Another characteristic of excessive arc length is a tendency for the arc to burn completely through. Fig. 27 illustrates this problem and suggests a repair procedure.

Faulty or unsuitable electrodes—Using an electrode not suited for the particular welding job can cause spatter, brittle welds (Fig.

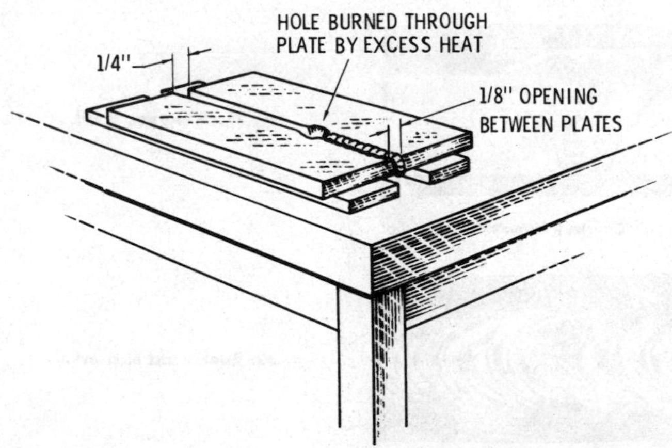

Fig. 27. Repairing a hole in a butt weld made by the arc burning through.

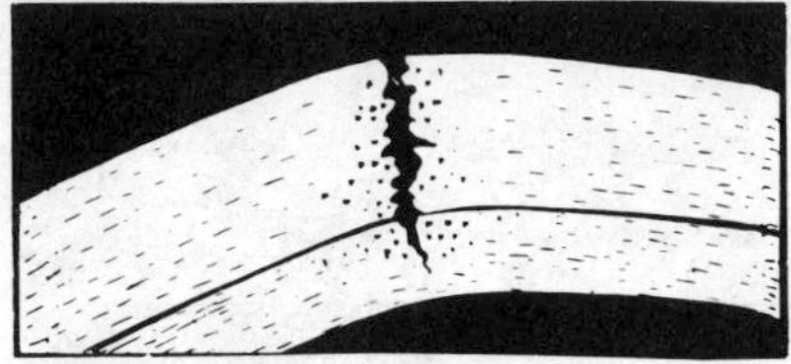

Fig. 28. A brittle weld.

Courtesy Hobart Brothers Co.

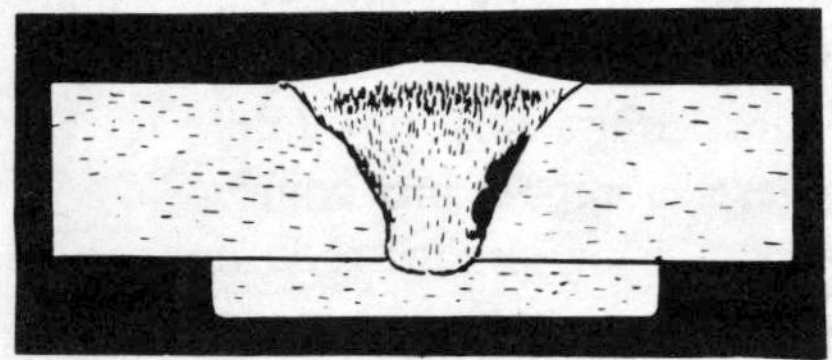

Fig. 29. A weld with poor fusion.

Courtesy Hobart Brothers Co.

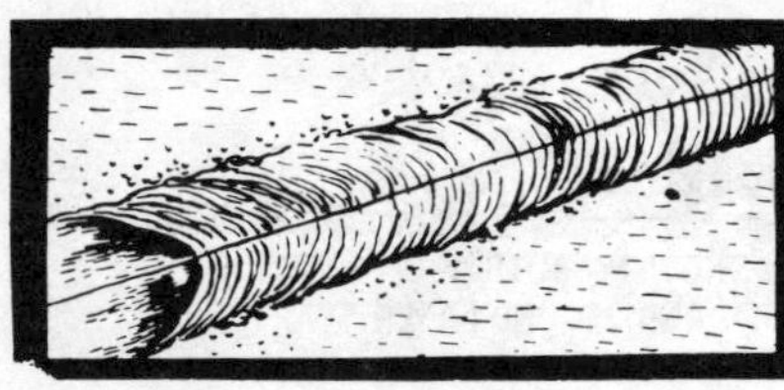

Fig. 30. A cracked weld.

Courtesy Hobart Brothers Co.

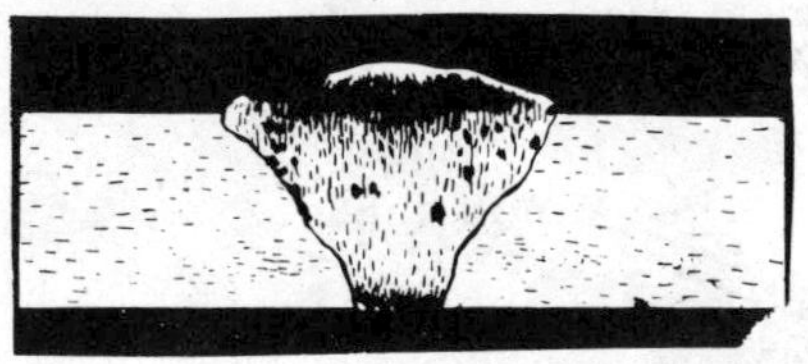

Fig. 31. A porous weld.

Courtesy Hobart Brothers Co.

Fig. 32. Poor weld appearance.

Courtesy Hobart Brothers Co.

28), poor fusion (Fig. 29), cracked welds (Fig. 30), undercutting, and poor penetration. A faulty electrode will produce a porous weld (Fig. 31) and a weld with poor appearance (Fig. 32). To avoid these conditions, a welder should inspect the electrode for defects and make certain that the most suitable electrode for the job has been selected before beginning to weld.

Electrodes exposed to damp atmosphere may pick up enough moisture to cause undercutting, rough welds, porosity, or cracking. Wet electrodes should be placed in a cabinet and heated at a temperature of 200° F for about one hour before using. Under most conditions, where only slight exposure to dampness is found, the electrodes can be dried at a temperature of 10° F above the surrounding atmosphere.

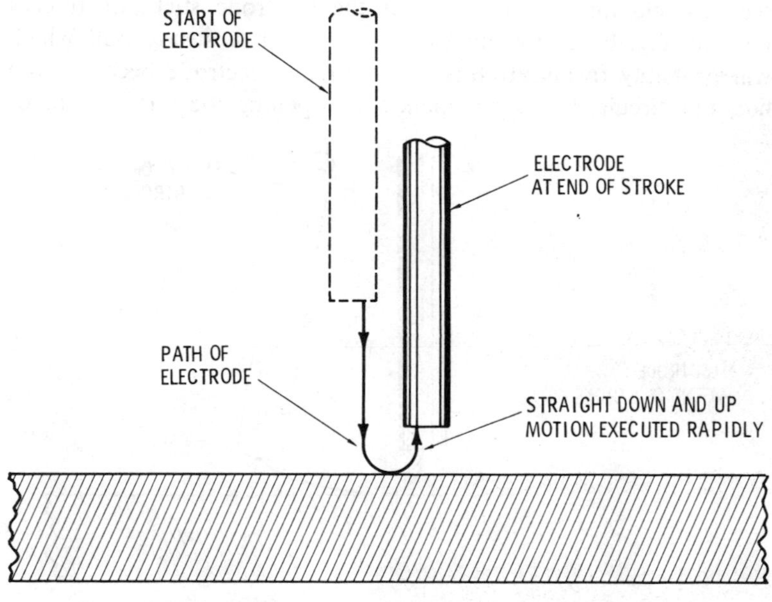

Fig. 33. The quick picking method of striking the arc.

The welder will encounter other problems in welding, but attention given to the six conditions described here will greatly reduce the chances of making unsatsifactory welds.

STRIKING THE ARC

Striking the arc is the operation of touching the work with the electrode and bringing it back from the work a proper distance in order to establish the arc.

The principal precaution to be observed in striking the arc is the speed in which the operation is performed. The electrode should be drawn back to the arc length somewhat more slowly than the movement in the first part of the action.

One method of striking the arc is the quick picking method (Fig. 33). The electrode is touched very lightly and quickly to the work by a motion of the wrist. The movement in touching the metal surface and just freeing the electrode should be quick. The electrode is then brought away more slowly, about ⅛ inch, or until the arc has the proper snapping sound. Hold the arc a few seconds and then snap it out. If the electrode sticks or freezes immediately, bend it from side to side with a steady pull which will probably free it. If this fails and the electrode becomes red hot, the circuit should be opened by opening the line switch, or

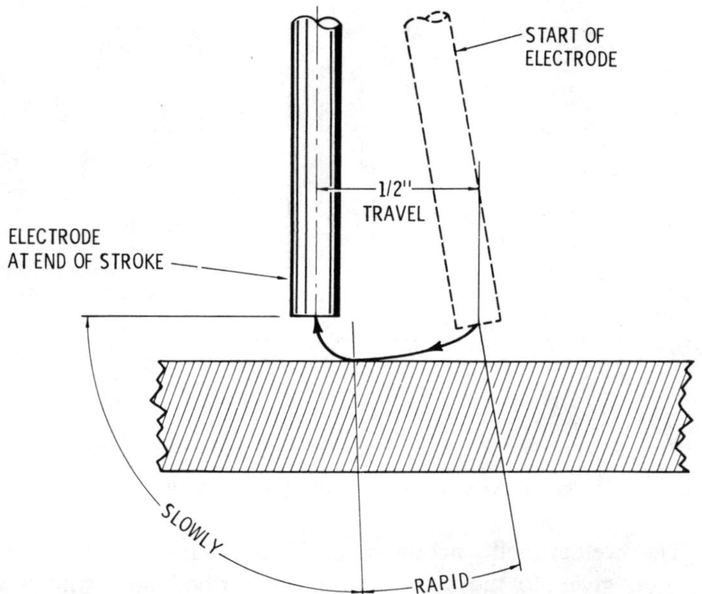

Fig. 34. The scratching method of striking the arc.

by freeing the electrode from the holder, or by lifting the metal plate or sheet from the bench. On cooling, the electrode can be broken away with a hammer.

The second method of striking the arc is designed to avoid sticking. It consists in lightly scratching the electrode on the surface of the metal. The withdrawal of the electrode should be slower than the rest of the motion. Fig. 34 illustrates the scratching method of striking the arc.

DEPOSITING THE METAL

In advancing the arc, care should be taken not to move the electrode faster than it is possible for the arc to melt a place on the surface for receiving the deposited metal. If the metal is moved too fast, the metal will be merely laid on the surface with no penetration. The operator should keep the arc traveling forward just fast enough to keep it at the forward edge of the crater.

Laying a Straight Continuous Bead

The beginning welder should first practice making straight continuous beads on a metal plate. Remember that when advancing the arc, care should be taken not to move the electrode

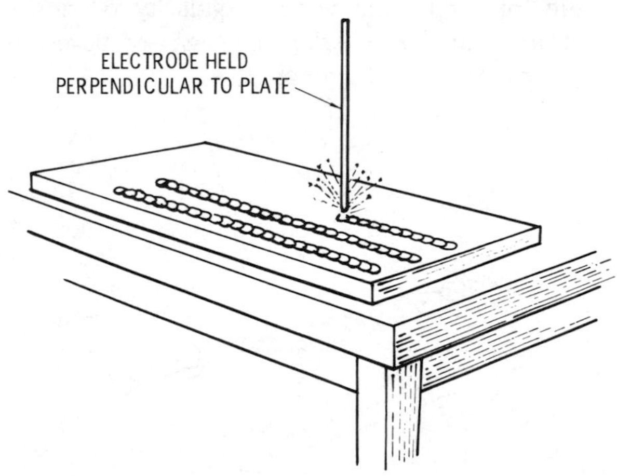

Fig. 35. Laying a straight continuous bead.

faster than it is possible for the arc to melt a place on the plate for receiving the deposited metal.

The procedure for laying straight continuous beads (Fig. 35) is as follows:

1. Select a steel plate with dimensions suitable for practice welding.
2. Hold the electrode perpendicular to the steel plate. The welder should try to assume an easy position in which the whole body is comfortable and braced so as to be steady without strain, leaving the right arm entirely free.
3. Strike the arc near the edge of the steel plate closest to the welder.
4. Do not let the arc go out while the bead is being run except to change electrodes.
5. Move the electrode slowly and steadily across the steel plate in a direction away from the welder.
6. Take care that the width and height of the bead is uniform. This will require a steady, continuous movement of the electrode. Varying the speed will result in a weld bead with uneven contours (Fig. 36).
7. Keep the length of arc constant. The deposited metal should meet all the requirements of a good weld—uniformity of height and width, regularity of ripples, good penetration, and no overlap nor signs of porosity.

Fig. 36. A weld bead with uneven contours.

Laying Three Parallel Beads

The beginning welder should now practice laying parallel beads 3/8 inch in width. These are wider than the straight, continuous beads described in the preceding paragraphs. Consequently, it will be necessary to spread the weld by weaving the electrode in a crescent motion from left to right and from right to left across the line of travel. The electrode will follow a path similar to that shown in Fig. 36.

The procedure for laying three parallel beads is similar to that described for laying a straight continuous bead with the following modifications:

1. The criss-cross motion required for the wider bead should not be too rapid or the weld will not penetrate.
2. The movement of the electrode should be governed by the same conditions laid down in the instructions for laying a straight continuous bead, allowing for the different motion. Note that the electrode is still held perpendicular to the steel plate.
3. In making this weld, it will be necessary to use more than one electrode per bead.
4. Each time an electrode is to be changed, clean the metal surface where the bead is to be deposited.

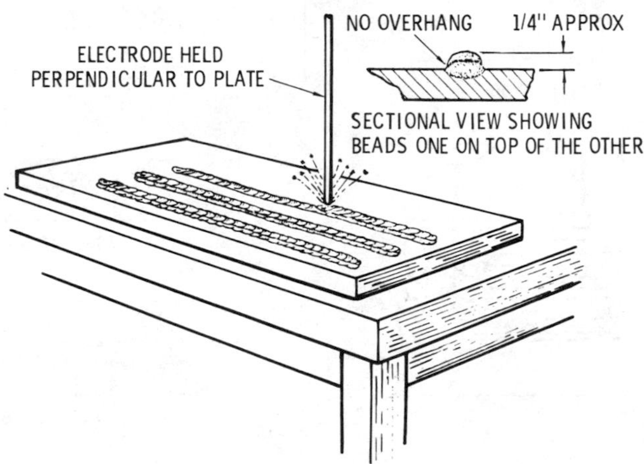

Fig. 37. Laying three parallel beads in two layers.

Laying Three Parallel Beads In Two Layers

The beginning welder should practice laying a second layer of bead over each of the previously described three parallel beads (Fig. 37). The total height of the double bead should be approximately ¼ inch. Do not let the metal from the second layer run over the edges of the first layer.

The current in the arc should be reduced to approximately 50 amperes. This current reduction is necessary because the conduction of heat into the plate is not as rapid, since the heat is applied to the top of a double layer of weld bead instead of the broad surface of the steel plate.

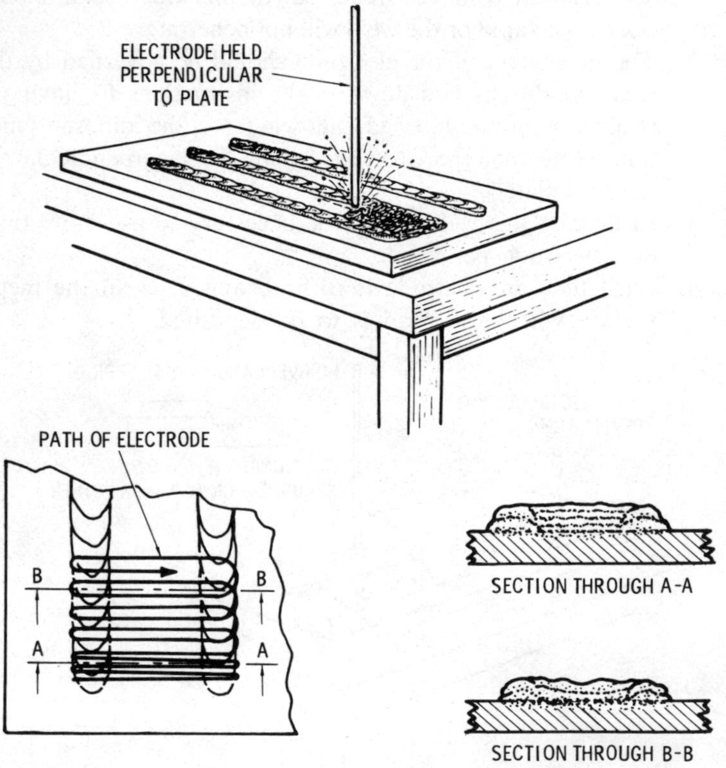

Fig. 38. Filling in the spaces between three parallel double-layered beads.

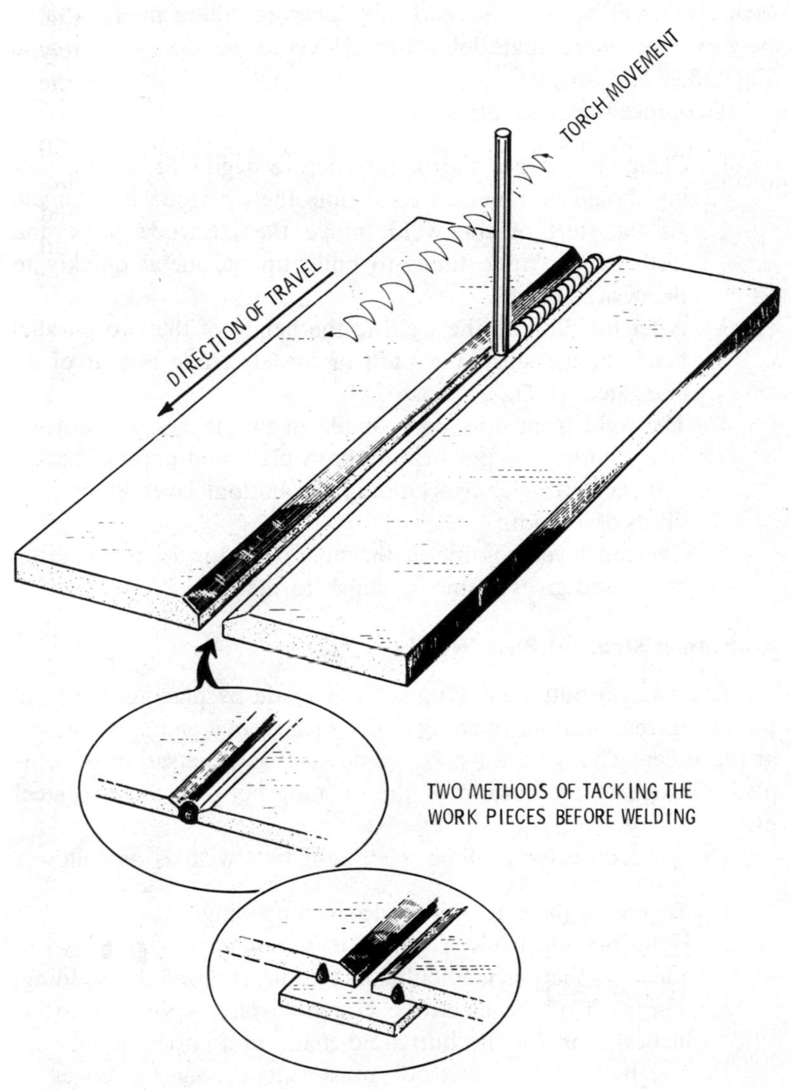

Fig. 39. Making a straight butt weld on a plate with beveled edges.

Filling the Spaces Between Parallel Beads

Once the beginning welder has successfully deposited two layers of weld bead, he should next practice filling in the spaces between the three, parallel, double-layered beads as shown in Fig. 38.

The procedure is as follows:

1. Clean the surface thoroughly before beginning. The cleaning should be repeated each time the electrode is changed.
2. At the start of the weld, move the electrode back and forth two or three times to build up the metal quickly to the desired height.
3. After building up the weld to the height of the two parallel beads at the start, the path of the electrode is that of an elongated spiral.
4. The weld from now on is made in two layers, the bottom layer thoroughly penetrating both plate and parallel beads, and the top layer overlapping the bottom layer about two thirds of its width.
5. The top layer completes the fill to the top of the parallel beads and gives a smooth finish to the weld.

Making a Straight Butt Weld

The straight butt weld (Fig. 39) is made by placing two steel plates together with an opening of ⅛ inch at one end and ¼ inch at the other. The weld must be made so that penetration is complete through to the bottom of the opening between the two steel plates.

The procedure for making a straight butt weld is as follows:

1. Tack weld the end with the ⅛-inch opening.
2. Hold the electrode so that it slants approximately 15° ahead of the perpendicular in the direction of the welding.
3. The path of the electrode is a bit wider (about ¼ to 5⁄16 inches) than for the butt weld made in a corner joint.
4. The path of the electrode must extend over the edges of the plates as shown in order that the deposited metal will be thoroughly fused with the plate.

Making a Corner Weld (Right-Angle Straight Butt Weld)

One method of producing a corner weld is to make a right angle straight butt weld (Fig. 40). This weld must extend through the bottom of the groove and show on the other side. The break should, in general, follow the middle of the weld. The grain of the metal should be uniformly fine and of a dull gray color.

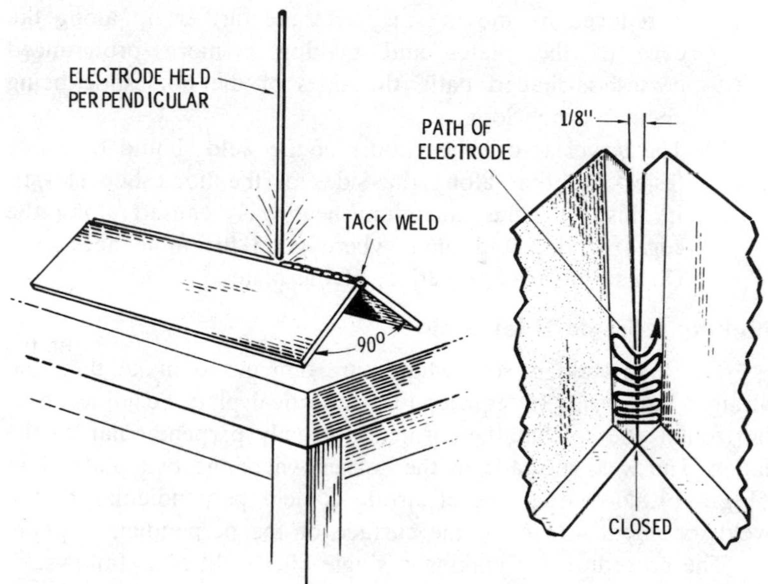

Fig. 40. A right angle straight butt weld.

The procedure for making a right angle straight butt weld is as follows:

1. Place the steel plates at an angle of 90°. One end should be closed, the other open at the end approximately ⅛ inch.
2. Tack the plates at the end where the edges touch.
3. Start the arc at the end at which the tack weld is located. The reason for this is that as the weld advances the shrinking of the deposited metal will gradually draw the plates together. Therefore as the arc reaches any point along the weld, the plates at that point will be spaced a slight distance apart.

4. The general rule for spacing the plates in welds of this type is ⅛ inch per foot length of the weld.
5. At the beginning, the path of the electrode will be simply a spreading motion to distribute the heat evenly.
6. As the plates become hot, the metal in the weld and at the bottom of the groove tends to fall or sag through. To prevent this the amount of heat in the middle of the weld is reduced by moving the electrode farther up along the edge of the plates and making a more pronounced horseshoe-shaped path, the sides of the horseshoe being about 3/16 inch long.
7. The travel across the middle of the weld should be made faster and that along the sides of the horseshoe slower. In this way, the maximum heating is caused along the edges of the cold plate where good fusion is necessary. Do not run over the edges of the plates.

Making a Single Fillet Weld

The fillet weld is somewhat more difficult to make than the straight butt weld. It requires that a vertical plate be joined to a horizontal one so that the former is exactly perpendicular to the latter. The weld is made in the corner where the two plates join (Fig. 41). Note that the electrode is held perpendicular to the weld, or about 45° from the surface of the perpendicular plate.

The procedure for making a single fillet weld is as follows:

1. Thoroughly clean the surface to be welded before laying the first bead.
2. Make tack welds at both ends of the vertical plate.
3. In making the first bead, the electrode should be moved slowly across the plate advancing as fast as necessary to keep the proper height of the deposited metal (which should be approximately ¼ inch).
4. Clean the surface of each bead layer with a hammer and a chisel and a steel brush before depositing a second layer.
5. Since the weld is being made in the middle of the horizontal plate, the heat will be conducted away in both directions by this plate and therefore at a greater rate than by the vertical plate.

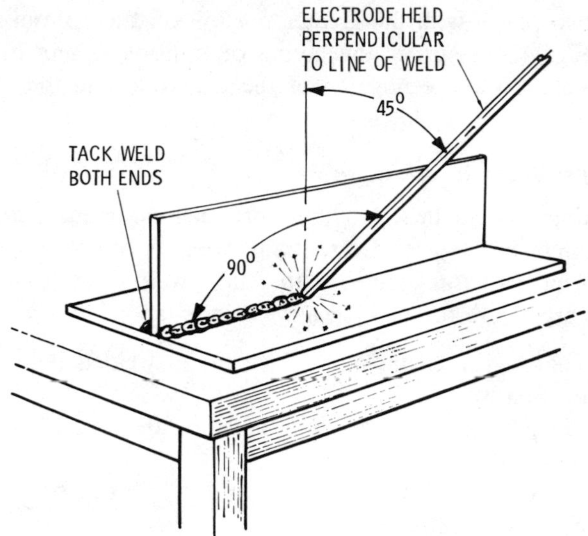

Fig. 41. Making a single fillet weld on a T-joint.

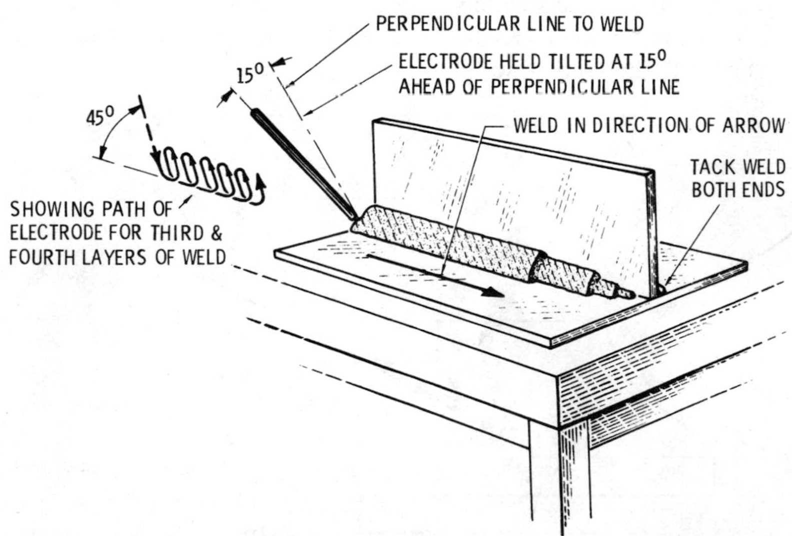

Fig. 42. Laying of multiple layers in a fillet weld.

6. The crater should be established at the junction of the two plates with the greater portion on the bottom plate.
7. Fig. 42 illustrates the laying of multiple layers in a fillet weld. Note the position of the electrode and the electrode path.

Making a Weld in a Lap Joint

Making a weld in a lap joint presents no great difficulty in arc welding. The lap joint essentially consists of two overlapping metal surfaces that are joined with a fillet weld (Fig. 43).

The procedure for making a weld in a lap joint is as follows:

1. Thoroughly clean the surface to be welded before laying the first weld bead.
2. Make tack welds at both ends of the overlapped edges (Fig. 44).
3. Make a fillet weld in the lap joint as illustrated in Fig. 43. Note the electrode angle.

CARBON-ARC WELDING

The carbon-arc welding process differs from metal-arc welding in that the additional metal required for the weld is obtained by melting a filler rod which is fed into the arc.

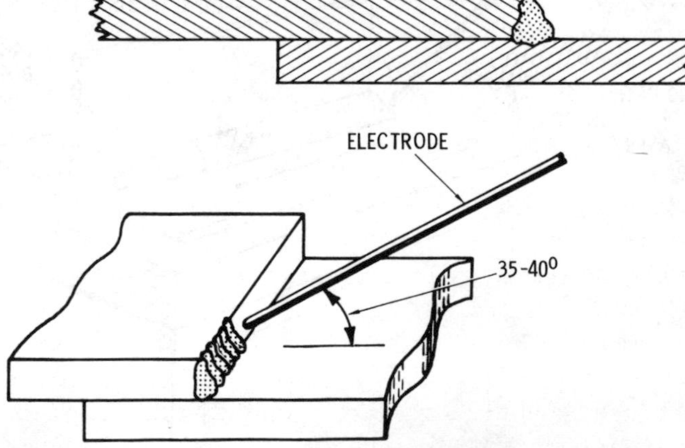

Fig. 43. Making a weld in a lap joint.

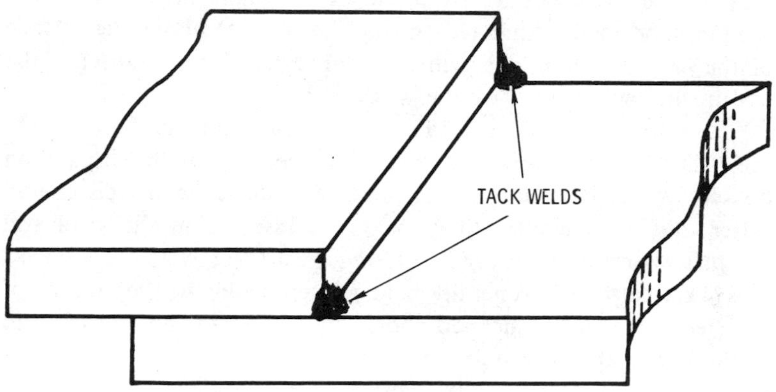

Fig. 44. Making a tack weld.

In carbon-arc welding, the arc is formed between the work and a carbon rod held in an electrode holder. The heat of the arc melts a small pool on the surface of the work to be welded. This pool is kept molten by playing the arc across it and extra metal for the weld is added by a filler rod.

Carbon-arc welding is a puddling process and is not applicable to vertical or overhead welding positions. The amperage required in carbon-arc welding is 15-30 amperes for common electrodes up to 300-500 amperes according to size.

The carbon arc is very stable and easy to maintain. The arc length can be varied over wide limits without causing the arc to go out. There is no tendency for the carbon electrode to freeze or stick, as is the case with electrodes used in metal-arc welding. Accordingly, the arc can be struck without difficulty at any point and rapidly moved over the surface of the work to the point where the weld is to be made.

In welding with the carbon electrode, a molten pool should be formed on the work and additional metal deposited in this pool. The arc should be kept at this point until the added metal is thoroughly melted and mixed with the original metal before more material is added.

The filler rod is held in the welder's left hand if he is right handed. The opposite is true should he be left handed. When the pool or the work is ready, the end of the filler rod is inserted in the pool and the arc directed against the rod just above the surface of the molten metal. This will melt through the rod and leave the end in the molten pool on the work.

The arc should be played about on the pool until the added metal is all melted down. At this point, because of the circulation caused by the heat, the molten metal will have been well mixed. After this, the filler rod may be again inserted in the pool and the process repeated advancing along the line of weld.

The electrode is generally held perpendicular to the surface of the steel plate but is inclined ahead about 15° to the line of weld to direct the arc back into the weld.

By welding from left to right, a right handed welder avoids awkward positions. The arc is directed backward into the weld and the position of the arms and hands is comfortable.

CHAPTER 7

TIG and MIG Welding

In any type of welding, the quality of a weld is determined by the success in which contaminants, particularly oxides, are prevented from accumulating on the metal surface. The cleaner the surface, the better the weld. To obtain such conditions, the molten weld metal must be protected from the atmosphere during the welding operation; otherwise, atmospheric oxygen and nitrogen, or other contaminants, will combine readily with the molten weld metal and result in a weak porous weld.

Before the development of the TIG and MIG inert gas shielded-arc welding processes, fluxes (chemical solutions) were relied upon to remove contaminants that could weaken the weld. The basic principle of the TIG and MIG welding processes is that a flow of inert gas (one that will not combine with other elements) is used to shield the flame and the area being welded from the surrounding atmosphere and all possible contaminants.

In addition to the need for some means of shielding the weld area, new metals and production techniques (particularly those being used in the aircraft industry) required greater welding speeds and easier more economical welding methods. It was to meet these needs that first TIG and then MIG welding were developed.

TIG welding was introduced to industry in the early 1940's and used primarily for the welding of aluminum, magnesium, and other hard-to-weld metals and their alloys. The principal shortcoming of TIG welding at its inception was its lack of suitability for welding metal thicknesses (particularly aluminum) greater than ¼ inch. MIG welding was developed and introduced toward the end of the same decade as a solution to this and other shortcomings. At first, limited to thicker sections of metal, further developments of the MIG welding process has made it suitable for all types of welding.

TIG WELDING PRINCIPLES

Gas tungsten-arc welding is the *American Welding Society's* official designation for this welding process. It is more commonly referred to as TIG (tungsten-inert-gas) welding, and it will be referred to by this abbreviated form in this chapter. *Heliarc* and *Heliweld* are examples of two manufacturing trade names for the TIG welding process.

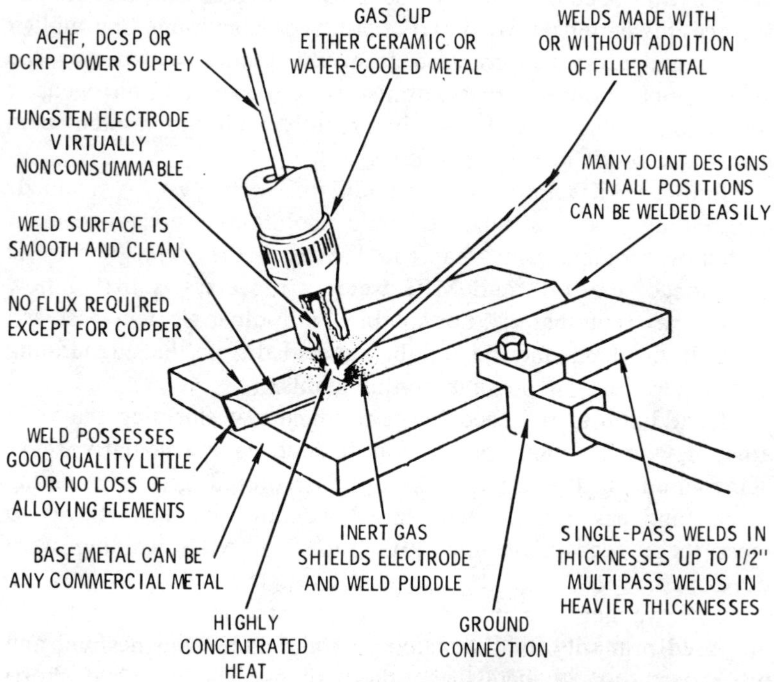

Fig. 1. The TIG welding process.

TIG welding is a shielded-arc welding process that represents a special and more complex form of standard metal-arc welding. The necessary heat for welding is provided by a very intense electric arc which is struck between a virtually nonconsumable tungsten electrode and the metal surface of the work piece. TIG welding differs from gas-metal arc (MIG) welding in that the electrode is not melted and used as a filler metal. On joints where filler metal is required, a welding rod is fed into the weld zone and melted with the base metal in the same manner used with oxyacetylene welding (Fig. 1).

The TIG welding process uses tungsten electrodes to conduct the electrical current to the arc. The arc and the molten puddle beneath the arc stream are shielded from the oxygen and other contaminants in the surrounding atmosphere by an inert gas (helium, argon, or a mixture of the two). Because the gas is inert, it will not combine with additional elements to form other compounds and thus provides an effective shield for the arc (Fig. 2).

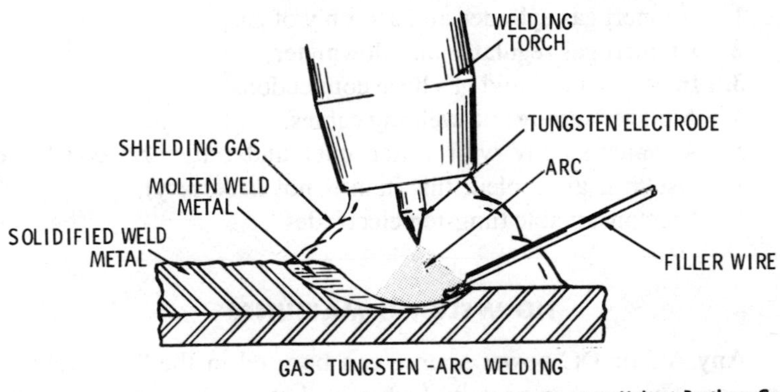

Fig. 2. The TIG welding process illustrating the inert gas shield.

TIG WELDING SYSTEM

A TIG welding system (Fig. 3) uses a special welding torch that permits the inert shielding gas, the water or air (for cooling), and the electrical current to flow through the handle. In addition to

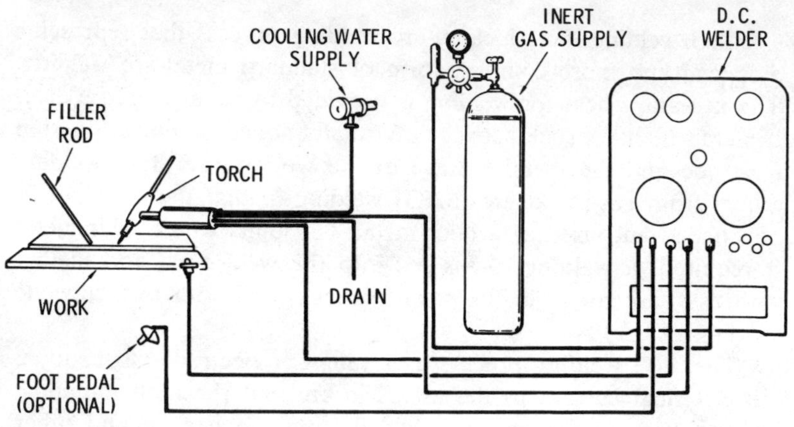

Fig. 3. The TIG welding system.

the welding torch, the TIG welding system requires the following equipment and supplies:

1. An inert gas cylinder and a supply of gas,
2. An inert gas regulator and flowmeter,
3. Inert gas hoses and gas hose connections,
4. A power supply and welding cables,
5. A water supply with water inlet and outlet hoses (if the system is air cooled, this item is not necessary),
6. Nonconsumable tungsten electrodes.

TIG WELDING MACHINES

Any AC or DC power source can be used in the TIG welding process. The most important factor is that the power source exhibit the ability to maintain a stable current at lower ranges. Many DC welders are constructed with this low range characteristic. An AC welder should be used only in conjunction with a high-frequency generator. A high-frequency generator will compensate for those "zero" periods in the alternating-current cycle. Several TIG welding power sources are illustrated in Figs. 4, 5, and 6.

The AC/DC high-frequency combination power source, illustrated in Fig. 4, is designed to provide constant output regardless

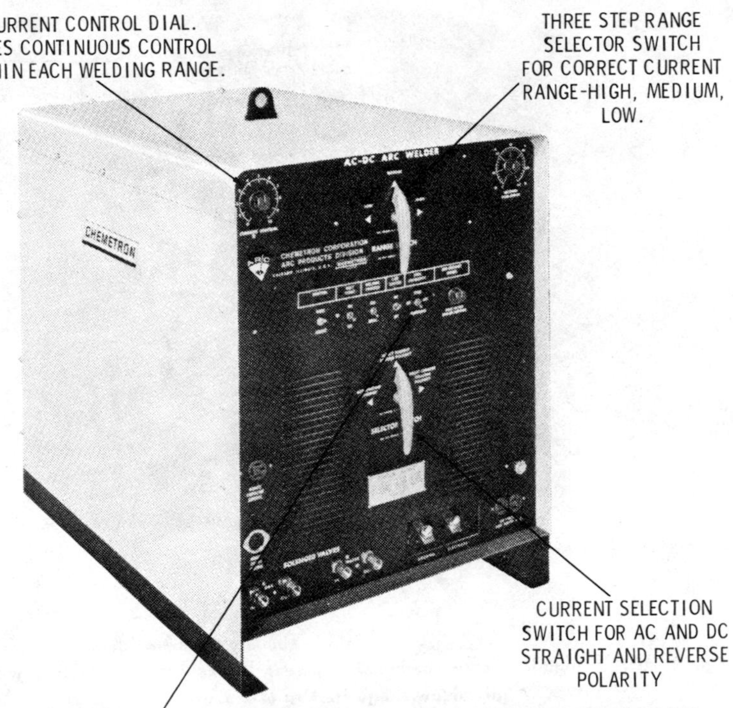

Courtesy Chemetron Corp.

Fig. 4. An AC/DC high-frequency combination power source for TIG metal-arc or spot-welding processes for which AC or DC is required.

of line voltage fluctuations, arc length, or warm-up, and is expected to produce a stable arc throughout the entire current range. The reactor design of this model produces wide current ranges and eliminates tungsten spitting throughout the entire range.

TIG WELDING TORCH

The TIG welding torch is designed to deliver both the electric current and the shielding gas to the point being welded as well as to provide passages within the torch for cooling purposes. The

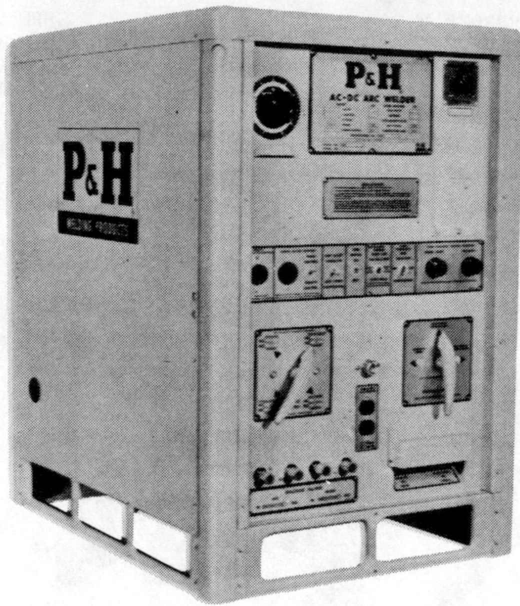

Courtesy Chemetron Corp.

Fig. 5. AC/DC high-frequency combination power source for TIG welding with a special low range starting at 2 amps.

tremendous heat of the arc and the high current often used (frequently in excess of 200 amperes) usually necessitates water or air cooling of the welding torch and power cable. Thus, adequate protection from heat is afforded, and the equipment is lightweight and flexible for easy handling. Although these torches may be cooled by either water or air, the former method is necessary when using currents of 200 amperes or more (Fig. 7). The cooling water must be clean; otherwise, restricted or blocked passages may cause excessive overheating and damage to the equipment. It is, therefore, advisable to use a suitable water strainer or filter at the water supply source.

Interchangeable ceramic cups control the rate of gas flow and provide direction for the flame. These are threaded and screw into the end of the torch. The nonconsumable tungsten electrode is firmly held in the torch by a chuck or collet that is also screwed

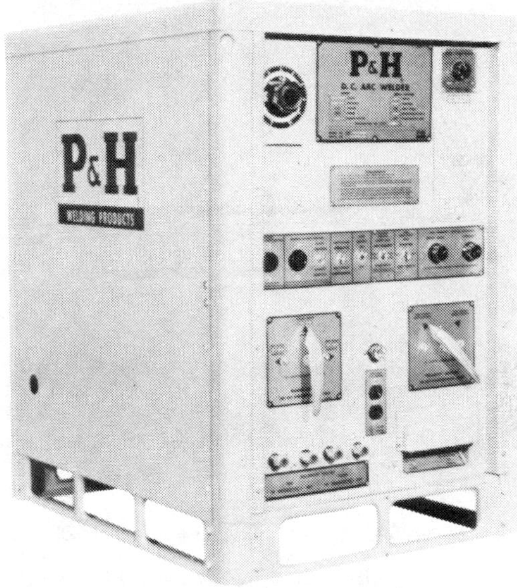

Courtesy Chemetron Corp.

Fig. 6. High-frequency DC power source for TIG welding available in either 200 or 300 amps.

into place. Both the chucks and collets differ in size according to the diameter of the electrode (Fig. 8).

TIG welding torches are available in a variety of sizes based on their amperage rated capacity. These sizes range from torches with a 100-amp rated capacity to 500 amps (Fig. 9). Other design characteristics involved in the different types of torches are:

1. Torch weight,
2. Torch length,
3. Electrode diameter capacity,
4. Electrode length capacity,
5. Chuck or collet size.

It is also possible to purchase flexible TIG welding torches (Fig. 10). The one illustrated in Fig. 11 is a water-cooled TIG welding torch with a flexible neoprene handle. The principal advantage of a flexible welding torch is that it enables the tip of the

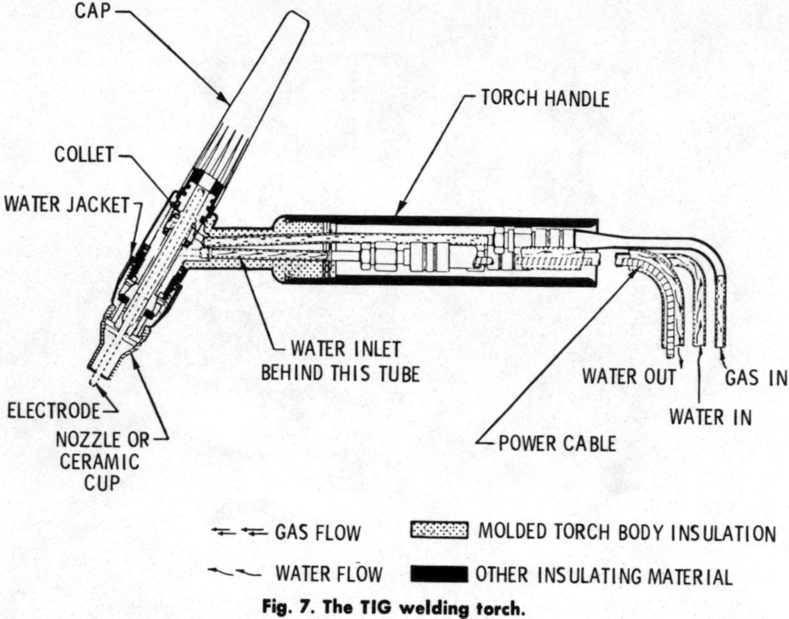

Fig. 7. The TIG welding torch.

torch to be extended into hidden recesses and other difficult to weld positions. In short, it permits an infinite number of operating angles (Fig. 12). As is shown in Fig. 11, the internal construction of these torches is similar to that of the nonflexible types.

NONCONSUMABLE TUNGSTEN ELECTRODES

Tungsten electrodes are available commercially in either pure tungsten or tungsten alloy types. The pure tungsten electrodes are designed for general purpose work where cost is a prime factor. The alloyed tungsten electrodes are recommended for work of a more critical nature. The alloyed tungsten electrodes include 1 or 2% thorium or zirconium. A 2% thoriated tungsten electrode would be used for work of a much greater critical nature than the 1% type. The addition of an alloy to the tungsten produces an arc that starts more easily and is more stable. The electrode life is longer than in the pure tungsten electrodes and the arc has a tendency to concentrate in the weld pool.

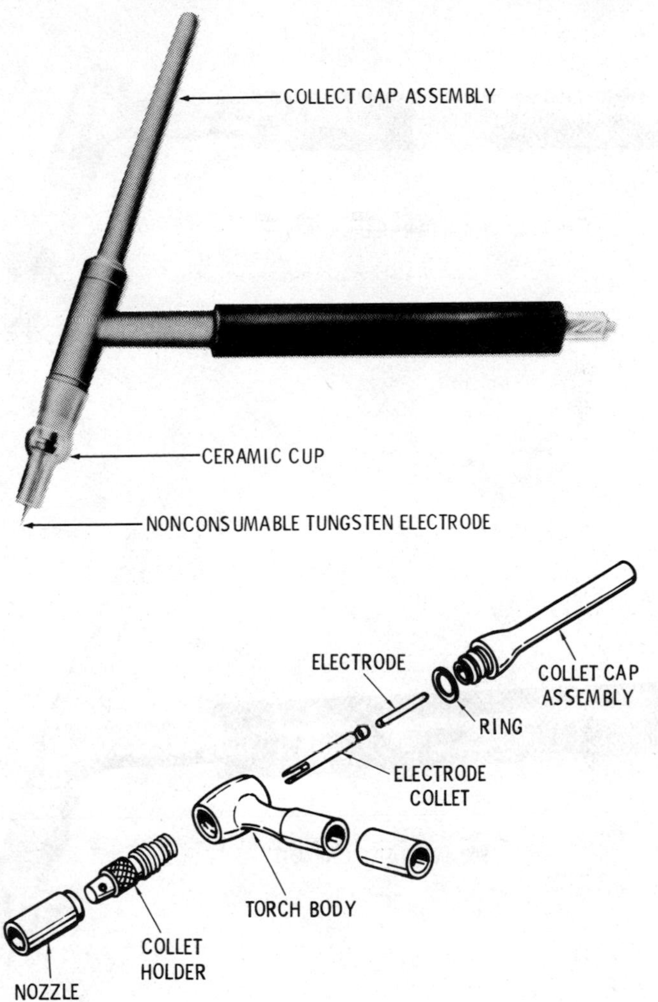

Courtesy Chemetron Corp.

Fig. 8. A TIG welding torch assembly.

Tungsten electrodes are available in a wide range of diameter sizes, including: 040, 1/16, 3/32, 1/8, 5/32, 3/16, and 1/4 inch. The standard lengths are 6, 7, and 18 inches. Special lengths and diameters can be manufactured to order.

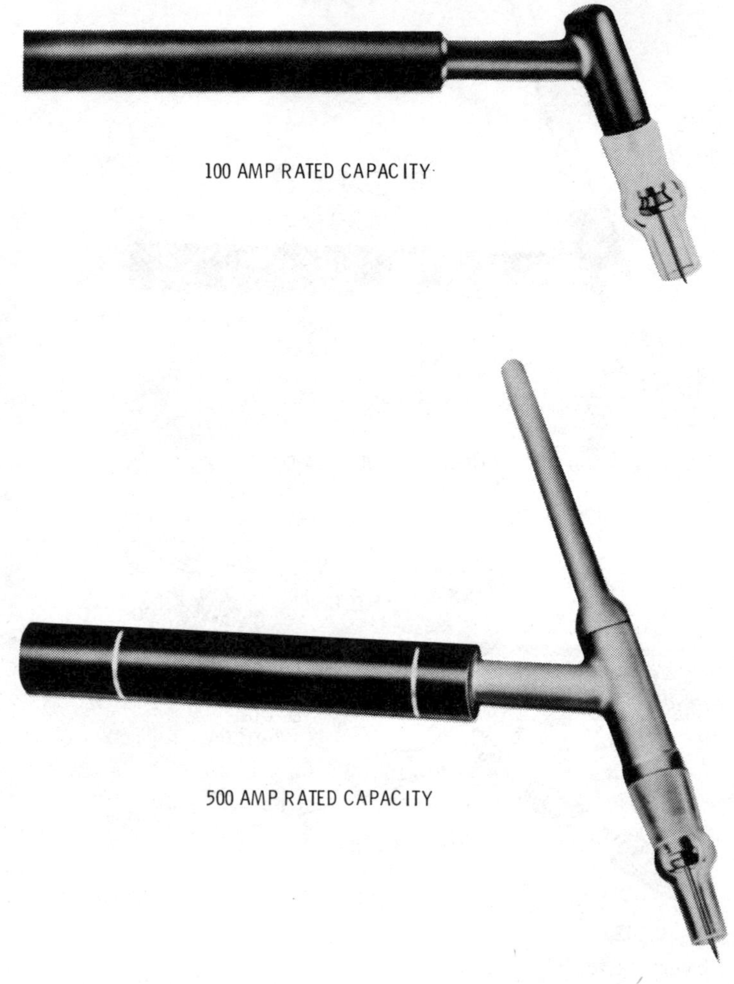

Courtesy Chemetron Corp.

Fig. 9. TIG welding torches rated at 100 and 500 amp capacity.

The shape of the electrode tip (Fig. 13) depends on the type of welding current used. A sharp pointed tip is used with DC and a rounded tip with AC.

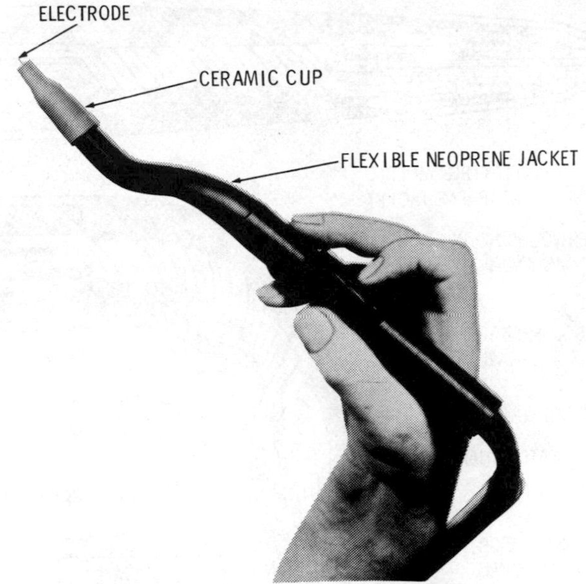

Courtesy Falstrom Co.

Fig. 10. A flexible TIG welding torch.

GAS SUPPLY AND EQUIPMENT

The pressure reducing regulators and flowmeters used in TIG welding are designed to reduce the gas pressure from the source to a level suitable for welding. A regulator and flowmeter designed for use with argon gas is shown in Fig. 14. The flowmeter is a device designed to precisely meter the flow of argon (or oxygen, helium, hydrogen, nitrogen, carbon dioxide, etc.) at regulated pressures. The flowmeter may be attached directly to the proper regulator or located some distance from the regulator at point of use.

Every piece of equipment that uses gas has a minimum delivery pressure (referred to as the *operating pressure*) designed into it. This operating pressure is the pressure necessary to force the required volumetric flow rate through it. The volumetric flow capacity in *cfh* for four different regulators is illustrated in Fig. 15.

According to the graph curve for regulator A (oxygen), a flow rate of 5000 cfh may be obtained at a delivery pressure of

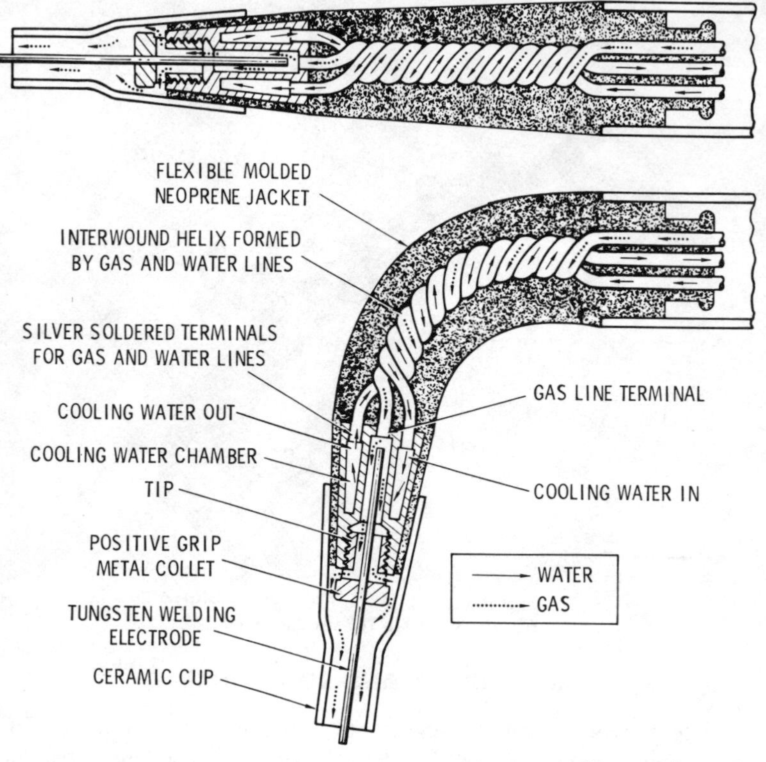

Courtesy Falstrom Co.

Fig. 11. A water-cooled TIG flexible torch.

50 psig. The curve also indicates that the same flow rate is obtainable at any pressure up to 180 psig. Regulator D (acetylene) indicates a flow rate of 1500 cfh may be obtained at a delivery pressure of 15 psig (the safety maximum for acetylene.)

The maximum flow rate obtainable at a given pressure is based on conditions that permit the flow of gas through a standard outlet bushing with no other restriction at any point further along the line. A given flow rate at a lower pressure than the corresponding pressure shown for the flow rate can not be obtained. However, it is possible to obtain a given flow rate at a higher pressure than the corresponding pressure shown for that flow rate. This is illustrated by the graph curve for regulator A in Fig. 15.

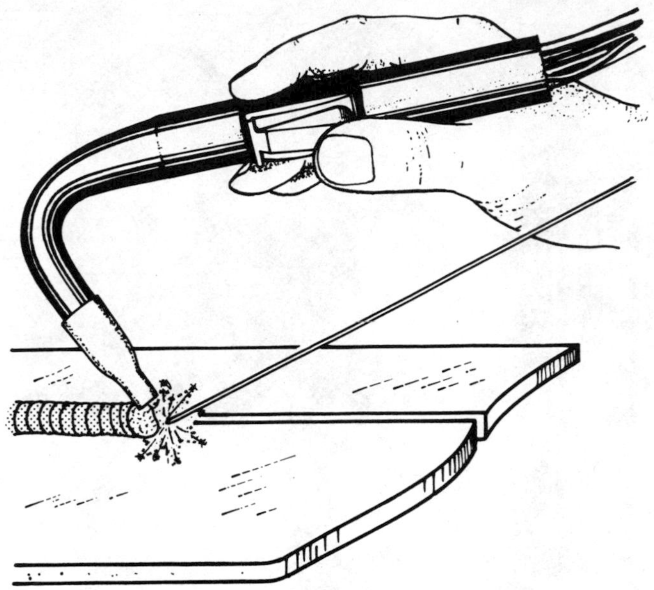

Fig. 12. Using a flexible TIG welding torch.

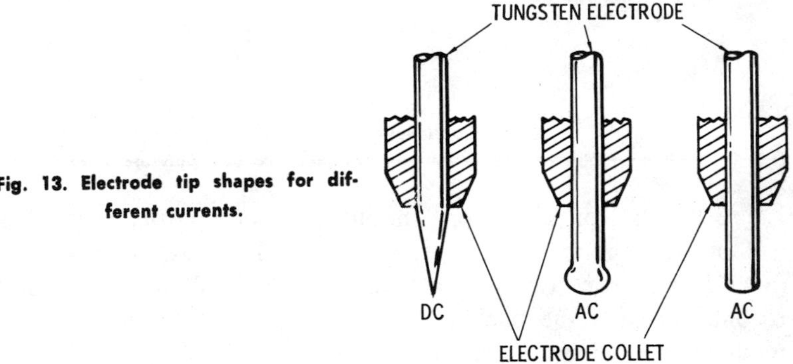

Fig. 13. Electrode tip shapes for different currents.

TIG JOINT PREPARATION

As with all other types of welding, the joint must be properly cleaned and beveled. All contaminants (dirt, grease, oxides, etc.) must be removed from the surface before the welding begins. Beveling is necessary on thicker sections of metal.

Courtesy National Cylinder Gas, Div. of Chemetron Corp.

Fig. 14. Welding regulator and flowmeter designed for use with argon or helium.

A backing of some sort should be used, particularly when welding the lighter gauge metals. The backing may consist of a metal bar, plastic, asbestos, or some other suitable material. Some recommend that the backing not come in contact with the immediate area being welded.

The joints suitable for joining by the TIG welding process are:

1. The butt joint,
2. The lap joint,
3. The T-joint,
4. The corner joint,
5. The edge joint.

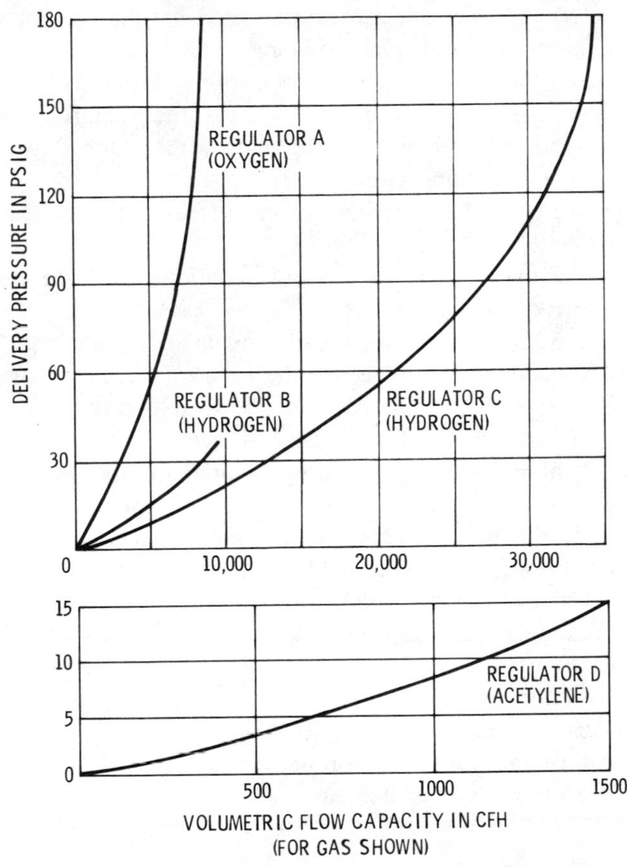

Courtesy Chemetron Corp.

Fig. 15. Volumetric flow capacity chart.

TIG WELDING PROCEDURE

A successful TIG welding operation depends on the correct and careful operation of the welding machine and other equipment. Because this is so important, the welder should familiarize himself with the machine he is using (thoroughly read the manufacturer's operating manual) and create an operating check list for the equipment.

The following items should be included in any such TIG welding check list:

1. Select the electrode size most suited for the type of metal, joint design, and current being used.
2. Insert the electrode in the collet of the torch. Make certain that it is firmly held and extends the required distance beyond the end of the nozzle. Lay the torch down *away* from the welding cables.
3. Move the polarity switch to the correct setting (e.g. to AC, DCRP, or DCSP).
4. Make the current adjustments recommended for the size electrode being used.
5. Make all other necessary adjustments on the machine. These will depend, of course, on the type of current being used. For example, a high-frequency switch may have to be switched on for AC welding. These will be described in the manufacturer's operating manual for the welding machine.
6. Turn on the cooling water (or air). Failure to do so will result in the destruction of the torch and the hoses by the extreme heat to which they will be subjected. Before turning on the welding machine, check the water outlet hoses to make certain that the water is flowing through the entire system.
7. Turn on the gas.
8. Adjust the flowmeter for the desired rate of flow.
9. Turn the power switch on.

TIG SPOT WELDING

Spot welding can be used with the TIG welding process to weld joints not easily accessible for resistance spot welding. It is particularly useful in joining light-gauge metals to one another, or joining light-gauge metals to heavier ones. The lap joint is the type most commonly welded by this process. A filler metal is used with TIG spot welding for the additional weld metal.

A specially designed TIG spot welding gun (Fig. 16) with a DC power supply produces small circular welds on the lap joint. The welding gun illustrated in Fig. 11 is water-cooled. The concentrated heat is such that the weld penetrates both overlapping

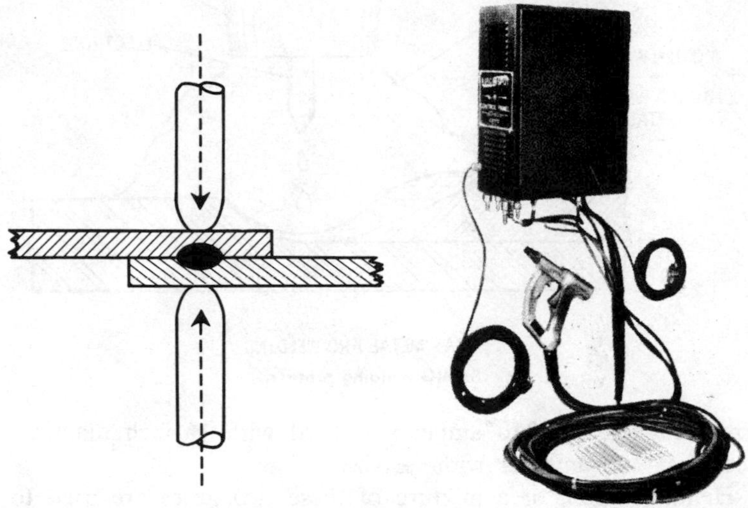

Fig. 16. A TIG spot welding gun with electrical control units.

Fig. 17. A TIG spot weld on a lap joint.

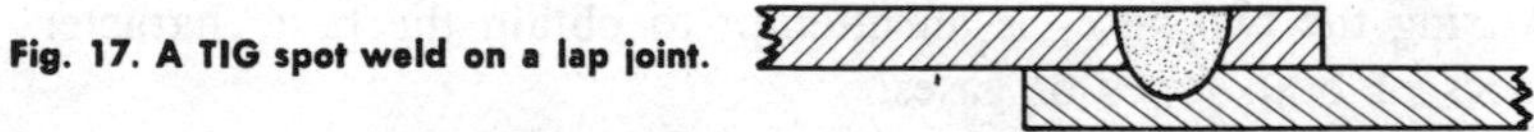

surfaces from one side (Fig. 17), fusing the two together. In this way, TIG spot welding differs from resistance spot welding which requires penetration from both sides of the lap joint. Electrode pressure is not required.

In TIG spot welding (as is the case in the resistance spot welding process), it is necessary to control the time variable with a timing device. The timer also controls the size of the spot weld and the amount of deposited metal. The timing control device on the power source, illustrated in Fig. 18, is adjustable for 0 to 360 cycles (0 to 6 seconds).

The nozzles for the TIG spot welding gun are available in a number of designs depending on the shape of the surface being welded. A nonconsumable tungsten electrode extends through the nozzle. These electrodes are available in 3/32 and 1/8-inch diameters, with the latter size being the most commonly used. A maximum

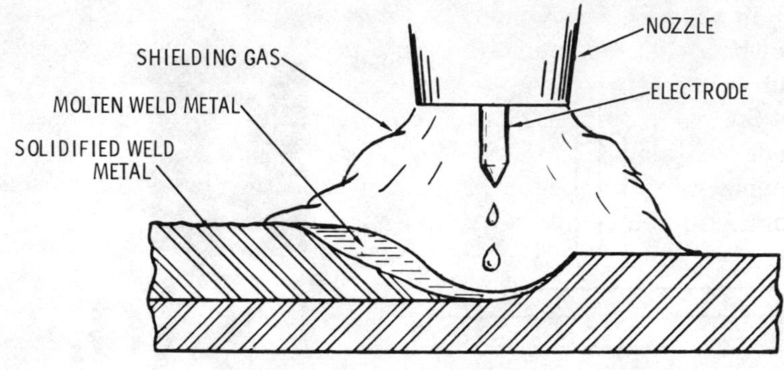

Fig. 18. MIG welding process.

current capacity of 250 amperes is used with ⅛-inch diameter electrodes; 200 amperes with 3/32 sizes.

Helium, argon, or a mixture of these two gases are used to shield the weld area in the TIG spot welding process. Helium produces a weld with deeper penetration than argon. On the other hand, argon results in a weld with a greater surface diameter. Mixing the two gases is an attempt to obtain the best characteristics of both types of gases.

Any drooping characteristic, DC power source capable of delivering up to 250 amperes with a minimum open circuit voltage of 55 volts can be used for TIG spot welding.

TIG WELDING APPLICATIONS

The TIG welding process can be used to weld almost all the metal alloys in use today. It is particularly effective and economical for welding light-gauge metals (under ⅛ inch thickness) and the hard-to-weld metals. Among the latter may be included the following:

1. Aluminum and aluminum alloys,
2. Copper and copper alloys,
3. Nickel and nickel alloys,
4. Magnesium and magnesium alloys,
5. Low-alloy and carbon steel.

In addition, TIG welding is suitable for joining Monel, Inconel, molybdenum, columbium, titanium, beryllium alloys, stainless steel, and some of the brasses.

Specific industrial applications of the TIG welding process include the manufacture of metal furniture and air conditioning equipment as well as widespread use in plumbing, sheet-metal work, and automotive body work.

MIG WELDING PRINCIPLES

Gas metal-arc welding (Fig. 18) is the *American Welding Society's* official designation for an inert-gas shielded arc welding process that uses a continuous consumable electrode wire. It is also frequently referred to as MIG (metal-inert-gas) welding, and this abbreviated form will be the one used throughout the remainder of this chapter. Commonly used trade names for the MIG welding process are *Aircomatic (Airco), Sigma (Linde),* and *Millermatic (Miller).*

This process consists of feeding a bare metal filler wire in conjunction with an inert shielding gas through a torch unit in which the welding wire picks up welding current supplied by a standard source of welding power. When the wire is fed from the unit to the work, it functions as a continuous consumable electrode. Thus, the filler wire and the electrode are one and the same.

MIG WELDING SYSTEM

A MIG welding system (Fig. 19) uses the following equipment and supplies:

1. An inert-gas cylinder and gas supply,
2. An inert-gas regulator and flowmeter,
3. Inert-gas hoses and connections,
4. A power supply and welding cables,
5. A MIG welding gun,
6. An electrode wire feeder,
7. A reel of electrode wire,
8. A remote control unit,
9. A supply of water (and water hoses) for cooling.

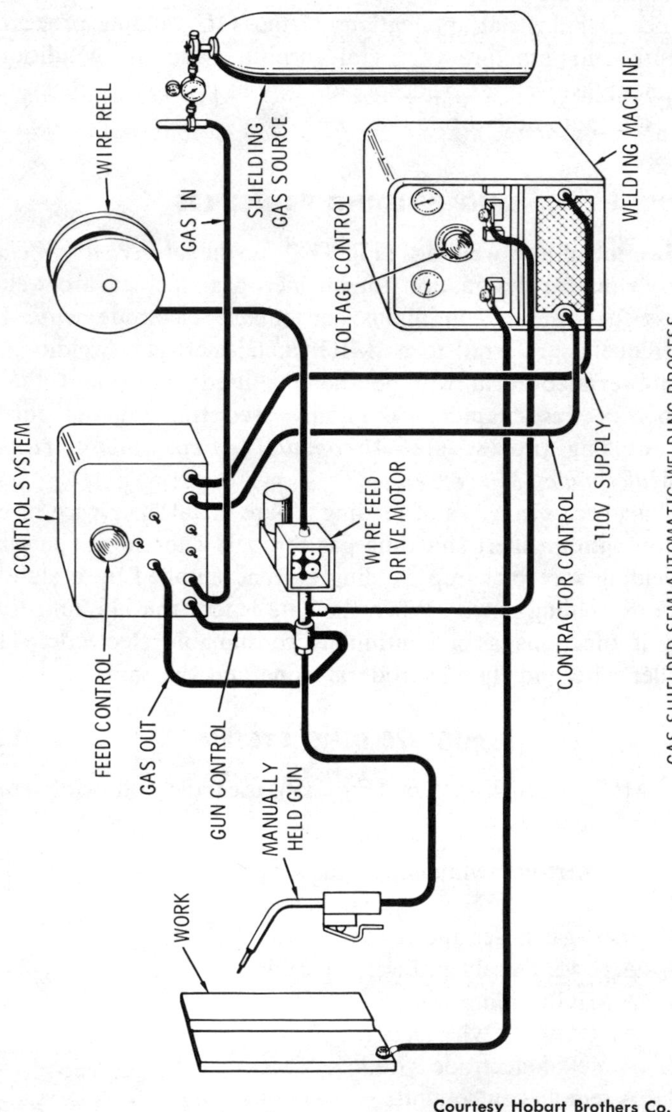

Courtesy Hobart Brothers Co.

Fig. 19. A MIG welding system.

MIG WELDING MACHINES

Direct current reverse polarity (DCRP) is recommended for use with the MIG welding process. Neither alternating current (AC) nor direct current straight polarity (DCSP) should be used. A DCRP current produces deeper penetration and a cleaner weld surface than the other types of currents. A MIG welding machine with an open circuit voltage range of 15-32 volts is illustrated in Fig. 20.

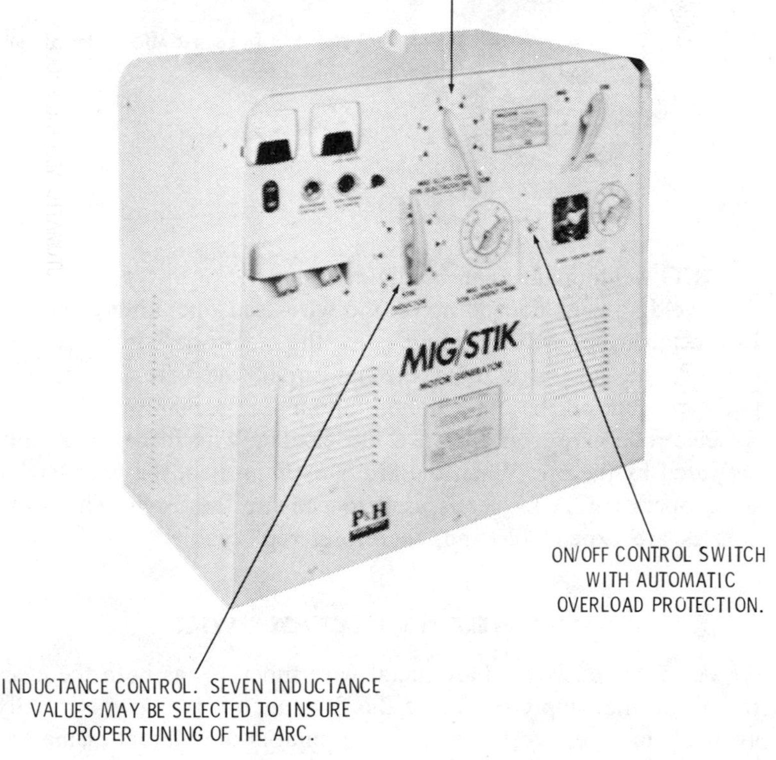

Courtesy Chemetron Corp.

Fig. 20. Electric motor driven MIG and conventional stick electrode welding machine.

MIG WELDING GUN

The MIG welding gun (torch) (Fig. 21) is essentially a specially designed electrode holder through which a continuous consumable electrode wire is fed to the arc. An inert gas flows through the nozzle and around the electrode to provide a protective shield for the area being welded. A supply of water or air is used to cool the torch.

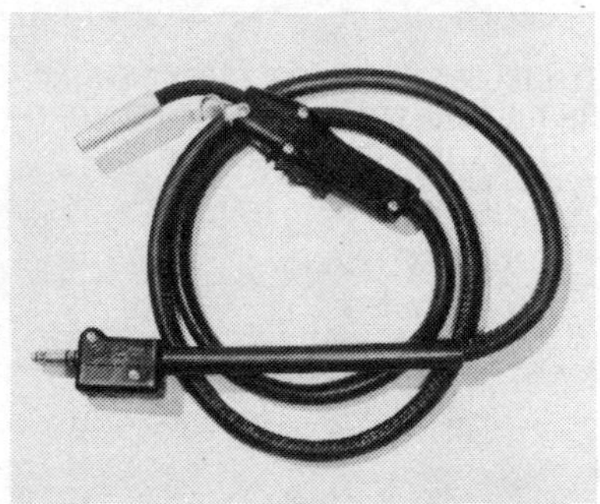

Courtesy Tweco Products, Inc.

Fig. 21. Air-cooled MIG welding gun.

MIG welding guns can be divided into the push-type and pull-type welding guns depending on the wire-feed operating principle. In the push-type MIG welding guns, the electrode wire is pushed through the gun by a roller mechanism in the wire feeder. The pull-type gun, on the other hand, contains a mechanism that pulls the electrode wire from the wire feeder. The wire feeder is usually activated by the arc. When the arc is extinguished, the wire feeder stops operating. A trigger is used to activate the torch. The torch nozzles are expendable and, therefore, replaceable.

MIG WELDING ELECTRODE WIRE

A continuously fed bare metal wire functions as both the electrode and the supply source of the filler metal. It is commercially available in spools, reels, or coils, in a number of different diameters, weights (e.g. 1, 5, 10, 25, 35, or 60 lb. spools), and metal compositions depending on the manufacturer and the type of equipment in which it is to be used.

AWS and *ASTM* specifications provide for three categories of electrode wire: (1) Nickel and Nickel-base-alloy Bare Welding Filler Metals (*AWS* A5.14, *ASTM* B304); (2) Corrosion-resisting Chromium and Chromium-Nickel Steel Welding Rods and Bare Electrodes (*AWS* A5.9, *ASTM* A371); (3) Copper and Copper-alloy Welding Electrodes (*AWS* A5.6, *ASTM* B225); and (4) Aluminum and Aluminum alloy Welding Rods and Bare Electrodes (*AWS* A5.10, *ASTM* B285).

The different types of electrode wire are identified by the letter E followed by numbers or chemical symbols that designate the major element or elements. For example, ECu would indicate a copper electrode wire; ECuNi, a copper and nickel, and so forth.

The electrode wire used in MIG welding is of relatively small diameter (generally ranging from 0.20 to ¼ inch) and is fed into the arc at high speed (over 400 ipm when welding thicker sections of aluminum plate). These factors contribute to high welding speeds and high metal deposition rates. As with other types of welding, the diameter size of the electrode wire is determined by such

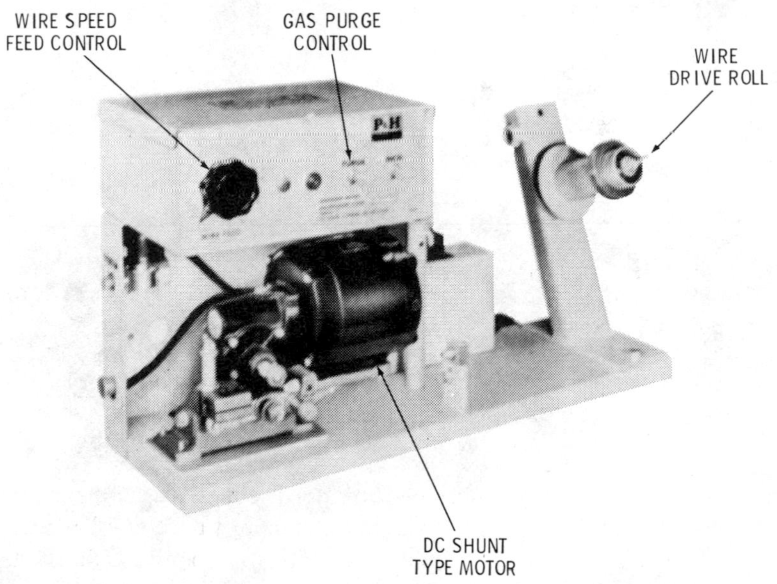

Courtesy Chemetron Corp.

Fig. 22. Wire feeder.

things as welding position, current, and the thickness of the metal. The welding current (amperage) will limit the speed at which the electrode wire can be fed to the arc.

Wire Feeder

MIG welding guns operate in conjunction with a wire feeder. These wire feeders are constructed so that they may be either attached to the welding machine or set independently some distance from the welding operation.

The wire feeders illustrated in Figs. 22 and 23 may be used with any constant-voltage power source and is designed to dispense continuous flux-cored or solid-wire electrode wire of 1/16, 5/64, 3/32, and 7/64 inch diameters.

Courtesy Chemetron Corp.

Fig. 23. Wire feeder for flux-cored or solid wire electrodes.

GAS SUPPLY AND EQUIPMENT

MIG welding uses an inert gas shield of helium, argon, carbon dioxide, or various mixtures of these gases (e.g. argon-helium, argon-carbon dioxide, argon-helium-carbon dioxide, etc.). The gas, or gas mixture, selected depends usually on the welding characteristics of the base metal. For example, an argon-carbon dioxide mixture is more efficient for welding certain classes of mild steel than either gas used alone.

Flowmeters (Fig. 24) calibrated for each type of inert gas are used to indicate the gas flow-rate. Their workings are described in the TIG welding section of this chapter.

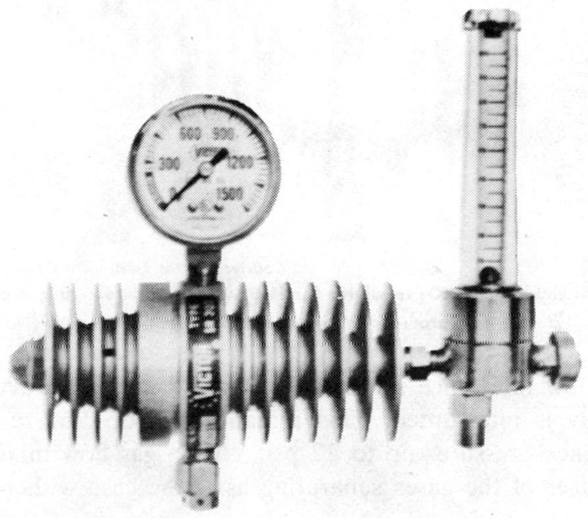

Courtesy Victor Equipment Co.

Fig. 24. Single-stage CO_2 regulator for MIG welding with a calibrated flowmeter.

Shielding Gas Flow Mixers

Gas flow mixers can be purchased that enable the operator to mix shielding gases according to the requirements of the job.

Gas flow mixers may consist of two or more modules. The gases are kept separated in the modules prior to mixing. There is a flow control knob and flowmeter for each module so that the operator can adjust the flow of the gases independently. This permits experimentation in obtaining the best weld.

The possibilities offered in gas mixtures may be illustrated with the three-module gas flow mixer (Fig. 25). For example, in a mixer designed to mix CO_2 and helium, the operator can also produce three mixtures of two gases: *argon,* CO_2, *argon-helium* and *helium-*CO_2.

Courtesy Victor Equipment Co.

Fig. 25. Single-stage CO_2 regulator for MIG welding where carbon dioxide provides the inert shielding gas.

The flow rate, once set, stays within 1% accuracy even if the gas supply is interrupted. This remains true of flow restrictions causing back pressures up to 22 psi. With a gas flow mixer, there is no danger of the gases separating as is the case with premixed gases.

Gas pressure regulators (Fig. 25) are necessary to reduce the line pressure from the cylinder to the welding gun (torch). The inert gas cylinders differ from those used in oxyacetylene or oxyhydrogen welding in the way that gas consumption is determined. Inert gas cylinders determine consumption in terms of cubic feet per hour (cfh).

The cables, hoses, and hose connections used in MIG welding are similar to those found in other types of welding. The same precautions necessary with other welding processes (e.g. selecting the correct hoses and connections, making certain that all connections are clean and tight, etc.) are also necessary with the MIG welding process.

Table 1 lists the various gas mixtures required by different welding jobs. The specific MIG welding method recommended for each gas mixture is given in the right-hand column.

Table 1. Various Gas Mixtures Required by Different Welding Jobs

Gas	Percentage Mixture	Metal	Method of Welding
CO_2	100%	Carbon Steel	Spray Transfer
	100%	Carbon and Low Alloy Steel	Flux-Cored Steel Wire
Argon	100%	Aluminum	Spray Transfer Short-Circuiting Metal Transfer
	100%	Magnesium	Spray Transfer
	100%	Copper	Spray Transfer
	100%	Monel	Spray Transfer
	100%	Titanium	Spray Transfer
	98% Oxygen 2%	Stainless Steel	Spray Transfer
	95% Oxygen 5%	Carbon Steel	Spray Transfer
	95% Oxygen 5%	Low-Alloy Steel	Spray Transfer
	95% Oxygen 5%	Stainless Steel	Spray Transfer
	75% CO_2 25%	Carbon Steel	Short Circuiting Metal Transfer
	75% CO_2 25%	Low-Alloy Steel	Short Circuiting Metal Transfer
	75% CO_2 25%	Stainless Steel	Short Circuiting Metal Transfer
	50% Helium 50%	Aluminum	Spray Transfer
	50% Helium 50%	Monel	Spray Transfer
Helium	90% Argon 7½% CO_2 2½%	Stainless Steel	Spray Transfer

MIG JOINT PREPARATION

The joint preparation (cleaning and beveling), the use of backing, and the types of joints suitable for welding are similar to those used in TIG welding (see TIG JOINT PREPARATION).

MIG WELDING PROCEDURE

MIG welding power sources should be operated with the same degree of caution as described for the types used in TIG welding (see TIG WELDING PROCEDURE). There are, however, certain aspects of the MIG welding process that are different and must be considered separately. These may be listed as:

1. The wire feeding unit,
2. A powdered flux feed unit,
3. The continuous electrode welding wire.

The electrode welding wire size best suited for the requirements of the welding job must be selected and placed on the wire feeder (some MIG welding guns are designed so that a small coil of electrode wire can be attached to the gun itself). Thread the electrode wire through the feed motor control unit and into the welding gun itself. The electrode wire extends into the arc through special tubes. These tubes must be the same size as the electrode welding wire.

Adjust the wire speed on the wire feeder to a speed suitable for the job. The deposition rate for the powdered-flux unit must also be within the recommended limits of the welding job. The operating controls and switches for the wire feeder are located on the front panel.

MIG WELDING METHODS

The two basic methods of MIG welding are:

1. Spray metal transfer,
2. Short circuiting transfer.

These essentially describe the way in which the filler metal is transferred to the weld. Other variations of MIG welding have been developed based on the distance of the arc from the work (MIG short arc welding) or the use of a flux with the gas shield (CO_2 MIG welding). Then, too, both standard MIG welding and CO_2 MIG welding have been adapted to the spot welding method.

SPRAY METAL TRANSFER METHOD

The DCRP welding current used in the MIG welding process results in the electrode filler wire melting off in the form of a spray or a steady stream of very small droplets of filler metal. This is referred to as the *spray metal transfer method.* The small droplets of filler metal do not interfere in any way with the stability of the arc.

SHORT CIRCUITING METAL TRANSFER METHOD

Another method of transferring the filler metal to the weld involves the temporary short circuiting of the arc. This is referred to as the *short circuiting metal transfer method.* A somewhat larger drop of filler metal than in the spray metal transfer method touches the metal surface before it detaches itself from the electrode. As a result, the arc is temporarily (for a mere fraction of a second) short circuited, but re-establishes itself as soon as the gap between the electrode wire and the work re-occurs.

CO_2 MIG WELDING

CO_2 MIG welding is similar in most respects to standard MIG welding except that it uses carbon dioxide as a shielding gas and either a magnetically coated or a flux-cored electrode. The *Dual Shield Process (Chemetron)* is a trade name for the latter. CO_2 MIG welding can also be done using electrodes coated with a deoxidizing agent. Except for the use of carbon dioxide as a shielding gas, this method bears the strongest resemblance to standard MIG welding.

CO_2 MIG welding with a flux-coated electrode requires that a magnetized granular flux be fed through the torch nozzle and into the arc where it attaches itself to the electrode. This coating provides protection to the electrode from contamination by impurities in the surrounding atmosphere. The carbon-dioxide gas provides a similar protection from these impurities in the weld area. The electrode wire is fed into the arc with a wire feeder unit in the same manner as in standard MIG welding. In fact, with the exception of the flux (either as a coating or as an electrode core), the entire welding system for CO_2 MIG welding is identical to that used in standard MIG welding. Fig. 26 illustrates a CO_2 MIG welding system that uses flux-cored electrodes.

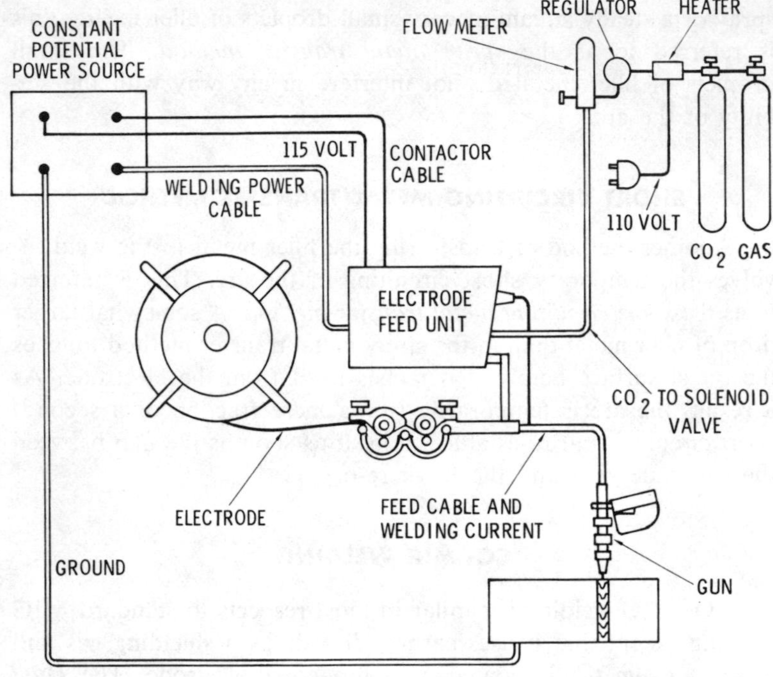

Courtesy Chemetron Corp.

Fig. 26. Components of the CO_2 MIG welding system.

Both the flux-coated and flux-cored CO_2 MIG welding processes allow the welder to observe the progress of the welding. This offers a distinct advantage over submerged-arc welding in which the arc (and welding progress) are hidden from view.

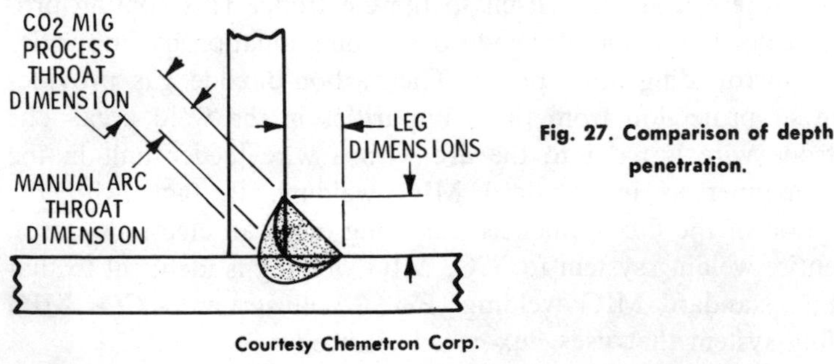

Courtesy Chemetron Corp.

Fig. 27. Comparison of depth penetration.

The principal advantages of flux-cored CO_2 MIG welding are:

1. The deep penetration of the arc (Fig. 27),
2. Edge preparation for butt joints is reduced to a minimum,
3. High deposition rate,
4. Less welding time.

All of these factors combine to produce cost savings.

MIG SPOT WELDING

Both standard MIG welding and CO_2 MIG welding have been adapted for spot welding. In addition to a special control panel and timing device. MIG spot welding requires a constant-potential power source capable of delivering currents as high as 500 amperes.

The MIG spot welding gun is equipped with removable nozzles constructed in a number of different designs. MIG spot welding is also used to weld joints not easily accessible for resistance spot welding. Its application, then, is similar to that of TIG spot welding.

CHAPTER 8

Submerged-Arc and Other Shielded-Arc Welding Processes

There are a number of shielded-arc welding processes that are either too limited in application, or their methods for shielding the arc and the molten weld metal are so radically different that it is next to impossible to justifiably include them in the description of the inert-gas shielded-arc welding processes contained in Chapter 7.

These various shielded-arc welding processes are as follows:

1. Submerged-arc welding,
2. Atomic-hydrogen welding,
3. Plasma-arc welding,
4. Electrogas welding,
5. Electroslag welding,
6. Underwater shielded metal-arc welding,
7. Vapor-shielded metal-arc welding,
8. CIG welding.

SUBMERGED-ARC WELDING

Submerged-arc welding (Fig. 1) is an automatic (or semiautomatic) welding process in which a granular flux blanket is

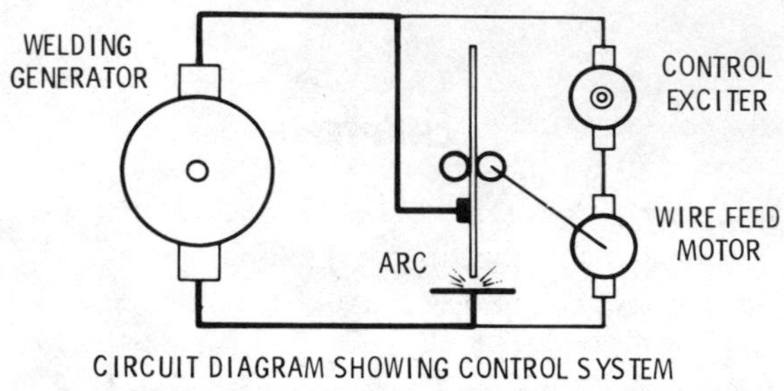

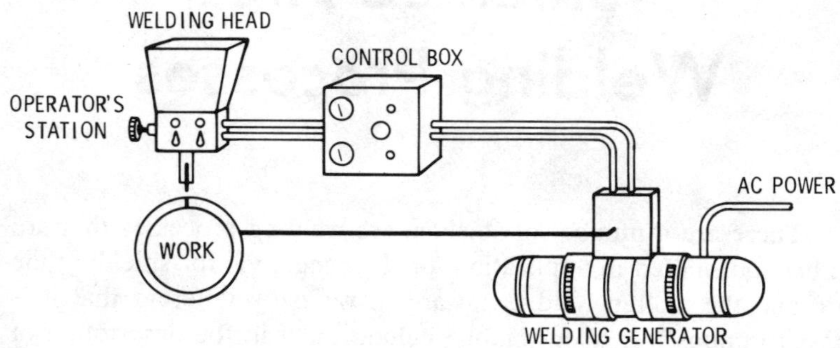

Fig. 1. Submerged-arc welding process.

used to completely cover the weld area while a consumable electrode is continuously and mechanically fed into the arc. It derives its name from the fact that the arc is hidden (submerged) beneath the flux blanket and not visible to the welder. This welding process is also referred to as *flux-covered arc welding, hidden-arc welding, subarc welding,* or *submerged-melt welding. Unionmelt welding* is a submerged-arc welding process developed by the *Linde Air Products Company. Unionmelt* is the trade name for their specially developed flux. This is an electric furnace product, carefully controlled as to composition and preparation.

The physical characteristics of welds made by the submerged-arc welding process are such that all types of inspection requirements such as full tensile, reduced tensile, free bends, nick bends, and Xray can be met easily if the procedures specified are followed.

The basic principle of this welding process centers on the concept of a gaseous shield around the arc and slag over the weld to protect both the arc and the molten metal from contaminants in the surrounding atmosphere. The results of this double shielding include deeper penetration, higher welding currents, faster welding, higher weld quality, and lower welding costs.

Instead of flux coated electrodes, granular flux is used in the latest development with automatic feed; it is in fact an automatic shielded metal-arc welding process.

Granular flux is deposited on the joint to be welded, being fed deep enough to cover the completed weld. A bare metal welding electrode wire is fed into the blanket of flux, its rate of feed controlled automatically for proper arc length. Direct current (DC) produces the arc between the electrode wire and the joint. The resultant arc heat fuses electrode and base metal, producing the weld.

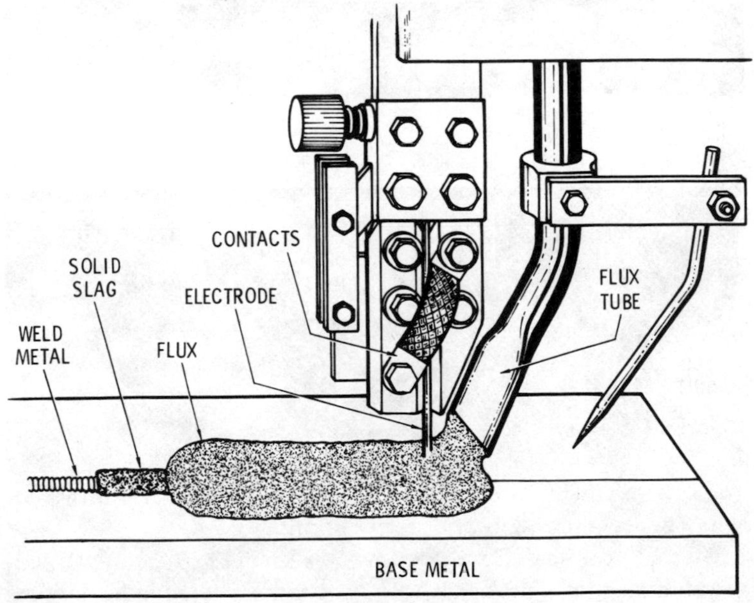

Fig. 2. Distribution of the flux.

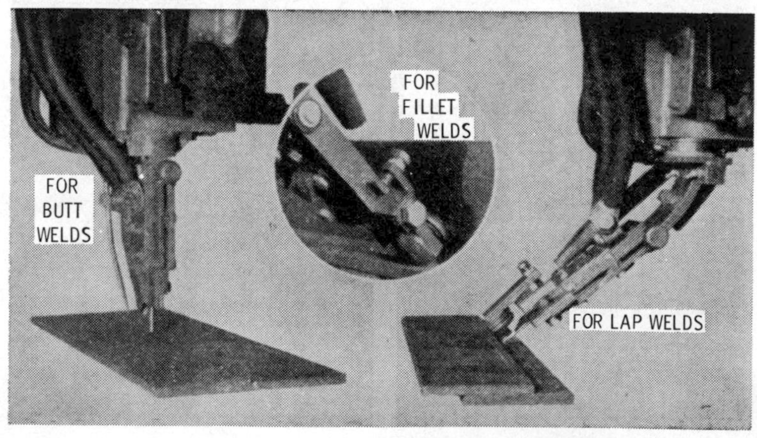

Fig. 3. Automatic shielded

Flux adjacent to the arc melts, floats on the surface of the molten metal, then solidifies as a slag on top of the weld. Since the arc and molten metal are blanketed by flux at all times, the metal is completely protected from contact with the air, assuming maximum quality of welds and making possible the use of very high amperage for faster welding (Fig. 2).

SUBMERGED-ARC WELDING EQUIPMENT

A submerged-arc welding system generally consists of the following basic components:

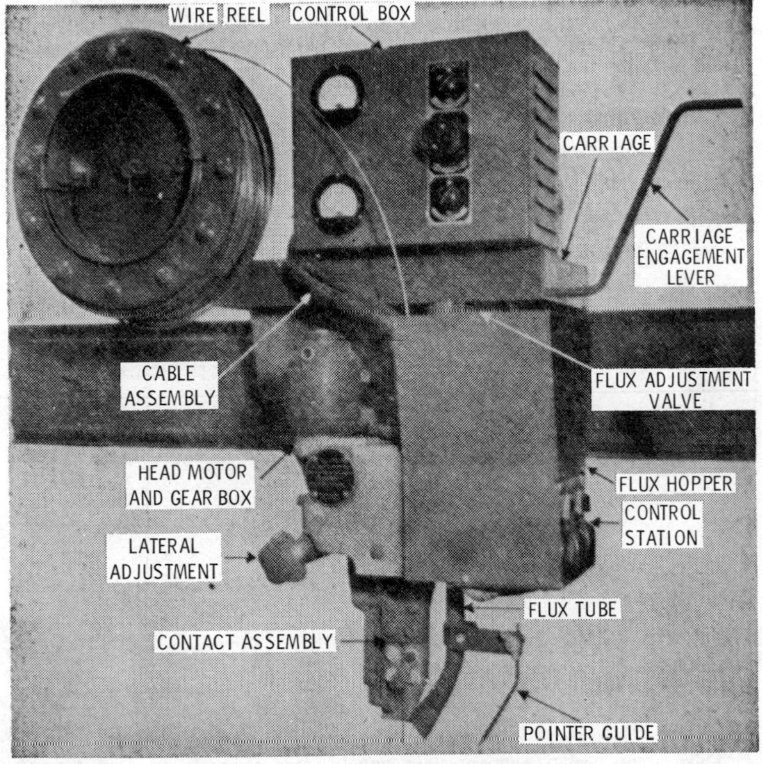

arc welding equipment.

1. An AC or DC power source,
2. A welding rod,
3. An electrode reel and motor to feed the electrode wire into the arc,
4. A flux hopper and a unit to recover unfused flux,
5. A control unit.

Almost any kind of standard DC generator can be used for submerged-arc welding. The heavy-duty AC transformers used in this welding process are usually those with 1000-ampere capacity (1000 amperes for one hour of operation, 750 amperes of continuous operation). However, both smaller and larger units are being used. Current higher than that obtainable from one transformer may be obtained in connecting two or more units in parallel. DC generators may also be connected in parallel in order to obtain sufficient amperage.

Both AC and DC power sources must employ a means of remote control for the welding current, so that welding can be started and stopped by a switch mounted near the welding head. This is provided by the control unit. If AC is used, a magnetically

Table 1. Approximate Maximum Currents

(*Unionmelt* Welding Process)

BUTT WELDS		FILLET WELDS	
Thickness, Inches	Current, Amperes	* Nominal Size, Inches	Current, Amperes
1/4	825	1/4 (3/16)	450
1/2	1175	3/8 (5/16)	650
3/4	1300	1/2 (7/16)	800
1	1600	5/8 (1/2)	900
1 1/2	2000	3/4 (5/8)	1000
2	2900	7/8 (11/16)	1200
2 1/2	3200	1 (13/16)	1500
3	4000	1 1/4 (1)	1600

* Nominal size is the leg dimension of ordinary arc-welded fillets. The figure in parenthesis is the approximate leg dimension of the equivalent UNIONMELT weld having approximately equal throat strength achieved through greater penetration.

operated contactor or heavy duty circuit breaker (a contactor is generally better for frequent on and off operation) should be installed in the primary supply leads to the transformers. The supply leads should have automatic cutout protection of sufficient capacity to protect the primary supply.

The control unit also indicates voltage and amperage levels, provides controls for current adjustments, the rate of electrode speed, and travel speed adjustments. See Fig. 3.

WELDING CURRENT

Either DC or single-phase AC may be used for submerged-arc welding, although the latter is the more frequently used of the two. An open-circuit welding voltage of 60 to 100 volts can be used. Table 1 gives approximate maximum currents required for various thicknesses of one-pass *Unionmelt* butt welds and fillet welds.

Sufficient adjustable reactance (resistance if direct current) to permit continuous stepless adjustment and control of the amperage during the welding operation should be provided. The reactance (or resistance) may be built into the power supply unit or it may be connected separately. The actual voltage at the welding zone will be adjusted through the special voltage control at approximately 25 to 50 volts, depending on the size and shape of the weld.

Submerged-arc welding uses a higher current than that in the manual arc welding processes. This results in such welding characteristics as:

1. Reduced weld shrinkage,
2. Minimum distortion of the welded structure,
3. A faster filler metal deposition rate,
4. A faster welding speed.

FLUXES AND ELECTRODES

The fluxes used in submerged-arc welding must be of a type that will not produce large amounts of gas. These fluxes are granular fusible substances available in a number of different grades. Each grade differs somewhat in chemical composition. The selection of an appropriate flux depends on a number of factors, including the type of metal to be welded, welding speed, and the thickness of the metal.

The flux is fed around the electrode during the welding process. Only enough flux should be used to submerge the arc. However, there will be no harmful effect on the weld, if the arc occasionally breaks out of the flux.

A special aspect of submerged-arc welding is that alloys can be added to the weld through the flux. A mild-steel electrode in conjunction with a special agglomerated alloy flux is used to do this. Fluxing and alloy materials are ground and mixed together in an agglomeration process so that neither can be separated from the other. The result is an agglomerated alloy flux that will ensure stable uniform welds. A wide variety of fluxes is possible by varying the alloy content. Because of the alloying elements that can be incorporated in the flux, the submerged-arc welding process can be applied to a wide range of welding jobs using only a small selection of standard electrodes.

The fluxes used in submerged-arc welding are not classified. Consequently, the welder should follow the manufacturer's recommendations for a particular job. The fluxes used in this process are relatively inexpensive. They should be stored in a dry place. If the flux becomes wet, dry it out by heating it to 500° F or 600° F.

Fused fluxes may be ground up and reused if they are mixed with at least 75% fresh flux. Grinding costs generally make this practice uneconomical.

Instead of stick electrodes, the submerged-arc welding process uses coils of bare metal wire that are continuously and mechanically fed into the arc. In this respect, submerged-arc welding strongly resembles MIG welding. The electrode wire diameters generally range from 0.20 to ¼ inch, and may be purchased in a number of different coil sizes (e.g. 1, 5, 10, 25, 35 lb., etc.).

UNIONMELT WELDING PROCESS

Unionmelt welding (Fig. 4) is a submerged-arc welding process developed by the *Linde Air Products Company*. The entire

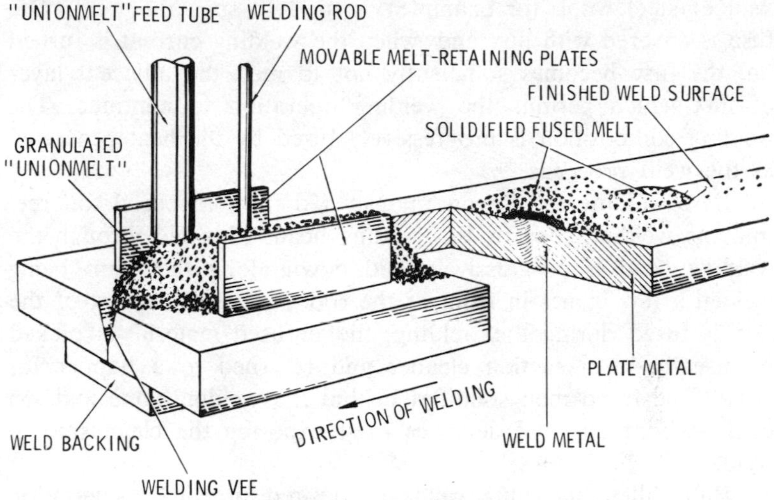

Fig. 4. *Unionmelt* welding process.

welding action of this process takes place beneath a flux cover. There is no open visible arc, a factor which prevents sparks and spatter. Within the layer of flux, an intense concentrated heat is generated by the electric current. As a result, the bare metal electrode and a portion of the edges being welded are melted and fused. Molten metal from the electrode is thoroughly mixed with the melted base metal to form the weld. When a weld is being made, a sub-surface layer of flux (*Unionmelt* in the *Linde* process) melts and floats as a liquid blanket over the molten weld metal.

The layer of flux functions as an excellent heat insulator and results in a concentration of the heat in a relatively small welding zone. It also makes possible the use of extra high current levels which permit a rapid generation of an intense heat.

Another function of the flux is to serve as a cleanser for the weld metal. It washes the metal which melts from the rod, and absorbs impurities from the fused metal in the base. It excludes the atmosphere and other gases from the weld metal. Special alloying elements can be introduced into the weld metal by means of the flux.

Since the granulated flux used in submerged-arc welding is not a conductor of electricity when it is cold, a special fuse (a wad of steel wool, for example) is used to start the weld. The fuse is covered with flux and, when the welding current is turned on, the fuse becomes sufficiently hot to melt the adjacent layer of flux which permits the welding operation to continue. The welding composition is progressively fused by the heat generated as the weld proceeds.

Bare welding rod is continuously fed from a special rod reel into the welding zone by the welding head. Flux fed through the welding head is progressively laid down along the seam being welded a few inches in front of the rod. Since only a part of the flux is fused during the welding, the unfused material is picked up usually by a suction cleaner and returned to a hopper for reuse. The fused melt solidifies behind the welding zone and, on cooling, contracts and detaches itself exposing the clean smooth weld.

Butt, fillet, and plug welds of desired practical penetration and contour in commercially used thicknesses of material can be made with the submerged-arc welding process. This process is especially suitable for production (repetitive) work and where welds can be made with the surface of the welding puddle approximately horizontal.

In order to make butt welds in one pass, with complete penetration, the weld must be backed to prevent the fused metal from falling through. A copper bar, a sliding copper shoe, or a trough of flux have all been successfully used for backing butt welds. Under some conditions, when flux backing is used, a smooth bead similar to that on the top surface can be obtained on the bottom surface of the weld. A fixed or rolling copper ring and a sliding shoe have been successfully used for backing circumferential butt welds in pipe and other cylindrical products. Steel backing pieces which may remain permanently attached to or may be cut from the under part of the finished weld are sometimes used. A partially penetrated unbacked weld made from one side of the joint is also extensively used to serve as backing for the main weld. Such two pass welds are made from opposite sides of the joint.

Fig. 5 illustrates the edge preparation recommended for joints welded by the *Unionmelt* welding process. These can be applied

Table 2. Butt Welds Welded from Both Sides

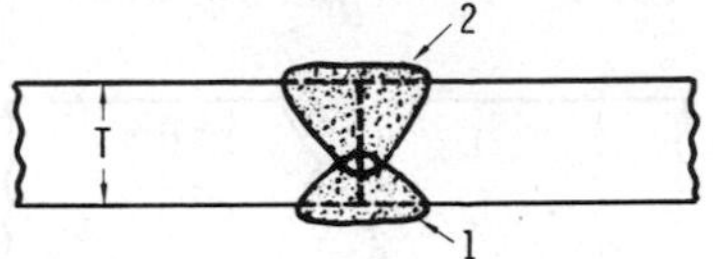

BUTT WELDS
Welded from Both Sides

Plate Thickness T	Pass	Amps.	Volts	Speed In/Min.	Wire Size
1/4*	1	475	29	48	5/32
	2	575	32		
3/8	1	500	33	32	7/32
	2	850	35		
1/2	1	700	35	27	7/32
	2	950	36		
5/8	1	800	35	24	7/32
	2	950	36		
3/4	1	900	36	22	7/32
	2	950	36		

*** Must be down flat on a platen.**
Seams must be butted tight. Seal all gaps on 1/2" or thicker with a fast seal bead (5/32" "Fleetweld 5") on the second pass side. Seal 3/8 and less on the first pass side.

to submerged-arc welding in general. More detailed specifications for submerged-arc welding joints are given in the following section.

JOINT PREPARATION

Tables 2 through 9 illustrated the type of joint preparation required where the work involves a flat horizontal surface. On a curved surface, such as small diameter girth welds, the speeds and currents are less than the equivalent seam on a flat horizontal surface. The reason for this is that the molten flux and steel are very fluid and it is necessary to limit the amount of molten material so it does not run off the curved surface. To control

Table 3. Butt Welds Welded from Both Sides

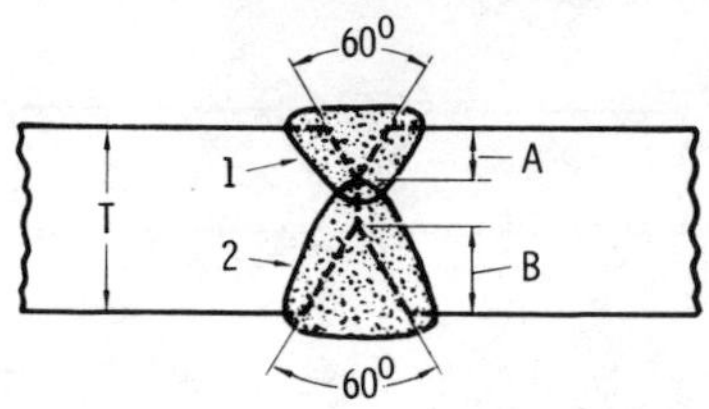

Plate Thickness T	Pass	A	B	Amps.	Volts	Speed In/Min.	Wire Size
3/4	1	1/8		700	35	22	7/32
	2		3/8	1000	36	16	
7/8	1	1/4		750	35	19	7/32
	2		3/8	1000	36	14	
1	1	3/8		900	35	14	7/32
	2		3/8	1050	36	12	
1 1/8	1	3/8		900	35	14	7/32
	2		1/2	1050	37	9	
1 1/4	1	1/2		1000	35	12	7/32
	2		5/8	1100	37	7 1/2	
1 3/8	1	1/2		1000	35	12	7/32
	2		3/4	1100	37	6 1/2	
1 1/2	1	5/8		1050	36	9	7/32
	2		3/4	1100	37	6 1/2	

Seams must be butted tight. Seal all gaps with a seal bead (1/4" "Fleetweld 5" or "9" on the first pass side.

this tendency to spill off, the current and the speed are reduced and the point of welding is ½ inch to 2½ inches off the vertical centerline in the direction opposite to the rotation of the work.

Where no gap is specified, the seam should be fitted tightly together. In butt seams, if the gap is 1/32 inch or more due to poor fitup, the seam must be sealed with a sealing bead.

SPECIAL SUBMERGED-ARC WELDING PROCESS

Several variations of the submerged-arc welding process have been developed to answer the needs of special applications. For

Table 4. Butt Welds Manually Welded from One Side

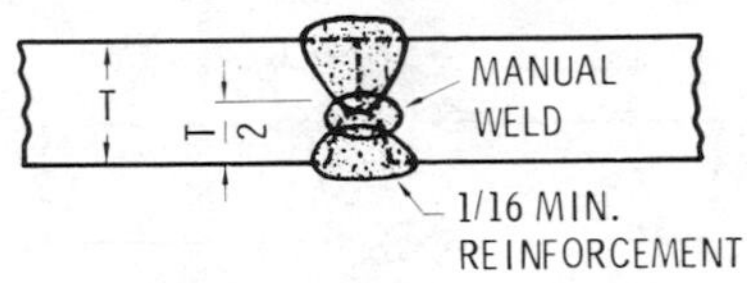

Plate Thickness T	Amps.	Volts	Speed In/Min.	Wire Size
1/4	575	33	48	7/32
3/8	800	35	30	7/32
1/2	950	36	27	7/32
5/8	950	36	24	7/32
3/4	950	36	22	7/32

Manual weld must be 50% penetration minimum. Make manual weld before automatic weld.

Table 5. Butt Welds with the Manual Weld Made Before the Automatic Weld

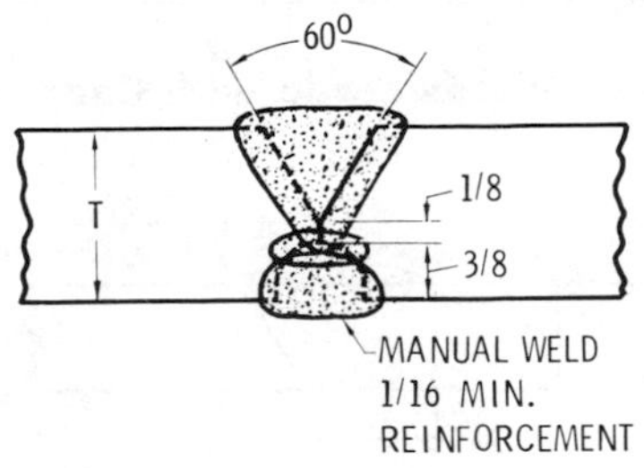

Plate Thickness T	Amps.	Volts	Speed In/Min.	Wire Size
3/4	900	35	18	7/32
7/8	1000	35	12	7/32
1	1050	37	9	7/32
1 1/8	1100	37	7 1/2	7/32

Make manual weld before automatic weld.

Table 6. Butt Welds Made with Steel Backing Bar

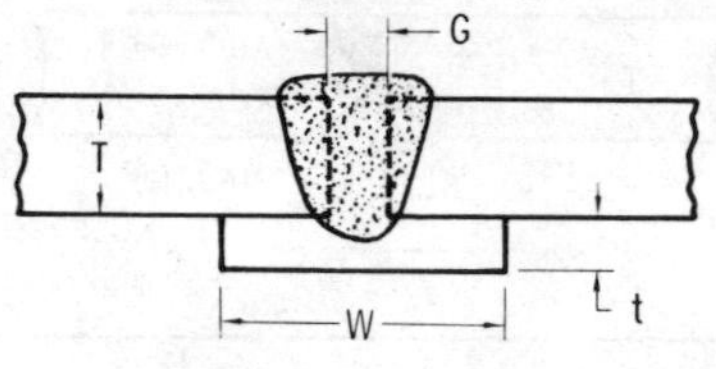

STEEL BACKING BAR

Plate Thickness T	Amps.	Volts	Speed In/Min.	Wire Size	G In.	t Min.	W Min.
16 Ga.	450	25	100-120	1/8	1/32	14 Ga.	3/8
14 Ga.	500	27	70-90	1/8	1/32	12 Ga.	3/8
12 Ga.	550	27	60	1/8	1/16	12 Ga.	1/2
10 Ga.	650	28	48	5/32	1/16	1/8	5/8
3/16	850	32	36	7/32	1/16	3/16	3/4
1/4	900	33	26	7/32	1/8	1/4	1
3/8	950	33	24	7/32	1/8	1/4	1
1/2	1100	34	18	7/32	3/16	3/8	1

Table 7. Butt Welds Made with Copper Backing

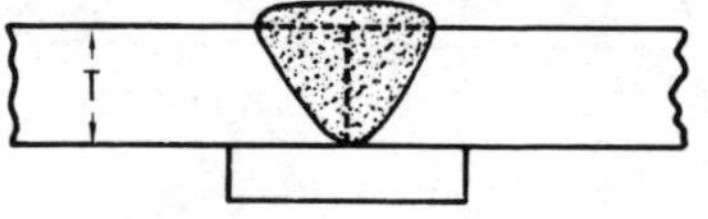

COPPER BACKING

Plate Thickness T	Amps.	Volts	Speed In/Min.	Wire Size
14 Ga.	400	24	100-120	1/8
12 Ga.	500	30	100-120	1/8
10 Ga.	650	31	80-100	1/8

Table 8. Butt Welds Having Specifications Not Acceptable For Codes

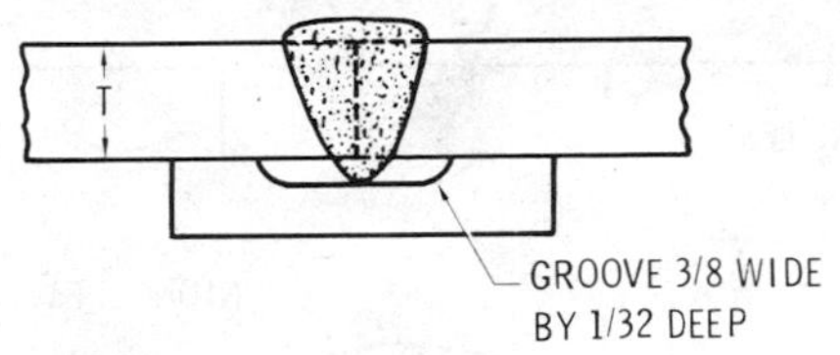

Plate Thickness—T	Amps.	Volts	Speed In/Min.	Wire Size
12 Ga.	500	30	100	5/32
10 Ga.	650	31	80	5/32
3/16	700	31	65	5/32
1/4	850	35	40	7/32
5/16	900	36	30	7/32

Table 9. Butt Welds

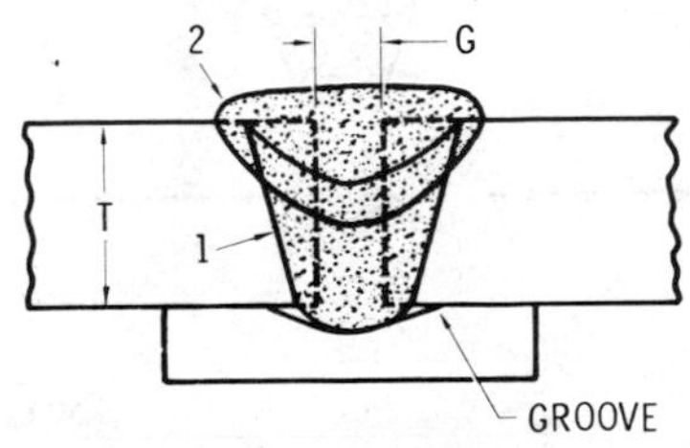

Plate Thickness T	Pass	Amps.	Volts	Speed In/Min.	G In.	Wire Size	Groove	
							Width	Depth
3/16	1	550	25	40	1/16-3/32	5/32	11/16	3/32
	2	500	30					
1/4	1	650	30	36	1/16-3/32	5/32	11/16	3/32
	2	600	35					
5/16*	1	650	30	30	3/32-1/8	5/32	11/16	1/8
	2	600	35					
3/8*	1	750	30	26	1/8-5/32	5/32	11/16	5/32
	2	700	35					

*** Fill groove in backing bar with flux before welding.**

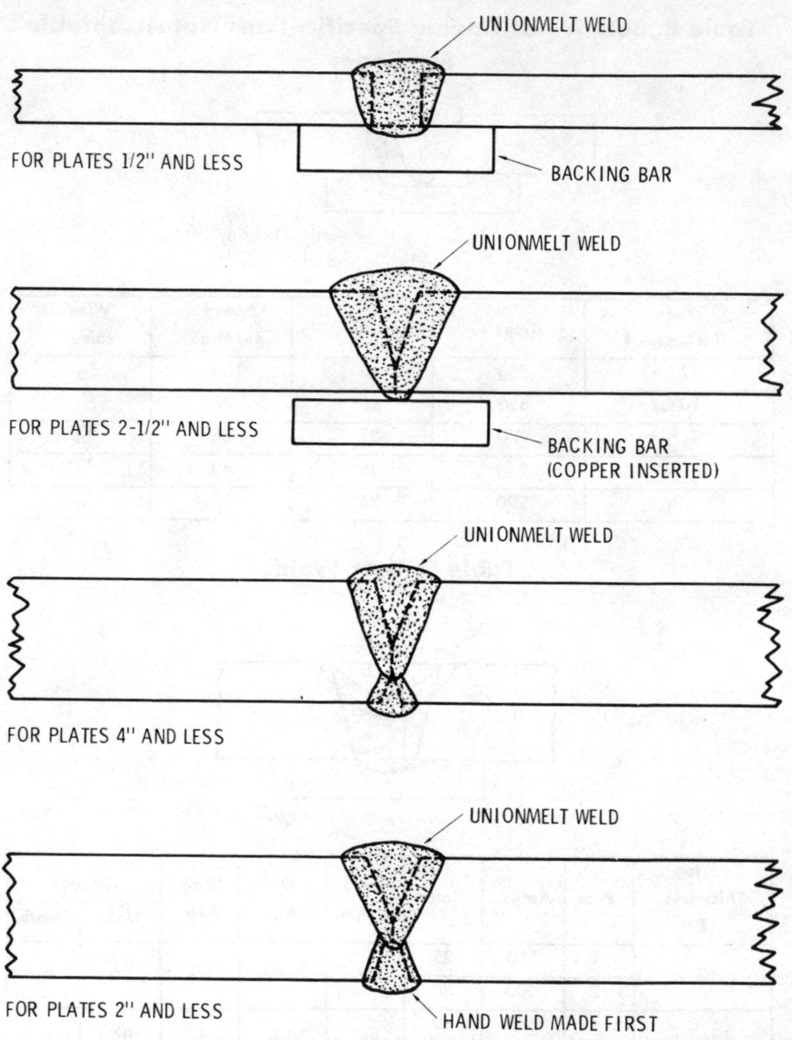

Courtesy Linde Air Products Co., Div. of Union Carbide Corp.

Fig. 5. Edge preparation for *Unionmelt* welds.

the most part, these variations are multiple-arc welding procedures (Fig. 6) and include:

1. Twinarc welding,

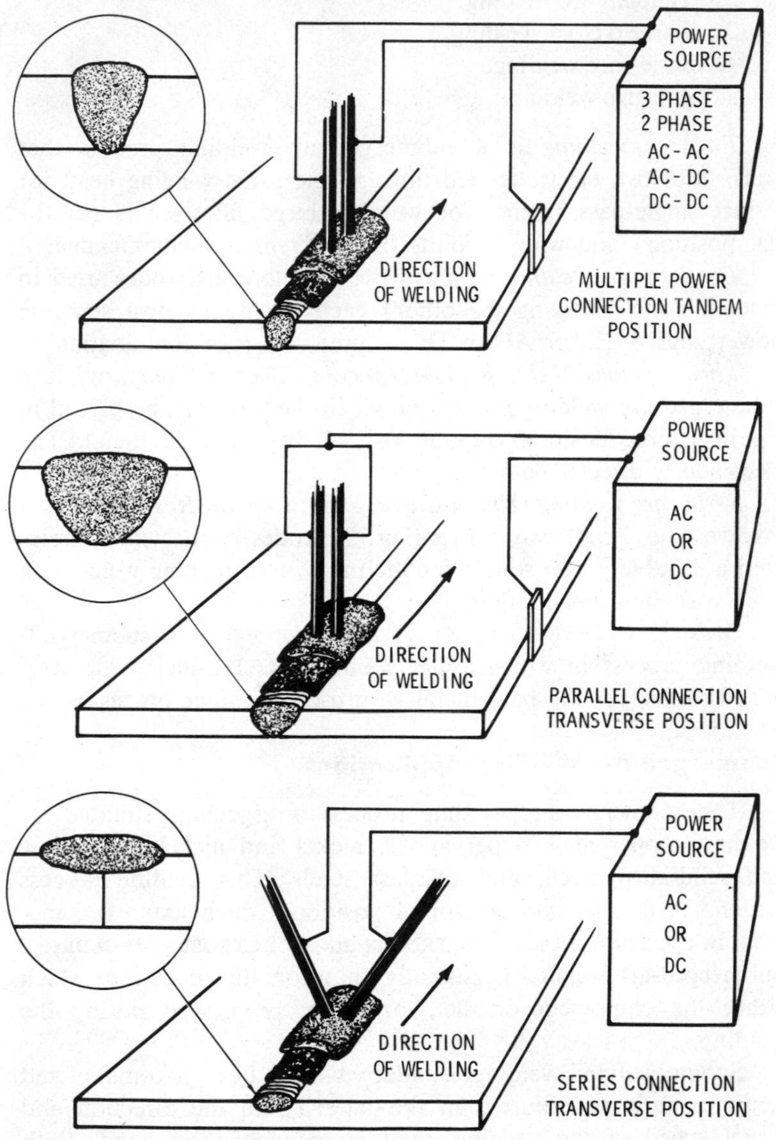

Fig. 6. Special submerged arc welding process.

2. Tandem-arc welding,
3. Three o'clock welding,
4. Series-arc welding,
5. Arcstrip welding.

Twinarc welding is a submerged-arc welding process that consists of two electrodes fed through the same welding head for a fast single-pass method of welding large fillet welds (in the flat position) and wide V-joints. DC current is recommended.

Tandem-arc welding employs two or more electrodes used in tandem (one following the other) each having its own separate power source. Either AC or DC is used, often in combination.

Three o'clock welding (The Lincoln Electric Company) is a submerged-arc welding process in which the joint can be placed in the horizontal position without loss of flux or weld metal. The electrode is directed into the joint.

Series-arc welding (Union Carbide and *Carbon Research Labs)* involves the use of two converging electrodes connected in series with a suitable power source for the purpose of making welds with a very shallow penetration.

Arcstrip welding (The *Arcos Corporation*) is a submerged-welding process in which a flat stainless steel 2-inch wide strip is substituted for the bare metal wire used by other processes.

Submerged-Arc Welding Applications

The submerged-arc welding process is especially suitable for welding copper and copper alloys, nickel and nickel alloys, low and mild-alloy steels, and stainless steels. This welding process is done in the flat and horizontal positions which limits it somewhat in application (see ELECTROGAS and ELECTROSLAG WELDING). Submerged-arc welding is basically an automatic process in which either the equipment or the work may be moved during the welding.

Submerged-arc welding is adaptable to both automatic and semiautomatic procedures. In the latter, both the direction and travel speed of the welding head is controlled by hand. Both procedures can be applied to a vast number of welding jobs ranging from the relatively minor operations of welding axles and spindles to such large scale ones as welding girders in major construction projects.

There is enough accessory equipment available in submerged-arc welding to make it extremely flexible and adaptable to almost any individual need. For this reason, submerged-arc welding has become almost synonymous with the terms automatic and semi-automatic welding.

ATOMIC-HYDROGEN WELDING

Atomic-hydrogen welding is an arc welding process in which an arc is maintained between two tungsten electrodes, and, at the same time, a stream of hydrogen gas is passed through the arc and around the electrodes.

A characteristic of an electric arc occurring in a stream of hydrogen gas is that much of the hydrogen is changed from a molecular state to an atomic state (whence the process gets its name) and thereby absorbs a considerable amount of energy.

In escaping the arc stream, these atoms recombine to form molecules again at the outer edge of the arc fan and the extra energy is released as heat. This extra heat, added to the normal heat of the arc, produces a considerably higher temperature than either the ordinary arc or gas torch. Since all of this heat is concentrated at high temperature in the small volume of the arc fan, excellent heat transfer to the work is obtained.

The work being welded does not form a part of the electric circuit and therefore does not need to be grounded.

Atomic-Hydrogen Arc Welding Equipment

Since the arc is independent of the work, and can be moved at will, atomic-hydrogen welding equipment provides an unusually mobile tool which is especially valuable in the fusion welding of delicate and exacting work, such as building up contours of molds and dies, welding thin sections, etc.

The equipment and supplies necessary for the atomic-hydrogen welding process include (Fig. 7):

1. A suitable AC transformer (constant potential),
2. An atomic-hydrogen electrode holder (torch),

3. Welding cables,
4. A hydrogen cylinder and a supply of hydrogen gas,
5. Gas hoses, gas connections, and a gas pressure regulator.

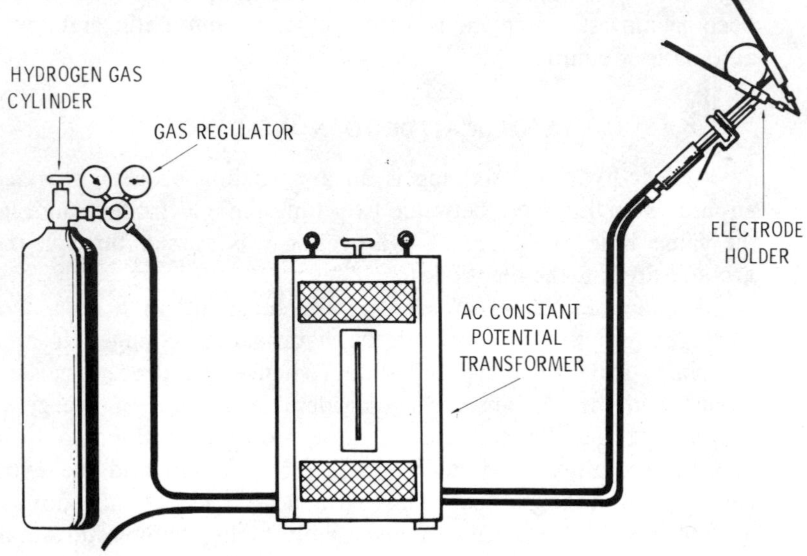

Fig. 7. The atomic-hydrogen welding system.

WELDING TORCH

The torch used in atomic-hydrogen arc welding (Fig. 8) consists of twin electrode clamps connected to the handle by twin tubes through which pass the electricity and hydrogen gas supplying the arc.

ELECTRODES

The current amperages, electrode, and filler rod sizes recommended for various metal thicknesses are given in Table 10.

The tungsten electrodes used in atomic-hydrogen welding do not enter the weld; they are used only as a means for establishing and maintaining the arc. They are, however, slowly evaporated by the heat.

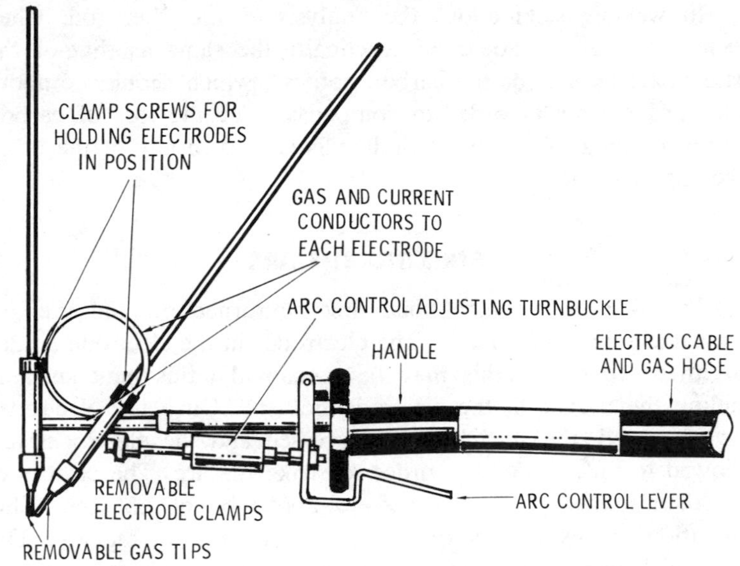

Fig. 8. The atomic-hydrogen welding torch.

Table 10. Current Amperage and Other Conditions for Atomic-Hydrogen Arc Welding

Thickness, inch	Electrode size	Current amperes	Filler rod diameter, inch
0.062 to 0.075	1/16	20-25	3/32
1/8 to 5/32	1/16	25-35	1/8
3/16 to 1/4	1/16 or 1/8	35-40	1/8 or 5/32
3/8 to 1/2	1/8	40-50	3/16
5/8 to 3/4	1/8	60-80	3/16 or 1/4

ATOMIC-HYDROGEN ARC WELDING FILLER ROD

The welding filler rod can be applied manually in much the same manner as gas welding. Accordingly, the process combines some of the flexibility of the gas welding operation with the concentrated heated zone of the metal or carbon-arc methods. The filler metal is added as needed at the forward edge of the arc during its continuing progress along the joint.

In welding steel alloys the analysis of the filler rod, when such rod is used, should be practically the same as that of the base metal except for the carbon content, which should be about one-third greater in order to compensate for the loss of carbon during welding. No appreciable loss is caused in any of the other alloying ingredients.

STRIKING THE ARC

The welding machine must first be turned on and the appropriate adjustments made. The electrodes in the electrode holder are then separated (this may be a manual adjustment such as pulling the trigger in the electrode holder). Once the electrodes are separated, the start button is pressed and the electrodes are allowed to touch briefly in order to strike the arc. The electrodes are then immediately separated to about 1/16 of an inch. This operation ignites the hydrogen as well as starting the arc. The flow of hydrogen is controlled automatically. The start button should be released as soon as the arc is established.

Viewed through a welding glass, the arc stream is seen to have a very definite circular outline (Fig. 9). The greatest concentration of atomic-hydrogen (and therefore heat) occurs at the

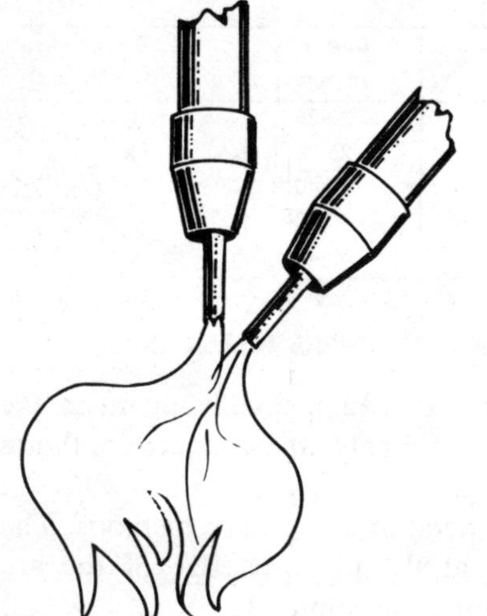

Fig. 9. Appearance of the atomic-hydrogen arc.

boundary of the arc stream, the edge of which should just touch the surface to be welded. The amount of heat available for welding depends on both the welding current and the size of the arc fan.

After striking the arc, the welder should adjust the current to a level suitable for the thickness of metal being welded. The amount of current flowing is indicated by the ammeter.

The arc may be stopped by depressing the stop button or by increasing the gap between the electrodes until the arc breaks.

ATOMIC-HYDROGEN WELDING METHODS

Automatic welding by machine operation can be obtained successfully with the atomic-hydrogen arc welding process and it can be profitably applied on production work involving straight or circular seams where the quantity is sufficient to warrant the investment. For such an installation, an atomic-hydrogen arc welding head is used in place of the hand operated electrode holder. In addition, a welding unit of proper size will be required for each arc; a thyratron panel for automatically controlling each arc; an operator's control station; and either a travel carriage for moving the welding head over the work or else a suitable travel mechanism for moving the work under the head.

As the atomic-hydrogen welding process involves an arc, the heated zone is controlled merely by moving the torch away from or toward the work. While the technique of making the weld resembles gas welding, some practice is required for the operator to become adept at holding the arc and controlling the weld puddle.

Joint preparation prior to welding is identical to that used on gas-welded joints. Surfaces to be welded must be cleaned of all dirt, grease, and other contaminants that could interfere with the formation of a good strong weld.

Sections ⅜ inch or more in thickness should be preheated to a temperature of 600° F to 700° F and maintained in this temperature range during welding. Furnace heating or heating with city gas and compressed air flame is recommended. If an open flame is used, local heating to the recommended heating range is usually adequate. This minimizes the extent of the annealed area when welding metals in the strain hardened condition. Sections lighter than ⅜-inch thickness are rapidly heated by the welding arc and preheating is not ordinarily advantageous.

Manual welding with this process is performed with a suitable gas welding flux. Both sides of the joint are painted with a water mixture of the flux and the filler rod is dipped into the mixture prior to welding.

The welding heat affects the flux more severely than in gas welding and after the joint solidifies a heavier scum than usual is present on the surface. The scum interferes with laying down subsequent weld beads so removal of the surface deposit is usually on parts where the thickness requires more than one weld bead to make the joint.

Atomic-Hydrogen Welding Applications

Atomic-hydrogen welding, although somewhat more costly than other arc welding methods, is used for the general welding of steel and ferrous alloys, thin sheet metals, and various nonferrous metals and alloys. This is a particularly desirable method for making fillet lap welds on alloys with less difficulty from cracking or overheating of the parts being welded. Good results are obtained from the standpoint of soundness and strength of welds in aluminum alloys with manual and automatic atomic-hydrogen equipment. The generally higher welding costs and the use of hydrogen (a reactive gas) are factors that limit the widespread use of this welding process.

PLASMA ARC WELDING

Plasma arc welding is a welding process in which a column of plasma-producing gas (e.g. argon, nitrogen, or hydrogen) is ionized by the heat of an electric arc and passed through a small welding torch orifice. The result is a plasma arc (an electric arc in combination with high energy ionized atoms) capable of delivering a high concentrated heat to the area being welded. The plasma arc causes deep penetration and produces welds with narrow beads and sharply limited heat-affected zones. *Thermal arc* is a trade name for a plasma arc welding process developed by the *Thermal Dynamics Corporation.*

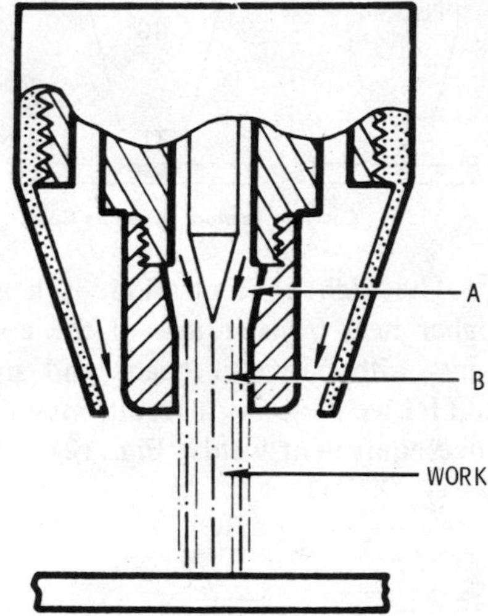

Fig. 10. Plasma arc welding process.

Courtesy Thermal Dynamics Corp.

Plasma refers to any gas that has been heated enough to become ionized. Fig. 10 illustrates the principle of the plasma arc welding process. Note that the gas flowing around the electrode at point (A) is unheated. Partial heating (i.e. ionization) occurs at point (B). Full ionization occurs before the ionized particles and the arc leave the torch in the form of a plasma arc stream.

The plasma arc welding process may be regarded as a further development of TIG welding over which it offers certain advantages, including:

1. Higher welding speeds,
2. Lower current requirement,
3. Cleaner welds,
4. Little or no effect from different arc stand-off distances.

The higher plasma arc welding speeds result from the controlled gas flow through the welding torch orifice. Instead of fanning out as is the case with the TIG arc, the plasma arc retains the form of a highly constricted arc. The higher arc current density results in concentrating the arc in a smaller surface area. This produces a higher welding speed (Fig. 11).

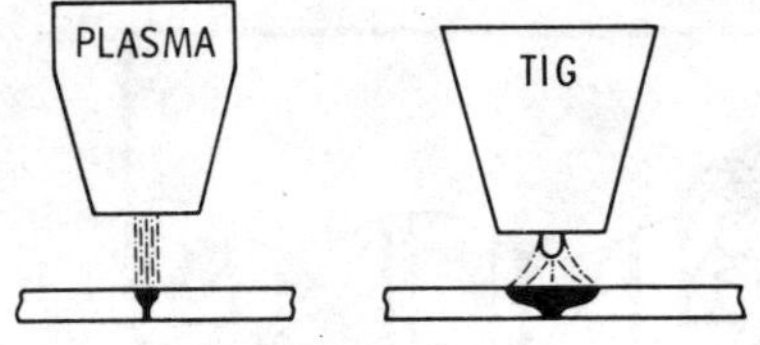

Courtesy Thermal Dynamics Corp.

Fig. 11. Comparison of plasma arc and TIG welding speeds.

The high concentration of heat, greater welding speed, and higher heat transfer rate of the columnated plasma arc produces joints with narrower beads and smaller heat-affected zones than in TIG welding. As a result, much less current is required to produce equivalent welds (Fig. 12).

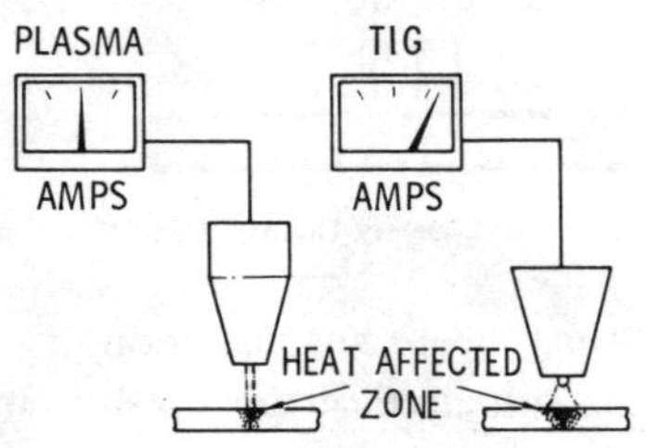

Courtesy Thermal Dynamics Corp.

Fig. 12. Comparative current requirements of plasma arc and TIG welding.

Because TIG welding has an open arc that fans out over the weld area, the distance of the torch orifice from the work surface will determine the width of the bead and the heat-affected zone. Varying the distance of the plasma arc from the work surface has very little effect on the width of the weld bead or the heat-affected zone because of its constricted nature (Fig. 13). A stand-off distance of ⅛ to ¼ inch at low currents is generally used with plasma arc welding.

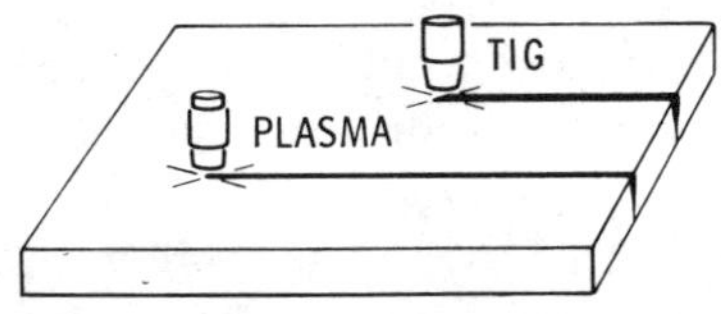

Courtesy Thermal Dynamics Corp.

Fig. 13. Comparative weld widths of plasma arc and TIG welding.

There is virtually no electrode contamination of the weld in plasma arc welding. This results from the electrode being recessed within the torch and receiving the protection of both the tip and the inert plasma gas (Fig. 14).

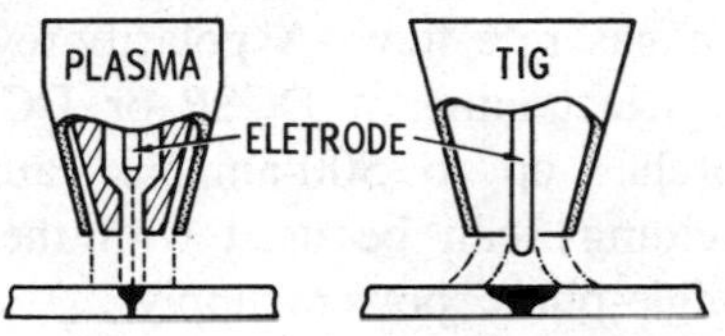

Fig. 14. Protection of the plasma arc electrode as compared to that of the TIG welding process.

Courtesy Thermal Dynamics Corp.

Plasma Arc Welding System

The plasma arc welding system (Fig. 15) consists of the following equipment and supplies:

1. DC power source,
2. Welding control console,
3. Recirculating water cooler,
4. Plasma welding torch,
5. Gas cylinders and a gas supply,
6. Gas pressure regulators,
7. Gas hoses and gas hose connections,
8. Water-cooled power cables.

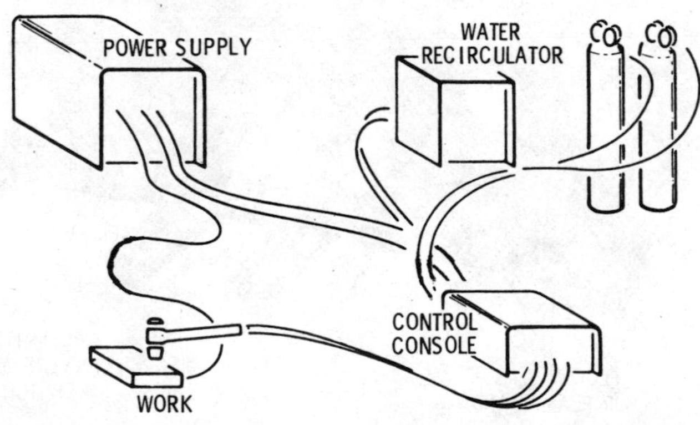

Courtesy Thermal Dynamics Corp.

Fig. 15. Plasma arc welding system.

Plasma Arc Welding Equipment

The plasma welding control console illustrated in Fig. 16 is usable with most DC power sources that meet the requirements of plasma arc welding. The control panel on the console contains flow meters (two to four) and fine control metering valves for the gas rate flow. A polarity-reversing switch enables the welder to select either a DCSP or DCRP welding current. Plasma arc torches up to 500-ampere capacity (for manual or automatic welding) can be used with the console. It requires a 115-volt single-phase power supply.

The high temperatures produced in plasma arc welding require a cooling system to prevent damage to the welding torch and other components. Water is used as the cooling agent, and it passes from

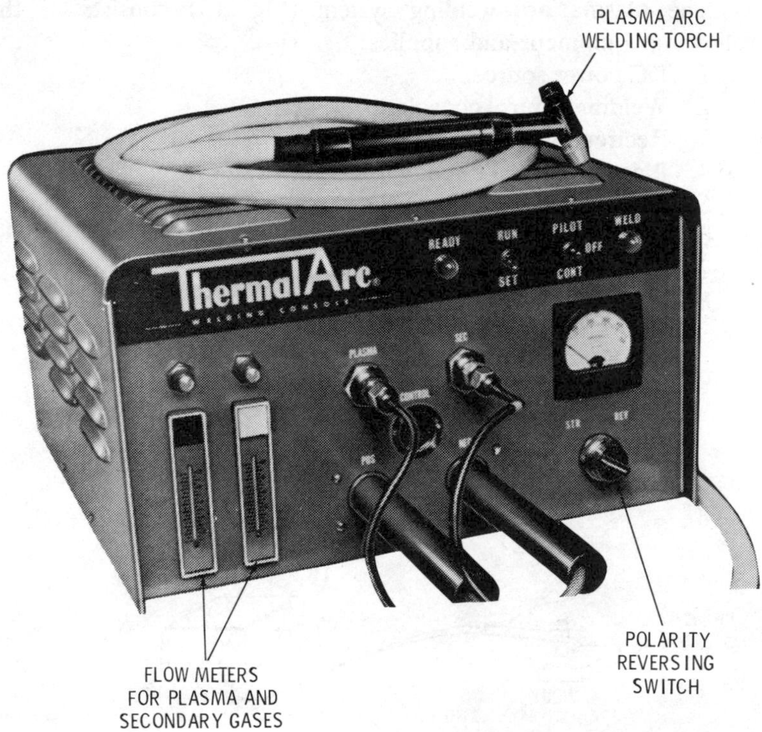

Courtesy Thermal Dynamics Corp.

Fig. 16. Plasma welding control console.

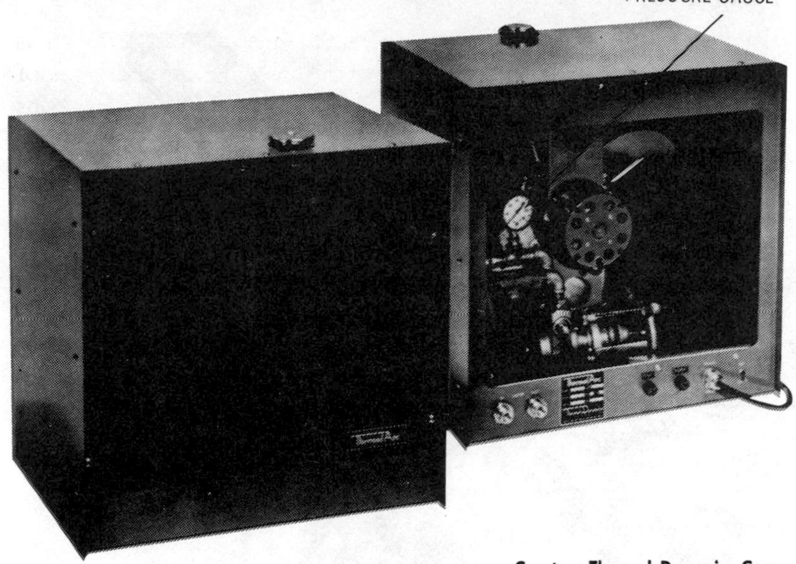

Courtesy Thermal Dynamics Corp.

Fig. 17. Water recirculator unit.

a specially designed water recirculator through the control console to the welding torch. The water recirculator illustrated in Fig. 17 operates with a pump capacity of 2.3 gpm at 125 psi and a heat exchanger capacity of 40,000 BTU/hr.

Plasma Arc Welding Torch

Both manual and machine-mounted torches can be purchased for plasma arc welding. The welding tips are removable, and are available in a wide variety of types for a number of different welding applications (Figs. 18 and 19). The torches illustrated in Figs. 18 and 19 are available in models using either DCSP or DCRP current. Argon is used as the plasma gas with argon, argon-helium, or helium functioning as the shielding gas. Water is used as the cooling agent with a water-flow capacity of ½ to ⅓ gpm at 50 psi.

A nonconsumable tungsten electrode is used in plasma arc welding and additional metal is added to the weld with a filler rod.

Courtesy Thermal Dynamics Corp.

Fig. 18. Plasma arc welding torch for manual operations.

Plasma Arc Welding Applications

Plasma arc welding finds a variety of applications in such widely differing areas as machinery and machine tools, aircraft, shipbuilding, metal fabrication, and electronics to mention only a few. The high temperatures of the concentrated plasma arc make it extremely suitable for welding such metals as carbon and stainless steel, copper and copper-nickel alloys, aluminum, titanium, brass alloys, as well as high-temperature and high-strength alloys.

ELECTROSLAG WELDING

Electroslag welding (Fig. 20) is a metal-arc welding process that may be considered to represent a further development of submerged-arc welding. The flux is continuously fed into the

Fig. 19. Plasma arc welding torch for automatic welding operations.

Courtesy Thermal Dynamics Corp.

area being welded where it forms a cover of molten slag over the weld metal. This covering of molten slag serves as a major source of heat for the electroslag welding process.

Electroslag Welding Equipment and Supplies

A solid filler wire (flux-cored electrode wire may also be used) is continuously fed under the thick flux cover. Up to three electrodes may be used to feed into the molten weld puddle, the number depending on the thickness of the metal being welded.

The direction of the welding is upward along a vertical path filling a 1 to 1$\frac{5}{16}$-inch spacing between two metal plates. A horizontal oscillation of the electrode is recommended for thicker metals. Special water-cooled copper shoes move vertically with the welding head. These copper shoes bridge the gap between the two plates and function as a mold for the molten weld metal.

Joint preparation is comparatively easy with flame cutting the plate edges usually being sufficient. If the electroslag welding

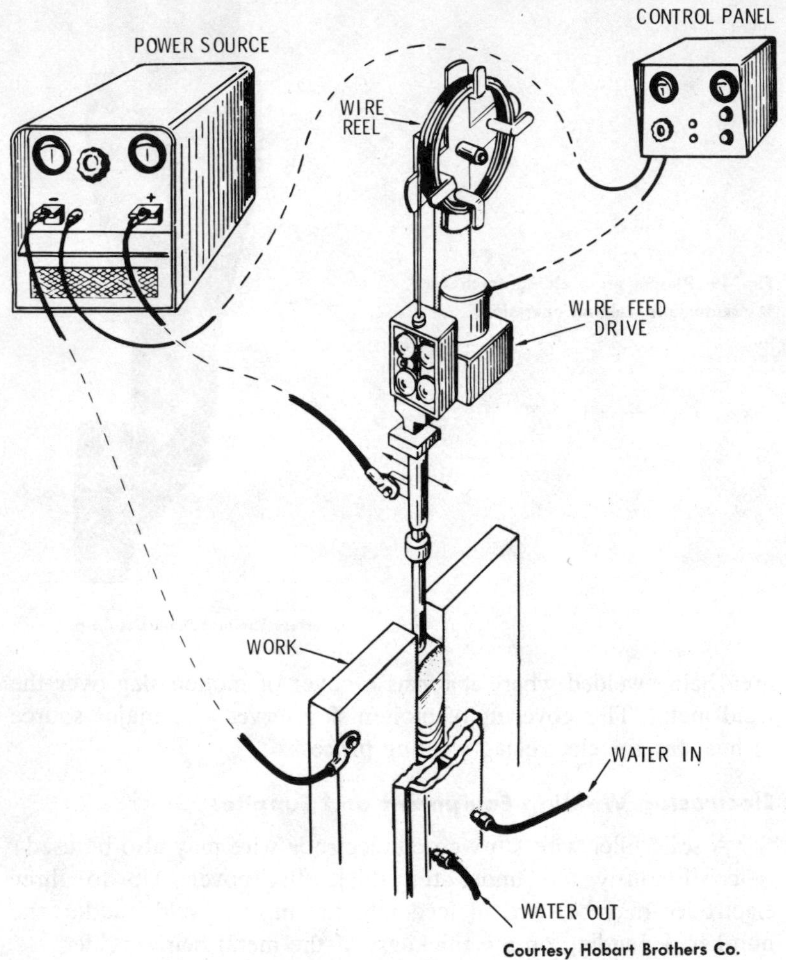

Courtesy Hobart Brothers Co.

Fig. 20. Electroslag welding process.

process is used correctly, only a single pass is needed to make a strong clean weld.

Electroslag welding requires a power source capable of delivering the AC welding current recommended for this process. The power source is stationary, but the rest of the equipment (copper shoes, flux hopper, wire feeder, etc.) ride vertical guide rails parallel to the area being welded. The flux is fed into the weld area, all but covering the ends of the electrodes. The welding heat is produced by the resistance of the flux to the passage of the flow of electricity from the electrode and by the layer of molten slag. There is no clearly defined arc as such.

Electroslag Welding Applications

The electroslag welding process is used for welding sections of steel plate ranging from 1 to 14 inches in thickness and occasionally in the production of ingots of refined metals. It is similar to the electrogas welding process in that it functions as a vertical form of submerged-arc welding.

ELECTROGAS WELDING

Electrogas welding (Fig. 21) is an automatic welding process that uses a carbon dioxide shielding gas to protect the arc and molten weld metal from contaminants in the surrounding atmosphere. A flux-cored electrode wire is continuously and mechanically fed into the arc and molten weld puddle. Because of the rapid deposition rate of electrogas welding, edge preparation is not necessary.

Electrogas Equipment and Supplies

Electrogas welding requires a suitable generator or transformer power source capable of delivering up to 700 amperes. A DCRP welding current is recommended.

Special copper shoes on either side of the steel plates function as a backing to the weld and prevent the molten weld metal from flowing out of the gap between the plates. These shoes move upward at a rate determined by the welder. Cooling water is fed into and circulated through the copper shoes. The carbon dioxide shielding gas is fed through the shoes and into the area being welded.

The electrode filler wire is purchased in coils or spools and is attached to feed unit that moves upward with the copper shoes and the welder's control panel.

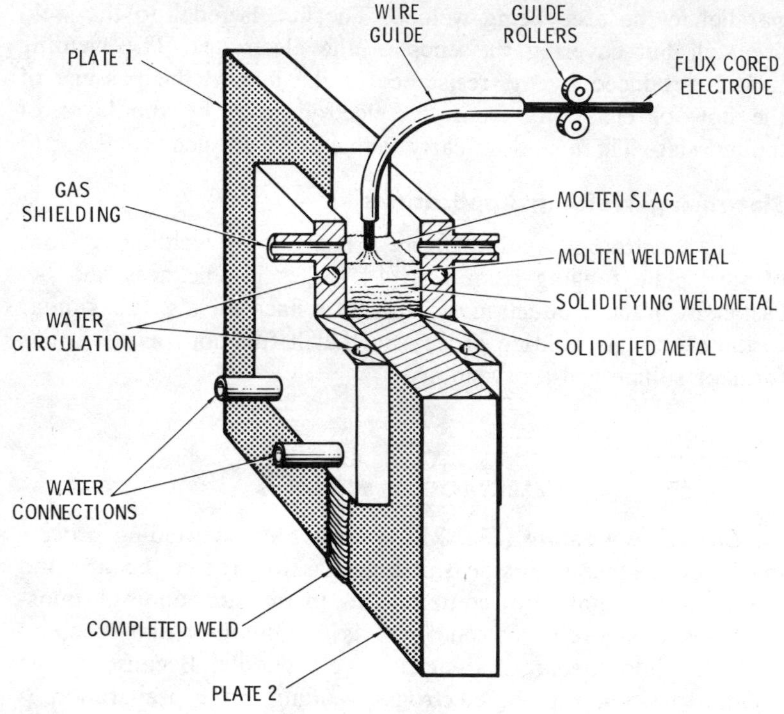

Fig. 21. Electrogas welding process.

Electrogas Welding Applications

The *electrogas welding* process is an automatic method for welding seams in steel plate in the vertical position. Consequently, it complements submerged-arc welding which is limited to the flat or horizontal position. It is suitable for welding metal under 1 inch in thickness.

UNDERWATER SHIELDED METAL-ARC WELDING

Underwater welds on mild steel can be made with practically the same form and generally the same properties as on the surface (about 50 percent of the ductility and 80 percent of the tensile strength of surface welds). But such results can be obtained only

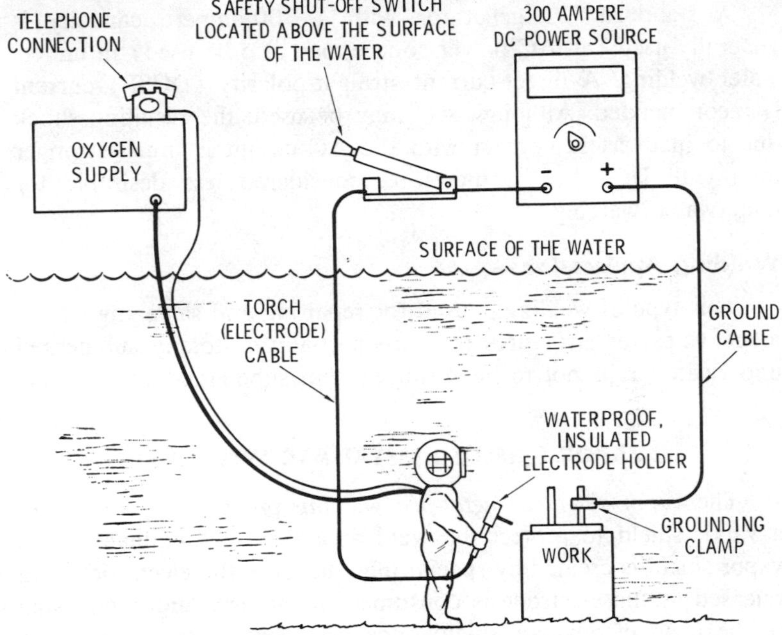

Fig. 22. Underwater shielded metal-arc welding process.

by skilled operators after adequate training. Because of the usual problems connected with all operations under water, the work requires more time and greater care and skill than on the surface.

Although underwater metal-arc welding has been known for a number of years, it attained the ranks of a standard process on mild steel joints as a result of development work carried on by the U.S. Navy in conjunction with manufacturers of welding equipment (Fig. 22).

Equipment and Supplies

Properly coated and fully waterproofed electrodes are used in underwater arc welding. The electrode holder must also be perfectly waterproofed with no bare surfaces exposed to contact with the water. The welding cables and all joints in the circuit must be properly insulated. The insulation should be carefully checked for current leakage at frequent intervals.

A standard DC generator with a 300-ampere capacity is generally used (an AC power source may also be used) in underwater welding. A direct-current straight-polarity (DCSP) current is recommended. Although AC may be used, the resulting shock due to inadvertent contact with the AC circuit is much stronger than with DC. AC is, therefore, considered less desirable for underwater welding.

Welding Applications

This type of welding is used for repairing and salvaging equipment, ships, or structures that are partially or totally submerged underwater. It is not to be confused with submerged-arc welding.

VAPOR-SHIELDED METAL-ARC WELDING

The *vapor-shielded metal-arc welding* process (Fig. 23) uses a vapor shield to protect the weld area from contaminants. The vapor shield is created by special ingredients in the electrode being released as the electrode is consumed in the arc, and condensing in the form of a vapor shield once they come into contact with the cooler surrounding atmosphere.

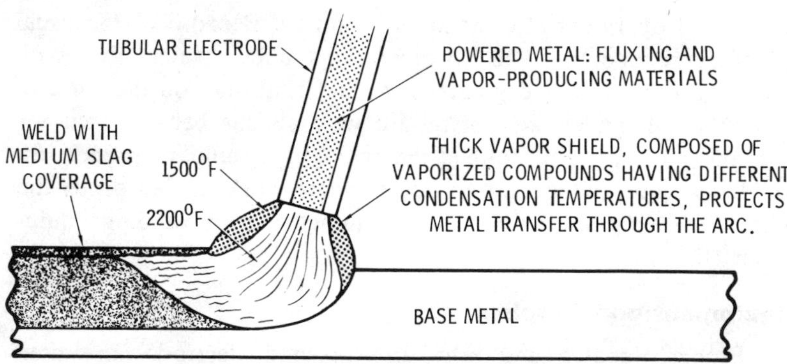

Fig. 23. Vapor-shielded metal-arc welding process.

Equipment and Supplies

Continuous tubular electrodes containing the ingredients for the vapor shield are fed mechanically through the welding gun and into the arc. The same types of power sources used in standard metal-arc welding are used with this welding process. A wire feeder mechanism similar to those in MIG and submerged-arc welding is necessary to feed the electrode wire into the arc. A control unit is also necessary to make adjustments in amperage and voltage.

Welding Applications

Vapor-shielded metal-arc welding is adaptable to both automatic and semiautomatic welding methods. Because the ingredients for the vapor shield are located in the core of the electrode, this welding process eliminates the necessity for any form of external shielding equipment and supplies (e.g. gas cylinders, gas supply, hoses, gas regulators, fluxes, etc.). This welding process is particularly suited for welding light-gauge mild steel.

CIG WELDING

CIG welding is the abbreviated form for carbon-inert-gas welding. The *American Welding Society's* official designation for this welding process is inert-gas carbon-arc welding.

The basic principle of CIG welding is that an electric arc is struck between a carbon electrode and the work within an inert gas shield. The gas shield protects the arc and weld area from contaminants, making the use of a flux unnecessary.

Equipment and Supplies

This process is identical to TIG welding in the equipment (similar power sources, etc.) and welding procedures used. It also uses similar shielding gases (helium, argon, etc.) to protect the arc and weld area from contaminants in the surrounding atmosphere. It differs in using carbon rather than tungsten electrodes.

The carbon electrode *must* be used with a DCSP welding current. These electrodes are tapered to a point with a predetermined diameter. This diameter is a controlling factor in the amount of current that can be used.

Welding Applications

CIG welding is used with particularly low currents, and is characterized by an easy quick starting arc. These two factors make CIG welding particularly applicable to light-gauge metals.

CHAPTER 9

Resistance Welding

Resistance welding is fundamentally a heat and squeeze process. When the parts to be welded are raised to the fusion temperature by the passage of a heavy electric current through the junction, pressure is applied mechanically to accomplish the welding. Examples of resistance welding include spot, seam, projection, and flash welding. Fig. 1 illustrates some of the many different resistance welding applications.

Most metals can be welded by at least one of the resistance welding methods. However, not all welding methods can be used with the same degree of success. The design of the joint, the heat resistance of the metal (or dissimilar metals), the pressure required, and other factors must be taken into consideration when selecting an appropriate resistance welding method.

Resistance welding has a limited application in general manufacturing because special equipment is usually required for each individual welding job. It is therefore practical chiefly for mass production. The machines are large, expensive, and individually designed to perform specific tasks.

RESISTANCE WELDING MACHINES

Most resistance welding machines use electric current directly from the power line. There are a few that store the electric energy (e.g. spike welding machines) for welding applications that require massive discharges of current during a brief interval

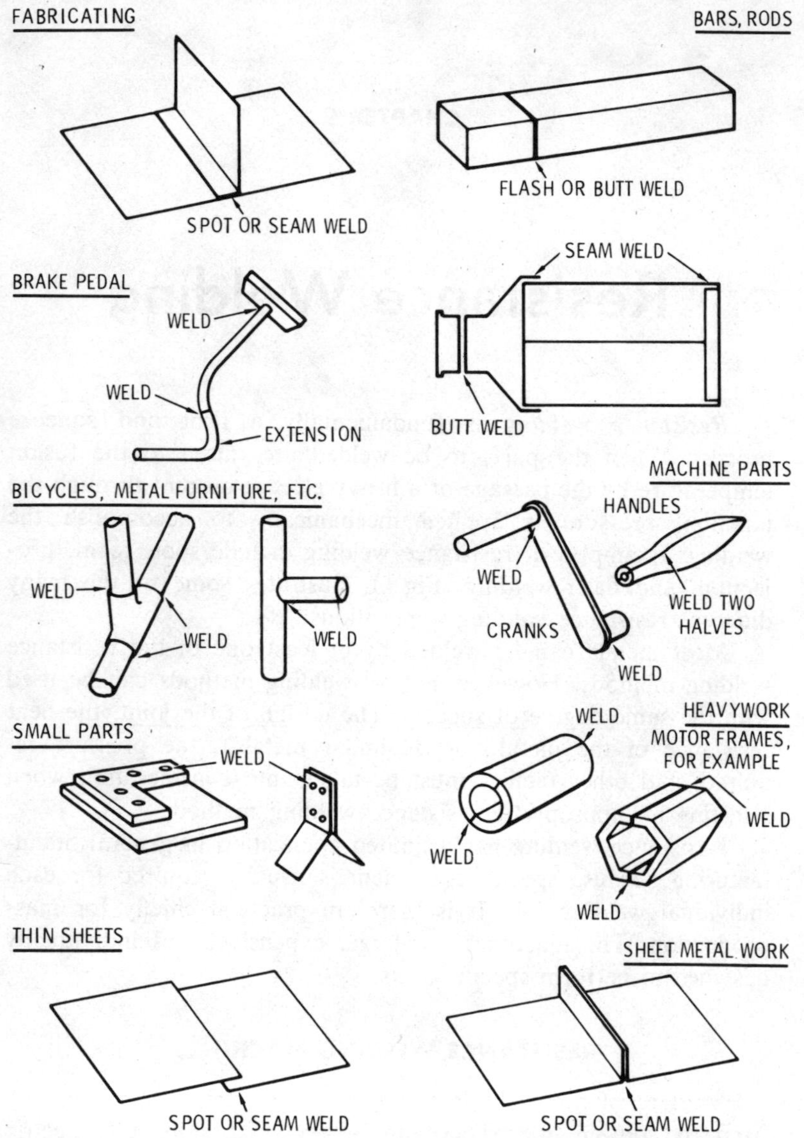

Fig. 1. Various resistance

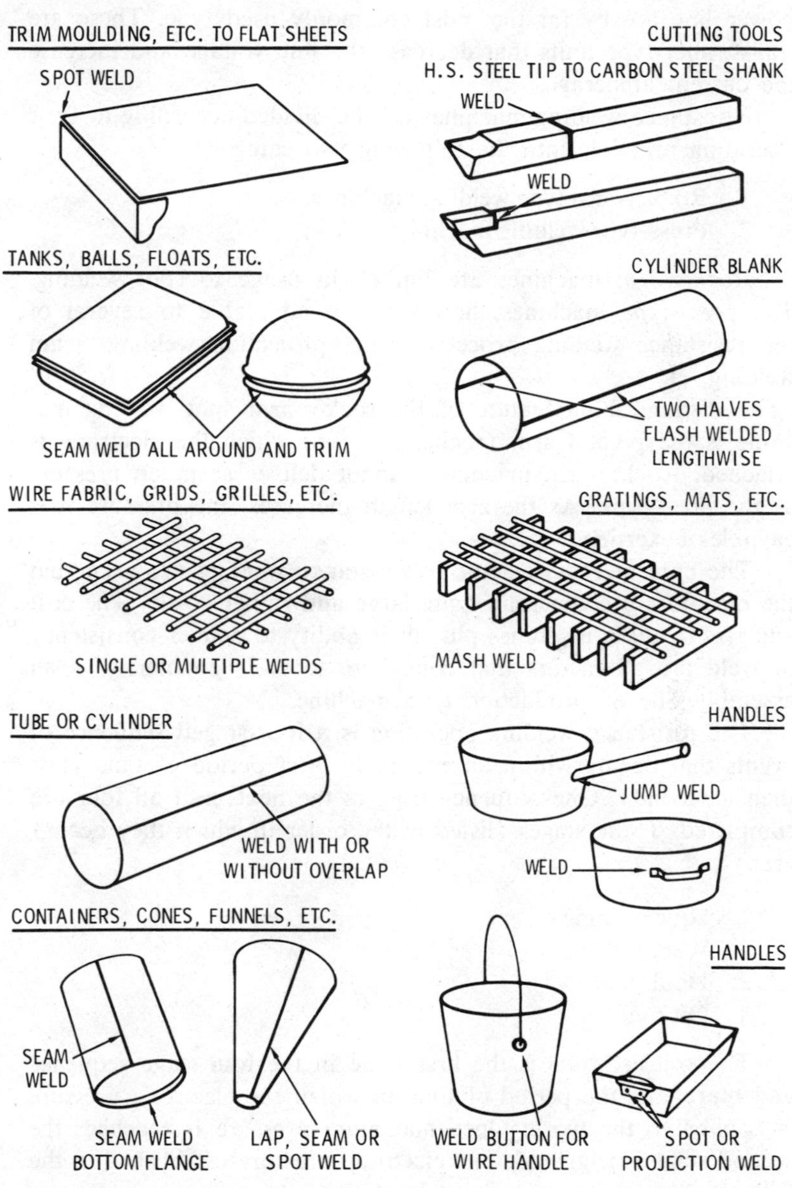

welding applications.

of time. However, machines that use the current directly from the power line are by far the most commonly used type. These are transformer-type units that decrease the line voltage and increase the current amperage.

Resistance welding machines can be divided according to their operating principle into the following two categories:

1. Rocker-arm type welding machines,
2. Press-type welding machines.

Rocker-arm machines are limited in usage to spot welding. The *press-type* machines, however, are adaptable to several of the resistance welding processes (e.g. projection welding, seam welding, etc.).

The distinguishing feature of the rocker-arm spot welding mechine is the pivotal arm (rocker arm) to which the electrode is attached. Rocker-arm machines cannot deliver as much pressure as the press-type. As the arm length increases, the pressure it is capable of exerting decreases.

The current requirements in resistance welding are such that the machines are generally quite large and rather costly. The cost and size of these machines plus their ability to deliver consistency of weld are all factors that contribute to their popularity as an assembly line or production type machine.

The resistance welding operation is a four-staged sequence of events that occurs within an extremely brief period of time (less than a second). One sequence triggers the next until all four are completed. These stages (listed in the order in which they occur) are:

1. Squeeze time,
2. Weld time.
3. Hold time,
4. Off time.

The *squeeze time* is the first stage in the four-stage sequence and represents the period of time in which the electrode pressure is applied to the work. Once maximum pressure is reached, the second stage is triggered. The electrode pressure established in the first stage is maintained throughout the next two stages. During the *weld time* (the second stage), the current flows through the work and welds the pieces of metal together. The current is interrupted again during the *hold time* (the third stage) to allow the weld to solidify. The electrode pressure is released during the *off time* (the fourth stage).

The press-type welding machines are distinguished by a vertical straight-line movement of the electrode. It is capable of exerting considerably more pressure than the rocker-arm machine. For this reason, it is favored when high quality spot welds are required. The press-type machine may be adapted to projection welding by using specially designed dies, or to seam welding by attaching copper alloy rollers.

STORED-ENERGY WELDING MACHINES

Some forms of resistance welding require more electrical energy than the power supply is capable of delivering. To compensate for this situation, electrical energy is stored in a capacitor bank (capacitor-discharge resistance welders) or a magnetic field (electromagnetic resistance welders) during the nonweld period of the resistance welding cycle. The energy is suddenly and explosively released during the weld time.

RESISTANCE WELDING MACHINE CONTROLS

Resistance welding machines are designed to control the three variables basic to most welding processes: time, heat (current), and pressure (force). The following are the principal control devices used to perform this function (Fig. 2).

1. The welding contactor controls,
2. The timing-sequence controls,
3. The heat controls,
4. The voltage and current regulating controls.

Welding contactor controls are switching devices designed to turn the large welding current on and off. This is necessary because of the requirement for brief weld time in resistance welding. A welding contactor may be operated electrically, magnetically, or mechanically depending on the design of the machine.

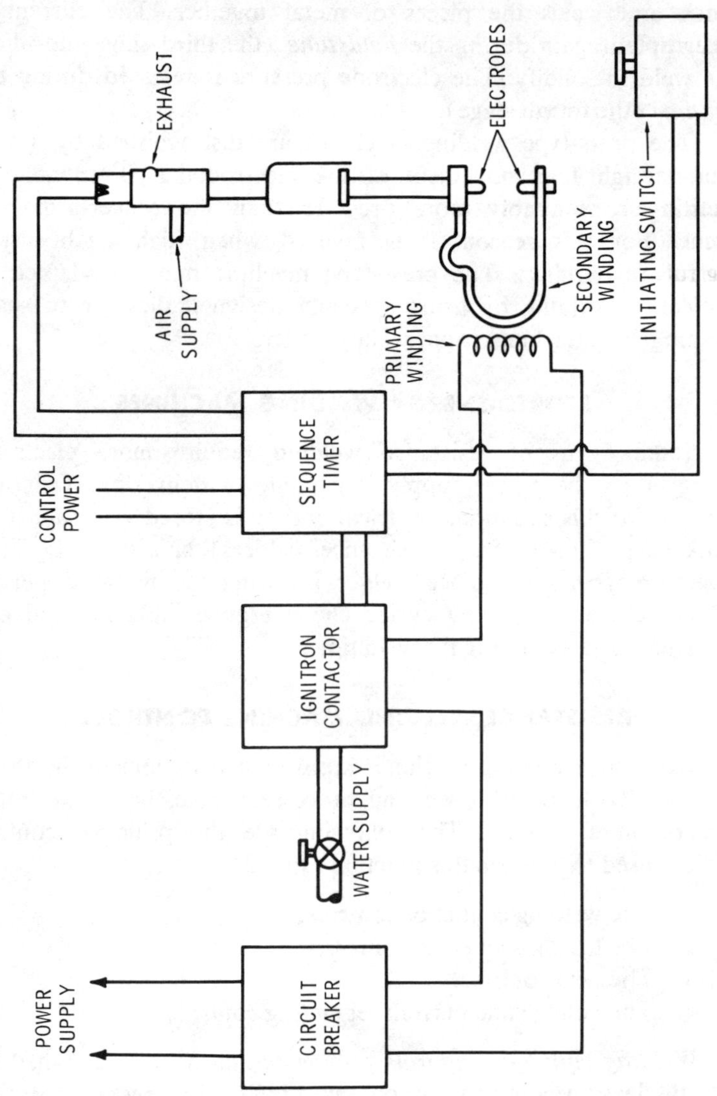

Fig. 2. Wiring diagram of a resistance welding machine illustrating the location of the ignitron contactor.

The mechanical-type welding contactor may be operated with a foot pedal or a built-in motor unit. The magnetic type can be designed to provide synchronous interruption of the current.

An ignitron welding contactor is designed to rapidly open and close the primary circuit of resistance welding machines. The ignitron contactor consists of two water-cooled ignitron tubes mounted on a suitable panel with other equipment for the protection and control of the tubes. These tubes are connected in such a way as to act as a single-pole single-throw switch, conducting full wave AC to the welding machine when the timing contacts are closed. When the timing contacts are open, the welding current ceases to flow.

The precise timing of synchronous electronic timers is required for practical application of short timing such as ½ to 3 cycles. If the timing mechanism is set for longer welding periods, the cycle is repeated over and over.

When used with an AC supply, an ignitron tube will pass current each half-cycle that the voltage on the anode is positive. A second ignitron tube is inversely connected to pass current during each of the other half-cycles.

The minimum current that can be satisfactorily controlled by ignitron tubes is approximately 40 amperes. For those types of resistance welding where less current is required, the addition of a shunt resistor that will demand 25 amperes (or a shunt reactor that will build up to 40 amperes the total current controlled by the ignitron tubes) will make possible the successful use of ignitron contactors on these low-current welding machines. Sometimes a welding machine will demand considerably more than 40 amperes except when the secondary circuit is open. No welding, of course, is done when the secondary circuit is open; but, if for testing and other purposes the primary circuit is closed and if the exciting current of the welding machine is less than 40 amperes a suitable shunt resistor or shunt reactor should be used.

A flow switch prevents operation of the control unless sufficient cooling water is flowing through the jackets of the water-cooled ignitron tubes. The normally open contacts of the flow switch are in series with the control circuit and close only when sufficient water is flowing.

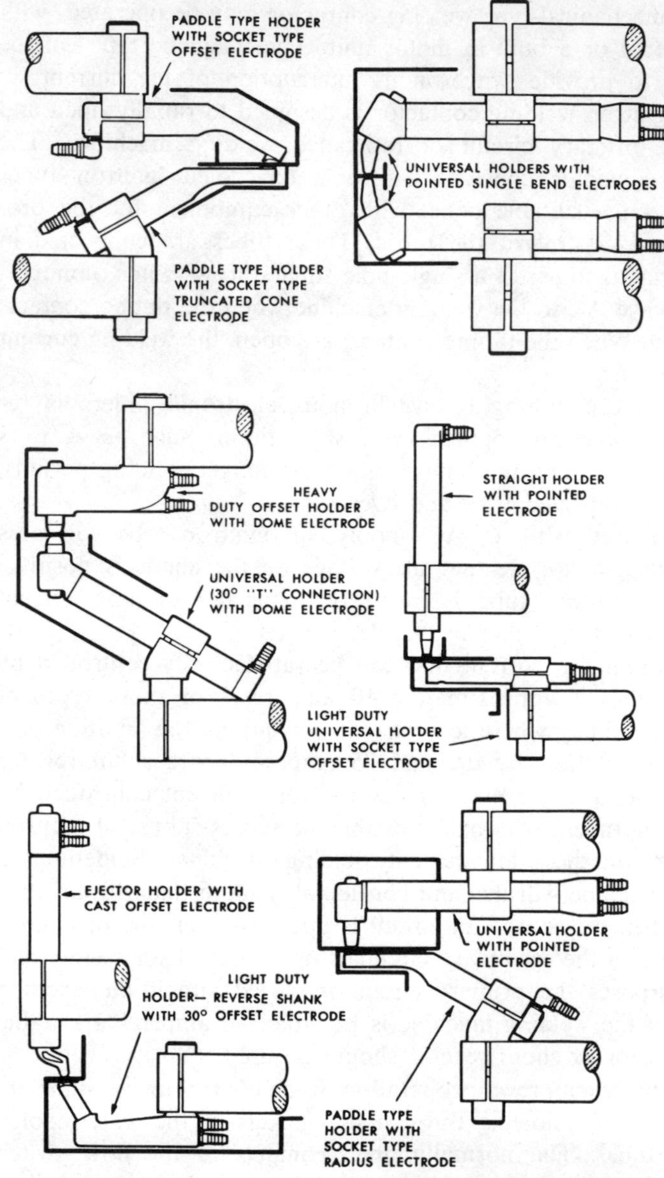

Fig. 3. Various combinations of standard holders, adaptors,

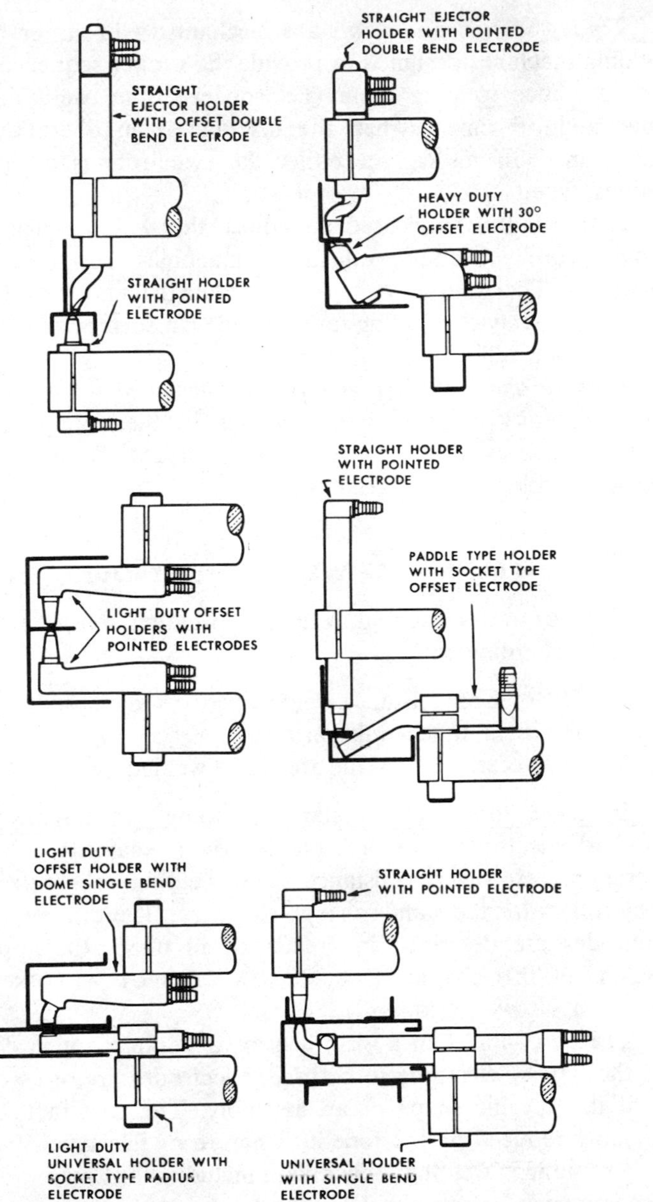

Courtesy P. R. Mallory and Co.

and electrodes used in resistance welding.

Timing-sequence controls are mechanisms in air or hydraulic welding machines designed to provide the proper sequence through the resistance welding cycle (i.e. squeeze time, weld time, hold time, and off time), when the machine is in operation. These timing mechanisms can be either the synchronous or nonsynchronous type.

Heat controls are used to adjust the welding heat and to provide preheat and postheat in machines designed for this provision. The welding heat (output voltage) may be adjusted in a variety of ways, including tapping the transformer or with series-parallel connections.

Voltage and current regulators are designed to maintain weld consistency by controlling fluctuations in the voltage from the power source as well as in the welding current. These fluctuations are kept within a 2% variation.

RESISTANCE WELDING ELECTRODES

The electrodes used in resistance welding must be carefully selected according to:

1. The type of resistance welding process to be used,
2. The welding schedule variations,
3. The accessibility of the area to be welded.

In most forms of resistance welding, various designs of standard electrodes are acceptable for a majority of welding operations. However, resistance seam welding substitutes copper alloy rollers for the standard type electrode. The different types of electrodes are described in greater detail under the appropriate sections of this chapter (see RESISTANCE SPOT WELDING, RESISTANCE SEAM WELDING, etc.).

The accessibility of a weld is also very important in determining the type of electrode to be used. Electrodes are often designed to fit the specific shape of an assembly. The cost factor is considerably reduced by the repetitive nature of this type of welding.

Resistance welding schedules include the many variables (recommended electrode size, on-off time, maximum weld speed, etc.) encountered in the various resistance welding processes.

These schedules are described in greater detail within the appropriate sections of this chapter.

The electrodes used in resistance welding must be carefully selected for the correct shape. The size of the weld is determined by the shape of the contact face of the electrode. The electrode must also be made of a metal that possesses a high degree of conductivity and resistance to wear. A copper and beryllium metal alloy possesses both these characteristics.

There are many different types of electrodes and electrode holders available in resistance welding. Fig. 3 illustrates a small portion of the many different combinations possible.

Electrode Dressers

Resistance welding electrodes are occasionally "dressed" to keep them both sharp and clean. The dressing operation results in the restoration of the electrode to its original contours. This operation can be performed with an electrode file (Fig. 4), manually operated electrode dressers (Figs. 5 and 6), and pneumatic

Courtesy P. R. Mallory and Co.

Fig. 4. An electrode file available in 2 and 8 inch radii.

Courtesy P. R. Mallory and Co.

Fig. 5. An electrode dresser manually operated with adjustable screw.

Courtesy P. R. Mallory and Co.

Fig. 6. Electrode dresser designed to restore spot welding electrodes to their normal contours.

Courtesy P. R. Mallory and Co.

Fig. 7. A pneumatic dresser which operates at a cutting speed of 1200 RPM.

or machine operated electrode dressers (Fig. 7). Electrode dressers have removable cutter blades that can be easily replaced to provide a wide variety of electrode contours.

Electrode Holders and Adapters

Electrode holders are specially designed extensions (or arms) that provide the electrodes with electrical current, pressure, and (in the case of water-cooled electrode holders) with a supply of cooling water. These electrode holders are made of a copper alloy, a composition which gives them a high degree of electrical conductivity.

Electrode holders may be either ejector or nonejector types (Fig. 8). In the latter, electrodes can be released (ejected) without damage to the electrode holder. Electrode holders are also classified according to design, such as:

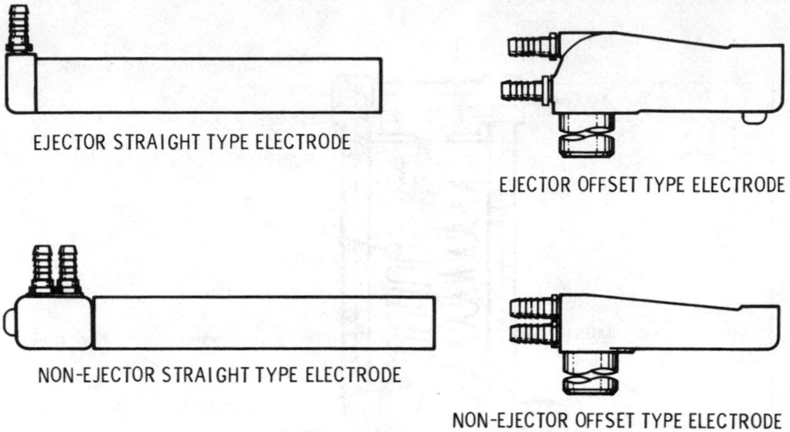

EJECTOR STRAIGHT TYPE ELECTRODE

EJECTOR OFFSET TYPE ELECTRODE

NON-EJECTOR STRAIGHT TYPE ELECTRODE

NON-EJECTOR OFFSET TYPE ELECTRODE

Courtesy P. R. Mallory and Co.

Fig. 8. Ejector and non-ejector type electrode holders.

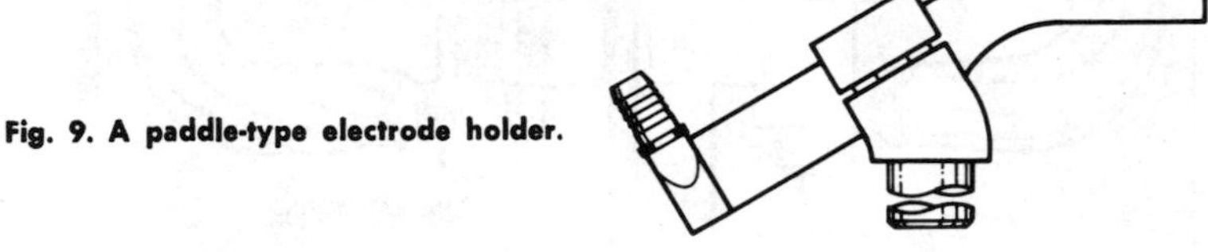

Fig. 9. A paddle-type electrode holder.

Courtesy P. R. Mallory and Co.

1. Straight,
2. Offset,
3. Paddle (Fig. 9).

Examples of water-cooled electrodes used in spot and projection welding are shown in Fig. 10.

Adapters (Fig. 11) are specially designed devices for holding one or more electrodes at the desired angle when making welds in series or parallel.

RESISTANCE SPOT WELDING

Resistance spot welding is a form of resistance welding in which the two surfaces are joined by spots of fused metal caused by heating small areas of the metal between suitable electrodes under pressure. A low voltage high amperage electric

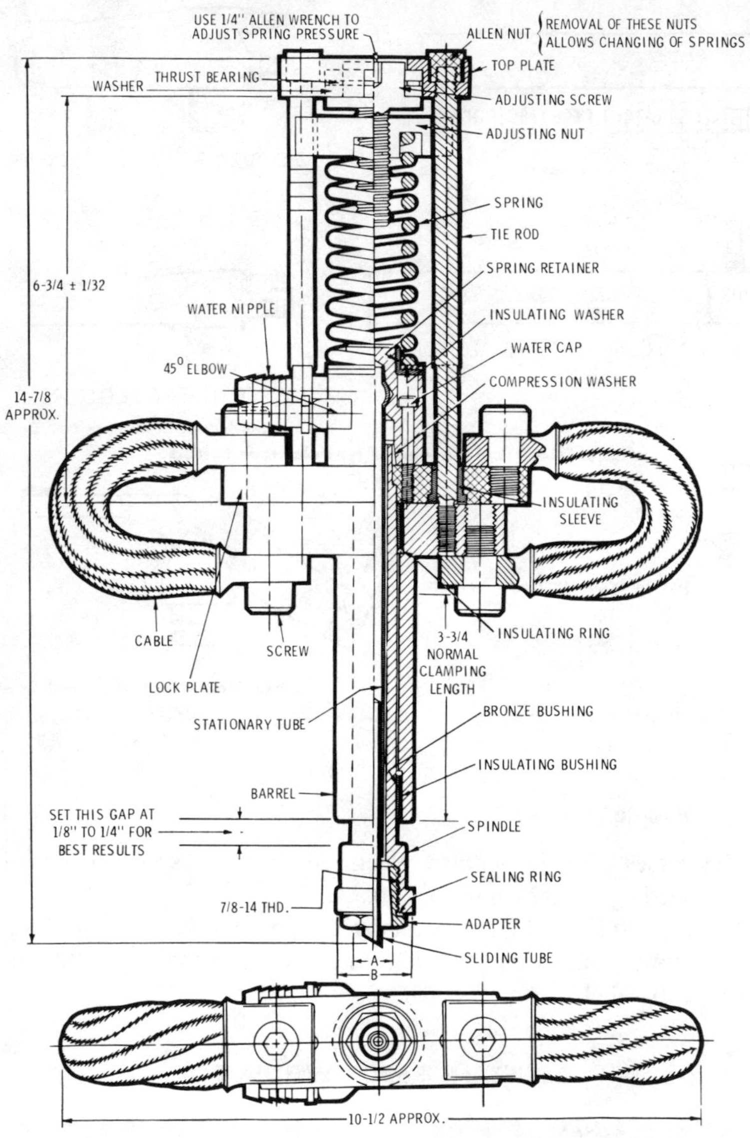

Courtesy P. R. Mallory and Co.

Fig. 10. Water-cooled low inertia electrode holder designed for the spot and and projection welding.

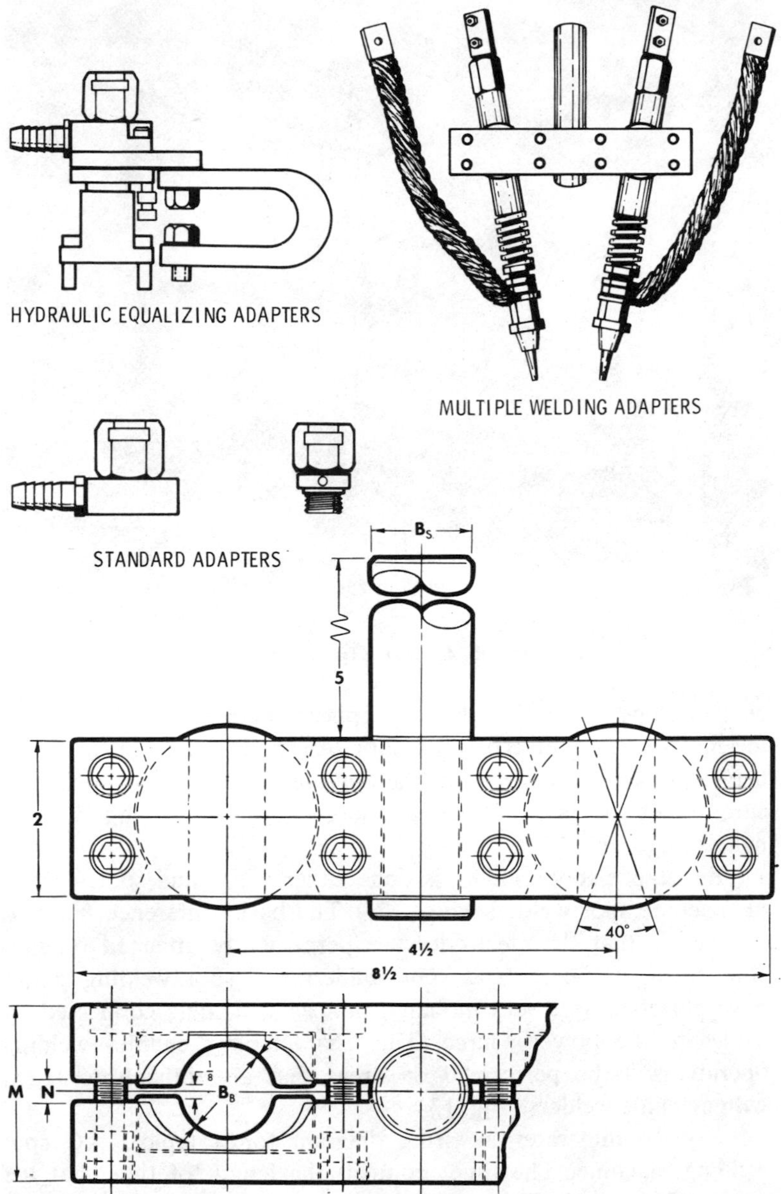

Courtesy P. R. Mallory and Co.

Fig. 11. Electrode adapters.

Courtesy Acme Electric Welder Co.

Fig. 12. A multiple spot welder.

current flows through a heavy copper conductor until it meets the metal surface placed in its path. An intense heat results from the resistance of the metal to the continuous flow of the electrical current. At the proper time, pressure is applied forcing the two metals together.

Resistance spot welding is done either with hand spot welders or machine spot welders (Fig. 12.) The basic difference between the two is that the electrodes are permanently attached on machine spot welders. Hand spot welders (or spot welding guns) have portable electrodes mounted in special holders connected by cables to the power source. This characteristic enables welding operations to be performed on areas that are difficult to reach with machine welders (Fig. 13).

Fig. 14 illustrates a wiring diagram for a typical AC spot welding machine. The timer controls the length of time that the current is applied. The number of turns on the primary winding can be controlled by the current regulator through a series of

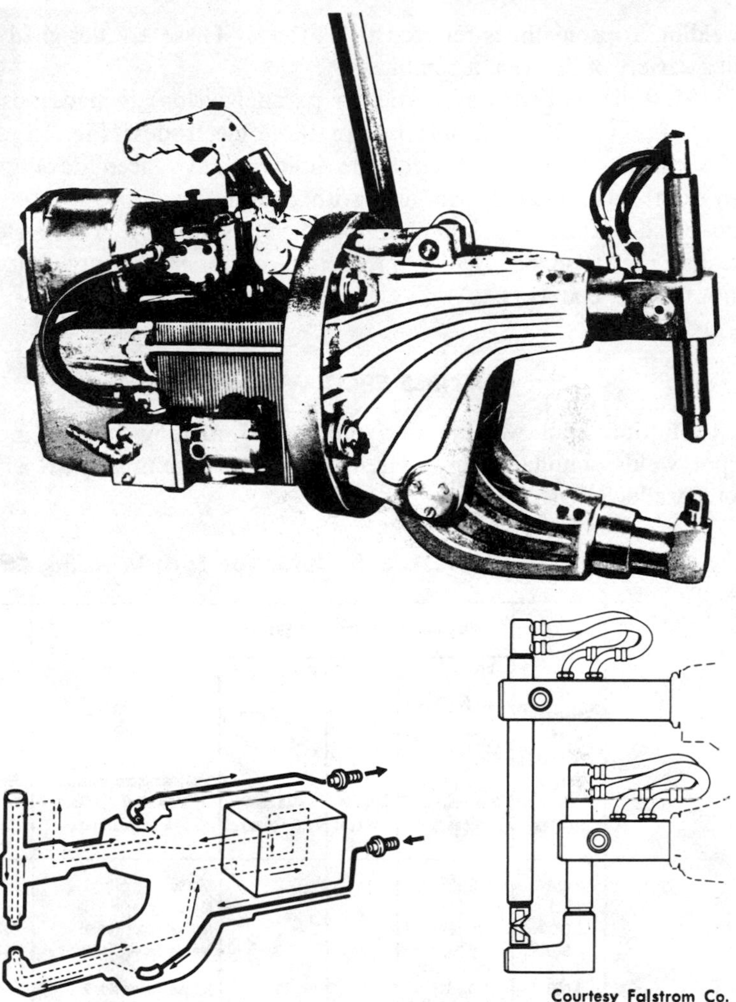

Courtesy Falstrom Co.

Fig. 13. Water-cooled spot welding gun.

tap settings. If a greater number of turns on the primary coil is selected, the secondary voltage decreases and the current increases. By decreasing the number of turns on the primary coil, the opposite effect is achieved.

Schedules for welding low carbon steel and stainless steel are shown in Tables 1, 2 and 3. The electrodes used in resistance spot welding are sometimes referred to as tongs. These are designed to fit a variety of different assemblies.

Multiple spot welding (series or parallel welds) is made possible by set ups that will hold two or three electrodes (Fig. 15).

Special water-cooled electrode holders have been developed by some manufacturers for use with metals having very little or no plastic range (e.g. aluminum and yellow brass). These electrodes may also be used in projection welding (see ELECTRODE HOLDERS AND ADAPTERS).

MULTIPLE SPOT WELDING

Multiple spot welding is a method of making two or more spot welds simultaneously. These welds may be made in series or parallel (Figs. 16 and 17).

Table 1. Schedules for Spot Welding Carb

Thickness of Thinnest Outside Piece (Inches)	Electrode Diameters and Shape			Recommended Minimum Standard Electrode Size	Weld Force (Lbs.)
	Flat Face (30°, d, D) Maximum d (Inches)	Min. D (Inches)	Radius Face (R, D) Radius R (Inches)		
0.010	0.125	½	2	Morse Taper No. 1	160
0.021	0.187	½	2	Morse Taper No. 1	244
0.031	0.187	½	2	Morse Taper No. 1	326
0.040	0.250	⅝	3	Morse Taper No. 2	412
0.050	0.250	⅝	3	Morse Taper No. 2	554
0.062	0.250	⅝	3	Morse Taper No. 2	670
0.078	0.312	⅝	3	Morse Taper No. 2	903
0.094	0.312	⅝	4	Morse Taper No. 3	1160
0.109	0.375	⅞	4	Morse Taper No. 3	1440
0.125	0.375	⅞	4	Morse Taper No. 3	1760
0.156	0.500	⅞	6	Male or Female Threaded	2500
0.187	0.625	1	6	Male or Female Threaded	3340
0.250	0.750	1¼	6	Male or Female Threaded	5560

The machines used for multiple spot welding are designed for specific resistance spot welding applications. An example of a multiple spot welding machine is shown in Fig. 18. These are transformer-type welding machines that control their electrodes in one of the three following ways:

1. Commutator method,
2. Hydraulic cylinders,
3. Individual transformers.

The commutator method employs a single transformer. The electrical energy is commutated in sequence to each of the electrodes at ultrahigh speed *after* all the electrodes are positioned against the work. Hydraulic cylinders may also be used to bring each electrode into contact with the work. Here, too, a single transformer is used. The electrodes contact the work one at a

on Steel (SAE 1010)—Optimum Conditions

Weld Time (Cycles) (60 Cycles per Sec.)	Hold Time (Cycles) Min.	Welding Current (Amps.) (Approx.)	Weld Shear Strength (For Steels Having Ultimate Tensile Strength of 90,000 psi and below) Min. Strength (Lbs./Weld)	Diameter of Fused Zone (Approx.) Dw (Inches)	Minimum Weld Spacing S (Inches)	Minimum Contacting Overlap L (Inches)
4	5	4000	130	0.113	¼	⅜
6	8	6500	300	0.139	⅜	7/16
8	10	8000	530	0.161	½	7/16
10	12	8800	812	0.181	¾	½
14	16	9600	1195	0.210	⅞	9/16
18	20	10600	1717	0.231	1	⅝
25	30	11800	2365	0.268	1⅛	11/16
34	35	13000	3054	0.304	1¼	¾
45	40	14200	3672	0.338	1 5/16	13/16
60	45	15600	4300	0.375	1½	⅞
93	50	18000	6500	0.446	1¾	1
130	55	20500	9000	0.516	2	1¼
230	60	26000	18000	0.660	4	1½

Courtesy R.W.M.A.

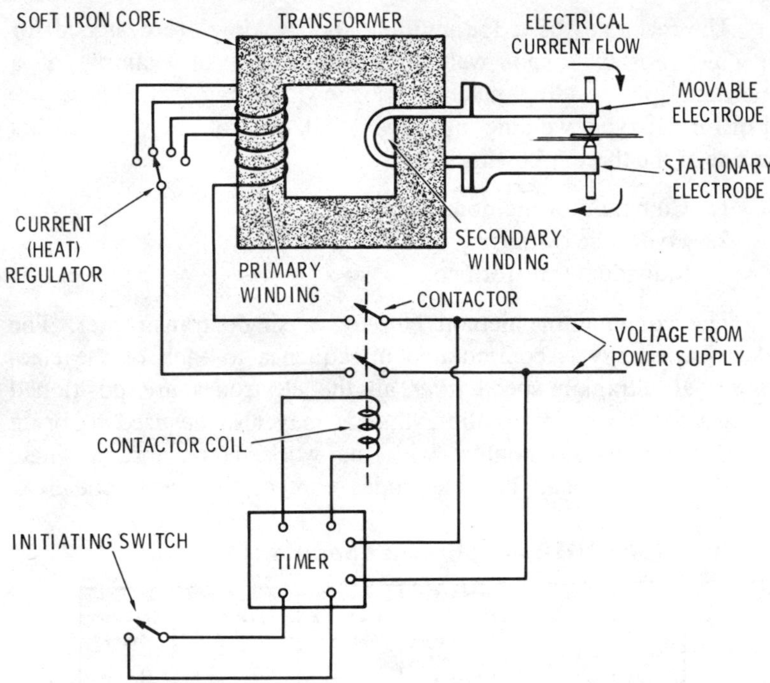

Fig. 14. Schematic diagram of a resistance spot welding machine.

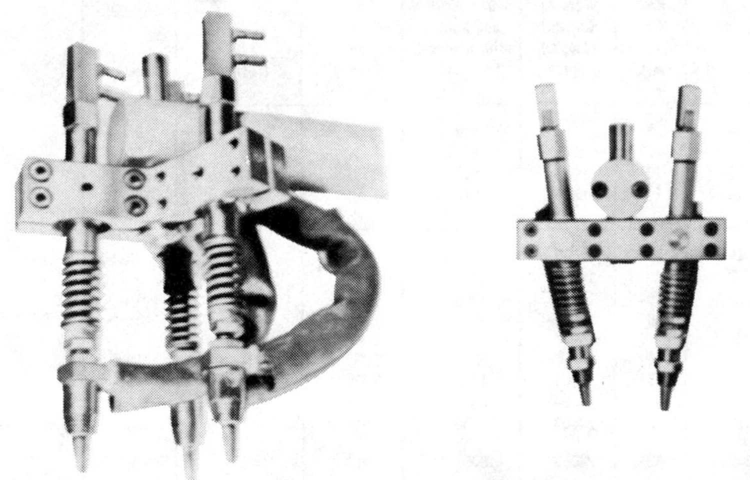

Courtesy P. R. Mallory and Co.

Fig. 15. Multiple spot welding.

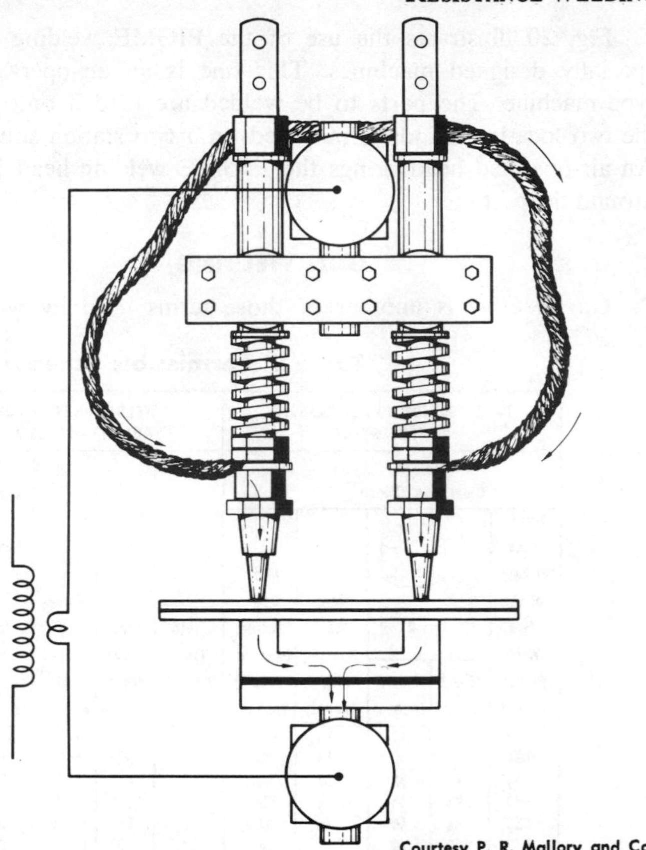

Courtesy P. R. Mallory and Co.

Fig. 16. Wiring diagram for parallel spot welds.

time in sequence. Some multiple spot welding systems use an individual transformer and controls for each electrode. The number of electrodes in this system can be increased by adding more transformers and control units.

PIGME WELDING PROCESS

The PIGME welding process uses special welding heads to produce the rapid firing of multiple electrodes in sequence. The welding head illustrated in Fig. 19 was designed to weld *I.B.M.* tape recorder cases.

Fig. 20 illustrates the use of the PIGME welding heads on specially designed machines. This one is an air-operated press-type machine. The parts to be welded are loaded onto either of the two locating mandrels mounted on a two station shuttle table. An air-operated head brings the PIGME welding head into place around the part.

GUN WELDING

Gas welding is another of those terms used by welders but

Table 2. Permissible Schedule Variat–

DATA COMMON TO ALL CLASSES OF SPOT WELDS					WELDING SET-UP FOR BEST QUALITY—CLASS A WELDS				
Thickness of Each of the Two Work Pieces Inches	Electrode Diam. and Shape: Min. D Inches	Electrode Diam. and Shape: Max. d Inches	Min. Weld Spacing (Note 4) Inches	Min. Contacting Overlap (Note 6) Inches	Weld Time (Note 7) Cycles	Electrode Force Pounds	Welding Current Amps.	Diam. of Fused Zone Inches	Average Tensile Shear Strength ±14% Pounds
.010	1/2	1/8	1/4	3/8	4	200	4000	.13	235
.021	1/2	3/16	3/8	7/16	6	300	6100	.17	530
.031	1/2	3/16	1/2	7/16	8	400	8000	.21	980
.040	5/8	1/4	3/4	1/2	10	500	9200	.23	1350
.050	5/8	1/4	7/8	9/16	12	650	10300	.25	1820
.062	5/8	1/4	1	5/8	14	800	11600	.27	2350
.078	5/8	5/16	1 1/8	11/16	21	1100	13300	.31	3225
.094	5/8	5/16	1 1/4	3/4	25	1300	14700	.34	4100
.109	7/8	3/8	1 5/16	13/16	29	1600	16100	.37	5300
.125	7/8	3/8	1 1/2	7/8	30	1800	17500	.40	6900

NOTES:

1. Low Carbon Steel is hot rolled, pickled, and slightly oiled with an ultimate strength of 42,000 to 45,000 PSI. Similar to SAE 1005—SAE 1010.
2. Electrode Material is Mallory 3.
3. Surface of steel is lightly oiled but free from grease, scale or dirt.
4. Minimum weld spacing is that distance for which no increase in welding current is necessary to compensate for the shunted current effect of adjacent welds.
5. Radius Face electrodes may be used:
   0.010 to 0.031—2″ Radius
   0.031 to 0.078—3″ Radius
   0.078 to 0.125—4″ Radius

not officially recognized as welding terminology by the *American Welding Society*.

Gun welding refers to the use of a gun welder and a portably designed spot-welding machine used to reach partially obstructed areas on irregular surfaces. With this advantage, the gun welder finds useful application in automotive body work and sheet metal fabrication.

The amount of current flow, pressure, and time is predetermined for the operator. The gun welder operator is only responsible for

ions for Spot Welding Low Carbon Steel

WELDING SET-UP FOR MEDIUM QUALITY—CLASS B WELDS					WELDING SET-UP FOR GOOD QUALITY—CLASS C WELDS				
Weld Time (Note 7) Cycles	Electrode Force Pounds	Welding Current Amps.	Diam. of Fused Zone Dw Inches	Average Tensile Shear Strength ±17% Pounds	Weld Time (Note 7) Cycles	Electrode Force Pounds	Welding Current Amps.	Diam. of Fused Zone Dw Inches	Average Tensile Shear Strength ±20% Pounds
5	130	3700	.12	200	15	65	3000	.11	160
10	200	5100	.16	460	22	100	3800	.14	390
15	275	6300	.20	850	29	135	4700	.18	790
21	360	7500	.22	1230	38	180	5600	.21	1180
24	410	8000	.23	1700	42	205	6100	.22	1600
29	500	9000	.26	2150	48	250	6800	.25	2050
36	650	10400	.30	3025	58	325	7900	.28	2900
44	790	11400	.33	3900	66	390	8800	.31	3750
50	960	12200	.36	5050	72	480	9500	.35	4850
60	1140	12900	.39	6500	78	570	10000	.37	6150

6. L L L

7. Weld time is indicated in cycles of 60 cycle frequency.
8. Tensile shear strength values are based on recommended test sample sizes.

Direction of Force	Thickness	Width	Length
	.000″ to .029″	5/8″	3″
	.030″ to .058″	1 ″	4″
	.059″ to .115″	1½″	5″
	.116″ to .190″	2 ″	6″

9. Tolerance for machining of electrode diameter "d" is ±.015″ of specified dimension.
10. Electrode force does not provide for force to press ill-fitting parts together.

Courtesy R.W.M.A.

keeping the electrodes in proper working condition and for applying them to the correct location when welding. If the electrodes are no longer suitable for efficient use, the operator must replace them.

SHOT WELDING

Shot welding is another term not officially recognized by the

Table 3. Schedule for Spot

THICKNESS "T" of THINNEST OUTSIDE PIECE (See Notes 1, 2, 3 & 4 Below)	ELECTRODE DIAMETER AND SHAPE (See Note 5)		ELECTRODE FORCE	WELD TIME	WELDING CURRENT (Approx.) AMPS
				CYCLES (60 Per Sec.)	Tensile Strength Below 150000 Psi
Inches	D, IN., Min.	d. IN., Max.	LB.		
0.006	3/16	3/32	180	2	2000
0.008	3/16	3/32	200	3	2000
0.010	3/16	1/8	230	3	2000
0.012	1/4	1/8	260	3	2100
0.014	1/4	1/8	300	4	2500
0.016	1/4	1/8	330	4	3000
0.018	1/4	1/8	380	4	3500
0.021	1/4	5/32	400	4	4000
0.025	3/8	5/32	520	5	5000
0.031	3/8	3/16	650	5	6000
0.034	3/8	3/16	750	6	7000
0.040	3/8	3/16	900	6	7800
0.044	3/8	3/16	1000	8	8700
0.050	1/2	1/4	1200	8	9500
0.056	1/2	1/4	1350	10	10300
0.062	1/2	1/4	1500	10	11000
0.070	5/8	1/4	1700	12	12300
0.078	5/8	5/16	1900	14	14000
0.094	5/8	5/16	2400	16	15700
0.109	3/4	3/8	2800	18	17700
0.125	3/4	3/8	3300	20	18000

NOTES:

1. Types of Steel—301, 302, 303, 304, 308, 309, 310, 316, 317, 321, 347 and 349
2. Material should be free from scale, oxides, paint, grease and oil
3. Welding conditions determined by thickness of thinnest outside piece "T"
4. Data for total thickness of pile-up not exceeding 4 "T". Maximum ratio between two thicknesses 3 to 1

American Welding Society, although it is commonly used in welding shops to refer to a special adaptation of spot welding. Here, the term is applied to the spot welding of aluminum alloys and stainless steel. Shot welding is characterized by: (1) a very brief period in the application of the current, and (2) a very careful control of the flow of the current. Thus, the metal to be joined is heated and cooled as quickly as possible. This has the advantage of reducing oxidation problems.

Welding Stainless Steel

Tensile Strength 150000 Psi and Higher	MINIMUM CONTACTING OVERLAP IN.	Minimum Weld Spacing (See note 6 Below) ℄ to ℄ IN.	DIAMETER OF FUSED ZONE IN. Approx.	MINIMUM SHEAR STRENGTH LB. Ultimate Tensile Strength of Metal: 70000 Up To 90000 Psi	90000 Up To 150000 Psi	150000 Psi and Higher
2000	3/16	3/16	0.045	60	70	85
2000	3/16	3/16	0.055	100	130	145
2000	3/16	3/16	0.065	150	170	210
2000	1/4	1/4	0.076	185	210	250
2200	1/4	1/4	0.082	240	250	320
2500	1/4	5/16	0.088	280	300	380
2800	1/4	5/16	0.093	320	360	470
3200	5/16	5/16	0.100	370	470	500
4100	3/8	7/16	0.120	500	600	680
4800	3/8	1/2	0.130	680	800	930
5500	7/16	9/16	0.150	800	920	1100
6300	7/16	5/8	0.160	1000	1270	1400
7000	7/16	11/16	0.180	1200	1450	1700
7500	1/2	3/4	0.190	1450	1700	2000
8300	9/16	7/8	0.210	1700	2000	2450
9000	5/8	1	0.220	1950	2400	2900
10000	5/8	1 1/8	0.250	2400	2800	3550
11000	11/16	1 1/4	0.275	2700	3400	4000
12700	3/4	1 3/8	0.285	3550	4200	5300
14000	13/16	1 1/2	0.290	4200	5000	6400
15500	7/8	2	0.300	5000	6000	7600

5. Electrode Material, Mallory 3, Mallory 100, or Elkonite 10W3
6. Minimum weld spacing is that spacing for two pieces for which no special precautions need be taken to compensate for shunted current effect of adjacent welds. For three pieces increase spacing 30 per cent.

Courtesy The American Welding Society

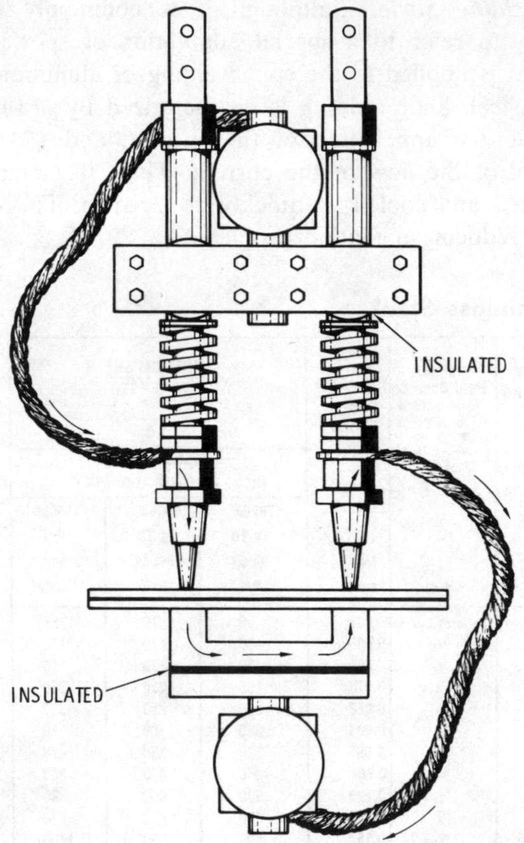

Courtesy P. R. Mallory and Co.

Fig. 17. Wiring diagram for series spot welds.

RESISTANCE SEAM WELDING

Although the term *seam welding* is generally used to indicate a special resistance welding process *(resistance seam welding),* it is also used at times in reference to the arc seam welding process.

The seam weld is similar to the spot weld, except that a circular rolling electrode is used to produce the effect of a continuous seam. In reality, it is a series of continuous overlapping spots produced by the rotating electrode and a regularly inter-

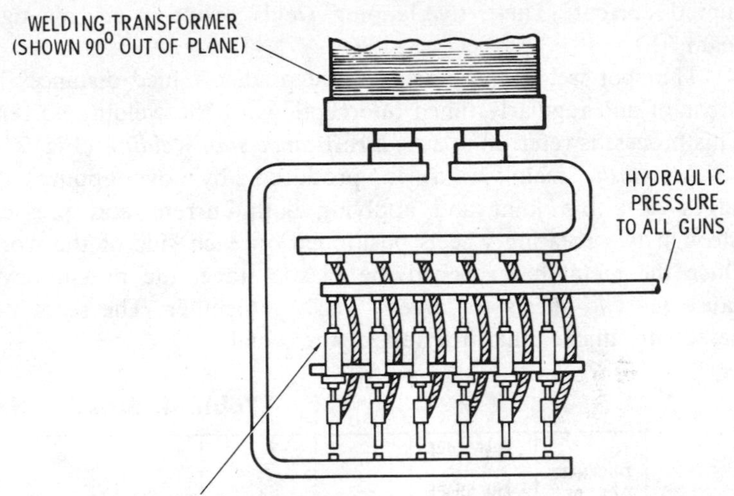

Fig. 18. A multiple spot welder.

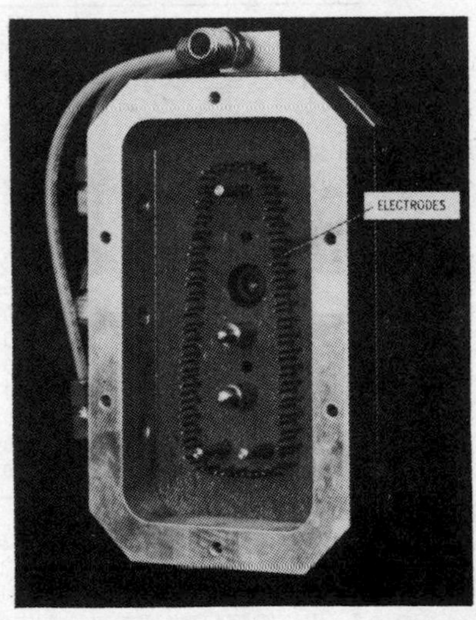

Fig. 19. A close up view of the PIGME welding head.

Courtesy Thomson Electric Welder Co.

rupted current. These overlapping welds result in an air tight seam.

The spot welds may be spaced at predetermined distances by constant and regularly timed interruptions of the welding current. This process is referred to as *roll resistance spot welding* (Fig. 21).

A *mash* seam weld is produced by overlapping the edges of a lap joint and applying both current and pressure through two rotating wheels positioned on each side of the work. Once the metal has reached the plastic stage, the pressure will cause the two sheets to forge ("mash") together. The seam will be slightly thicker than the rest of the metal.

Table 4. Schedule for

THICKNESS "T" OF THINNEST OUTSIDE PIECE (See Notes 1, 2, 3 & 4 Below) INCHES	ELECTRODE WIDTH AND SHAPE (See Note 5 Below) R = 3" W, IN., Min.	ELECTRODE FORCE LB.	ON TIME CYCLES (60 Per Sec.)	OFF TIME FOR MAXIMUM SPEED (Pressure-Tight) CYCLES 2 "T"	4 "T"
0.006	3/16	300	2	1	1
0.008	3/16	350	2	1	2
0.010	3/16	400	3	2	2
0.012	1/4	450	3	2	2
0.014	1/4	500	3	2	3
0.016	1/4	600	3	2	3
0.018	1/4	650	3	2	3
0.021	1/4	700	3	2	3
0.025	3/8	850	3	3	4
0.031	3/8	1000	3	3	4
0.040	3/8	1300	3	4	5
0.050	1/2	1600	4	4	5
0.062	1/2	1850	4	5	7
0.070	5/8	2150	4	5	7
0.078	5/8	2300	4	6	7
0.094	5/8	2550	5	6	7
0.109	3/4	2950	5	7	9
0.125	3/4	3300	6	6	8

NOTES:

1. Types of Steel—301, 302, 303, 304, 308, 309, 310, 316, 317, 321, 347 and 349
2. Material should be free from scale, oxides, paint, grease and oil
3. Welding conditions determined by thickness of thinnest outside piece "T."
4. Data for total thickness of pile-up not exceeding 4 "T." Maximum ratio between

Seam welders (Fig. 22) are resistance welding machines using specially designed roller (disc)-type electrodes. The electrodes can be set so that the welded seam moves either perpendicularly or longitudinally to the welding arms. A schedule for seam welding stainless steel is shown in Table 4.

PROJECTION WELDING

Resistance projection welding is a resistance welding process in which the position of the weld is established by the predetermined design of a projection on one of the two metal surfaces

Seam Welding Stainless Steel

MAXIMUM WELD SPEED IN. PER MINUTE		WELDS PER INCH		WELDING CURRENT (Approx.)	MINIMUM CONTACTING OVERLAP (See Note 6)
2 "T"	4 "T"	2 "T"	4 "T"	AMPS.	IN.
60	67	20	18	4000	1/4
67	56	18	16	4600	1/4
45	51	16	14	5000	1/4
48	55	15	13	5600	5/16
51	46	14	13	6200	5/16
51	50	14	12	6700	5/16
55	50	13	12	7300	5/16
55	55	13	11	7900	3/8
50	47	12	11	9200	7/16
50	47	12	11	10600	7/16
47	45	11	10	13000	1/2
45	44	10	9	14200	5/8
40	41	10	8	15100	5/8
44	41	9	8	15900	11/16
40	41	9	8	16500	11/16
36	38	9	8	16600	3/4
38	37	8	7	16800	13/16
38	37	8	7	17000	7/8

thicknesses 3 to 1.

5. Electrode material, Mallory 100
6. For large assemblies minimum contacting overlap indicated should be increased 30 per cent.

Courtesy The American Welding Society

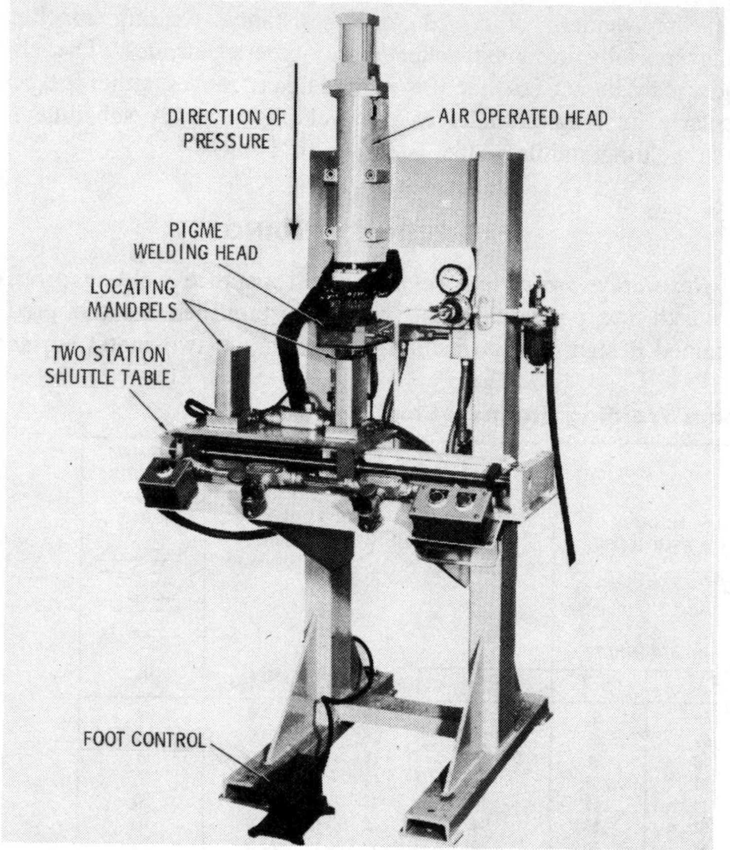

Courtesy Thomson Electric Welder Co.

Fig. 20. PIGME welding machine.

to be joined. The two surfaces are held together under pressure by the electrodes. When the electric current flows through the electrodes, it causes the projecting metal to melt and fuse the two metal surfaces together. This is referred to as a projection weld (Fig. 23).

The three types of projections used in resistance projection welding are:

1. The cone type,
2. The spherical type,

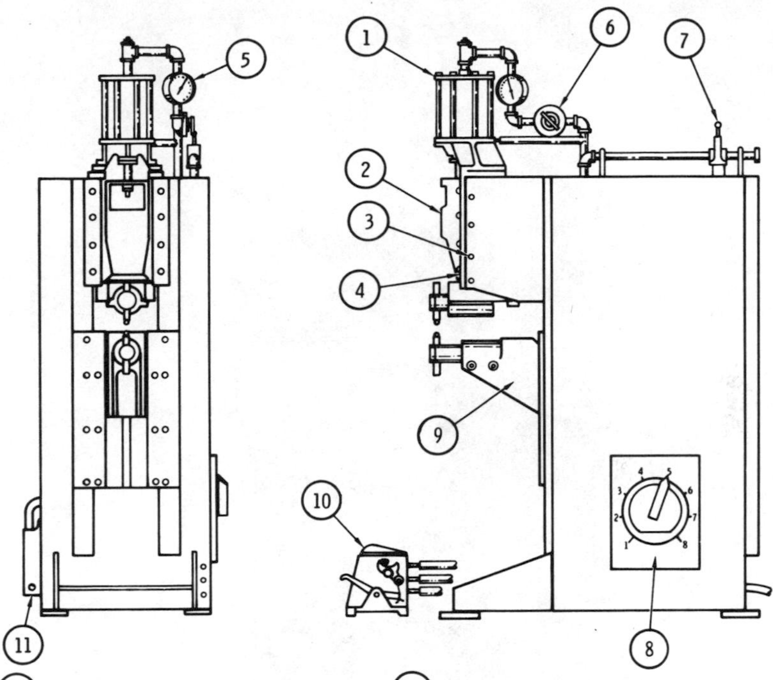

(1) WELDING HEAD, EQUIPPED WITH AIR CYLINDER, CUSHIONED WITH OIL PROOF BUMPERS

(2) HEAD SLIDE KEEPS INERTIA OF MOVING PARTS AT A MINIMUM, PERMITTING EFFICIENT "FOLLOW UP" OF ELECTRODES WHEN WELDING.

(3) GIBBED WAYS FOR THE HEAD SLIDES

(4) PEEL OFF SHIMS.

(5) GAUGE FOR READING WELDING PRESSURE

(6) WELDING PRESSURE REGULATING VALVE

(7) AIR CYLINDER AND VALVE

(8) WELDING HEAT REGULATOR

(9) LOWER ARM

(10) FOOT VALVE AND LIMIT SWITCH

(11) DRAIN FOR COOLING CIRCUITS

Fig. 21. The spot or projection welder.

3. The button type.

The *button-type* projection is used for joining the thinner gauge flat sheets (24 to 13 gauge). The *cone-type* projection is used on 12 to 5 gauge metal sheets, and the *spherical-type* on

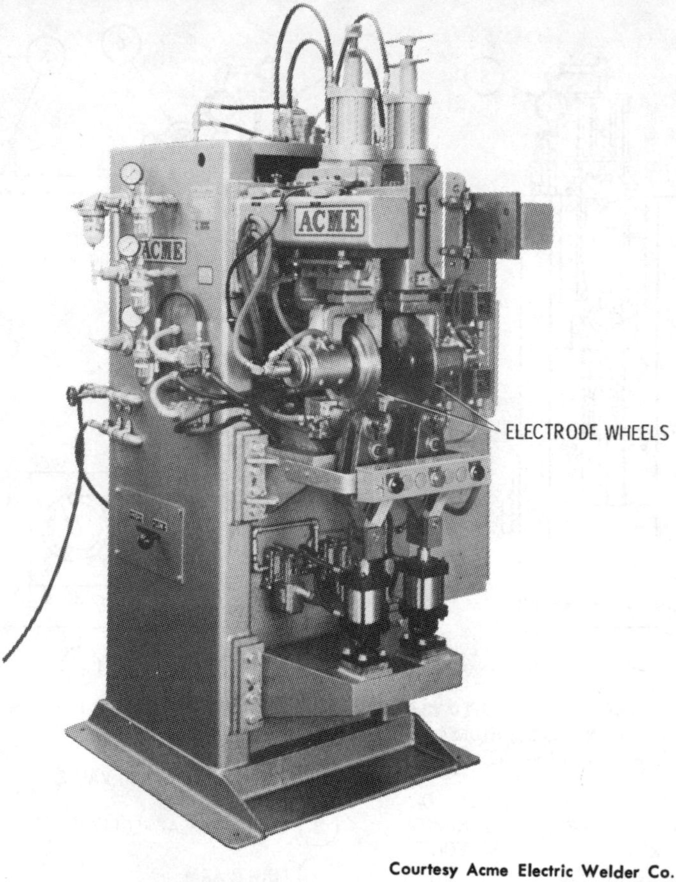

ELECTRODE WHEELS

Courtesy Acme Electric Welder Co.

Fig. 22. A seam welding machine.

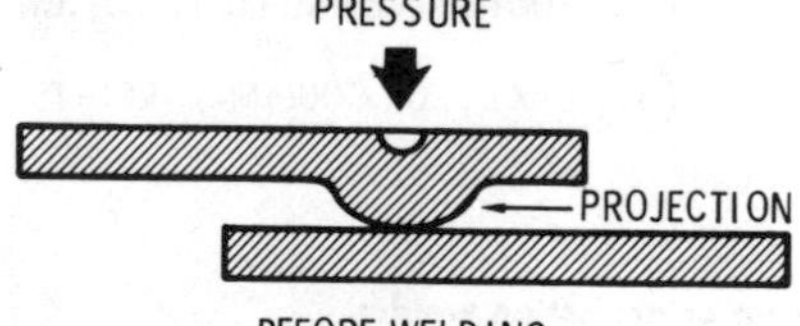

Fig. 23. Projection welding.

heavier gauges. Only the spherical-type projection has a cavity running along either side of the projection. The cavity receives the displaced metal during the welding process.

Poor projection welds are indicated by separated surfaces, partially sheared projections, and the expulsion of metal. Satisfactory projection welds, on the other hand, are indicated by the absence of such metal expulsion at the weld. Both surfaces must also be in direct contact with no space between them. The projection must be large enough to bring the contacted metal surface to the welding temperature. Otherwise, an unsatisfactory weld will occur.

Projection welding machines are of the press-type construction, and are also used for spot welding.

A schedule for the projection welding of low carbon steels is illustrated in Table 5. Table 6 illustrates a punch and die design for forming welding projections and lists the pertinent data.

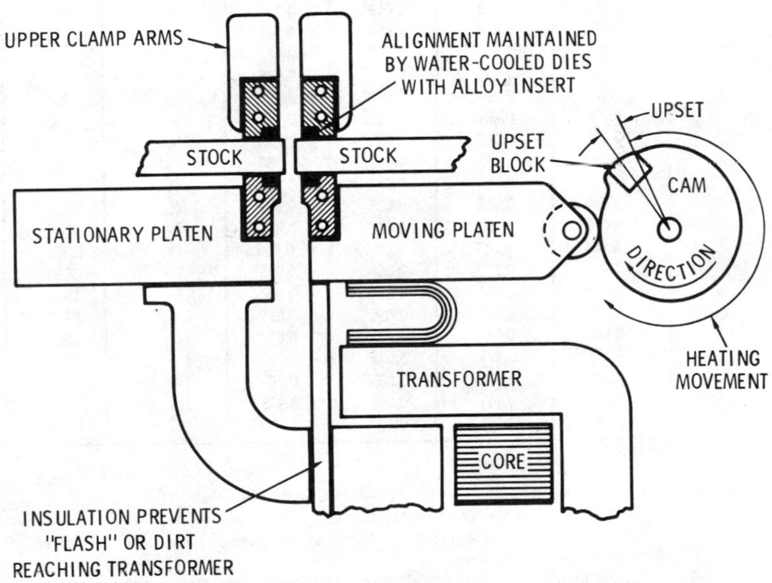

Fig. 24. The flash welding process.

FLASH WELDING

Flash welding (Fig. 24) is a resistance welding process in which the heat is applied *before* the two parts are pressed together. When the pressure is applied, molten metal is squeezed out of the weld. The period during which this activity takes place is known as the *flashing time*. The expelled molten metal is referred to as *flash*. Once it has cooled, the flash solidifies around the weld

Table 5. Design and Welding Data for

Thickness of Thinnest Outside Piece Inches	PROJECTION DESIGN		ELECTRODE DIAMETERS (d = 2 x Projection Diameter)		Electrode Force Pounds
	Base Diameter of Projection Dp Inches	Height of Projection H Inches	Minimum d Inches	Minimum D Inches	
0.010	0.055	0.015	0.125	½	50
0.012	0.055	0.015	0.125	½	80
0.014	0.055	0.015	0.125	½	100
0.016	0.067	0.017	0.187	½	115
0.021	0.067	0.017	0.187	½	150
0.025	0.081	0.020	0.187	½	200
0.031	0.094	0.022	0.187	½	300
0.034	0.094	0.022	0.187	½	350
0.044	0.119	0.028	0.250	⅝	480
0.050	0.119	0.028	0.250	⅝	580
0.062	0.156	0.035	0.312	⅞	750
0.070	0.156	0.035	0.312	⅞	900
0.078	0.187	0.041	0.375	⅞	1050
0.094	0.218	0.048	0.500	⅞	1300
0.109	0.250	0.054	0.500	⅞	1650
0.125	0.281	0.060	0.500	⅞	1900
0.140	0.312	0.066	0.625	1	2300
0.156	0.343	0.072	0.625	1	2800
0.171	0.375	0.078	0.750	1	3300
0.187	0.406	0.085	0.750	1	3900
0.203	0.437	0.091	0.875	1¼	4500
0.250	0.581	0.110	1.000	1¼	6600

NOTES:

1. Type of Steel—Low Carbon SAE 1010—0.15% Carbon Maximum.
2. Material free of scale, oxide, paint, dirt, etc.
3. Size of projection determined by thickness of thinnest piece and projection should be on thickest piece.
4. Data is based on thickness of thinnest sheet for two thicknesses only. Maximum ratio between two thicknesses=3 to 1.

(flash weld). Fusion occurs simultaneously over the entire contact area, making warpage almost non-existent. A slight upset (Fig. 25) remains after the flash welding process has been completed. If appearance is important, the upset bulge can be mechanically removed.

This is an automatic welding process in which resistance welding variables are machine controlled. Operator error is therefore

Projection Welding Low Carbon Steels

Weld Time (Cycles) 60 Cycles per Sec.	Hold Time (Cycles) Minimum	Welding Current Amperes (Approx.)	Diameter of Fused Zone Dw Inches	Minimum Shear Strength (Single Projection Only) (For Steels Having Strength of 100,000 psi and below) Pounds	Minimum Contacting Overlap L Inches (S = 2Dp MIN)
3	3	2800	0.112	150	1/8
3	3	3100	0.112	200	1/8
3	3	3400	0.112	250	1/8
4	4	3600	0.112	285	5/32
6	6	4000	0.140	380	5/32
6	8	4500	0.140	525	3/16
8	8	5100	0.169	740	7/32
10	10	5400	0.169	900	7/32
13	14	6500	0.169	1080	9/32
16	16	7100	0.225	1500	9/32
21	20	8400	0.225	2100	3/8
24	24	9200	0.281	2550	3/8
26	30	10,500	0.281	2950	7/16
32	30	11,800	0.281	3700	1/2
38	36	13,300	0.338	4500	5/8
45	40	15,000	0.338	5200	11/16
60	45	15,700	0.437	6000	3/4
80	50	17,250	0.500	7500	13/16
105	50	18,600	0.562	8500	7/8
125	50	20,000	0.562	10,000	15/16
145	55	21,500	0.625	12,000	1
230	60	26,000	0.687	15,000	1 1/4

5. Contacting overlap does not include any radii from forming.
6. Projection should be located in center of overlap.
7. Tolerance for Projection Dimensions:

Dimension	Thickness Up to 0.050″	Thickness Over 0.050″
Diameter "D"	±0.003″	±0.007″
Height "H"	±0.002″	±0.005″

Courtesy The American Welding Society

Table 6. Punch and Die Design for Forming Welding Projections

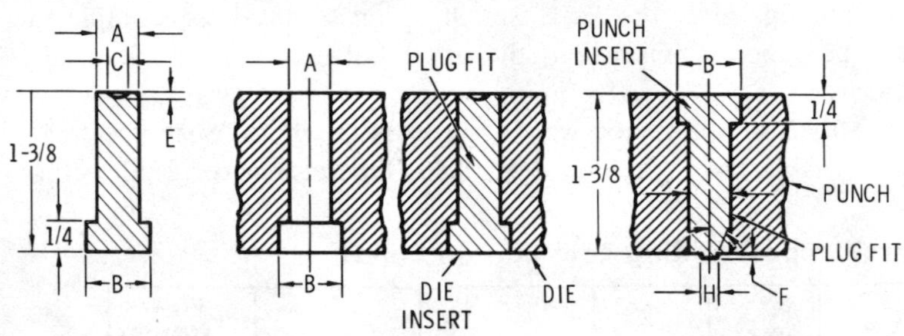

Mat Thickness	Pt. No.	A	B	±.002 C	Dr	±.001 E	±.001 F	±.001 H	Jr
0.010-0.015	1	3/8	9/16	.055	.033	.015	.015	.035	.005
0.016-0.021	2	3/8	9/16	.067	.042	.017	.020	.039	.005
.025	3	3/8	9/16	.081	.050	.020	.025	.044	.005
.031	4	3/8	9/16	.094	.062	.022	.030	.050	.005
.034	5	3/8	9/16	.094	.062	.022	.030	.050	.005
.044	6	3/8	9/16	.119	.078	.028	.035	.062	.005
.050	7	3/8	9/16	.119	.078	.028	.035	.062	.005
.062	8	3/8	9/16	.156	.105	.035	.043	.081	.005
.070	9	3/8	9/16	.156	.105	.035	.043	.081	.005
.078	10	3/8	9/16	.187	.128	.041	.055	.104	.010
.094	11	1/2	11/16	.218	.148	.048	.065	.115	.010
.109	12	1/2	11/16	.250	.172	.054	.075	.137	1/64
.125	13	1/2	11/16	.281	.193	.060	.085	.154	1/64
.140	14	1/2	11/16	.312	.217	.066	.096	.172	1/64
.156	15	5/8	13/16	.343	.243	.072	.107	.191	1/64
.171	16	5/8	13/16	.375	.265	.078	.118	.210	1/64
.187	17	5/8	13/16	.406	.285	.085	.130	.229	1/64
.203	18	11/16	7/8	.437	.308	.091	.143	.240	.020
.250	19	13/16	1	.531	.375	.110	.175	.285	.025

Material: Tool Steel.
Finish all over and harden to 65-68 Rockwell "C" scale.
Note: All working surfaces of die unit must be polished.

Courtesy The American Welding Society

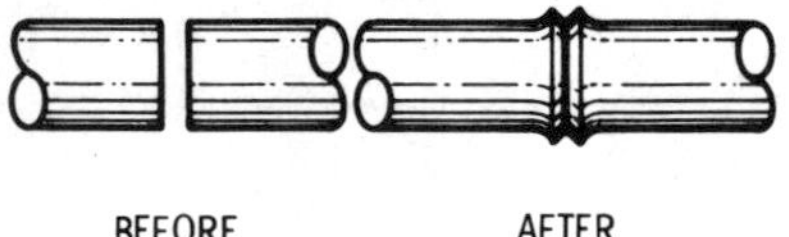

Fig. 25. The upset bulge after the completion of the flash welding process.

reduced to a minimum. A welding machine designed specifically for flash welding is illustrated in Fig. 35. The electrodes function as clamps which hold the two pieces to be welded. These electrode-

Courtesy Omega Machinery Corp.

Fig. 26. Electric wire butt welding machine equipped with a re-upsetting and burr removing device.

clamps are moved toward one another until slight contact is made, or until only a very short distance remains between the edges of the two pieces being welded. An arc is then established between the two surfaces.

This welding process is applied primarily in the butt welding of metal sheets, tubing, bars, and rods, and is referred to as *flash butt welding* or *resistance flash butt welding*.

Figs. 26, 27 and 28 illustrates welding machines constructed for the butt welding of wire, wire frames, wire rings, and similarly

Courtesy Omega Machinery Corp.

Fig. 27. Electric wire butt welding machine with pneumatic-hydraulic operation.

designed items. The welding current is adjusted by a rotary selector switch to different wire diameters. The entire welding cycle is automatically controlled. The welding current switches on automatically immediately after the clamping jaws have closed over the wire. Once the welding temperature has been reached and the correct amount of upset pressure has been established, the welding current automatically shuts off. The latter action occurs either by means of an electronic switch or an upset switch. The clamping jaws open after a pre-set cooling period is completed.

Courtesy Omega Machinery Corp.

Fig. 28. Electric wire butt welding machine with foot operated clamping control.

UPSET WELDING

Upset welding is a resistance welding process in which the heat is applied after the two metal pieces are brought together under continuously applied pressure. In this way, it is distinguished from flash welding in which the heat application precedes the pressure. Pressure is maintained throughout the entire period during which the heat is applied and afterwards while the metals are cooling. The pressure exerted at the weld is known as the *upsetting force.* This process has been largely replaced by the flash welding process. Upset welding machines do not differ essentially from those designed for flash welding.

PERCUSSION WELDING

Percussion welding (or *percussive welding*) is a resistance welding process in which the parts to be welded are prepositioned and clamped in place, heated to the plastic stage by the explosive release of stored up electrical energy, and suddenly brought together with percussive force. The sudden percussive force is produced by a pneumatic cylinder or spring action. The arc is extinguished as soon as contact is made between the two surfaces. The electric current is stored in a stored-energy resistance welding machine usually of the electrostatic (capacitor-discharge) or electromagnetic type.

SPIKE WELDING

Spike welding is another of those welding terms not officially recognized by the *American Welding Society*. It is used to refer to a type of resistance welding in which large amounts of electricity are stored up and then released rapidly through the two metals to be joined. Spike welding has the advantage of being suitable for almost any kind of metal or metal alloy, and can be applied through different thicknesses of metal. Spike welding is also very suitable for welding dissimilar metals. Spike welding machines are of the stored-energy type (see STORED-ENERGY WELDING MACHINES).

CHAPTER 10

Thermit Welding

Thermit welding is a term used to designate a group of welding processes that achieve coalescence by pouring a superheated liquid metal (thermit) around the parts to be welded. *Thermit* is a trade name for a mixture of finely divided metal oxide and a metal reducing agent.

The thermit welding process was discovered by Goldschmidt in 1895 or 1896. He believed that a cold mixture of metallic oxide and finely divided aluminum (the reducing agent) could be ignited in one spot and that the reaction of this spot would furnish sufficient heat to propagate reaction throughout the entire mass. These mixtures were ignited in refractory-lined crucibles and produced masses of molten metal and aluminum oxide slag, which separated of their own accord because of the wide differences in weight. At the same time, it was noted that the reaction would, in most cases, proceed at a rate making possible the production of any quantity up to a ton or more of material in a single reaction lasting only about thirty seconds.

PRINCIPLES OF THERMIT WELDING

The heat necessary for coalescence in the thermit welding processes is obtained from a chemical reaction that takes place between a metal oxide (usually iron oxide) and a metal reducing agent (almost always aluminum). The high chemical affinity of aluminum for oxygen is the basis for the thermit process. The igniting temperature of thermit is approximately 2200° F (1204° C). When ignited in one spot, the reaction spreads throughout the mass—the aluminum uniting with the oxygen of the iron oxide setting free the iron which comes down into the mold as highly superheated liquid steel. Due to the chilling effect of the crucible, liquid steel has a slightly lower temperature when it is poured into the mold. Despite this, superheated liquid steel is still approximately twice as hot as ordinary molten steel.

The thermit reaction is a nonexplosive one requiring about thirty seconds for completion. Before the reaction can begin, a portion of the thermit must be raised to its ignition temperature of 2200° F (1204° C). A special low-ignition-temperature thermit powder is commonly used for this purpose. Other low-ignition powders, particularly those used in the fireworks industry, are also used.

Other materials may be added to the aluminum and iron oxide mixture to form thermit. In designing such mixtures many variables controlling both the time and the temperature of the reaction as well as the required chemical analysis of the resultant weld metals are taken into account. For example, through the addition of metallic elements, either by means of metallic pieces which are melted during the reaction, or in the form of combinations of oxides of elements with aluminum, a wide variation in the analysis of thermit steel is provided.

As was mentioned previously, the most common thermit reaction is one using aluminum in combination with iron oxide.

Table 1. Typical Thermit Reactions

$3Fe_30_4 + 8A1$	$\rightarrow$	$9Fe + 4A1_20_3$	(5590°F/3088°C)	719.3 Kcal
$3Fe0 + 2A1$	$\rightarrow$	$3Fe + A1_20_3$	(4532°F/2500°C)	187.1 Kcal
$Fe_20_3 + 2A1$	$\rightarrow$	$2Fe + A1_20_3$	(5360°F/2960°C)	181.5 Kcal
$3Cu0 + 2A1$	$\rightarrow$	$3Cu + A1_20_3$	(8790°F/4865°C)	275.3 Kcal
$3Cu_20 + 2A1$	$\rightarrow$	$6Cu + A1_20_3$	(5680°F/3138°C)	260.3 Kcal
$3Ni0 + 2A1$	$\rightarrow$	$3Ni + A1_20_3$	(5740°F/3171°C)	206.6 Kcal
$Cr_20_3 + 2A1$	$\rightarrow$	$2Cr + A1_20_3$	(5390°F/2977°C)	546.5 Kcal
$3Mn0 + 2A1$	$\rightarrow$	$3Mn + A1_20_3$	(4400°F/2427°C)	403 Kcal
$3Mn0_2 + 4A1$	$\rightarrow$	$3Mn + 2A1_20_3$	(9020°F/2771°C)	1041 Kcal

Courtesy Thermex Metallurgical, Inc.

Aluminum is almost always used as the reducing agent, although magnesium has been employed on occasion as a substitute. However, magnesium is extremely limited in application due to the nonfluid nature of its slag. The use of different metal oxides (other than iron oxide) in the thermit reaction is a more common practice. Table 1 illustrates a number of different thermit reactions, the temperatures obtained, and the type of metal oxide used.

EQUIPMENT, MATERIALS, AND SUPPLIES

The thermit welding process requires an adequate supply of thermit mixture, a thermit ignition powder, and a device (hot iron, flint gun, etc.) for igniting the ignition powder. Many thermit molds can be purchased ready-made, or they can be designed and constructed to job specification.

If preheating is necessary, there are a number of fuel burning (propane, gasoline, kerosene, etc.) heating torches commercially available.

THERMIT MIXTURES

The most commonly used types of thermit for welding the various ferrous metals are:

1. Plain thermit,
2. Forging thermit,
3. Cast iron thermit.

Other thermit mixtures are available for special welding purposes such as welding rails, welding electrical connections, and building up worn wobbler ends of rolls and pinions in steel mills.

Plain thermit is a mixture of finely divided aluminum and iron oxide which is the basis for most thermit mixtures. This mixture yields one of the highest temperatures for thermit welding.

Forging thermit (or *mild-steel thermit*) is plain thermit with the additions of manganese and mild-steel punchings, and is used in welding steel. Manganese is added to adjust the chemistry of the thermit mixture (carbon may also be added); the mild-steel punchings are used to augment the metal content.

Cast iron thermit consists of plain thermit with additions of ferro-silicon and mild-steel punchings and is used for welding cast iron. The mild-steel punchings augment the total metal content. Unfortunately, the machinability of the weld metal is very poor, a factor that limits the general use of this thermit mixture. A post-heat treatment of the weld area is required to improve its machinability.

Thermit mixtures for welding rails and building up worn steel-mill wobblers consist of plain thermit with additions of manganese and carbon. The amount of manganese and carbon added to the mixture is adjusted in accordance with: (1) the desired hardness of the deposited metal, and (2) the need for controlling resistance to abrasion and functioning as grain refiners.

Thermit mixtures for welding electrical connections consist of copper oxide and aluminum. The principal advantage of this method of joining electrical connections over soldering and brazing them is the relatively low input of the thermit welding process.

THERMIT IGNITION POWDER

The ignition temperature of a thermit mixture is in excess of 2200° F (1204° C). Therefore, a low temperature ignition powder is necessary to bring the thermit mixture to the ignition point.

There are a number of different ignition powders available for this purpose. A barium peroxide is frequently used as an oxidizing agent and aluminum dust as the reducing agent in these low-ignition-point thermits. Essentially, then, a thermit ignition powder is a thermit mixture that ignites and burns at lower temperatures.

The thermit ignition powder is ignited directly with a match. This should be done by partly burying the match head in the mixture and igniting the match with a thin, red hot rod. This avoids the danger of burn injuries to the fingers or hand due to the sudden flare in igniting. The thermit ignition powder can also be ignited with a spark from a flint gun, or by using a burning magnesium ribbon.

MOLDS FOR THERMIT WELDING

The type of mold used for thermit welding depends on both the size and the purpose of the weld. Another factor to be taken into consideration when selecting the type of mold to be used is whether or not the surfaces being welded will require preheating.

All thermit molds consist of two principal sections. The upper section, or crucible, is the portion of the thermit welding apparatus in which the thermit reaction takes place. It is here that the superheated liquid steel is formed. The crucible may be constructed as a single unit with the lower section (Fig. 8) or it may exist as an independent unit suspended over the lower section or mold (Fig. 1). The lower section, or mold proper, receives the liquid steel from the crucible and forms the weld.

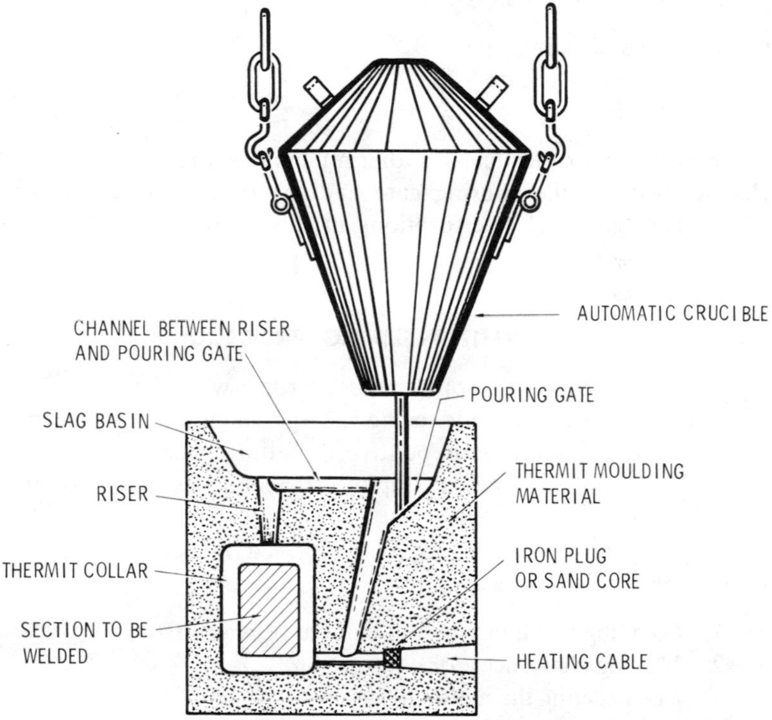

Courtesy Thermex Metallurgical Inc.

Fig. 1. Thermit mold and an independent crucible.

The construction and use of molds designed for repair or large single-purpose welds will be described in the following paragraphs. Other types of thermit molds are described in the sections concerned with welding reinforcing bars and rails.

Molds used for repair or large single-purpose welds generally require preheating as a part of the welding procedure. The principal parts for such molds are:

1. A mold box,
2. A crucible,
3. Molding material,
4. A wax pattern.

In construcing the mold proper, wood patterns are made for the principal ducts required. These ducts are known as:

1. The preheating gate,
2. The pouring gate,
3. The riser.

In addition to the ducts, a thermit mold should have: (1) a channel between the pouring gate and the riser, (2) vent holes, (3) a slag basin, (4) perforations in the mold box, and (5) a preheating gate plug.

THERMIT WELDING PROCEDURE

The following paragraphs are concerned with the procedure involved in constructing and using a large single-purpose thermit mold. These molds are nonrepetitive in nature, generally used for repair welds, and commonly employ preheating as a part of the welding process.

The various operations to be performed in constructing and using such a mold are as follows:

1. Cleaning and lining up the parts to be welded,
2. Making allowances for contraction,
3. Constructing the mold,
4. Ramming the mold,
5. Preheating the mold,

6. Preparing the crucible,
7. Plugging the preheating gate,
8. Applying the ignition powder,
9. Removing the mold.

CLEANING AND LINING UP THE PARTS TO BE WELDED

Regardless of the type of thermit weld being used, the parts to be welded must be properly cleaned and lined up before the mold is applied.

The metal surfaces must be thoroughly cleaned in order to ensure a strong weld. The adjacent ends are cleaned by a sand blast or other suitable means. Clean the ends thoroughly for a space of at least 5 inches in order to expose good bright metal on each side. Also, remove all grease or dirt over a space reaching to the limits of the mold box. This precaution is taken to prevent any foreign matter from burning out and leaving a space between the mold and the parts to be welded.

The parts to be welded are lined up so that there is a space between the ends. The extent of the space depends on the size of the parts to be joined.

MAKING ALLOWANCES FOR CONTRACTION

Proper allowance should be made for contraction during the alignment and placement operations by setting the parts away from each other a sufficient distance to make up for the contraction of the thermit steel in cooling. This should be varied from 3⁄16 inch to 5⁄16 inch depending on the size of the weld and the length of time of the preheating.

In many cases, it is necessary to obtain this increased space by forcing the sections apart with a jack or other mechanical means. In other cases, it will be necessary to heat an opposite member to obtain the desired results.

It is important that the expansion previously obtained by jacking or by heating a parallel member should be maintained until the weld has solidified sufficiently to enable it to resist the tendency of the frame to upset it. On the other hand, it is probably even more dangerous to resist shrinkage of the metal in the weld during solidification, so the operator should use careful judgment in this matter.

Where the expansion has been obtained by preheating a parallel member, the preheater can be removed when or shortly after the weld is poured. If properly timed in this way, the two parallel members will contract together satisfactorily. The operator should be certain that the contraction of the weld is not resisted by the jack; otherwise, at these high temperatures, the weld material has little strength and may be damaged by this resistance.

CONSTRUCTING THE MOLD

This type of mold requires the forming of a wax pattern around the area to be welded. The wax pattern provides a space around the parts to be welded for the flow of the superheated liquid steel. The wax burns out completely during the preheating leaving a space surrounding the ends to be welded.

The wax pattern is made by melting yellow wax in a pan and allowing it to cool until it becomes plastic. Then it is shaped around the parts to be welded. The space between the ends is also filled with wax. A hot rod is used to provide a vent duct through the wax extending from the preheating gate to the riser.

The weight of the wax pattern is determined by subtracting the weight of the wax and leftover from the weight of the wax taken in making the pattern. The mold itself consist of a mold box and the molding material, which is packed around the inside of the box.

The selection of a proper molding material is very important. It should exhibit both the necessary refractory qualities and the very necessary porosity after preheating. An example of a molding material which meets these requirements is one composed of 3 parts clean sharp silica sand (100% of which should pass through a screen having a .03-inch square opening), mixed with 1 part Welsh Mountain plastic clay. These parts are first thoroughly mixed in the molder together with 1/40 part glutrin by volume and sufficient water (1/2 part) to bring the mixture to the proper consistency.

If mixed by hand, the sand and clay must be dried before mixing (being careful not to subject the clay to a temperature higher than 400° F) and thoroughly mixed with water before it is added to the sand and clay.

The mold should be provided with the necessary number of pouring gates, heating gates, and risers (depending on the size of the weld). The risers communicate with the highest points of the wax pattern, and the heating gates and the pouring gates with the lower point.

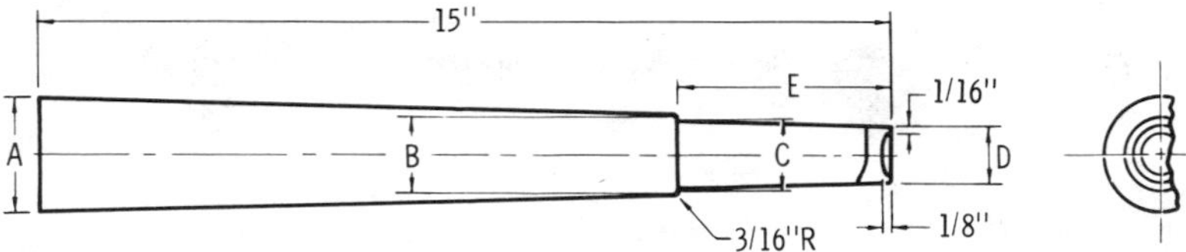

HEATING AT GATE PATTERN

SIZE	A IN.	B IN.	C IN.	D IN.	E IN.
NO. 1	2-1/2	1-1/2	1-1/8	1	3
NO. 2	2-3/4	1-3/4	1-3/8	1-1/4	4
NO. 3	3	2	1-3/8	1-1/2	5

Courtesy Thermex Metallurgical Inc.

Fig. 2. Heating gate pattern.

The preheating gate pattern (Fig. 2) is placed so that it projects through a hole in the box provided for it and set against the lowest point of the wax pattern.

The pouring gate pattern (Fig. 3) is placed so that the lower end is set against the preheating pattern about midway between the wax and the preheating pattern shoulder. This should be a neat joint to prevent any fins of sand which might break off and mix with the thermit steel.

The riser pattern (Fig. 4) is placed so that it is set against the highest point of the wax pattern. If there are two or more high points, a riser is provided for each.

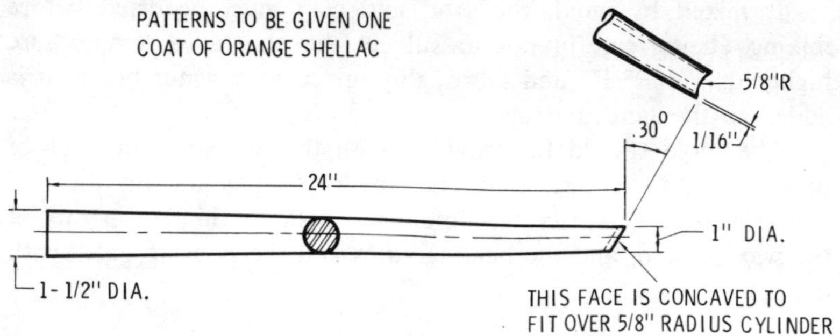

Courtesy Thermex Metallurgical Inc.

Fig. 3. Pouring gate pattern.

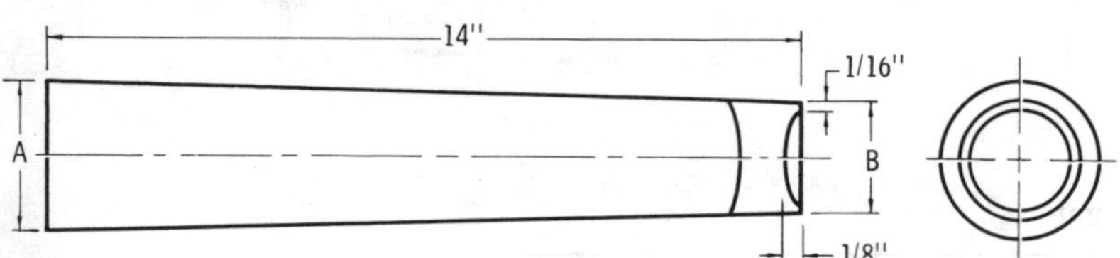

RISER PATTERN

SIZE	A IN.	B IN.
NO. 1	1-1/4	1
NO. 2	1-3/4	1-1/2
NO. 3	2-1/4	2

Courtesy Thermex Metallurgical Inc.

Fig. 4. Riser pattern.

The riser is designed and constructed to hold a supply of thermit steel that will provide for shrinkage in cooling and to form a reservoir for the collection of any foreign matter. It also functions as a means of equalizing pressure in pouring.

The slag basin (Fig. 5) prevents the slag from overflowing the mold box. It is constructed by hollowing out the top of the mold after the ramming operation has been completed. The channel (Fig. 5) is cut in the top of the mold, extending from the pouring

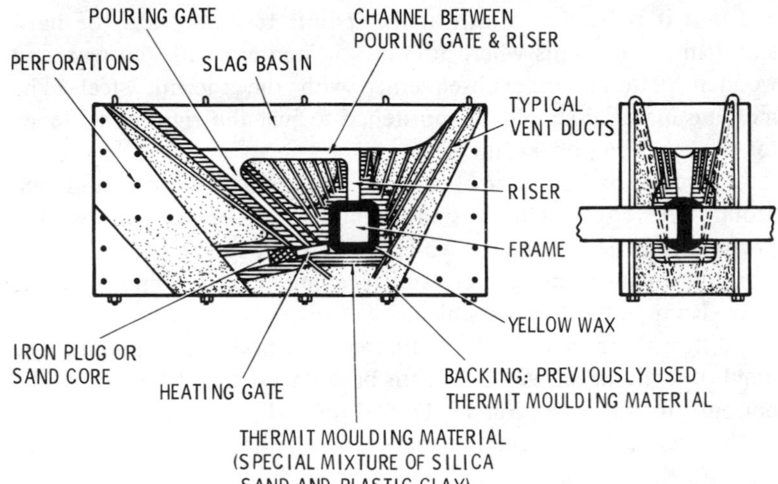

Fig. 5. Method employed in constructing molds for making thermit welds.

gate to the riser. The channel allows the first slag overflowing on top of the mold to flow across quickly to the riser and equalize the pressure of the pouring gate.

The vent ducts (Fig. 5) are made with a vent rod of 8 to 10 gauge steel wire. The rod is forced down through the molding material to within 3 or 4 inches of the bottom. Care should be taken that the end of the rod does not touch the wax pattern, because in such cases the thermit steel will flow into the vent ducts rendering them inoperative. An ample number of vent ducts should be provided.

The vent ducts provide a means of escape of any air that may have been trapped in front of the incoming thermit steel and for the escape of any gases formed by the high temperature of the molten metal.

RAMMING THE MOLDING MATERIAL

It is important to thoroughly ram hard the molding material in the construction of the mold, because the safety of the entire welding operation depends upon it. The operator must be certain that the material is well rammed underneath the pattern and that it is hard and firm at this point. It should also be hard and firm at all points where it comes in contact with the wax and wooden patterns and subsequently with the thermit steel. The molding material should be moistened to just the right consistency for the ramming operation.

Place a small amount of molding material in the box and ram around the edges with a small rammer, or better still, with a pneumatic bench rammer (which will save a great deal of time) working toward the center and keeping the molding material level. Ram as hard as possible.

Different operators adopt different methods for ramming the mold, but the usual way is to ram backing up to a little above the bottom of the wax collar. Trowel out the 1-inch space next to the wax collar where the facing should be and ram this hard with the facing sand.

In the same way, the backing and facing are rammed hard to a point above the heating gate. A channel is troweled out for the heating gate and the heating gate pattern pounded or tapped down with the rammer until it has sunk into the molding material to its proper position.

More facing sand is then rammed on top of this heating gate, completely burying it; and on either side of this more backing is rammed even against the wax collar itself. When a depth of 2 inches or 3 inches above the heating gate has been reached, the approximately 1 inch next to the wax collar is again troweled out and facing put in and rammed hard.

In a similar way, a round hole is dug down to expose the top of the heating gate pattern an inch or more away from the wax collar, the pouring gate pattern carefully placed, and the space around the pouring gate then filled with facing to the level to which the mold has already been rammed. This method of ramming in, backing and cutting away those portions where it must be replaced with facing, is then followed until the mold is completely filled, the riser pattern placed practically in the same manner as was the pouring gate pattern and the top of the mold hollowed out to form a basin.

The molding material should be about 4 inches thick between the wax pattern and the mold box at all points. This thickness is considered sufficient to hold the intensely hot molten thermit steel.

In those cases where a wooden riser pattern cannot be withdrawn, use a wax riser pattern that will melt out during preheating with the rest of the wax pattern.

After the ramming operation has been completed, the operator should lightly rap the gate, riser, and pre-heat opening patterns and draw them out carefully. Wipe away any loose sand that might tend to fall into the holes. A molder's slick, trowel, and lifter are used for this operation.

After removing the pouring and preheating gate patterns, carefully smooth off any sharp edges or fins at the top of the pouring gate and at the junction of the pouring and heating gates. This latter point can best be reached by means of a ¼-inch rod slightly curved at its end. After preheating, the sharp edges and fins at the junction of the heating gate and wax collar and at the base of the riser can be similarly smoothed off. The mold is now ready for preheating.

PREHEATING THE MOLD

The principal reasons for preheating the mold are:

1. To melt the wax so that it will drain out,
2. To thoroughly dry the mold,
3. To bring the parts to be welded to a proper temperature.

In the preheating process, the flame of a compressed air liquid fuel (gasoline, kerosene, etc.) heater is directed into the heating gate melting the wax pattern and leaving a space between and around the ends to be welded. This space is later occupied by the molten thermit steel. The heat is continued until the parts to be united have been brought to a good, red, workable heat.

A single or double burner preheater is used to heat the mold. These are designed to burn a mixture of air and liquid fuel in such a manner that the combination takes place in the interior of the mold where it is required to heat the parts to be welded.

The burner of the preheater should be set so as to point into the heating gate of the mold about 1 inch from the opening and the blast applied.

On completion of preheating, remove the burner and direct it down the riser so as to blow out any sand or dirt which may be in the mold. If the riser is difficult to reach, the burner may be directed down the pouring gate. The operator should now examine, as carefully as possible, the interior of the mold to locate any possible loose sand particles and to be certain that these are blown out before the heating gate is plugged.

PREPARING THE CRUCIBLE

The thermit reaction takes place in a conical shaped vessel (crucible) having a magnesia tar lining. (Fig. 6). This lining is

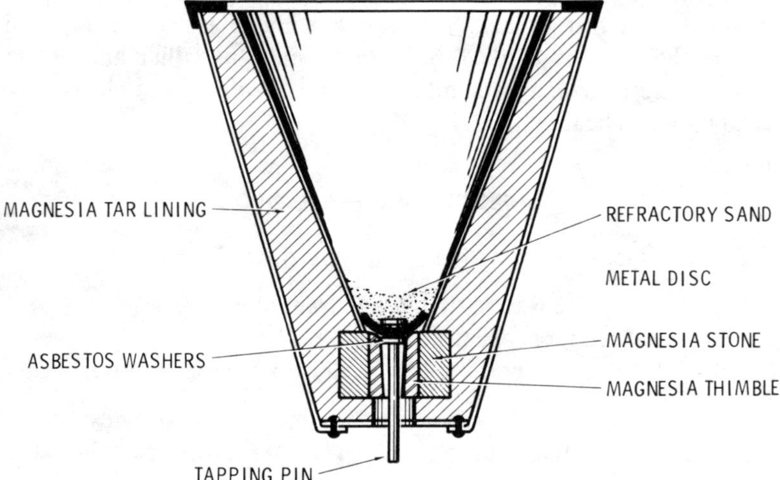

Fig. 6. Thermit automatic crucible showing the arrangement of plugging material.

thickest at the bottom or zone (of the internal heat) of the molten thermit steel. At the bottom of the crucible is a hard burnt magnesia stone. This latter has a tubular opening into which a small magnesia stone or thimble of conical form is made to fit. This thimble provides a channel through which the liquid thermit steel is poured. To replace, it can be easily removed by gently knocking upward and a new thimble inserted. It is safer to use a new thimble for each reaction.

The steel formed by the thermit reaction settles to the bottom of the crucible, separating itself as it does so from the slag. For this reason, the opening to the crucible is placed on the bottom. That way, the steel is placed between the parts to be welded and the slag.

The crucible is closed at the bottom by first inserting the thimble with one thickness of uncreased paper. This is plugged with plugging material as follows:

1. Suspend the tapping pin through the thimble,
2. Place the asbestos washers over the tapping pin,
3. Place the metal disc over the washers,
4. Cover the aforementioned with the entire contents of a package of refractory sand.

The magnesia tar lining is supported by a sheet iron shell, the bottom of which forms the opening through which the thermit steel flows. The lining should be heated until it becomes plastic. A few handfuls should then be in the bottom of the crucible shell and a magnesia stone imbedded in this material and centered over the hole. More magnesia tar should then be rammed around the cone to hold it firmly in place. The cast-iron crucible cone should then be placed in position with the small projecting teat at the lower end set in the hole of the magnesia stone (the teat on the cone has the same dimensions as the thimble previously described). The upper part should then be centered inside the shell by means of wedges inserted at equal distances along the circumference. The magnesia tar may then be rammed into the space between the cone and the shell, a little at a time and tamped hard. The life of the crucible depends upon the hardness or density of the lining. Special iron tools should be made up for this tamping operation and should have flat ends. Good hard blows should be struck with a hammer on the upper end of the tool when ramming in the lining, or better still, use a pneumatic bench rammer for this purpose.

When completely filled and tamped, a mark should be made with a piece of chalk on the cone, and the point opposite to it on the lining, so that when the cone is withdrawn it may be replaced exactly as before. Remove the cone, exercising care not to disturb the lining, place a layer of wrapping paper or newspaper over the tar lining, then replace the cone carefully so that the marks previously made appear opposite to each other. After this, put on the crucible ring and lute carefully around the top with fire clay to protect the upper part of the lining from the heat in baking. It is also a good practice to place damp fire clay around the bottom of the crucible and inside the stone for the same purpose.

The crucible is now ready for baking and for this purpose it should be placed in a suitable oven. The heat should gradually be raised until the cast-iron cone becomes red hot and should be held at that temperature until all fumes are removed from the tar, after which it can be allowed to cool gradually before removing from the oven.

Before placing the crucible in position, the various openings on the top of the mold should be covered so that nothing will fall into them.

Lift the crucible into position securing it firmly with its bottom directly over and about 3 inches above the center of the pouring gate, so that the metal, when tapped, will fall into the hole itself and not on the sloping sand surrounding the mold. If the crucible cannot be placed directly over the pouring gate it will be necessary to construct a runner, to lead the thermit steel into the pouring gate.

The bottom of the crucible is closed before charging by first inserting the thimble wrapped with one thickness of uncreased paper. This is plugged with plugging material as follows: the tapping pin is suspended through the thimble and over this the asbestos washers, and then the metal disc is pressed lightly down on the asbestos washers. All this is then covered with about ½ inch of refractory sand.

While the preheating is in progress, and after having made certain that the crucible has been thoroughly dried out and possibly is even still warm from this drying operation, the charge of thermit and additions should be placed in the crucible.

The amount of thermit required for a given mold is determined by the weight of the wax pattern. The usual allowance is twenty pounds of thermit mixture for each pound of wax in the pattern.

Before charging, each bag of thermit should be thoroughly mixed by dumping it into a pan and turning the pile over a number of times before putting it into the crucible. The thermit cannot be properly mixed after it is in the crucible. It is important to put in a few handfuls of thermit first, before dumping in the rest of the charge, so as not to disturb the plugging material. The ignition powder should be added only when the thermit charge is ready to be ignited.

PLUGGING THE PREHEATING GATE

The preheating gate may be plugged with a sand core or iron plug. Care must be taken to see that the iron plug is clean and not rusted, tin-galvanized, etc.

Many operators make sand cores by wrapping a piece of heavy paper around the proper part of the heating gate pattern, sliding the paper off carefully and then ramming the molding material into the form so made and baking slowly and carefully over the mold during preheating.

The "proper" part of the heating gate pattern is the enlarged section adjoining the shoulder. The shoulder is designed to prevent the plug from being forced too far into the heating gate. Otherwise, the heating gate would interfere with the bottom of the pouring gate.

When working with a short plug, care should be taken that it fits neatly in the heating gate. Such a short plug could then be backed up with molding material, lightly tamped into the heating gate, and this in turn backed up with more molding material rammed in between the mold box and the steel plate provided for this purpose. This will prevent any possibility of the thermit steel running out through the preheating opening.

APPLYING THE IGNITION POWDER

Mix the contents of the ignition powder and then place one-half teaspoonful of ignition powder on top of the thermit in the crucible (thermit will not ignite from the heat of the preheater and the reaction cannot be started without the use of the ignition powder). It is advisable to use the heated-rod method of igniting the ignition powder.

It is important that ample time be allowed for the completion of the reaction. The crucible should be tapped when it is heard that the reaction is over, which usually takes from 25 to 35 seconds. The operator can very easily hear when the reaction has quieted down.

All heating apparatus should be removed to a safe distance, while the thermit reaction is in progress. The reason for this is that the moisture in the crucible lining, or in the thermit itself, will cause some of the slag to splash out of the crucible during the reaction. Consequently, it is well to ensure that the riser and all openings other than the pouring gate are covered with small pieces of sheet iron to prevent this slag from dropping into the mold and adhering to the sections being welded.

REMOVING THE MOLD

The mold should be allowed to remain in place as long as possible, preferably overnight, so as to anneal the steel in the weld, but (in case of necessity) it can be removed after two or three hours in light frame welds, or in about four hours in heavier frame welds.

After removing the mold, cut off the metal in the riser and pouring gate with an oxyacetylene torch, or drill through the sections and knock them off.

THERMIT WELDING OF REINFORCING BARS

Many modern building codes require the use of continuous reinforcing bars in concrete construction. These reinforcing bars may be either overlapped and wired together or they may be welded at the end to form one continuous length. The end welding of these reinforcing bars is done with the thermit welding process. There is no overlapping of the bars as is the case when they are tied together (Fig. 7).

The thermit welding of reinforcing bars is economical and simple. No special skills are required by the worker. The reinforc-

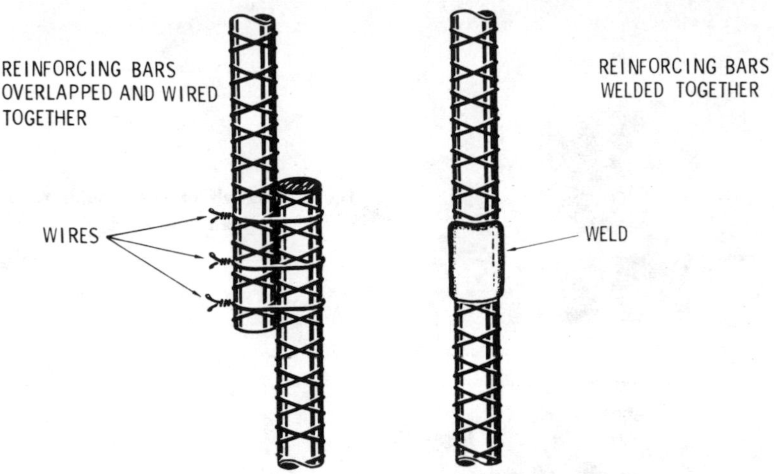

Fig. 7. Two different methods of joining reinforcing bars.

ing bars may be welded together in horizontal or vertical positions, or they may be welded to plates. Disposable molds can be purchased for each position and for specified bar diameters.

The procedure for welding reinforcing bars with the thermit welding process is as follows:

1. Clean the ends of the bars to be joined. They must be free of dirt, rust, and other contaminants in order to produce a strong weld.
2. The ends must provide a flat square surface. This can be accomplished by cutting with a torch. They may also be sawed or sheared. Trimming is usually necessary and should be done with a cutting torch.
3. A final cleaning is generally required to remove any slag or loose scale deposited during torch cutting.
4. Align the two bars so that the mold halves will fit freely. Misalignment will result in reducing the tensile strength of the weld. There should be approximately a ⅜ inch gap between the bar ends. The pouring channel of the mold will not be centered over the gap. It is not necessary for the ribs to line up when the bars are aligned (Figs. 8, 9 and 10).

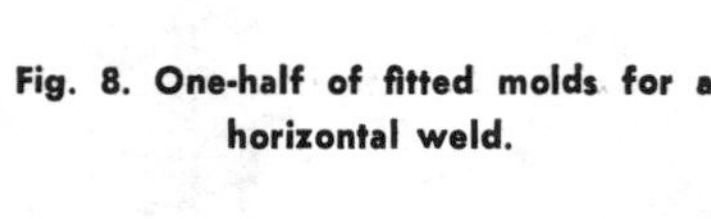

Fig. 8. One-half of fitted molds for a horizontal weld.

Courtesy Thermex Metallurgical Inc.

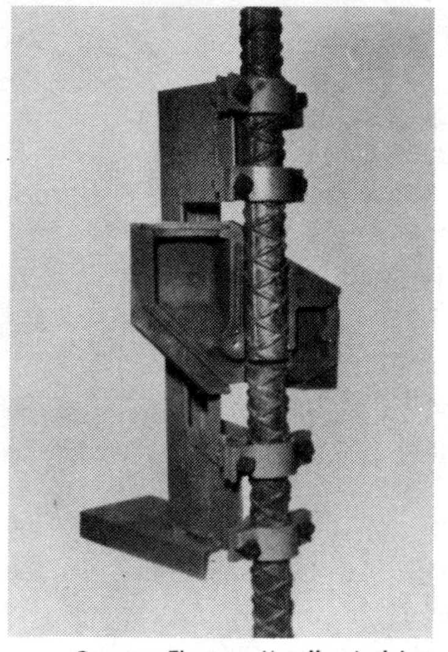

Fig. 9. One-half of fitted molds for a vertical weld.

Courtesy Thermex Metallurgical Inc.

5. The bars only need to be preheated in temperatures below 40° F. It is also advisable to preheat in order to remove moisture or grease.
6. Fasten the two halves of the mold together around the reinforcing bars. Make certain that the pouring channel is directly over the gap between the two bars. Seal the mold with sand to prevent molten metal escaping from between the bars and the mold (Fig. 11).

Fig. 10. One-half of fitted molds for a bar-to-plate weld.

Courtesy Thermex Metallurgical Inc.

Fig. 11. A closed and sealed horizontal weld.

Courtesy Thermex Metallurgical Inc.

7. Place the tapping discs in the mold. Add the thermit and cover with the igniting compound.
8. Ignite the mixture with a flint lighter or other appropriate means. Upon igniting, an exothermic reaction occurs lasting for about 20 seconds. The reaction produces a quantity of superheated molten steel. The molten steel then flows through the tapping discs into the bottom half of

the mold where it preheats and welds the reinforcing bars together.

9. Cooling takes approximately ten minutes, although longer periods improve the possibility that solidification is complete.
10. Remove the two halves of the mold (Figs. 12 and 13).

Courtesy Thermex Metallurgical Inc.

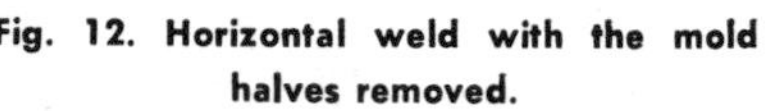

Fig. 12. Horizontal weld with the mold halves removed.

Courtesy Thermex Metallurgical Inc.

Fig. 13. Vertical weld with the mold halves removed.

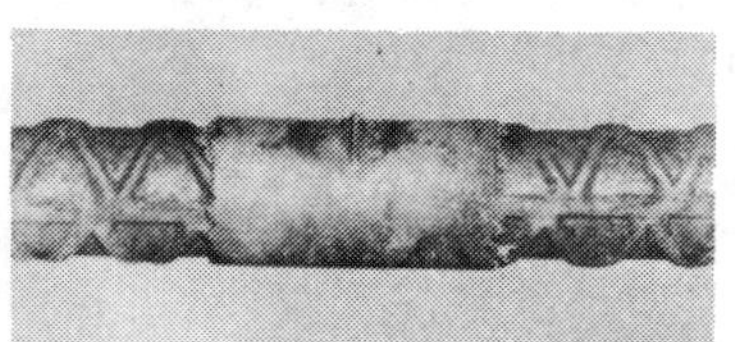

Courtesy Thermex Metallurgical Inc.

Fig. 14. Complete horizontal weld.

11. Break off the risers and gates. A hammer can be used to remove them (Figs. 14 and 15).

The *Thermit Rebar Weld* manufactured by *Thermex Metallurgical, Inc.* provides a complete thermit welding kit consisting

Fig. 15. Complete vertical weld.

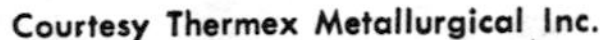

Courtesy Thermex Metallurgical Inc.

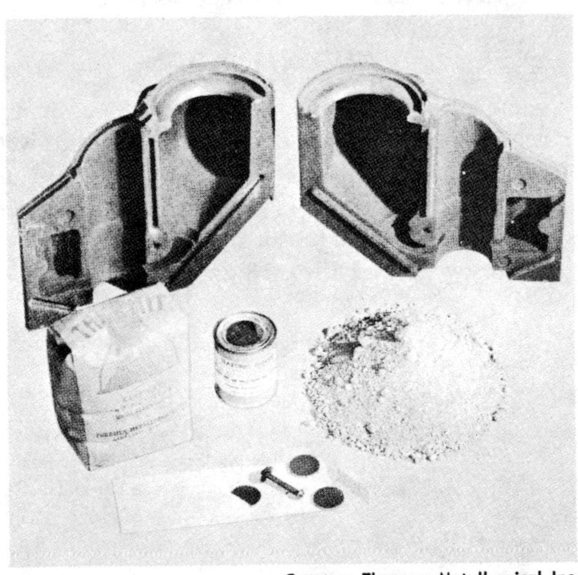

Courtesy Thermex Metallurgical Inc.

Fig. 16. Thermit *Rebar* weld kit.

of a disposable mold, tapping discs, molding sand, thermit welding compound, and ignition compound (Fig. 16). The size of the mold is determined by the size of the reinforcing bar (Fig. 17).

A. HORIZONTAL WELD:

BAR SIZE	A	B	C
9	1-7/8	6-1/2	1-1/2
10	1-7/8	6-3/8	1-1/2
11	1-7/8	6-1/4	1-5/8
14S	2-3/4	6-1/8	1-3/4
18S	3	8-3/8	2-3/16

B. VERTICAL WELD:

BAR SIZE	A	B
14S	2	3-1/8
18S	2	3
14S/18S	2	3-1/8

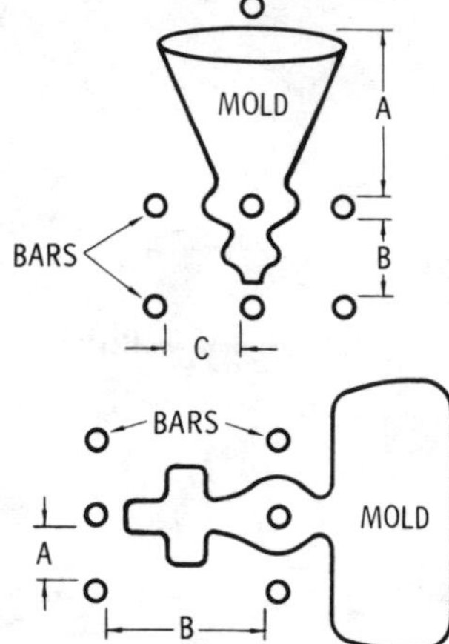

Courtesy Thermex Metallurgical Inc.

Fig. 17. Bar clearance required for *Rebar* welds.

WELD INSPECTION

Either gamma-ray or X-ray inspection methods may be used to determine the quality of reinforcing bar thermit welds. The X-ray sound welds are generally the most commonly used with this type of welding. Etched cross sections of thermit welds illustrate the fusion of filler metal with opposing surfaces to form a strong satisfactory weld. See Figs. 18 and 19.

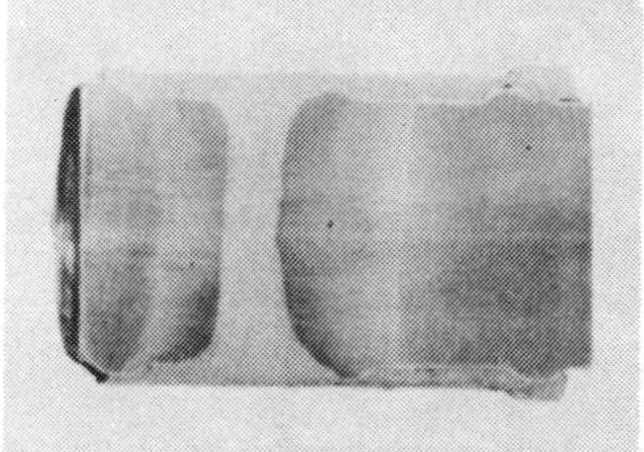

Courtesy Thermex Metallurgical Inc.

Fig. 18. Etched cross section of a joint showing how the filler metal fuses with the bar ends.

Fig. 19. Etched cross section of a joint showing how the filler metal fuses with the bar and plate.

Courtesy Thermex Metallurgical Inc.

Unsatisfactory welds may be cut apart and rewelded following the procedure outlined for welding reinforcing bars with the thermit welding process.

WELDING RAILS

The standard lengths of rail used by railroads, rapid transit lines, and industrial tracks (e.g. crane runways, marine railways, etc.) are welded into longer continuous stretches in order to reduce the number of joints and thereby lower maintenance costs. Thermit welding is the method commonly used to join these rails (Fig. 20).

A thermit rail weld is made by pouring superheated steel obtained by the thermit reaction into a mold surrounding the rail ends at the joint. This thermit steel melts all parts of the rail section with which it comes into contact and, on cooling, solidifies with them in a homogenous weld.

Thermex Metallurgical, Inc. manufactures disposable thermit welding molds for joining rails. These can also be used for welding crane runway rails. The molds are moisture-free, nonhydroscopic, and can be designed to fit compromise joints.

These factory produced molds result in a welding process that differs in several important ways from other methods. These special characteristics are as follows:

1. The preheating of the rail ends is automatic and accomplished by a portion of the superheated liquid steel produced

(A) Thermit mold being used.

(B) Rails welded.

Courtesy Thermex Metallurgical Inc.

Fig. 20. Transit rails being welded.

in the thermit reaction. This eliminates the slow manual and sometimes uneven preheating done with a torch in other thermit welding methods.

2. All welds are uniform.
3. Tapping of the molten metal into the weld section of the mold is automatic. The possibility of late or premature tapping is eliminated.
4. A shorter preheating period which reduces damage to adjacent rail steel.

CHAPTER 11

Stud Welding

Stud welding is an arc welding process in which coalescence occurs as a result of an electric arc drawn between a stud (or similar metal part) and the work. The stud functions as the electrode. Once the proper fusion temperature has been reached, the stud and work surface are brought together under pressure to form a weld. A diagram of a typical stud-welding circuit is shown in Fig. 1.

Stud welding is a low cost method of fastening extensions (studs) to a metal surface. It is generally cheaper than such conventional methods as drilling and tapping, bolting, spot welding, and the fusion of metals. Table 1 illustrates the typical weld stud types and their various applications.

The stud-welding process uses the following equipment and supplies:

1. A power source,
2. A timing control unit,
3. A portable or mounted stud-welding gun,
4. Welding cables,
5. A ground cable with ground clamp,
6. A generator cable,
7. Studs,
8. Ferrules.

The various types of cables used in stud welding are discussed in Chapter 5 (ARC WELDING EQUIPMENT AND SUPPLIES). About 50

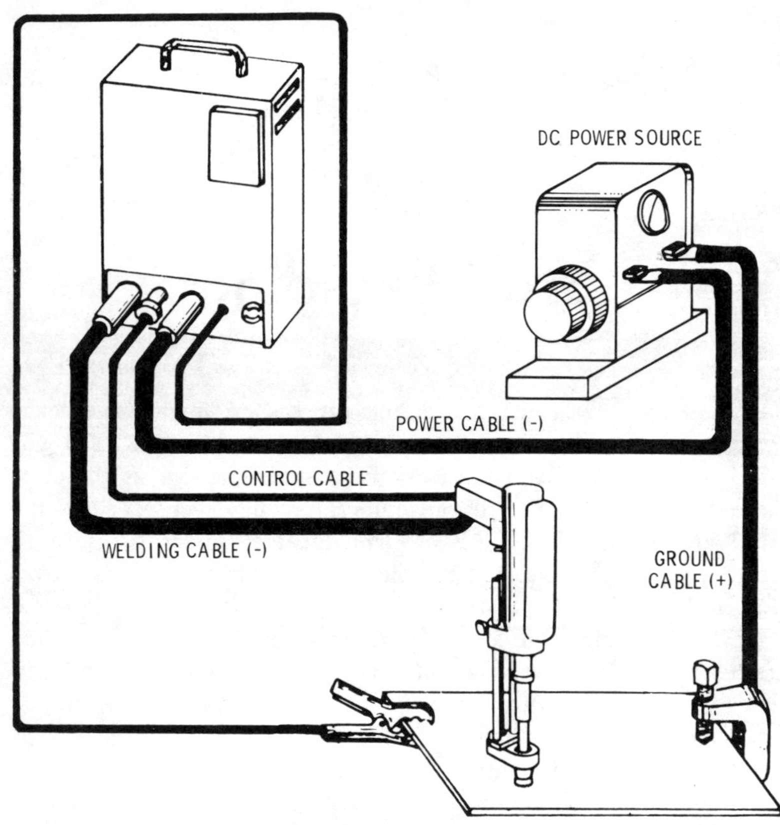

Courtesy Tru-Weld, Div. of Tru-Fit Screw Products Corp.

Fig. 1. Typical stud welding circuit.

feet of 4/0 welding cable is needed for the circuit illustrated in Fig. 1. In addition, approximately 25 feet of 4/0 ground cable, 25 feet of 4/0 generator cable, and 50 feet of wire for the control cable are required.

The power source used in stud welding should be a DC generator capable of delivering a current rated at 300 amps or more. Note the positioning of the DC power source with respect to the control unit and welding gun in the stud-welding circuit shown in Fig. 1.

Table 1. Typical Weld Stud Types and Applications Table

	SHIPBUILDING							RAILROAD									
Typical Weld Stud Types and Applications / APPLICATIONS / WELD STUD TYPES	wood decking	insulation	electrical wireways & control panels	furniture	sheet metal coverings	piping	gage lines	woodliners, furring	wood flooring	steel liners, auto cars	cab linings	wire conduit, cables	coverplates, diesel engines	braking resistors	side sheets, (Diesel)	insulation, tank & box cars	hatch covers, studs, refrigerator cars
Threaded (Pitch Dia. Weld Base)	●	●	●	●	●	●	●	●	●	●	●	●	●	●	●	●	●
Fully Threaded	●	●	●	●	●	●	●	●	●	●	●	●	●	●	●	●	●
Threaded (Full Weld Base)	●	●	●	●	●	●	●	●	●	●	●	●	●	●	●	●	●
Threaded (Reduced Weld Base)	●	●	●	●	●	●	●	●	●	●	●	●	●	●	●	●	●
Threaded (Full Reduced Weld Base)	●		●	●	●	●	●	●	●	●	●	●	●	●	●	●	●
No Thread		●															
Shoulder			●		●							●					
Collar			●		●							●					
Tapped			●		●	●	●					●					
Bent																	
Concrete Anchor																	
Shear Connector																	
Double Pointed Insulation Pin	●									●	●					●	
Set Lock Stud					●												

CONTROL UNIT

The control unit is a device designed to control the flow time of the current. Its components should be capable of handling the large momentary surges of current during the weld cycle. The control unit illustrated in Fig. 2 has a sequence controller and a timer. The welding time is marked in cycles and can be set within a range of 7 to 40 cycles. It can operate on DC current up to 1200 amps at approximately 50 cycles.

Fig. 3 illustrates still another type of stud-welding circuit in which the control mechanism and power source are contained in a single portable unit. This control and power unit can be plugged

Table 1. Typical Weld Stud Types and Application Table (Contd.)

Typical Weld Stud Types and Applications / WELD STUD TYPES	CONSTRUCTION (Bridge & Multi-story Bldg.): shear connectors composite construction	concrete anchors composite construction	stair forms	truck docks	door framing	supports	elevator shafts	column or cover guards	highway curbing	expansion joints	GENERAL INDUSTRIAL EQUIPMENT: metal furniture	kitchen equipment	appliances	household furniture	food processing & beverage equipment	textile machinery	furnaces	tanks & pressure vessels	electrical equipment	conveyors	lift trucks	farm equipment
Threaded (Pitch Dia. Weld Base)	●		●	●	●	●	●	●			●	●	●	●	●	●	●	●	●	●	●	●
Fully Threaded											●	●	●	●	●	●	●	●	●	●	●	●
Threaded (Full Weld Base)	●		●	●	●	●	●	●			●	●	●	●	●	●	●	●	●	●	●	●
Threaded (Reduced Weld Base)											●	●	●	●	●	●	●	●	●	●	●	●
Threaded (Full Reduced Weld Base)											●	●	●	●	●	●	●	●	●	●	●	●
No Thread			●	●	●	●	●	●														
Shoulder															●	●	●		●	●	●	●
Collar															●	●	●		●	●	●	●
Tapped											●	●	●	●	●	●	●		●	●	●	●
Bent	●	●	●	●	●	●	●	●	●	●												
Concrete Anchor	●	●	●	●	●	●	●	●	●	●												
Shear Connector	●	●	●	●	●	●	●	●	●	●												
Double Pointed Insulation Pin																						
Set Lock Stud																						

into any 115-volt 60-hertz 10-amp AC outlet. It will develop a voltage output of 0 to 175 volts DC with 65,000 μF capacitance available.

STUD-WELDING GUN

Portable stud-welding guns (Fig. 4) are available in light, medium, and heavy duty types. The type of gun selected by the welder depends on the nature of the job. Mounted stud-welding guns are found in semiautomatic and automatic welding equipment. Both the portable and mounted stud-welding guns use

Table 1. Typical Weld Stud Types and Application Table (Contd.)

	AUTOMOTIVE				TRAILER				BOILERS				OTHER		
Typical Weld Stud Types and Applications / WELD STUD TYPES	passenger cars bumper & bumper guards	shock absorbers	sway bar attachments	trim	rub bars	flooring	rope hooks	flanges for tank trucks	cover plates	clean out or access doors	studded boiler tubing	steel heater coil openings	jet engine forging blanks	armored vehicles	tanks
Threaded (Pitch Dia. Weld Base)	●	●	●	●	●	●		●	●	●		●		●	●
Fully Threaded	●	●	●	●	●	●		●	●	●		●		●	●
Threaded (Full Weld Base)	●	●	●		●	●		●	●	●		●		●	●
Threaded (Reduced Weld Base)	●	●	●		●	●		●	●	●		●		●	●
Threaded (Full Reduced Weld Base)	●	●	●		●	●		●	●	●		●		●	●
No Thread				●							●		●	●	●
Shoulder														●	●
Collar														●	●
Tapped														●	●
Bent							●								
Concrete Anchor															
Shear Connector															
Double Pointed Insulation Pin															
Set Lock Stud															

Courtesy Tru-Weld

DCSP current for all metals except aluminum. With aluminum, a DCRP current is used with an argon or helium shielding gas.

WELDING STUDS

Welding studs are commercially available in many different sizes and shapes. In addition to the standard designs, most manufacturers will produce to individual job specifications. Some examples of stud types showing their before and after weld shapes are illustrated in Tables 2 and 3.

Table 2. Examples of Nonthreaded Weld Studs and Their Specifications

B	L	HD	HT	FD	FH
1/2	2	1	5/16	11/16	5/32
5/8	2 1/2	1 1/4	1/2	7/8	3/16
3/4	3	1 1/4	1/2	1 1/16	1/4
3/4	4	1 1/4	1/2	1 1/16	1/4
3/4	5	1 1/4	1/2	1 1/16	1/4
3/4	6	1 1/4	1/2	1 1/16	1/4
3/4	7	1 1/4	1/2	1 1/16	1/4
3/4	8	1 1/4	1/2	1 1/16	1/4
7/8	3 1/2	1 3/8	1/2	1 1/8	5/16
7/8	4	1 3/8	1/2	1 1/8	5/16
7/8	5	1 3/8	1/2	1 1/8	5/16
7/8	6	1 3/8	1/2	1 1/8	5/16
7/8	7	1 3/8	1/2	1 1/8	5/16
7/8	8	1 3/8	1/2	1 1/8	5/16

The length of a stud is reduced approximately 1/8 to 3/16 inch during the welding operation. Consequently, studs are supplied correspondingly longer than the length specified in the manufacturer's catalog. The catalog length is the *after weld* length of the stud.

In addition to the many different sizes and shapes, studs are also commercially available in a number of different metals including stainless steel (304, 305, and other grades in the 300 series) and some nonferrous metals. The most common type is one composed of a low-carbon steel with a tensile strength of 60,000 psi.

When required, studs can be supplied annealed by the manufacturer to the following specifications: (1) low-carbon steel at

Table 2. Examples of Nonthreaded Weld Studs and Their Specifications (Contd.)

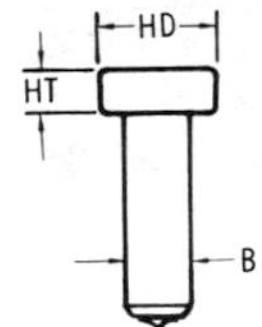

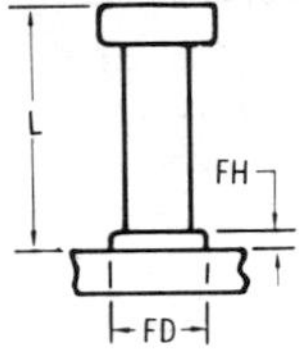

B	L	HD Head Diameter	HT Head Thickness	FD	FH
1/4	2 5/8	1/2	3/16	23/64	7/64
1/4	4	1/2	3/16	23/64	7/64
3/8	4	3/4	9/32	1/2	1/8
3/8	6	1	9/32	1/2	1/8
1/2	4	1	5/16	11/16	5/32
1/2	5 3/16	1	5/16	11/16	5/32
1/2	6	1	5/16	11/16	5/32
1/2	8	1	5/16	11/16	5/32
5/8	6 3/8	1 1/4	5/16	7/8	3/16
3/4	7 5/16	1 1/2	3/8	1 1/16	1/4
7/8	8 7/16	1 3/4	7/16	1 1/8	5/16

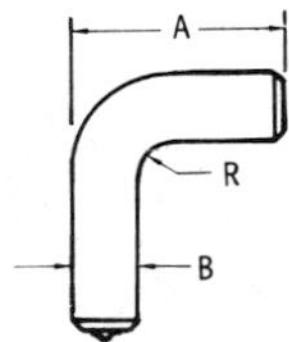

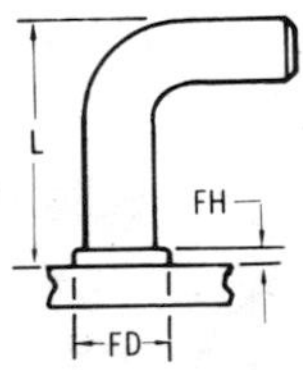

B Diameter	L Min. Length After Weld	A Minimum Length	R Radius	FD Weld Fillet Diameter	FH Weld Fillet Height
1/4	1	1 5/16	1/8	23/64	7/64
5/16	1 1/8	1 1/2	7/32	7/16	7/64
3/8	1 3/8	1 17/32	7/32	1/2	1/8
7/16	1 1/2	1 5/8	1/4	19/32	9/64
1/2	1 5/8	1 11/16	1/4	11/16	5/32
5/8	1 3/4	2	5/16	7/8	3/16
3/4	2 5/8	2 13/16	1/2	1 1/16	1/4

Table 2. Examples of Nonthreaded Weld Studs and Their Specifications (Contd.)

B Diameter	L Min. Length After Weld	A Overall Length
$\frac{7}{16}$	$1\frac{21}{32}$	$2\frac{1}{4}$
$\frac{1}{2}$	$1\frac{21}{32}$	$2\frac{1}{4}$

Courtesy Tru-Weld

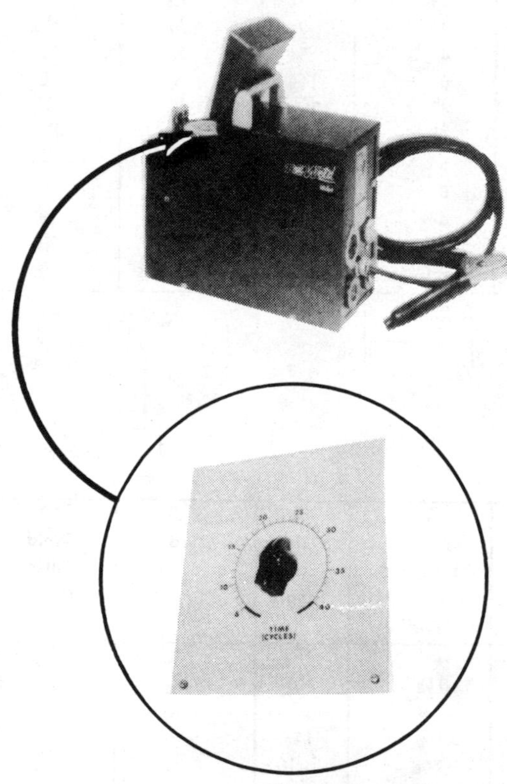

Courtesy Tru-Weld, Div. of Tru-Fit Screw Products Corp.

Fig. 2. Control unit and weld timer.

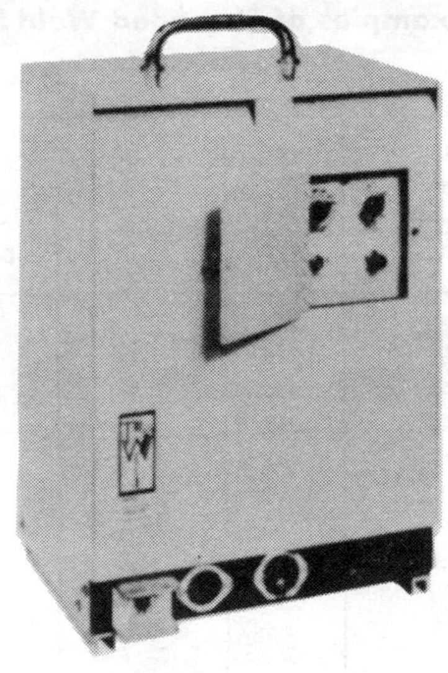

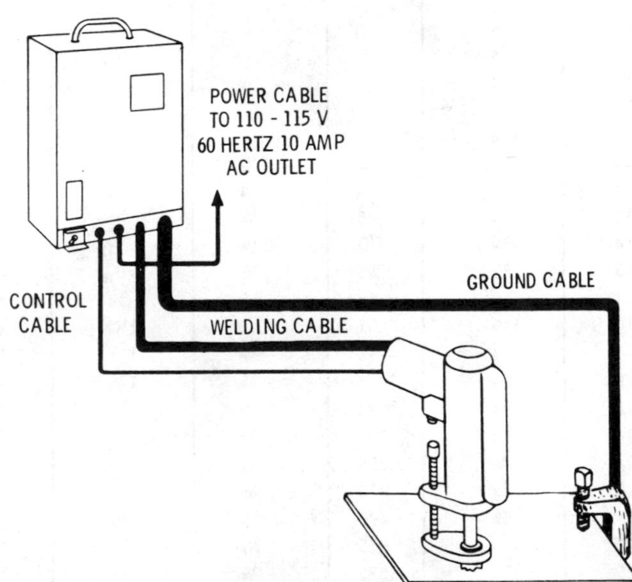

Courtesy Tru-Weld, Div. of Tru-Fit Screw Products Corp.

Fig. 3. A stud welding unit which houses the control unit and power source in one container.

Table 3. Examples of Threaded Weld Studs and Their Specifications

D Diameter and Thread	L Min. Length After Weld	B	U	FD Weld Fillet Diameter	FH Weld Fillet Height
1/4—20	5/8	.215	3/8	5/16	3/32
5/16—18	5/8	.275	3/8	13/32	7/64
3/8—16	5/8	.330	13/32	7/16	7/64
7/16—14	3/4	.387	7/16	1/2	1/8
1/2—13	3/4	.448	1/2	19/32	5/32
5/8—11	7/8	.562	5/8	3/4	3/16
3/4—10	1 1/8	.680	13/16	7/8	1/4
7/8— 9	1 3/8	.798	7/8	1	5/16
1— 8	1 7/8	.915	15/16	1 1/8	5/16
10—24	3/4			9/32	3/32
1/4—20	3/4			23/64	7/64
5/16—18	3/4			7/16	7/64
3/8—16	3/4			1/2	1/8
7/16—14	7/8			19/32	9/64
1/2—13	7/8			11/16	5/32
5/8—11	1 1/8			7/8	3/16
3/4—10	1 1/8			1 1/16	1/4
1/4—20	3/4	1/4	3/16	23/64	7/64
5/16—18	3/4	5/16	1/4	7/16	7/64
3/8—16	3/4	3/8	1/4	1/2	1/8
7/16—14	7/8	7/16	1/4	19/32	9/64
1/2—13	7/8	1/2	1/4	11/16	5/32
5/8—11	1 1/8	5/8	1/4	7/8	3/16
3/4—10	1 1/8	3/4	1/4	1 1/16	1/4
7/8— 9	1 3/8	7/8	3/8	1 1/8	5/16
1— 8	1 5/8	1	3/8	1 3/8	3/8
5/16—18	5/8	.225	3/8	5/16	7/64
3/8—16	5/8	.280	3/8	3/8	7/64
7/16—14	3/4	.325	3/8	7/16	1/8
1/2—13	3/4	.370	3/8	1/2	1/8
5/8—11	7/8	.490	7/16	5/8	5/32
3/4—10	1	.615	9/16	51/64	3/16
7/8— 9	1 1/4	.675	3/4	7/8	1/4
1— 8	1 1/2	.790	13/16	1	5/16

Courtesy Tru-Weld

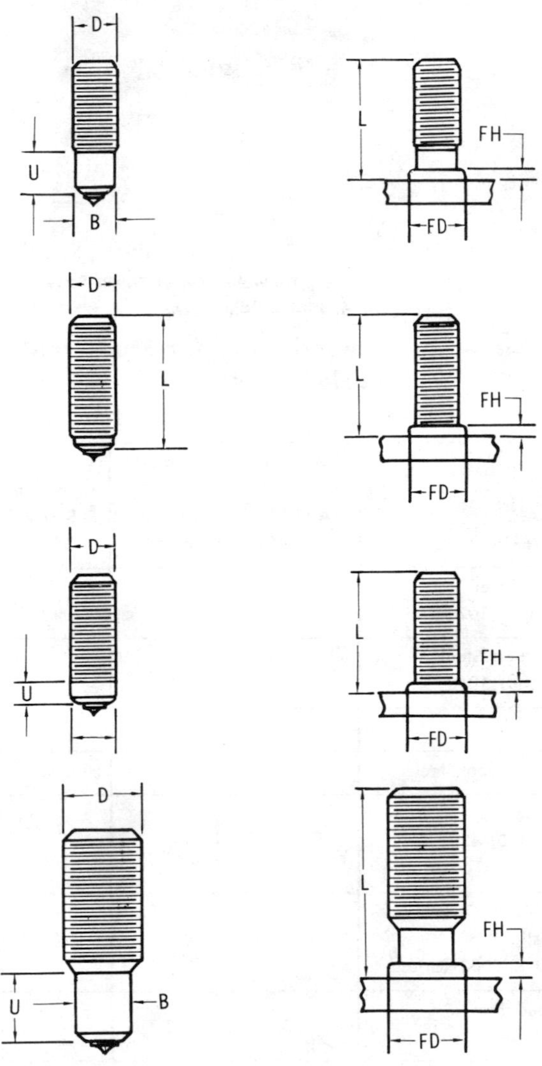

75 Rockwell B. Max., and (2) stainless steel at 90 Rockwell B. Max. Some recommendations for stud welding various materials is listed in Table 4.

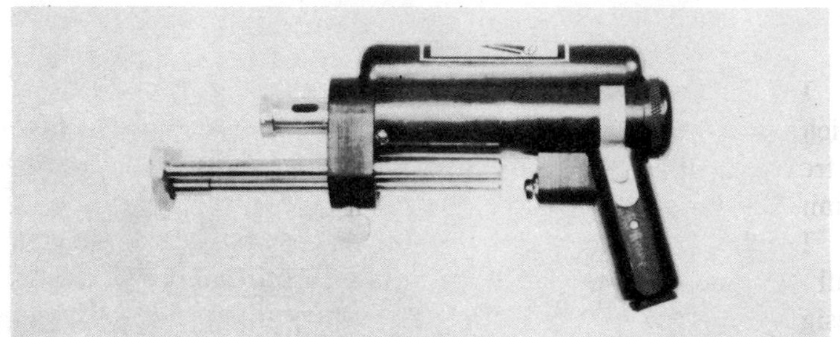

Courtesy Tru-Weld, Div. of Tru-Fit Screw Products Corp.

Fig. 4. Stud welding gun.

Table 4. Recommendations for Stud-Welding Various Materials

	WELD STUD MATERIAL			
BASE MATERIAL	Low Carbon Steel AISI 1008 thru 1020	Stainless Steel Type 304 and 305	Aluminum Series 1100	Brass 65-35, 70-30
Low Carbon Steel AISI 1008 thru 1020	1	1		1
Medium Carbon Steel AISI 1020 thru 1050	2	2		
Structural Steel	1	1		2
Galvanized Carbon Steel	1	1		
Stainless Steel Types 405, 410, 430, 302, 304, 305, 316	1	1		1
Aluminum—1000 Series and most alloys of 3000, 5000 & 6000 Series			3	
Copper—Lead Free Brass—Lead Free	2	2		1

1—Will develop full weld strength under normal welding conditions.

2—Weld strength limited by stud size and base material characteristics.

3—For sizes larger than 3/16″ diameter, use of inert gas atmosphere at weld is recommended for maximum weld strength.

Courtesy Tru-Weld

STUD-WELDING FERRULES

The stud-welding ferrule is a device that functions as an arc shield. As such, it concentrates the heat of the arc and protects it from contamination. The ferrule also serves as a dam or containing wall for the molten metal.

These ferrules used in stud welding are specially designed of ball clay and talc and are manufactured in a number of different designs (Fig. 5). Their composition enables them to withstand

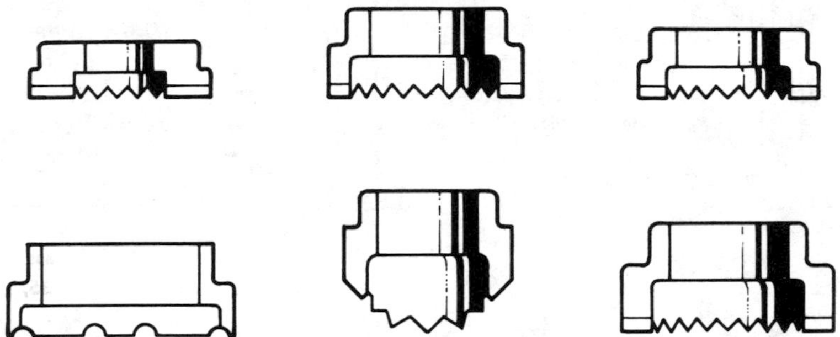

Courtesy Tru-Weld, Div. of Tru-Fit Screw Products Corp.

Fig. 5. Various types of stud welding ferrules.

heat shock as well as reasonable high welding temperatures without melting or breaking. They are fitted over the stud prior to the beginning of the stud welding operation. They may be removed afterwards by cracking them open with a hammer.

WELDING FLUX

A welding flux is used in stud welding to produce a more stable arc and to prevent oxides from forming on the surface during welding. Most studs are produced with the flux packed into a cavity at the end of the stud. However, studs with a weld base under 5/16 inch in diameter are supplied non-fluxed.

STUD-WELDING CYCLE

The complete stud-welding cycle (Fig. 6) takes place in a mere fraction of a second. It may be manually, semiautomatically,

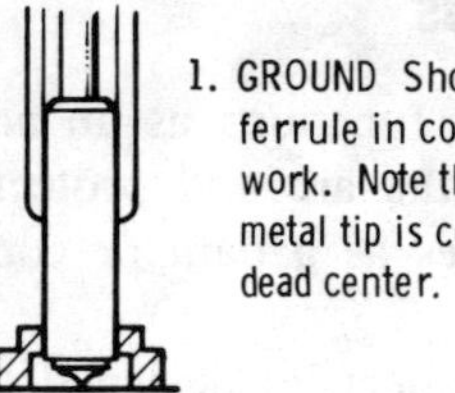

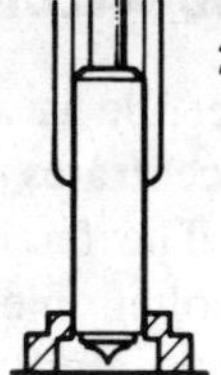

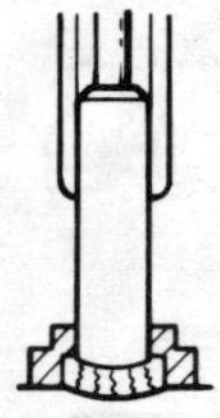

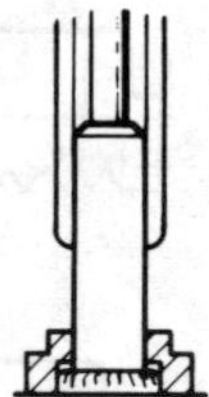

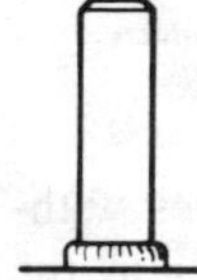

Courtesy of Tru-Weld Div. of Tru-Fit Screw Products Corp.

Fig. 6. The five stages of the stud welding cycle.

or automatically operated. In the manual operation, the operator positions the stud-welding gun against the work. The manual stud-welding procedure is as follows:

1. Place the stud in the welding gun. It should be firmly held by the chuck.
2. Place the ferrule over the stud.
3. Position the stud-welding gun against the surface to which the stud is to be attached.
4. Depress the gun trigger.

The stud will be forced back into the gun when the trigger is depressed. The retraction of the stud will create an arc between

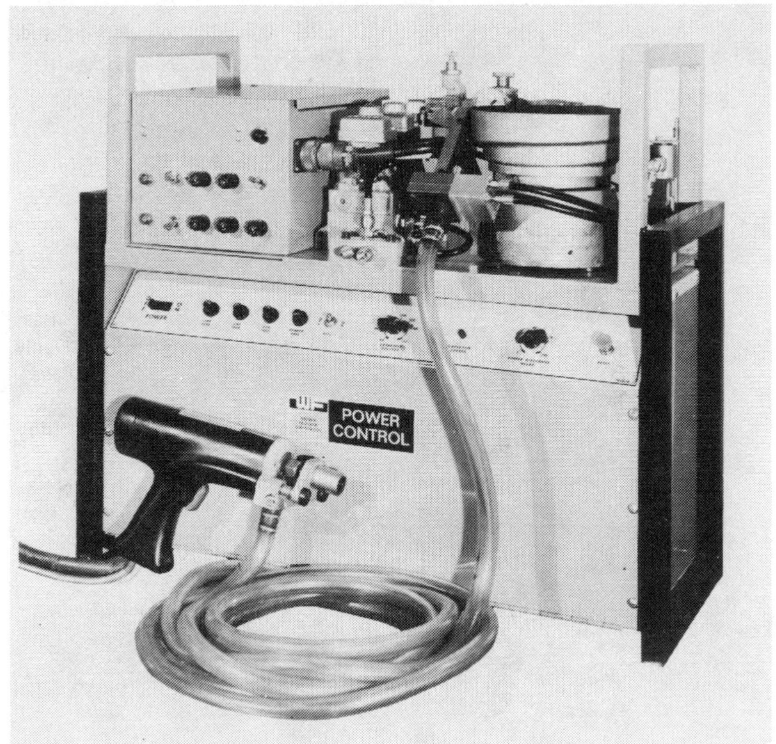

Courtesy Warren Fastener Corp.

Fig. 7. Automatic production stud welder.

the stud and the work surface. The arc will automatically shut off after a predetermined and pre-set period of time has been completed. The shutting off of the arc activates a spring in the gun which drives the molten stud end against a similarly heated point on the work surface. The impact force results in a fillet forming around the base of the stud.

AUTOMATIC AND SEMIAUTOMATIC STUD WELDERS

Automatic stud welders (Fig. 7) are commercially available for work on a production line basis. These welders are capable of producing up to ten welded pieces per minute. The operator needs

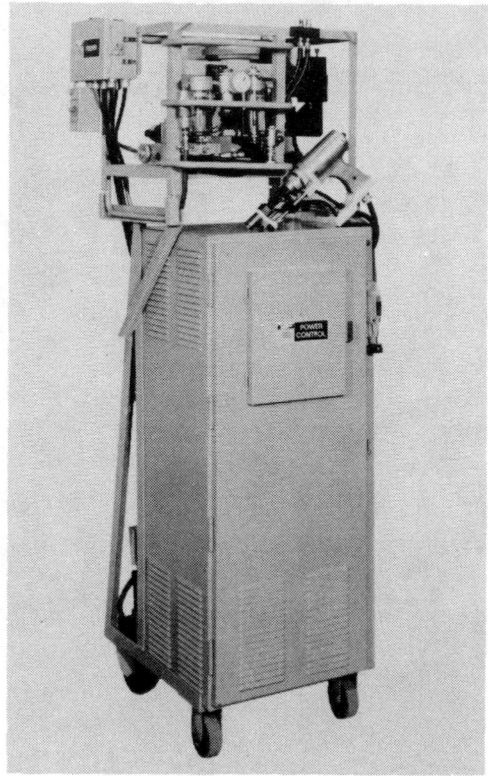

Fig. 8. Semiautomatic stud welder.

only to observe the unit's operations and maintain the supply of ferrules, studs, and workpieces.

The automatic stud welder shown in Fig. 7 operates either on a single cycle or on an automatic repeat cycle. The sequences followed in a single cycle are as follows:

1. Position the workpiece,
2. Position the ferrule,
3. Position the stud,
4. Weld the stud to the workpiece,
5. Unload and retract,
6. Repeat the cycle.

The semiautomatic stud welder illustrated in Fig. 8 can also be set for manual operation. The gun arm can be moved into a wide variety of positions depending on the size of the work. The plate covering the top of the work table is grounded. This unit contains electrically controlled pneumatic equipment. Air power positions the gun for welding and retracts it after welding. This necessitates an air supply with 60 to 70 psig.

CHAPTER 12

Other Welding Processes

This chapter is concerned with a brief discussion of nine welding processes not previously described. These nine welding processes are:

1. Induction welding,
2. Forge welding,
3. Friction welding,
4. Pressure welding,
5. Flow welding,
6. Cold pressure welding,
7. Laser welding,
8. Ultrasonic welding,
9. Electron beam welding.

Each of these welding processes is limited in its application. None can be classified with any of those welding processes discussed in Chapters 3 through 11.

INDUCTION WELDING

Induction welding (Fig. 1) is a fusion welding process which involves the heating of metal by exposing it to a strong alternating electromagnetic field. It represents a further development of the induction heating process which had previously been limited to applications in hardening, brazing, and soldering. Filler metals

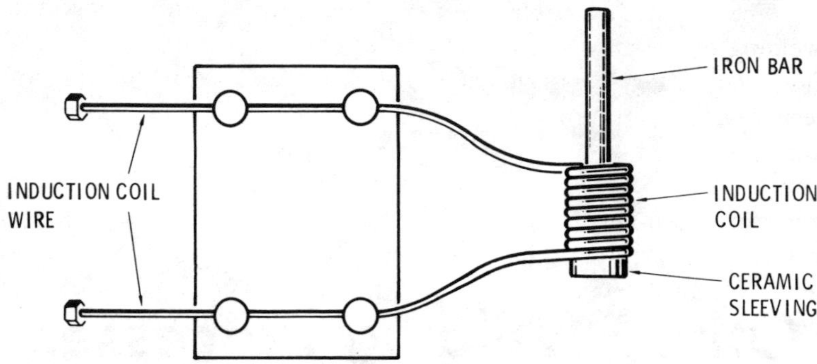

Fig. 1. Induction welding process.

are added to the joint by capillary action. Pressure may or may not be used.

Induction heating can be divided into low- and high-frequency types. The latter traces its development to research in the nineteenth century when the theory of high-frequency current advanced from its hypothetical stage to practical realization with the discovery (1842) by Henry that the discharge of an electric capacitor across a gap was oscillatory. Years later when Marconi was making use of this important discovery in developing wireless telegraphy, Egbert von Lepel, a noted scientist, recognized the shortcomings of the spark gaps used by Henry and Marconi and invented what he called the "quenched" spark gap. The quenched spark gap principle permitted the use of frequencies of 100,000 to 500,000 hertz (cycles per second). The ultrahigh frequencies (those approaching 500,000 hertz) used in conjunction with the continuous tube mill increased the rate at which a metal surface could be brought to a fusion temperature and concentrated the heat in a very narrow area.

Low-frequency heating is not a new process. For many years, ordinary 60-hertz current and currents up to a few thousand hertz have been employed in furnaces for melting metal. Alternating current, because of its inductive effect, tends to concentrate on the surface of a conductor. This tendency is low with 60-hertz current but at frequencies required in induction heating it becomes very pronounced.

The alternating electromagnetic field necessary to achieve the welding temperatures in induction welding is produced by a high-frequency alternating current flowing through a coil. As an alternating current flows in a wire or conductor, the amount of current (expressed in amperes) constantly changes. It fluctuates continually from zero to a positive maximum and back again to zero; then to a negative maximum and back to zero. This complete reversal of current is called a cycle. The number of these cycles occurring in a second is known as frequency and is expressed as hertz.

Induction welding involves high-frequency heating of the metal. High-frequency heating simply means that the number of cycles per second (hertz) of the current normally supplied from the power lines (60 hertz) has been stepped up many times by a motor generator set, spark gap, or by various types of vacuum-tube equipment.

Almost any conductive object can be heated if it is exposed to a strong electromagnetic field. Metals are, in general, good conductors of electricity, and their surface resistance is such that an appreciably high current (eddy current) can flow on it. These eddy currents depend on the frequency of the equipment, the surface resistance of the metal, and the power applied. The higher either or all of these are, the more heat that is produced in the metal. Low frequencies require very close coupling between coil and work to transfer energy; high frequencies permit a wider spacing between coil and work.

When an electric current flows in a conductor, magnetic lines of force are formed around the conductor. If this wire is formed into a loop or coil, the magnetic field is concentrated.

The strength of the magnetic field depends on the following three factors:

1. The amount of current flowing in the wire,
2. The number of turns or loops in the coil,
3. The type of material inside or surrounding the coil.

A single loop carrying a large current may produce the same effect as a coil with many loops or turns carrying a small current. When the medium around, or within the coil, is air or some poor conductor of magnetic lines of force, the field within the coil is weak. By inserting a bar of soft iron in the coil carrying a current, a path for the magnetism or lines of force is provided and an electromagnet is created.

The electric currents are produced in the bar through electromagnetic induction. Only when there is a magnetic field of varying intensity surrounding the coil can a current be induced in the bar (or any other conducting object) placed in the coil. With an alternating current flowing in the coil, fluctuating from zero to a maximum, the magnetic lines of force expand. When the current reverses, they contract. If these magnetic lines of force are cut by an object as they expand and contract, a voltage is induced in that object. In both the induction brazing and induction welding processes, objects to be brazed or welded are placed within coils designed to approximate their shapes.

As previously mentioned, high frequencies often range between as high as 100,000 and 500,000 hertz. High-frequency induction heating reduces the time required for the hardening, annealing, stress relieving, brazing, soldering, and forging of metals from hours to minutes and even seconds. In the hardening of metals, the heat can be applied where it is needed, as hot as it is needed, and as long as it is needed.

The surface of metal parts such as crankshafts, tools, pins, gears, valve stems, valve seats, and the internal surface of engine cylinders can be hardened without affecting the toughness of the metal under the surface. High-frequency brazing gives a smooth, tight, strong union of the parts. In forging, the advantage of high-frequency heating is in the short heating cycle.

Quenched spark gap induction heat treating units are manufactured in sizes ranging from 4 to 30 kW input. Larger size units are expected to be available in the near future. These units are provided with a device which corrects the power factor to practically unity. The high-frequency energy is generated in the unit and is applied to the object being heated through "work coils." These work coils are made of copper tubing formed into a coil of few or many turns. They are connected to the outside of the unit and are water cooled (water flowing through the copper tubing).

The object or parts being heat treated are placed within the turns of the induction coil and are thus exposed to the field of the high-frequency current. This produces the heat energy required to heat the object or the particular section where heating is required.

The heat energy is induced in the area lying within the turns of the work coil. Since the heating takes place with such extreme rapidity (almost instantaneously) it is a relatively simple matter to apply the coils to one section of a part, or simultaneously to two or more sections of the part, and to heat those sections in such a manner that no distortions or structural changes occur in any section of the heated part. The heat that the high-frequency current induces is concentrated on the outer surface of the metal part which is being heated, and the higher the frequency the more pronounced is this "skin effect." Another advantage to working with high frequencies is that the work coils are not required to fit closely around the object being treated. This enables the same coil to heat many different objects of different shapes and sizes and, in fact, obviates the necessity of requiring odd shaped coils to heat irregular shaped objects.

The procedure for operating an induction heat treating unit (manual control) is as follows:

1. Determine the temperature required for the particular heat treating operation involved as well as the time cycle required to produce such temperature.
2. Step on the foot control pedal which switches on the current (this will result in the heat being almost instantly applied to the metal part).
3. Maintain current for the number of seconds determined in step one.
4. Switch off the current.
5. Remove and quench the part.

There are some units equipped with automatic timing devices which take the place of manual control by the operator. The quenching medium is usually water, brine, or oil, although others can be used.

FORGE WELDING

Forge welding may well represent the oldest known welding process. Essentially it is an application of the blacksmith's method of joining metals. The metals to be joined are heated in a furnace or some other device to the plastic stage (i.e. just below the molten stage). The metals are then withdrawn from the heat source, superimposed one on the other in the position to be welded, and hammered or pressed together until a weld has been created. Forge welding is generally used to join low-carbon steel and wrought iron.

The heating must be uniform (along the entire area to be joined) and neither too much nor too little. If too much heat is applied, the metal acquires a spongy rough appearance—too little heat results in a weak incomplete weld. Fluxes are used to prevent or delay oxidation.

There are generally three basic forge welding processes which can be used:

1. Hammer welding,
2. Die welding,
3. Roll welding.

All three heat the metals in the same general way (i.e. in a furnace or similar device). It is in the method of applying the pressure that these processes differ.

As the name suggests, the *hammer-welding process* involves the repeated blows of a hammer (manually or mechanically operated) against the surface to be joined until a weld has been formed. The time-honored example of this method is work performed by the blacksmith. It may also be found in the manufacture of hardware. Hammer welding is rapidly finding fewer and fewer applications.

The *die welding process* involves bringing pressure by means of tube rollers, a bell, or a mandrel. Its principal application is in the manufacture of large diameter water pipes.

In the *roll welding process,* pressure is applied by pressing the metal surfaces between plate rolls. This process is principally used for the welding of cladding metal to steel plate.

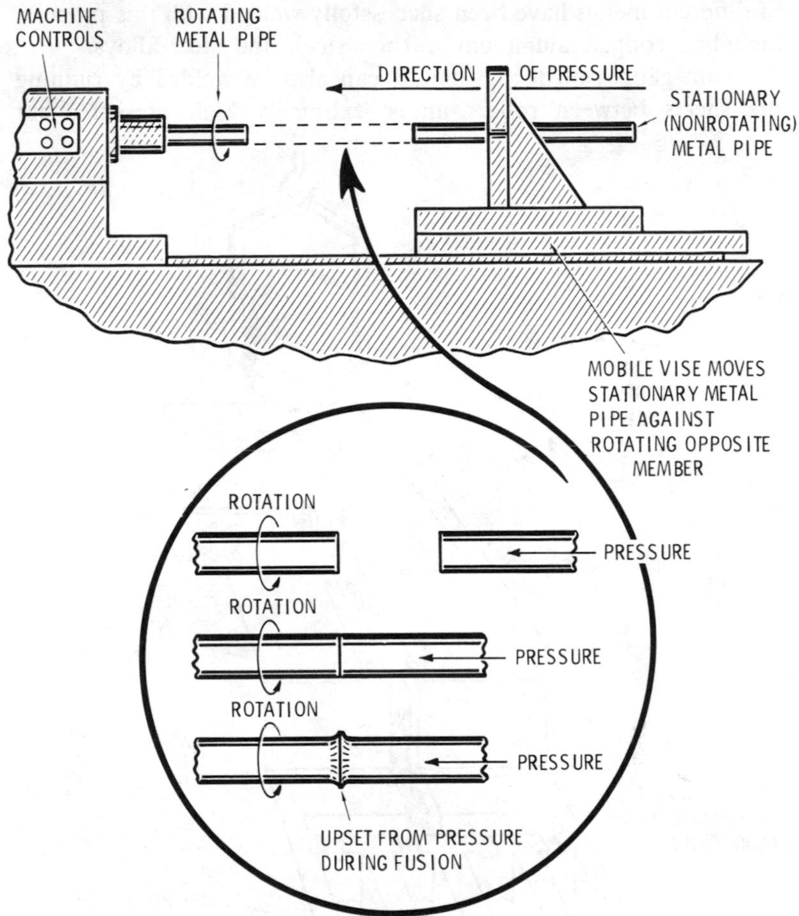

Fig. 2. Friction welding.

FRICTION AND PRESSURE WELDING

Friction welding (Fig. 2) is a welding process limited to the joining of tubes, pipes, or rods. Fusion is created by rotating one of the parts to be joined against the stationary surface of the other part in a lathe-like friction welding machine. The rotation is done at high speeds at low pressure until fusion temperatures are reached. Greater pressure is then applied until the two parts are welded together. Some upset results from the pressure. A number of different metals have been successfully welded with this process, including copper, aluminum, carbon steel, and steel alloys.

Thin gauges of ductile metals can also be welded by running the sheets between rollers under extremely high pressure, but

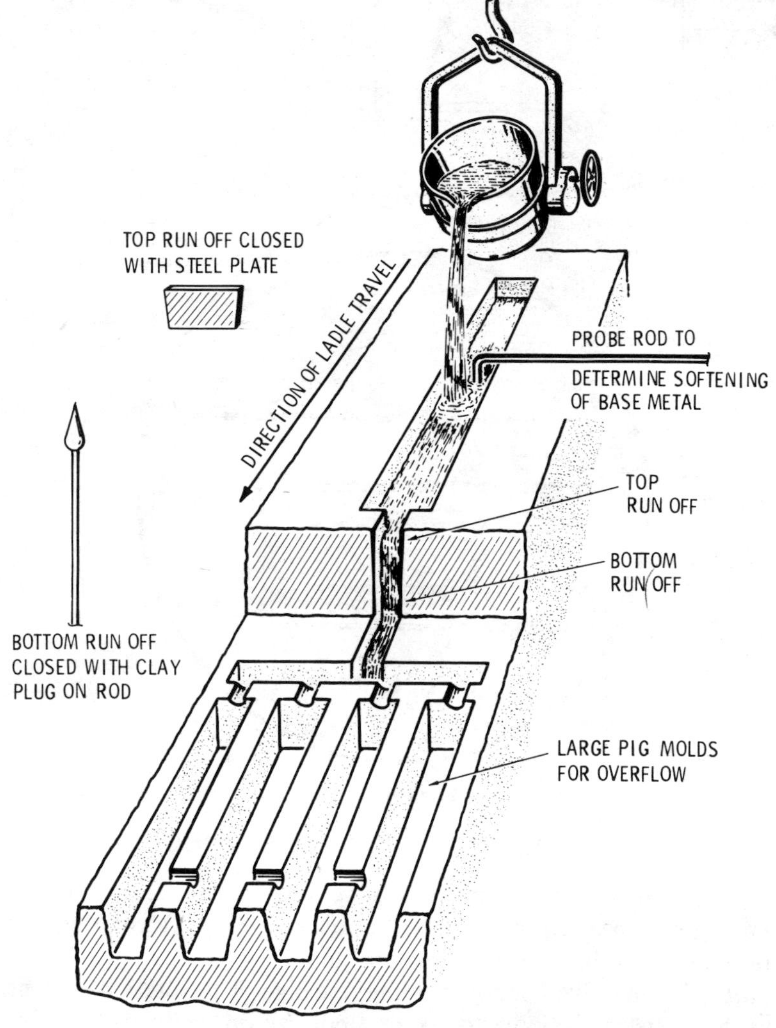

Courtesy American Welding Society

Fig. 3. Pouring operation during flow welding.

without the use of heat. The high pressure causes the metal surfaces to deform and flow together in a bond. The greatest application of *pressure welding* is in the making of lap and butt welds on thin gauges of aluminum and aluminum alloys.

FLOW WELDING

Flow welding (Fig. 3) is a welding process in which the heat is derived from molten filler metal poured over the area to be joined. The flow of molten filler metal is continued until (1) fusion of the base metal occurs, and (2) the required amount of filler metal has been deposited. Excess filler metal runs off into large pig molds. Runoff channels are blocked by specially constructed steel plate gates as soon as the weld has formed.

A probe rod is used to determine the condition of the base metal. This would be impossible to detect otherwise since it remains beneath the surface of flowing filler metal.

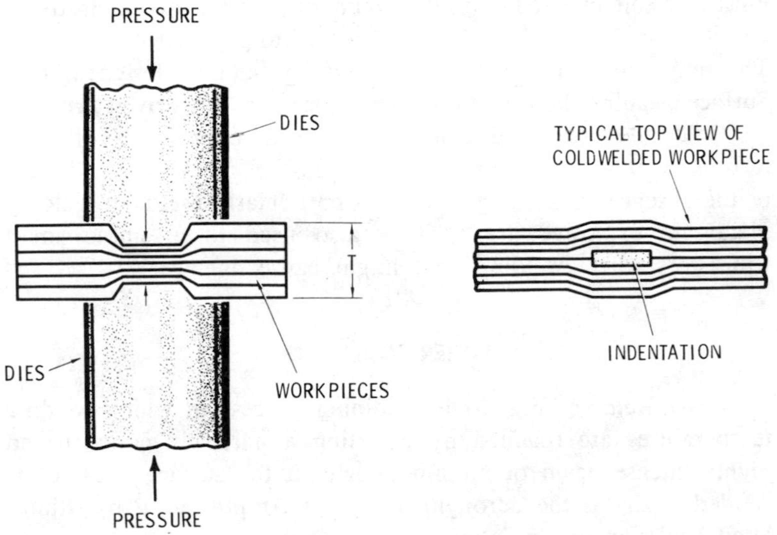

Fig. 4. Cross-sectional view showing a pair of specially constructed dies used for cold welding a pair of overlapping metal strips shown at the end of a cold-welding operation.

COLD PRESSURE WELDING

Cold pressure welding (Fig. 4) is a welding process that does not require heat. Ductile metals can be welded with this process at room temperatures by applying severe pressure to the area being welded. Specially constructed dies are used to apply a controlled deformation of the surface until an unrestricted free flow of metal occurs.

The pressure may be applied manually or with power-driven sources. The amount of pressure to be used depends upon three factors:

1. The thickness of the metal,
2. The type of metal,
3. The surface area of the die.

The length of time pressure is applied is of no particular importance in cold pressure welding.

Just as in other welding processes, the surfaces to be welded must be clean and free of all contaminants. This is a comparatively easy low-cost process that lends itself well to production line work. The major problem consists in obtaining suitably designed dies. Surface cleaning should be done with special motor-driven brushes.

Cold pressure welding is applicable to most kinds of nonferrous metals, especially higher grades of aluminum. Because of the absence of welding heat, it is particularly useful in welding metals in explosive areas. It is also used to assemble small transistors where welding heat might cause damage.

LASER WELDING

Laser welding (Fig. 5) is a joining process in which welding temperatures are reached by directing a narrowly concentrated highly intense beam of amplified light at the surface area to be welded. *Laser* is the acronym for Light Amplification by Stimulated Emission of Radiation.

A major advantage of the laser welding process is the narrow size of the heat-affected zone. This is due primarily to the fact that the laser beam is highly concentrated and narrowly defined.

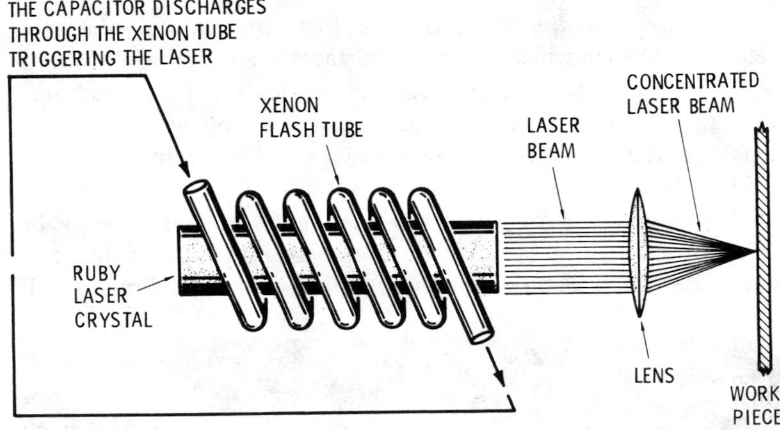

Courtesy Sonobond Corp.

Fig. 5. Laser welding process.

An equally important factor is the high degree of control over heat input in this welding process. Laser welding is used in the space, aircraft, and electronics industries for lighter gauge metals.

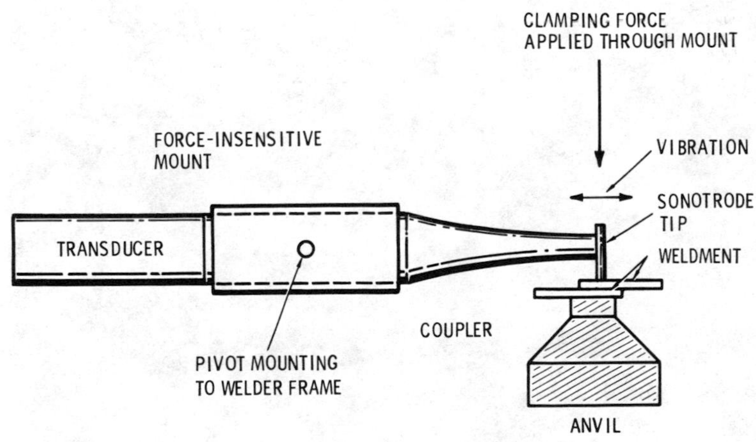

Fig. 6. Principles of ultrasonic welding.

ULTRASONIC WELDING

Ultrasonic welding (Fig. 6) is a low temperature solid-state welding process in which two metal surfaces are joined while vibrating under pressure. The friction that causes the two metal surfaces to become welded together is produced by acoustic vibrations operating in a frequency range extending from 10,000 to 175,000 Hz (cycles per second).

The pressure provided when clamping the surfaces to be welded should only be enough to prevent slippage of the two pieces. Excessive clamping pressure is unnecessary and may result

Courtesy Sonobond Corp.

Fig. 7. Ultrasonic frequency converter.

in deformation of the surface. On the other hand, insufficient clamping can result in damage to the transducer-coupler.

Ultrasonic welding equipment is designed to convert electrical energy to mechanical (vibratory) energy. The mechanical (vibratory) energy is transferred to the surfaces to be joined by a sonotrode tip oscillating approximately parallel to the plane of the interface between the two metal surfaces. This is basically a shear-type vibratory force.

An ultrasonic welding system will generally consist of the following components:

1. A frequency converter,
2. A transducer-coupling system,
3. An anvil,
4. A clamping mechanism.

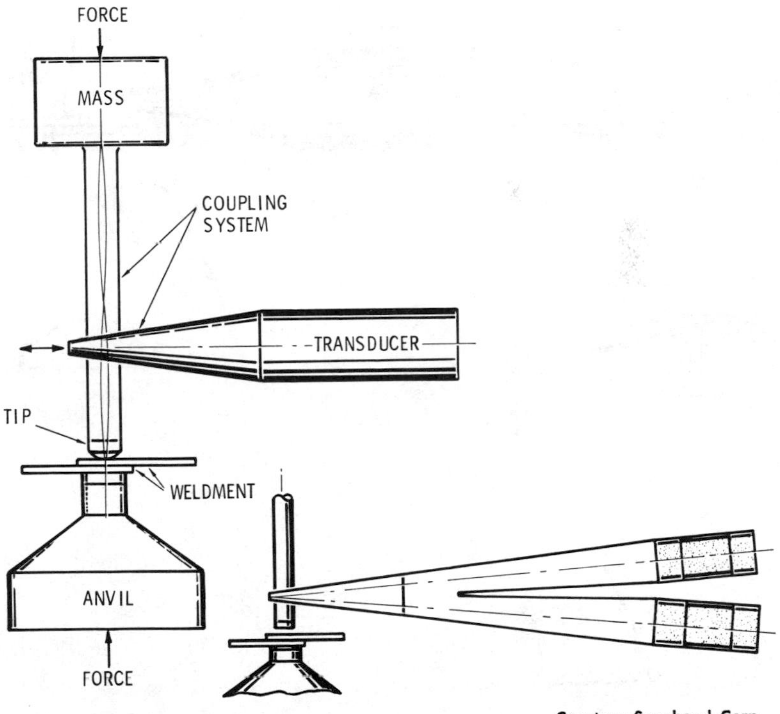

Courtesy Sonobond Corp.

Fig. 8. The wedge-reed transducer-coupling system.

The frequency converter is designed to convert 60-Hz power from the line to a high-frequency electrical current. The high-frequency current then feeds into a transducer where it is converted to energy high-frequency vibrations. Simply stated, an ultrasonic welding system converts standard (60-Hz) electrical current to high-frequency electrical current and then to vibratory power.

Fig. 7 shows a frequency converter designed for use in ultrasonic welding. It can be used with the ultrasonic spot-type welding machine shown in Fig. 12. This particular frequency converter is a 750-AW unit. It is capable of delivering 750 acoustical watts to the workpiece.

The transducer-coupling system is located in the welding machine. The purpose of the transducer is to convert high-frequency

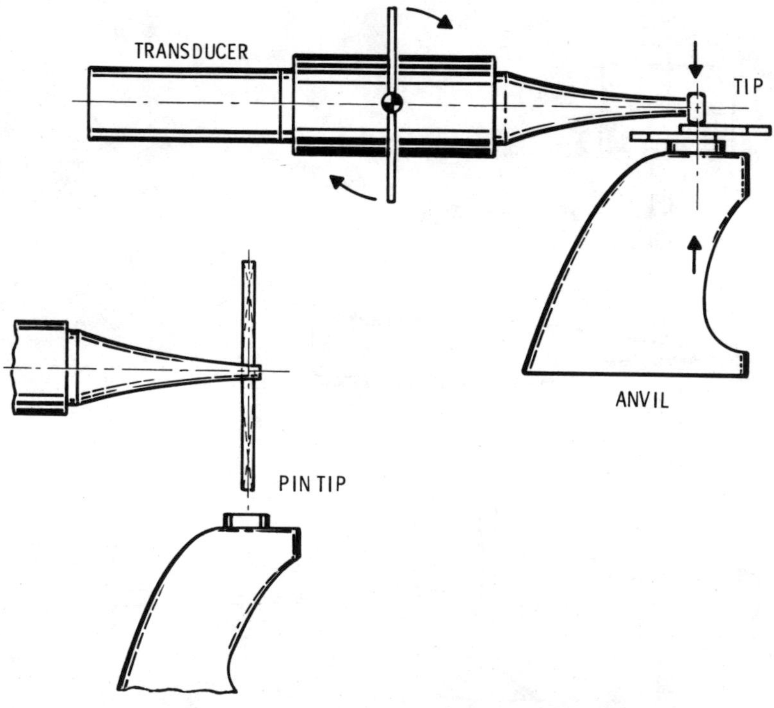

Courtesy Sonobond Corp.

Fig. 9. A lateral-drive transducer-coupling system.

electrical current into a mechanical vibratory power. The coupling system is designed to carry the vibratory energy to the work surface. The sonotrode tip is a part of the coupling system.

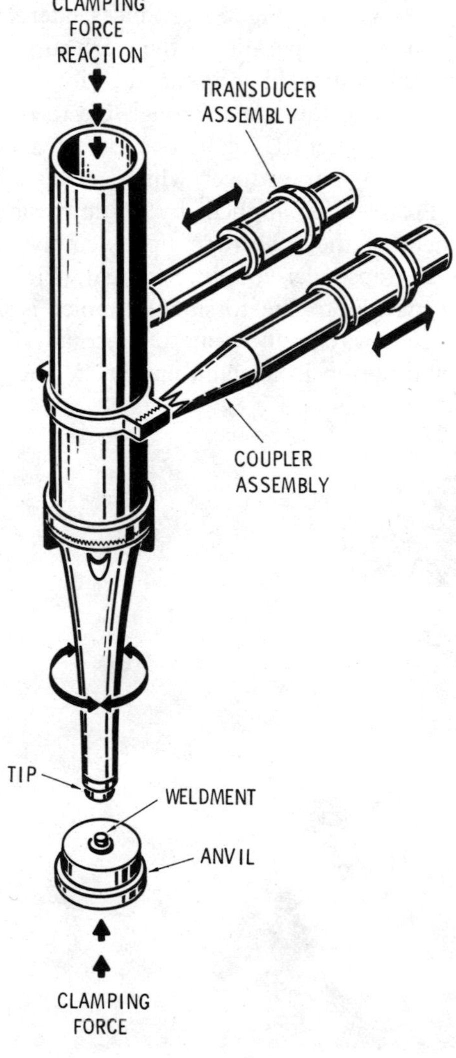

Courtesy Sonobond Corp.

Fig. 10. A torsional transducer-coupling system.

Three different types of transducer-coupling systems are available for use in ultrasonic welding machines. These are: (1) the wedge-reed system, (2) the lateral-drive system, and (3) the torsional system.

The *wedge-reed system* (Fig. 8) produces lateral vibrations of the sonotrode tip in a plane parallel to the work surface. The wedge is driven by the transducer in longitudinal vibration. Pressure is applied perpendicular to the work through the reed.

The lateral-drive system (Fig. 9) consists of a tip and lateral sonotrode attached to a transducer which moves parallel to the work surface. Pressure is applied by a downward forcing or bending movement of the sonotrode tip assembly.

The torsional transducer-coupling system (Fig. 10) is found in ring welding machines. The torsional coupler is driven by the transducer in such a way as to create a torsional vibratory movement of the sonotrode tip in a plane parallel to the work surface.

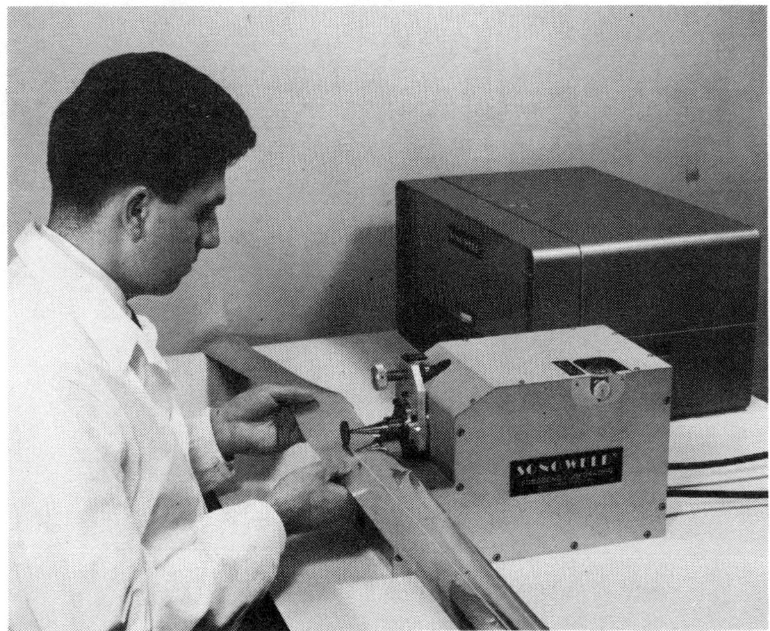

Courtesy Sonobond Corp.

Fig. 11. Ultrasonic spot-type welding machine.

The anvil provides a counterreactive force to the pressure exerted by the tip against the surface. It can also serve as a transporter for the work in some kinds of ultrasonic welding machines (e.g. the continuous-seam type).

Clamping force may be applied by: (1) pneumatically actuated systems, (2) spring systems, and (3) hydraulic systems. The last named are generally found in the larger welding machines.

Most ultrasonic welding machines will operate at 15,000 Hz ($\pm$150 Hz) with a 750 to 4500 acoustic watts rating. The very-high-frequency ranges (up to 175,000 Hz) are found in smaller machines designed for welding fine wire.

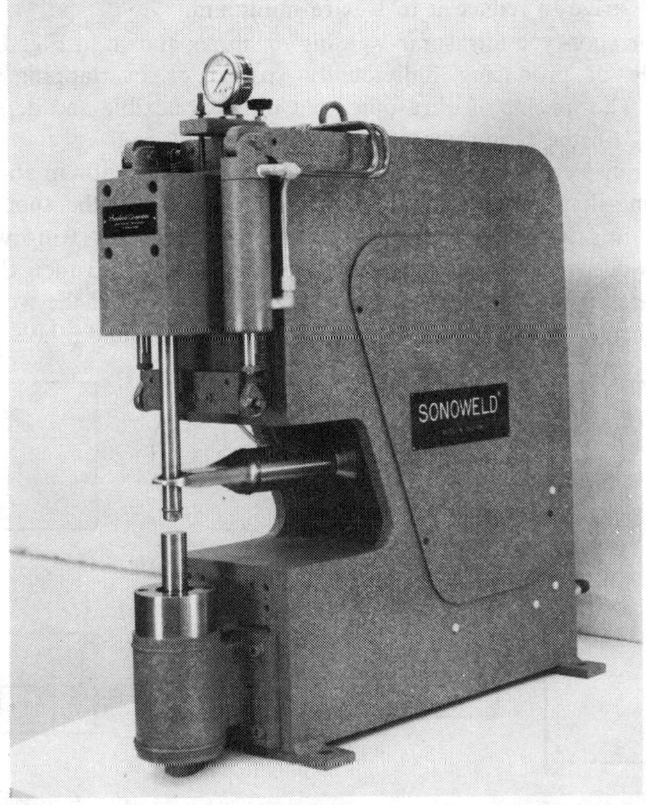

Courtesy Sonobond Corp.

Fig. 12. Ultrasonic spot-type welding machine.

Ultrasonic welding machines can be used to produce the following four types of welds: (1) spot, (2) ring, (3) line, and (4) continuous-seam.

Fig. 11 shows an ultrasonic welding machine equipped with a rotating circular sonotrode tip for continuous-seam welding. Disc-shaped tips are also used to produce the continuous-seam weld. There are several methods of clamping the work and bringing it in contact with the welding head. For example, the anvil may be fixed in position with the sonotrode tip rotating across the work surface or the anvil can be designed to transport the work under a fixed welding head. A number of other combinations are also possible. Although some slippage can be tolerated, the operator should strive to reduce it to a bare minimum.

The spot-type ultrasonic welding machine, shown in Fig. 12, is capable of producing individually spaced or overlapping spot welds. The spacing of ultrasonic spot welds is flexible and depends entirely on the requirements of the joint design.

A ring weld is produced by a machine equipped with an annular (i.e. ring-shaped) sonotrode tip. As is the case with the spot-type welds, ring welds can be made so that they overlap. Ring welds are produced in several different designs (Fig. 13) and their diameters are limited only by the size and power capacity of the welding machine. The stress is evenly distributed around the weld,

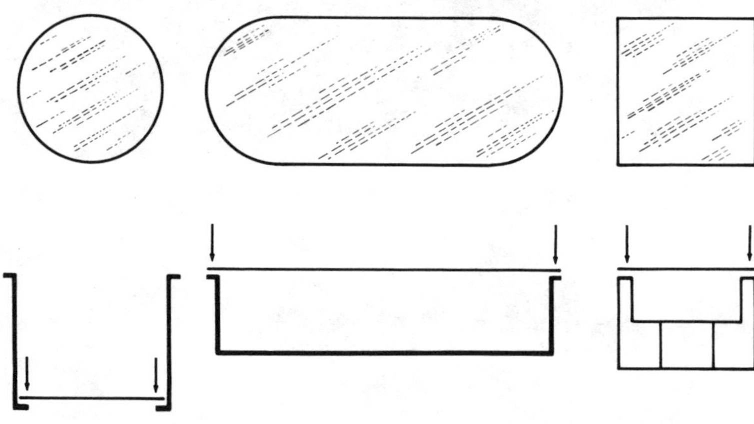

Courtesy Sonobond Corp.

Fig. 13. Typical ring welds.

thereby eliminating the tendency to crack. This is a major advantage of ring welds over the spot-type welds.

Line welds approximately 6 inches or more in length are produced by a welding machine equipped with an elongated sonotrode tip. The power interval is determined by preset machine controls. The work is usually clamped between the elongated welding tip and an anvil. The welding tip oscillates perpendicular to the weld line and therefore in a plane that will be parallel to the plane of the work surface.

A thorough cleaning of the base metal is not as important in ultrasonic welding as it is with other welding processes. Oxides and other surface contaminants are usually broken up by the vibratory force. However, any possible problems can be avoided

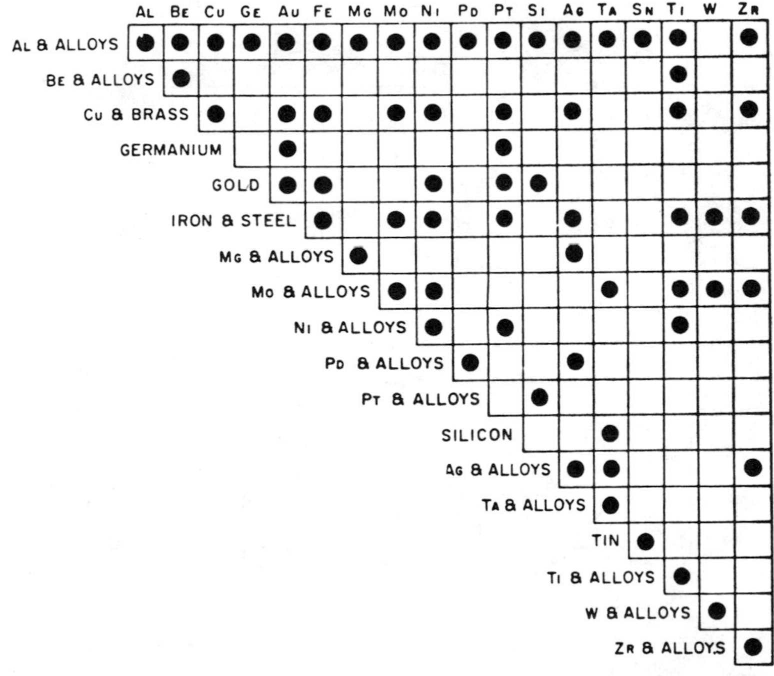

Courtesy Sonobond Corp.

Fig. 14. Metals and metal alloys successfully joined by the ultrasonic welding process.

by simply taking the same precautions normally employed in preparing the surface for gas or arc welding.

Ultrasonic welding has found useful applications in such widely differing areas as the production of aluminum foil and aluminum products, the hermetic sealing of volatile substances (explosives, reactive chemicals, etc.), and the electronics industry. There is no metal or metal alloy that cannot be joined by the ultrasonic welding process, although some can be done more economically than others. Fig. 14 shows those most successfully joined by this process (indicated by the black dot).

ELECTRON-BEAM WELDING

Electron-beam welding (Fig. 15) is an automatic welding process performed in a vacuum without a shielding gas. Neither

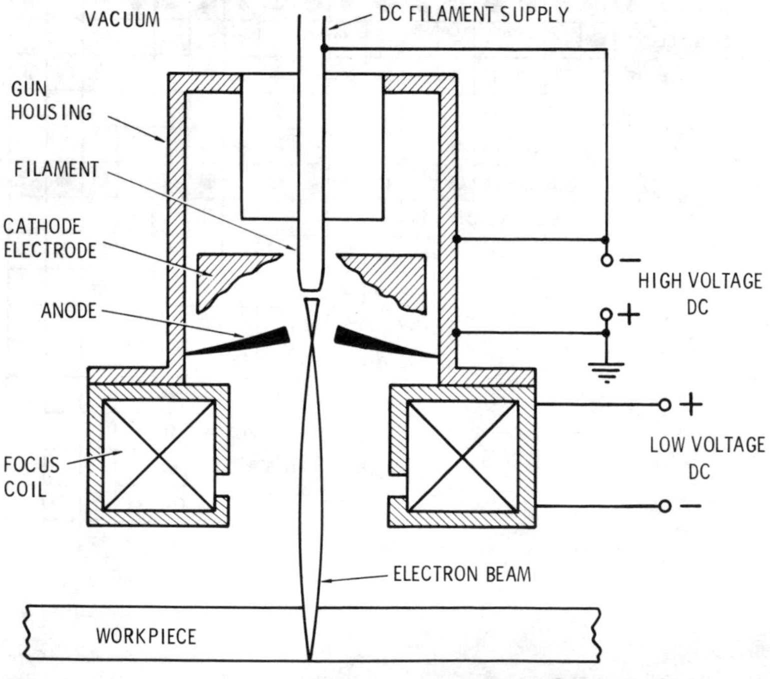

Courtesy Sciaky Bros., Inc.

Fig. 15. The electron-beam welding process.

an electrode nor a filler rod is used. Both seam (continuous) and spot welding are possible with the electron-beam welding process.

Basically, the electron beam consists of negatively charged particles (electrons) which are propelled from a negative cathode to a positive anode across a vacuum. The velocity of the electrons is accelerated as they pass across the vacuum, reaching a maximum rate at their point of contact with the anode. The velocity of the electrons is accelerated by increasing the voltage to the electron-beam gun. The welding heat is directly proportional to the velocity of the electrons: the greaer the velocity of the electrons, the higher the welding temperatures become. The metal work surface functions as the anode. The electron beam is focused against the anode by means of an electromagnetic coil within the gun.

Fig. 16 illustrates the differences between welds produced by the electron beam and standard fusion welding processes. Note that the latter results in greater angular distortion caused by shrinkage

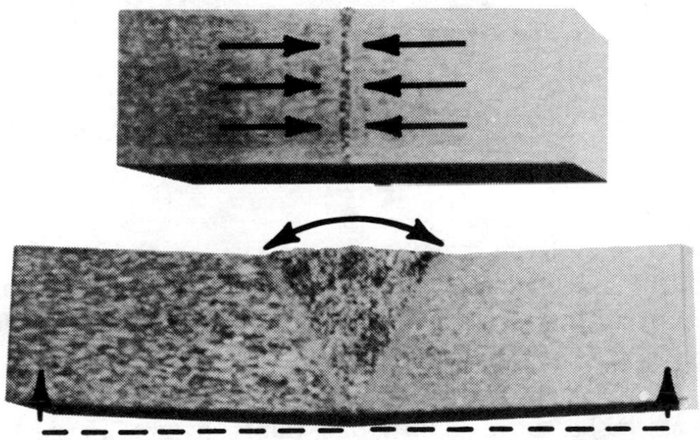

Courtesy Sciaky Bros., Inc.

Fig. 16. Comparison of electron-beam welds with standard fusion welds.

at the top surface. Note also the narrowness of the electrom-beam weld in comparison to the fusion weld. In the electrom-beam weld, parallel sides result in linear shrinkage rather than angular distortion.

A high deposition rate produces welds of excellent quality with only a single pass. Because of the concentrated nature of the electron beam, arc length may be varied from very short to very long without changing the quality of the weld. As a result of its variable arc length, electron-beam welding can be used to weld or repair points that are virtually inaccessible for other welding processes. By way of example, Fig. 17 shows a ring gear welded to a gear housing by the electron-beam welding process. The gears are fully machined and hardened.

Courtesy Fusion Labs, Inc.

Fig. 17. A ring-gear welded to a gear housing by the electron beam welding process.

The electron beam is highly concentrated, a characteristic that enables the welder to make deep narrow welds, and to work in close proximity with portions of an assembly that would be damaged using other welding processes. The intense concentration of the beam is shown in Fig. 18, which illustrates the welding of piston assemblies on a multistation rotary fixture. Fig. 19 illustrates the welding of a bearing thrust washer to a fully hardened and ground gear shaft assembly. The weld is ⅛ inch deep and only .050 inch from the carburized teeth.

An electron-beam welding machine consists basically of the following components:

1. An electron-beam gun,
2. A vacuum chamber,
3. A transport system.

Courtesy Fusion Labs, Inc.

Fig. 18. The electron-beam welding of piston assemblies on a multistation rotary fixture.

Most electron-beam guns used today are the triode type. The general arrangement of the principal elements of a triode electron-beam gun is shown in Fig. 20.

The vacuum chamber is operated in conjunction with pumps and a pumping system to control the vacuum environment. The vacuum environment may be a rough vacuum, soft vacuum, or hard vacuum, depending on the type of pumping system used to evacuate the chamber (mechanical piston pumps, mechanical blower pumps, and oil diffusion pumps respectively).

The size and design of the vacuum chamber used in electron-beam welding will depend on the dimensions of the weldment. The fact that they can be quite large is illustrated in Fig. 21 by

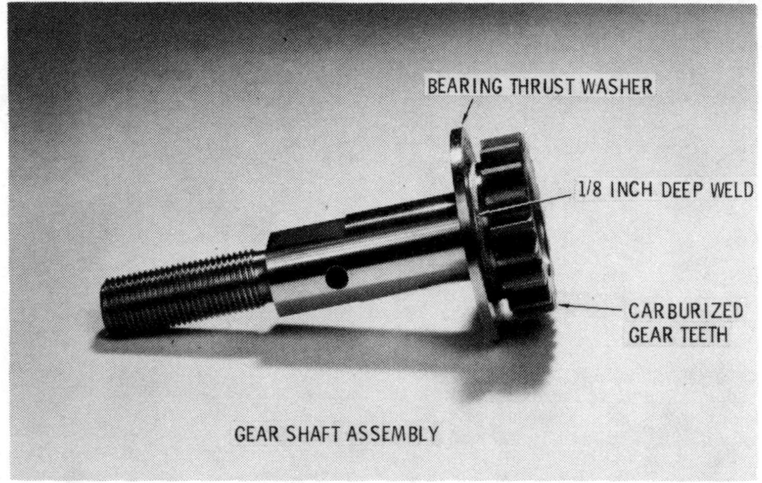

Courtesy Fusion Labs, Inc.

Fig. 19. Electron-beam welding of a bearing thrust washer to fully hardened and ground gear shaft assembly.

the clamshell style machine used for fabricating the Navy F-14A jet fighter. This particular machine is 32 ft. long, 10½ ft. wide and 8 ft. tall, and is very useful for welding assemblies that are long and of intermediate height. The clamshell vacuum chamber provides maximum accessibility to the work and will pump to the vacuum level necessary for welding in a little less than 20 seconds. This is one of the largest electron-beam welders in the world.

Taller assemblies can be welded in rectangular machines with tunnel-type vacuum chambers. Fig. 22 shows an electron-beam welder with a high rectangular tunnel style chamber. It is somewhat shorter in length than the clamshell machine (approximately 7 ft.), but it is 3 ft. taller.

Each electron-beam welding machine is equipped with some form of transport system to provide movement for the gun and work. These systems are usually driven by DC motors. Because of the nature of electron-beam welding, they must be extremely precise in their location and control of the positions and velocities of both the gun and the work.

Incorporated with the transport system is some means of observing the work. This may be done by visual observations (aided

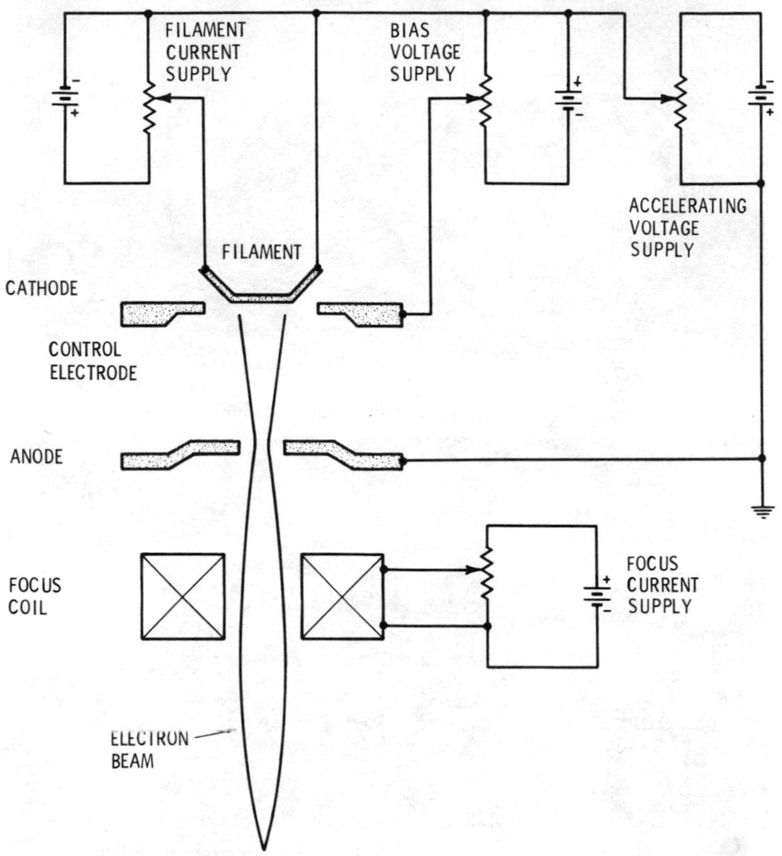

Courtesy Sciaky Bros., Inc.

Fig. 20. General arrangement of the principal elements of a triode electron beam gun.

by monocular or binocular means) through a window in the machine. However, an electronic scanning device is a more accurate and consistent means for identifying the weld joint and determining accurate alignment of the electrode beam relative to the weld joint while the chamber is closed. Figs. 23 to 25 illustrate the operating principle of just such a device and a typical scope pattern for a butt weld with beam aligned to the weld joint and one with the beam aligned to the right of the joint.

Courtesy Sciaky Bros., Inc.

Fig. 21. Clamshell style welding machine.

Courtesy Sciaky Bros., Inc.

Fig. 22. Rectangular machine with tunnel-type vacuum chamber.

Recent engineering developments have resulted in the design of electron-beam welders capable of making two parallel beads simultaneously with one gun. One of the first major applications of these machines has been in the mass production of typewriter carriages. The machine developed for this purpose (a *Sciaky* soft vacuum EB welder) is shown in Fig. 26. This particular ma-

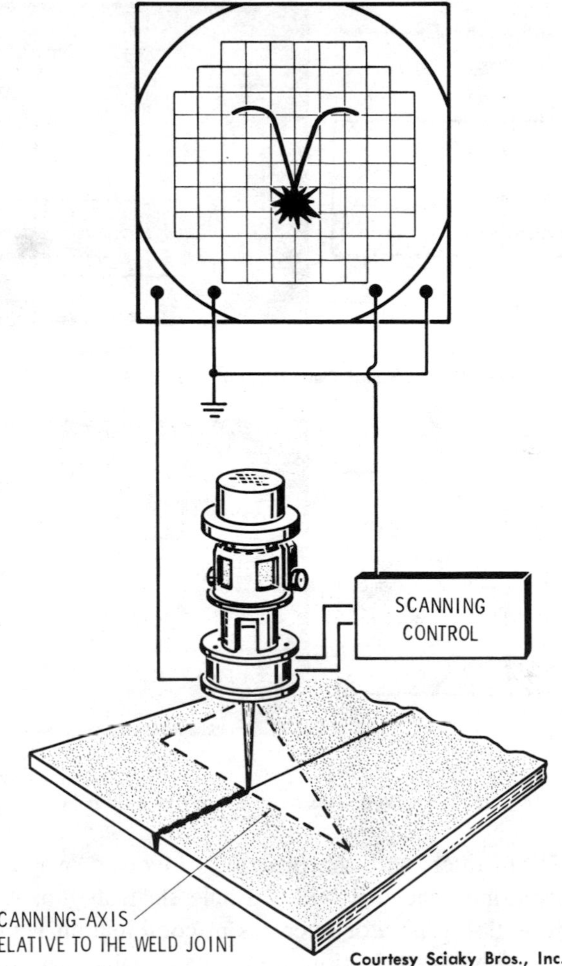

Courtesy Sciaky Bros., Inc.

Fig. 23. Operating principles of an electronic scanning device.

chine has six welding stations on a rotating fixture capable of holding twelve typewriter carriages (two per station). When the rotating fixture moves a station into position under the electron gun, the tooling plate is raised and the stations (and workpiece) are sealed in a soft vacuum chamber. Since the rotating fixture is trunnion-mounted, the carriage can be turned over 180° to allow welding on the other side.

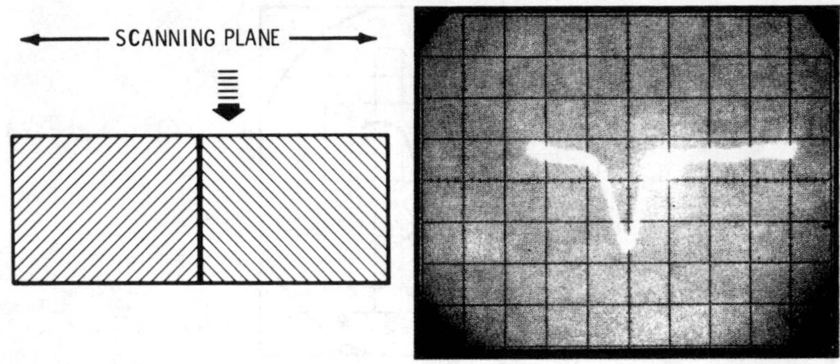

Courtesy Sciaky Bros., Inc.

Fig. 24. Square butt weld aligned to right of joint.

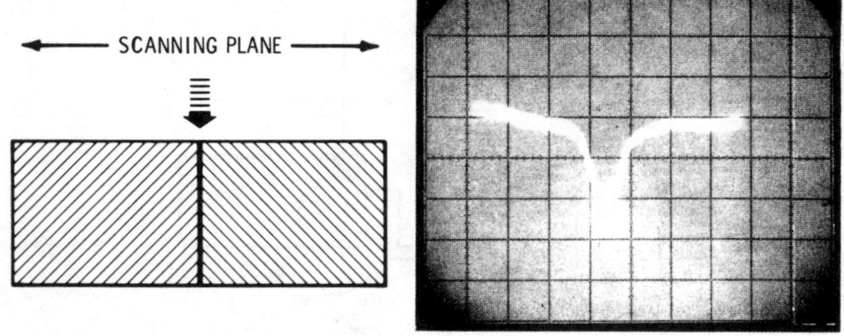

Courtesy Sciaky Bros., Inc.

Fig. 25. Square butt weld beam aligned to weld joint.

Fig. 27 illustrates the basic principles involved in making two parallel beads simultaneously with a single electron gun. A square-wave beam motion generator operates in combination with special deflecting devices. A deflection coil deflects the electron beam equal distances along each side of the workpiece as it is positioned under the gun. The electron gun is activated and deflected when a square-wave deflection signal is fed to the deflection coil. Welding will occur *only* during the dwell of the square wave. These machines are capable of making intermittent or continuous-seam welds.

Fig. 28 illustrates typical joint designs for electron-beam welding. Note that a backing strip is used for the square butt weld in Fig.

29. When a backing strip (or rod) is necessary, it *must* be metallurgically compatible with the base metal. The welder who wishes to obtain strong high quality welds should also pay particular attention to the following recommendations:

1. Keep gaps to a maximum of .005 inch for most joints,
2. Do not use lap joints where high stress or fatigue will be encountered,
3. Use lap joints for thin gauge metals and sheet metal,
4. Properly vent parts having internal cavities.

Courtesy Sciaky Bros., Inc.

Fig. 26. Soft vacuum electron beam welder.

Much more extensive guides to the operator are provided by manufacturers of electron-beam welding equipment.

Joint preparation requires near perfect fitup. The surface must be thoroughly cleaned prior to welding. Acetone or an equivalent

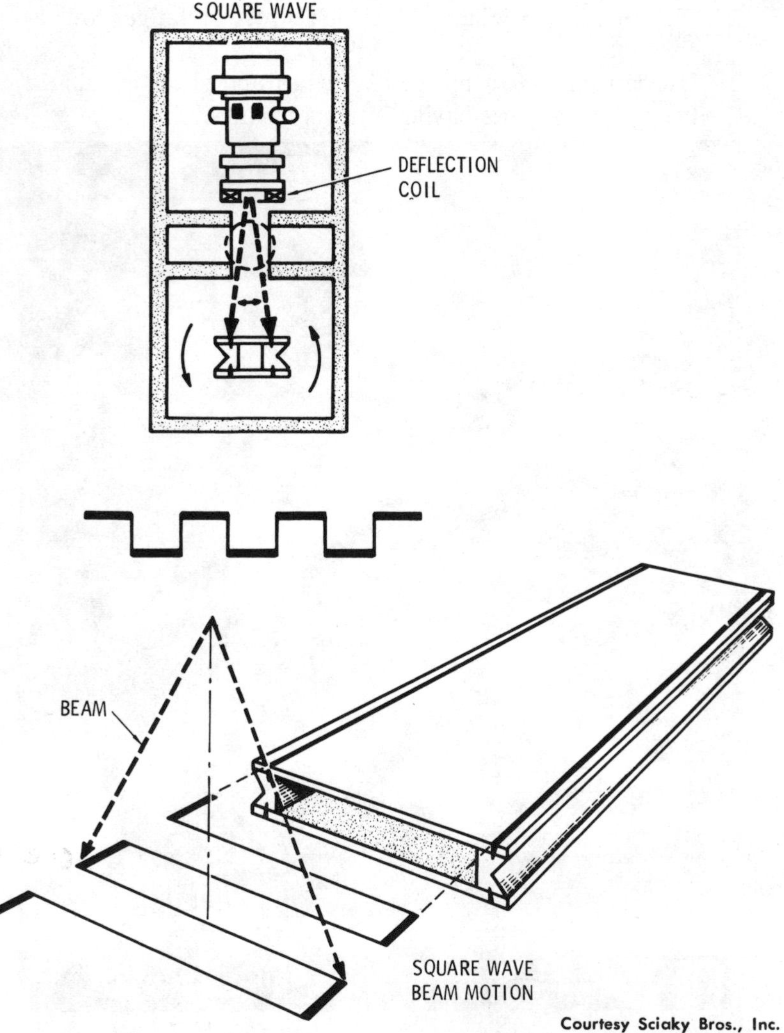

Courtesy Sciaky Bros., Inc.

Fig. 27. Operating principles in making two parallel beads simultaneously.

solvent is recommended for cleaning. Solvents containing oil should be avoided when machining nonferrous metals or cast materials (dry machining is recommended for both). Solvents containing chlorides should also be avoided. All contaminants (oil, dirt, grease, heavy scale, etc.) should be removed. If optimum weld quality is desired, then *all* traces of scale and surface oxides should be removed. Irregular surfaces can be cleaned by pickling. It is not recommended that nonferrous metals be prepared by grinding or wire brushing.

Because the electron beam generates small amounts of radiation, care should be taken to protect the welder from any danger. An adequate lead shield between the welder and the source of radiation will provide sufficient protection.

Electron-beam welding has no minimum thickness limitations; consequently, it is very useful in welding the thin gauges of metals.

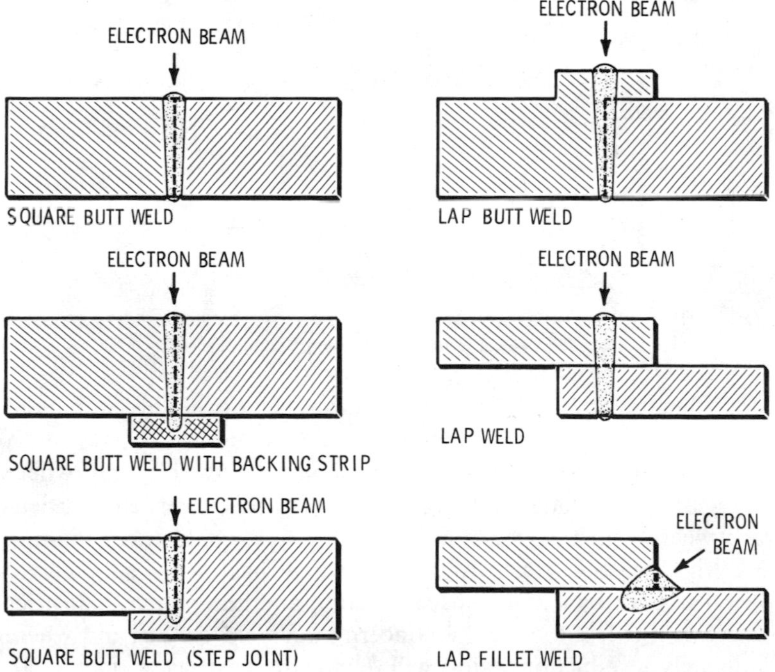

Courtesy Sciaky Bros., Inc.

Fig. 28. Typical joint designs for electron beam welding.

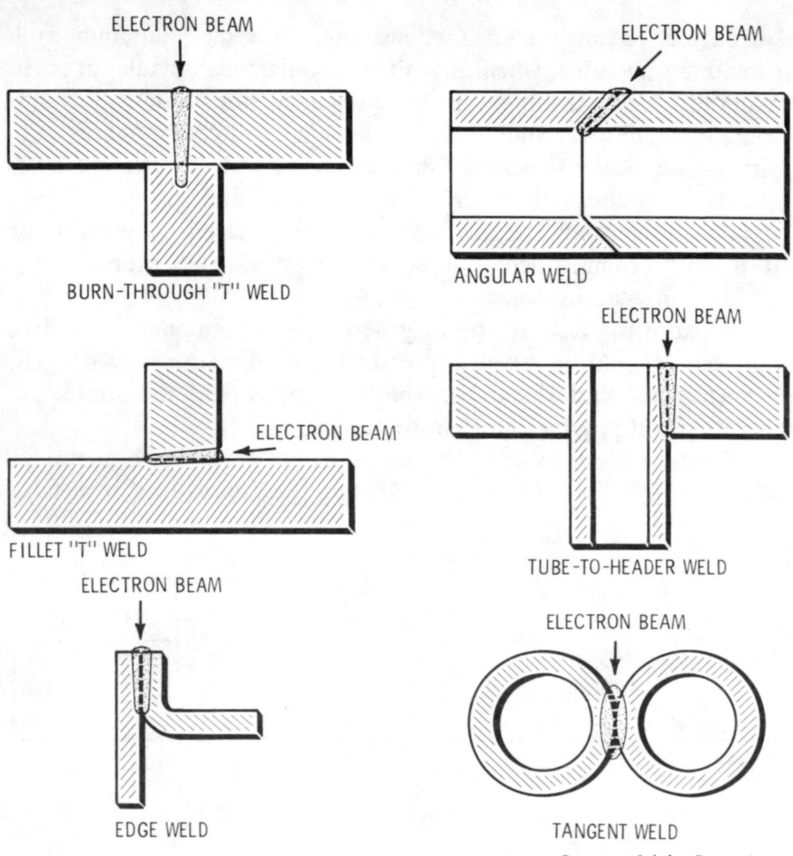

Courtesy Sciaky Bros., Inc.

Fig. 29. Typical joint designs using backing strips.

Unfortunately, the initial cost of electron-beam welding equipment is relatively high when compared to other welding processes. As a result, electron-beam welding is economically feasible only where large-scale, automatic welding operations are concerned, particularly where a high degree of accuracy and weld quality are required. For this reason, electron-beam welding has been adapted to welding operations in the space, aircraft, and automotive industries where cost is less a factor of concern than weld quality and where ultimately production rates will eliminate cost as a factor altogether.

CHAPTER 13

Metal Cutting, Flame Hardening

The welder frequently encounters situations in which heat is applied to metal not for the purpose of joining two or more surfaces together but to cut, flame harden, stress relieve, or other related functions. This chapter concerns itself primarily with the various metal cutting processes and principally with gas flame cutting.

Gas flame cutting is a chemical process in which the oxygen present in the air surrounding the cut is absorbed by the flame. The different types of gas flame cutting described in this chapter include:

1. Oxyacetylene cutting,
2. Oxyfuel gas cutting,
3. Semiautomatic and automatic flame cutting,
4. Underwater hydrogen cutting.

Also included are such specially developed cutting processes as oxygen lance cutting, powder cutting, and flux injection cutting.

Arc cutting is a melting process in which the metal is heated until it reaches the flowing point. A fundamental difficulty with arc cutting is the removal of the molten metal from the cutting zone. A number of arc cutting processes have been developed, many as attempts to solve the problem of removing the molten metal. This chapter is concerned with the following:

1. Carbon-arc cutting,
2. Air carbon-arc cutting,
3. Metal-arc cutting,
4. Oxyarc cutting,
5. Underwater arc cutting,
6. TIG and MIG cutting,
7. Plasma-arc cutting.

The electric arc and gas flame are also used to cut or shape the metal in preparation for welding or working (Fig. 1). Such operations include gouging, scarfing, and piercing. Other types of surface preparation operations in which the gas flame functions exclusively are flame machining, flame hardening, flame strengthening, stress relieving, flame softening, and flame cleaning. All of these operations are described in this chapter.

Courtesy Harris Calorific Sales, Inc.

Fig. 1. Using a heavy duty hand cutting torch.

CHOOSING THE CUTTING PROCESS

Choosing the most suitable cutting process is dependent on a number of different factors. The cost of the cutting operation is always an important consideration and is closely related to the ease and speed of cutting. The quality of a cut is another important factor. A cut that does not require extensive finishing and cleanup contributes to reducing cutting costs. Finally, the type of metal to be cut should also be taken into consideration. This last factor will be the one that concerns us in this section.

The ferrous metals (wrought iron, cast iron, low-alloy steels, etc.) are characterized by a high degree of nonresistance to oxidation. When the surfaces of ferrous metals are heated to temperatures of 1500°F to 1600°F and subjected to contact with the oxygen in the surrounding atmosphere or a stream of cutting oxygen, the iron in their composition tends to oxidize (i.e. burn) rapidly. This exothermic oxidation of iron and other ferrous metals increases the temperatures of the cutting operation. Thus, the most suitable cutting process for the ferrous metals is one that uses oxygen in combination with a fuel gas, preferably acetylene. Low-alloy steels, manganese steels, the various types of iron, and the carbon steels (up to .80% carbon) are examples of the types of ferrous metals and metal alloys that can be cut with the oxyacetylene cutting process.

The higher the alloy content in a particular metal, the more difficult it will be to cut with an oxyacetylene flame. A waster (backing) plate is sometimes used to compensate for this difficulty. The heat produced by the waster plate as it is being consumed increases the heat of the cutting operation. A higher alloy content requires a higher cutting heat, particularly if the alloy is one of the refractory metals (those having melting points above 3500°F). Waster plates have been used successfully with both the oxyacetylene and powder cutting processes.

Standard arc cutting produces a rougher cut than oxyacetylene cutting. The latter requires little cleanup or finishing work. Automatic machine flame cutting produces an even higher quality cut, often necessitating no finishing at all. Unfortunately, a cutting process cannot be selected solely on the merits of appearance. In

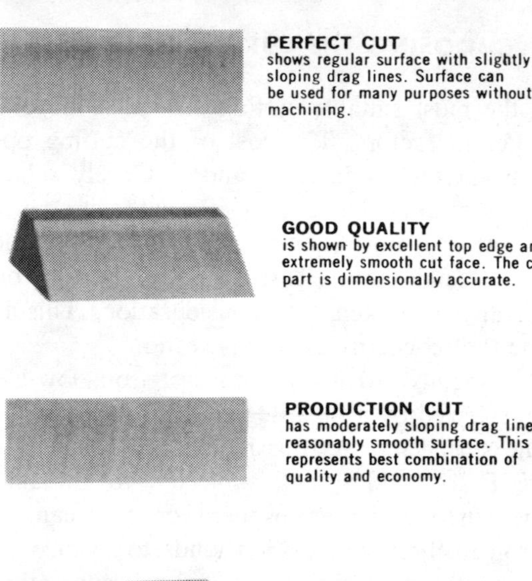

Courtesy Harris Calorific Sales, Inc.

Fig. 2. Correct cutting techniques.

some cases, a particular cutting process may be the only one that can be used.

Examples of correct cutting techniques are shown in Fig. 2. By way of contrast, Fig. 3 illustrates many common cutting faults.

Oxyacetylene Cutting

Cutting metals with an oxyacetylene flame is purely a chemical process; that is, a very rapid form of rusting (oxidation) and should not be confused with mere melting. The metals that can be cut with the oxyacetylene cutting torch are all commercial forms of iron, such as steel, wrought iron, and cast iron.

The principle of oxyacetylene cutting is quite simple. After a small spot of metal has been heated red hot, a jet of oxygen is shot against it. The iron and oxygen combine to form iron oxide, thus burning a narrow slit or *kerf* in the metal surface.

CUTTING SPEED

EXTREMELY FAST
Not enough time is allowed for slag to blow out of the kerf. Cut face is often slightly concave.

SLIGHTLY TOO FAST
makes drag lines incline backwards, but a "drop cut" is still attained. Quality is satisfactory for production work.

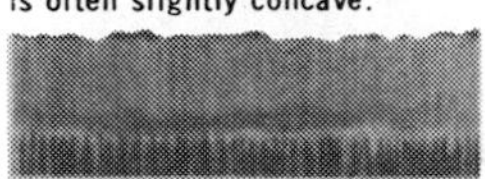

EXTREMELY SLOW
produces pressure marks which indicate too much oxygen for cutting conditions.

SLIGHTLY TOO SLOW
produces high quality cut, although there is some surface roughness caused by vertical drag lines.

TIP DISTANCE

TOO CLOSE
Grooves and deep drag lines caused by unstable cutting action. Part of preheat cone burns inside kerf where normal gas expansion deflects oxygen cutting stream.

TOO HIGH
Top edge is beaded or rounded, cut face is not smooth and often is slightly beveled when preheat effectiveness is partially lost due to the tip being held too high. Cutting speed is reduced because of the danger of losing the cut.

GAS ADJUSTMENT

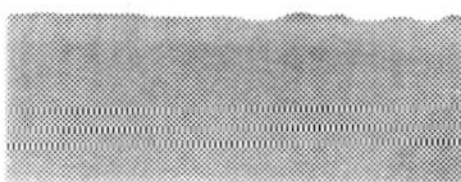

TOO HOT PREHEAT
Rounded top edge caused by too much preheat. Excess preheat does not increase cutting speed. It only wastes gases.

TOO MUCH CUTTING OXYGEN
Pressure marks are caused by too much cutting oxygen. Correct this fault by lowering cutting oxygen pressure, increasing speed, or using a smaller tip. As oxygen volume nears correct proportion, pressure marks appear closer to the bottom edge until they finally disappear.

DIRTY TIP

DIRT
or scale in the tip will deflect oxygen stream and cause excess slag, pitting and undercutting.

Courtesy Harris Calorific Sales, Inc.

Fig. 3. Common cutting faults.

A cutting torch differs from a welding torch in that in addition to having an oxyacetylene flame, it also has another gas stream of pure oxygen under high pressure which does the cutting after the metal has been raised to the melting point by the heating flame (Fig. 4).

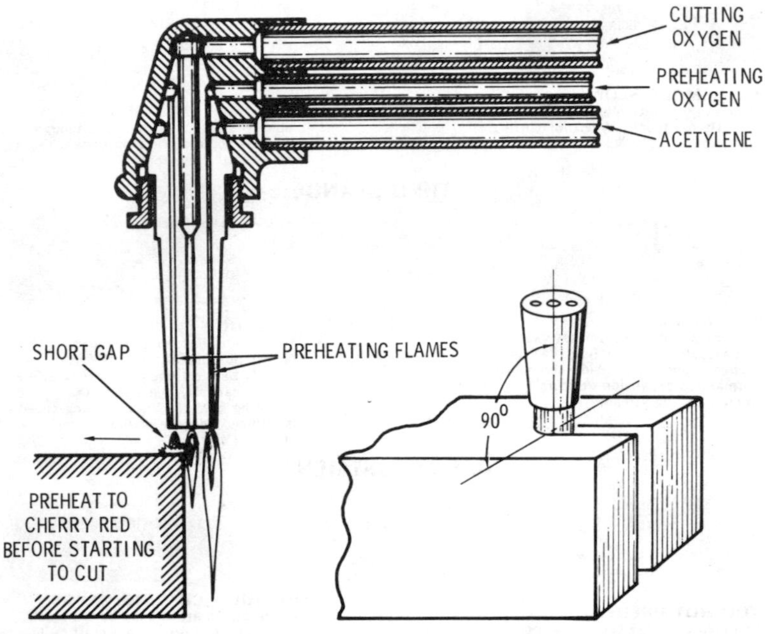

Courtesy Airco Welding Products

Fig. 4. A cutting torch showing the preheating and cutting oxygen tubes.

The oxygen cutting jet is adjusted by maintaining it under sufficient pressure to blow away the oxide as fast as it is formed so that the slot being cut is kept open.

A typical oxyacetylene cutting installation will consist of the following components:

1. An oxyacetylene cutting torch,
2. An oxygen cylinder,
3. A heavy-duty oxygen regulator,
4. Oxygen hose and connections,

5. An acetylene cylinder,
6. A standard acetylene regulator,
7. Acetylene hose and connections.

With the exception of the oxyacetylene cutting, this equipment was described in Chapter 3. The cutting torch is distinguished from the standard oxyacetylene welding torch by having a device (usually a lever) to control the oxygen cutting stream (Fig. 5).

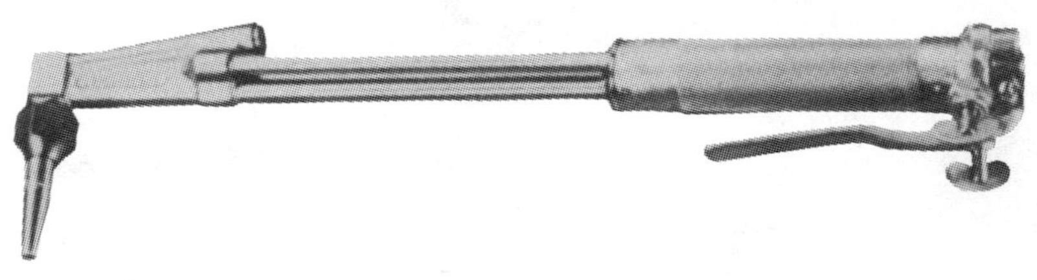

Courtesy Harris Calorific Sales, Inc.

Fig. 5. A typical hand cutting torch.

Cutting torches are available with the oxygen control lever located in a number of different positions depending on the preference of the operator (Fig. 6).

Cutting Torch—The oxyacetylene cutting torch is designed to provide an oxygen cutting stream to the surface area being cut. It is in this way that cutting torches are distinguished from standard welding torches. The provision for an oxygen cutting stream also influences the design of the cutting tip. A typical cutting tip will have a centrally located orifice with a number of others located around it in a circular pattern. The oxygen cutting stream comes from the orifice in the center of the tip; the heating flame from those located around the outer edge (Fig. 7).

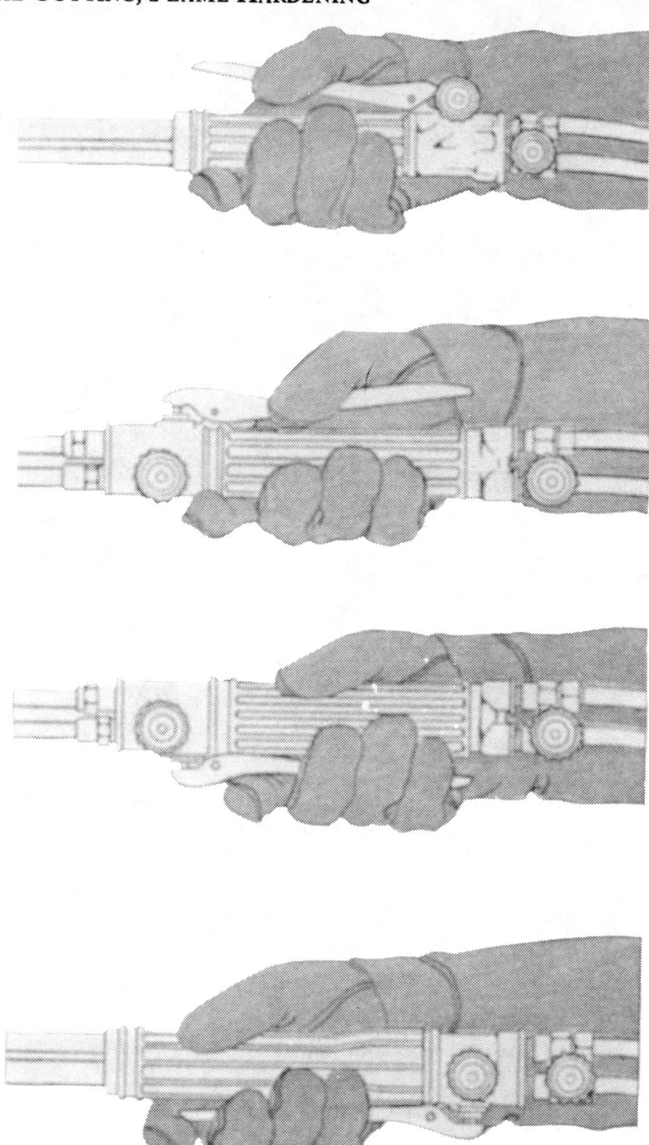

Courtesy Victor Equipment Co.

Fig. 6. Various oxygen control lever positions.

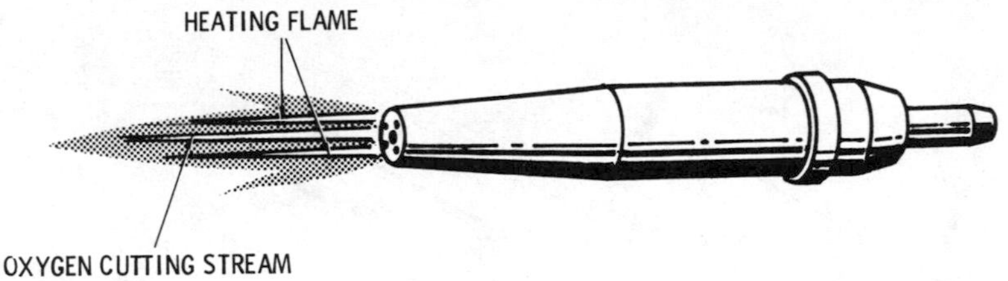

Fig. 7. Operating principles of a typical oxyacetylene cutting tip.

The two basic types of oxyacetylene cutting torches are: (1) the injector type, and (2) the equal pressure type. In the injector-type, the acetylene is delivered to the torch at pressures below 1 psig. The gases in the equal pressure type torches, on the other hand, are delivered at pressures above 1 psig. The operating principles of these two types of torches are described in greater detail in Chapter 3.

Insofar as construction and design are concerned, an oxvacety-lene cutting torch may be either a specially designed cutting torch or the body of a welding torch with a cutting attachment fastened to it.

Fig. 8 illustrates the construction of a typical hand cutting torch. This model has the cutting oxygen valve lever located on the bottom of the torch handle. Note that different head angles (straight, 45°, and 75° are available.

Hand Cutting Attachments

The hand cutting attachment has been developed to provide the standard welding torch with greater operating flexibility. These hand cutting attachments are connected to the body of the welding torch after removing the welding tip assembly.

Hand cutting torch attachments are available in models similar to standard cutting torches. These models include such design variables as: (1) the position of the oxygen control lever, (2) the angle of the cutting head (straight, 75°, and 90° angles), and (3) the fuel gas they are to use (acetylene or MAPP; natural gas or propane; etc.).

The procedure for connecting a hand cutting attachment to the body (handle) of a welding torch is as follows:

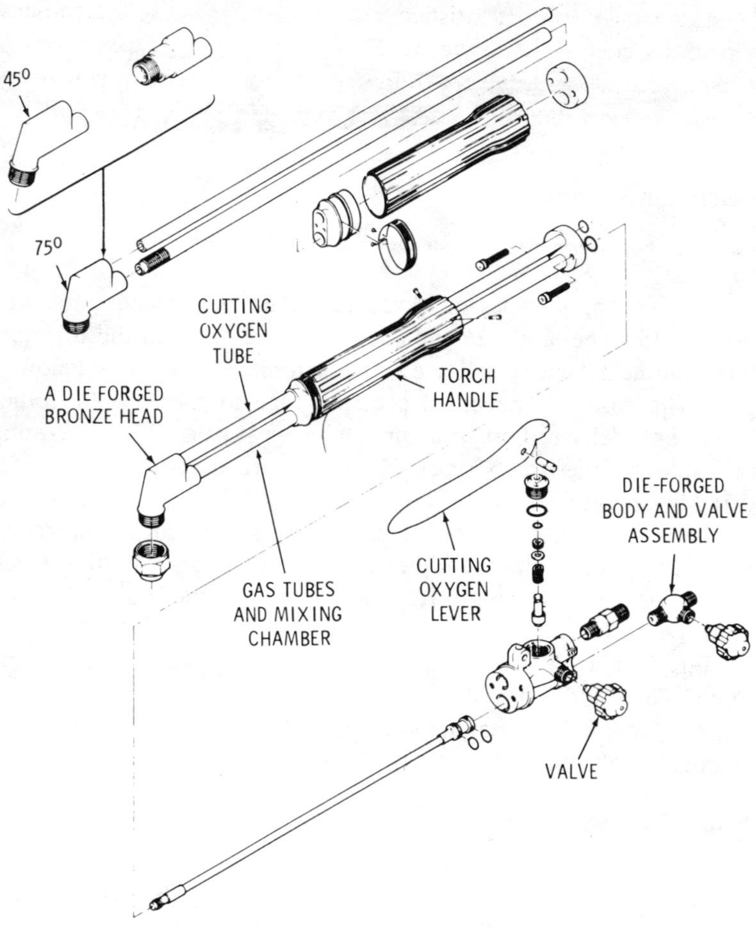

Courtesy Victor Equipment Co.

Fig. 8. The construction of a typical hand cutting torch.

1. Remove the welding tip and mixing assembly,
2. Connect the hand cutting attachment. Tighten securely enough to prevent any possibility of gas leakage. Do *not* force it.
3. Select an appropriate tip and insert it into the hand cutting attachment. Tighten the tip according to the recommendations of the manufacturer.

As soon as you are satisfied that the hand cutting attachment is properly connected to the welding torch handle, you are ready to make the necessary valve adjustments for cutting. These may be described as follows:

1. Open wide the oxygen needle valve located on the welding torch handle and *leave* it open. All oxygen adjustments will be made with the oxygen needle valve located on the hand cutting attachment.
2. Close the oxygen needle valve located on the hand cutting attachment.
3. Open the fuel gas needle valve approximately one full turn.
4. Adjust the fuel gas regulator for the pressure recommended by the manufacturer for the cutting tip size you have selected for use.
5. Close the fuel gas needle valve.
6. Open the cutting oxygen valve.
7. Adjust the oxygen regulator for the pressure recommended by the manufacturer for the thickness of the metal being cut.
8. Close the cutting oxygen valve. The torch is now ready to light. The remainder of the flame cutting procedure is described in a later section.

Types of Cutting Tips—A good cutting tip should demonstrate a high degree of heat resistance and long-wear characteristics. Such tips are usually made from a copper alloy to reduce pocking or burning the tip when slag splatters from a hot cut.

Better cutting tips will produce sharper cuts and cleaner faces with a minimum kerf. The gas delivery should be continuous and free of turbulence. The seating surface of the cutting tip should be precision machined for a tight leakproof fit in the torch head.

Cutting tips are available in a wide variety of shapes, lengths and designs. This variety is due basically to the need to design the cutting tip with a specific job in mind. Thus, cutting tips used for thin gauge metal will differ in design from those used for thicker gauges. Gouging, scarfing, and rivet-washing tips, will differ in design from standard cutting tips.

Cutting tips are generally of the straight-bore type tapering slightly toward the cutting orifice. However, curved-bore designs for special cutting jobs are also available (Fig. 9).

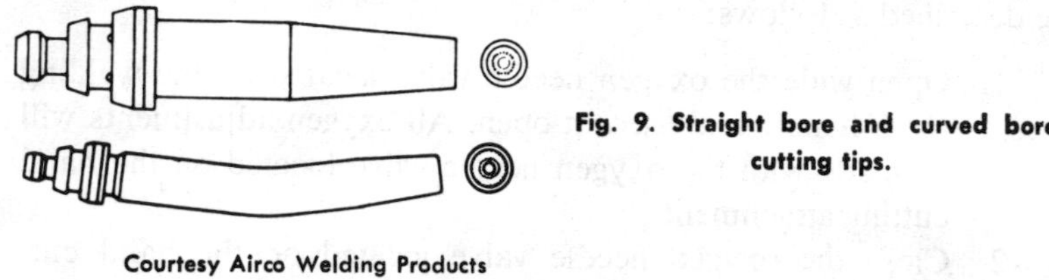

Fig. 9. Straight bore and curved bore cutting tips.

Courtesy Airco Welding Products

Cutting tips designed for use with natural gas or propane differ from those intended for acetylene gas. The principal differences are shown in Fig. 10. Note that the natural gas and propane cutting tips have an outer shield (1) into which the inner structure (2) fits. The shield protects the flame from being blown away from the cut and increases the stability of the flame.

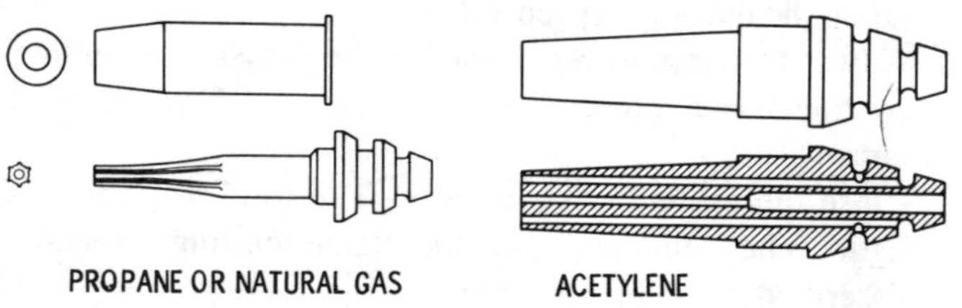

Courtesy Airco Welding Products

Fig. 10. Propane or natural gas cutting tips compared with one designed for use with acetylene.

Flame Adjustment—Correct flame adjustments are very important in ensuring the quality of a cut. Fig. 11 shows the types of oxyacetylene flame adjustments recommended for hand cutting metals. Note that the oxidizing flame is not included in this group. The reason for this is that the oxidizing flame is used in machine cutting torches rather than the hand type.

The type of metal and its thickness also determines the nature of the flame adjustment. For example, cast iron is most successfully cut with a carburizing flame. However, most steels are cut with a neutral flame and thick steel castings require an oxidizing flame.

CUTTING FLAME	EFFECT ON METAL
 f. Acetylene burning in air	Not suitable for cutting
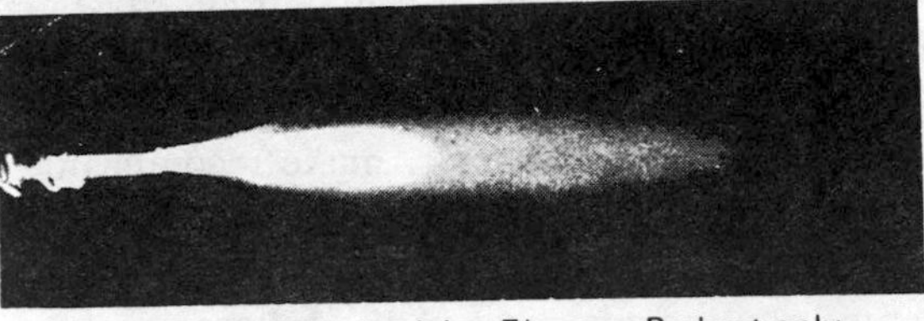 g. Strongly Carburizing Flame — Preheat only  h. Strongly Carburizing Flame — Cutting Oxygen flowing	Excess acetylene helps to get heat down to the bottom of material being cut, this is especially suitable for cutting cast iron.
 i. Neutral Flame — Preheat only  j. Neutral Flame — Cutting Oxygen flowing	Standard adjustment for cutting steel.

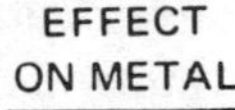

Courtesy Airco Welding Products

Fig. 11. Cutting flame adjustments.

Never allow the oxygen cutting stream to interfere with the preheating flames. Always release the cutting oxygen *after* the preheating flame has been adjusted.

Cutting Tip—The selection of a cutting tip is often determined by the type of fuel gas. For example, an oxyacetylene flame will require a type of cutting tip different from the one recommended for use with oxypropane or oxy-natural gas. The reason for this, of course, is that different fuel gases have different flame characteristics. Using the wrong tip can result in an unstable flame, uneconomical fuel gas consumption, and many other problems. Welding equipment manufacturers will usually make recommendations for using their cutting tips, and these recommendations should be followed as closely as possible. The information given to the operator will generally include most of the following:

1. The manufacturer's tip size,
2. Oxygen cutting orifice drill size,
3. Plate thickness,
4. Oxygen pressure (psi),
5. Fuel gas pressure (psi),
6. Hand cutting speed (inches per minute),
7. Machine cutting speed (inches per minute),
8. Kerf width.

Table 1 is an example of how the manufacturer will present this data. These particular charts describe operating data for *Harris* cutting tips. The drill size (oxygen cutting orifice drill size) is more important than the tip size. The latter is essentially a manufacturer's identification number and will not necessarily correlate from one manufacturer to another. However, the size of the oxygen cutting orifice is equated to a specific drill size. This can prove very helpful when an operator is confronted with the problem of matching the cutting characteristics of tips from different manufacturers.

The type of fuel gas being used is not the sole determining factor in the selection of the cutting tip. Equally important is the *pressure* required by the cutting operation. The nature of the operation (thickness of the metal, type of metal, etc.) will determine the amount of oxygen and fuel-gas pressure needed for

Table 1. Cutting Tip Data

6290-VVCM, 6290-NHM, 2490-VVCM and 2490-NHM
MACHINE CUTTING TIP CHARTS

Plate Thickness	Tip	Cutting Orifice Drill Size	Cutting Oxygen Pressure	Fuel Gas Pressure	Width Kerf in Inches
1/16″	5/0VVCM	#75	40	4 oz. or more	.035
1/8″-3/16″	5/0VVCM	#75	60	4 oz. or more	.035
1/4″	4/0VVCM	#68	40	4 oz. or more	.06
3/8″	000VVCM	#64	75	4 oz. or more	.065
1/2″	00VVCM	#62	75	4 oz. or more	.075
3/4″	0VVCM	#60	90	4 oz. or more	.075
1″	0 1/2VVCM	#58	100	4 oz. or more	.075
1 1/4″	0 1/2VVCM	#58	100	4 oz. or more	.075
1 1/2″	1VVCM	#56	100	4 oz. or more	.08
2″	1VVCM	#56	105	4 oz. or more	.08
2 1/2″	1VVCM	#56	90	4 oz. or more	.08
3″	1 1/2VVCM	#54	95	4 oz. or more	.09
4″	2VVCM	#53	95	4 oz. or more	.09
5″	2VVCM	#53	105	4 oz. or more	.09
6″	2 1/2VVCM	#51	95	4 oz. or more	.09
7″	3VVCM	#49	100	4 oz. or more	.11
8″	4VVCM	#45	95	4 oz. or more	.13
9″	5VVCM	#41	90	4 oz. or more	.15
10″	5 1/2VVCM	#39	90	4 oz. or more	.17
10″	5NHM	#35	60	4 oz. or more	.23
11″	6NHM	#31	60	4 oz. or more	.26
12″	7NHM	#29	60	4 oz. or more	.27
15″	8NHM	#25	60	4 oz. or more	.29

Courtesy Harris Calorific Sales, Inc.

successful cutting. This information is also necessary in the selection of a suitable cutting tip.

As you can see, selecting a cutting tip is not as simple as it would first seem. Careful consideration must be given to a number of different aspects of the cutting operation before the final decision is made. Experience ultimately will be the best guide for making these decisions.

Hand Cutting Procedure—The procedure for oxyacetylene cutting may vary depending on the requirements of a specific job. However, there are certain fundamental procedural steps that can be considered common to all cutting operations. In the interest of cutting efficiency and safety, the operator should establish for himself a set cutting procedure. The cutting procedural steps listed below are arranged in the sequence they would most likely occur in a cutting operation. This format essentially follows the *NCG* cutting instructions provided by *Chemetron* for its welding equipment.

1. Using the thickness of the metal to be cut as a guide, select an appropriate cutting tip size for the torch.
2. Insert the cutting tip in the torch and tighten it with the wrench recommended by the welding equipment manufacturer. Do not attempt to force the tip. If it will not tighten properly, check the heads for wear or stripping. It may need to be replaced.
3. Open the fuel gas valve on the torch and purge the lines of any air.
4. Adjust the fuel gas regulator to the working pressure recommended for this size tip.
5. Close the fuel gas valve on the torch.
6. Open the torch cutting oxygen valve.
7. Adjust the oxygen regulator to the working pressure recommended for this size tip.
8. Close the cutting oxygen valve. At this point, both the fuel gas and oxygen regulators have been set at their recommended working pressures. The fuel gas and cutting oxygen valves on the torch should be closed. Fuel gas and oxygen will have filled the lines down to the torch needle valves. The torch is now ready to light.
9. Begin the flame adjustment by opening the torch fuel-gas needle valve slightly and igniting the gas with a spark lighter. Adjust the flame so that it is steady and well-defined.
10. Open the oxygen needle valve (*not* the cutting oxygen valve) on the torch until a bluish-white cone is established.

11. Open the cutting oxygen valve.
12. Readjust the oxygen needle valve for the correct preheat flame.
13. Position the torch so that the cutting tip is placed at a right angle (perpendicular) to the surface.
14. Allow the flame to play on one spot on the surface until it sweats and changes to a red color.
15. As soon as the red color is obtained, gradually open the high-pressure oxygen cutting valve. The operator is now ready to begin cutting.
16. The cutting of different types of metals and metal alloys is described in the appropriate sections of this chapter.
17. The fuel-gas needle valve *must* be closed first when shutting off the torch. As soon as this is done, close the oxygen needle valve. The regulator pressure adjusting screws should be released last. These adjustments are for temporary shut-downs of the torch. When long periods are involved, both cylinder valves should also be closed.

STRAIGHT CUTTING OF STEEL

For practice in cutting, it is best to start with a piece of ½-inch steel plate about 6 inches wide. Rule a chalk line about ½-inch from one edge of the plate and place it so the line just clears the far side of the welding table.

Connect a cutting torch to a cylinder of oxygen and one of acetylene, insert the proper size nozzle and adjust pressures for ½-inch metal. Before proceeding further, the learner should put on goggles, light the torch, adjust the preheating flame to neutral, and switch off the cutting oxygen.

Hold the torch vertical to the plate so that the preheating flame just touches the end of the chalk line on the edge of the steel plate. Hold the torch steady until the spot is heated to a bright red. The next step is to open the valve that controls the cutting oxygen and move the torch slowly, but steadily along the chalk line.

If the cut has been properly started, a shower of sparks will fall from the underside of the plate. This action indicates that the cut is penetrating entirely through the plate. The cutting torch should be moved just fast enough so that the cut continues to penetrate the plate completely. If the torch is moved too fast the cutting jet will fail to penetrate entirely through the plate and the cutting will stop.

If the cutting stops, immediately close the cutting valve and preheat the point where the cut stopped until it is a bright red. On opening the cutting valve there should be no difficulty in restarting the cut.

The heat of the preheating flame will tend to melt the edges of the cut if the torch is moved too slowly. This will produce a very ragged appearance or at times fuse the metal together again.

CUTTING HOLES IN STEEL

To cut a small hole in steel, support a piece of ½-inch steel plate between two fire bricks on top of the welding table. Hold the torch over the center of the plate until the preheating flame has heated a round spot bright red (Fig. 12).

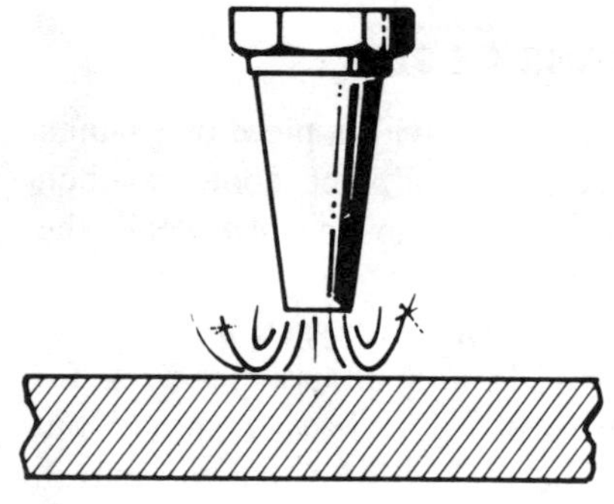

Fig. 12. Preheating the surface in preparation to cutting a small hole.

Courtesy Airco Welding Products

Raise the torch approximately ½-inch (Fig. 13) and open the cutting valve very gradually to avoid blowing slag back into the torch tip. The molten slag can also be removed by tilting the tip or moving it side ways and rotating it around the cut (Fig. 14).

Cutting larger holes does not present a problem. Mark the outline of the hole with chalk. At any point within the chalk outline, cut a hole (as previously described). Start a cut from this

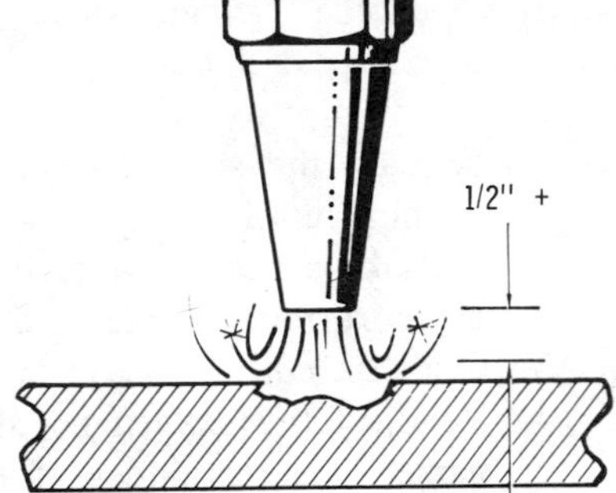

Fig. 13. Lifting the torch.

Courtesy Airco Welding Products

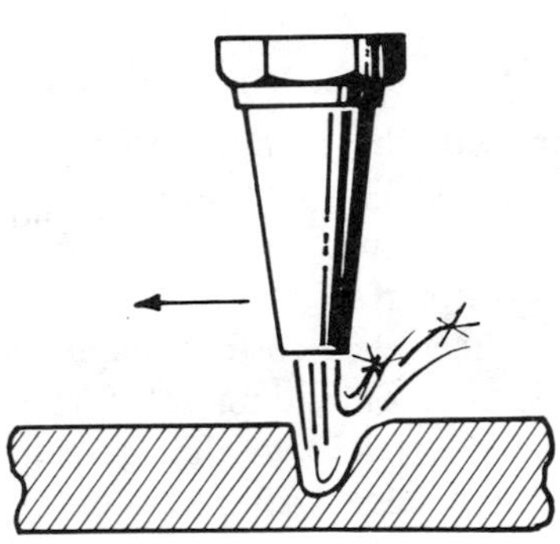

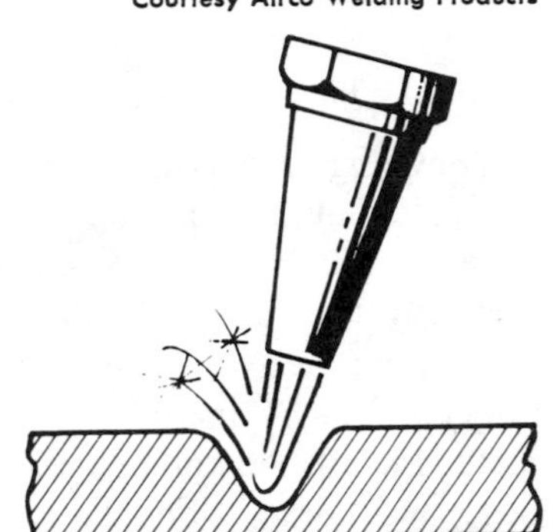

Courtesy Airco Welding Products

Fig. 14. Tilting the torch.

point, working toward and then following the outline that has been drawn on the plate. In this way the hole will have clean sharp edges all around.

Large round holes can be made accurately with the aid of a radius bar or a compass, one leg of which is fixed at the center and the other is made by the torch nozzle.

OXYFUEL GAS HAND CUTTING PROCEDURE

The procedure for oxyfuel gas (propane, natural gas, LPG, etc.) cutting is similar in most respects to the procedure previously described for oxyacetylene cutting. The principal differences between the two involve flame adjustment. The procedure is as follows:

1. Open the fuel gas valve very slightly and ignite the gas with a spark lighter.
2. Open the oxygen valve very slightly and continually increase the size of *both* the fuel gas and oxygen valve opening until enough preheat flame has been established to make a suitable adjustment. The cone of the flame will be short, fairly stable, and bluish white in color.
3. Open the cutting oxygen valve.
4. Readjust the oxygen needle valve for the desired preheat flame. The remainder of the cutting procedure is similar to that described for oxyacetylene cutting.

BEVEL CUT

Beveling is a common operation with the cutting torch and one that should be mastered by all welders. This operation is performed by holding the torch head at an inclined angle instead of vertically (i.e. perpendicular to the work surface).

Place a piece of ½ inch or thicker steel plate so that one edge projects 3 or 4 inches over the side of the welding table steel plate. Incline the torch head at an angle of 45° to the top surface, and cut a triangular prism off the edge. After the cut is finished, the piece of plate remaining will have one edge beveled at an angle of 45°.

Beveling can also be done by holding the torch perpendicular to the surface. However, when beveling, the preheat holes are aligned so that they are parallel to the kerf. Contrast this with the

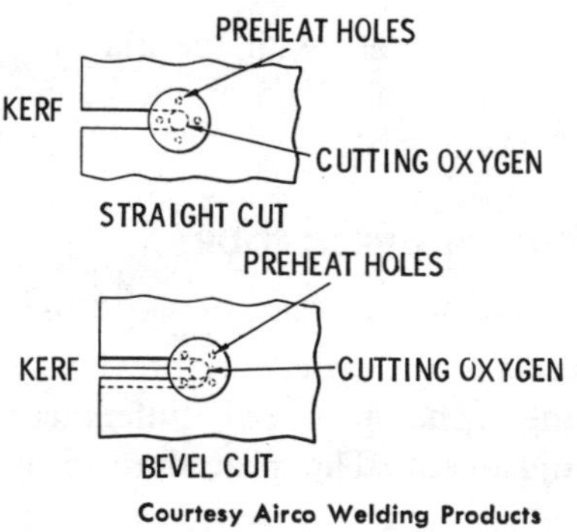

Courtesy Airco Welding Products

Fig. 15. Position of the preheat holes for a bevel cut.

position of the preheat holes when making a straight cut as shown in (Fig. 15).

CUTTING CAST IRON

In spite of the fact that steel and cast iron each contain well over 90% iron, the two metals behave quite differently under the cutting torch. There is no difficulty in cutting cast iron, once the operator has become familiar with its somewhat different actions.

Cast iron requires the following adjustments during cutting as opposed to steel:

1. The preheating flames should be adjusted with an excess of acetylene,
2. The nozzle must be held farther away from the metal,
3. Longer preheating is necessary (cast iron must be almost molten before the cut will start),
4. The nozzle must be oscillated continually across the line of cut.

The ease of cutting cast iron depends on the quality of the metal. The grade of castings known in the trade as "good grade iron castings" is easier to cut then the poorer material used for such parts as grate bars or floor plates.

OXYACETYLENE PROCEDURE FOR CUTTING CAST IRON

The operations involved in cutting cast iron should include the following steps:

1. Adjust the oxygen cutting regulator so as to give the correct pressure for the thickness of the cut. This information is furnished by manufacturers of cutting equipment.
2. With the cutting valve open, adjust the heating flames to give the correct excess of acetylene, as noted in the instructions supplied by the manufacturer of the torch. The amount will vary for different makes of torches.
3. The oxygen pressure must be increased with the thickness of the metal to be cut. The exact pressure of oxygen to be used in each case is given by the manufacturer.

4. Close the oxygen cutting valve. Then begin with a preliminary preheating with the nozzle along the edge to be cut from top to bottom. Preheat what will be the first "bite" of the cutting torch through the entire thickness of the metal. If this part of the casting is warmed, the cut will start quicker and more easily. In general, the warmer the whole mass of iron, the easier it is to cut (Fig. 16).

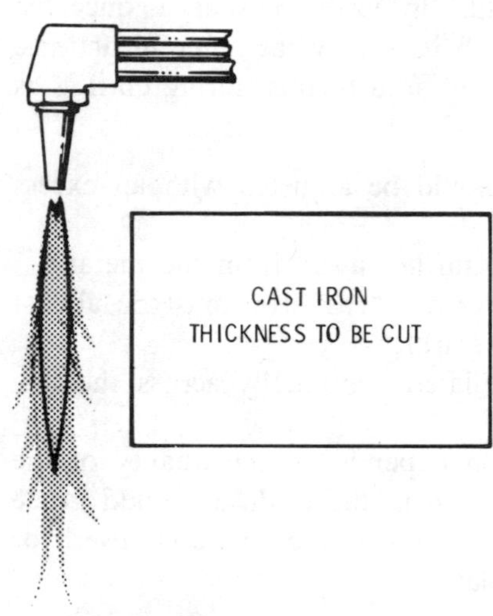

Courtesy Airco Welding Products

Fig. 16. Preheating the flame so that there is an excess fuel streamer equal to the thickness to be cut.

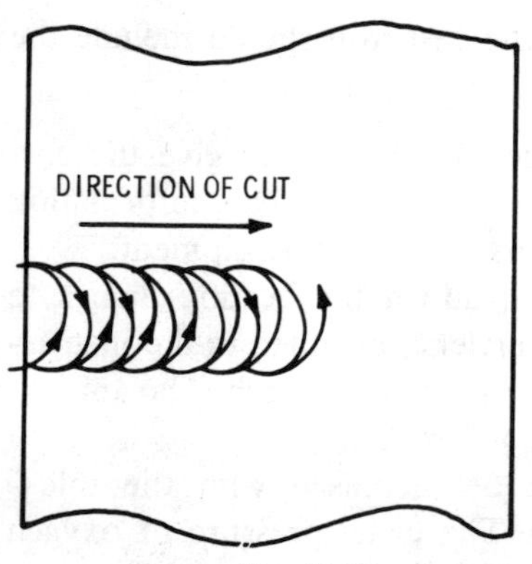

Courtesy Airco Welding Products

Fig. 17. The torch is moved in semicircular motions ½ inch to ¾ inch as required to clear cut in heavy sections.

5. After the preheating is accomplished, hold the torch so that the tip points backward at an angle of 75°. The inner cones of the flames should be ⅛ to ¼ inch above the surface.
6. To start the cut, give the torch tip a swinging motion describing semicircles across the line of cut (Fig. 17).
7. Heat a semicircular area about ½ to ¾ inches in diameter until the metal is actually molten. When the metal begins to "boil" open the cutting valve for an instant to blow off the slag.
8. Move the tip just off the heated edge, open the cutting valve quickly and then move the torch along the line of cut, with the tip at an angle of 75° (Fig. 18). Use the same swinging motion described previously and keep the metal hot.

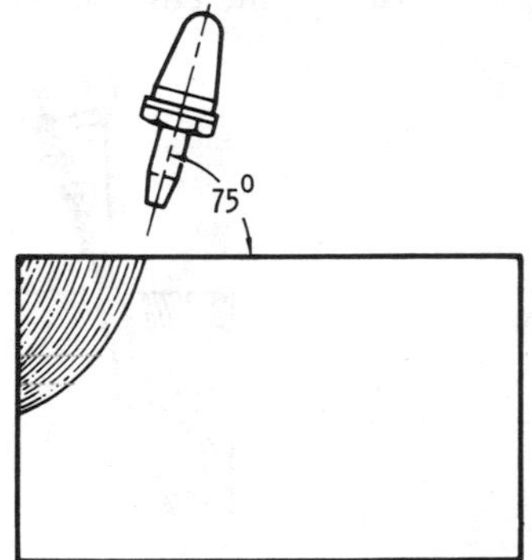

Fig. 18. Cutting tip at the 75° angle.

Courtesy Airco Welding Products

9. As the cut progresses, gradually straighten up the torch until it is at an angle of about 90° (Fig. 19). On heavy material, the metal will be sufficiently preheated by the cutting reaction and the cut should go forward regularly.
10. The torch should be swung from side to side, describing semicircles across the line of cut, as shown in Fig. 17 for the entire length. The diameter of these semicircles will depend on the thickness of the metal and the proficiency of the operator. As the operator becomes experienced

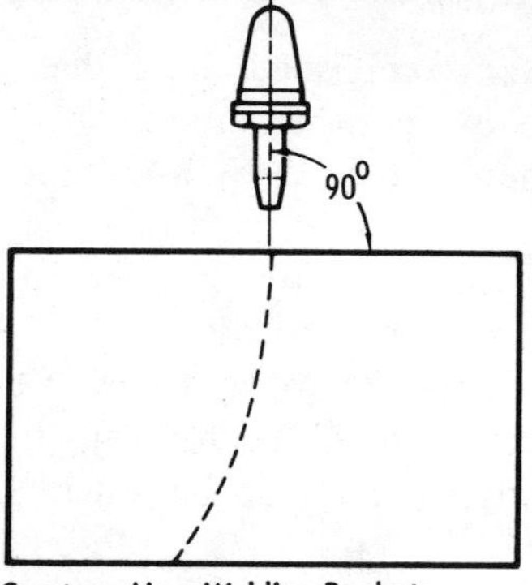

Fig. 19. Cutting tip held at a 90° angle.

Courtesy Airco Welding Products

with this work he will be able to reduce the diameter of the swing, and therefore the width of the kerfs. The cutting speed should be such that the cutting oxygen jet just sweeps the edge of the cut (Fig. 20).

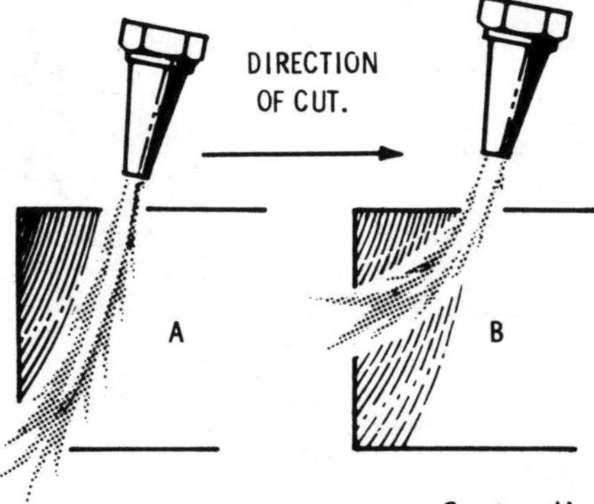

Courtesy Airco Welding Products

Fig. 20. Cutting jet should just sweep edge of cut as shown in A and not advance too deeply as shown in B.

11. The cutting temperatures can be increased by sticking a steel flux rod into the cut as shown in (Fig. 21).

OXYGEN LANCE CUTTING

The oxygen lance cutting process is used to cut heavy thick sections of low carbon steel and other metals. The oxygen lance

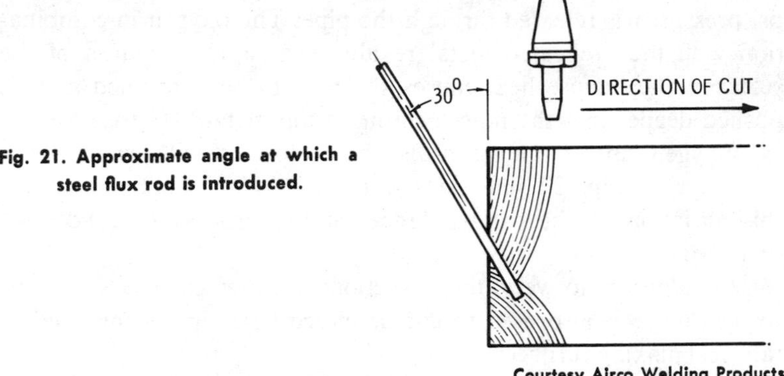

Fig. 21. Approximate angle at which a steel flux rod is introduced.

Courtesy Airco Welding Products

(a length of pipe) operates in conjunction with a preheating torch. This combination results in a much deeper cut than is capable with the ordinary oxyacetylene cutting process. Oxygen lance cutting is also refered to as *lancing, oxygen lancing,* and *lance cutting.*

The oxygen lance is basically a length of black iron pipe with an internal diameter of 1/8 or 1/4 inch. Because of the very high temperatures involved in oxygen lance cutting, much of the lance is also consumed.

When cutting low carbon steel, a section of the metal is first preheated to a cherry red. The lance is brought against this pre-

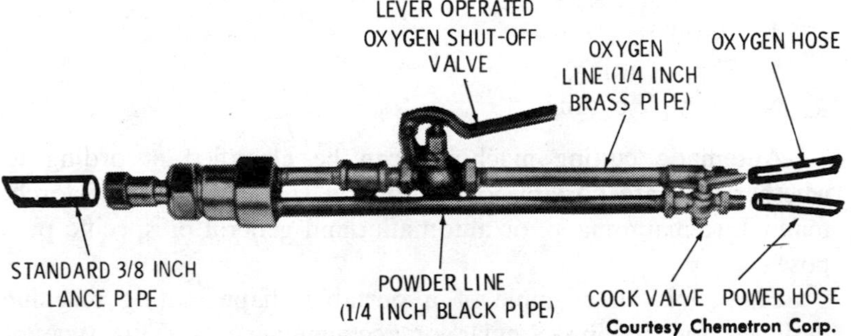

Courtesy Chemetron Corp.

Fig. 22. A lance valve used in the power-oxygen variation of lance cutting. The Ferrojet® lance valve is used in the steel industry for piercing, skull-cutting and lance cutting when the material thickness exceeds the capability of hand cutting torches.

heated section, and a stream of oxygen (at approximately 40 to 50 psi pressure) is released through the pipe. The oxygen in combination with the preheated metal results in a rapid oxidation of the surface. The intense heat causes the pipe to be consumed as it is pushed deeper into the hole forming in the metal.

Oxygen lance cutting can also be performed with iron powder and oxygen. Fig. 27 illustrates a *Ferrojet®* lance attachment for this variation of the oxygen lance cutting process (see POWDER CUTTING).

In addition to very thick sections of low carbon steel, the oxygen lance is also used to cut or pierce holes in castings and to tap steel making furnaces.

AUTOMATIC MACHINE FLAME CUTTING

Many different types of automatic flame cutting machines have been developed for a wide variety of applications. These flame cutting machines are designed and constructed to hold the torch (or torches) in specific predetermined positions for repeated and rapid high quality cuts. The types of cutting operations for which automatic flame cutting machines are generally designed for include:

1. Straight line cutting,
2. Circle cutting,
3. Pipe cutting,
4. Beveling,
5. Shape cutting,
6. Multiple cutting.

Automatic cutting machines can be classified according to whether they are portable or stationary; single or multiple torch; manual, semiautomatic, or automatic; and general or specific purpose.

Fig. 23 is an example of a portable shape cutting machine designed for making circular or rectangular cuts. This type of machine uses steel templates to make identical cuts in production quantities. The maximum diameter of the circles that can be cut is limited by the length of the cutting arm.

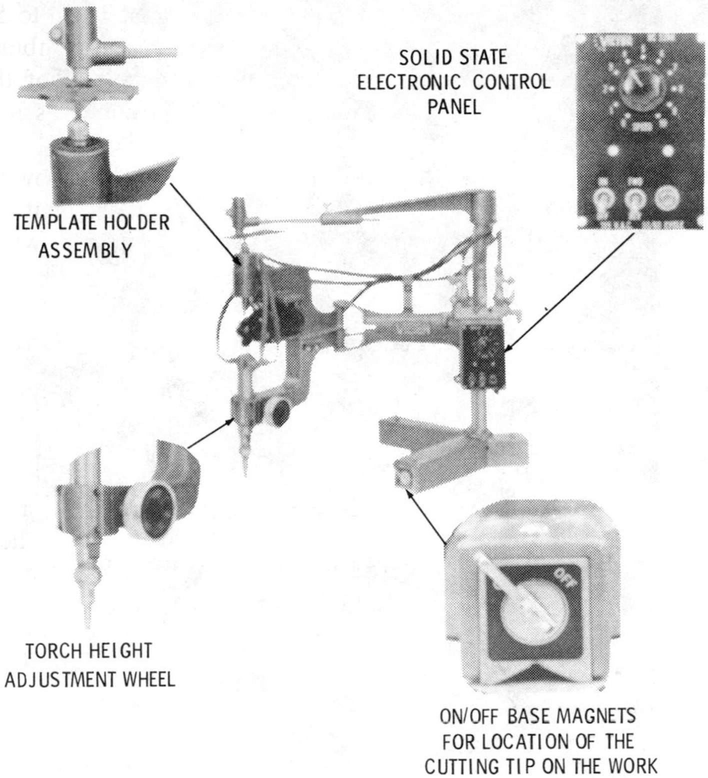

Courtesy Victor Equipment Co.

Fig. 23. A portable cutting machine.

Portable pipe cutting and beveling machines are generally limited to a standard pipe diameter range (e.g. 4 to 30 inches). These are lightweight machines with no more than one or two torches.

The tractor type machines (Fig. 24) are guided by steel tracks (for straight cuts) or a radius rod and a center point (for circles, etc.). These are small lightweight machines mounting one or more torches. Guided tractor cutting machines can be adapted to flame, plasma-arc, or carbon-arc cutting operations. A built-in protractor gives a complete range of bevel cutting angles. The controls on the tractor control panel will include:

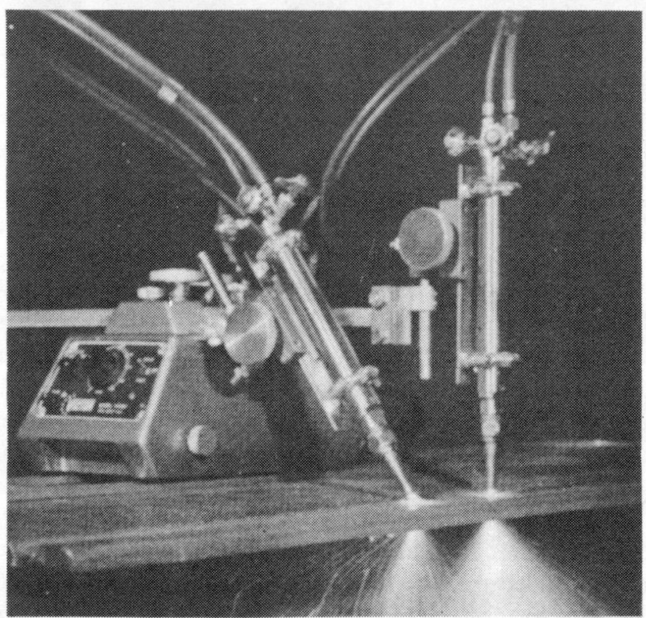

Courtesy Victor Equipment Co.

Fig. 24. Portable tractor-type cutting machine.

1. A calibrated speed control dial,
2. A motor control switch (for drive and free-wheeling settings),
3. A direction control switch (forward and reverse).

The wide range of cutting applications for guided tractor cutting machines is shown in Figs. 25 to 28.

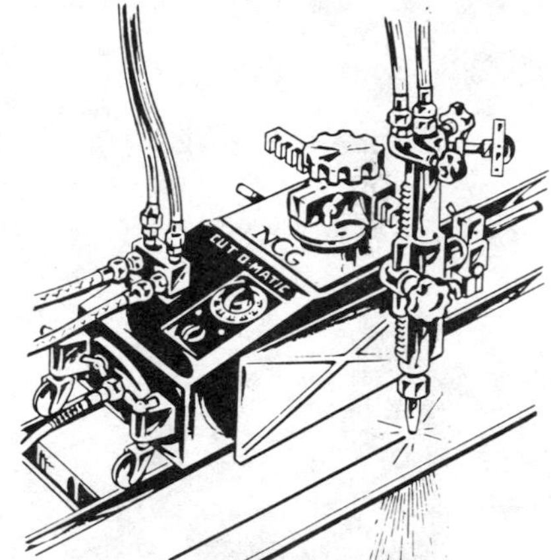

Fig. 25. Straight line cutting.

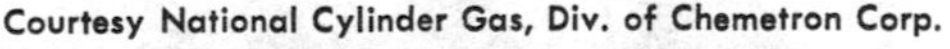
Courtesy National Cylinder Gas, Div. of Chemetron Corp.

Fig. 26. Strip cutting.

Courtesy National Cylinder Gas, Div. of Chemetron Corp.

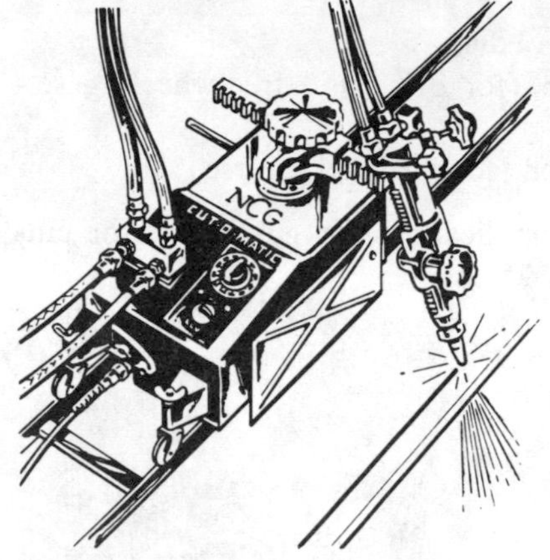

Fig. 27. Automatic bevel cutting.

Courtesy National Cylinder Gas, Div. of Chemetron Corp.

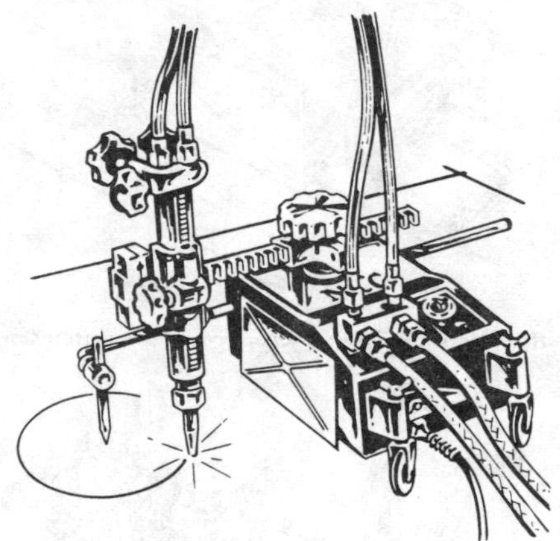

Fig. 28. Circle cutting.

Courtesy National Cylinder Gas, Div. of Chemetron Corp.

The large, stationary-type shape cutting machines operate on the pantograph principle. A table adjoining the cutting machine contains the template design (Fig. 29). An electronic tracing device follows the lines of the template. The movements of the tracing device are coordinated by means of a long bar with those of the pantagraph assembly. The latter, complete with one or more torches, is mounted on tracks directly over the work. This gantry-type design is capable of reproducing large regular and irregular shapes with extreme accuracy.

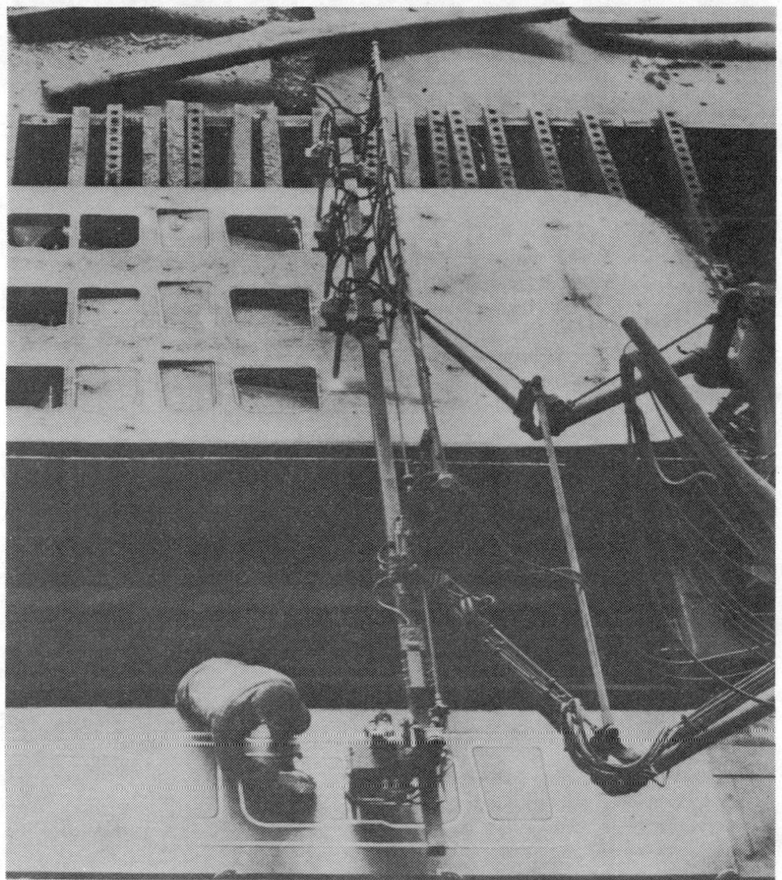

Courtesy Airco Welding Products

Fig. 29. A large stationary-type cutting machine.

Gantry-type, stationary, automatic flame cutting machines can be photo-tracer controlled or numerically controlled. The photo-tracer cutting machine, shown in Fig. 30, contains a photoelectric noncontact type tracer for use with both line and edge templates. An indicator signals when the tracer fails to follow the outline of the template and the cutting oxygen automatically shuts off.

The numerically controlled (n/c) cutting machine, shown in Fig. 31, is radically different from the photo-tracer type machine.

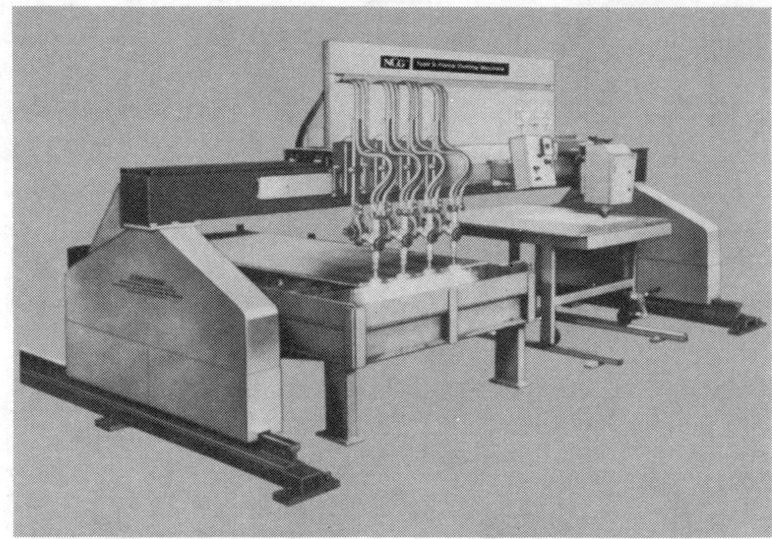

Courtesy National Cylinder Gas, Div. of Chemetron Corp.

Fig. 30. Dual drive gantry flame cutting machine.

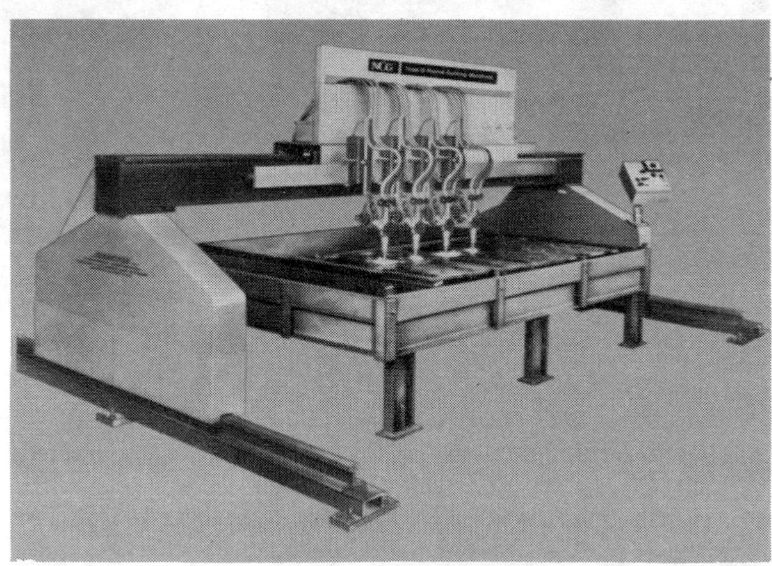

Courtesy National Cylinder Gas, Div. of Chemetron Corp.

Fig. 31. A numerically controlled gantry flame cutting machine.

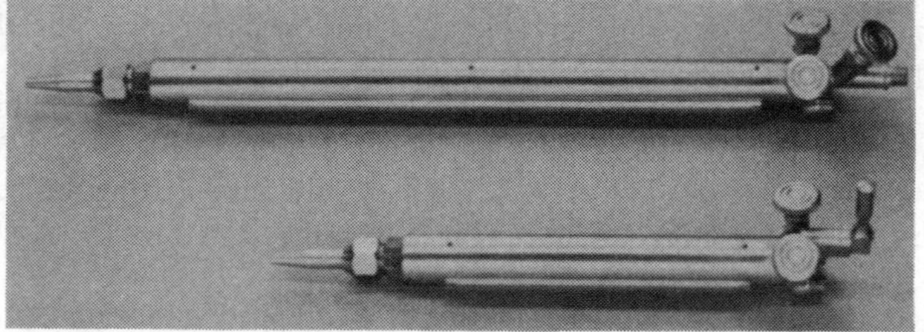

Courtesy Harris Calorific Sales, Inc.

Fig. 32. Heavy-duty type machine cutting torches.

The cutting is controlled by: (1) an eight-channel punched-tape input with 300 characters per second, (2) a photoelectric tape reader, and (3) tape reels. Because templates are not used, off-line cutting errors are eliminated.

Machine cutting torches (Fig. 32) are available in a number of different designs. The selection of a suitable torch depends on the type of fuel gas being used, the thickness of the metal, the nature of the job, and other variables. The welding equipment

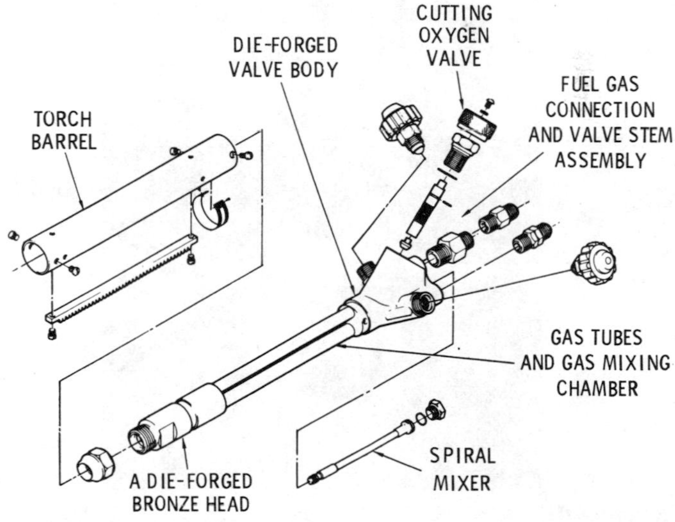

Courtesy Victor Equipment Co.

Fig. 33. The construction of a typical machine cutting torch.

manufacturers generally give detailed information concerning the operating characteristics of their torches.

The construction of a typical machine cutting torch is illustrated in Fig. 33. The gas tubing is made from stainless steel for maximum heat resistance. As is the case with hand cutting torches, the machine cutting torch can be either the injector-type or the equal-pressure type.

Some machine torch attachments are equipped with water spray nozzles to minimize warping, especially when cutting the thinner sections of steel plate (one inch thick or less) (Fig. 34). The rate of flow is established by needle valves on the individual torch assemblies. This arrangement enables the operator to turn off a specific torch in a multiple torch operation.

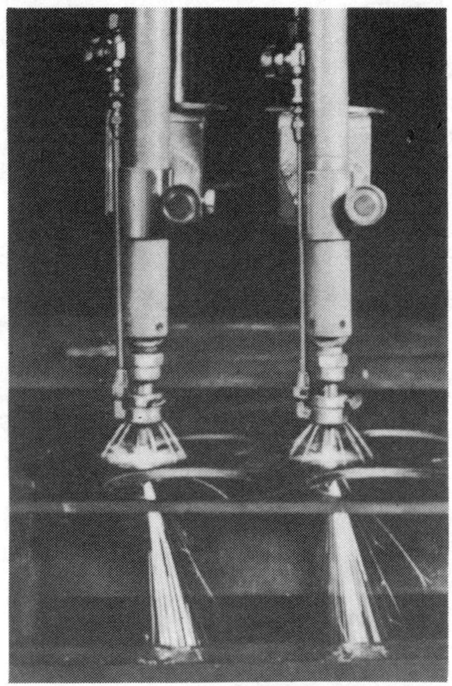

Courtesy Chemetron Corp.

Fig. 34. Water spray attachment for machine cutting torches.

Using an automatic flame cutting machine is almost mandatory when production-type precision cutting is required. Practical experience has shown that automated cutting procedures provide greater speed and, in the long run, reduced costs.

POWDER CUTTING

The iron-powder cutting process was developed for metals that exhibit a tendency to develop oxides when in contact with oxygen. This is a particular difficulty when using the oxyacetylene cutting process, because the oxygen cutting stream causes a rapid built up of oxides in some metals.

A finely divided iron powder is injected into the oxygen cutting stream of the torch and immediately finds its way into the cutting area. The iron particles are rapidly oxidized resulting in a sudden increase of heat on the metal surface. As a result of the intense supplemental heat, the refractory oxides that form on the surface are melted and flushed from the cutting area. This permits the cutting flame of the torch to cut the metal without interference.

The iron-powder cutting process was originally developed to cut stainless steel, and is now successfully applied to a whole range of oxidation-resistant metals (nickel-chrome, nickel and chrome alloys, cast iron, aluminum, bronzes, steels containing more than 5% alloy, etc.). *Ferrojet* is a manufacturer's trade name for this cutting process which was developed by the *Chemetron Corp.*

A typical iron-powder cutting installation (Fig. 35) consists of the following components:

1. A hopper and a supply of iron powder,
2. Cylinders and supplies of oxygen, acetylene, and nitrogen,
3. Three gas regulators,
4. A powder control regulator,
5. A powder control valve and ejector device,
6. A cutting torch,
7. Oxygen and gas fuel hoses,
8. Two gas cylinder trucks.

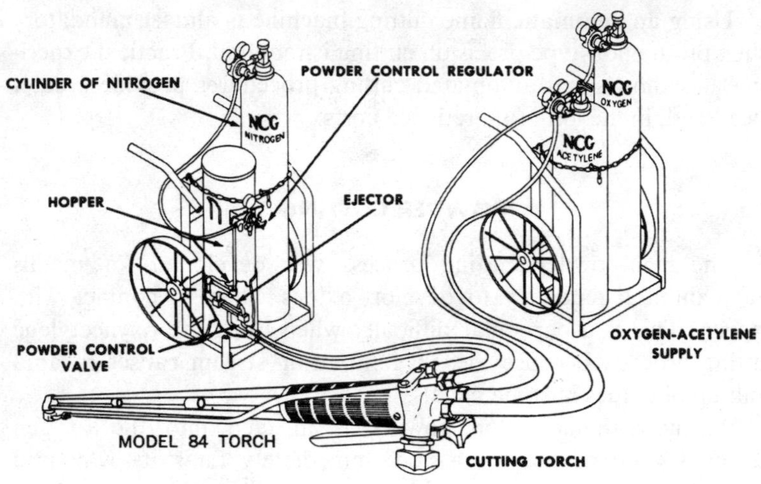

Courtesy National Cylinder Gas, Div. of Chemetron Corp.
Fig. 35. Iron powder cutting equipment.

The iron-powder hopper illustrated in Fig. 36 has a capacity of 150 lbs. It is designed to provide an automatically uniform flow of powder and operates in conjunction with the cutting oxygen valve on the torch. The iron powder is fed under pressure into the ejector unit at the bottom of the hopper where it is mixed with a converting gas and fed in the form of an atomized powder to the torch tip.

The powder control regulator is connected externally to the hopper. Other control components, the ejector mechanism and the powder control valve, are located internally in the bottom half of the hopper.

FLUX INJECTION CUTTING

Another method for preventing the interference of refractory oxides in the cutting process is through the use of a flux (Fig. 37). A flux (a nonmetallic compound) is introduced into the cutting area. This material unites with the oxides and forms a slag which

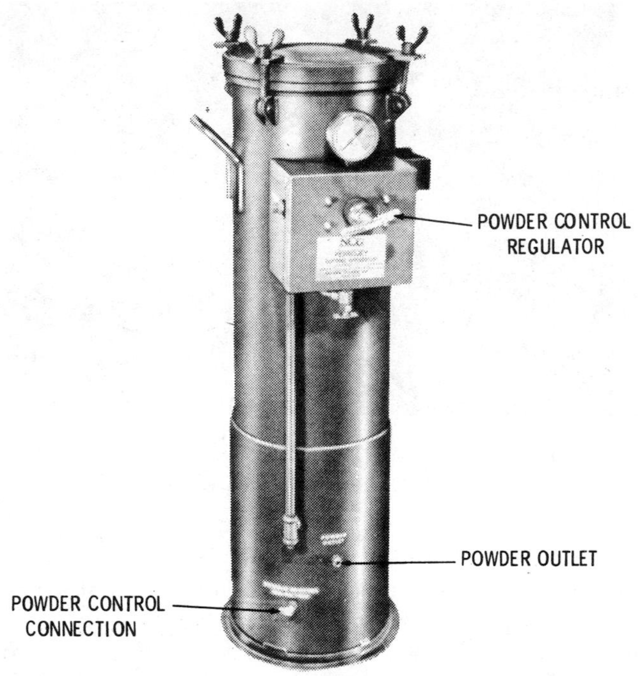

Courtesy National Cylinder Gas, Div. of Chemetron Corp.
Fig. 36. Iron-powder hopper.

is washed from the cut. The equipment used is quite similar to that employed in the iron-powder process.

UNDERWATER OXYHYDROGEN CUTTING

Underwater cutting of metals with gas torches has been known in theory for almost as long as surface cutting. But it remained dormant until 1925, when Captain Edward Ellsberg, USN, developed the first practical torch and used it in the raising of the U.S. Submarine S-51 the following year. The equipment and technique were first made available on a commercial basis in 1927 by *Craftsweld.*

The two principal methods for gas cutting metal underwater are: (1) the oxyhydrogen method, and (2) the oxyacetylene

Courtesy Airco Welding Products

Fig. 37. Flux injection cutting torch.

method. Oxyacetylene cutting underwater is limited to a depth of less than 20 feet because acetylene becomes unstable when used at a pressure over 15 lbs. per square inch. Acetylene gas is generally not recommended for underwater cutting. Oxyhydrogen cutting, on the other hand, may be used for any thickness at any depth. Table 2 gives the various working depths and the recommended pressures (air, hydrogen, and oxygen) for each.

A complete underwater gas cutting outfit will generally include the following:

1. An underwater torch,
2. A supply of oxygen gas in cylinders manifolded together.
3. A suitable oxygen regulator,
4. High-pressure oxygen hose,
5. A supply of hydrogen gas in cylinders manifolded together.
6. A suitable hydrogen regulator,
7. High-pressure oxygen hose,
8. A supply of compressed air of adequate pressure and volume,
9. High-pressure compressed air hose,
10. A complete diving outfit with loudspeaker telephone.

Table 2. Recommended Pressures (Gauge Pressure at Surface) for Oxyhydrogen Underwater Cutting Torch

Working depth, feet	Water Pressure pounds (Approx.)	Length of hose feet	Pressures		
			Air lbs.	Hydrogen lbs.	Oxygen lbs.
10	4	100	55	55	75
20	9	100	60	60	80
30	13	100	65	65	85
40	17	150	75	75	95
50	22	150	80	80	100
60	26	200	90	90	110
70	30	200	95	95	115
80	35	250	100	100	120
90	39	250	105	105	125
100	43	300	115	115	135
125	54	300	125	125	145
150	65	300	140	140	160
175	76	400	155	155	175
200	87	450	170	170	190
225	97	450	185	185	200

Courtesy United States Navy

The basic principles of an underwater gas cutting torch are very simple. The fuel gas used, whether acetylene or hydrogen, requires two parts of oxygen for complete combustion. One part is supplied in the form of pure oxygen, delivered to the torch under proper control from the oxygen cylinder or a similar source. The other part of oxygen is obtained from the atmosphere surrounding the flames at the end of the tip (Fig. 38). Since there is no air under water, the deficiency is made up by creating an artificial atmosphere around the tip by means of an air jacket and delivering to it a supply of compressed air from a suitable source on the surface (Fig. 39).

For efficient operation, the discharge end of the air nozzle must be designed to supply the compressed air without creating turbulence and without contaminating the purity of the oxygen delivered through the central orifice of the tip. Turbulence will interfere with preheating, while a reduction of the purity of the oxygen by contamination with compressed air will reduce the cutting speed, if carried far enough, may even make any cutting impossible.

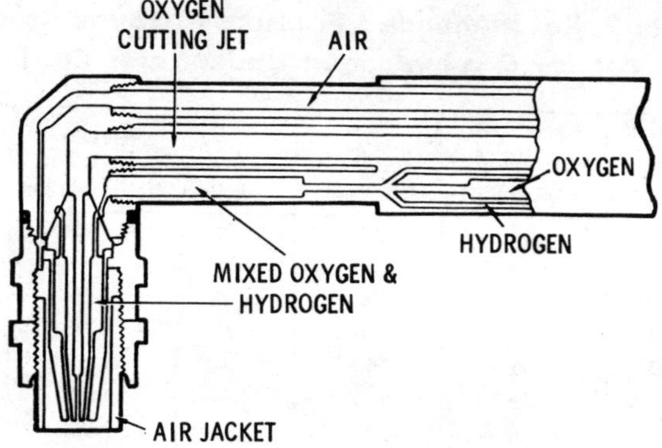

Courtesy United States Navy

Fig. 38. Cross section of an underwater cutting torch.

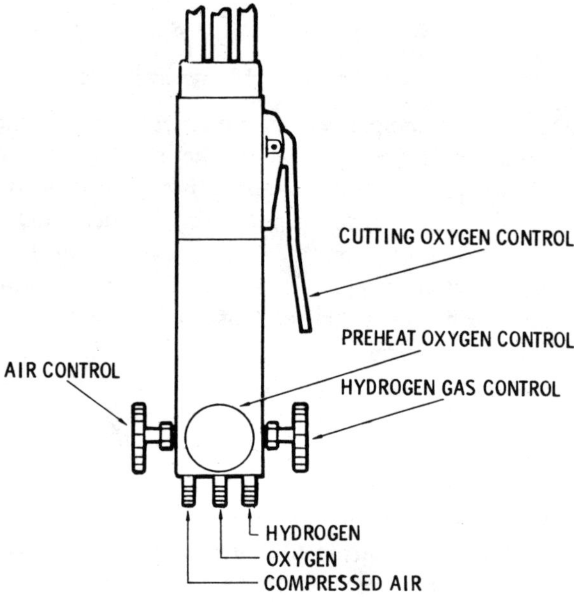

Courtesy United States Navy

Fig. 39. Underwater torch controls.

An underwater cutting torch is subject to more rigorous conditions than a surface torch because of its exposure to the erosive and corrosive action of mud, sand, oil, and similar chemicals frequently present below the surface. When the valves are closed before returning the torch back to the surface, the water and its impurities will back into the torch.

An underwater cutting torch must, therefore, be made of tough, noncorrosive metals exclusively and, in addition, must be so constructed as to allow easy and direct access to all valves and seats for cleaning without dismantling adjacent parts.

For efficient operation, the torch should also be free from attachments and similar projections since they are too easily damaged and even broken off when the torch is lowered to the diver or raised to the surface.

Fig. 40 shows an example of one of the several different types of cutting tips designed for use in underwater cutting. This particular one can be used with hydrogen or acetylene.

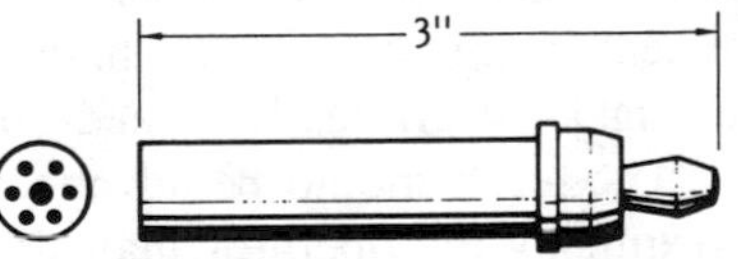

Fig. 40. Underwater cutting tip designed for use with either hydrogen or acetylene.

Courtesy Victor Equipment Co.

Underwater gas torches are most frequently ignited on the surface, adjusted for the depth of water at which cutting is to be done, and taken down by the diver or lowered to him under his instructions. On locations where this is not practical, the diver may use an electric spark created between two carbon tips of an insulated holder connected to a supply of current on the surface. A pneumatic type spark lighter or a chemical capsule has also been used.

Fig. 41 illustrates two procedures for starting a cut underwater. One involves starting the cut at the edge of a plate, whereas the other starts it somewhere in the middle. Both involve pre-

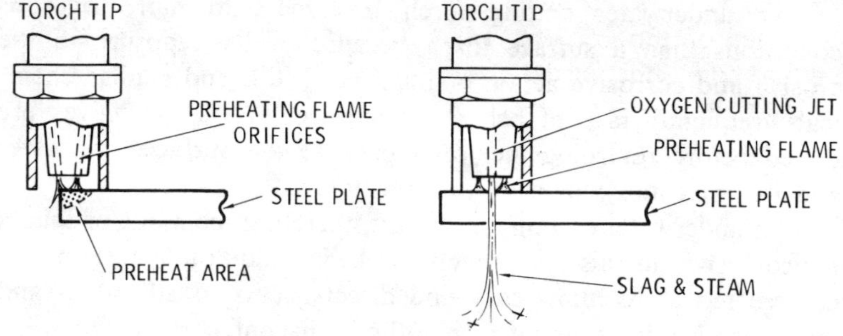

Courtesy United States Navy

Fig. 41. Starting an underwater cut.

heating the surface until it reaches suitable cutting temperatures. On an edge cut, this will be indicated by sparks flying out from the metal. When the metal is ready for cutting, press the cutting jet lever and direct the cutting stream against the preheat spot. Beginning a cut in the middle of a plate results in blowing the molten slag and steam through the plate and out the back. The torch is held against the plate until a spot has reached suitable preheat temperatures. Then, it is withdrawn slightly while the cutting jet is gradually turned on.

The torch should be advanced steadily and evenly. By moving too quickly the operator may skip over spots leaving them uncut. Too slow a speed may result in cooling the surface below a suitable temperature for cutting. Consequently, the cut will be lost and will have to be started again.

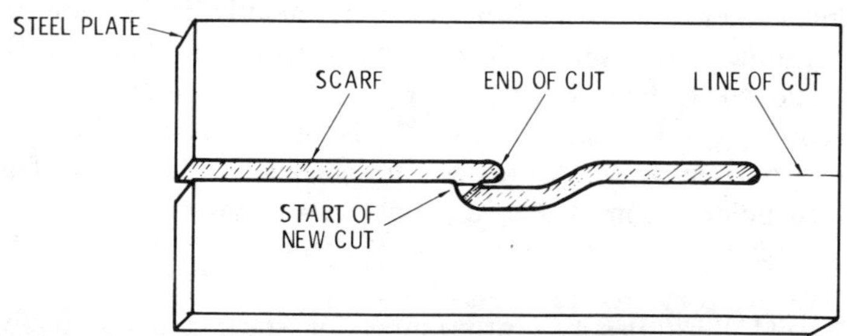

Courtesy United States Navy

Fig. 42. Re-starting an underwater cut.

Fig. 42 shows the method for restarting a cut. Note the new cut is started on the side of the old cut and then returned to the old cutting path. This is done to assure a complete cut.

UNDERWATER ARC CUTTING PROCESSES

The two principal methods of cutting metal under water with an electric arc are:

1. The arc-oxygen underwater cutting process,
2. The metallic arc underwater cutting process.

The latter is basically a melting process employed when an oxygen supply is not available.

The basic principle of the arc-oxygen underwater cutting process consists of combining an electric arc as a source of heat with pure oxygen under pressure as a means of rapidly oxidizing the molten metal.

A typical arc oxygen underwater cutting installation will generally include the following:

1. A fully insulated underwater arc cutting torch with a short length of cable attached,
2. A supply of oxygen gas in cylinders manifolded together,
3. A suitable oxygen regulator,
4. A standard DC welding machine (AC may also be used) with a current capacity suited to the type of electrode used,
5. Standard welding cable of suitable size for the required length connected and probably insulated to the lead cable of the torch,
6. A standard welding cable with clamp for ground connection,
7. A shut-off switch (fused),
8. A supply of coated electrodes thoroughly waterproofed and fully insulated over the entire length projecting out of the torch.

A fully insulated arc cutting torch (Fig. 43) consists of a head and directly connected power cables. Midway on the underside of this head is attached a dielectric coupling which insulates

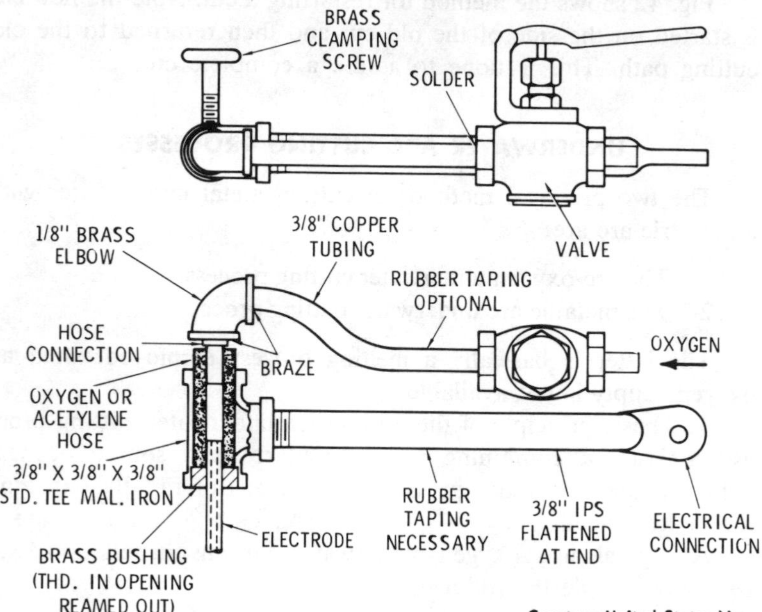

Courtesy United States Navy

Fig. 43. Arc-oxygen underwater cutting torch.

the head from the rest of the torch. To the coupling is attached the oxygen control valve, lever operated, and the oxygen supply hose connected at the inlet.

At the front end of the torch head is a collet-type electrode grip with provisions for sealing the inlet end of the tubular electrode to the oxygen supply with the same insulated locknut which locks the collet around the electrode.

The interior of the torch head is equipped with a screened flash arrester cartridge (inserted from the front) which absorbs the force and heat of the flash, dissipates its effect, and protects the torch from the damage. The cartridge can be readily removed for cleaning or replacement by the diver under water. The collet lock nut, the torch head, and the torch insulator coupling are fully insulated with dielectric jackets made of materials not affected by water, oil, acids, alkalis, or spatter.

The three basic types of electrodes that have been used in arc-oxygen underwater cutting are:

1. Tubular carbon electrodes,
2. Tubular ceramic electrodes,
3. Tubular steel electrodes.

The tubular steel electrodes have largely replaced the others in usage.

The tubular (cored) carbon electrode is about ⅜ inch by ¾ inch in cross section and has one or more metal tube cores along its length for transmitting the oxygen under pressure to the arc. It is 12 inches long and uses from 600 to 1000 amperes. It lasts about 20 to 30 minutes in operation. After it is consumed, the torch must be returned to the surface where the insulator is cut away, a new electrode screwed into position, and the electrode and front end of the torch carefully taped with several layers of rubber and friction tape before the torch is returned to the diver. The tendency of tubular carbon electrodes to be brittle contributed to their replacement by tubular steel electrodes.

Tubular ceramic electrodes consist of hollow rods of silicon carbide or some other ceramic material. They are generally ½ inch in diameter and 8 inches long. In general, they are used in the same way as tubular steel electrodes and with a similar type of torch or holder except that the torch must be equipped with a larger collet to accomodate the ½-inch diameter rod. Each electrode will burn for a period of about ten minutes, so that not as much time is lost in changing electrodes as with the more rapidly consumed steel rods. Also, because of the larger cross section, heavier currents may be used. However, careful handling is required to avoid breaking the electrode when changing rods or working in awkward places.

Tubular steel electrodes consist of a steel tube with an extruded and waterproofed coating. These electrodes are available in 5⁄16-inch diameters (OD) and 14-inch lengths. Changing tubular steel electrodes under water is easily accomplished. The electrode is inserted into the torch head and locked and sealed by taking a quarter turn on the locknut. When the electrode is consumed, the lock nut is given a reverse quarter turn and the stub is blown out by depressing the oxygen valve lever.

Tubular steel electrodes can be employed under water not only for cutting steel, but also for cutting cast iron and nonferrous

Table 3. Electrodes Recommended for Underwater Cutting

ELECTRODE	Size	Position	Current (Amperes) (X)	Time for 12" Burnoff (Seconds)
Westinghouse Flexarc SW	5/32"	H	170-210	56-44
		V	170-210	56-44
		O.H.	170-190	56-50
	3/16"	H	220-260	59-50
		V	220-260	59-50
		O.H.	190-210	66-61
Lincoln Fleetweld 37	3/16"	H	220-260	60-49
		V	220-260	60-49
		O.H.	200-220	66-60

SUBSTITUTE ELECTRODES

ELECTRODE	Size X	Position	Current (Amperes) (X)	Time for 12" Burnoff (Seconds)
A. O. Smith Smithweld 15	3/16"	H	220-260	55-45
		V	220-260	55-45
	1/4"	H	250-290	86-77
Metal and Thermit Murex Alternex	3/16"	H	210-250	55-45
		V	210-250	55-45
		O.H.	190-210	60-55
	1/4"	H	230-270	90-73
Hollup Sureweld "C"	3/16"	H	200-240	83-73
		V	200-240	83-73
Reid-Avery Raco 7	1/4"	H	210-230	74-65
		V	210-230	74-65
Metal and Thermit Murex Type A	3/16"	H	200-240	55-49
		V	200-240	55-49
		O.H.	170-190	61-57
General Electric G.E. W-25	3/16"	H	180-220	61-51
		V	180-220	61-51

(X)—For D.C. only; add 10% for A.C.

X —These electrodes shall not be tested in 5/32 inch size. In most cases 5/32 inch will be satisfactory.

Courtesy United States Navy

metals. It may also be used in the open air for the cutting of difficult high alloy steels such as the stainless group, some tool steel, and manganese high-alloy steel.

Electrodes recommended for underwater cutting are listed by trade name in Table 3. The *Westinghouse Flexarc SW* and the *Lincoln Fleetweld 37* are the most commonly used for this type of cutting.

The metallic arc underwater cutting process is basically a melting process. Solid steel electrodes with extruded and waterproofed coatings may also be used for underwater cutting. Such electrodes are available commercially in a complete range of sizes; the 3⁄16 and 1⁄4 inch diameter will operate successfully in cutting plate up to 3⁄4 inch in thickness.

An arc-ogygen underwater cutting torch may be used by removing the collet used with the tubuluar steel electrodes and replacing it with a collet of correct dimensions. Standard underwater welding equipment is also used.

TIG AND MIG CUTTING

Tungsten inert gas (TIG) cutting uses equipment similar to that found in the TIG welding process, but with certain important differences. The most important difference is found in the positioning of the tungsten electrode which is recessed in the ceramic cup of the torch. The gas within the cup is ionized by an electric arc. This ionized gas stream subsequently functions as a vehicle for the arc. Both the gas stream and arc are tremendously constricted as they leave the torch so that a narrow highly concentrated column of heat cuts into the metal surface. TIG cutting, then, is actually an entirely new process characterized by the ionization of the inert gas. The name *plasma-arc cutting* has been given to this new process.

Metal inert gas (MIG) cutting generally uses the same equipment as found in the MIG welding process. The principal difference between the MIG welding and the MIG cutting is that higher amperages are required for the latter (as high as 1800 amperes).

PLASMA-ARC CUTTING

Plasma-arc cutting is a high-speed cutting process capable of attaining temperatures as high as 50,000° F. It is used to advantage .in cutting magnesium, brass manganese, nickel, *Monel, Inconel,* copper, aluminum, cast iron, clad steels, and many other metals.

Plasma-arc cutting is not an officially recognized term by the *American Welding Society,* but its use has become so widespread in the welding equipment industry that it is commonly known by this name. The *American Welding Society* has applied to this cutting process the term gas tungsten-arc cutting (that is to say, TIG cutting).

The operating principle of plasma-arc cutting is based on the observation that when a high-frequency electric arc is passed through a stream of inert gas (usually nitrogen) the latter is ionized (changed to plasma). Both the ionized gas column and the arc are forced through a small orifice in the torch nozzle. The continuous effort of the column of ionized (plasma) gas to return to its original gaseous state causes the high temperatures that

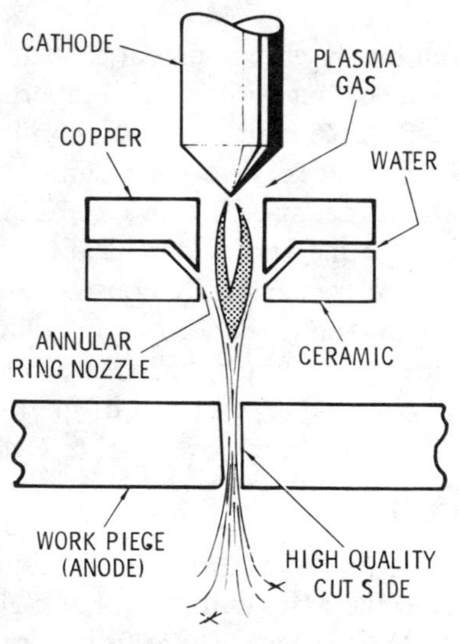

Courtesy Chemetron Corp.

Fig. 44. The plasma-arc cutting process.

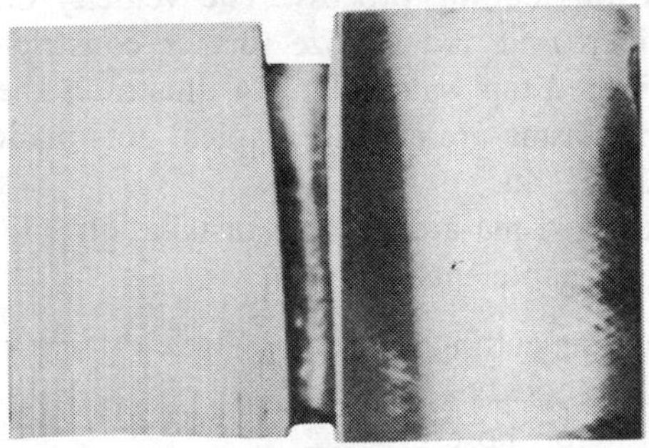

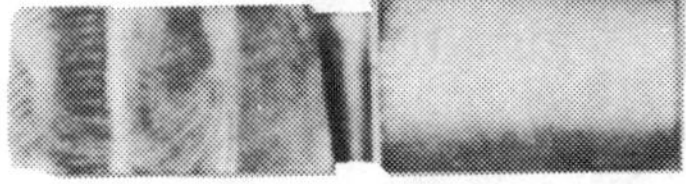

Courtesy Thermal Dynamics Corp.

Fig. 45. Examples of cuts made with a plasma-arc cutting torch.

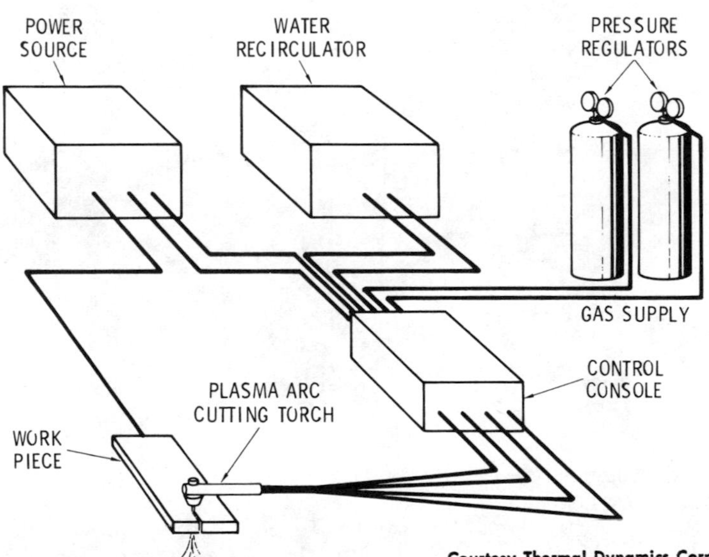

Courtesy Thermal Dynamics Corp.

Fig. 46. Plasma-arc cutting system.

characterize this cutting process. The velocity of the plasma gas column is increased and its size further constricted by a sprayer stream of filtered tap water. Fig. 44 illustrates the basic operating principles of plasma-arc cutting. Typical cuts made by this process are shown in Fig. 45.

A typical plasma-arc cutting installation (Fig. 46) includes the following components:

1. A plasma cutting torch,
2. Water-cooled power leads,
3. Gas supply hoses,
4. A cutting control console,
5. A recirculating water cooler,

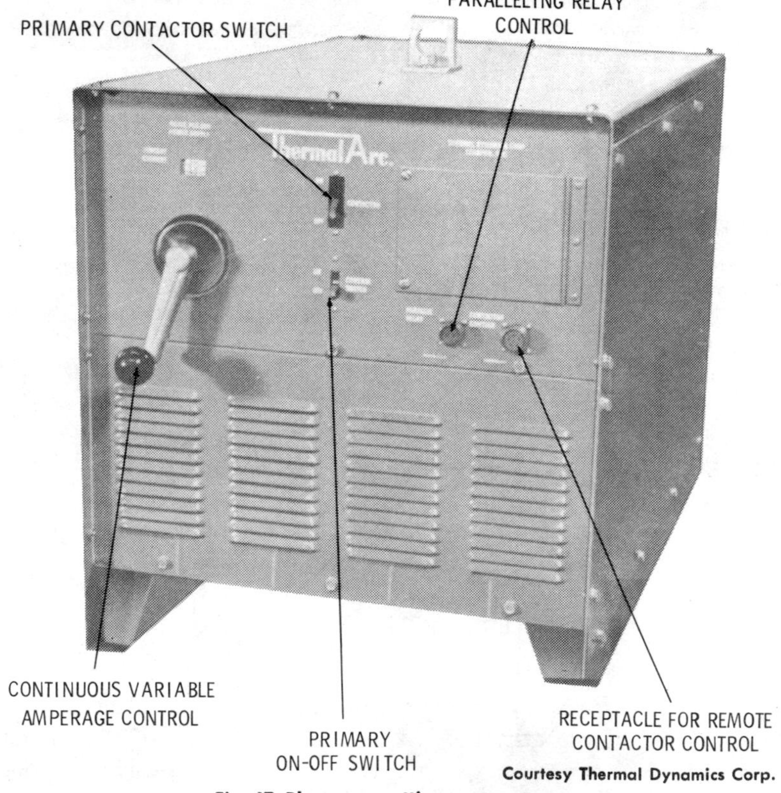

Courtesy Thermal Dynamics Corp.

Fig. 47. Plasma-arc cutting power source.

Fig. 48. Two plasma-arc cutting sources in combination with a pilot console unit.

Courtesy Thermal Dynamics Corp.

6. Gas cylinders and a gas supply,
7. A power source.

The plasma-arc cutting power source illustrated in Fig. 47 has input current characteristics of 230/460 volts, 3-phase, 60 Hz. It is capable of delivering the following output:

amperes	*load volts*	*open circuit volts*
250	200	400
500	100	200
1000	50	100

This power source is equipped with a silicon rectifier with a moving core reactor. Continuous variable amperage control permits major current adjustments. Fig. 48 illustrates a conventional method of combining more than one power source with a pilot console unit.

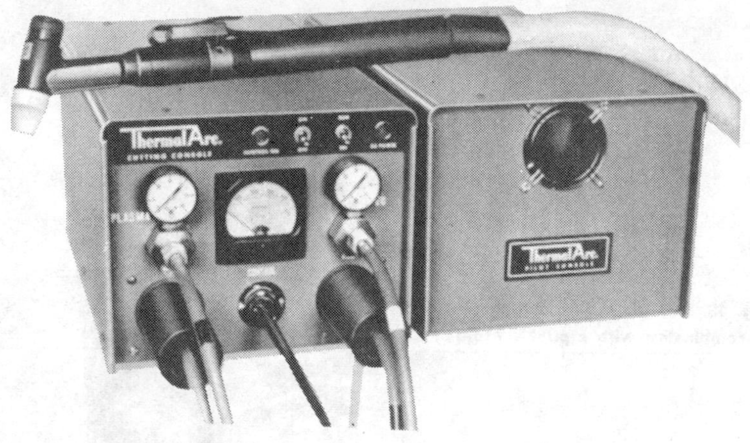

Courtesy Thermal Dynamics Corp.

Fig. 49. Plasma-arc cutting console and pilot console.

Fig. 49 illustrates both the plasma-arc cutting console and the plasma-arc pilot console. These two consoles in combination permit the performance of other welding operations with the same standard power source when not being used for plasma-arc cutting.

Machine-mounted plasma-arc cutting torches (Fig. 50) are available in several types depending on the thickness of metal to be cut. The front end components are replaceable according to the gas used or whether or not bevel cutting is to be performed. The plasma gas is either nitrogen (140 SCFH maximum) or, in the case of the heavy-duty type, a mixture of 65% argon and 35% hydrogen (175 SCFH maximum). The water flow for these torches operates at 5 gpm at 125 psi (130°F maximum input temperature). Table 4 lists some typical operating parameters for plasma-arc cutting.

CARBON-ARC CUTTING

The carbon arc is used to cut metals that do not readily oxidize and are therefore difficult to cut with a gas flame. This is a high temperature melting operation. The fact that cutting depends on melting rather than chemical action distinguishes the carbon-arc cutting process from cutting with the gas flame. The carbon-arc cutting process is particularly suited to cutting cast iron.

Table 4. Operating Parameters for *Water-Arc*® Plasma Cutting

STAINLESS, MILD AND ALLOY STEEL

Nozzle Size	Power (kw)	Current (amp)	Material Thickness (inches)	Speed (ipm)	Water (gpm)*	Nitrogen at 150 psig (scfh)
.166	20	160	1/8	80-120	.33	140
.166	30	195	1/4	60-100	.33	140
.166	40	245	3/8	50-100	.33	140
.166	60	350	1/2	50-100	.33	140
.166	70	400	3/4	35-70	.33	140
.187	80	450	1	30-60	.38	165
.187	90	500	1 1/2	10-20	.38	165
.187	100	550	2	5-15	.38	165
.187	100	550	3	5-10	.38	165
.220	150	750	4	6-8	.55	240

ALUMINUM

Nozzle Size	Power (kw)	Current (amp)	Material Thickness (inches)	Speed (ipm)	Water (gpm)*	Nitrogen at 150 psig (scfh)
.166	20	160	1/8	80-200	.33	140
.166	30	195	1/4	60-150	.33	140
.166	40	245	3/8	50-150	.33	140
.166	50	300	1/2	50-140	.33	140
.187	60	350	1	40-90	.38	165
.187	70	400	1 1/2	20-40	.38	165
.187	100	550	2	10-30	.38	165
.187	100	550	3	5-15	.38	165
.220	150	750	4	10-15	.55	240

*For constricting the arc.

Courtesy Chemetron Corp.

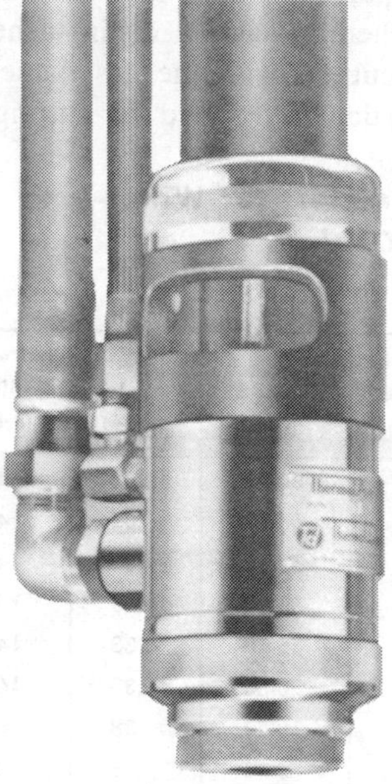

Courtesy Thermal Dynamics Corp.

Fig. 50. Machine-mounted plasma-arc cutting torch.

Carbon, carbon-graphite, graphite, and shielded metal-arc electrodes can all be used, although the carbon electrodes are the most popular. A graphite electrode permits the use of higher currents than the other two.

Because carbon-arc cutting is a melting operation, the work must be positioned (or the cut started) in such a way that the molten metal can flow out of the cutting area. Generally, cuts are made on the under side of a metal surface and begun on the lowest corner or side.

The carbon-arc cutting process uses a standard arc-welding power source (adjusted for DCSP), an electrode holder, and the equipment found in standard arc-cutting stations.

AIR CARBON-ARC CUTTING

The air carbon-arc cutting process (Fig. 51) combines an electric arc with a jet of compressed air. The air stream passes through a specially designed torch (electrode holder) parallel to the electrode and is used to remove the molten metal while the

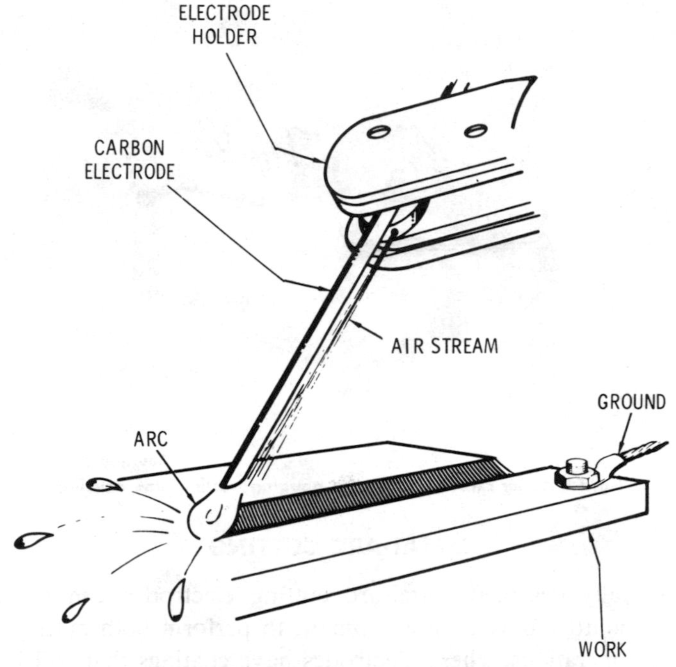

Fig. 51. Air carbon-arc cutting process.

cutting is in progress. As a result, the work can be cut in any position. Some of the advantages of air carbon-arc cutting include little or no distortion and high travel speed.

The air carbon-arc cutting process uses the same equipment and supplies as standard carbon-arc cutting except for the addition of the following:

1. A supply of compressed air,
2. A specially designed air carbon-arc electrode holder,
3. Compressed air hoses.

An air carbon-arc gun is shown in Fig. 52. It operates on the principle described above.

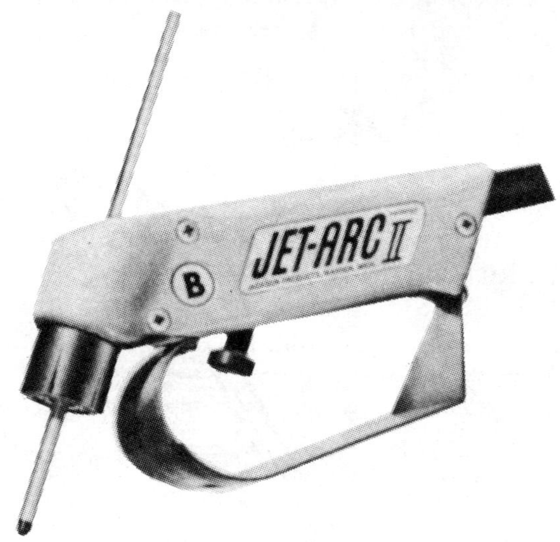

Courtesy Chemetron Corp.

Fig. 52. An air carbon-arc gun for gouging, cutting, and beveling.

METAL-ARC CUTTING

Specially designed metal-arc cutting electrodes can be used with standard arc-welding equipment to perform both cutting and gouging operations. These electrodes have coatings that will intensify the cutting action and blow away the molten metal.

A compressed-air stream can be combined with the arc to create a cutting process similar to air carbon-arc cutting. It uses the same equipment and supplies as in the air carbon-arc cutting process.

OXY-ARC CUTTING

The oxy-arc cutting process uses an electric arc in combination with a stream of oxygen to cut metals. Along with air carbon-arc cutting, this represents another approach to cutting metals that prove difficult for the gas flame processes.

Oxy-arc cutting employs flux coated (or non-coated) tubular

steel electrodes and a specially designed electrode holder (to carry both the arc and the oxygen stream). A suitable source of oxygen is also necessary along with oxygen hoses. The power source may be either AC or DC. This process was originally developed for underwater cutting, but is also used as an above-surface cutting process.

GOUGING, SCARFING, AND FLAME MACHINING

Gouging is a cutting or machining procedure used to remove excess metal (particularly from round objects rotated in a lathe), to cut (gouge) out defective welds, to remove defects from plate metal and metal castings, and to cut curved grooves along the edge of a metal plate in preparation for welding. The amount of metal that can be removed by gouging is far in excess of the amount removed by a lathe. The object of gouging is not to penetrate completely through the metal, but rather to gouge out a specifically desired contour. It is important to maintain a steady travel speed (approximately 30 to 50 inches per minute), a correct angle, and low oxygen pressure (25 to 35 lbs. psi) to ensure uniform contour. Both manual and machine gouging is possible.

Gouging can be performed by using a gas flame, an electric arc, or an arc in combination with a stream of compressed air (air carbon-arc gouging). In the steel industry, flame gouging is known as *scarfing* and finds useful application in the removal of seams, cracks, and other defects from billets and blooms. The term *flame machining* is often applied to mechanized flame gouging in which gouging speeds of 100 feet per minute are common. Flame machining is commonly used to cut grooves in steel plate.

In flame (oxyacetylene) gouging, the oxygen stream flows at a much slower rate than the one used for cutting. This results in a slower rate of oxidation which permits greater accuracy and control. For those metals that resist oxidation (e.g. cast iron, stainless steel, etc.), arc gouging (using a carbon or graphite electrode) is preferred to flame gouging. Air carbon-arc gouging represents a further development of the arc gouging process. The excess molten metal is removed by a special high velocity stream of air after the metal has been melted by the arc.

The tips (Fig. 53) used in flame gouging are designed for an even preheat treatment of the metal and accuracy in cutting. These tips can also be used for other metal removing procedures and preparations, including deseaming, rivet washing, and rivet cutting. Standard cutting tips can also be used for gouging, but the specially designed gouging tip is recommended. Many tips (as well as torches) are designed to accept most fuels used for preheating. The type of fuel used will also influence the tip selection. A gas mixer is commonly incorporated in these tips—a design factor that enables fast starts and prevents flashbacks or backfire.

Table 5. Scarfing Tip Selection Data

SCARF WIDTH AND DEPTH PER TIP SIZE

Tip Size	Scarfing O_2 Pressure psig	A Appr. Width of Scarf Inches	B Appr. Depth of Scarf Inches
12	70	1⅜	⅛
12	80	1½	⅛
12	90	1⅝	⅛
12	90	1½	3/16
12	70	1½	⅛
12	80	1⅝	⅛
12	90	1¾	⅛
12	90	1⅝	3/16
16	80	2¼	⅛
16	85	2½	⅛
16	90	2½	3/16
16	100	2¾	⅛
16	80	2½	⅛
16	85	2⅝	⅛
16	90	2¾	3/16
16	100	3	⅛
20	80	2½	⅛
20	85	2½	3/16
20	90	3	⅛
20	95	3	⅛
20	100	3¼	⅛
20	80	2¾	⅛
20	85	2¾	3/16
20	90	3	⅛
20	95	3	⅛
20	100	3½	⅛

The torch illustrated in Fig. 54 can be used for either heavy-duty cutting operations or scarfing. In the latter capacity, it is essentially designed for scarfing blooms, billets and slabs, carbon steel and alloy steel ingots. The scarfing tips are replaceable and are manufactured in orifice sizes (diameters) ranging from ¼ to ⅝ inch. The lever mounted on top of the torch activates both the scarfing oxygen and the wire starter feed mechanism. The wire is fed to the tip in ⅝-inch lengths until the lever is released.

The scarfing operator usually walks forward so that he faces the scarf, adjusting the torch to the desired angle. Various angles

Selection Data

C Appr. Depth of Scarf Inches	Appr. Angle of Tip to surface of billet	Appr. Angle of Tip to line of scarf	Position of Billet	
1/16	10°	10°	Flat	
1/16	10°	10°	Flat	
1/16	10°	10°	Flat	
⅛	20°	10°	Flat	FLAT
1/16	10°	20°	45° Angle	
1/16	10°	20°	45° Angle	
1/16	10°	20°	45° Angle	
⅛	20°	10°	45° Angle	45° ANGLE
1/16	10°	10°	Flat	
1/16	10°	10°	Flat	
⅛	20°	10°	Flat	
1/16	10°	20°	Flat	FLAT
1/16	10°	10°	45° Angle	
1/16	10°	10°	45° Angle	
⅛	20°	10°	45° Angle	
1/16	10°	20°	45° Angle	45° ANGLE
1/16	10°	10°	Flat	
⅛	20°	10°	Flat	
1/16	10°	20°	Flat	
1/16	10°	10°	Flat	
1/16	10°	10°	Flat	FLAT
1/16	10°	10°	45° Angle	
⅛	20°	10°	45° Angle	
1/16	10°	20°	45° Angle	
1/16	10°	10°	45° Angle	
1/16	10°	20°	45° Angle	45° ANGLE

Courtesy Chemetron Corp.

TIP STYLE	SERIES	TIP* SIZE	TYPE OF GAS	NO. OF PREHEATS	DESCRIPTION
	720 720 720	12 16 20	ACETYLENE	10 14 16	These tips are 5 inches long. They are of one-piece construction. A heavy alloy wear ring and tip nut for extreme heavy service are supplied.
	723 723 723	12 16 20	NATRUAL GAS SCARFING	14 16 16	These tips are 5 inches long. Sizes 16 and 20 are of one-piece construction. Size 12 is a two piece tip. A special alloy wear ring and tip nut for extra heavy service are supplied.

*DESIGNATES THE SIZE OF THE SCARFING ORIFICE IN 32NDS OF AN INCH.

Courtesy Chemetron Corp.

Fig. 53. Scarfing tips.

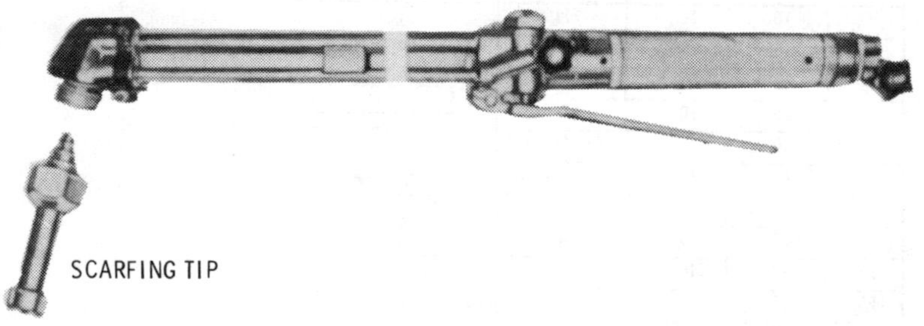

Courtesy Chemetron Corp.

Fig. 54. Heavy-duty hand scarfing torch.

for different scarf widths and depths are shown in Table 5. Note that the dimensions of a scarf depend upon several interrelated factors, including (1) tip size, (2) scarfing oxygen pressure, (3) tip angle, and (4) work position. Speed of travel, though not listed, is also a factor influencing the width and depth of the scarf.

Scarfing is also used synonymously with the term *deseaming*. This stems from the use of the scarfing torch to remove surface seams from ingots, bars, billets and blooms prior to their entering the finishing cycle at the mill.

FLAME HARDENING AND STRENGTHENING

Flame hardening is a process in which the surface of a quench hardening ferrous metal is locally heated by means of an oxyacetylene flame followed by a suitable quench. In other words, the surface of the part to be hardened is rapidly raised to the required temperature with the intense heat of the oxyacetylene flame and then just as rapidly cooled, usually by a stream of water (Fig. 55).

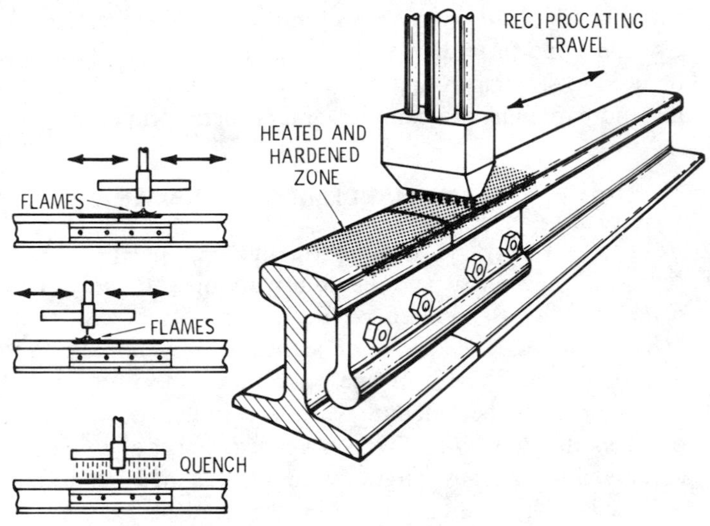

Courtesy Airco Welding Products

Fig. 55. Flame hardening process.

The major advantage of flame hardening is that parts are hardened only where the hardness is desired, such as the teeth of gears. The depth to which the surface is hardened may be controlled within reasonable limits. There are also applications where the core of some machine part must be of an alloy steel not suited to case carburizing because of desired physical properties. The flame-hardening process provides a method of obtaining the desired hard surface without changing the properties of the core material. Because the part is briefly above the critical point and then promptly quenched, flame-hardened parts are scale free. Generally speaking, distortion is well within manufacturing tolerances.

Another feature of flame hardening lies in the fact that the case depth may be varied within reasonable limits. Ordinary, requirements call for case thicknesses from 1/16 to 1/4 inch. In addition, case depth is uniform because the rate of heat input is under exact control. Flame hardening is adaptable to a wide variety of sizes and shapes of parts. This is illustrated by the flame hardening of such varied articles as rail ends, pump liners, crane wheels, gears, tractor shoes, sheave wheels, machine tool ways, crank shafts, cam shafts, and many other parts.

The same procedure used to flame-harden a surface can also be used to strengthen that portion of the surface. Both strengthening and hardening are complementary metal properties.

FLAME-HARDENING EQUIPMENT

The oxyacetylene flame-hardening equipment should be of sufficient ruggedness and flexibility to operate under the rather severe conditions imposed on it. It is necessary that such equipment be water cooled and of sufficient gas capacity to treat an area of reasonable size in one operation.

In order to accomodate various widths, as well as irregular profiles, the tips are of the threaded removable type and are obtainable in various lengths with various size tip orifices. Plugs are used so that only a portion of the head may be operated if so desired. In addition to all-purpose heads, it has been necessary to design heads for specialized applications such as gear tooth hardening.

For progressive hardening it is often convenient to mount flame hardening equipment on one of the standard oxyacetylene cutting machines. It so happens that the flame-hardening speed falls within the range of speeds obtainable with cutting machines and they thus become ideal traversing devices.

FLAME-HARDENING PROCEDURE

Due to the very high temperature of the oxyacetylene flame, it is possible to raise the temperature of the surface so rapidly that the surface can be hardened without affecting the core. As a result, the hardening process can be restricted to the surface. A neutral flame is, in nearly all cases, best for flame hardening.

In the usual methods of hardening, the heating and quenching of thin sections, say ½ inch thick, must be avoided because the slow speed of heating makes it impossible to keep the entire section from heating to a hardening temperature that will destroy desirable core properties. On the other hand, the speed of heating by the oxyacetylene flame makes possible the surface hardening of sections as thin as ⅜ inch with proper technique.

The hardness obtainable by the flame-hardening process is at least equal to that obtainable on steel of the same analysis by furnace hardening methods. In fact, on heavier sections the unheated metal greatly assists the quenching medium and permits the use of a less drastic external quench than would be required in furnace hardening.

The cooling rate which controls the hardening process is in turn controlled by the quenching medium used, and depends on its quantity and certain other physical factors as well as to some extent on its initial temperature. It is the use of a quenching step that distinguishes flame hardening from flame softening.

Water is widely used as a quenching medium and is generally acceptable as a flame-hardening quench. Water sprayed under pressure upon the metal provides a quicker cooling method than water as a bath. On certain types of steel, an air quench may be sufficiently drastic to give the desired hardness. In other cases, oil or oil and water are most effective although it is necessary to observe certain precautions when using oil.

Flame-hardening quenching arrangements are easy to set up. In many operations, a small stream of water is all that is necessary to cover the area to be hardened. Where a wider path must be covered by water, a fan-shaped nozzle or spray may be used. Either of these is simple to make or obtain. Best results are usually obtained when the torch cooling water and the water for quenching are separately controlled.

In general, any steel that may be hardened by simple heating and quenching may be treated by the flame-hardening process. In addition, plain or alloy cast iron and malleable iron may be flame-hardened.

The ability of plain carbon steels to quench-harden is dependent on the carbon content of the steel. To obtain a maximum degree of hardening, the steel should contain at least .40% carbon. As the carbon increases, the hardness obtainable also increases. The general range for plain carbon steel is from .40% to .70% carbon. Generally steels with greater carbon percentages can be flame hardened but greater care is required to prevent surface checking.

The most desirable steels for flame hardening are straight carbon or low alloy steels. These usually harden to a good degree and, except for certain types, are better able to withstand heating and quenching without checking or cracking.

The higher alloy steels present a more difficult problem from a heating and quenching standpoint and must be individually considered. However, certain high alloy steels can be sufficiently hardened by an air quench.

Practically all flame-hardening operations can and should be mechanical. Hand operations, although occasionally desirable, are not conducive to uniform results.

There is usually sufficient mechanical equipment available in most shops to eliminate the necessity of special equipment. For example, a lathe is very easily adapted to flame hardening. The torch is mounted on the tool carrier for smooth horizontal motion along the surface to be hardened. The part to be hardened is mounted on or alongside the lathe. The lathe is also adaptable to hardening of circular pieces, in which case the part to be hardened is rotated.

The various methods of flame hardening may be classified as follows:

1. Stationary,
2. Progressive,
3. Spinning,
4. Combination.

The *stationary* method includes those operations wherein the torch and the work are motionless during flame hardening. An example is the flame hardening of automotive valve stems, sometimes referred to as *spot hardening*.

In the *progressive* method, the torch and the work move with respect to each other. That is, the torch may move with the work stationary or the work may move with the torch stationary. In the former variation of the progressive method, the lighted torch, with a head having sufficient flame area to cover the path to be hardened, is directed along the surface to be hardened at the maximum speed which will heat the steel to the hardening temperature. In most cases, immediately behind the flame is a stream or spray of water which progressively quenches the heated surface.

The speed of flame travel is determined by considering flame intensity, type of steel being treated, the temperature desired, and the depth of case desired. The rate of speed may vary from 4 to 10 inches per minute.

The torch should be placed so that the ends of the inner tips (cones) of the flames are 1/16 to 1/8 inch from the surface being hardened. In no case should the inner tips touch the work. On large circular work, the most desirable method is to heat and quench during one revolution of the work. The work revolves at a speed of 4 to 10 inches per minute. A spray or stream of quenching water is directed against the heated surface immediately following the torch flames.

The type of work in which a circular path is to be hardened can be done by spinning the part under the flames of one or more stationary torches, and quenching the heated portion while it is still spinning. The tips used in machine flame hardening are sometimes designed to incorporate both the quenching and heating jets in the same unit (Fig. 56). The rotation speed is usually about 100 rpm. The time required for the spinning method varies from a few seconds to a minute, depending on the diameter of the piece treated, the number of torches, and other factors. It is generally undesirable to take more than one minute for hardening. Any piece that requires more time than this should have more heat, or be treated by the single revolution method.

For the *spinning* method, it is advisable to use a quenching in which a large volume of water under low head can be released

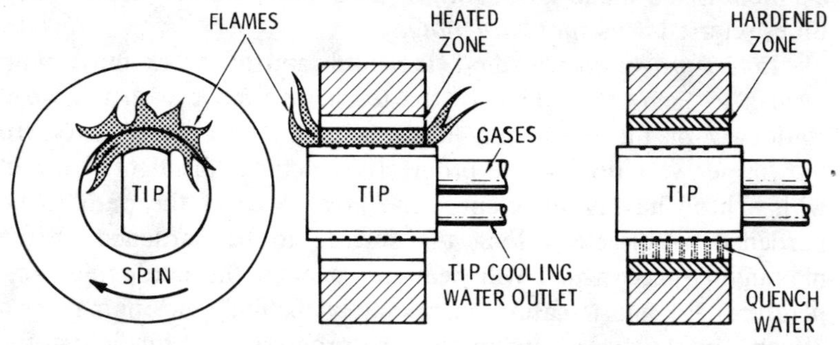

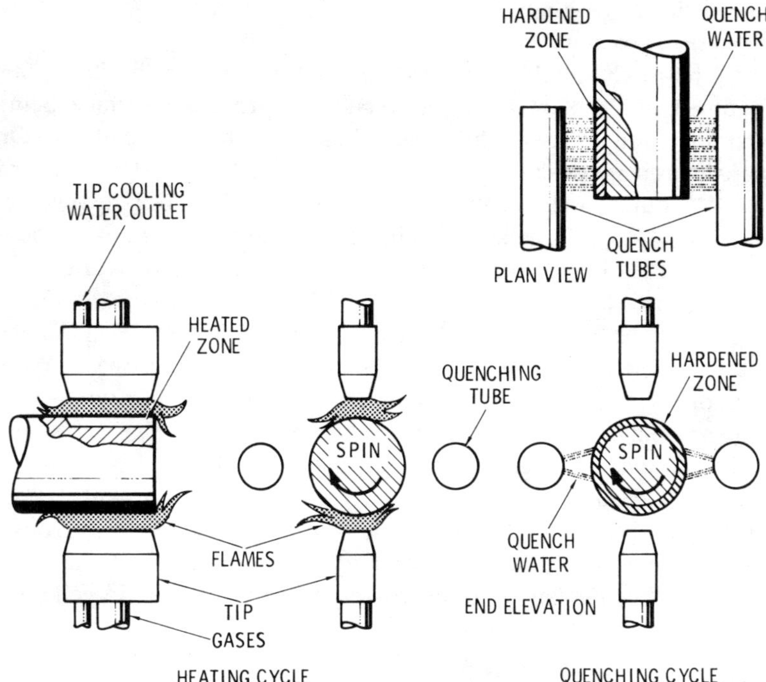

Courtesy Airco Welding Products

Fig. 56. The spinning method of flame hardening.

to cover the entire part at once and flow over it in a solid stream for the necessary time. The flame-hardening operation should be followed by a stress relieving draw shortly after quenching.

The spinning method is often applied to the hardening of local areas on shafts, small gears, pinions having small teeth, lathe spindles, pump shafts, crank-shaft bearings, and objects of similar design.

The *combination* method of hardening combines the progressive and spinning methods and is applied to circular objects of such length that the torch may traverse while the part is being spun. An example of this application is the combination method used in the flame hardening of shafts by spinning the shaft in a lathe while the torch, mounted on the tool rack, traverses the shaft.

STRESS RELIEVING

Flame hardening should be followed by a stress relieving draw shortly after quenching. Stress relieving is simply a heating operation conducted in an oil bath or furnace. The drawing prevents

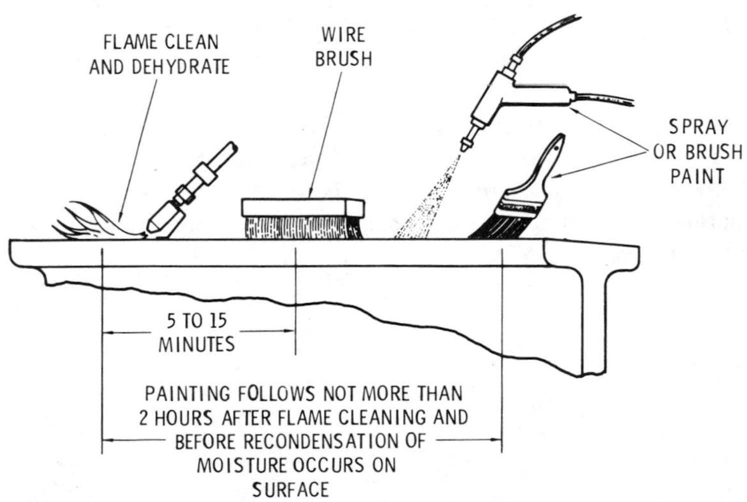

Courtesy Airco Welding Products

Fig. 57. Using the gas flame to clean the surface.

checking, minimizes distortion, and, if conducted at temperatures of 300°F to 400°F it will have little if any effect on hardness.

The drawing should be performed within a half hour after the hardening. The part should remain in the oil bath or furnace for a sufficient time to relieve the stresses set up by quenching. Except for large masses of metal, this drawing period need not exceed an hour. The article may be cooled in still air.

FLAME SOFTENING

Flame softening is a heating process used to reduce the tendency of a steel to become brittle and hard after rapid cooling from a red hot condition. This is particularly important after cutting operations because high temperature of the torch flame or arc can cause the edges of the cut to become very hard and brittle. By softening these localized areas with a gas flame, the metal is restored to an acceptable level of softness. Flame softening can be done by hand or it can be mechanized.

CLEANING AND OTHER SURFACE PREPARATIONS

The heat of a torch flame is frequently used for a variety of metal surface preparations designed to clean the surface and improve appearance. For example, the scale that can form on the surface of castings, blooms, and billets can be removed by heating the area slowly with a gas flame. The scale will expand and crack off. A gas flame can also be used to dry a surface in preparation for painting, or to remove paint from a previously finished surface (Fig. 57).

CHAPTER 14

Solders and Soldering

Soldering is a joining process used to fasten two metal surfaces together. This joining process is also referred to as *soft soldering,* thereby distinguishing it from silver brazing, which is sometimes called *hard soldering.*

Soldering is similar to brazing in that a nonferrous filler metal is used to join the two metal surfaces. It differs from brazing in its use of a nonferrous metal that melts at temperatures below 800°F (in brazing the filler metal melts at temperatures above 800°F, but below that of the base metal). In both soldering and brazing, the base metals are not melted. This distinguishes these two joining processes from welding in which the base metals are heated until they fuse or flow together. A comparison of the heating temperatures required by these various joining processes is shown in Fig. 1.

Soldered joints are weaker than those produced by either brazing or welding. On the other hand, brazed joints are stronger than soldered ones, but weaker than welded joints. Thus, it can be seen that a hierarchy of strength exists among these three types of joining processes with the soldered joint being the weakest. Because of their weakness, soldered joints are seldom used where strength is a requirement. If undue pressure or stress is to be placed on a soldered joint, rivets, bolts, or some other means of reinforcing it are recommended.

PRINCIPLES OF SOLDERING

The correct temperature for soldering must be determined for each particular job largely by experiment. It varies with the size of the work, the type of work, the solder composition and the nature of the flux. Usually soldering ranges are between 500°F and 700°F. The operator should not attempt to solder at temperatures close to 800°F. At approximately this temperature the drosses (the scum thrown off from molten metal) become soluble in the solder and the solder is said to be "burnt" and will behave badly.

When a piece of soft solder is heated, it starts to melt at a definite temperature. The melting temperature will depend on the type of solder. As additional heat is applied, the temperature

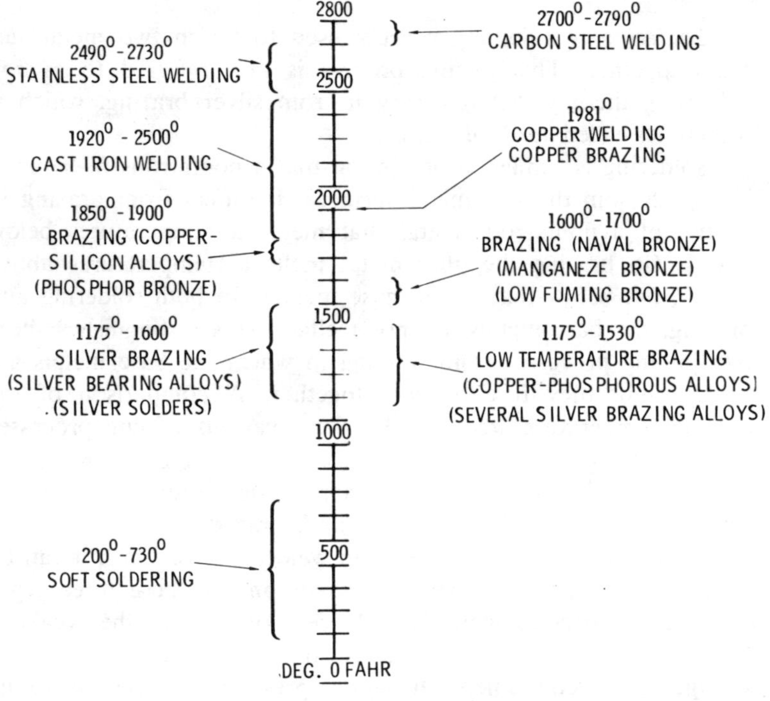

Courtesy American Welding Society

Fig. 1. Temperature ranges of various joining processes.

increases, and more and more solder melts until it is all melted. The first temperature is called the *melting point (solidus* temperature) and the second temperature the *flowing point (liquidus* temperature). The latter temperature is important because the metal must be heated to a temperature higher than the flowing (liquidus) temperature, or the solder will not flow.

The ability of a solder to flow over a surface is one determination of the solderability of that metal. However, the flow characteristics will vary from one solder to the next. For example, tin-lead solder of, say, a 5 to 95 composition will flow quite a bit differently from a 95 to 5 tin-silver solder. Only practice will enable an operator to become accustomed to these differences.

As a soft-soldered joint cools, the solder solidifies at the melting point. However, until the joint has cooled to a temperature below the melting point no strain can be applied to the joint, since solder has no strength, while it is partially solidified. Between the flowing and melting points is the so-called plastic range in which

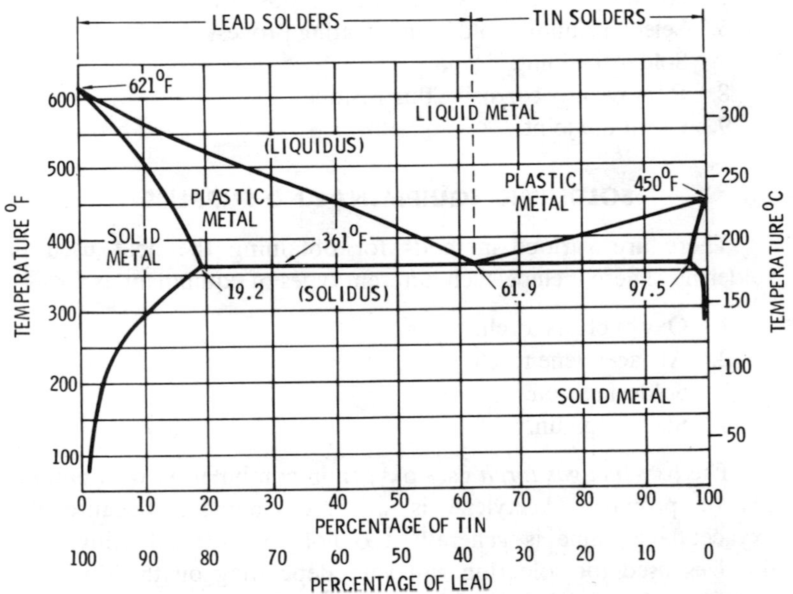

Courtesy American Welding Society

Fig. 2. Temperature ranges of the tin-lead solders.

there is a mixture of solid and liquid solder. Fig. 2 illustrates the relationship of the various temperatures of tin-lead solder.

A few solders melt and flow at the same temperature. These are referred to as *eutectic* solders and are suitable for making bead joints, but are not suitable for running into lap joints. It helps to know the melting and flowing points and the plastic range when changing from a solder of known composition to a substitute solder.

Soldered joints of high quality can be assured if the operator will carefully adhere to a number of basic soldering principles. These principles (in approximately their sequential order) are as follows:

1. Determine initially the solderability of the metal and make certain that soldering is the most efficient joining process.
2. Select an appropriate joint design.
3. Carefully and thoroughly clean the surface.
4. Select a suitable flux and apply it to the surface.
5. Align the parts and secure them in position.
6. Select a suitable solder and heating process.
7. Solder the joint.
8. Remove any corrosive flux residue.
9. Clean the joint.

SOLDERING EQUIPMENT AND SUPPLIES

There are various methods for obtaining the heat used in soldering. These include such different types of equipment as:

1. Oxy-fuel gas torch,
2. Air-acetylene torch,
3. Soldering iron,
4. Soldering gun.

The *oxy-fuel gas torch* uses oxygen in combination with natural gas or propane. Acetylene is not recommended because the oxyacetylene flame is generally too hot for soft soldering. The tip sizes used for soldering will vary depending on the thickness and type of metal being soldered. These tips will also frequently be designed for use with other types of surface work such as heating or silver brazing. Fig. 3 illustrates a multiuse torch tip of this type.

Fig. 3. Torch tip used for soft soldering.

Courtesy Victor Equipment Co.

Some welding equipment manufacturers have also developed torches specifically designed for soft soldering. Fig. 4 illustrates a torch that can be used for both soft soldering (large areas) and heating. This particular torch has a pilot light that can be retained during down time. The full flame is restored by pressing the lever.

The *air-acetylene torch* with a single hose connection (Fig. 5) is also used in soft soldering. The hose carries the fuel-gas (in this case, acetylene) to the torch. The flow of the fuel-gas under pressure draws in (i.e. aspirates) sufficient air for combus-

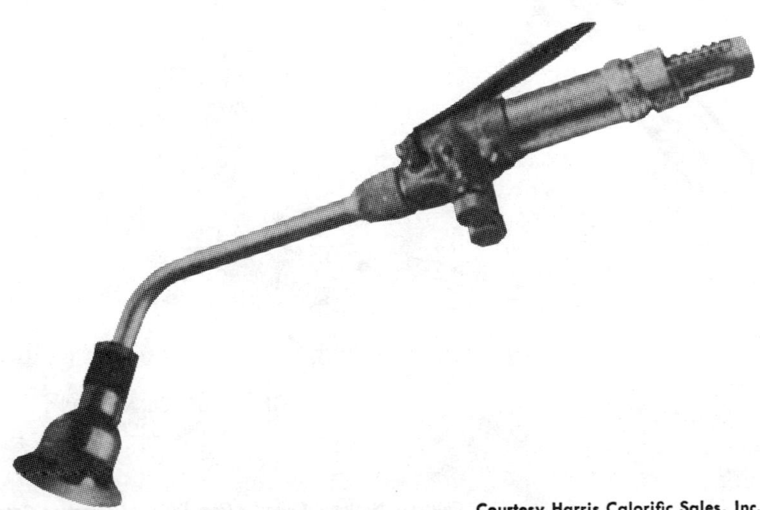

Courtesy Harris Calorific Sales, Inc.

Fig. 4. A heating and soldering torch.

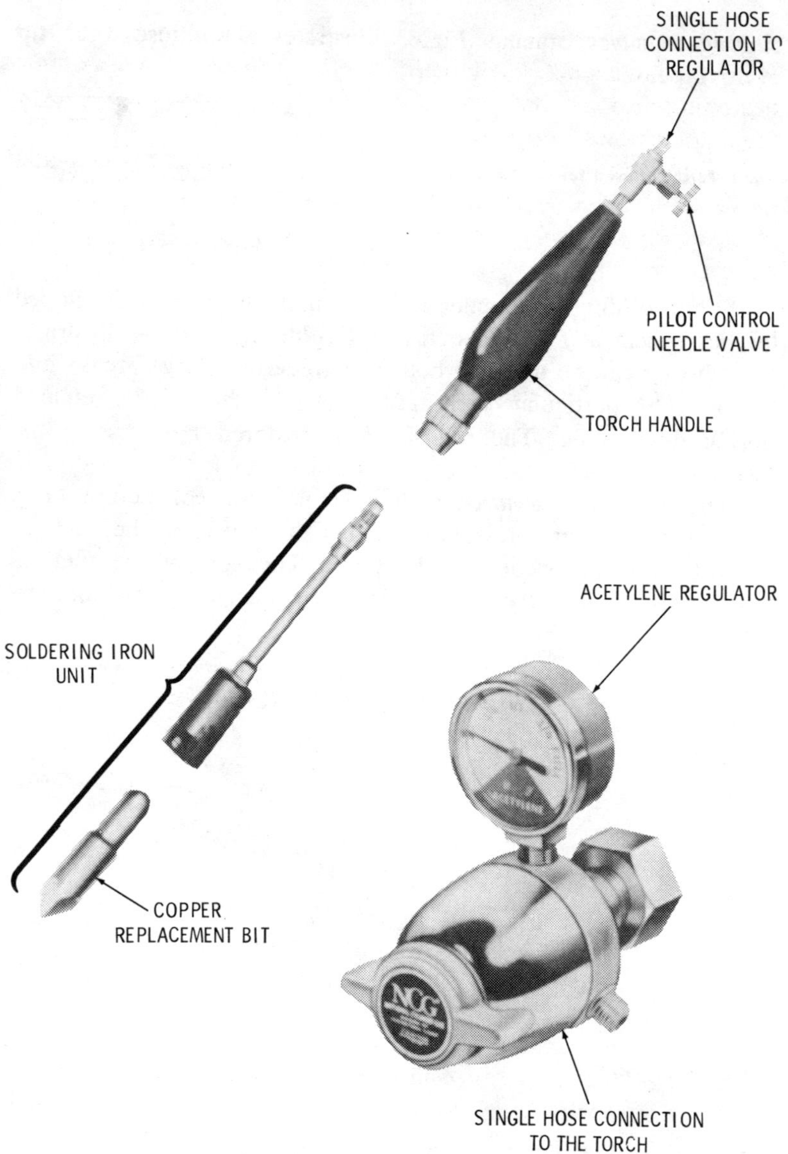

Courtesy National Cylinder Gas, Div. of Chemetron Corp.

Fig. 5. An air-acetylene regulator and soldering iron.

tion. Pressure reducing valves or regulators are required on the fuel-gas cylinder. These torches have replaceable tips for maximum flexibility. Propane gas can also be used, but a different type regulator is necessary.

A gasoline torch with a soldering iron attachment operates on the same principle as the air-acetylene torch. However, in this case, the fuel is contained in the torch itself and there are no hose connections with an external supply. Fig. 6 shows such a device for soldering.

Fig. 6. Gasoline operated torch.

Courtesy Turner Corp.

A *soldering iron* (also referred to as a soldering bit) consists of a large piece of copper, drawn to a point or edge and fastened to an iron rod having a handle grip. These soldering irons are available in sizes ranging from the small lightweight types to the large heavy duty models found in the industrial plants.

Soldering irons may be either internally or externally heated. External heating is usually done in ovens (furnaces) with a gas flame or other means. Internal methods of heating involve either gas or electricity, with the latter being the most predominant form.

The electric, internally heated, soldering iron is available in two basic types:

1. The hatchet iron,
2. The pencil iron.

The *hatchet iron* (Fig. 7) enables the operator to solder in areas that are not easily accessible. A wide variety of tips are available. The elements are interchangeable, thereby providing the operator with a wide wattage range.

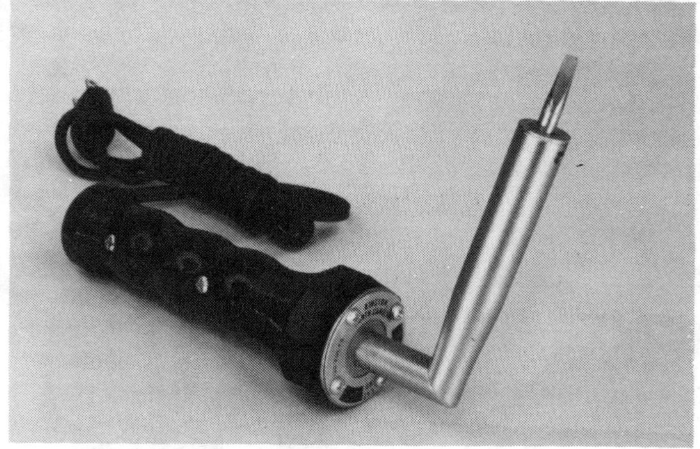

Courtesy Wall Manufacturing Co.

Fig. 7. Heavy-duty hatchet soldering iron.

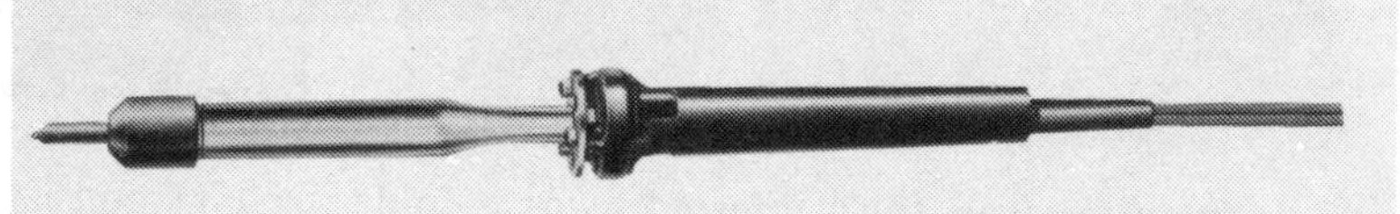

Courtesy Wall Manufacturing Co.

Fig. 8. A typical pencil-type soldering iron.

The *pencil iron* (Fig. 8) receives its name from its slim, pencil-like appearance. It is the type most commonly used in soldering and is available in a variety of different sizes.

Soldering irons can be fitted with tips of many different sizes and shapes. The type selected will depend on the work to be performed. The latter will also determine such factors as voltage and wattage rating, and the temperatures to be used. A heavy bit is

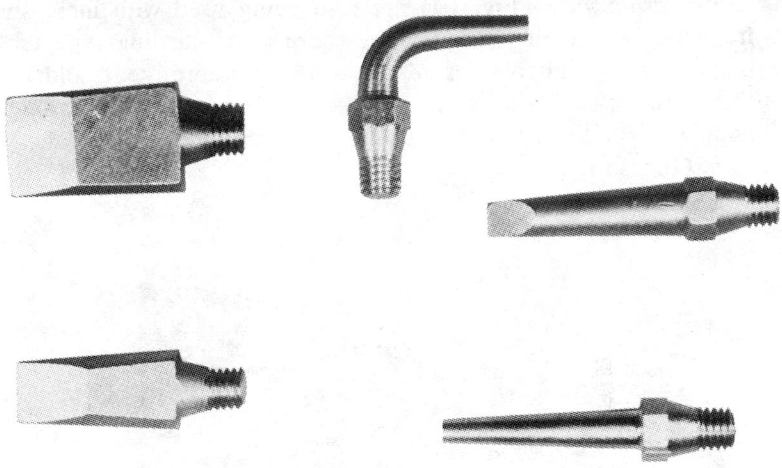

Courtesy Wall Manufacturing Co.

Fig. 9. Various types of soldering iron tips.

Courtesy Wall Manufacturing Co.

Fig. 10. Soldering gun with replaceable tip and element.

generally preferable to one weighing less than 2 lbs. because it will retain the heat longer. Fig. 9 shows some examples of typical soldering bits.

Soldering guns (Fig. 10) are also being used with increasing frequency for joining metals. These are quick-heating type tools that are more suitable for repair work or intermittent soldering than production type work. These are becoming increasingly popular with the home craftsman and are now available in kit form (Fig. 11).

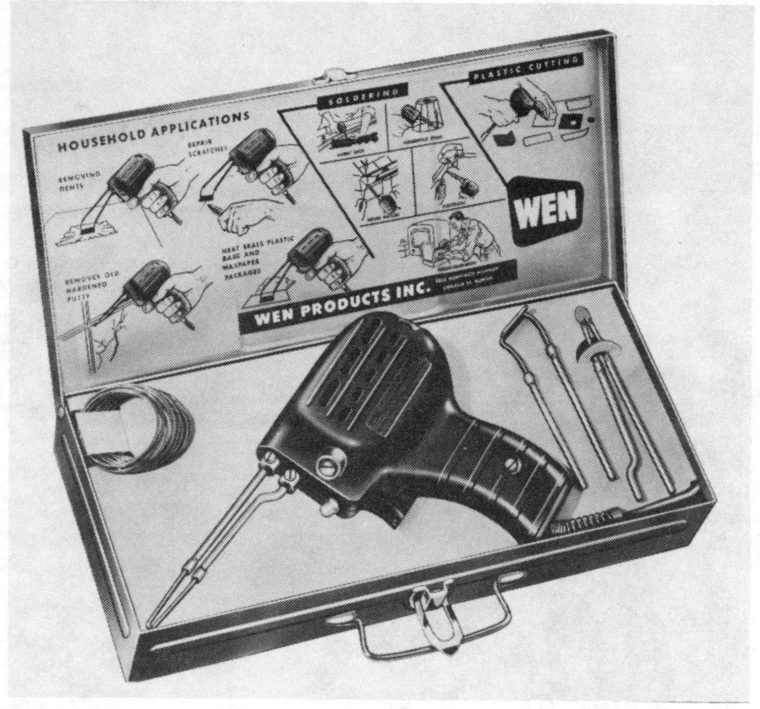

Courtesy Wen Products Inc.

Fig. 11. A typical soldering gun kit.

Another useful piece of equipment is the desoldering tool (Fig. 12). This tool is designed to melt and remove the solder by first heating it and then removing the molten solder by suction. The vacuum bulb located above the heating iron sucks the molten solder off of the surface. The desoldering tool is operated with one hand, leaving the other one free to remove the compo-

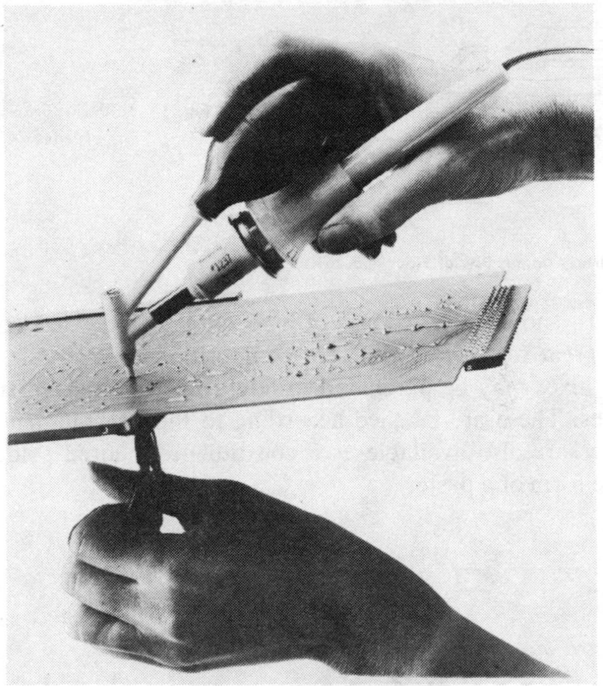

Courtesy Ungar, Div. of Eldon Industries

Fig. 12. A desoldering tool.

nents. The tips are heat-resistant and replaceable. The tip orifices are available in a number of different sizes.

An iron holder is a particularly useful device for the soldering workbench. It is designed to hold the iron during down-time, and thereby reduce the possibility of burning the cord with a hot tip, burn injuries to the operator, or dropping the soldering iron on the floor.

"Hot-knife" attachments have been developed for special tasks such as cutting the epoxy coatings on printed-circuit boards. The blades shown in Fig. 13 attach to 27 watt, 35 watt, and 45 watt *Ungar* irons. The blade and collet holder are purchased together.

Solders are available as wire coils (flux cored or solid), bars, ingots, preforms, foil, powder, paste, and a variety of other forms.

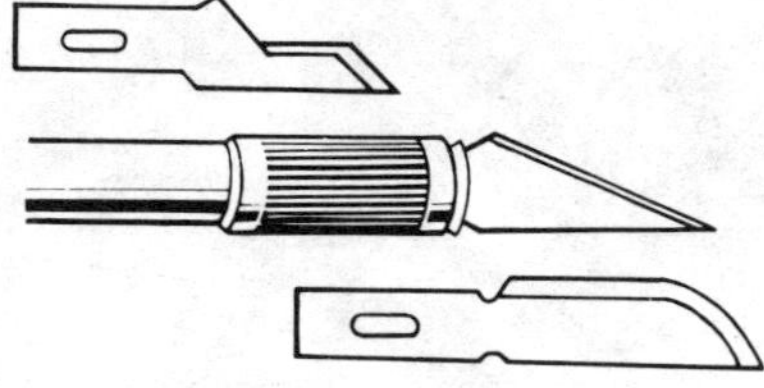

Fig. 13. Soldering iron attachment blades.

Courtesy Ungar, Div. of Eldon Industries

Fig. 14 shows a typical coil of wire solder. Solder is also available in a great number of different compositions.

Fluxes may be purchased in liquid form in convenient squeeze bottles. These are labelled according to their recommended usage. Fluxes are also available as a constituent of cored solder wire or in the form of a paste.

Fig. 14. Solder in wire coil form.

Courtesy Lenk Manufacturing Co.

The operator may wish to mix his own flux. Chemicals (zinc chloride, ammonium chloride, hydrochloric acid, etc.) are commercially available for this purpose as well as the necessary formulas. Furthermore, special flux compositions can be ordered from manufacturers of soldering equipment and supplies. Information is readily available from the various manufacturers.

SOLDERS

Solders are nonferrous filler metals used to join metal surfaces. These soldering metals are formed from fusible alloys. There are a number of different types of solders and the type used depends on the composition of the metals being joined. Not all metals and metal alloys can be soldered. Among those that will not take a soldered joint are:

1. Titanium,
2. Chromium,
3. High-tensile magnesium-bronze,
4. Beryllium,
5. Cobalt,
6. Silicon.

Some classifications of solders are based on the ASTM specification B32-58T. One category receives its designation from the silver content of the solder. For example, a 2.5S solder is one containing 97.5% lead and 2.3 to 2.7% silver with 2.5% being the *desired* level. These are the so-called lead-silver solders.

The largest classification category is designated according to its tin content (for example, a 5A solder represents one having 5% tin and 95% lead). There are three letter suffixes (A, B, and C) added to the numeral that indicates the tin percentage. These denote the following three composition classes:

1. A—0.12% maximum antimony content allowed for solders containing 35% tin or more,
2. B—0.50% maximum antimony content allowed for solders containing 35% tin or more,
3. C—the antimony content may not exceed 6% of the tin content in the so-called tin-antimony solders (those containing 20 to 40% tin and an appropriate percentage of antimony).

By way of illustration, a 40A solder will consist of 40% tin, 60% lead and a maximum of 0.12% antimony. The percentage given for tin is the *desired* percentage, the one for lead a *nominal* percentage. These amounts will vary to some degree. A 40B solder will contain 40% tin, 60% lead and a maximum of 0.50%

antimony. Finally, a 40C solder will contain 40% tin, 58% lead and a maximum of 2.4% antimony (6% of the tin content).

Solders can be divided into a number of different groups according to their composition. The largest group and most commonly used are the tin-lead alloy solders.

The strongest tin-lead alloy is one composed of 42% tin and 58% lead (in other words, a 42-58 solder). It works well and has sufficient viscosity to fill commutator lead slots and other closely spaced openings automatically. It is therefore not surprising that the nearest commercial composition 40% tin and 60% lead (also known as "commercial" or 40-60 solder) is the most popular composition for soldering commutators and similar work. It is cheap, works easily and does not throw out of hot commutators as readily as solders containing a higher proportion of tin.

A solder containing too much lead makes a weak joint because the lead does not transfuse with brass. A solder containing too much tin becomes brittle.

The use of 60-40 solder (60% tin, 40% lead) or 50-50 solder (50% tin and 50% lead) makes it possible for even the novice to do acceptable soft soldering. The high tin content of these solders makes them easy to work. With solders in which the maximum tin content is limited to 30%, however, even experienced workers encounter difficulty when soldering.

In addition to the very large tin-lead alloy group of solders, the following groups of solder alloys are also represented:

1. Tin-antimony-lead solders,
2. Tin-antimony solders,
3. Tin-zinc solders,
4. Tin-silver solders,
5. Lead-silver solders,
6. Cadmium-silver solders,
7. Cadmium-zinc solders,
8. Zinc-aluminum solders,
9. Lead-bismuth-tin solders,
10. Indium solders.

Each of the above groups represents the addition of specific alloys to produce solders with certain required characteristics.

For example, antimony is added for extra strength (under specified conditions) and increased electrical conductivity. Silver added to lead will improve wetting characteristics, particularly on copper or steel. Cadmium, silver, and zinc combinations produce a solder particularly suitable for joining aluminum or dissimilar metals.

Sometimes it is necessary to solder at a temperature below that of the tin-lead alloys (i.e. below 361° F). A solder containing lead, tin, and bismuth was developed to meet this requirement. These are frequently referred to as *fusible alloys*.

The Indium solders were also developed for special purposes. Some show a high degree of corrosion (alkaline) resistance. They are available in 50-50 or 48-52 tin-Indium compositions or with additions of varying amounts of bismuth, lead, and cadmium.

CLEANING THE SURFACE

Thoroughly cleaning the surface is, perhaps, the single most important step in soldering. The solder will not alloy to the surface of the base metal unless it is perfectly clean. Any surface contaminants (e.g. dirt, grease, etc.) will obstruct the alloying or wetting process of the solder.

Cleaning can be done mechanically or chemically. The former may be accomplished by machining, grinding, blasting (with sand, grit or shot), or hand abrading. Generally, hand abrading is the method most commonly used in the small workshop and can be done in a number of ways, including:

1. Rubbing with steel wool (fine grade),
2. Filing,
3. Brushing with a wire brush or scraping.

The metal surface is hand abraded until it has a bright shiny appearance. The only exception to this rule is a tin surface for which a wiping with a solvent will usually suffice or surfaces on which a noncorrosive flux is to be applied.

A chemical cleaning can be accomplished by wiping the surface with a clean cloth dipped in a solvent or detergent. This will remove most forms of grease, oil or dirt. The chemical cleaning agent should be removed with hot water before applying the flux.

SOLDERING FLUXES

Metal surfaces are covered with thin, invisible oxides and other impurities. These must be removed before the solder is applied or it will not adhere to the surface. The removal of these impurities is accomplished by applying a flux.

A *flux* is a chemical that not only removes surface contaminants, but also forms a thin film over the surface preventing contact with the air. Moreover, the flux contributes to the free flowing characteristic of the solder. The operator should be cautioned at this point *not* to skip the cleaning operation that precedes the application of the flux. Although the flux does remove oxides and other impurities, in most cases, it is not a complete cleaning operation.

No one flux can be assigned to any one metal as being peculiarly suited to that metal for all purposes. The nature of the solder often determines the selection of the flux. In electrical work, for example, the best flux is a pine amber rosin because it does not cause corrosion. A corrosive flux, such as one containing zinc chloride solution should be strictly excluded from any electrical work.

The *Underwriter's Code* permits the use of a flux composed of chloride of zinc, alcohol, glycerine, and water. This preparation permits the solder to flow freely and is not highly corrosive. This flux is prepared as follows:

1. 5 parts zinc chloride,
2. 4 parts alcohol,
3. 3 parts glycerine.

Anhydrous zinc chloride crystals should be used dissolved in alcohol. The glycerine makes the flux adhesive. To prevent the alcohol igniting, the mixture may be diluted with water.

Fluxes play an important part in the soldering operation. By shielding the cleaned metal from the air, fluxes retard the formation of an oxide film. A flux will also remove any oxide that might be on the surface when the solder is applied.

Some flux manufacturers add wetting agents to their fluxes, having determined the amount to be added by experimental work.

Table 1. Flux Recommendations for Various Metals and Metal Alloys

RELATIVE SOLDERABILITY OF METALS, ALLOYS AND COATINGS

Base Metal, Alloy or Applied Finish	Flux Requirements			Soldering Not Recom-mended
	Non-Corrosive	Corrosive	Special Flux and/or Solder	
Aluminum			X	
Aluminum-Bronze			X	
Beryllium				X
Beryllium Copper		X		
Brass	X	X		
Cadmium	X	X		
Cast Iron			X	
Chromium				X
Copper	X	X		
Copper-Chromium		X		
Copper-Nickel		X		
Copper-Silicon		X		
Gold	X			
Inconel			X	
Lead	X	X		
Magnesium			X	
Manganese-Bronze (High Tensile)				X
Monel		X		
Nickel		X		
Nichrome			X	
Palladium	X			
Platinum	X			
Rhodium		X		
Silver	X	X		
Stainless Steel			X	
Steel		X		
Tin	X	X		
Tin-Bronze	X	X		
Tin-Lead	X	X		

Table 1. Flux Recommendations for Various Metals and Metal Alloys (Contd.)

RELATIVE SOLDERABILITY OF METALS, ALLOYS AND COATINGS

Base Metal, Alloy or Applied Finish	Flux Requirements			Soldering Not Recommended
	Non-Corrosive	Corrosive	Special Flux and/or Solder	
Tin-Nickel	X	X		
Tin-Zinc	X	X		
Titanium				X
Zinc		X		
Zinc Die Castings			X	

Courtesy The American Welding Society

Wetting agents are penetrants that cause the solder film to thin out and cover a larger area. Fluxes with wetting agents are not necessary for 50-50 or 60-40 solders, but they are not detrimental. However, it is advisable to use fluxes with wetting agents for the lower tin solders. This is particularly necessary for long joints on copper or for all joints on brasses and bronzes.

Fluxes may be divided into two basic classes:

1. Noncorrosive fluxes.
2. Corrosive fluxes.

Some authorities recommended listing mild fluxes (slightly corrosive fluxes) as a third class. However, because they result in some corrosion, they will be grouped with the corrosive fluxes. Using the two-class distinction as a basis, Table 1 lists various metals and metal alloys with a suitable flux (i.e. noncorrosive or corrosive) for each.

Rosin and rosin alcohol are noncorrosive fluxes. Although these fluxes will not cause corrosion, their fluxing action is very weak. Moreover, the noncorrosive fluxes leave an unsightly brown stain on the surface. This noncorrosive flux residue is not electrically conductive and must be removed after soldering is completed.

Zinc chloride is the principal element found in most corrosive fluxes. Other ingredients include ammonium chloride, sodium chloride, hydrochoric acid, hydrofluoric acid, and water. None of these ingredients are present in every corrosive flux, but every corrosive flux does contain mixtures of some kind of salts and inorganic acids.

The corrosive fluxes have a highly active fluxing action, are relatively stable over different temperature ranges, and are very suitable for solders that have the higher melting temperatures. All flux residue must be removed immediately after soldering or it will corrode the metal.

The so-called mild or intermediate fluxes are slightly corrosive, but weaker in their fluxing action than the corrosive fluxes. The mild fluxes contain organic acids (e.g. citric acid, glutamic acid, etc.), whereas the acids found in the corrosive fluxes are inorganic in origin.

Mild fluxes are used for quick spot soldering jobs. Longer usage results in burning or other forms of breakdown caused by their reaction to prolonged heating. The residue of these fluxes can be easily removed with water.

CARE AND DRESSING OF TIPS

The life expectancy of a soldering iron is often considerably reduced by one or more of the following:

1. Line voltage fluctuations,
2. Improper care and handling,
3. Normal usage.

Line voltage fluctuations are a problem in production work and involve current surges or drops. A surge of current can damage the heating element of the iron. On the other hand, drops in voltage result in reduced heat output which indirectly affect the bit. Line voltage fluctuations can be guarded against by constantly checking the voltage.

Improper care and handling probably contributes more to shortening the life expectancy of soldering irons than line voltage fluctuations. Some examples of improper handling are: (1) damaged cords (usually by allowing the cord to come into contact with the hot bit, (2) cracked handles (often by dropping), (3) damaged tips (again by dropping), and (4) overheated irons (generally by prolonged idling periods).

Most of the problems associated with care and handling can be eliminated simply by being more careful. This, of course, is a problem for the individual. The worker should try to learn proper work habits. Every soldering station should also have a soldering iron holder. These not only prevent breakage, burned cords, and other mechanical problems, they also reduce the possibility of the operator burning himself.

Over a period of time, the soldering tip will deteriorate as a result of normal usage. Proper care can prolong its life expectancy beyond what would normally be the case.

The three most common tip malfunctions that occur with normal usage are: (1) tip erosion, (2) tip corrosion, and (3) tip freezing.

Plain copper tips (i.e. those without special coatings) are those most subject to tip erosion. During the soldering process, small bits of the tip break off and are carried away. This eventually destroys the effectiveness of the tip. Unfortunately, little can be done about this and, for this reason, plain copper tips are not used for extensive production operations.

Copper tips with coatings resistant to erosion are designed to prolong the life of the tip. This erosion-resistant coating is a thin plating of metal that can be tinned. It is also soluble in solder, but to a much lesser extent than a plain copper tip. Oxides will form on the surface of a coated tip and must be removed. A wire brush is recommended for this purpose. The operator must not brush beyond the exposure of the original plating or the tip will be ruined. The tip should be re-tinned immediately after brushing.

The formation of oxides on the surface of the tip is known as tip corrosion. A wire brush or file is recommended for their removal. These oxides should be frequently removed, care should be taken not to remove the original coating, and the tip should be immediately re-tinned.

Frequently while soldering scale will collect inside the core of the soldering iron, if the scale is not removed, it will solidify and freeze the tip making it almost impossible to remove. The chance of this occuring can be reduced by removing the tip after soldering and cleaning off any scale that may have formed inside the core. A frozen tip will frequently result in the destruction of the heating element itself. Consequently, it is a wise move to prevent freezing from occurring in the first place.

TINNING THE BIT

A properly tinned bit is extremely important to a successful soldering operation. This consists of coating the bit with solder before beginning the work. The bit is tinned by heating it in a fire or gas flame until hot enough to rapidly melt a stick of solder when the bit is pressed against it. The solder covering the surface of the tip must be replaced frequently in order to maintain a clean coating. Some operators find that by wiping the tip frequently on a damp sponge and then immediately re-coating it with solder, the transfer rate of the soldering heat at the joint can be greatly increased.

When the bit is heated to the correct temperature, the face of the copper should be cleaned up with an old file. If the bit is too hot, the copper surface will tarnish immediately. To correct this, allow the bit to cool slightly and repeat the cleaning.

When the surface only tarnishes slowly, a little flux is sprinkled on it and then rubbed with a stick of solder. After the molten metal has spread over the surface to be tinned, the superfluous solder is wiped off with a clean damp rag. The surface should then present a bright silver appearance when properly tinned.

It is impossible to solder with an untinned or badly tinned bit, because the oxidized film of copper on the surface prevents the ready transmission of heat.

Tinning (Precoating) The Surface

Tinning is a term used to describe the covering of a metal surface with a thin permanent precoating of molten solder, pure tin, copper, nickel or some other suitable metal. This term is sometimes incorrectly used synonymously with *wetting* which refers to the spreading characteristics of the solder itself. In tinning, a strong semi-chemical attraction occurs between the atoms of the precoating metal and those of the base metal. This tinning film, in turn, functions as a base coat for the flux and the second layer of solder. Tinning greatly facilitates the flow of solder on small parts and assemblies.

Wiping

Wiping is a technique used for sealing lead joints. The solders used for wiping contain 30 to 40% tin, 58 to 68% lead, and the remainder antimony. The solder is first melted and then poured in the molten state over the joint.

Sweating

Sweating (Fig. 15) is a term sometimes used to refer to a procedure for temporarily holding together work which has to be turned or shaped, and which could not be so conveniently held by other methods. After having been turned or shaped, the separation of the parts is readily effected by the aid of heat.

In this operation the surfaces are cleaned, fluxed, heated, and covered with a film of solder. The soldered surfaces are then placed together and heated by passing the bit over the outside surface until the solder melts and unites the two surfaces.

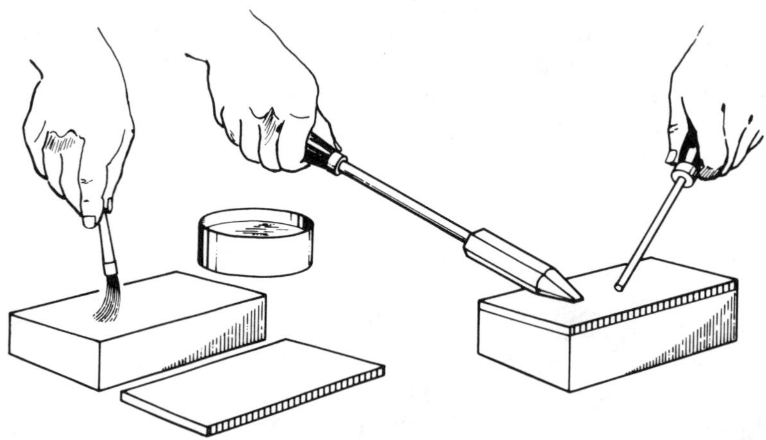

Courtesy Branson Sonic Power Co.

Fig. 15. The sweating procedure.

The procedure for the sweating together of two metal surfaces is as follows:

1. Clean the two surfaces thoroughly,
2. Apply a flux to both surfaces,
3. Put a piece of tinfoil over one of the surfaces,
4. Place the other surfaces on top of the tinfoil,
5. Clamp the two surfaces together,
6. Heat the combined surfaces with a hot bit or torch (if the combined metals have considerable thickness) until the solder melts,
7. Allow the solder to cool. The two surfaces should be firmly united.

SOLDERING METHODS

This chapter concentrates primarily on the soldering-copper method of joining metals. The emphasis placed upon this method is due to the fact that it is the soldering method most commonly used both in the home workshop and industry.

Other soldering methods described in this chapter are:

1. Torch (gas-flame) soldering,
2. Dip bath soldering,
3. Carbon-arc soldering,
4. Induction heating soldering,
5. Ultrasonic soldering,
6. Resistance heating soldering,
7. Furnace heating soldering,
8. Automatic soldering systems and modules.

Torch (Gas-Flame) Soldering

An oxyacetylene torch is generally not recommended for soldering because the temperature of the flame is too high. It usually results in destroying the protective action of the flux. This can be avoided by directing the full force of the flame away from the spot being soldered. This form of indirect heating can also be used with other types of gas flame soldering when the temperatures become too high for the metal or metal alloy.

The air-acetylene torch should show a bright, sharply defined inner cone and a pale blue outer flame. A yellow flame would indicate that the acetylene pressure is inadequate, that the needle valve is not opened sufficiently, or that the soldering bit is clogged. The temperatures with this form of gas flame soldering are somewhat lower than those obtained from the oxyacetylene flame.

Still lower flame temperatures can be obtained by using propane, natural gas, or butane. Oxygen burned with a fuel gas will produce higher temperatures than air burned with the same gases. The former will also produce a sharper, more well-defined flame than a torch operating on the bunsen burner principle. The latter are generally gasoline filled torches, and they produce a widespread flame that is somewhat bushy in appearance.

Dip Bath Soldering

Dip bath soldering (or *dip soldering*) consists of dipping the parts to be soldered into a bath of molten solder. The joint must be firmly held in position by jigs or other clamping devices until the solder has completely solidified. The pots containing the solder bath are available in a wide variety of sizes (from several lbs. to over 10,000 lbs.) depending on the requirements of the job. They may be either electrically or gas heated, with the latter method of heating being reserved for the larger size pots.

Applications of the dip bath method are found in the soldering of automobile radiator cores, tin can seams, printed circuits and other types of electronic equipment that would ordinarily require numerous individually soldered joints.

The steps in the dip bath method of soldering are as follows:

1. Thoroughly clean the parts to be soldered by removing any dirt, grease, or other contaminants from the surfaces.
2. Select a suitable joint design and clamp the parts in position.
3. Heat the solder to a suitable temperature (it is important not to overheat the solder).
4. Preheat the parts (to approximately 240 to 345° F) and dip them into a suitable flux.

5. Allow the parts to dry.
6. Dip the fluxed parts into the bath of molten solder.
7. Remove the parts as soon as the wetting action has been completed.

Carbon-Arc Soldering

A low amperage carbon-arc can be used for soldering under certain conditions. It is particularly important not to allow an arc to be drawn between the carbon electrode and the work when soldering. Otherwise, the heat will be too intense for soldering and may fuse or cut the base metal. Either an AC or DC welder can be used with an amperage setting of no higher than 25 amperes. The surface must be thoroughly cleaned and a flux must be used.

Soldering Copper Method

The *soldering-copper method* (also referred to as the *soldering-iron* method) is a term used to describe soldering done with the copper-tipped soldering iron. This is the oldest and most commonly used method of joining metals with solder.

The soldering-copper method is described in considerable detail in the following sections of this chapter:

1. Tinning the bit,
2. Wetting,
3. Tinning the surface,
4. The soldering procedure.

Induction Heating Method of Soldering

When an electric current flows in a conductor, magnetic lines of force are formed around it. If the conductor is a wire formed into a loop or coil, the magnetic field is concentrated. With an alternating current flowing in the coil, fluctuating from zero to a maximum, the magnetic lines of force expand. When the current reverses, they contract. If these magnetic lines of force are cut by an object as they expand and contract, a voltage is induced in that object. Thus, the object functions as a short-circuited secondary coil, and begins to build up heat on its surface. The amount of heat will depend on whether welding, brazing, or soldering temperatures are required.

The induction heating method of soldering consists of placing the surfaces to be joined in this type of strong alternating field. As was mentioned in the previous paragraph, this field is produced by an alternating electric current flowing through a coil. The amount of heat is determined by: (1) the frequency of the equipment, (2) the surface resistance of the metal, and (3) the power applied. Low frequencies require very close coupling between the coil and work to transfer energy; high frequencies permits a wider spacing between coil and load. Power sources are varied and may include specially designed induction heaters or generators.

Because induction heating is expensive, it finds its most widespread application in production-type work where repetition eventually reduces cost. Most joint designs can be soldered since irregular shapes simply require specially designed coils. However, the size of object does prove to be a limitation in induction heating. The larger the object, the less efficient this method becomes.

Other Soldering Methods

Some metals, such as aluminum, develop an oxide film on their surfaces which makes soldering by the standard methods very unsatisfactory. Ultrasonic soldering was developed to solder such metals. Its success is attested to by its growing use in the production soldering of nonferrous metals (which were formerly considered difficult to solder), dissimilar metals, and materials long considered unsolderable (e.g. ceramics, glass, plastic, fiber glass, quartz and ferrites).

An ultrasonic soldering system designed for production work consists of the three following basic components:

1. A power supply,
2. A transducer,
3. A soldering unit.

The power supply is a solid-state unit capable of delivering 20-kHz ultrasonic energy to a lead zirconate titanate transducer. The transducer converts the electrical energy fed to it by the

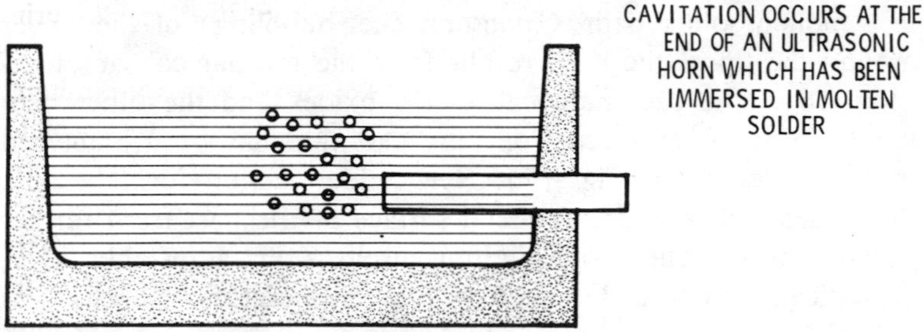

CAVITATION OCCURS AT THE END OF AN ULTRASONIC HORN WHICH HAS BEEN IMMERSED IN MOLTEN SOLDER

Fig. 16. Cavitation occurring in a molten solder bath.

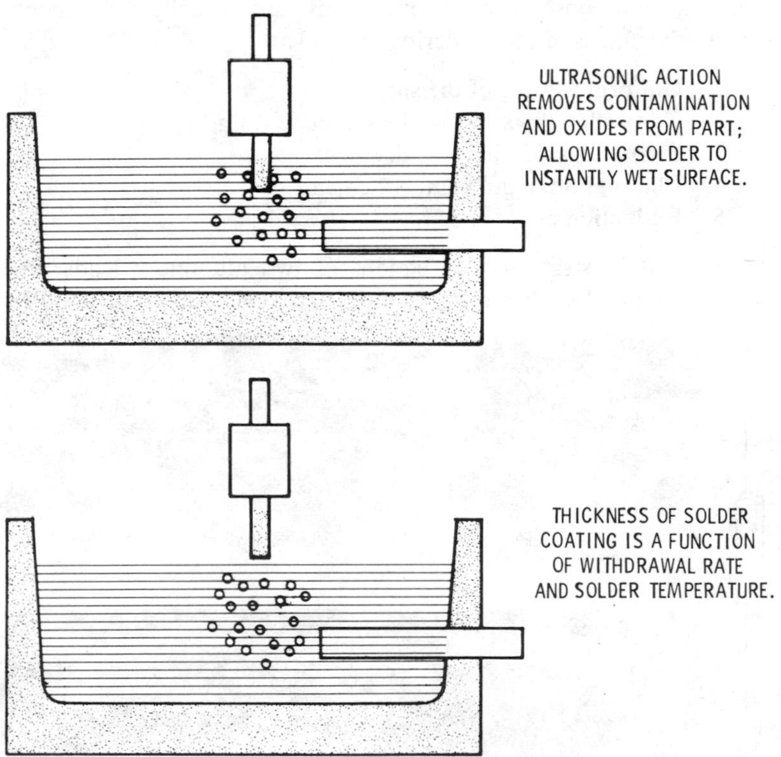

Courtesy Branson Sonic Power Co.

Fig. 17. Immersion and removal of parts from the cavitation zone.

power supply unit into 20-kHz mechanical vibrations. The vibrations are then transmitted to a specially designed horn immersed in a molten solder bath. Cavitation (i.e. bubbling) of the solder around the tip of the horn results from the mechanical vibrations (Fig. 16). The cavitation causes the oxides and the other contaminants to be removed from any part that has been immersed in the solder bath (Fig. 17). New oxides cannot form because the solder bath shields the metal surface of the part from further contamination. The soldering unit itself is an adjustable, thermostatically controlled heater with a temperature range up to 800°F.

Ultrasonic soldering does not require a flux. Moreover, precleaning and post-cleaning operations are usually eliminated. Other advantages of this soldering method are:

1. Uniform wetting of the surface,
2. Controlled thickness of the solder coating,
3. Increased soldering speed,
4. Elimination of gold embrittlement,
5. Reduction of cost.

Fig. 18 illustrates cross sections of two aluminum leads, one soldered with the ultrasonic method and one without it. In Fig.

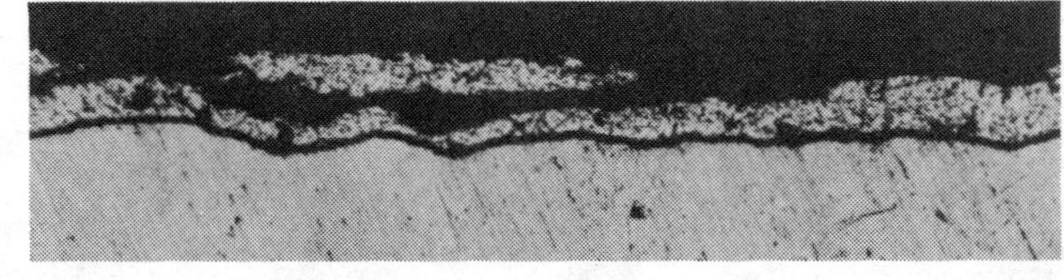

Without ultrasonics, considerable oxides are present at interface, resulting in poor bond. (Immersion @ 500° F for 1.5 seconds.)

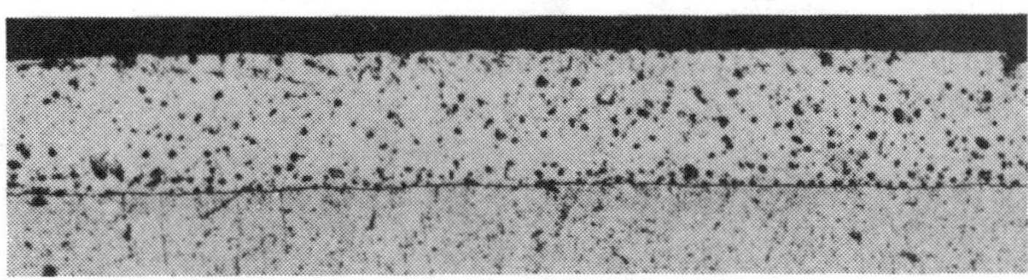

With ultrasonics, excellent bonding occurs since oxides have been scrubbed away as the solder wets the surface. (Immersion @ 500° F for 1.5 seconds.)

Courtesy Branson Sonic Power Co.

Fig. 18. Microscopic cross sections of aluminum axial leads soldered without flux.

19, the cross section of a joint between two sections of aluminum tubing clearly shows the excellent diffusion of solder and aluminum possible with ultrasonic soldering.

Ultrasonic vibrations can also be applied to the tip of a soldering iron with a unit consisting of a power supply, a foot-switch, and a soldering iron (Fig. 20). Tips for ultrasonic soldering irons are available in several sizes and shapes.

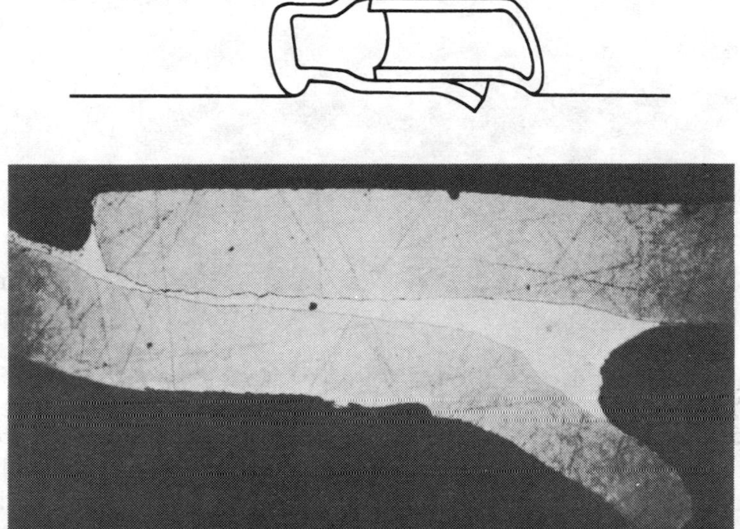

Courtesy Branson Sonic Power Co.

Fig. 19. These two sections of aluminum tubing are fitted together in a bell joint configuration and solder ultrasonically without any precleaning or fluxing.

Resistance soldering employs the same heating principle used in resistance welding. The heat is produced by the resistance of the metal to the passage of an electrical current. The electrodes are also similar in design, although currents of a lower voltage are used. A major problem in resistance soldering is that the electrode usually cannot be tinned. Therefore, the operator must exercise special care in positioning both the flux and solder exactly where it is to be used.

Special furnaces or ovens are also used for producing the heat for soldering. These may be either gas or electrically heated, and some are equipped with devices for rapid cooling. This method of soldering is recommended for complicated assemblies. The size of the assembly is restricted by the size of the furnace.

AUTOMATIC SOLDERING SYSTEMS AND MODULES

A number of different types of automatic soldering systems have been developed for use in industry. One broad application

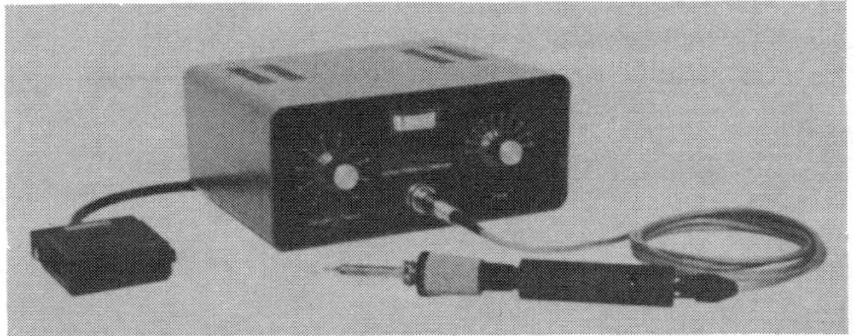

Courtesy Branson Sonic Power Co.

Fig. 20. Ultrasonic soldering iron and power source.

of these systems is in the soldering of PC (printed-circuit) boards and computer back plane boards. Those systems based on modular design provide maximum flexibility through an interchangeability of major components.

A basic component of any automatic soldering system is the solder applicator. Two types of solder applicators used in industry are the cascade wave type and the flat wave type. A variation of the flat wave type is the flat wave applicator with extenders (Fig. 21).

Standard PC board soldering systems are capable of handling work ranging from small PC boards to the larger, more complex computer back plane boards. These systems are modular in design and generally consist of the following components:

1. A flux applicator,
2. A preheating unit,

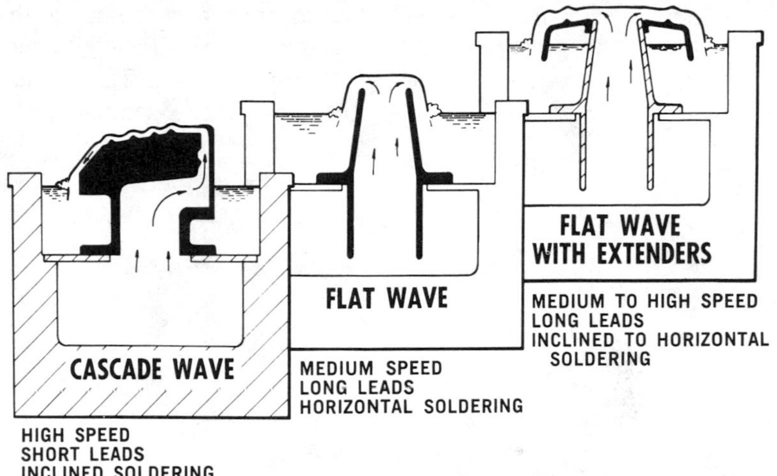

Courtesy RCA Industrial and Automation System

Fig. 21. Three types of solder applicators.

3. A soldering module,
4. A control panel,
5. A conveyor.

Washing and exhaust stations also form a part of these automatic soldering systems. Multiple rotary brush cleaning systems have been developed that employ high pressure hot water sprays with a deionized water rinse.

The soldering module may be mounted on a roll-out base to facilitate maintenance. The one illustrated in Fig. 22 is of this type. This particular model has a cast-iron solder pot with a capacity of 1200 lbs. The flux applicator is limited to 24-inch wide boards or smaller.

Smaller bench soldering machines have also been developed to solder PC boards. These are designed to handle maximum widths of either 7 or 10 inches. Like the larger types, these, too, are modular systems.

Manufacturers of automatic soldering systems are frequently called on to design machines for a specific use. Each of these machines will conform to the various sequential steps in the soldering process.

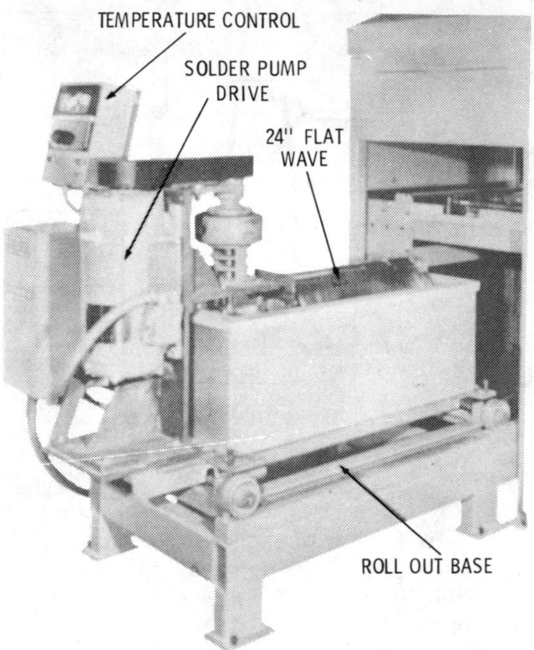

Courtesy RCA Industrial and Automation System

Fig. 22. A soldering module mounted on a roll-out base.

JOINT DESIGNS FOR SOLDERING

A basic requirement for a soldered joint is that it be designed to compensate for the weaknesses of solder as a joining material. A soldered joint does not have the strength of a welded joint, because the base metal does not melt and fuse with the filler metal. Consequently, a suitable joint design for soldering is determined by the stresses or loads to which it will be subjected. The operator should consider the degree to which the joint must withstand vibration, impact, tension, compression, and shear stress. The more pronounced these factors become, the greater is the need for such additional support as bolts, screws, or other fastening devices.

The electrical conductivity of a soldered joint is also an important consideration. Because so many soldered joints serve as electrical connections (e.g. on printed-circuit boards), one possessing poor electrical conductivity would hardly be acceptable. Both the composition of the solder and the finished joint must be conducive to a high degree of electrical conductivity. The operator must produce a soldered joint without points and avoid bridging electrical connections.

The lap and butt joints are the most common type of joint used for soldering. Of the two, the former is recommended and preferred by most operators. Fig. 23 illustrates many of the typical joint designs used for soldering.

SOLDERING PROCEDURE

Soldering is both a skill and an art. Although soldering gives the appearance of being easy, it requires a great deal of practice before even a passing skill can be acquired. The following procedure for soldering wire joints provides a basic step-by-step outline for the soldering process.

1. Clean and tin the bit thoroughly.
2. Heat the bit until it reaches the right temperature. Do not try to solder a joint with a bit so cool that it only melts the solder slowly, nor with one so hot that it gives dense clouds of smoke when in contact with rosin. Burned rosin must be regarded as dirt.
3. Remove the bit from the fire and hold it, or preferably support it on a brick or block of other material which does not conduct heat readily.
4. Wipe the surface clean with a rag. Apply solder until a pool remains on the flat surface, or in the groove, if a grooved bit is used.
5. Sprinkle with rosin, lay the joint in the pool of solder and again sprinkle with rosin.
6. Rub the joint with a stick of solder so that every crevice is thoroughly filled.
7. Remove the bit and lightly brush superfluous solder from the bottom of the joint. See that no sharp points of solder remain which may afterwards pierce the insulation.

When the bit is first placed on the joint, the solder should run up into the joint. This will occur only when the joint is well

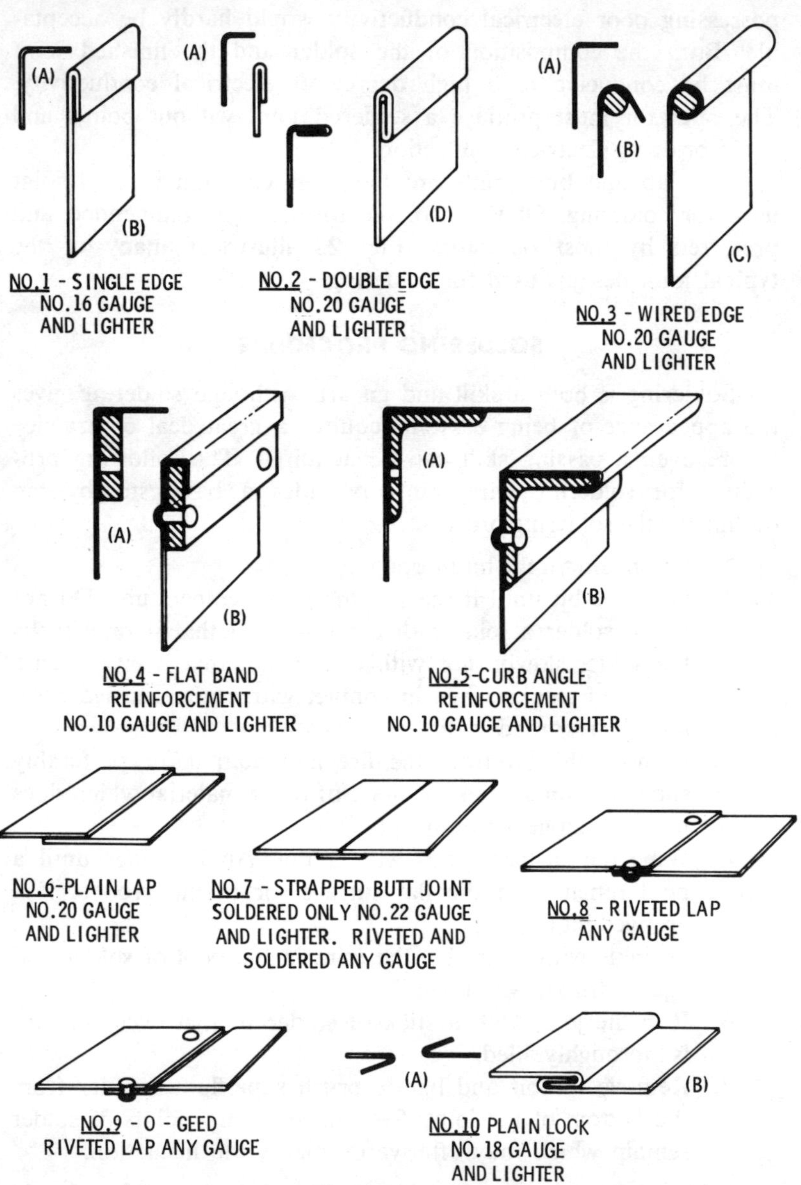

Fig. 23. Typical designs

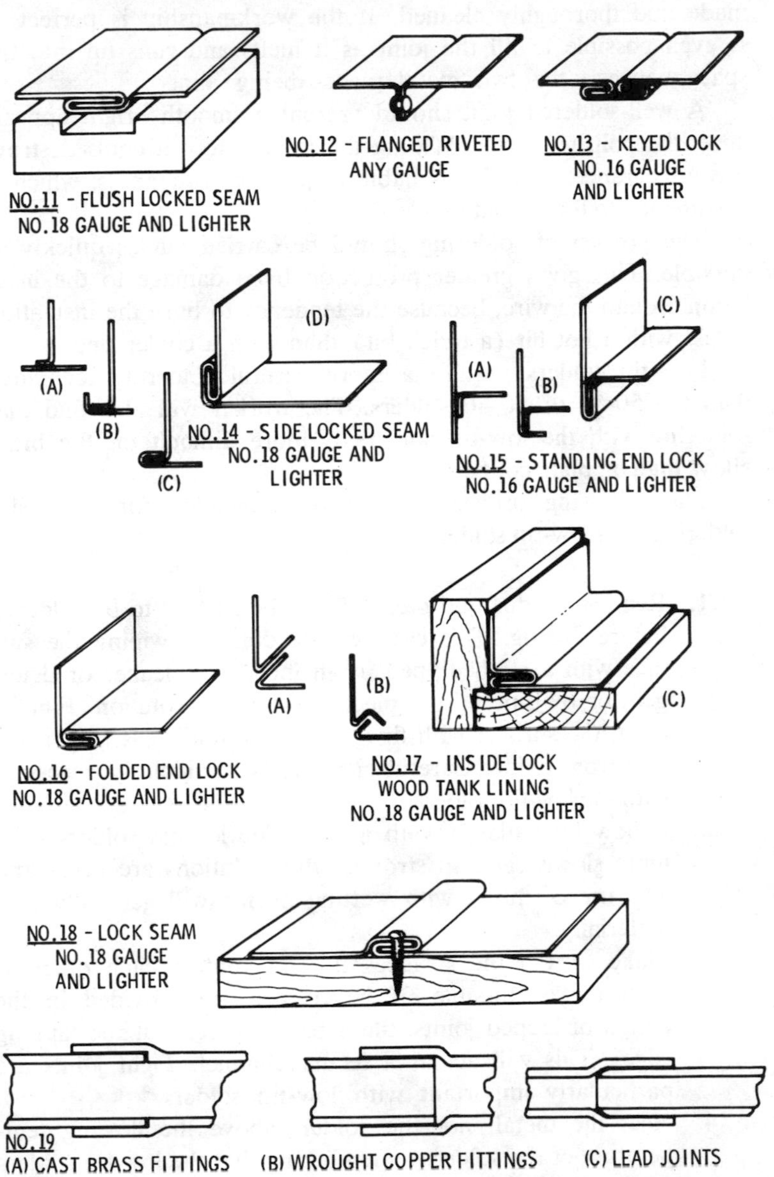

Courtesy American Welding Society

for soldering.

made and thoroughly cleaned. If the workmanship is perfect, it is even possible to fill the joint as it melts and runs up into the space between the two metal pieces being joined.

A well-soldered joint should present a smooth bright appearance like polished silver. Wiping the joint before it cools destroys this appearance, and also is liable to produce roughness, which is detrimental to the insulation.

The process of soldering should be carried out as quickly as possible. This gives greater protection from damage to the insulation around the wire, because the tendency to burn the insulation is less with a hot bit (a quick bit) than with a cooler one.

Low-tin solders require a more careful soldering technique than the 50-50 or 60-40 solders. The worker will also find that soldering with the low-tin solders is more difficult on the brass alloys than on plain copper.

The following procedures are recommended for successful soldering with low-tin solders:

1. Remove all dirt and grease from the surface to be soldered before fluxing. This can best be done by wiping the surface with a cloth dipped in an inorganic cleaner or detergent or by dipping the piece itself in the solution. Finally, clean the surface with fine steel wool until it is bright and free from oxide. Careful cleaning is particularly necessary with the low tin solders.
2. Follow immediately with a flux. The low-tin solders alloy more slowly, and so stronger flux solutions are necessary. The use of fluxes with wetting agents will generally help materially.
3. Make certain the surfaces fit and contact each other as perfectly as possible. If bevels are not provided in the design of lapped joints, file a bevel or bell out the lapping edges. This will act as a feeder channel. Tight joints are particularly important with low-tin solders.
4. Heat the metal, *not* the solder, above the flowing temperature of the solder, remembering that the low-tin solders require higher metal temperatures. Also, remember that just filling the feeder channel with solder is not enough.

Heat the lapped joint until the solder drops disappear from the bevel and continue to apply solder until the joint is filled.

5. Use a flux while soldering, applying it by dipping the solder in the flux from time to time.
6. Do not touch or disturb the soldered joint until it has cooled below the melting point and is therefore fully set. After the temperature has dropped below the melting point, wet cloths can be used to decrease the time normally required to cool the joint to room temperature.
7. The stronger the flux, the more corrosive it is, so be certain to clean off all traces of flux from the completed joint. This is best done by using water.
8. If a smooth job is desired, finish the joint by removing the high spots with a medium file.

If the joint has a small enough diameter to be heated all over, such as in water service tubing, there should be no trouble with low tin solders if the foregoing procedure is used. However, on large diameter joints, pretinning both sides of the joint, after proper cleaning and fluxing, is an additional help.

The same procedure applies to the high-lead silver solders except that a temperature of about 250°F higher than that for 50-50 solder is required to make the solder flow. The high-lead silver solders can be used where higher temperatures of application are not a handicap. However, because of the higher temperature required, the same flux that is used for 50-50 solder cannot be used for the lead silver solders. Special fluxes for this purpose are available, and flux manufacturers should be consulted.

LEAD BURNING

Lead burning (sometimes referred to as *autogenous soldering* or *lead welding*) is the process of joining pieces of lead together simply by placing the edges to be joined close to, or overlapping each other. They are melted so that they flow and intermingle with each other, forming one piece, and retaining the same condition of unison on solidifying.

A lead burn joint can generally be reinforced by melting a strip of lead along the joint at the same time as the edges are joined.

For joining lead sheets together, it is essential that the pieces touch or overlap each other when in horizontal position and overlap when in either slanting upright or overhead position. It is not necessary to soil the sides of the seams, because the lead will flow only where it is directed by the flame jet.

A hydrogen flame in combination with a stream of air was formerly the most common method of applying heat in lead burning. The oxyacetylene flame has largely replaced the use of hydrogen in recent years, because it is faster and cheaper.

The oxyacetylene torch used for lead burning is equipped with a valve block for regulating the gases (there are no valves on the torch handle). The gas regulators are similar (though smaller) in design to those used in oxyacetylene welding and operate on the same principle. There are a number of different tip sizes, and their use will depend on the nature of the job.

For flat butt burning, the end of a stick of lead should be held on the seam so as to be melted at the same time as the other lead. During the process of burning, the sheet lead will be expanded when the heat is applied. Being a poor conductor, the heat is not distributed to the adjoining sides of the seam, hence the heated parts will rise up and leave hollow spaces underneath. When this happens, leaving places where the lead does not rest on the board, the lead melts more readily with the result that a hole is made through which the molten metal will flow. To prevent this, the lead should be held down with the end of the stick of feeding lead, which is held in the left hand.

The following recommendations are suggested for improving the quality of lead burning:

1. Hold the flame perpendicular to the work (never allow the flame to strike the work from an angle or be held flat against the work),
2. Move the flame as rapidly as the work will permit,
3. Keep the flame as small as possible (that is, no larger than required by the work).

SOLDERING COPPER AND COPPER ALLOYS

Copper is a very easy metal to solder, but the copper-base alloys differ in this respect according to the type of alloy used. The copper-tin, copper-zinc, and copper-nickel alloys can all be soldered with relative ease. However, those containing such alloying metals as silicon, chromium, aluminum, or beryllium present varying degrees of difficulty to the operator.

Copper is characterized by a very high degree of thermal conductivity. Consequently, the heat received from the soldering iron is rapidly dissipated over a wide area of the metal surface. A high rate of heat input will be necessary to counteract this effect in cases where a localized heat application is necessary.

A 50-50 solder (50% tin, 50% lead) with a temperature range of 361°F (solidus) to 421°F (liquidus) is commonly used to solder copper and copper-base alloys. A 95% tin, 5% antimony is a quick-setting solder frequently used to join these metals. However, it should not be used with a copper-zinc alloy because it tends to increase joint brittleness.

Any of the fluxes used in soldering can be used on copper and the copper-base alloys. That is not to say that each flux works equally well on every one of these metals. Many of the copper-base alloys (e.g. the aluminum bronzes) require special fluxes. Generally speaking, corrosive type fluxes can be used on all the copper-base alloys, and the noncorrosive types on the coppers.

SOLDERING ALUMINUM AND ALUMINUM ALLOYS

Solder for aluminum should preferably be one which is self fluxing and should be applied with a nonoxidizing flame. Lead solders are not recommended for soldering aluminum. A soldering torch or a gasoline blow torch may be used. Soldering bits are not recommended, because the degree of heat required is far greater than any soldering bit is capable of producing.

All metals or combinations of metals used for aluminum soldering are subject to corrosion unless they are protected by a paint or varnish. Since all metals or combinations of metals are electro-positive to aluminum, a soldered joint when exposed to moisture is rapidly attacked and disintegrated.

The corrosive fluxes are recommended for soldering aluminum because they can counteract the tendency of the aluminum surface to form an oxide film during the application of heat. Since those fluxes recommended for use with aluminum contain fluorides, great care must be taken to ensure proper ventilation during the soldering operation.

SOLDERING MAGNESIUM AND MAGNESIUM ALLOYS

Magnesium and the magnesium alloys are difficult to solder because of the refractory oxide film that forms on the surface of these metals. No effective flux has been developed to chemically remove this oxide film. Consequently, the surface must be cleaned with steel wool, a wire brush, or some other mechanical means.

Like copper, magnesium is characterized by high thermal conductivity. Therefore, any soldering operation will require a high heat output.

A 70% tin, 30% zinc solder (as a precoating) followed by a 60% cadmium, 30% zinc, and 10% tin solder is generally recommended for soldering magnesium and the magnesium alloys.

Because of the difficulty in removing the oxide film from metals, soldering applications are largely limited to repair-type operations (filling surface dents, cracks, or other defects).

SOLDERING OTHER METALS

Other metals that can be joined by soldering with varying degrees of success include the following:

1. Cast irons,
2. Tin,
3. Lead,
4. Nickel,
5. Steel,
6. Stainless steel,
7. Precious metals.

The cast irons are not easily soldered, but they can be joined by this method if proper precautions are taken.

It is the presence of carbon in cast iron that presents the greatest difficulty (particularly the graphitic carbon flakes in gray cast iron) because it resists the wetting action of the solder. If a surface cannot be wetted, the soldered joint may fail. The task, then, is to prepare the surface in such a way that this tendency of carbon will be drastically reduced or eliminated. One method is to use an oxidizing flame to sear the surface. Chemical cleaning techniques are also used. Another method is to chip the surface until it shines brightly and then brush it with a stiff wire brush in the direction it was chipped.

Corrosive fluxes are recommended for soldering cast irons. Any type of heating method is acceptable, and most solders will do the job well. Soldering cast irons is basically a repair procedure.

Tin and the tin alloys are as easy to solder as copper. The surface preparation is even easier, because it is only necessary to remove any surface contaminants (dirt, grease, oil, etc.) before soldering. A rosin flux and a 63%-37% lead solder is recommended for soldering tin and its alloys. A torch flame is usually preferred for heating.

The basic problem involved when soldering lead is its low melting temperature. Lead and the various lead alloys melt at temperatures ranging from 450°F to 621°F. Every precaution should be taken to avoid melting the base metal. The solder used for joining these metals should be selected with this consideration in mind.

A rosin flux is recommended for soldering tin and the tin alloys. The several solders used for joining these metals include:

(1) 50% tin -50% lead,
(2) 38% tin -60% lead -2% antimony,
(3) 34.25% tin -64.39% lead -1.25% antimony, and 0.11% arsenic.

A 60% tin -40% lead or 50% tin -50% lead solder is generally recommended for soldering nickel and the high-nickel alloys (Inconel, Duranickel, Monel, etc.). In the case of the high-nickel alloys, the selection of a suitable flux depends on the metal used as an alloy with the lead. There is no restriction placed on the type of heating method employed.

The ease by which a steel can be soldered depends on the carbon content of the metal. As the percentage of carbon increases, the steel becomes more difficult to solder.

A 40% tin 60% lead flux is generally recommended for soldering steel and its alloys. A strongly corrosive flux with a suitable heating method (all types apply to steel) will produce a satisfactory soldered joint.

Stainless steels contain elements (e.g. chromium) that increase their resistance to oxidation and corrosion. These same elements make stainless steel somewhat more difficult to join than the standard steel types.

A 50% tin 50% lead solder is generally preferred. A corrosive flux is absolutely necessary and accompanies an extensive thorough cleaning of the metal surface. After removing all traces of the corrosive flux, a rosin flux is used.

The precious metals are commonly represented by gold and silver. Others (rhodium, palladium, etc.) are also used but with far less recognition on the part of the public.

A rosin flux and a 50% tin -50% lead solder is usually recommended for soldering the precious metals. A suitable soldering iron is selected for the heat source.

It is sometimes necessary to join metals that are dissimilar in chemical composition. In most cases, one of the higher temperature joining processes (e.g. welding, braze welding, or brazing) can be used with satisfactory results. However, in those instances where the differences between the melting point temperatures of each metal exceeds 50°F, soldering is the recommended joining method. The joint designs for soldering dissimilar metals are the same as those used to solder other metals.

The most efficient production-type soldering method for joining dissimilar metals is the ultrasonic soldering process. Because the solder bath shields the metal from contamination, no flux is necessary with ultrasonic soldering. This soldering process is described in greater detail in another section of this chapter.

INSPECTING SOLDERED JOINTS

The inspection of a soldered joint is generally a visual one whether it be on the production line or elsewhere. As a matter of fact, visual inspections are more likely to occur on the production line than elsewhere. This is due largely to the lack of sufficient time to apply electrical or mechanical tests.

The types of soldered joints that fail to meet visual inspection standards are as follows:

1. Joints with excess solder,
2. Joints with no solder at all,
3. Joints with insufficient solder,
4. Flux residue in the joint,
5. A joint in which solder functions as a ground or causes a short,
6. A joint with points protruding from the solder,
7. Indications that the joint was made with a poor flow of solder,
8. Indications that the joint was moved before it was completely cooled.

Movement of the soldered joint before it has completely cooled will cause it to be materially weakened. Premature movement all too often causes the solder to fracture which, in turn, may result in joint failure.

The amount of solder applied to a joint accounts for the greatest number of soldered joints that meet rejection. For example, some operators apply entirely too much solder to the joint. This can be corrected before final inspection by removing the excess solder. G.R. Nichelson of the *Wall Manufacturing Company* recommends dipping the scrap end of unsoldered stranded conductor into flux and pressing between the terminal and the hot iron tip. This will cause the excess solder to be drawn off into the strand.

Joints that have insufficient solder are weak and will usually fail if subjected to heavy vibrations. These joints, and those to which no solder has been applied, are regarded as not being repairable and are rejected.

Sometimes excess flux will remain in the joint after soldering. If the soldered parts (wire, tube, sheet, etc.) are loose, the joint is defective and must be rejected. However, if no movement is possible, the flux can be removed by reheating.

If the soldered joint has resulted in the shorting or grounding of electrical equipment, it should be rejected. This also applies to the formation of points on the surface of the solder, for these can disrupt a high-voltage circuit.

Sometimes the solder does not have the opportunity to flow evenly through the joint. Re-heating will correct this defect prior to the final inspection.

CHAPTER 15

Brazing and Braze Welding

Brazing and *braze welding* join metals by applying heat to a nonferrous filler metal that melts above 800°F but below the melting point of the base metal.

Both brazing and braze welding are distinct joining processes and one should not be confused with the other. Unfortunately, careless individuals, or those ignorant of the distinction, will use "brazing" to mean both brazing and braze welding. Actually, braze welding is a gas welding technique; whereas, brazing represents a whole group of welding processes.

The principal ways in which brazing and braze welding differ are:

1. Joint design,
2. Joint preparation,
3. The method of filler metal distribution.

Since braze welding is essentially a gas welding technique, the joint designs will be similar to those used in gas welding. The joint edges are often beveled and this is an edge preparation procedure not found in brazing. Because of the method of filler metal distribution in brazing, the joints are fitted very closely together (within thousandths of an inch). Joint clearances for braze welding, on the other hand, are no different from those for gas welding.

In brazing, the filler metal is distributed by capillary attraction. The filler metal flows freely over the surface and into the recesses of the joint. The force of capillary attraction is such that the filler metal will even flow against gravity (e.g. vertically in the pipe joints). The filler metal in braze welding is fused where it is deposited. In other words, it is not distributed by capillary attraction.

Brazing and braze welding offer certain advantages over other joining processes. Most of these advantages stem from the lower heating temperatures used (below the melting point of the base metal) and include the following:

1. The elimination of warping,
2. Increased welding speed,
3. Reduction of the effects of expansion, contraction, and distortion,
4. Little chance of locked-in stress,
5. Reduced expense.

The reduction in operating expense, logically enough, results from the lower heat requirement and faster welding speed. The joints produced in brazing and braze welding are stronger than soldered joints, but not quite as strong as those produced by other welding processes (e.g. plasma arc, oxyacetylene, etc.).

Courtesy Airco Welding Products

Fig. 1. Braze welding a locomotive piston head.

Fig. 2. Braze welding a cracked cast-iron pressure cylinder.

Brazing and braze welding are highly suitable methods for joining dissimilar metals, repairing gray iron castings, and joining pipes, thin-gauge metals, and small assemblies. They are extensively used in such industries and occupations as refrigeration, heating, air conditioning, electronics, and automotive (Figs. 1 and 2).

Finally, a word about terminology and the confusion to which it can lead. In the first paragraph of this section, it was stated that braze welding is sometimes erroneously referred to as "brazing." Another misnomer for braze welding is "bronze welding." This stems from the fact that bronze (brass) welding rods are used as filler metal in braze welding. Brazing, in turn is sometimes referred to as "hard soldering" (the filler metals are essentially hard solders), "spelter soldering" (spelter was at one time widely used as a filler metal), and "silver soldering" (from the use of a silver-based alloy as a filler metal).

BRAZING AND BRAZE WELDING EQUIPMENT

Braze welding is essentially a gas welding technique. As such, it can be performed with a standard torch using oxyacetylene, oxyhydrogen, air-fuel gas, or other oxy-fuel gas combinations. Carbon-arc torches have been successfully used in braze welding, too.

Brazing also uses standard welding torches with a number of different air- or oxy-fuel gas combinations to obtain the required brazing temperature. However, many welding equip-

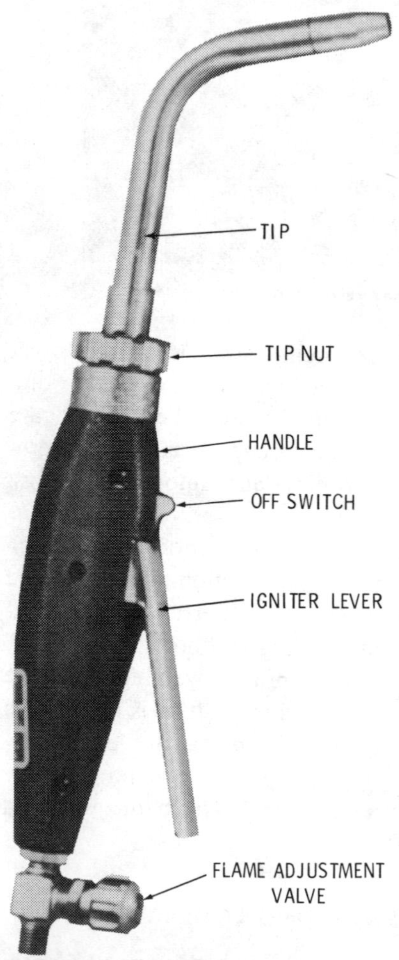

Courtesy Harris Calorific Sales, Inc.

Fig. 3. A torch designed for silver brazing, soldering, and other uses in metal work.

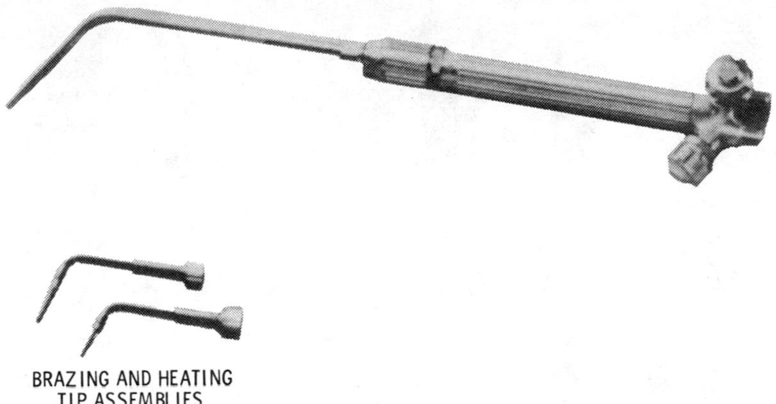

Courtesy Harris Calorific Sales, Inc.

Fig. 4. A torch designed for cutting, welding, and brazing, depending on the type of tip used.

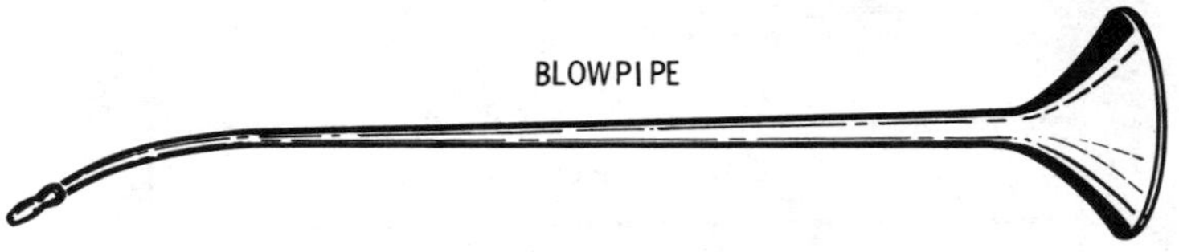

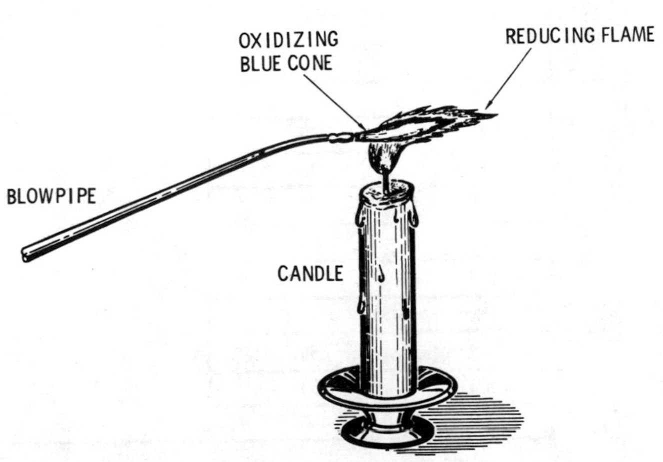

Fig. 5. A typical mouth blowpipe designed for small work.

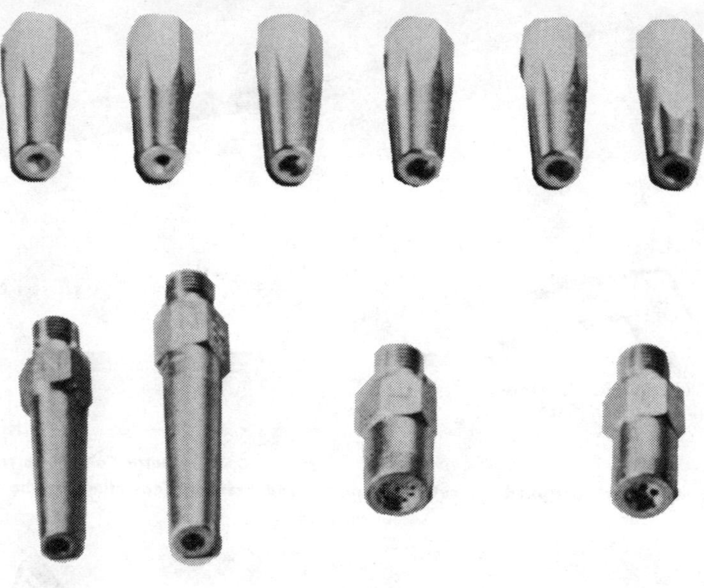

BRAZING TIP CHART		
TIP SIZE	OXYGEN PSI	FUEL GAS PRESSURE
4N	20	4 OZ OR MORE
5N	25	
6N	25	
7N	30	
8N	30	
10N	35	
13N	40	
15N	45	
20N	50	
30N	50	
80N	60	

Courtesy Harris Calorific Sales, Inc.

Fig. 6. Various types and sizes of brazing tips.

ment manufacturers have created specially designed torches for brazing, soldering, and other essentially low temperature joining processes. Examples of some of these torches are shown in Figs. 3 and 4.

One of the oldest "torches" used in brazing and soldering is the blowpipe. The blowpipe is simply a tube that may or may not be bent at the end. The operator blows through the tube in order to increase the intensity of the flame. An alcohol lamp is sometimes used to concentrate and direct the flame. A typical blowpipe is illustrated in Fig. 5.

Brazing and braze welding tips and tip assemblies are generally the same size as those used in other welding processes for a particular job. Here, again, welding equipment manufacturers have produced a special line of tips. Some examples of these are illustrated in Figs. 6 and 7.

Specially designed furnaces and pots are also used in soldering and some brazing processes. The furnaces should be equipped with protective devices, because there is always the danger of explosion from the accumulation of excess gas. This can be avoided by carefully following the instructions of the furnace manufacturer for discharging the excess gas. This equipment should only be used in a well ventilated area.

Some furnaces provide vacuums rather than shielding gases for the protection of the metal surface from atmospheric contamination (oxidation, etc.). These are designed for use with the refractory metals (e.g. zirconium, beryllium, etc.).

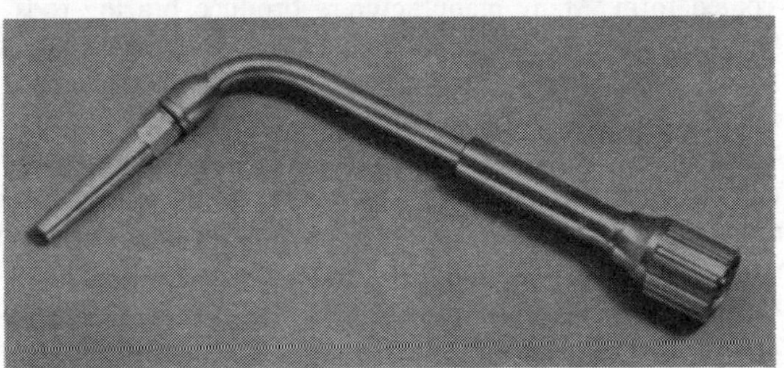

Courtesy Harris Calorific Sales, Inc.

Fig. 7. Brazing tip assembly.

A high-frequency AC generator provides the necessary heat in induction brazing. Because these are expensive pieces of equipment they are used only in production work.

BRAZING AND BRAZE WELDING FLUXES

Flux is a substance added to the metal surface for the purpose of eliminating any oxides or similar contaminants that may have formed, and thereby insuring a stronger bond between the surfaces to be joined.

Braze welding and most brazing processes require the use of a flux to prevent oxidation during the application of heat. Some heating methods, such as furnace brazing which is performed within a shielding atmosphere of inert gas or a vacuum, do not require their use. Brazing and braze welding, then, do not differ greatly from standard welding processes in their need to protect the metal surface against contamination from gases and other contaminants in the surrounding atmosphere. In addition to providing a protective coating against oxidation, a flux also has two other important functions:

1. It contributes to the capillary attraction of the brazing alloy by enabling it to run more freely.
2. It dissolves any oxides that may form *during* the brazing or braze welding operation.

Flux is commercially available in powder, paste, liquid, or rod-coated form. Many manufacturers produce brazing rods that are given a coating of flux at the factory. Brazing flux is generally available either in jars or cans in 4 oz., 8 oz., and 1 lb. weights.

The flux should be carefully selected for each particular brazing or braze welding operation. There is no all-purpose flux for all operations, although some exhibit wider application than others. Generally, the manufacturer's label will specify the particular metal with which the flux may be used. For example, *Chemetron Corporation's* NCG No. 2 brazing flux is designed to be used when brazing copper, brass, bronze, and steel. Their stainless steel flux, on the other hand, is used when brazing nickel-chrome alloys, Inconel nickel, and Monel metal. Aluminum brazing fluxes (regardless of the manufacturer) are limited in usage to aluminum brazing operations.

BRAZING AND BRAZE WELDING ALLOYS (FILLER METALS)

Almost any metal (or dissimilar metals) can be brazed or braze welded using a suitable filler metal (brazing alloy) and flux. Brazing filler metals generally include the following eight categories:

1. Silver-based alloys (BAg),
2. Aluminum-silicon alloys (BA1Si),
3. Copper (BCu),
4. Copper-zinc (brass) alloys (BCuZn),
5. Copper-phosphorous alloys (BCuP),
6. Copper-gold alloys (BCuAu),
7. Nickel-chromium alloys (BNiCr),
8. Magnesium alloys (BMg).

Classifications and requirements for the various brazing filler metals are established by the *American Welding Society* and the *American Society for Testing Materials* in their jointly prepared specification: Brazing Filler Metals, AWS A5.8, ASTM B260. Table 1 lists representative types of brazing filler metals, giving their specifications and uses.

The AWS-ASTM designation for brazing filler metal alloys begins with a prefixed letter (B-) which indicates it is a filler metal used for brazing. This is followed by a group of letters indicating the principal alloys. Thus, BAg indicates a silver base alloy (Ag) used for brazing (B). A number usually follows, separated from the letters by a hyphen (e.g. BCuP-1, BA1Si-3, etc.). This number does not refer to any aspect of alloy composition. It is simply a convenient means of numerically listing the various brazing alloys of a particular group (e.g. BCuP-1, BCuP-2, BCuP-3, etc.)

Some of the compositions of these brazing and braze welding alloys are given in the sections of this chapter on SILVER BRAZING, COPPER-ALLOY BRAZING, and ALUMINUM BRAZING.

Table 1. AWS-ASTM Classification: Brazing Filler Metals

Class	Cu	Ag	P	Percentage			Al	Other	Solidus Temp.	Liquidus Temp.
				Zn	Cr	Ni				
BCu	99								1980	1980
BCuP-1	95		5						1305	1650
BCuP-3	89	5	6						1195	1500
BCuP-4	87.5	6	7.25						1185	1360
BCuP-5	80	15	5						1185	1300
BCuZn-1	60			40					1650	1660
BNiCr-2					7	86		4 Si, 3 B	1750	1825
BAg-1a	15	45		16				24 Cd	1125	1145
BAg-1b	15½	50		16½				18 Cd	1160	1175
BAg-2	26	35		21				18 Cd	1125	1295
BAg-3	15½	50		15½		3		16 Cd	1170	1270
BAg-4	30	40		28		2			1220	1435
BAg-5	30	45		25					1250	1370
BAg-Mn		85						15 Mn	1760	1778
BAlSi-1	4						95	5 Si	1070	1165
BAlSi-3	4						86	10 Si	970	1085
BAlSi-4	4						88	12 Si	1070	1080

PRINCIPLES OF BRAZING AND BRAZE WELDING

The process of brazing is based on the fact that a molten nonferrous metal will form a strong bond with metal surfaces that are clean and properly fluxed, and that have been heated to the proper temperature. Simply stated, then, brazing involves the melting of a low fusing metal against the metals to be united while they are in such a condition of cleanliness and temperature that the metal welds itself to them.

It is not necessary to melt the base metal. This means that the weld will progress more rapidly than it could if fusion of the base metal were required. In the case of castings, it also means that repairs can, with a few exceptions, be made without extensive preheating, that frequently repairs can be made in places which would otherwise require dismantling.

Braze welding is another bonding process that does not melt the base metal. Brazing and braze welding operate under essentially the same basic principles, but differ in application and procedure.

Both brazing and braze welding use nonferrous filler metals that melt above 800°F but below the melting point of the base metal. Because a brazing or braze welding filler metal is composed of more than one alloy, it will not solidify (freeze) at one given temperature. Consequently, two temperatures—the *solidus* (melting point temperature) and *liquidus* (flowing point temperature) and *liquidus* (flowing point temperature)—are given for each brazing alloy. These are expressed in degrees Fahrenheit. In other words, the filler metal solidifies *below* the *solidus* temperatures and flows freely at the *liquidus* temperature or above. The brazing temperature range (also expressed in degrees Fahrenheit) extends approximately 200° F to 300° F above the *liquidus* temperature. For example, BAg-3, a silver brazing alloy, has a liquidus temperature of 1270° F, and a brazing temperature range of 1270°-1500° F.

HEATING METHODS USED IN BRAZING

There are a number of different methods used for applying heat in brazing, and it is on this basis that the different brazing processes are distinguished from one another. Because of the higher temperatures required in brazing, a flame is generally used instead of a heat bit. For small work, a blowpipe or torch is often sufficient. However, the torch is not limited to small assemblies. It is commonly and successfully applied to larger production and repair work. Forges (a smith's forge) and furnaces are frequently used, the furnaces being either electrically or flame heated. The work in a smith's forge is being held high up in the forge, so that it does not rest on the heat source, being suspended between banks of incandescent fuel so that the heating will be as near uniform as possible. Charcoal is the best fuel, although bituminous coal may also be used. If the latter is selected, the sulphur in the soft coal should be removed by coking. Otherwise, the brazing will be of very poor quality.

A gas furnace is very desirable for brazing. An air blast is necessary as in the forge, but a comparatively small blower will suffice. Dip baths of molten metal or chemicals have also been successfully employed in brazing.

Sometimes a charcoal bed is used to increase the brazing heat. This is usually done by building around the work with charcoal which becomes incandescent from the heat of the torch flame, and also gives off heat from its own combustion. Small articles are brazed by placing them in a hole scooped in the charcoal bed (Fig. 8).

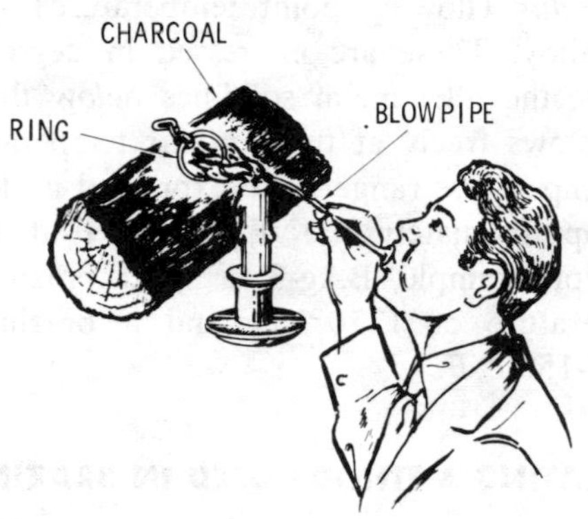

Fig. 8. The use of a charcoal bed.

On the basis of the method of heating, brazing can be divided into the following processes:

1. Torch brazing,
2. Furnace brazing,
3. Induction brazing,
4. Dip brazing,
5. Resistance brazing,
6. Carbon-arc brazing,
7. Twin carbon-arc brazing,
8. Flow brazing,
9. Block brazing,
10. Automatic brazing.

Torch Brazing

Torch brazing uses the heat of an oxygen (or air)-fuel gas flame to raise the base metal to the temperature required for brazing. Both a suitable flux and a filler metal are necessary. As is the case in most other brazing processes, the joint to be brazed must first be thoroughly cleaned and then correctly positioned. The filler metal (brazing alloy) is generally added next, although it can be added while the heat is being applied. The heat is applied until the filler metal melts into the joint by capillary attraction.

The torch brazing process finds wide application in industry in both fabrication and repair work. Consequently, this chapter will concentrate primarily on this brazing process in the sections concerned with brazing specific metals or metal alloys (see JOINING COPPER AND COPPER ALLOYS, etc.).

The various torch brazing methods are classified according to the type of filler metal (brazing alloy) used in the operation. The most commonly used are:

1. Silver based alloy,
2. Copper-zinc based alloy,
3. Aluminum based alloy.

These are described in greater detail further on in this chapter (see SILVER-ALLOY BRAZING, COPPER-ALLOY BRAZING, and ALUMINUM BRAZING).

Furnace Brazing

A suitable furnace is necessary for furnace brazing. The suitability of a furnace depends on its ability to raise the heat to the required brazing temperature and on the inclusion of any necessary safety devices.

The surfaces to be brazed are thoroughly cleaned and aligned. The brazing metal is then applied, and the complete assembly is then placed in the furnace. A flux is sometimes used to counteract oxidation, although some furnaces are so designed as to make this unnecessary. The heat in the furnace is increased until brazing temperatures are reached.

Induction Brazing

Induction brazing uses a high-frequency electric current as its source of heat. A flux or special gaseous atmosphere is necessary to prevent oxidation during brazing. Induction brazing is characterized by rapid heating. Distortion can be controlled by changing the current.

Dip Brazing

Dip brazing is a brazing process in which the parts to be brazed are dipped or immersed in a molten metal or chemical bath. Oxidation is prevented by the shielding action of the bath. The high heat conductivity of the molten metal or chemicals results in a rapid attainment of brazing temperatures. The molten bath may consist of either molten salt or a brazing filler-metal alloy. Molten salt baths are used for the dip brazing of larger assemblies.

A brazing bath can be constructed for the workshop by using an iron pot and a coal fire, gas furnace, or other suitable means of heating. The brazing alloy is melted in the pot over a fire, as

Fig. 9. The dip brazing process.

shown in Fig. 9, flux being placed on top of the brazing alloy. In brazing, first hold the object in the flux a little while to heat and coat the article with a film of flux. Then, when it is lowered into the molten brazing alloy, the latter will flow in the joint and firmly attach itself to the surface. Before dipping, the article to be brazed is coated with a special antiflux graphite covering all the surface except that which is to be brazed. The layer of flux in the pot may be kept from ½ to 2 inches deep.

Resistance Brazing

Resistance brazing can be performed on a standard projection or spot welder. Water-cooled electrodes are necessary, because of the high currents used. Oxidation is prevented with a suitable flux (usually the BCuZn or BCuP types), although specially designed gaseous atmospheres have also proven to be effective. Resistance brazing uses lower pressures than the welding process.

Carbon-Arc Brazing

Carbon-arc brazing resembles the welding process in that the arc formed between the carbon electrode and the work provides the necessary heat. A phosphor bronze or silicon bronze metal filler rod is used. This brazing process is characterized by relatively high operating speeds.

Twin Carbon-Arc Brazing

Twin carbon-arc brazing uses a twin-electrode torch, carbon electrodes, and alternating current (AC) to obtain suitable brazing temperatures. The temperatures are considerably higher than those used in carbon-arc brazing. Unfortunately, these high temperatures result in problems of control for the operator, a factor that limits twin carbon-arc brazing to thin gauge sheets.

Flow Brazing

Flow brazing operates on the same principle as flow welding. The molten brazing filler metal is allow to flow over the joint until the base metal is heated to a suitable temperature for brazing.

Block Brazing

In block brazing, the base metal is raised to brazing temperatures by applying heated blocks to the joint. Various means (fuel-gas flame, electricity, etc.) are used to heat the blocks.

Automatic Brazing

Automatic brazing consists basically of continuous flux feeding devices and specially designed filler metal forms. In the former, a liquid flux is mixed with the torch-brazing fuel gas and continuously applied to the surface during brazing. As a result, the fluxing step is eliminated from the brazing procedure. The specially designed filler metal forms are used in furnace brazing.

BRAZING JOINT DESIGNS

In brazing, the joints are designed to fit within thousandths of an inch. A joint clearance of 0.002 to 0.006 inch is recommended when copper alloy and silver alloy brazing. A separation of 0.006 to 0.015 inch is generally suggested for aluminum brazing. The molten filler metal will distribute throughout the joint surface by capillary attraction.

Many types of joints can be brazed, but the lap joint probably gives the most satisfactory results in terms of optimum strength requirements. The length of the lap should be equal to at least three times the thickness of the thinnest member of the joint. Fig. 10 illustrates some basic types of joints used in brazing.

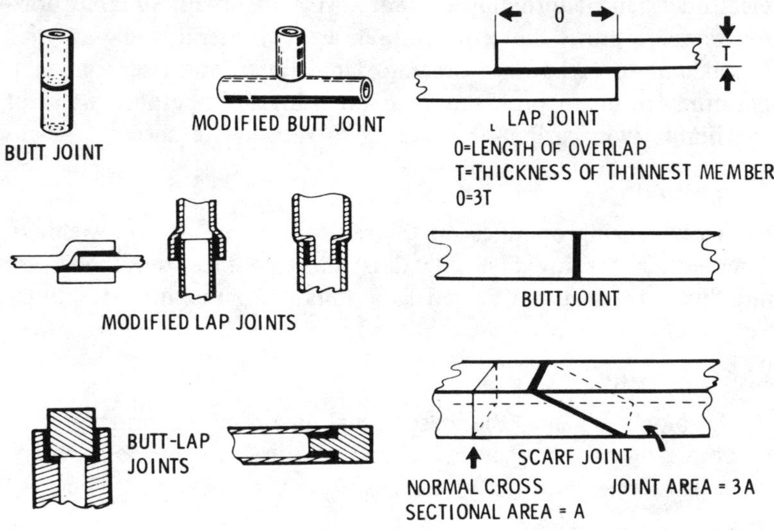

Courtesy Airco Welding Products

Fig. 10. Basic types of brazed joints.

An important consideration in joint design for brazing is the type of stress to which the joint will be subjected. The joint should be designed so that the brazing alloys are subjected to a shearing stress (rather than direct stress) across the deposited metal. Fig. 11 illustrates both good and bad joint designs.

BRAZING PROCEDURE

The procedure for brazing consists essentially of the following steps:

1. Cleaning the surface,
2. Applying a suitable flux,
3. Aligning the parts to be joined,
4. Applying the heat,
5. Applying the appropriate brazing alloy,
6. Cooling the surface,
7. Cleaning the surface.

The work must first be thoroughly cleaned both mechanically and chemically. All dirt, grease, oil, and other surface contaminants must be removed or the capillary attraction so important to the brazing process will not function properly. The result will be a weak bond or no bond at all. Fluxes are *not* designed for this purpose, and should not be used for this. Steel wool, emery cloth, or other suitable means can be used for the mechanical cleaning. If necessary, the surface can then be chemically cleaned by pickling in acid. The acid should be removed by washing the surface with warm water.

Select a suitable flux for the brazing operation. Remember that fluxes differ in their chemical compositions and the brazing temperature ranges within which they are designed to operate. Select the one that meets the requirements of the work. The flux is applied to the joint surface of the parts to be brazed. This is usually applied as a paste with a brush.

Align the parts to be brazed and secure them in position by clamping or other suitable means. Tongs may be used for small pieces. The parts must be held together until the brazing alloy solidifies.

Preheat the surface to the required brazing temperature. The correct brazing temperature will be indicated first by a gradual drying out of the flux and finally by its becoming fluid.

As soon as the flux has reached the fluid stage, the brazing alloy should be added. If all conditions (cleaning, aligning, etc.) in the brazing procedure have been properly met, the brazing alloy (filler metal) will spread over the metal surface and into the joint by capillary attraction. Stop heating as soon as the brazing alloy has completely covered the surface of the joint. Never overheat when brazing as this will tend to increase the porosity of the brazing alloy and weaken the bond.

Allow the surface to cool and the brazing alloy to completely solidify before removing the clamps or other devices used to hold the parts in position.

Remove any excess flux. Some fluxes tend to corrode the surface, and all fluxes (or other substances) will interfere with painting or any other finishing methods. Scrubbing with hot water is

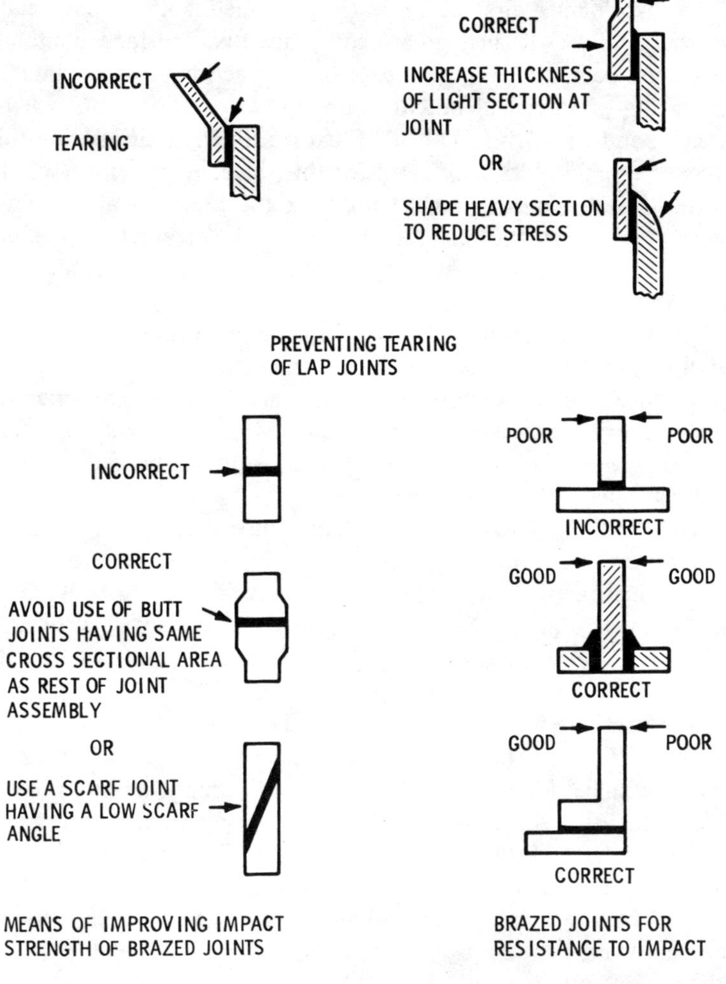

Fig. 11. Examples of good

generally sufficient for the post-cleaning operation. Excess metal around the joint can be removed with a file.

The various methods (techniques) of brazing include the following:

1. Butt brazing,
2. Lap brazing,

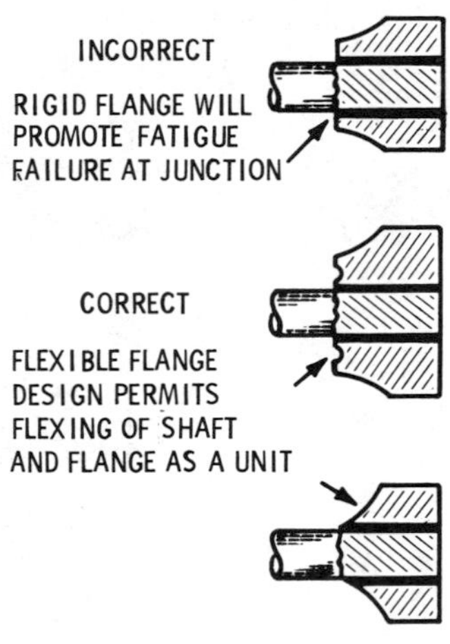

BRAZED JOINTS FOR SHAFTS AND FLANGES

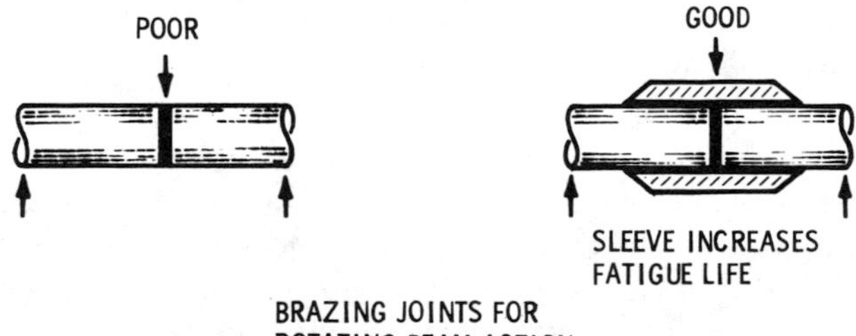

BRAZING JOINTS FOR ROTATING BEAM ACTION

Courtesy Airco Welding Products

and bad brazing joint design.

3. Muffle brazing,
4. Brazing with filler metal inserts.

Butt Brazing

Butt brazing is a brazing method frequently used to join lengths of pipe. This method is illustrated in Fig. 12. After cleaning the ends to be brazed and fluxing, they are clamped in position, butt to butt, using a vise and clamp (as shown in Fig. 12) or other means. A little brazing alloy is placed on the joint and heat is applied. When the pieces are hot enough to melt the brazing

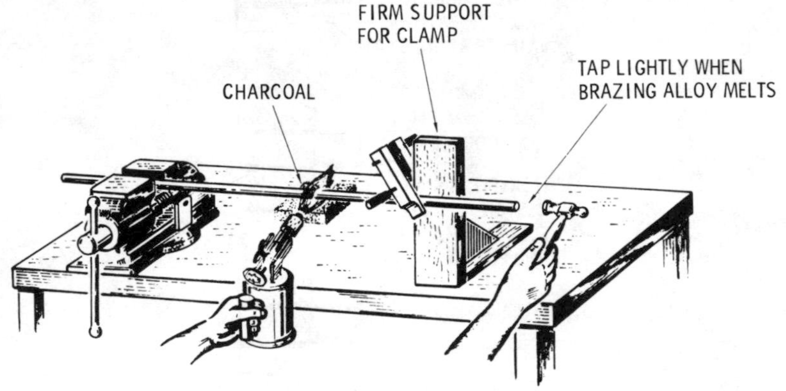

Fig. 12. Butt brazing.

alloy it will flow into the joint, butt brazing the two pieces. By giving one of the pieces a slight tap on the end, when the brazing alloy melts, the surplus brazing alloy is squeezed out, making a good and firm joint. If the pipes are large or of considerable length, the heat is quickly conducted away, necessitating a charcoal banking or more adequate means of heating.

Lap Brazing

The lap brazing method is illustrated in Fig. 13 with the joining of two sections of a band saw. Note that in making the lap, the two ends are beveled to make an accurate, close fitting joint. A file can be used for this purpose. The brazing alloy is either placed between the two surfaces, or the surfaces are coated with a suitable flux and the brazing alloy is allowed to flow into

Fig. 13. Lap brazing.

the joint from the edges. After firmly clamping the parts in position, the brazing alloy is laid over the joint, or it may be placed between the two pieces to be united. When the heat is applied, the brazing alloy melts and the two pieces are squeezed tightly together to force out surplus brazing metal.

Muffle Brazing

In muffle brazing, a tube or muffle is used for enclosing the parts to be brazed. The object of the muffle is to ensure uniform heating. It is especially adapted to brazing alloys, the melting temperatures of which are rather close to that of solder.

Courtesy Airco Welding Products

Fig. 14. Filler metal inserts.

Brazing With Filler Metal Inserts

Filler metal (brazing alloy) inserts are produced by some welding equipment manufacturers for use in brazing. These are preformed and prefitted before applying the heat. Fig. 14 illustrates both the seal and the spud type inserts.

SILVER-ALLOY BRAZING

Silver brazing has been used for years in the manufacture of silverware and jewelry because it makes a strong joint and because the color of the base metal can be matched. For industrial purposes, it has many additional advantages, several of which are due to comparatively recent new developments and improvements in formulas by silver brazing wire manufacturers. This process is also referred to as *silver soldering* or *hard soldering*.

Many operations, difficult if not impossible to perform by welding, are readily and economically done by silver brazing. While the silver brazing wire is relatively expensive per unit of weight, only a comparatively small quantity is needed for any one joint and consequently its cost does not materially affect the cost of the finished product.

The chief components of standard silver brazing alloys are silver, copper, and zinc. The silver content ranges from 10% to as much as 80%. The melting point varies with the composition, generally ranging from 1100° F to 1500° F which is approximately midway between the melting point of ordinary solder and that of bronze welding rod. Aluminum brazing alloys will have a somewhat lower working temperature, usually in the 1000° F to 1100° F range. The standard specifications for silver brazing alloys are given in Table 2.

Silver brazing is best and most economically done with the oxyacetylene torch, although for some classes of light work an air-acetylene torch may be satisfactory.

The parts to be joined by silver brazing must be thoroughly cleaned—first mechanically and then chemically. Apply a fine file, steel wool, or emery cloth until bright metal is exposed. Then coat with flux. The edges and surfaces should be smooth and fitted tightly together. The merest film of silver brazing alloy is sufficient

Table 2. A List of Silver Brazing Alloys

AWS-ASTM Class	% Silver	% Copper	% Zinc	% Cadmium	% Others	Solidus °F	Liquidus °F	Brazing Temperature Range °F
BAg-1	45	15	16	24	—	1125	1145	1145-1400
BAg-1a	50	15.5	16.5	18	—	1160	1175	1175-1400
BAg-2	35	26	21	18	—	1125	1295	1295-1550
BAg-3	50	15.5	15.5	16	3 (Ni)	1195	1270	1270-1500
BAg-4	40	30	28	—	2 (Ni)	1240	1435	1435-1650
BAg-5	45	30	25	—	—	1250	1370	1370-1550
BAg-6	50	34	16	—	—	1270	1425	1425-1600
BAg-7	56	22	17	—	5 (Sn)	1145	1205	1205-1400
BAg-8	72	28	—	—	—	1435	1435	1435-1600
BAg-9	65	20	15	—	—	1280	1325	1325-1550
BAg-10	70	20	10	—	—	1335	1390	1890-1600
BAg-11	75	22	3	—	—	1365	1450	1450-1650
BAg-Mn	85	—	—	—	15 (Mn)	1760	1778	1780-2100

Courtesy Airco Welding Products

to effect a good bond in a tight fitting joint, and any greater thickness of the alloy is wasteful.

Good fluxing is extremely important in silver brazing. The flux dissolves any oxide that may form on the surface of the metal while heated, blankets the heated surface from atmospheric oxidation and facilitates easy tinning.

There are several good silver brazing fluxes commercially available. These should be applied according to the directions of the manufacturer. Since the alloy will flow over the base metal wherever flux is present, materials and time will be saved if care is taken to put the flux only on those parts where silver alloy is supposed to adhere.

Since most silver brazing is done on relatively small work, the heat of an oxyacetylene lead burning blowpipe or small sheet-metal-welding blowpipe is usually sufficient. For heavier work, however, particularly on metals that have a high heat conductivity a large full size blowpipe may be required.

After the parts have been evenly fluxed, heat the joint with a soft neutral flame. Keep the blowpipe moving in circles, with the inner cone a couple of inches above the surface so that a spreading rather than a localized heat is applied.

When the bare metal has reached the melting point of the alloy, withdraw the flame and touch the end of the brazing wire to the surface. It will quickly melt and flow (due to capillary attraction) to all parts that have been cleaned and fluxed. The silver alloy flows so easily that it penetrates quickly and deeply into all parts of the joint leaving no pinholes.

In order to obtain the full advantage of the low temperature of silver brazing, it is, of course, most desirable not to heat the joint beyond the temperature required for free flowing and bonding of the brazing alloy.

If the operator overheats the joint in order to be sure that he has a surplus of heat to melt the alloy, he will lose a great deal of the benefit of this low-temperature brazing process. If there is insufficient heat for the alloy to flow freely, it will tend to "ball up." If this happens, do not try to remedy the difficulty by applying more heat. It will only make matters worse. Instead, withdraw the flame until the joint has cooled sufficiently so that more flux can be applied and then it will be found that the operation can be resumed in the normal manner. The heat should be applied only to the work, and never to the wire or the alloy in the joint itself.

Many operators have never tried silver brazing because they have the mistaken idea that it is difficult to do. The fact is that any operator who has mastered the standard welding procedures or who knows how to handle a blowpipe will have no difficulty whatever. Some operators actually find it so simple that they are sometimes inclined to become a little careless. However, with ordinary care, it can be done entirely satisfactorily and at great speed. It should be remembered that silver brazing is not a substitute for welding—it is a valuable supplement to welding.

COPPER-ALLOY BRAZING

Copper-alloy brazing represents one of the three major categories of brazing and receives its name from the use of copper

Table 3. Approximate Compositions of Various Alloy Filler Metals

Name	CU	ZN	SN	Fe	Mn	Si	Ni	P	Solidus (Melting Point) °F	Liquidus (Flow Point) °F
Brass Brazing Alloy, Muntz Metal	60	40	—	—	—	—	—	—	1650	1660
Naval Brass, Tobin Bronze, Yellow Bronze,	59-61	39.25	0.50-1.00	—	—	—	—	—	1630	1650
Phosphor Bronze, Silicon Bronze	98.2	—	1.5	—	—	—	—	—	1850	1980
Low Fuming Bronze	56-59	40-41	.9	1.0	.03	.09	—	—	1570	1595
Ordinary Refractory Spelter	50	50	—	—	—	—	—	—	1585	1610
Nickel Silver, White Brazing Rod	46-50	38-44	.15	.25	—	.15	9-11	—	1690	1715
Manganese Bronze	58.5-59.5	39.25	0.75-0.10	0.40-0.80	0.01-0.09	0.04-0.14	—	—	1590	1630
Phosphor-Copper	90.0-96.0	—	1.0	—	—	—	—	4.0-8.0	1300	1530

Courtesy Airco Welding Products

and copper base alloy filler metals. It can be used to braze both ferrous and nonferrous metals and employs some of the highest temperature ranges in brazing. These filler metals can also be used in braze welding.

A copper-zinc alloy is the most common filler metal used in this type of brazing. Most of the copper-zinc alloys fall within the 60-40 range (60% copper and 40% zinc) or very close to it. Some brazing alloys (e.g. phosphor bronze or copper silicon) will use alloys other than zinc. Tin, iron, manganese, silicon, nickel, phosphorous, and zinc, are all used in varying amounts in copper and copper base alloy filler metals. Table 3 lists a few of these alloy filler metals and gives their approximate compositions. Remember, however, that this is only a representative listing and that there are many other brazing filler metals of the copper alloy composition.

ALUMINUM BRAZING

Aluminum brazing represents a third category of brazing, and like the others take its name from the basic composition of the filler metal it uses. Aluminum filler metals (brazing alloys) were developed for brazing aluminum and aluminum alloys. However, unlike other filler metals, aluminum brazing alloys are restricted in usage to this particular metal and its alloys. The aluminum brazing alloys are also distinguished by having the lowest brazing temperature range (815° F to 1220° F).

A suitable flux and brazing alloy filler metal are necessary when brazing aluminum. Manufacturers will usually recommend the flux and brazing alloy to be used with a particular type of aluminum or aluminum alloy. For example, *Airco Welding Products* recommends their *Airco* 4043 drawn silicon aluminum rods with an *Airco* No. 60 aluminum brazing flux when brazing aluminum of the 1100 and 3003 compositions. Other manufacturers will make similar recommendations for what they believe to be the most suitable use of their products. Most catalogs include data on chemical analysis and recommendations for usage. In the final analysis, trial and error will prove to be the best guide when brazing aluminum.

The number of aluminum brazing alloys are not as numerous as those used in copper-alloy brazing. Their composition generally ranges from 85% to 95% aluminum with silicon (5% to 12%) being the other major alloying element. They are available in several forms including rod, wire, strip, and sheet. Aluminum brazing alloys can also be used to braze weld aluminum and its alloys.

Aluminum brazing (and braze welding) require much the same surface preparation as does welding aluminum. A *thorough* cleaning of the base metal is a particularly important step in aluminum brazing, especially when brazing the 52S, 53S, 61S, and 63S compositions. Both chemical and mechanical means are used, but the chemical cleaning is recommended because it is generally more thorough. A fine steel wool and a file are adequate for a mechanical cleaning of the surface. A chemical cleaning should consist of the following steps:

1. Dip the aluminum or aluminum alloy surface into a caustic bath,
2. Rinse the surface,
3. Dip into an acid bath,
4. Rinse the surface.

The chemical cleaning should be repeated until the surface is completely clean.

THE BRAZE WELDING PROCEDURE

The procedure for braze welding will generally consist of the following steps:

1. Edge preparation,
2. Cleaning,
3. Alignment and clamping,
4. Preheating,
5. Flame adjustment,
7. Applying the flux and filler metal,
8. Cooling and cleaning.

Edge preparation is necessary in braze welding. It must be remembered that braze welding is essentially a gas welding technique. The edge preparation, then, is similar to that employed in gas welding. Beveling with a 90° included vee is recommended for pieces ¼ inch thick or more. This should be done by hand chipping with a chisel. The chips can then be removed from the surface by sandblasting or other suitable means. If the edge is under ¼ inch thick, the surfaces should be chipped until bright metal shows. This will insure a good bond between the base metal and the filler metal.

Clean the surface of the edge and the area extending approximately one inch back from the top and bottom of the vee. This will increase the "tinning" activity of the filler metal (i.e. the spreading of the filler metal in a thin layer over the surface of the base metal). Both mechanical (steel wool, emery cloth, etc.) and chemical (cleaning solutions, etc.) means are used to clean the surface prior to braze welding.

Once the joint edges have been prepared and cleaned, the two pieces to be joined must be correctly aligned and clamped in position. Tack welds are used as an added precaution. Tack welding is described in considerable detail in Chapter 4 (GAS WELDING PROCESS).

Preheating is frequently necessary in braze welding to reduce the possibility of expansion and contraction stresses caused by the heat of the braze welding process. The extent of preheating will depend upon the size of the casting or assembly. Naturally, larger pieces will require more extensive preheating than smaller ones. Preheating also reduces operating costs, because less heat is required when depositing the brazing alloy.

The flame adjustment for the torch is also very important and will depend upon the type of metal being braze welded. A slightly oxidizing flame is recommended for braze welding cast iron. The flux is generally applied by dipping the brazing rod into the flux and applying it to the surface. A high quality flux is particularly necessary for removing oxides that may form on the surface. More flux is used during the tinning operation than during subsequent passes to fill the vee.

An even rate of cooling is necessary. One method ensuring this is to heat the area around the joint at least twice the distance back from the edge. Excess flux and other particles may be removed by using a stiff wire brush or steel wool.

JOINING COPPER AND COPPER ALLOYS

The brazing of copper and copper alloys uses filler metals that melt within a 1100° F to 1350° F temperature range. Brazing within this temperature range is referred to as low temperature brazing, and the filler metals as low temperature alloys.

Fig. 15 illustrates typical joint designs used in low temperature brazing. These are basically tightly fitted joints with a clearance of 0.003 to 0.006 inch. The lap joint is preferred but other types (e.g. scarf, shear, and butt) may also be used. Fig. 15 also illustrates an insert type joint in which the brazing alloy is prepositioned in the form of a washer, ring, or a specially designed shape.

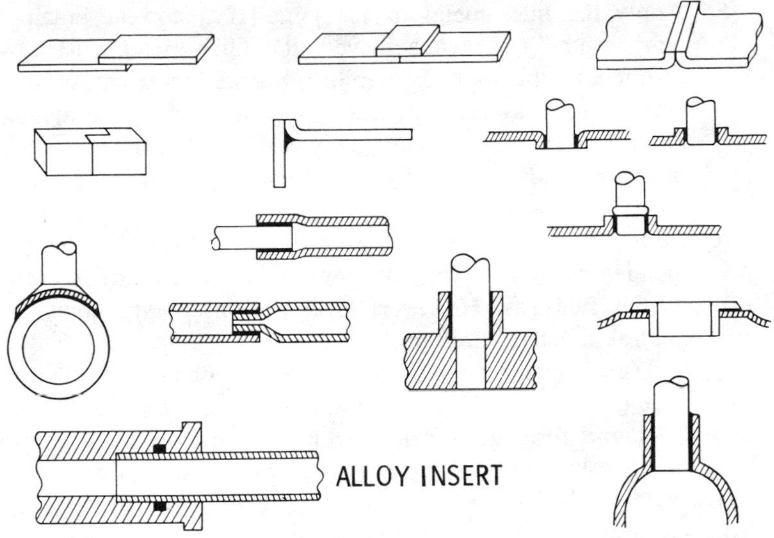

Courtesy Airco Welding Products

Fig. 15. Joint designs used in low temperature brazing.

The procedure for brazing copper and copper alloys is as follows:

1. Select a suitable joint design. If a lap joint design is selected, provide enough overlap to give maximum strength. A butt joint design may require machining the two surfaces until close clearance is obtained. Both scarf and shear type joints require grinding until the desired angle is obtained.
2. Thoroughly clean the surfaces to be joined. Begin with a mechanical cleaning (using a grinder, steel wool, emery cloth, etc.) and finish with a chemical cleaning (a suitable cleaning solution).
3. Coat the surfaces to be joined and the surrounding area with a suitable flux.
4. Assemble the joint and preheat the base metal until the flux becomes fluid. The joint is now ready for the application of the filler metal (brazing metal). Do *not* overheat the joint.

5. Apply the filler metal to the joint. If a correct brazing temperature has been achieved, the filler metal will flow throughout the joint by capillary attraction. Remove the heat as soon as this operation is completed. Once again, do not overheat. Overheating gives no advantage to the operator, and will undoubtedly result in a weak porous joint.
6. In torch brazing, a slightly carburizing flame is recommended for most types of joints when acetylene is used as the fuel gas. However, with the insert-type joint, a neutral flame is preferred.
7. Only the outer envelope (not the inner cone) of the flame should be allowed to touch the surface. Play the flame back and forth across the surface in an even continuous motion. In joints having members of unequal size, apply more of the heat to the larger member. This will insure uniformity in heat distribution.

Flanges that are to be brazed to copper pipes must be of copper or what is known as brazing metal (98% copper and 2% tin), as gun metal flanges would melt before the brazing alloy ran.

The hole in the flange is slightly tapered, and the end of the pipe, also, to form a clearance in which the brazing alloy may flow. A countersink is also formed in the face side of the flange, and the pipe slightly opened to fit it.

After the mixture is placed in the joint and the parts put together, the countersink is stopped with clay to retain the brazing alloy. The pipe is then slung vertically over the source of heat, with the flanges underneath and the previous process carried out. It will frequently be necessary to close the pipe with a clay tamping or other suitable means (wooden plug, etc.) to prevent the heat from going up. Projections from flanges are protected from the fire by means of a covering of clay.

Braze welding is frequently used to join copper pipe, copper sheet, and copper plate. Beveling is recommended for the edges of copper plate exceeding 3/16 inches in thickness.

As with all brazing and braze welding operations, a thoroughly clean surface is absolutely essential. Apply the flux, heat the base metal to the 1150° F to 1350° F temperature range, and add the filler metal. A flux coated rod is recommended. The type of welding rod selected will depend upon the composition of the base metal.

JOINING IRON AND STEEL

Iron and steel are both classified as ferrous (iron bearing) metals. Iron is the basic metal in any of the various types of iron and steel produced for use in industry or elsewhere. Steel differs from iron in that the former uses small amounts of carbon or carbon in combination with other metal alloys (e.g. tungsten, vanadium, chromium, etc.) to provide the different characteristics required by the manufacturer.

A detailed description of the different types of iron and steel is given in Chapter 19 (WELDING CAST IRON) and Chapter 22 (WELDING STEEL AND STEEL ALLOYS).

The metals and metal alloys chosen for the descriptions of brazing and braze welding techniques in the following sections were chosen on a selective basis only. This was done to provide examples of techniques used with the more commonly joined metals and metal alloys.

Gray cast iron can be brazed, but only with extreme difficulty. The major problem confronting the operator is the presence of a number of surface impurities (e.g. graphitic carbon, silicon, etc.) which will prevent a satisfactory bond unless removed. One successful method of removal consists in suspending the castings in a bath of catalyzed molten salts and passing an electric current through them. If the impurities can be removed, a silver base alloy is recommended as a filler metal.

Braze welding is a far more successful method for joining gray cast iron than brazing. Its principal application is in repairing castings where the lower temperatures of braze welding avoid the formation of hard martensitic cast iron during heating and contraction during cooling. A copper base alloy is recommended.

Braze welding is also recommended for repairing castings of malleable iron (a specially treated form of white cast iron). Suitable fluxes and filler metals are available.

One way in which braze welding is distinguished from brazing is the greater amount of filler metal required by the former. This is due to the fact that in joint preparation (particularly beveling) is very similar to that in fusion welding. Similarly, a joint formed by braze welding may require several passes. Figs. 16 and 17 illustrate cast iron joints made by multiple passes.

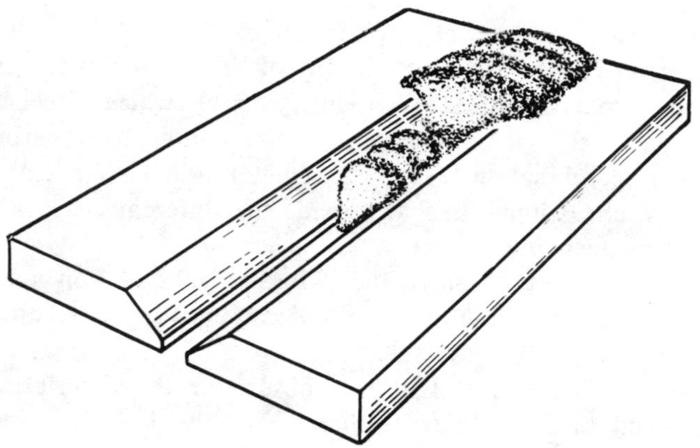

Courtesy Airco Welding Products

Fig. 16. Braze welding with two passes.

The bevel should be formed by chipping followed by sand-blasting. This method tends to decrease the amount of exposed carbon flakes on the joint surface. Thoroughly cleaning the surfaces to be joined. A flux (usually a dry one) can then be applied. Position the parts to be joined according to the desired joint design, and tack weld them so that they will not slip out of position.

Preheating to a low red heat is recommended. For smaller assemblies, a localized preheat is sufficient. As the assemblies increase in size, the extent of preheating also increases. A manganese-bronze fluxed rod is recommended for braze welding cast iron. In braze welding, the filler metal will fuse where it is deposited. The capillary attraction of a filler metal is characteristic only of brazing.

Stainless steel should be silver brazed rather than welded unless the stress to which the joint will be subjected is too great.

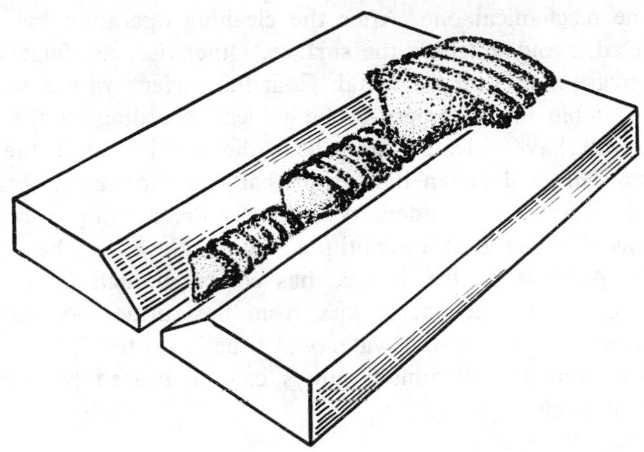

Courtesy Airco Welding Products

Fig. 17. Braze welding with three passes.

The lower temperatures used in silver brazing greatly reduce the chances of warping the base metal.

JOINING OTHER METALS

Brazing aluminum is another low temperature brazing process similar to the one described in the section on copper and copper alloys (see BRAZING COPPER AND COPPER ALLOYS). Here, too, the filler metal is distributed throughout the joint by capillary attraction. The two main distinguishing features of brazing aluminum are:

1. The flux is specifically designed for use with aluminum and no other metal,
2. It operates within a narrower and somewhat lower temperature range (1160° F to 1185° F) than copper brazing.

Both lap and T-joint designs can be used when brazing aluminum. Lap joint clearances range from 0.006 to 0.010 inch for smaller laps (generally under ¼ inch) to 0.010 to 0.250 inch for larger ones.

The surface must be thoroughly cleaned. In the case of aluminum, the chemical cleaning will be even more important than the mechanical one. After the cleaning operation has been completed, avoid touching the surface. Otherwise, the fingers will leave grease marks on the metal. Coat the surface with a suitable flux, assemble the members to be joined according to the joint design you have selected, and heat the surface until the flux liquifies. This will be an indication that the filler metal (brazing alloy) is ready to be added. As soon as the aluminum brazing alloy has filled the joint, discontinue heating. Never overheat when brazing. As soon as the surface has cooled enough (at least to 800° F), remove the excess flux from the surface. A wash of nitric acid followed by hot water will usually do the job.

Aluminum and aluminum alloys can be brazed by the following methods:

1. Torch brazing,
2. Dip brazing,
3. Furnace brazing.

Torch brazing is the quickest and most economical method of the three. A neutral flame is recommended when torch brazing aluminum. The filler metal rods belong to the BT1Si AWS classification and are suitable *only* for brazing aluminum and aluminum base alloys. The fluxes used for brazing aluminum and aluminum alloys usually contain chlorides and fluorides.

The same methods used to braze aluminum (i.e. torch brazing, dip brazing, and furnace brazing) can also be applied to magnesium. The filler metal used in brazing magnesium is generally an AZ92 alloy (89.0% magnesium, 9.0% aluminum, and 2.0% zinc) with a brazing temperature range of 1130° F to 1160° F.

As is the case when brazing aluminum and aluminum alloys, a chemical cleansing of the magnesium base metal is generally more effective than a mechanical one. All other aspects of brazing magnesium are similar to those encountered when brazing aluminum.

JOINING DISSIMILAR METALS

Brazing and braze welding are particularly suited for joining dissimilar metals having widely differing melting points. The joint designs do not differ from those used to join similar metals, and are secondary in importance to metal properties (e.g. thermal conductivity, etc.) that may influence the strength of the weld. Silver-based filler metals are particularly suited to joining dissimilar metals.

BRAZING AND BRAZE WELDING PROBLEMS

The problems that the operator will encounter in brazing and braze welding are in no way insurmountable. However, he should be made aware of them before beginning the work so that complications can be avoided.

Health and safety are always of primary importance to the operator, and for that reason one should be cautioned about using brazing and braze welding fluxes that contain fluorides, cadmium, and other potentially harmful substances. All of these substances contribute to the effectiveness of the flux, but can produce fumes that may irritate the eyes, nose, and throat. In some cases they could prove injurious to the operator's health. These fluxes should only be used where adequate ventilation is available. Read the manufacturer's instructions before using the flux, and follow these instructions carefully.

Unclean metal surfaces offer another problem to the operator. Unless the surface is thoroughly cleaned, the joint formed by brazing or braze welding will be too weak to withstand the stresses placed upon it. A flux will not remove dirt, grease, or oil from the surface. This must be done both mechanically and chemically. The surfaces of the two members being joined should be first mechanically cleaned by grinding, filing, or another suitable method. This is followed by a chemical cleaning.

CHAPTER 16

Surfacing and Metallizing Processes

In industry, special protective surfaces are frequently applied to parts that are subject to heat, impact, abrasion, corrosion, and other forms of wear. It is also frequently necessary to build up surfaces that have been worn down, incorrectly cast, or mis-machined. Both of these approaches to the problem of metal wear are economically more desirable than replacing the worn part. Down-time is eliminated and the cost of a new part considerably reduced by extending the service life of the old one. Surfacing and metallizing are the processes used to apply special surfaces or coatings to metal. The basic differences that distinguish these processes will be examined in this chapter.

SURFACE CONDITIONS

The various service conditions to which surfacing and metallizing deposits are subjected are classified in three groups as follows:

1. Abrasion,
2. Impact,
3. Corrosion.

Abrasion is a grinding action due to a material such as rocks or sand caught against the surface of the part; or a sliding, rolling or scrubbing action of one metal part against another. Abrasion will not only scratch or grind away metal surfaces that come into contact with one another in this manner, it will also generate enough heat to seriously reduce the hardness of metals that are hardened by heat treatment methods.

Impact is the result of abnormal loads placed against a metal surface. This may be either light or heavy and tends to deform the surface (usually with a gouging effect) which will cause cracking or shipping.

Corrosion is a deformation of the surface caused by the action of various chemicals (ordinary water causing rust) and also oxidation or scaling at elevated temperatures.

The surface can be protected from all of these service conditions by selecting and depositing a suitable surfacing or metallizing alloy metal.

PRINCIPLES OF SURFACING

Surfacing is the process of applying a metal coating edge or point onto surfaces or parts that are subject to wear or corrosion. This is usually done either to build up a metal surface to its former dimensions or to create a protective coating over metals that are lower in resistance to wear or corrosion.

The special surface forms an integral layer with the metal over which it is applied. It may be applied with the same equipment and supplies used in the various welding processes (though with a somewhat different technique). It is in the method of application and in the manner by which bonding occurs that surfacing can be distinguished from metallizing.

Surfacing is also referred to as *metal-surfacing, hard-surfacing,* or *hard-facing* (sometimes written as one word). Unfortunately, the last two terms seem limited to describing those surfaces designed to resist impact or abrasion, while excluding those that may be designed specifically to resist different forms of corrosion. The *American Welding Society* uses the term "surfacing" to describe any protective or built-up surface of metal applied by one of the welding processes commonly used and suited for that purpose. This is the term that will be used in this text.

Surfacing, as a process, consists of a number of steps that should be carefully considered and followed in the *approximate* order given below:

1. Identification of the base metal,
2. Determination of the surface problem,
3. Determination of the welding position,
4. Estimation of the area to be surfaced,
5. Selection of an appropriate surfacing metal alloy,
6. Selection of an appropriate surfacing process,
7. Preparation of the surface,
8. Application of the surfacing metal alloy,
9. Application of finishing procedures.

Most of these steps in the surfacing process are self-explanatory and will be explained in greater detail in the appropriate sections of this chapter. Some of these steps, such as the identification of the base metal, may appear to be almost too self-evident to be included in this listing. However, determining the exact composition of the base metal can be quite tricky and the correctness of its identification will strongly influence the success and quality of the surfacing process.

Surface problems will include such variables as the types and degrees of wear that must be met. These variables in combination with the welding position of the job will determine the selection of the surfacing alloy. Many manufacturers of welding equipment and supplies will give specific information concerning the suitability of each of their surfacing alloys for various types of jobs. This information will include recommended welding (i.e. surfacing) positions, typical analyses, and suggested applications.

Surfacing Applications

Any type of equipment, tool, or part that is subject to the various types of corrosion and wear will have its useful service life effectively prolonged by surfacing. A brief and very incomplete listing of surfacing applications would include the build-up or surfacing of hammer-mill parts, grader-end bits, railroad tamper bars, tractor grousers, exhaust valves, pulverizer hammer, coal augers, and dipper lips (Fig. 1) to mention only a few. This process

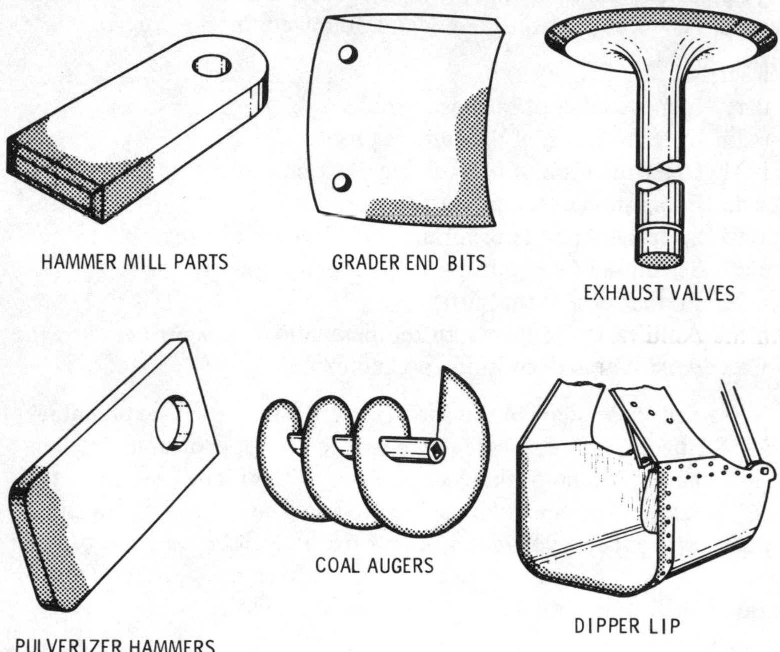

Courtesy Airco Welding Products

Fig. 1. Typical surfacing applications.

is particularly directed toward rebuilding or protecting cutting edges and gear teeth.

Surfacing Processes

Surfacing can be done by any one of the following welding processes:

1. Oxyacetylene welding,
2. Shielded metal-arc welding,
3. Carbon-arc welding,
4. TIG welding,
5. Submerged-arc welding.

In this chapter, the greater emphasis will be placed on oxyacetylene and shielded metal-arc welding as surfacing methods.

Oxyacetylene Surfacing

Oxyacetylene surfacing employs principles similar to those described for tinning or sweating; to flow the surfacing metal or alloy onto the surface to be covered. The oxyacetylene flame causes a thin film of low-melting constituents to form on the surface of the base metal. This thin film acts as a bonding surface for the surfacing alloy that is subsequently flowed onto the base metal. It also prevents deep penetration of the base metal by the surfacing alloy. As the surfacing process progresses, the thin film of low-melting constituents is absorbed by the surfacing overlay resulting in the bonding of the latter to the base metal. Fusion between the surfacing alloy and the base metal seldom (or desirably) reaches a depth greater than 0.01 inches. Because of this minimal penetration, dilution of the surfacing alloy by the base metal is kept at a minimum. Excess dilution will weaken the protective characteristics of the surfacing overlay.

Surfacing Rods and Fluxes

Standard welding fluxes are recommended for the various surfacing alloy rods. Sometimes a rod will already be coated with a flux and using a separate one will not be necessary. In any event, the use of a flux is mandatory with any of the bronze alloys.

There are a wide variety of metals and metal alloys available for surfacing. Each must be carefully selected according to the conmposition of the base metal it is to cover and the particular protective characteristics the operator wishes to impart to the base metal. Manufacturers of welding equipment and supplies will generally provide sufficient data to make a suitable selection.

OXYACETYLENE SURFACING PROCEDURE

A knowledge of general gas-welding procedures is a prerequisite to acceptable surfacing. The area to be surfaced must be thoroughly cleaned. All dirt, grease, or other foreign matter must be removed. This can be accomplished by grinding or machining the surface. All sharp corners should be rounded off. If the surface is not completely clean, blow holes or pockets will form and thus impair the efficiency of the surfacing procedure. If a corner or

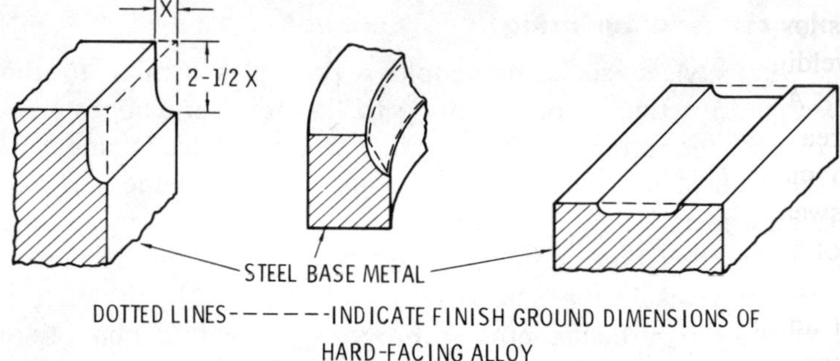

Fig. 2. Sectional views showing method of machining steel edges and corners preparatory to surfacing.

edge to be surfaced is subject to particularly strong shock, it might be wise to create machined recesses on the surface, as shown in Fig. 2. This will give additional support to the surfacing deposits. If it is impossible to grind or machine the surface, it can be filed. In any event, the surface on which the surfacing alloy is to be deposited must be clean.

Most medium size parts can be preheated with a neutral welding flame before surfacing. This should be done with a neutral flame of about 800° F (a very faint red heat, visible only without goggles in a darkened room). For large parts, preheat in a furnace.

It is economical to preheat in a furnace when such equipment is available unless the part can be preheated quickly and thoroughly with a torch flame. Do not, however, preheat to a temperature higher than the critical temperature of the metal to be surfaced or such a high temperature that scale will be formed.

Adjust the flame so that it has an excess of acetylene. The total length of the acetylene feather should be about three times

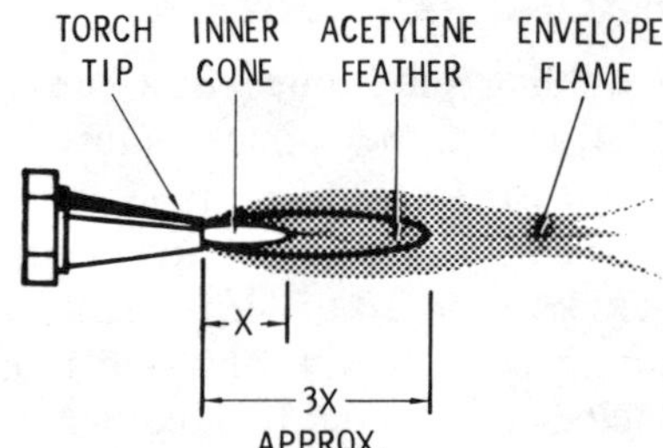

Fig. 3. An excess acetylene flame used for all surfacing operations.

as long as the flame tip (inner cone) measured from the end of the welding tip (Fig. 3).

An excessive amount of acetylene in the flame prepares the area for surfacing by causing an extremely thin layer of the surface to melt. This produces the watery glazed appearance known as "sweating." Unless an excess acetylene flame is used, the alloy will not spread and may become porous.

Hold the torch so that the excess acetylene flame is directed at an angle from 30 to 60 degrees to the surface. The end of the inner cone of the flame should be about ⅛ inch from the surface. Keep the torch in this position until the surface under the flame suddenly becomes glazed indicating that an extremely thin surface layer has melted. This is the condition known as "sweating." (Fig. 4)

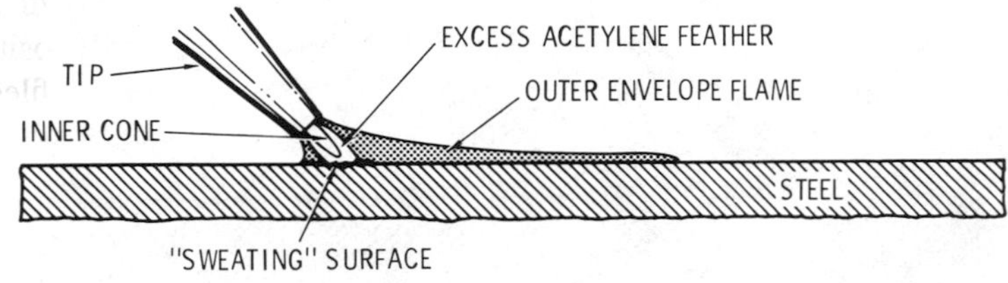

Fig. 4. Condition of the flame during sweating.

The sweating area will vary according to the size of the welding tip, but for a medium size tip the surface should "sweat" for a distance extending about ¼ inch around the excess acetylene feather.

After sweating, withdraw the flame just enough so that the end of the welding (surfacing) rod can be brought between the inner tip (cone) of the flame and the hot steel. The end of the inner cone should just about touch the rod and the rod should lightly touch the "sweating" area. The end of the rod will then melt and form a puddle on the "sweating" steel surface.

If the first few drops from the welding rod form a puddle or do not spread uniformly, it is an indication that the base metal surface is too cold. Should this be the case, bring the base metal to the full sweating temperature once again. Spread the molten alloy over the sweating area by removing the rod from the flame and directing the flame into the puddle. Return the rod to the flame and melt off more alloy into the puddle, as required.

Next, direct the flame so that part of it plays only on the edge of the puddle to keep it molten and part of it plays on the base metal surface adjoining the puddle. As the base metal reaches the sweating heat, the puddle of surfacing alloy will spread over the sweating area.

Continue these steps until the desired area is coated. With a little practice, the correct amount of surfacing alloy can be added to make the deposit the desired thickness. It is better to do this in one operation than to go back over the whole job to add another layer or alloy. Figs. 5 and 6 illustrate the approximate relationships between tip, rod, puddle, and base metal, for two methods of applying the surfacing deposit.

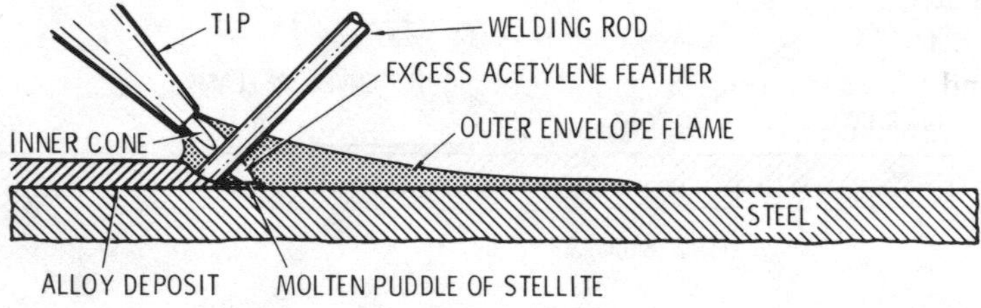

Fig. 5. Approximate relationship of the tip, rod, puddle, and base metal during surfacing.

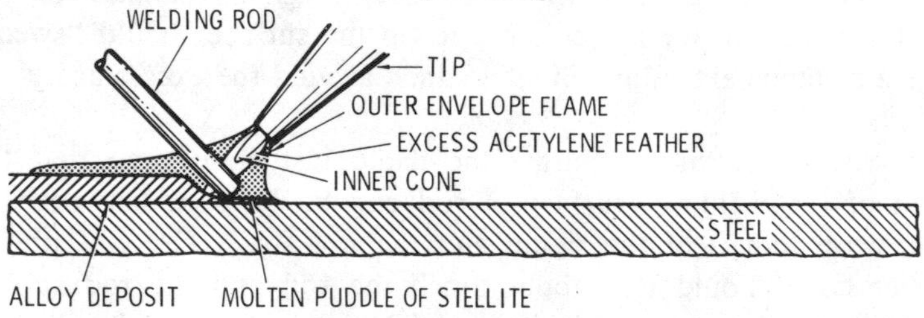

Fig. 6. The backhand method of surfacing.

During the operation, the flame can be repeatedly moved back to melt just a thin surface layer of the deposited alloy in order to smooth out high spots just behind the molten puddle. This should be done quickly without allowing the front edge of the puddle to solidify and without interrupting the steady forward progress of the work.

If desired, a second pass can be made with the flame to smooth off the surface and to minimize grinding. If this second pass is made, care should be taken to melt only the surface of the deposited surfacing material without melting down to the base metal. This precaution is necessary to avoid diluting the surfacing alloy with the base metal. The puddle should be flowed in the direction desired by means of the pressure of the flame, not by stirring with the rod.

The work generally progresses in the direction of the hand holding the rod. Sometimes the backhand method (Fig. 6) is used for base metals that scale badly (e.g. steels) or on very thin sections. If dirt or scale should appear, float it to the surface with the flame. If it will not float, dislodge it with the rod.

Once the surfacing overlay has been built-up to the desired size and thickness, remove the flame slowly from the puddle to prevent the formation of shrinkage, cracks, or blow holes. If these are present, remelt the deposit in their immediate vicinity, flick any particles of scale or oxide from the pool with the rod and add a little more surfacing alloy from the rod. If the holes still appear, it may be necessary to grind off the alloy deposit down to the base metal at that point, warm the area gradually with the flame and deposit additional surfacing alloy.

The surfacing overlay should be allowed to cool slowly. This is absolutely essential in order to produce a deposit free from cracks and internal stresses. Parts which show a strong tendency to crack, such as large gate-valve wedges and seat rings, pump-shaft sleeves or parts on which the deposit is circular or large in area, should preferably be placed in a heat-treating furnace while still hot from welding, brought slowly to a low red heat (about 1150° F) and then allowed to cool overnight in the furnace with the door closed and the heat turned off. *Never* cool a part by dipping it in water.

Finishing is the last stage in the surfacing process. After the surfaced part has cooled, it may be necessary to remove the high spots or grind the surfacing overlay to an exact size. This is done by hand, or on a grinding machine.

SHIELDED METAL-ARC SURFACING

Arc welding is extensively used in surfacing. Its ability to fuse new metal to worn down parts provides an efficient method of repairing damage caused by abrasion, impact, and corrosion. In coping with wear and tear of equipment, the welder has only to choose the proper arc welding electrode. Apply the proper procedure to replace worn and broken metal surfaces with new metal, no matter whether the equipment is damaged by grinding or sliding abrasion, repeated impact, or plain everyday rust. Examples of such repair include the restoration of worn parts of excavating equipment such as teeth; lips and bottoms of power shovels; bulldozer blades, blades of grinding machines; teeth of rooters; etc.

SHIELDED METAL-ARC SURFACING PROCEDURE

It is important that the operator be familiar with standard arc welding procedures, since shielded metal-arc surfacing is essentially the application of techniques found in the arc welding process.

Once again, surface preparation is very important. All dirt, grease, or other foreign matter must be removed prior to surfacing or there will not be adequate bonding between the overlay and the base metal.

All nine steps of the surfacing procedure outlined in the section PRINCIPLES OF SURFACING will apply here. There are, however, certain modifications of these procedural steps that apply only to shielded metal-arc surfacing. These modifications are as follows:

1. Allow the beads to cool slowly before adding the next layer (this will tend to increase the hardness of the deposit),
2. Remove all slag after depositing each layer,
3. Avoid undercutting,
4. Make certain that there is sufficient penetration of the underlying and adjoining beads,
5. Do *not* overheat. Use only enough amperage to keep the arc from extinguishing,
6. Use a straight bead to make the deposit.

SURFACING ELECTRODES

A major problem encountered when surfacing with an electric arc is the selection of a suitable electrode. The electrodes are melted during the surfacing process and are deposited as a relatively thin layer over the base metal. The selection of a suitable electrode depends on whether the base metal is to be protected from one of the various kinds of wear or corrosion, or built up to its proper dimensions.

There are a wide variety of different electrodes commercially available from which the operator may select the one most suited for his particular surfacing problem. For example, the high-carbon electrodes provide excellent protection against both abrasion and impact. Tungsten-carbide electrodes provide even better protection against abrasive wear, but they lack resistance to impact and cost more than high-carbon electrodes. Chromium-carbide electrodes share approximately the same characteristics as tungsten carbide electrodes. Stainless steel electrodes provide the best resistance to impact, but have very little resistance to abrasion.

In choosing the proper electrode for repairing worn or broken equipment, it is necessary to bear in mind the quality of the metal being welded and the conditions under which the equipment operates. Failure to consider either of these factors will be reflected in the quality of the results to be obtained. For instance, if the equipment is made of high-manganese steel and subject to repeated impact, an electrode must be used which will provide a deposit with the qualities of the base metal and at the same time be resistant to batter. If, on the other hand, the equipment being surfaced must possess high resistance to abrasive action, then another electrode should be used. Then again, the particular application may require a deposit which will resist both impact and abrasion. Here, too, a particular electrode will provide the desired result. However, where both impact and abrasion are present, it is important to determine which condition is most active. Having determined this, the welding operator employs a particular electrode which provides the most effective deposit. Manufacturers of welding equipment and supplies will generally provide detailed information on their surfacing electrodes. An example of a surfacing-electrode guide is illustrated in Table 1.

Table 1. Example of a

Hobart Type	Identification end color	A.W.S. class	Size		Amperes*	Deposition rate lb/hr DCR current
TUFANHARD 150 (BUILD-UP)	Silver	EFeMn-A	1/8 x 14	DCR or AC	80-140	2.3 lb. @ 120-amps
			5/32 x 14		110-180	3.7-lb. @ 170-amps
			3/16 x 14		160-270	5.2-lb. @ 270-amps
			1/4 x 18		240-370	8.3 lb. 370-amps
TUFANHARD 160 (BUILD-UP)	Red	None	1/8 x 14	DCR or AC	75-160	2.6-lb. @ 130-amps
			5/32 x 14		120-210	3.3-lb. @ 170-amps
			3/16 x 14		140-260	5.2-lb. @ 240-amps
TUFANHARD 250 (BUILD-UP)	Black	EFe5-C	5/32 x 14	DCR or AC	130-150	2.7-lb. @ 150-amps
			3/16 x 14		170-190	7.0-lb. @ 230-amps
			1/4 x 18		210-240	10.7-lb. @ 350-amps
TUFANHARD 320 (BUILD-UP)	Green	EFe5-C	1/8 x 14	DCR or AC	90-125	2.4-lb. @ 110-amps
			5/32 x 14		110-180	3.0-lb. @ 150-amps
			3/16 x 14		150-250	4.3-lb. @ 200-amps
			1/4 x 18		250-380	7.0-lb. @ 320-amps

Portion of an Electrode Guide

Typical hardness	Typical all weld deposit chemistry						Typical Applications
	C	Mn	Si	Cr	Ni	Mo V	
14—18 Rc as welded Work hardens to 45—50 Rc	.7	14.0	.5	4.0	4.0		Build up on Manganese Steel, Dredge Pump Shells & Cutter Heads, Re-building Crusher Rolls & Mantles, Tractor Rollers, Rails, & Sprockets, Re-building Impact Breaker Bars & Hammer Mill Hammers, Dipper Repair & Build-up Wheel Excavator Teeth, Dragline Buckets & Lips, Power Shovel Teeth.
15—20 Rc as welded Work hardens to 50 Rc	.75	4.0	.5	20.0	9.0		Manganese Track work, Frogs, Switches, Crossovers, Rail ends, Railroad Car Castings, Shovel Tracks.
Layer Mild Steel 1 23 Rc 2 24 Rc 3 26 Rc	.20	.12		1.05			Build-up on mild steel where good impact resistance is required. Widely used for build-up of Power Shovel and Tractor parts. Also used on certain pipe line parts, such as ball joints, swivels, elbows & valves.
Layer Mild Steel 1 26 Rc 2 28 Rc 3 34 Rc	.20	1.0	.40	.90	.30	.40	Carbon steel frogs, Sow Blocks, Rail ends, Crane Wheels, Mine Car Wheels, Roll Wobblers, & Couplings, Dredge Pump Doors, Power Shovel Rollers & Drive Tumblers. Build-up on where low-hydrogen Mild & Alloy Steels quality & maximum machineable hardness is required.

Courtesy Hobart Brothers Company

In some instances, two electrodes may be employed in order to meet all requirements of the job at hand. The deposit may be built up part way with one and finished off with another. It is usually preferable, however, to build up the entire deposit with the best electrode available for the purpose.

Electrodes for surfacing are commonly divided into martensitic and austenitic types depending on the deposit they produce.

Martensitic deposits show practically their full hardness in the state as deposited and little or no increase in hardness on cold working. Generally speaking, the higher the hardness of the martensitic deposit, the better the resistance to wear and deformation and conversely the lower it is to resistance and to impact or chipping.

Austenitic deposits are relatively soft as deposited and have the property of hardening at the surface when deformation takes place as in cold working, while the rest of the deposit remains relatively soft and tough.

CARBON-ARC SURFACING

The carbon-arc welding process is recommended for surfacing when it is necessary to build up or surface small parts or pieces, or when an extremely thin surfacing layer is required.

Essentially the same types of equipment and materials are used in surfacing as are employed when welding with this process. Carbon-arc welding can also be adapted to the powder-spray form of surface application.

TIG SURFACING

TIG surfacing consists of applying the principles of inert-gas metal-arc welding to surfacing. In this process, an arc is maintained between a nonconsumable tungsten electrode and the work surface. The surfacing metal is added in the form of a surfacing alloy rod that is fed into the arc. The arc and immediate work area are shielded by an inert gas (usually argon or helium). TIG surfacing is especially recommended when surfacing the high-nickel alloys.

SUBMERGED-ARC SURFACING

Production-type, or volume, buildup of large surface areas is being accomplished in industry with the submerged-arc welding process. This can be either an automatic or semiautomatic technique depending on the requirements of the job.

The equipment and materials used when surfacing with this process are identical to those employed when welding. Suppliers will generally recommend a suitable flux (this will be an agglomerated alloy flux) and electrode for the job at hand. Usually, the electrode will supply enough surfacing metal. Supplementary welding rods can be used in cases where additional filler metal is required.

It is recommended that DCSP (direct current, straight polarity) in combination with high voltage and a *slow* rate of deposition be used when surfacing with the submerged-arc welding process.

PRINCIPLES OF METALLIZING

Metallizing is a coating or impregnating process that essentially consists of spraying extremely small particles of molten metal onto a specially prepared surface. The molten metal particles bond themselves to the base metal surface. In almost all cases the base metal is not fused.

A sprayed metal coating will generally have somewhat greater porosity than in its other forms. This is duc primarily to thc fact that the metal particles are surrounded by a thin oxide film. When the spherical particles strike and adhere to the base metal surface, they take the form of flattened, overlapping shingle-type deposits. Apparently, the oxide film remains undisturbed and assumes the shape of the flattened particle.

All the properties of a sprayed metal coating must be carefully considered before application. Otherwise, the results could be both costly and structurally unacceptable. Foremost in consideration should be the cooling and contraction of the sprayed metal coating. If it is excesively greater than the base metal, it can result in serious warping of the latter. An important point to remember is that the mechanical and physical properties of a sprayed metal coating *do not* equate with those of the same metal in other forms. The operator must determine the properties of the metal coating. This may be done through information from manufacturers of welding equipment and supplies or such reference works as the *AWS Welding Handbook* (3rd Edition).

The metallizing processes consist of basically the same procedural steps that were listed for surfacing, but with certain differences in technique that should be noted. For example, the surface preparations for metallizing require the same thorough cleaning standards that surfacing does. However, the base metal must be roughened to provide an anchor for the sprayed metal particles. This is necessary for all metallizing processes in which the base metal is not fused. Once again, the *Sprayweld* process stands as an exception to this rule.

Dovetail keying is the most common method of roughening a surface for metallizing. In this method, grooves or threads (usually on cylindrical objects) are etched or scratched into the base metal surface by blasting or with special tools or electrodes. The sprayed metal flows into and under the projections on the roughened surface, thereby providing a secure anchor for the deposited layer.

When metallizing, the molten metal is generally applied in several successive thin layers. The operator should not attempt to apply the coating in a single pass as adequate bonding cannot always be guaranteed.

METALLIZING APPLICATIONS

The principal application of metallizing is in the repair of worn machinery parts or the correction of parts that have not been produced or machined to the correct specifications. The advantage of using metallizing for this purpose is that the lower heat involved reduces or eliminates the possibility of overheating. Consequently, distortion and other effects of overheating on the base metal do not become a troublesome factor. The heat levels are significantly lower than those used in surfacing. Basically, then, metallizing is the mechanical bonding of a sprayed metal alloy to the surface of a base metal. No fusion occurs between the sprayed metal alloy and the base metal (an exception being the *Sprayweld* process, developed by *Wall Colmonoy Corp.*

Metallizing has been used for decorative purposes, but the relative high cost of the process has limited its use in this area. Most metallizing applications are similar to those used for surfacing (e.g. the creating of corrosion-resistant surfaces, etc.). However, the build up or coating of parts by the metallizing process presents one distinct advantage over surfacing. A metallized surface is far more porous, a factor that causes lubricants to be trapped and maintained over a considerably longer period of time.

Metallizing Processes

Sprayed metal coatings can be applied to a metal surface in several different ways. A basic distinguishing factor among these different metallizing processes is the method by which the heat is obtained for the process. Accordingly, this chapter will be concerned with the following four metallizing processes:

1. Oxy-fuel gas spraying,
2. Plasma-arc spraying,
3. Electric-arc spraying,
4. *Sprayweld* metallizing.

OXY-FUEL GAS SPRAYING

The various metallizing processes in which an oxy-fuel gas gun is used to supply the sprayed metal coating may be distinguished from one another by the method in which the metal coating is supplied to the spray gun. For example, in some, the metal is supplied to the gun in the form of a wire or rod, whereas in others a metal powder is fed into the gun. In either type, the molten particles are sprayed onto the base metal surface by means of a stream of compressed air.

Oxygen is used in combination with one of the several commercially available fuel gases to feed the gun. Although acetylene is generally preferred, other fuel gases may also be used.

A typical oxy-fuel gas spraying setup (Fig. 7) will include the following equipment and supplies:

1. A spray gun,
2. A control console with flowmeters for air and gas,

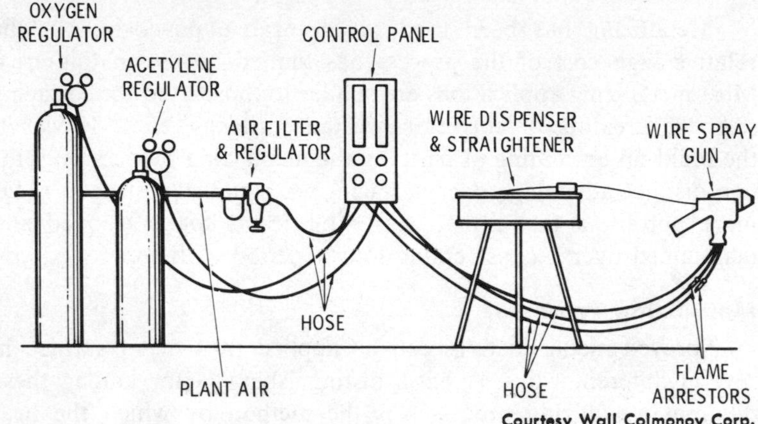

Courtesy Wall Colmonoy Corp.

Fig. 7. A semiautomatic metalizing system equipped with a wire spray gun.

3. An oxygen cylinder and regulator,
4. An acetylene cylinder and regulator,
5. An air filter and regulator,
6. A wire dispenser and straightener,
7. Air and gas hoses,
8. Flame arresters,
9. A supply of oxygen,
10. A supply of acetylene,
11. A plant air supply.

The cutaway view of a typical wire-type spray gun is shown in Fig. 8. Wire (or rod) up to 3⁄16 inch in diameter is fed into the gun through the wire guide atomized and sprayed at a rate varying from one to twenty-five feet per minute. The variations in the rate of spraying are made possible by interchangeable air motors located in a unit at the back of the gun. One motor is designed for the higher range of speeds; the other one for the lower range. All controls (speed adjustment dial, on-off valve, gas valve, and wire feeder) are located on the gun.

The control console permits manual, semiautomatic, and automatic operation of the spray gun. It also contains controls for regulating the air pressure, oxygen and fuel-gas flow, and automatically lighting the spray gun. Most oxy-fuel gas spraying systems also provide means for remote control.

The oxygen and fuel-gas (usually acetylene) regulators are of the two-stage type. Air and gas flowmeters are mounted on the panel of the console. Each of the gas flowmeters has an integral needle valve for flow adjustment.

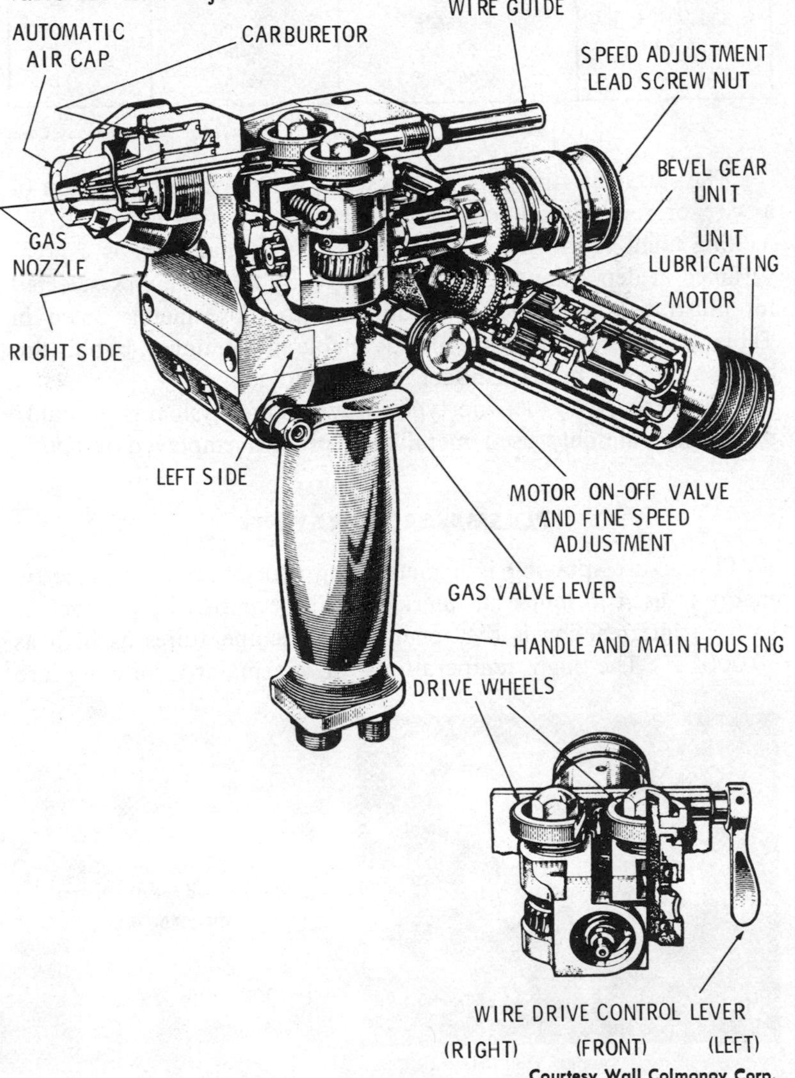

Courtesy Wall Colmonoy Corp.

Fig. 8. Working mechanism of a wire-type spray gun.

Table 2. Examples of Deposition Rates

Metal	Melting Point	Wire Dia.	Lbs. per Hr.
Lead	620°F	1/8-in.	85
Zinc	786°F	3/16-in.	45
Steel (.10)	2700°F	5/32-in.	7
Molybdenum	4532°F	1/8-in.	6

Courtesy Wall Colmonoy Corp.

The surfacing alloy for these systems is available in the form of a wire or rod. The deposition rates will vary, depending on the melting point of the surfacing alloy. The fact that there is a great variance in deposition rates is illustrated by the examples selected for illustration in Table 2. In addition to those metals listed in Table 2, surfacing alloys are also available in aluminum, babbit-tin, brass, aluminum bronze, monel, nickel, and copper.

Metallizing with the wire-type spray gun and system is probably the most commonly used metallizing process employed today.

PLASMA-ARC SPRAYING

Plasma-arc spraying is a metallizing process in which electric energy is used to ionize an inert gas. The eventual by-product of such an interreaction is the generation of temperatures as high as 30,000° F. The high temperatures of plasma-arc spraying are

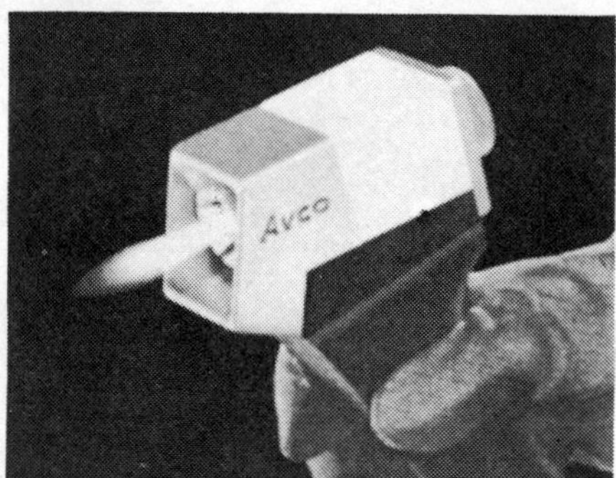

Courtesy Wall Colmonoy Corp.

Fig. 9. Gun used for plasma-arc spraying.

created in the plasma spray gun (Fig. 9) without combustion. The absence of combustion eliminates or reduces, to minimal levels, the posibility of oxides forming on the surface.

The high temperatures used in this metallizing process causes the small particles of metal to reach the molten stage much more quickly than in other metallizing processes (e.g. oxy-fuel gas spraying). As a result, the molten metal particles flow more quickly through the plasma-arc reaching the surface being coated with less distortion and greater impact.

Among the many advantages of plasma-arc spraying can be included the following:

1. High bond strength,
2. No limitation on the types of materials that can be joined,
3. Dense structure to the coating,
4. Minimal heat distortion problems.

Plasma-arc spraying can be used to coat inexpensive base metals with coatings that are selected to impart certain desired surface characteristics (higher corrosion resistance, greater wear resistance, etc.). Usually the sprayed coating consists of expensive metal. By spraying the more expensive coating onto a comparatively inexpensive base or core, the desired physical and mechanical

Courtesy Wall Colmonoy Corp.

Fig. 10. Spraying a part with a coating of chromium oxide.

characteristics of the former are combined with the economy of the latter. Fig. 10 shows an operator spraying an injection molding ram with chromium oxide. The part is held in a rotating fixture.

Plasma-arc spraying can also be used to fabricate complete parts. This is possible because plasma-arc coatings are stronger, denser, and can be applied more thickly than other metallizing coatings. The beryllium nose cone, shown in Fig. 11, is an example

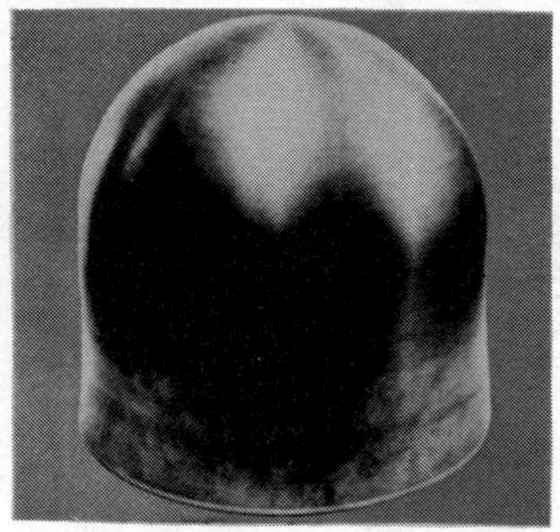

Courtesy Wall Colmonoy Corp.

Fig. 11. Fabricating a beryllium nose cone.

of the fabrication of a complete part. This particular part was fabricated as a "free standing shape." It is about 12-inches in diameter and 13-inches high with a wall thickness of .070 inch. It was formed by spraying a coating of metal over a shaped mandrel, which is later removed.

ELECTRIC-ARC SPRAYING

Electric-arc spraying is a metallizing process in which twin electrodes in the form of continuous wire coils are fed into an arc, melted, atomized by a jet of compressed air and propelled onto the surface of the metal. Electric-arc spraying is also referred to as *arc metallizing* or by one of the trade names (e.g. *Colmonoy Electrospray*).

This is the cheapest metallizing process used in production-type work. The operating temperatures are higher than those used in oxyacetylene spraying (approximately 7000° F as compared with about 6000° F) and the application rate is faster (3 to 5 times greater). In addition, surface preparation need not be as thorough as when metallizing with the oxy-fuel gas process.

Fig. 12 shows a typical electric-arc spraying setup. The principal components of this metallizing system are:

1. DC power source,
2. Dual wire feeder,
3. Electric-arc spray gun.

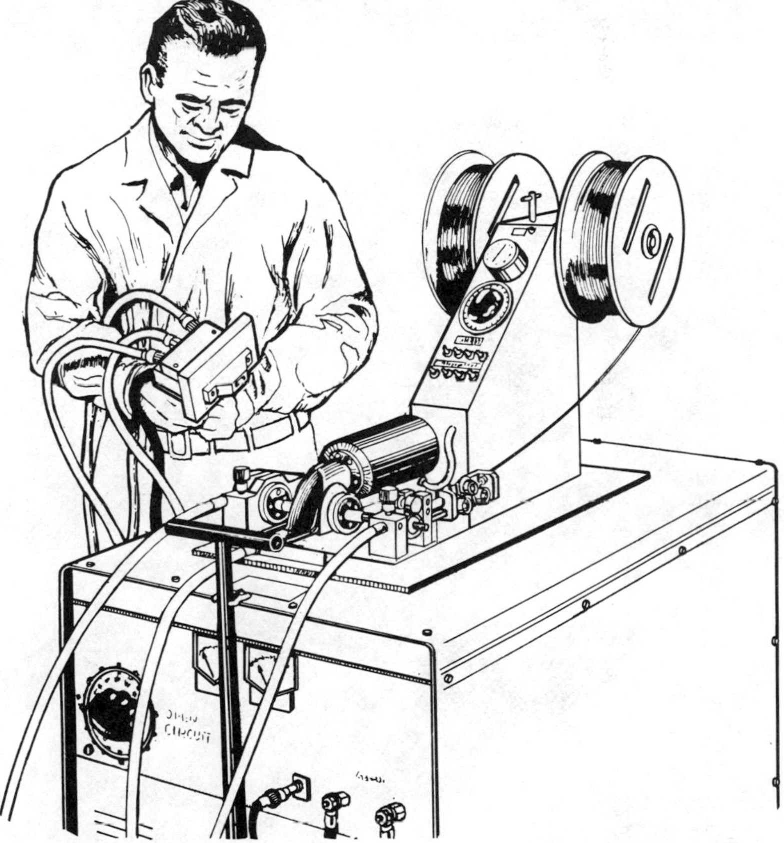

Courtesy Wall Colmonoy Corp.

Fig. 12. A typical electric-arc metal spraying system.

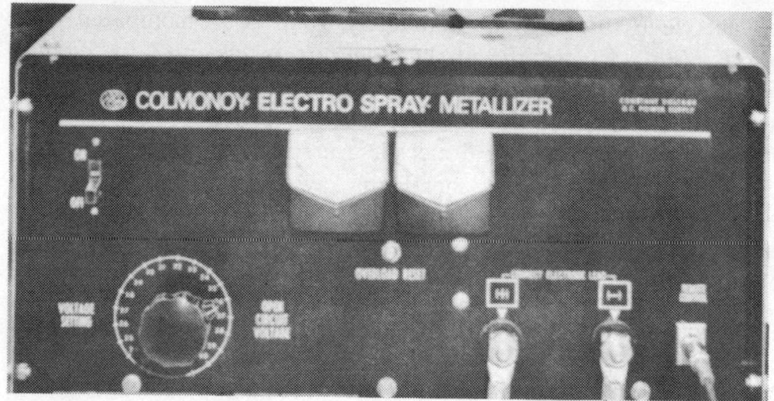

Courtesy Wall Colmonoy Corp.

Fig. 13. DC power source used in electric-arc metallizing.

Courtesy Wall Colmonoy Corp.

Fig. 14. A machine mounted wire feeder.

A typical constant-voltage, silicon-rectifier DC power source commonly used in electric-arc spraying is shown in Fig. 13. This is a solid-state power source available in either 300-amp or 500-amp continuous duty models. Note the voltage setting control for varying the open-circuit voltage.

The dual wire feeding unit (Fig. 14) can be mounted on the power source but is controlled separately. Independent adjustment is necessary to obtain proper burn-off rate of the wire electrodes. The wire-feeder is powered by an electric motor contained in the unit. The wire feed controls are located above the motor on the portion of the unit supporting the two reels of electrode wire. Larger wire-feeders are available for production work that requires greater volume and longer operating periods. (Fig. 15).

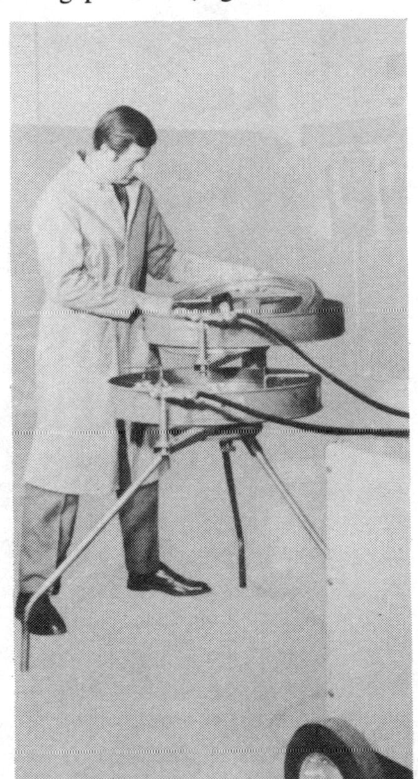

Fig. 15. Large volume wire feed unit.

Courtesy Wall Colmonoy Corp.

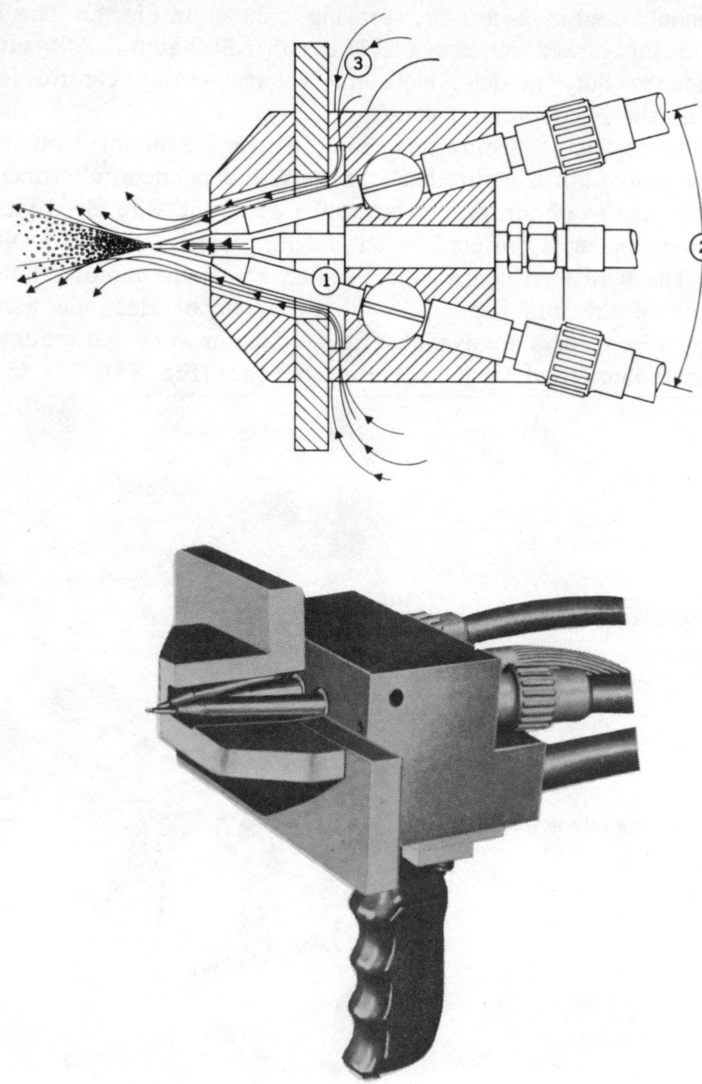

Courtesy Wall Colmonoy Corp.

Fig. 16. An electric-arc spray gun.

The spray guns used in this metallizing process are typically lightweight with an absence of moving parts. Fig. 16 shows a gun with: (1) straight electrode tips, (2) a particularly small included angle of 30°, and (3) openings behind the arc shield that permit free air to be pulled in and around the spray stream by the jet of compressed air. This last feature creates a protective envelope of air that prevents the loss of metal particles on the periphery of the metal spray. Control switches on the gun enable the operator to turn the arc current on or off and to start or stop the wire feed mechanism.

SPRAYWELD METALLIZING PROCESS

The *Sprayweld* metallizing process is a method of applying surfacing alloys in which the sprayed molten particles are fused to the base metal. This characteristic distinguishes *Sprayweld* from standard metallizing processes in which simple bonding occurs. On the other hand, it differs from surfacing processes in that the method of application does not make use of the gas or arc welding techniques. The *Sprayweld* process was developed by engineers at *Wall Colmonoy,* and is a trade name of that company.

A spray gun (Fig. 17) is used to apply surfacing alloys up to .060 inch thick to the base metal. After the spraying is com-

Courtesy Wall Colmonoy Corp.

Fig. 17. Gun used in the Sprayweld process.

pleted, the surfacing deposit is fused to the base metal with an oxyacetylene welding torch or a controlled atmosphere furnace (the torch method is usually preferred).

The gun used in the *Sprayweld* process can also be used to apply nonfusible metallized overlays. However, it is more economical to use standard equipment (oxy-fuel gas, electric-arc, etc.) when metallizing.

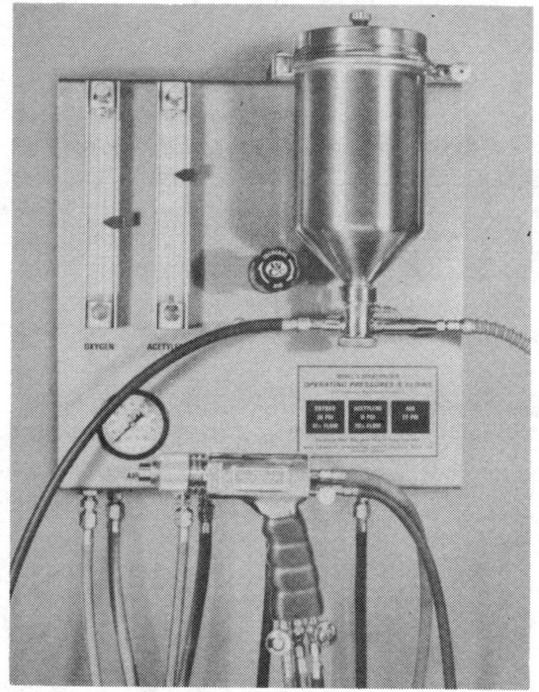

Courtesy Wall Colmonoy Corp.

Fig. 18. Major components of a Sprayweld set-up.

A *Sprayweld* metallizing setup (Fig. 18) will consist of the following basic components:

1. A spray gun,
2. A control panel with oxygen and acetylene flowmeters,
3. A powder hopper,
4. A carburetor,

5. An air filter and regulator,
6. Oxygen and acetylene hoses,
7. Oxygen and acetylene regulators.

The powder carburetor and powder hopper are located on the control panel unit along with the oxygen and acetylene ffowmeters. It is possible to regulate the flow rate of both the oxygen and acetylene by integral needle valves incorporated in the flowmeters. The powder hopper has a 15 lb. capacity. High-capacity *Spraywelders* can spray over 13 pounds of powder per hour.

The surfacing alloys used in the *Sprayweld* process have excellent "wetting" properties. The sprayed molten particles have a wide plastic range and will wet and fuse to the base metal without losing their shape. The high speed impact of the particles tends to eliminate porosity in the overlay deposit. The finished deposit can be made in any thickness up to .060 inch per side. Finishing should result in removing not more than .015 inch of the overlay deposit.

The *Sprayweld* process offers certain advantages over other surfacing methods. Chief among these are: (1) an increased speed of application (almost 4½ times faster than hand welding), (2) easily adapted to surfacing cylindrical shapes or small shapes with unusual contours, and (3) capable of applying a controlled-depth overlay.

The procedure for surfacing with the *Sprayweld* process is as follows:

1. Undercut the base metal surface by blasting with angular chilled iron grit. This provides an anchor for the overlay coating prior to fusion.
2. Spray the base metal with the selected surfacing alloy.
3. Fuse the sprayed overlay with the heat of an oxyacetylene welding torch.
4. Finish the overlay coat by grinding. This should be done wet when possible, and with light fast cuts.

There are four stages in the *Sprayweld* process on a metal part. From left to right they are:

1. Undercut and grit blasting,
2. The sprayed overlay,

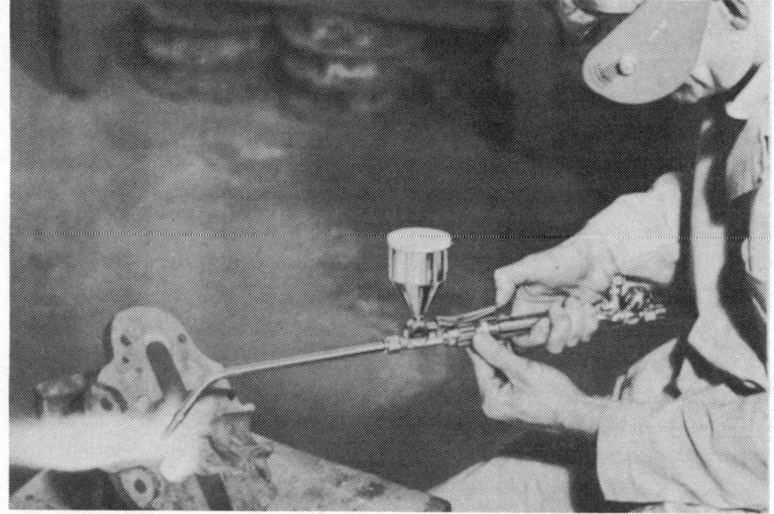

Courtesy Wall Colmonoy Corp.

Fig. 19. Using a torch designed for both spraying and fusing simultaneously.

3. The overlay after fusion,
4. The surface as a result of grinding.

A specially designed welding torch (Fig. 19) is now commercially available that combines the two spraying and fusing operations necessary in the *Sprayweld* process. This torch both sprays and fuses buildup and surfacing overlay metals. It can also be used for braze welding.

The surfacing and buildup metal alloy powders are contained in a powder hopper located on top of the torch. The torch proper is a standard welding torch body. These torches are produced by *Wall Colmonoy* under the trade name *Fusewelder*.

Surfacing applications with these dual purpose torches are generally limited to small irregularly shaped objects (worn camshaft lobes, stainless steel heat exchangers, etc.). Its capacity is much more limited than the *Sprayweld* process and should not be used as a substitute for the latter.

CHAPTER 17

Welding Pipes, Tubes, and Pressure Vessels

Entire books have been written about the very complex and highly specialized skill of pipe welding. Consequently, the inclusion of sections on tube and pressure vessel welding, plus the space limitations of a single chapter, require that this section be a very concise summary of these welding specialties.

A significant portion of our work force is either directly or indirectly engaged in the manufacture of pipe. The product that results from their efforts finds widespread application in both industrial and nonindustrial sectors of our economy.

Pipe functions basically as a conveyor of raw materials to a central point of consumption and as a conveyor of the products of that consumption to points away from it. The raw materials include a tremendous variety of things including gas, slurries of coal and water, and chemicals. The products of consumption include such diverse forms as waste and steam. All of this suggests that pipes must be fabricated to meet a wide variety of specifications and job requirements.

Pipe usage, then, dictates the need for high levels of quality in its manufacture. A defective pipe would be very dangerous in a pipe line carrying volatile high pressure gas. The welder is also governed by these same requirements. All welding must be of consistently high quality. A weak weld would be just as dangerous as a defective pipe. A pipe welder, then, must be able to show that he is certified to do this kind of work.

TUBING AND PIPE MANUFACTURING METHODS

Tubing differs from pipe in the thickness of its walls, being generally thinner than pipe for equal diameters. Pipes can also differ from one another according to the thickness of their walls. Fig. 1 shows the three weights of wrought iron pipe: standard, extra

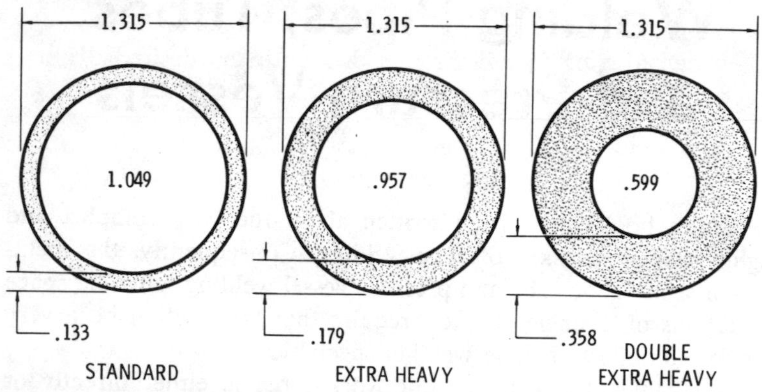

Fig. 1. The three weights of wrought pipe.

heavy, and double extra heavy. The illustration is approximately actual size, thickness, etc. of the three grades of 1-inch wrought iron pipe.

Pipe is made from both ferrous metals (e.g. wrought iron pipe, steel pipe, etc.) and nonferrous ones (e.g. copper). As a result, the welder *must* be able to identify the type of metal he will be working with.

A smith may take two pieces of metal of unknown composition and join them by some kind of weld, but the apparent ease with which he secures the result is no measure of the quality of the work.

In welded pipe, the strongest and most efficient weld is the one produced by the most thorough amalgamation of the joined portions. Such a weld represents skill in metallurgy as well as in the mechanical arts of rolling and welding.

The work must not fail under the internal hydrostatic pressure tests given by the pipe manufacturers, varying from 450 to 3000 lbs. per square inch according to the size of the pipe or tubing and the service for which it is intended. The standard hydrostatic test pressures to which pipes are subjected at the mills is very much greater than any pressure stress likely to occur in ordinary practice. For example, test pressures for a standard 2-inch pipe are: (1) butt welded 700 lbs. per square inch, and (2) lap welded 1000 lbs. per square inch. The bursting pressures are much higher.

In the manufacture of pipe, billets rolled into narrow strips of skelp are formed into pipe in special pipe-welding machines. The billets can also be forced through dies until the desired diameter and gauge is reached.

The various methods of manufacturing pipe and tubing are as follows:

1. Butt welding,
2. Lap welding,
3. Seamless forming,
4. Electric welding,
5. Hammer welding.

Butt Welding Method

The butt welding method (Fig. 2) is used in making pipe sizes ⅛ inch to 3 inches. Sheets of metal called *skelp* used in making butt weld pipe come from the rolling department of the steel mills with a specified length, width and thickness, according to the size of pipe to be made.

The edges are slightly beveled with the face of the skelp so that the surface of the plate which is to become the inside of the pipe is not quite as wide as that which forms the outside; thus, when the edges are brought together, they meet squarely.

One end of the skelp is trimmed to a vee shape for a short distance and slightly turned up to facilitate the grip of the welding tongs. When heated to the proper welding temperature, the vee-shaped ends of the skelp are gripped by heavy tongs and drawn from the furnaces, one at a time, through funnel shaped dies or bells.

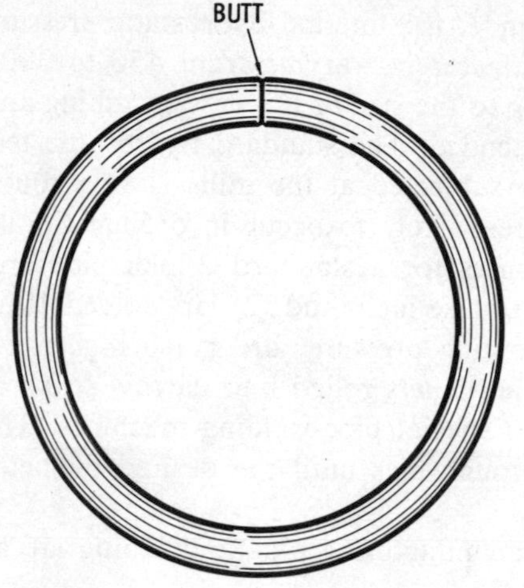

Fig. 2. The butt weld method.

The inside of the welding bell is so shaped that the plate is gradually turned or formed into the shape of a pipe, the edges being forced squarely together and welded.

After the skelp has been welded into pipe, it passes through a jet of sizing rolls, where it is reduced slightly in size and elongated. From the sizing rolls, the pipe passes across a cooling table to another set of specifically designed rolls which again slightly reduce the size and elongate the pipe, giving it the correct finished diameter and circular contour. Any heavy mill scale, or welding scale, which may be present is removed by this rolling operation, leaving the pipe walls smooth and clean.

The pipe then passes across a second cooling table and then to the cross rolls where any straightening necessary is done, and then to a tank of water where any loose scale is washed out (for certain sizes, the loose scale is blown out by a blast of compressed air).

The pipe is then conveyed to a cold saw where the rough ends are removed, after which each length is carefully inspected.

Lap Welding Method

The lap welding (Fig. 3) method of pipe manufacture is used on sizes 2 inches to 24 inches in diameter. The skelp used in making lap-weld pipe is rolled to the necessary width and thickness for

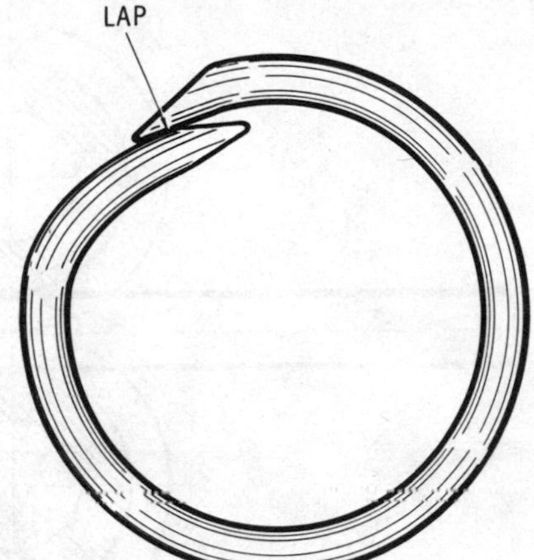

Fig. 3. The lap weld method.

the size of pipe to be made. The edges are then scarfed and overlapped, so that when the skelp is bent into shape a comparatively large welding surface is provided, compared to the thickness of the pipe.

The skelp is first heated to redness in a "bending furnace" and then drawn from the front of the furnace through a die, the inside of which gradually assumes a circular shape so that the skelp is bent into the form of a pipe with the edges overlapping.

Next, the bent skelp is heated to the proper welding temperature and pushed through an opening in the front of the furnace into the welding rolls, consisting of two rolls set one above the other, each having a semicircular groove so that the two together form a circular pass.

Between these rolls a mandrel called a *welding bell* (Fig. 4) is held in position inside the pipe so that the lapped edges of the skelp

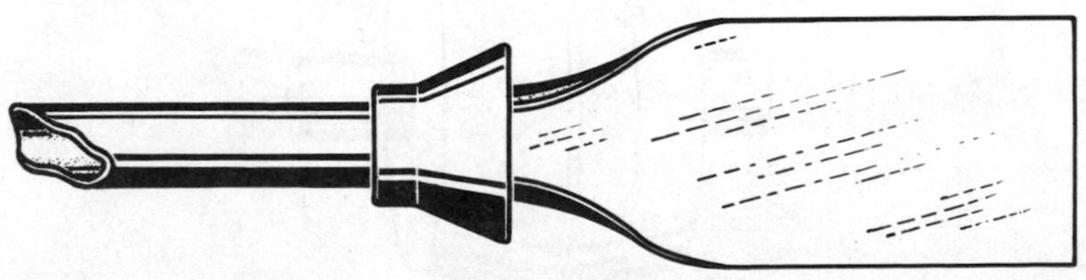

Fig. 4. The welding bell.

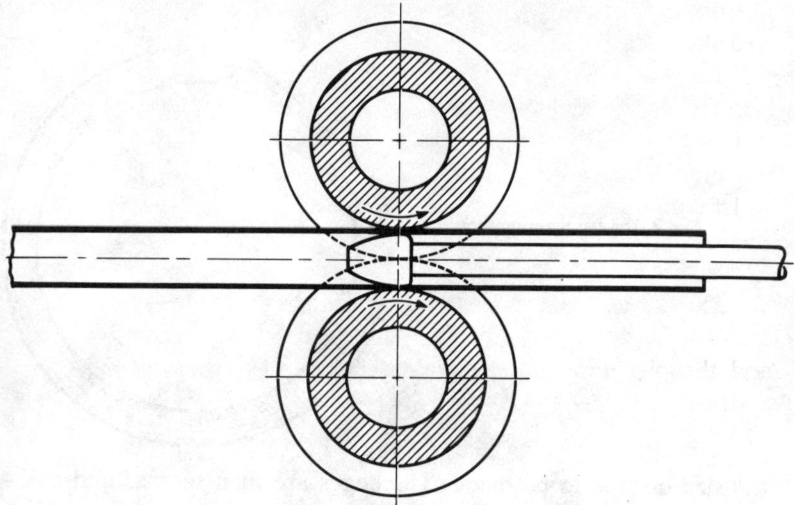

Fig. 5. Welding rolls to correct any irregularities in the pipe.

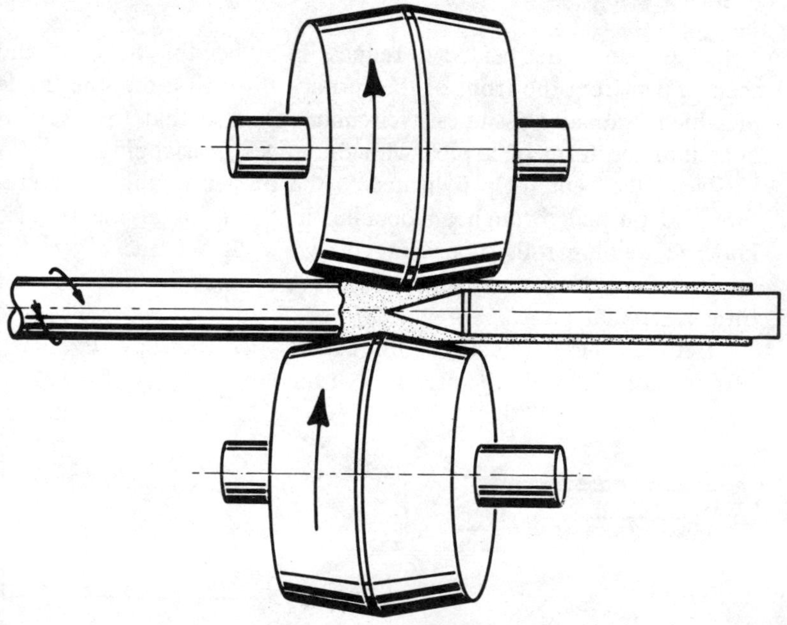

Fig. 6. The piercing operation.

are firmly pressed together at a welding heat between the mandrel and the rolls.

The pipe then passes through a similarly shaped set of rolls to correct any irregularities and to obtain the outside diameter required (Fig. 5).

Finally, the pipe is passed through the straightening or cross rolls, consisting of two rolls set with their axes askew. The pipe is made practically straight by the cross rolls and is also given a clean finish with a firmly adhering film of blue oxide.

The pipe is then rolled up an inclined cooling table so that the metal will cool off slowly and uniformly, without internal strain, and the pipe will remain straight. When cool enough, the rough ends are removed by cold saws or in a cutting-off machine, after which the pipe is ready for inspection, threading, and testing.

Seamless Forming Method

Seamless pipe and tubing can be made either by a piercing operation or a reeling operation. In the piercing operation (Fig. 6), sizes up to 14 inches of steel are delivered to the heating furnace in the form of solid cylindrical billets, center punched at one end

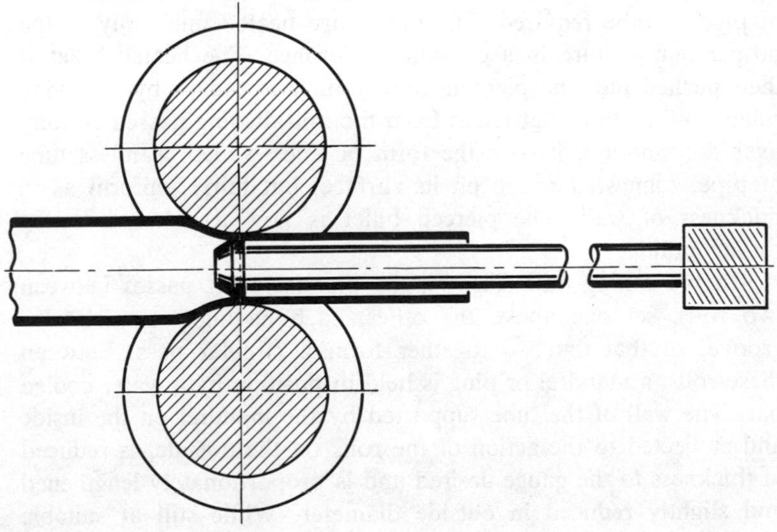

Fig. 7. The rolling mill operation.

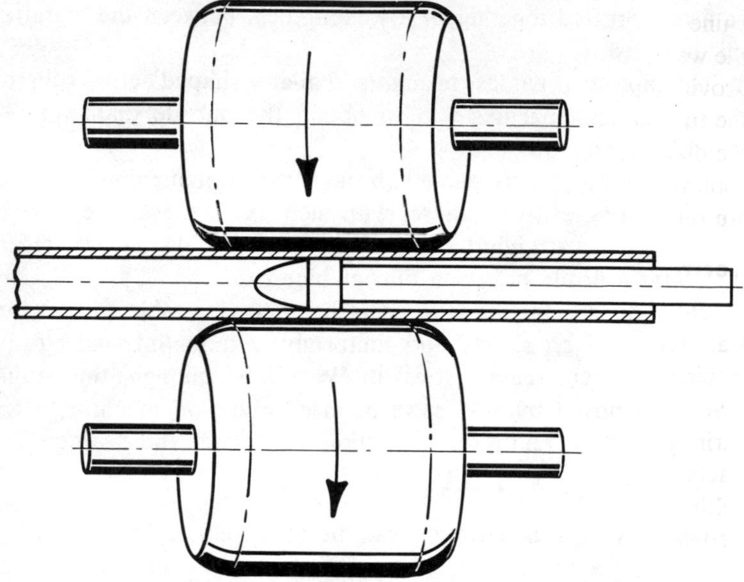

Fig. 8. The reeling operation.

and of proper diameter and length to make the size and length of pipe or tube required. The billets are heated uniformly to the proper temperature in a continuous furnace. The heated billet is then pushed into the piercing mill until it is caught by revolving rollers. When the billet issues from the mill, having passed entirely over the mandrel, it is in the form of thick-walled seamless tube or pipe, somewhat rough on its surface, but fairly uniform as to thickness of wall. The pierced billet is now transferred to the rolling mill.

In the rolling mill (Fig. 7), the pierced billet passes between two rolls, set one above the other, each having a semicircular groove, so that the two together form a circular pass. Between these rolls, a mandrel or plug is held in position by a water cooled bar. The wall of the tube supported by the mandrel on the inside and subjected to the action of the rolls on the outside, is reduced in thickness to the gauge desired and is proportionately lengthened and slightly reduced in outside diameter. While still at suitable working temperature, the tube passes on through the reeling machine (Fig. 8) between barrel shaped rolls and over a plug where the wall is brought to uniform thickness, the tube is rounded up and provided with a smooth, burnished surface. From this machine the tube passes to the sizing or finishing rolls, the tubes travel to a the diameter required. From the finishing rolls, the tubes travel to a cooling table and after being sorted and inspected, the rough ends are removed and the tubes cut to proper length.

Electric Welding Method

Both the arc and resistance welding processes can be used in the manufacture of pipe. Resistance welding is used in the manufacture of pipe sizes ranging from 30 inches to 96 inches. The skelp is beveled, crimped, bent, and tacked to maintain a cylindrical shape during welding. The tacked pipe is then placed on the welding machine and held stationary. A number of automatic electric welding heads are placed along the pass, each unit covering a predetermined distance.

Hammer Welding Method

The hammer welding method is used in the manufacture of pipe exceeding 30 inches in diameter. The flat plate is curled on specially designed rollers, heated to a suitable temperature and a weld is forged on the overlapping edges of plate by blows from a pneumatic hammer. An anvil is inserted in the pipe to absorb the hammer blows and prevent distortion of the cylindrical shape.

PIPE LINE WELDING

In the construction of large pipe lines, a number of different operations are performed. This includes clearing the right of way, stringing the pipe, and ditching. Line-up and tacking crews precede the major construction force. Aided by a tractor and hoist, they put the pipe lengths on ball-bearing dollies for rolling. The tack welder tacks the adjacent lengths of pipe together usually at four points in the circumference. He joins as many lengths as the nature of the line and the contour of the country will permit. The long tack-welded section is then left on the dollies ready for the firing line welders who follow close behind.

The three principal methods employed for welding large pipe lines are:

1. Roll welding,
2. Bell-hole welding,
3. Stove pipe welding.

In *roll welding,* the operators weld at the top of the pipe while a helper turns the pipe by means of a chain pipe wench. The principal advantage of this method is that it permits welding to be done in a flat downward position and by using large sized electrodes (¼ inch, 5/16 inch and sometimes ⅜ inch) and heavy welding current, high welding speeds are obtained.

After the completion of each weld, the helper cleans off all the slag. Each weld is inspected for any imperfection before the welder proceeds to the next joint. As soon as a section is completed, the pipe is rolled off the dollies and the pipe sections are painted and then joined into a continuous stretch by bell-hole welding.

Bell-hole welding may be defined as the placing of a welded section of pipe over the open ditch on skids and welding it to the preceding section. It is generally necessary to dig out the sides of the ditch forming a bell shaped hole or space in the ditch to permit easy access to the entire circumference of the joint, hence the name.

A major problem with bell-hole welding is that overhead welding is necessitated due to the impossibility of turning the pipe line.

In bell-hole welding, the operator first makes the burning-in bead in the top half of the bell-hole joint. After this portion of the joint is welded, the helper cleans the bead while the operator continues to weld the lower half of the joint. The operator then stands by until the cleaning of the entire burning-in bead is finished. He then proceeds to weld the final or finish bead.

All bends in a section are made, where possible, in the middle of a pipe line. Bending of a section to conform to line of ditch is usually done immediately after the section is welded to the pipe line. Where expansion joints are not welded into a line, slack is put into the line by forcing the line into the ditch, so that it will lie on the bottom of the ditch, not in a perfectly straight line, but weaved from wall to wall.

In the construction of oil and gas lines, in rough terrain or very muddy or sandy country, the *stove pipe* method is frequently employed. Here, all joints (usually plain end without liner) are position welded like bell-hole joints. One joint at a time is added to the line, making it possible to reduce the size of the crew and amount of equipment and keep operations bunched together under one supervisor.

During the alignment and tack welding process the joint is generally held in place by a line up clamp. After tacking, two bell-hole welders work simultaneously on both sides of the joint, making the complete first bead—each welder coming down from the top.

The wall thickness of the pipe for river crossings is generally 50% to 100% greater than that used in the rest of the line. The pipe is roll-welded and bell-hole welded into a section long enough to cross the river. The joints may be reinforced by a welded-on sleeve. Special river clamps 4 to 10 feet long and weighing 1000 to 3000 lbs. each are often bolted to the pipe at intervals, serving as a means of anchoring.

The pipe section is attached by cable to tractors and pulled onto pontoons. It is then guided by stakes into a dredged out ditch in the river bottom where it is usually buried at a minimum depth of 8 feet.

PIPE LINE WELDS

A number of different types of welded joints for pipe lines have been designed and employed in the installations to date. Some of these designs call for recessed and raised bead backing strips; others involve up-setting of the pipe ends to permit recessing of the ring in the pipe wall without reducing the cross section at the weld; and still others require machining of pipe edges to form U-grooves and bevels with lips of varying dimensions. Good welds may be produced and entirely satisfactory results have been obtained with any of these designs. All things considered, however, the primary requisite of any weld design is weldability and, as a general rule, the more simplified the design the easier it becomes for operators to produce sound and satisfactory welds.

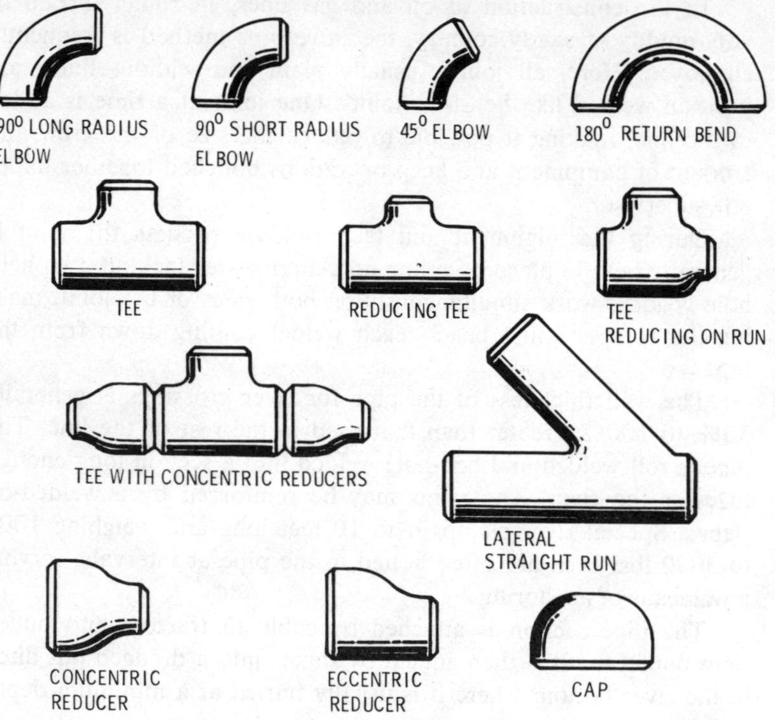

Courtesy The American Welding Society

Fig. 9. Typical manufactured welding fittings.

PIPE FITTINGS FOR WELDED JOINTS

Various kinds of standard fittings have been designed for welded joints. Standard dimensions have been adopted for each fitting. The fittings consist of elbows, tees, offsets, reducers, crosses, manifolds, saddles, swage nipples and other designs. Fig. 9 illustrates some examples of typical welding fittings available from manufacturers.

Pipe fittings are used to provide any kind of elbow, offset, branch or compound turn in piping, headers and manifolds, or wherever pipe welding might be used. The 180° return fitting is used in coils, retorts, tube stills, etc. The 90° types and straight pipe are used to make up expansion loops.

Pipe fittings are made in two types; the long radius and standard radius. Long radius pipe fittings have, in every size, a radius of curvature equal to 1½ times the nominal pipe size. Standard radius pipe fittings are formed to a radius equal to the nominal pipe size.

Long radius pipe fittings (radius = 1½ D) are made in 45°, 90° and 180° (return bend) types in standard pipe thickness, in iron pipe sizes from ¾-inch to 18-inch inclusive. They are made in the same types, 45°, 90° and 180° in extra strong pipe thickness sizes from 1 inch through 18 inches.

Standard radius pipe fittings (radius = 1 D) are made in 90° and 180° return bend types only, except in sizes 20-inch (OD) and 24-inch (OD) which are furnished in the 45° type. The 90° fittings, both standard and extra strong, are furnished in sizes 2 inches through 24 inches. The 180° return fittings are made in both standard and extra heavy pipe thicknesses in sizes from 2-inch through 24-inch (OD) except the 18-inch (OD) which is available only in the long radius.

Pipe fittings are available in materials other than steel, and with walls as light as sixteen gauge, or as heavy as 1 inch in thickness.

PIPE FLANGE

The use of welded pipe joints has been further facilitated by the introduction of specially designed forged flanges. These are

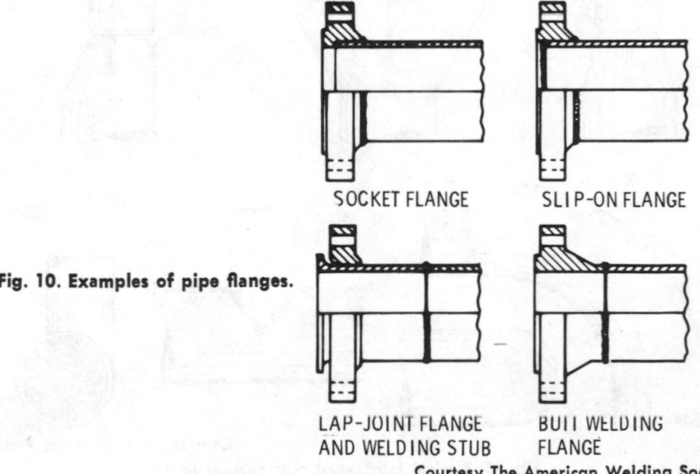

Fig. 10. Examples of pipe flanges.

Courtesy The American Welding Society

available in several types, including butt, lap, socket and slip-on (Fig. 10). Flanges for butt and lap joints can be used under all service conditions. Socket and slip-on types are limited to working pressures under 300 psi.

In a welded pipe assembly, flanges are required at the connections with valves and similar fittings. They are also required

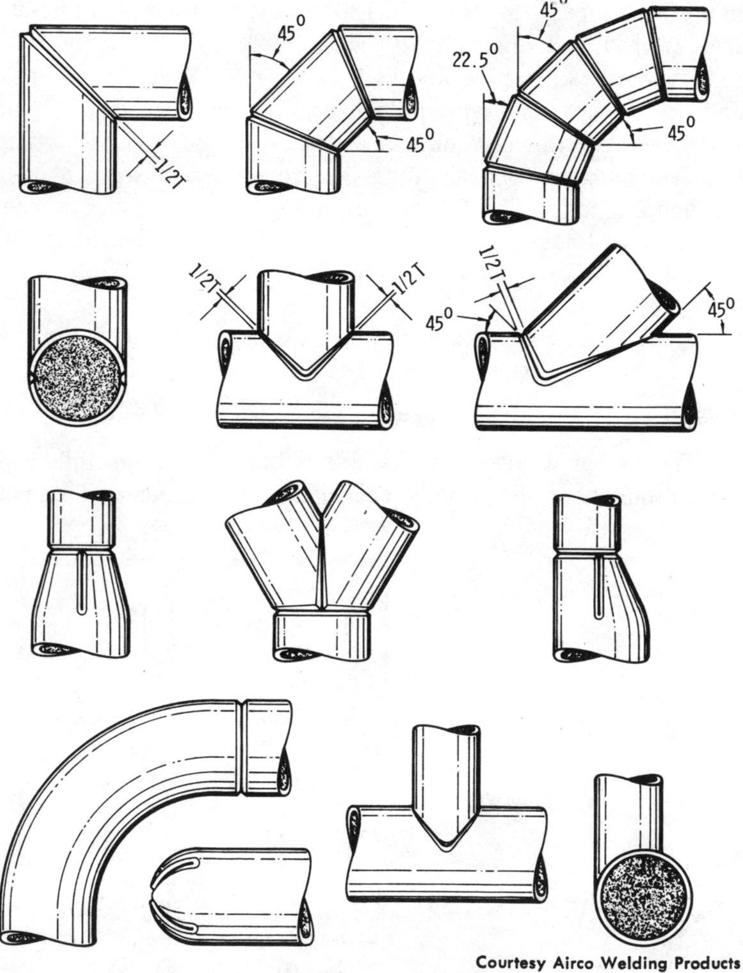

Courtesy Airco Welding Products

Fig. 11. Typical fabricated pipe connections.

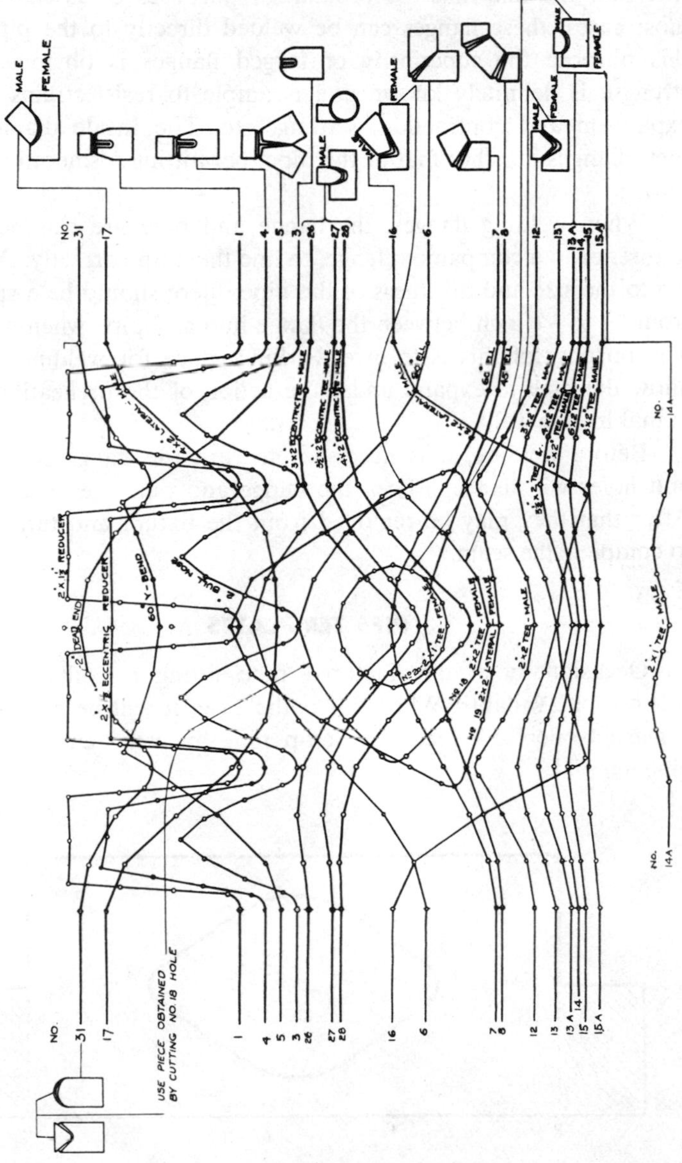

Courtesy Airco Welding Products

Fig. 12. Example of a manufacturer's pipe template.

wherever the line must be broken for purposes of assembling. In most cases, these flanges can be welded directly to the pipe. For this purpose the superiority of forged flanges is obvious. Their strength is definitely known and is ample to resist strains due to expansion and contraction, settling, etc. The inside diameter of these flanges matches that of the pipe, permitting a smooth internal flow.

When welding flanges, the flange and pipe are clamped in a fixture, or to a companion flange, to line them up correctly. According to the size and thickness of the pipe, there should be a space of from 1/16 to 1/4 inch between the flange hub and pipe when welding. The purpose of this is to give working space for welding, and to allow the pipe to expand under the action of the preheating flame in making the weld.

Before welding, it is necessary to turn the flange so that the bolt holes will line up. Then the flange and pipe are tack welded. After that they may be removed from the fixture and turned over to complete the seam.

PIPE TEMPLATES

Occasionally a manufactured pipe fitting of suitable design will not be available. When this is the case, it will be necessary to construct one from pieces of scrap pipe by using an appropriate pipe template.

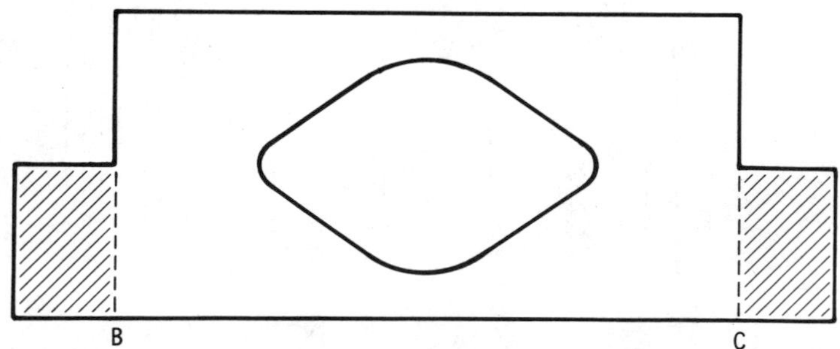

Courtesy Airco Welding Products

Fig. 13. Prepared template for run of a tee.

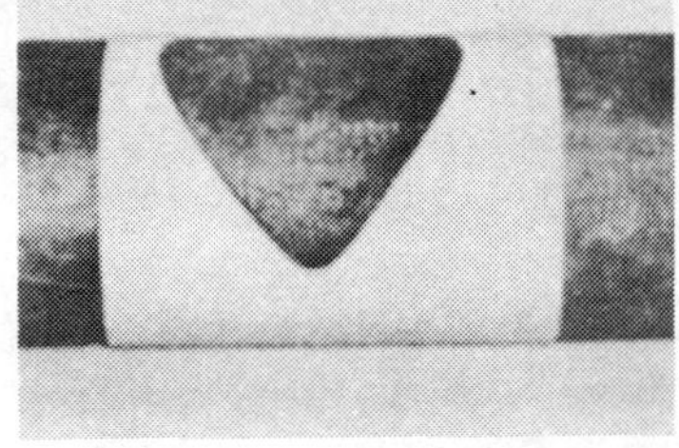

Fig. 14. Template wrapped around pipe ready for marking.

Courtesy Airco Welding Products

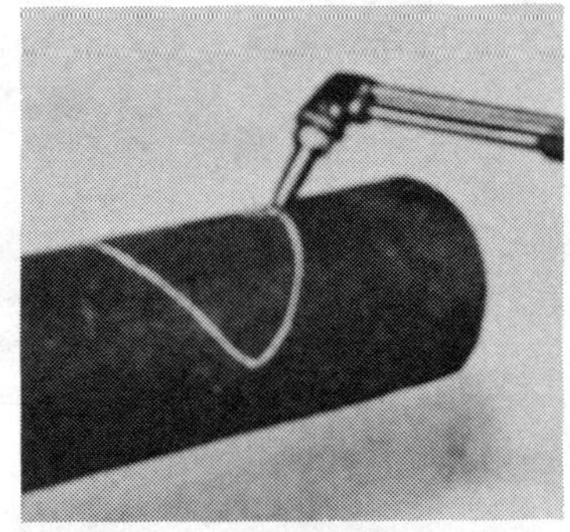

Fig. 15. Cutting end of branch for tee connection.

Courtesy Airco Welding Products

Fig. 16. Starting cut of hole in run of a tee connection.

Courtesy Airco Welding Products

Pipe templates are available in a variety of sizes (generally ranging from 2 to 16 inches) from manufacturers of welding

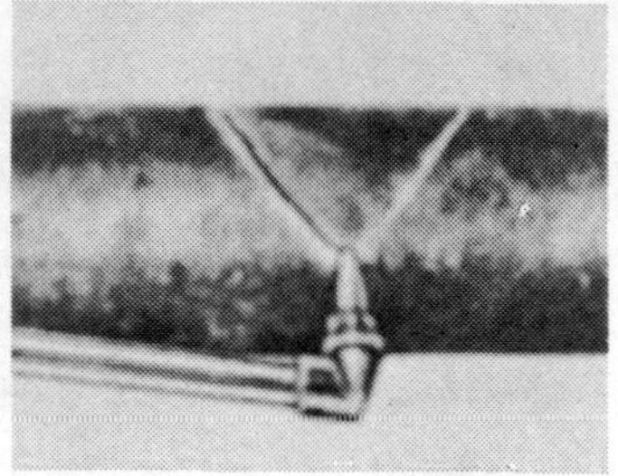

Courtesy Airco Welding Products

Fig. 17. Cut for hole in run of a tee connection partially completed.

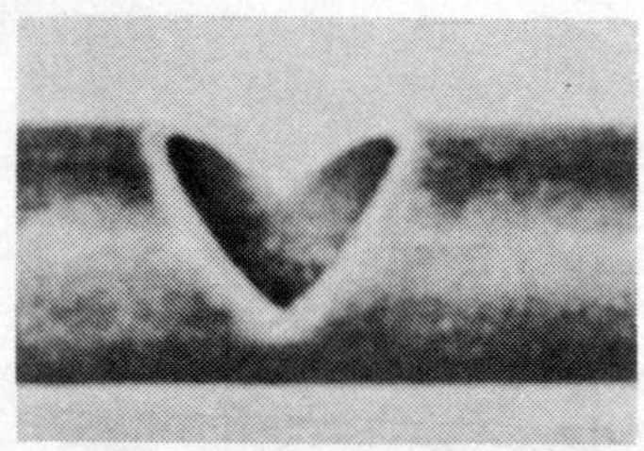

Courtesy Airco Welding Products

Fig. 18. Complete cut.

Courtesy Airco Welding Products

Fig. 19. Tee connection tack welded.

equipment and supplies. Fig. 11 illustrates the types of pipe connections that can be made from templates with oxyacetylene welding and cutting. A typical pipe template is shown in Fig. 12. This is a template for 2-inch pipe, and is not drawn to scale. However, the template received from the manufacturer will be drawn to scale, and should be used as a tracing guide for making working templates.

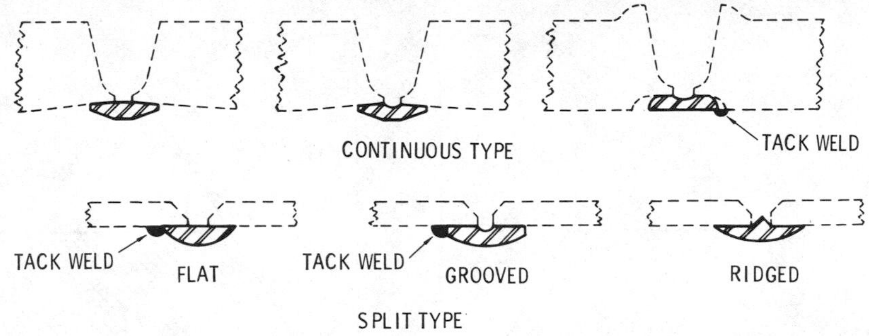

Courtesy The American Welding Society

Fig. 20. Various type backing rings.

Figs. 13 to 19 illustrate the use of a template for constructing a T-joint.

BACKING RINGS

A backing ring is a ring-like device sized to fit the bore of the pipe and cover the gap between the two pipe ends to be joined. The backing ring prevents formation of icicles and weld spatter inside the pipe and provides complete fusion at the root of the weld. These rings are usually made of the same metal as the pipe being welded or they may be of plain carbon steel.

Since pipe walls may vary slightly in thickness and pipe is seldom perfectly round, it is frequently necessary, when using backing rings, to machine the pipe ends in order to provide true round inside contours and assure a good fit for the rings.

There are many sizes of backing rings in use. Fig. 20 shows some typical examples of backing rings. Split-type backing rings are cut at one point in their circumference to permit adjustment to the diameter of the pipe. They represent one of the two categories of backing rings. The other category consists of continuous-type backing rings, and are so named because the circumference of the ring remains unbroken.

ARC WELDING PROCEDURE

Fig. 21 illustrates a basic joint design for welds in plain carbon steel or carbon-molybdenum steel piping 6 inches or more in diameter in the horizontal rolled and the horizontal fixed position.

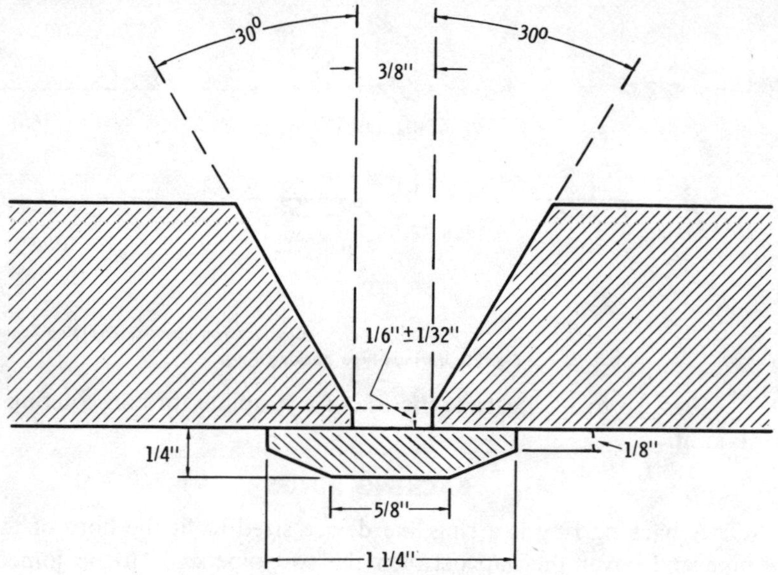

Fig. 21. Joint design for welds in plain carbon steel or carbon-molybdenum steel.

The larger gap (⅜ inch) permits better manipulation of the electrode during welding, makes easier the laying down of the first few beads (which are usually the most difficult) and results in better welds. For smaller pipe, a 3⁄16 inch or ¼ inch gap is generally recommended.

A design for welds in the vertical fixed position is shown in Fig. 22. Note that the bevel of the lower pipe edge is reduced to approximately 7°, while that of the upper pipe edge is increased to about 45°. The gap remains at ⅜ inch. This design for vertical welds provides a nearly horizontal shelf for the operator to build on and makes every bead deposited practically a fillet weld. The welding technique thus becomes simply that of continuous fillet welding.

Pipes with thick walls require multilayer welds for satisfactory results. In this procedure the deposition of a number of thin layers of weld metal by means of a multiplicity of passes is preferable to the use of a few heavy layers and a small number of passes. Both

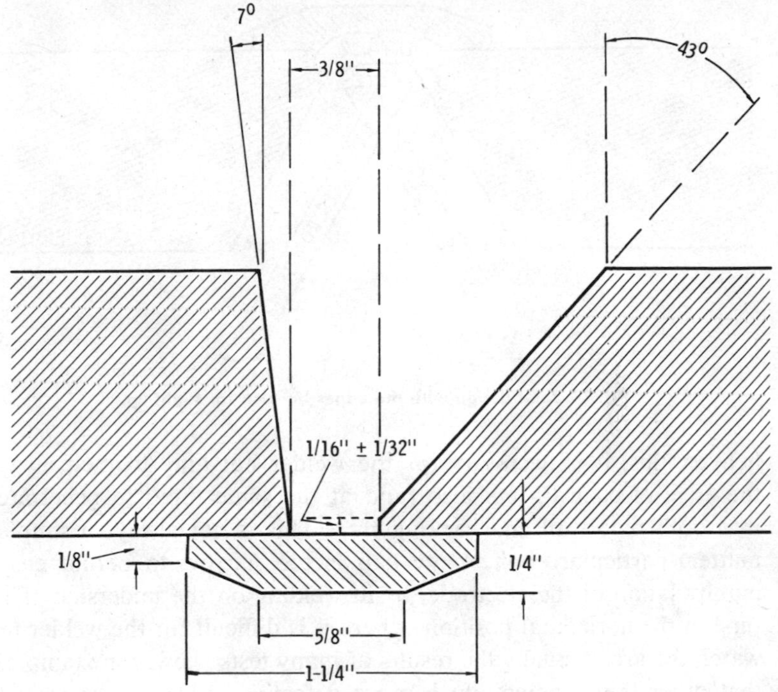

Fig. 22. Design for welds in the vertical position.

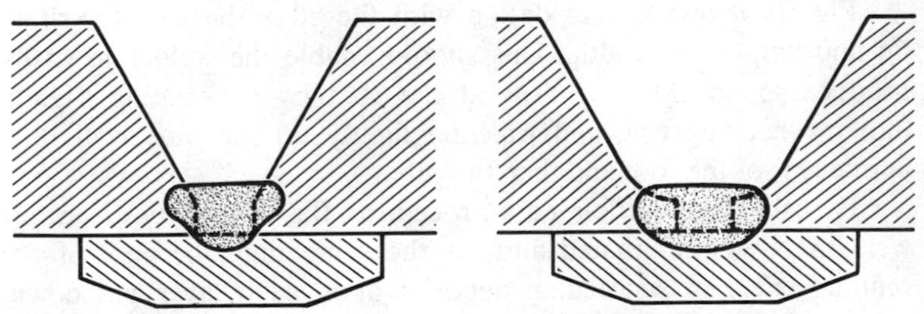

Fig. 23. Vee- and U-groove joints with large lips.

the vee- and U-groove are used in multipass welding. However, effective use of these joints require that the edge be removed or greatly reduced before welding. The difficulties that can occur with a large lip are illustrated in Fig. 23. When depositing the first layer of weld metal at the root of a U-groove, or in a vee with

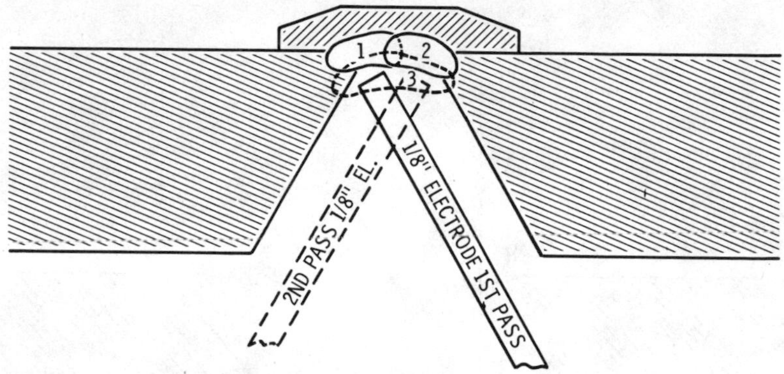

Fig. 24. Joint design with the edges beveled for slight lip.

heavier lips, it is necessary for the welder not only to melt away the edge of each lip but also (and at the same time) to penetrate into both pipe wall and backing ring. This is not always a simple matter, particularly where the gap is too narrow to permit easy manipulation of the electrode, or in welding on the underside of a pipe in the horizontal position, where it is difficult for the welder to watch the arc. Actually the results of many tests show, for example, that other things being equal, more defective welds can be traced directly to this particular condition than to any other source.

Fig. 24 shows a joint design with the edges beveled for slight lip and ample gap width. This should enable the welder to easily produce sound clean deposits at the root by penetrating slightly into the backing ring and washing lightly up the pipe wall, first at one side of the root and then the other.

By using the multiple pass procedure, the possibility of porous weld metals is eliminated and, at the same time, because of the refining effect of succeeding deposits upon each layer put down, a better grain structure throughout the weld is provided.

When welding in the horizontal fixed position, the ⅛-inch electrodes are used at approximately 90 to 100 amperes and the 5⁄32-inch electrodes at 130 to 160 amperes. Where the pipe is rotated during welding, or, when welds are being made in the vertical position, larger electrodes and higher current values are employed.

The weld deposit is built up approximately 1/16 inch beyond the outside of the pipe before the cover layer is applied. Proper grain refinement in the weld for the full thickness of the pipe wall is thus assured.

The backing ring is inserted in position in one pipe end and tack welded at several points along its inside edge. The end of the adjoining pipe valve or fitting is then brought into position, slipped over the free end of the ring and tack welded in place, leaving 3/8 inch space between the beveled edges at the root of the gap.

Tack welds in the vee are made with a 1/8 inch electrode at 115 amperes. The arc is struck and the electrode withdrawn slowly, leaving flat "shoe button" tack welds which do not require chipping out later on but are readily washed away when the first welding bead is put down. The advantage of this method of tacking is that it avoids any possibility of injury to the backing ring, which might otherwise be caused by the chipping hammer.

In shop work, where pipe (in horizontal position) is rotated as welded, the first layer is deposited by weaving across this entire gap at the root, penetrating well into the backing ring and washing lightly up the pipe walls at each side. The second layer is deposited directly on top of the first layer and is also made the full width of the gap. Successive layers follow the same pattern (Fig. 25).

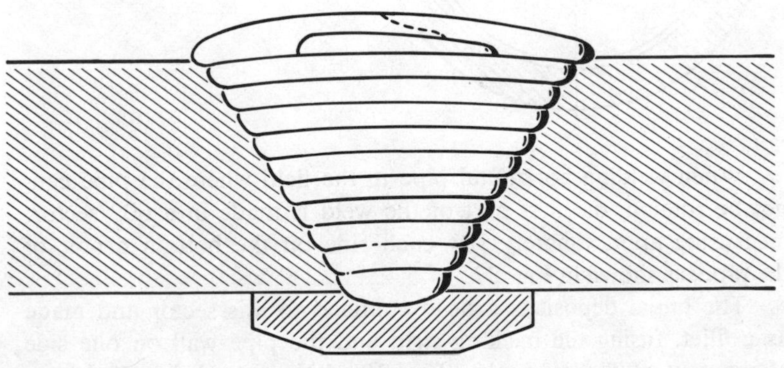

Fig. 25. Method of putting down suceeding weld layers.

The electrode may be held perpendicular to the pipe, or inclined toward the welder at about 30° off the vertical and slightly ahead of the top center of the pipe, allowing the slag to flow back of the deposited metal as the pipe rotates. In general, the perpendicular position is employed when high amperages are being used, while the inclined position has advantages when using relatively low amperages. The electrode positions are shown in Fig. 26.

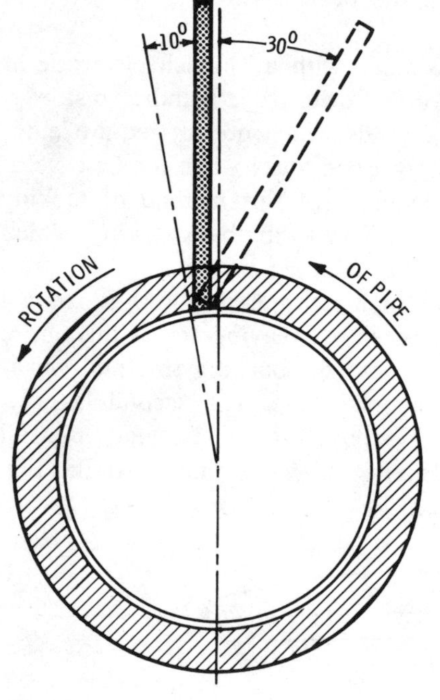

Fig. 26. The electrode positions.

When welding horizontal pipe in the field where the pipe cannot be rotated, the first part of the weld is made in a 60° sector, measuring approximately 30° each side on the bottom center of the pipe, as shown in Fig. 27.

The initial deposit is a bead confined to this sector and made as a fillet, fusing the backing strip and the pipe wall on one side of the root of the gap only. (Fig. 28). The second deposit is run both ways from the bottom center of the pipe and is put down in

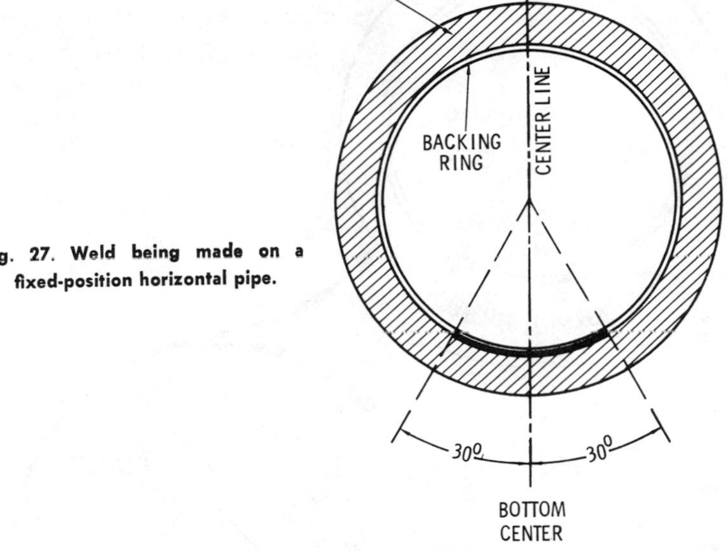

Fig. 27. Weld being made on a fixed-position horizontal pipe.

the same manner as the first bead, but on the opposite side of the gap and up to the 30° point only. From this point, it is continued as a full width layer up to the top center of the pipe on each side. (Fig. 29).

The third deposit is also run both ways from the bottom center of the pipe and covers the first two beads, while succeeding layers are made as shown in Fig. 30.

This method of welding eliminates the usual difficulties encountered on the underside of pipe in the horizontal fixed position, where frequently it is difficult for operators to watch the arc and where, therefore, a majority of defects ordinarily occur.

A somewhat different procedure is used for making welds in the fixed vertical position (Fig. 30). The first deposit is a bead, penetrating the backing ring and lower pipe wall. A second bead penetrates the ring and upper pipe wall and fuses with the first bead. Succeeding deposits are put down as illustrated in Fig. 30, each bead being nearly a perfect convex fillet weld.

Care must be taken to thoroughly remove the slag and to clean each bead or layer before the succeeding deposit is put down. This

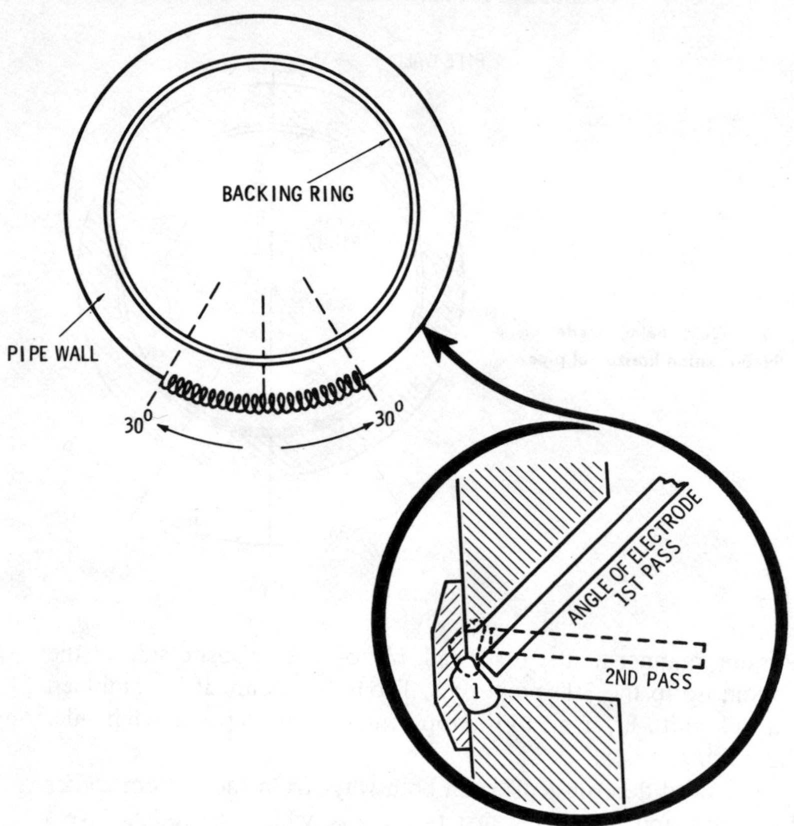

Fig. 28. Beginning the weld on a horizontal fixed position pipe.

can be done with a wire brush or a light hammer, following which the weld metal is smoothed with a chipping tool.

The chipping operation may result in a slight peening effect but is purely a means of cleaning and smoothing and possesses no value as far as stress relief and grain refinement are concerned. The final or cover layer may be chipped smooth and peened lightly to give a finished appearance similar to knurling.

Preheating is not necessarily an essential part of the arc welding procedure. Its purpose is to reduce the rate of heat conduction away from the freshly deposited metal, and to aid in refining the deposit so that coarse grain formations in the weld are avoided.

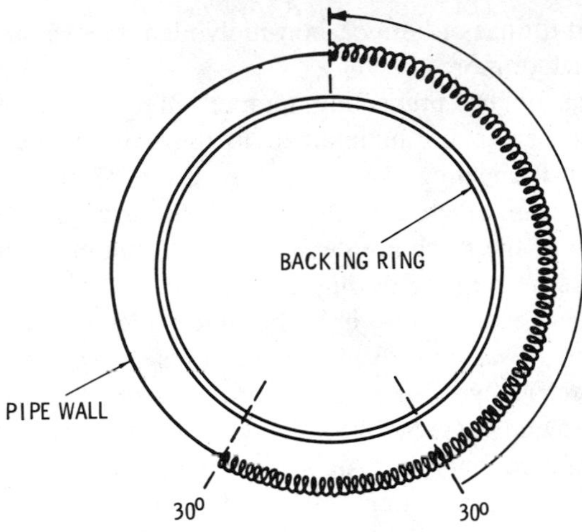

Fig. 29. Completing the weld.

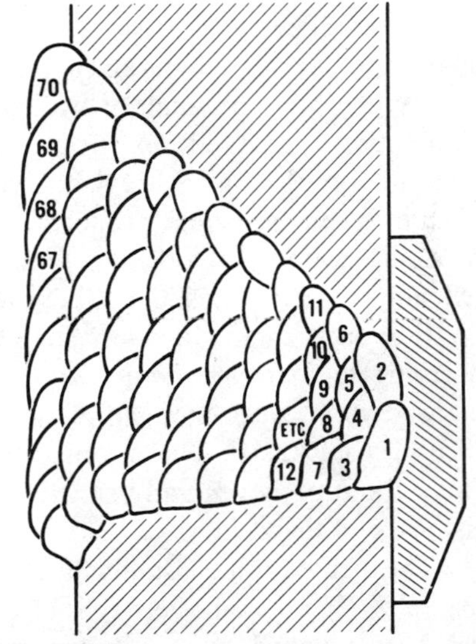

Fig. 30. Laying multiple deposits for welds in the vertical fixed position.

The same objectives may be attained under ideal conditions, without preheating when highly skilled operators, using a well developed technique, are employed. Nevertheless under ordinary circumstances, it is usual to preheat the welding ends of all pipes, valves, and fittings of both carbon-molybdenum steel and mild steel ½ inch thick or more.

The customary preheating temperatures are from 400°F to 600°F, the heat being maintained in the work while the welding is going on. Preheating of welds in low pressure lines may be done with torches, in which case the heat of the arc and an occasional application of the torch are depended on to maintain the necessary temperature in the pipe during welding.

On high pressure lines, in carbon-molybdenum steel pipe, however, where closer control of temperatures is required, preheating is done electrically. Pipe ends are enclosed in resistance or induction heaters and covered with a heavy coating of plastic insulation, leaving exposed only the surfaces to be welded. This procedure assures uniform preheating and maintenance of heat during welding. The insulation not only helps the parts to retain heat, but also protects the operator and permits using the same equipment for stress relieving without change or interruption.

Stress-relieving consists in bringing the welded joint slowly up to the proper temperature and holding it there for approximately one hour per inch of wall thickness of pipe. To ensure the elimination of locked up stresses all welds in material, ½ inch thick or more, require stress relieving. As a general rule, because of the creep resisting properties of carbon-molybdenum steel, welds in carbon-molybdenum pipe require somewhat higher stress relieving temperatures than those in mild steel pipe.

For carbon-molybdenum steel pipe, the recommended stress relieving temperatures range from 1200°F to 1250°F, while for plain carbon steel pipe ample stress relief will be obtained at 1150°F to 1200°F. The distance that the preheating zone extends along the pipe is generally six times the wall thickness of the pipe each side of the weld.

The heated zone usually is not allowed to cool more rapidly than 250°F an hour between 1200°F and 600°F. When 200°F has been reached, the insulation may be removed and the pipe is allowed to cool in still air to atmospheric temperature. Stress-relieving of joints which have been preheated electrically is conveniently done immediately after completion of welding. Additional insulation is placed around the weld and the current is

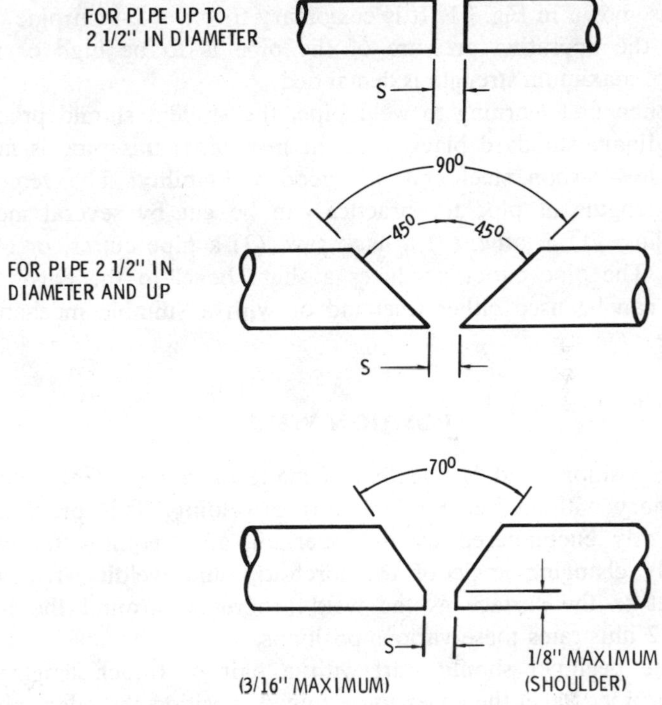

Fig. 31. Recommended joint preparation for butt welding standard pipe.

stepped up to provide the proper temperature. Where electrical preheating has not been employed, electric induction or resistance stress relieving equipment also is usually used. The weld is brought up to proper temperature and allowed to cool at the usual rate of not more than 250°F an hour.

OXYACETYLENE WELDING PROCEDURE

In pipe welding, the edges to be joined must be beveled whenever the metal is more than 3⁄16 inch thick so that the weld will penetrate to the inside wall of the pipe. If the walls of the pipe are less than 3⁄16 inch thick, the edges may be cut square and welded without beveling.

Recommended joint preparation for welding standard steel pipe is shown in Fig. 31. It is customary to bevel 2-inch pipe only when the operating pressure of the pipe is to be high or if a joint of maximum strength is demanded

When first learning to weld pipe, the student should practice on ordinary standard black wrought iron pipe; this pipe is made from low-carbon steel and has good weldability. The required short lengths of pipe for practice can be cut by several means including: (1) a lathe, (2) a hack saw, (3) a pipe cutter, or (4) a torch. The pipe cutter produces a slight bevel to the edge. The torch can be used either freehand or with a suitable mechanical device.

POSITION WELDS

A position weld is one that is made on a pipe that remains stationary without being rolled during welding. This problem is frequently encountered by the operator and requires the continually changing angle of the torch tip and welding rod with respect to the surface as the weld progresses around the pipe. Fig. 32 illustrates these various positions.

The beginner should start with a pair of 6-inch lengths of 6-inch pipe. Bevel the edges to be joined, position the edges about ⅛ inch apart, and make four tack welds at appropriate locations around the pipe.

Place the tack welded lengths of pipe in a piece of 3-inch channel iron and then insert an iron bar inside the pipe and clamp the bar and the channel to the welding table in such a manner that the pipe is held in rigid position.

Start the weld at the bottom and weld around one side of the pipe to the top, then go back to the bottom and weld around to the top on the other side. The beginner should take particular care to study the various angles of the torch tip and the welding rod for making position welds, illustrated in Fig. 32.

It will be noted that the torch tip is first held in an inverted position with the flame pointing upward until one-quarter of the weld has been made to position E. The torch tip is then turned over from the inverted to the normal position. At position E, the bent

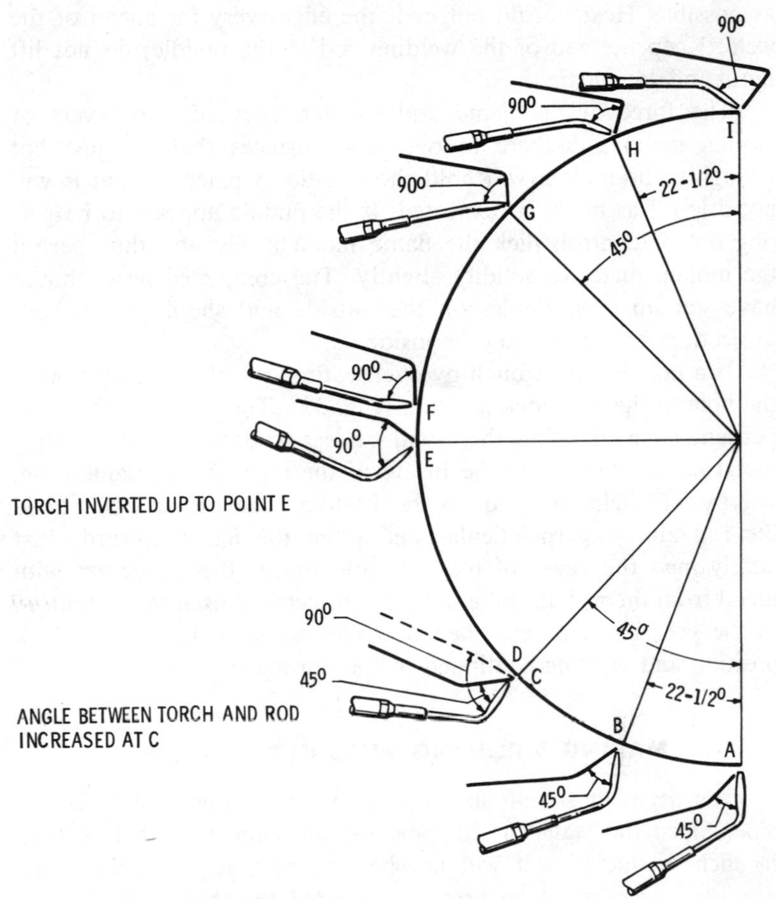

Fig. 32. Various welding positions.

part of the torch tip points slightly upward. It is almost horizontal at position F and as the weld progresses toward the top of the pipe, the angle of the head gradually approaches that of the normal downhand weld as at position I.

The operator should start at the bottom with the torch flame pointing almost straight up and with the welding rod inclined at a slight angle. First, close the bottom of the vee and then fill up the entire vee with one pass. Keep the molten puddle as small as possible. Heat but do not melt the edges very far ahead of the weld. Keep the end of the welding rod in the puddle; do not lift it out and do not stir.

The force of the flame and the tendency of thin layers of molten metal to adhere to overhead surfaces that are just hot enough to be molten will hold the puddle in place so that it will not drip off as might be expected. If the puddle appears to be getting out of control, flick the flame momentarily and thus permit the molten metal to solidify slightly. The completed weld should have smooth even ripples on the outside and should be without projections of "icicles" on the inside.

If a hole burns through by overheating, roll the pipe and bring the hole to the 3 o'clock position (Fig. 32). The edges will then be perpendicular and thus there will be less tendency for the molten metal to fall through to the inside of the pipe. At the same time, gravity will help to build up the bottom edge of the hole. Hold the rod almost perpendicular and point the flame upward. Just barely melt the edges of the hole and bridge them together with metal from the rod until the hole is completely closed in the bottom of the vee. The pipe can then be rolled back to the 1:30 o'clock position and welding continued in the normal manner.

MAKING A ROLLING WELD IN 2-INCH PIPE

The beginner should start by making a rolling weld in 2-inch pipe. Since the walls of the pipe of this diameter are less than 3⁄16 inch in thickness it will not be necessary to bevel the ends. The two pieces must be first tack welded together to hold them in the proper position. Place the two pieces of pipe in the valley of a 2-inch angle iron as in Fig. 33. Space the ends ⅛ inch apart and make tack welds at three points equally spaced around the pipe (a ⅛-inch welding rod is recommended).

The tack welds are made by heating a spot on the two edges and just as they melt and flow together introduce the rod into the puddle and melt off enough to make the tack weld. Hold the torch and the rod so that they form an angle of about 90° with each other. Each will thus be at an angle of about 45° with the surface of the pipe.

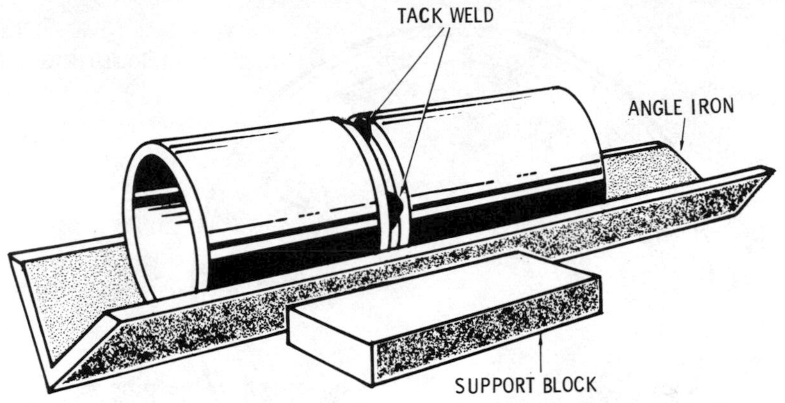

Courtesy Airco Welding Products

Fig. 33. Position the pipe sections in an angle iron and make the tack weld.

The tack welds should be about ¼ inch long. It is not necessary to try to get fusion through the entire thickness of the pipe wall as the tack weld will be remelted when the weld proper is made.

After the tack welds are completed, remove the angle iron, place the pipe on the rollers of the rig and proceed to make the weld proper. (Fig. 34). Begin the weld about halfway between any two tack welds. Roll the pipe so that the starting point is 1:30

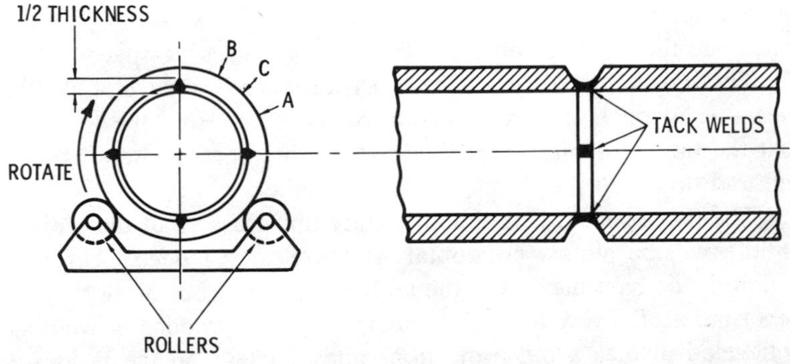

Fig. 34. Placing pipe on rollers. Welding begins at point C and moves to point B. Rotate pipe so that point B is at point A. Now weld from point A to point B and continue until pipe is completed.

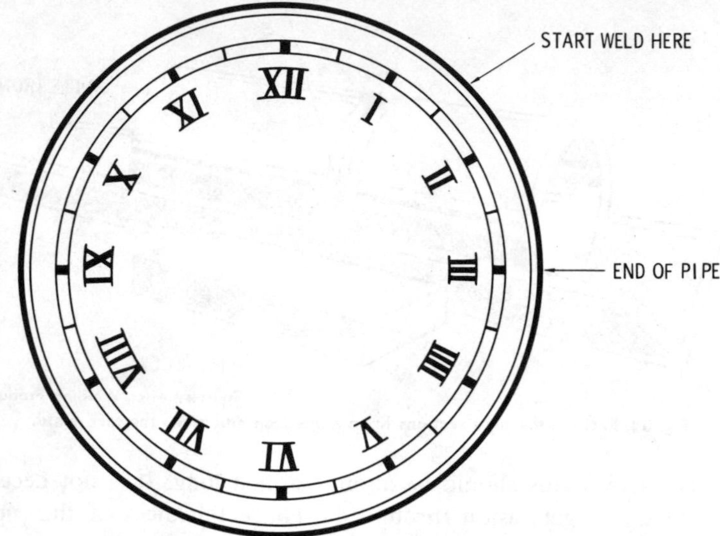

Fig. 35. The clock face and its relationship to pipe welding.

o'clock when comparing the end of the pipe with the face of a clock (Fig. 35).

Heat the general area of the weld with a circular motion of the flame adjusted to neutral, until a visible red color is obtained, gradually narrowing the circles to a small spot. Melt the two edges for a distance of about ⅝ inch. Just as the metal starts to flow, introduce the welding rod and fill up the intervening space.

The rod should be held so that it is practically vertical and is in the same plane as the cross section of the pipe. Hold the torch so that the tip and flame are approximately horizontal and so that the rod and tip are 90° with respect to each other.

The flame should be pointed slightly upward so that the molten puddle will be almost horizontal. Be sure to get deep, thorough fusion of the base metal and the added rod metal but do not try to penetrate all the way through. This is common practice in welding unbeveled pipe to avoid projections and "icicles" on the inside of the pipe.

After the first ⅝-inch section of pipe has been welded, roll the pipe clockwise ⅝ inch. Then start over as before and weld the next section. The welds should be reinforced by building them up about ⅜ inch higher than the surface of the pipe.

When a tack weld is reached, remelt it into the weld as the weld progresses; otherwise the metal under the shallow tack weld will not be thoroughly fused and a weak spot will result. Do not use it as a finished part of the weld.

The finished weld should have a smooth even appearance. If test specimens are required, they should be cut out from the point at which welding started and finished. With this in mind, it is recommended that the beginner mark the beginning and finishing point with a piece of soapstone.

MAKING A HORIZONTAL WELD IN 2-INCH PIPE

Take two more lengths of 2-inch pipe and tack weld them together in three places in the same manner as previously described for making a rolling weld, but this time stand them on end. The line of weld will then be horizontal. In order to prevent the tack welded pipes from being overturned during the welding operation, tack weld them to a plate as shown in Fig 36.

A bronze welding rod is recommended for making the tack welds since it will be easier to melt loose after the weld in the pipe has been completed. Also, it will not deface or mar the plate for future use.

Start the weld about half way between the tack welds. Hold the torch so that the portion of the tip assembly below the bend is

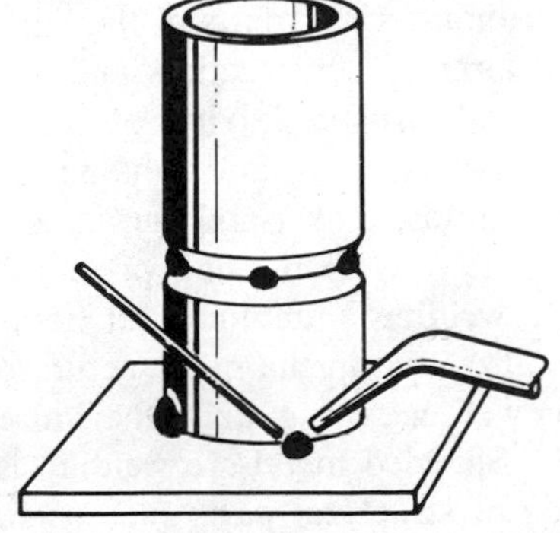

Fig. 36. Vertical pipe tack welded to a plate.

horizontal and at about 45° to the pipe surface. Heat an inch or a little less of the two edges, playing the heat in between them so that they melt and flow together. Go back and fill in with the rod, which should also be horizontal and, almost but not quite, parallel with the line of weld. Here, again, the rod and the torch tip should form an angle of about 90°.

When adding filler metal, lower the torch so that the flame points slightly upward. By doing this the force of the flame will help keep the molten puddle in place. Be sure that both the rod and base metal are melted but keep the puddle as small as possible. If it seems that the puddle is about to get out of control and run down the side of the pipe, flick the flame away momentarily and the metal will solidify slightly.

AIRCRAFT WELDING

Welding is widely used in the construction and repair of aircraft. In fact, most of the joining processes described in the previous chapters can be used in the aircraft industry for either fabrication or repair work. Many of these joining processes, of course, are strictly automatic or semiautomatic in nature and require production line application for their effective use. Some, such as oxyhydrogen welding, are generally restricted to specific welding problems (e.g. welding aluminum and the nonheat-treatable aluminum alloys). On the other hand, oxyacetylene or shielded metal-arc welding are almost unlimited in their application.

Oxyacetylene welding is adaptable to the welding of ferrous and non-ferrous parts of the lightest and the heaviest sections used in aircraft construction and repair. It is used for joining the several structural parts of the aircraft, such as wings, fuselage, control surfaces and landing gears; for engine jackets, exhaust manifolds, tanks, and surfacing tail skids and engine valves.

As was mentioned previously, oxyhydrogen welding is used for welding aluminum and nonheat-treatable aluminum alloys. It is also used in the manufacture of fuel tanks, oil and water tanks, as well as fairing, and other miscellaneous low stressed parts.

Shielded metal-arc welding is applied principally to the welding of structural parts such as fuselages, landing chassis and control surfaces. It has been used satisfactorily for metal as thin as .031 inch but only by the most skillful welders.

Resistance welding is used in the form of flash welding for simple butt joints, and spot welding for corrosion resistant steels and aluminum and aluminum alloys. Flash welded joints have been used in engine mounts and tubular axles for butt joints of tubes and forgings.

Other joining processes that have been used in the aircraft industry include pressure gas welding (landing gear and other heavy structures), TIG and MIG welding (high speed welding of almost any kind of joint), brazing (limited use in nonstructural applications), and soldering (restricted to connections in the electrical system and similar applications).

Light gauge metal in the form of tubing is still used in the construction of fuselages. Tubular structure was, at one time, quite widespread in the aircraft industry and a knowledge of the welding procedures used for its construction and repair can prove quite useful to the beginning welder.

This section will concentrate primarily on describing the oxyacetylene welding of light gauge metal, and particularly metal in the form of tubing used in aircraft construction. The fundamentals of oxyacetylene welding were fully described in Chapter 3 and 4, and will apply to the welding operations described in this chapter. The student is urged to review these chapters before proceeding further. The student of aircraft welding should also have a knowledge of the metals used, how they act in welding and their distinguishing characteristics.

TYPES OF JOINTS

There are basically five types of joints used in aircraft welding (Fig. 37). These joints are determined by tubular structure design or their repair function, and may be listed as follows:

1. Butt joint,
2. T-joint,
3. Lap joint,
4. Angle joint,
5. Composite joint.

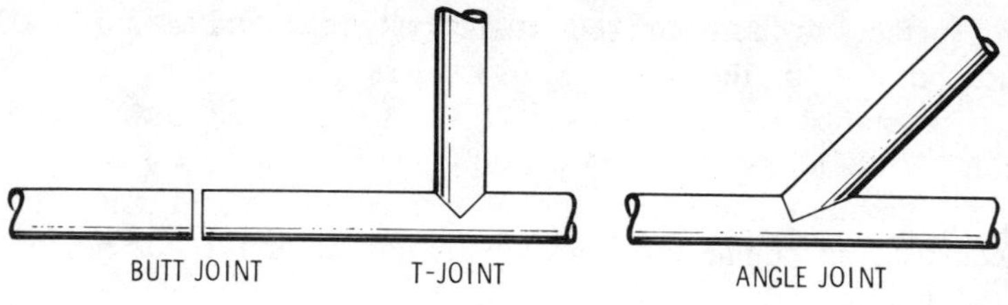

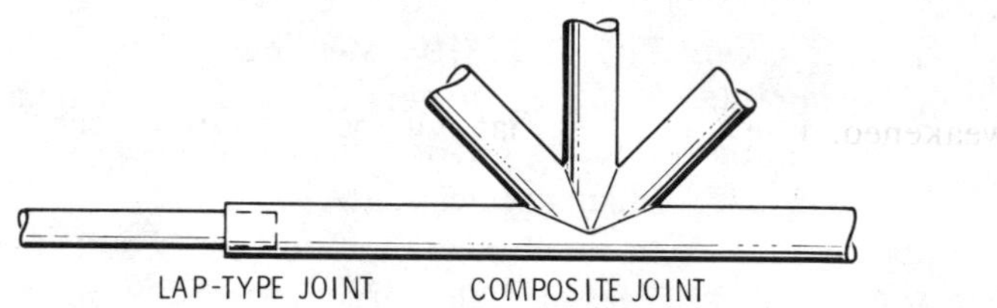

Fig. 37. Five types of joints used in aircraft welding.

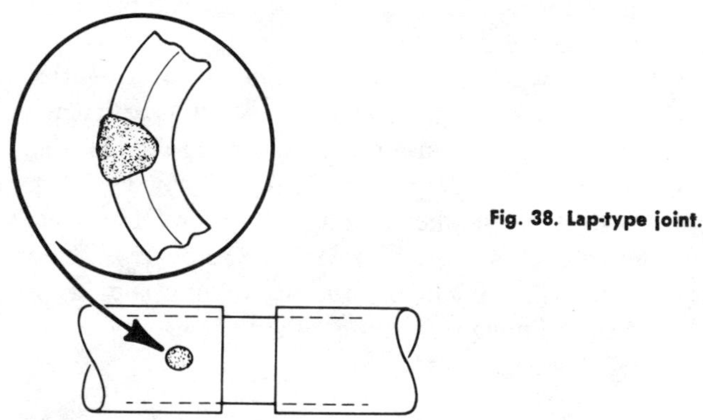

Fig. 38. Lap-type joint.

The *butt joint* and *T-joint* have already been described in the introductory chapters of this book, and will receive further attention in other sections of this chapter.

The *lap joint,* as it is used here, refers to the practice of using inserted tubing or overlapping sleeves for purposes of reinforcement (Fig. 38). A plug weld is used to join the two surfaces.

The *angle joint* is similar in design to the lateral connections used in pipe welding. The *composite joint* represents a combination of a T-joint and one or more angle joints.

OXYACETYLENE WELDING PROCEDURE

Once again, the fundamentals of oxyacetylene welding described in previous chapters will apply to aircraft welding. The importance of clean metal surfaces, proper fit-up, and correct identification of the base metal cannot be overstressed. Low-carbon steel welding rods are recommended for use with most alloy steel tubing. Nonferrous tubing (e.g. aluminum) require welding rods of similar composition to the base metal.

A soft neutral flame should be used. Over-heating and particularly burning should be avoided or the structure may be seriously weakened. Use a tip size that will permit rapid welding without skipping. The thickness of the metal will determine the size of the tip.

The beginner should start on straight horizontal seams, using short pieces of scrap tubing split longitudinally or small pieces of steel sheet equivalent in thickness to the tubing wall. It is recommended that the beginner follow the procedure outlined as follows:

1. Select a welding torch tip of suitable size,
2. Set the oxygen and acetylene regulators at correct working pressure with the welding torch valves open,
3. Light the welding torch and adjust the flame to neutral,
4. Incline the torch tip in the direction the weld is to progress. This will result in preheating the metal ahead of the puddle,
5. Apply the flame to the end of the seam, keeping the tip of the inner cone about 1/16 inch away from the metal,
6. Watch the metal closely and as the edges of the seam melt and fuse together, move the torch along the seam,
7. Continue along the seam until the weld is completed. The completed weld should be clamped in a vise and bent back and forth until it breaks through the weld. The weld fracture should then be examined for defects.

The beginner should take particular care in establishing and maintaining a proper welding speed. If the torch is moved too slowly too much metal will be melted and it will run through the bottom of the sheet, leaving a hole. On the other hand, if the torch is moved too quickly the edges will not be properly fused and the result will not be a weld at all. Experience will soon develop the correct speed. Almost unconsciously, the student will learn to alternately dip and raise the blowpipe slightly describing a series of overlapping circles and will thus control the pool of molten metal under the flame.

The beginner should also practice welding with a welding rod, because it will frequently be necessary to add filler metal during a welding operation. The end of the welding rod should be heated in the flame so that it is about ready to melt when the molten puddle begins to form at the seam. Insert the heated end of the rod in the molten puddle and move it from side to side. The motion of the rod is opposite to that of the torch tip and it will require a little practice to coordinate the two correctly.

MAKING BUTT WELDS IN TUBING

It is frequently necessary to make butt welds when welding tubing and the beginner should become familiar with the technique. Fig. 39 illustrates a typical butt weld in tubing and includes suggested positions for the tack welds.

The procedure for making butt welds in tubing is as follows:

1. Cut the ends of the tubing to be joined so that they are square,

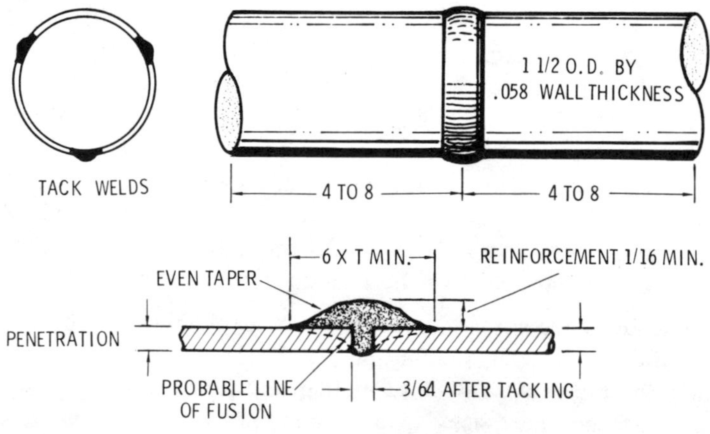

Fig. 39. A butt weld in tubing.

2. Clean inside and out beyond the width to be welded,
3. Make tack welds 3⁄16 inch to 1⁄4 inch long at three equidistant points on the tubing,
4. Begin welding and rotate the tubing as the weld progresses,
5. Melt away each tack weld as it is reached by the flame and fill the space with filler metal,
6. Continue along the seam until the point at which the weld began is reached.

The weld reinforcement for the tubing shown in Fig. 39 should be 1⁄16 inch minimum and 1⁄8 inch maximum. The width of the finished weld should be not less than six times the tube wall thickness. Make the contour of the weld so that it tapers gradually to the base metal on either side of the weld. The penetration should extend to the inside wall of the tube without melting holes through and without having excessive weld metal in the shape of icicles inside the tube.

As with any type of welding with the oxyacetylene flame, the surface must be thoroughly cleaned preparatory to welding. This can be done with either a stiff wire brush, a file, emery paper, or by grinding if necessary. Proper cleaning is very important.

The welding tip should be clean and the torch adjusted to a soft neutral flame. There should be even fusion of the welding rod and base metal with a steady movement of the torch, and removal of all oxides or dirt in the welding puddle.

The welding rod should be fed evenly into the puddle and the weld itself should be regular in appearance with either a smooth ripple or a level flat surface, according to the method of welding desired. Some firms prefer the ripple type of weld, while others find the flat weld more satisfactory.

Particular care should be taken that the weld metal is not deposited only on one side. This is a common fault with beginners, the tendency being to allow the metal to build up at one side of the joint, leaving a corresponding lack of reinforcement at the other side. The welder should occasionally check the flame during the operation to see that it is always neutral. Make certain that the tack welds are thoroughly melted out and refused with the base metal as the weld puddle advances.

Several butt weld specimens should be made under various conditions. The first one should be made with the tubing rotated in a horizontal position; another with the tubing vertical; a third at an angle; and a fourth should be made in a fixed horizontal position so that the welder will have not only horizontal welding, but vertical and overhead welding to perform.

In making a butt weld, the two members are spaced a distance equal to the wall thickness of the tubing. In this way, penetration to the inside wall of the tubing is secured.

After welding, tests for tensile strength and ductility should be made. Tensile tests should result in strengths of at least 50,000 lb. per square inch for mild-steel tubing and 80,000 lb. per square inch for chrome-molybdenum tubing. For ductility, tests strips cut from tubing across the weld can be tested by bending or breaking in a vise.

MAKING WELDS IN T-JOINTS

Some T-joints will be made with tubing of the same diameter; in others, the branch tubes will be of smaller diameter than the main tube (Fig. 40).

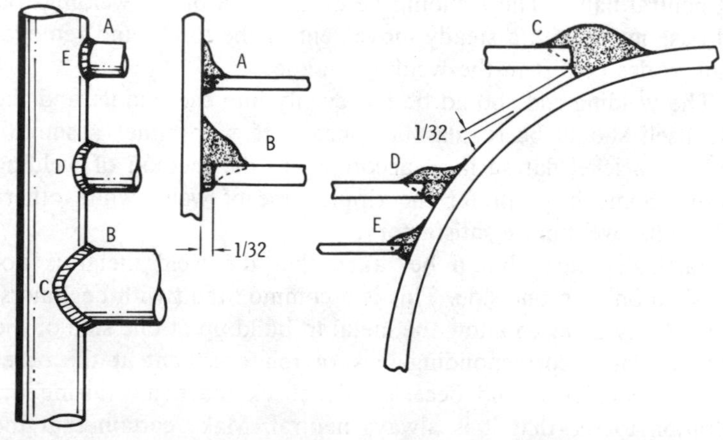

Fig. 40. Types of T-joints.

The penetration of the weld at the T-joint is shown in the cross sections in Fig. 40. The extent of instruction and types of joint depends on the actual work in which the plant is engaged or will be engaged. Lattice joints are very common in aircraft fuselages but their variety makes thorough instruction in all types somewhat difficult.

MAKING FILLET WELDS

The fillet weld shown in Fig. 41 is used to join a flat plate to a section of tubing. The edge of the plate abutting the tube should be square cut. In addition, all surfaces to be joined (including the tubing) must be thoroughly cleaned of oxides and any foreign matter.

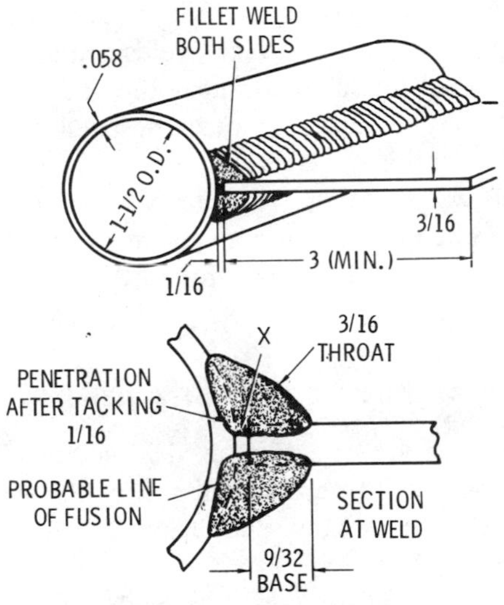

Fig. 41. A fillet weld made on tubing.

The tack welds should be 3/16 inch to 1/4 inch long on both sides near the ends of the plate. It is recommended that the tube be clamped to the table or bench so that welds can be made in a normal horizontal position. Penetration should be made to the root of the fillet, as shown in Fig. 39.

MAKING INSERT PLATE WELDS

Fig. 42 shows a typical insert plate weld. Note that the flat plate is inserted *through* the tubing rather than abutting on an exterior surface.

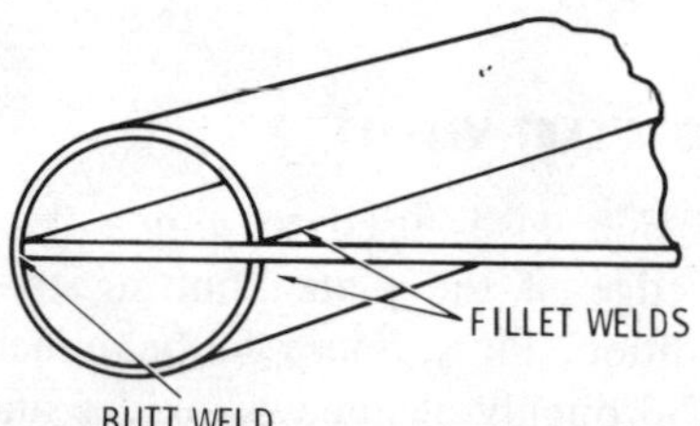

Fig. 42. Insert plate weld.

Once again, surfaces to be welded including the surface of the inside joints must be thoroughly clean and free from oxides and any foreign matter. Four tack welds should be made. Two should be positioned on each side near the end of the plate at the point of the fillet welds, and two near the ends of the plate at the point of the butt weld.

Three welds will be made (one butt weld and two fillet welds). The work should be placed in a convenient position for each of the three welds. Welding is to be done forward or backward, with the latter recommended.

MAKING COMPOSITE JOINTS

Composite joints are of a type used in actual production work in the factory and should be the final test of the beginner's ability in airplane welding. Side and end views of a typical composite joint are shown in Fig. 43.

The nature of the design and the position of the specimen for welding should be such that the beginner will be required to make welds in several unhandy positions, and also to improve welds on previously made joints.

The penetration and fusion of the weld metal with previously made weld metal are important details to be carried out by the welder in this operation. The several welds must be uniformly and

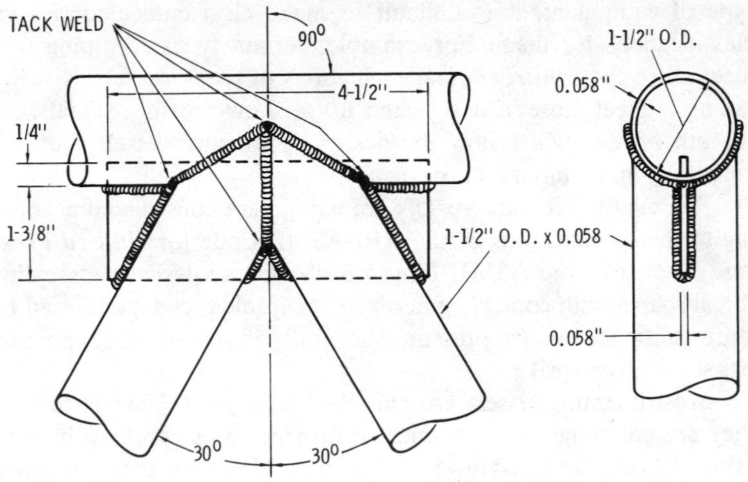

Fig. 43. A composite joint.

completely fused at all points of contact. The intersections of the composite joint must be carefully matched. Surfaces to be welded (including surfaces inside the joints) must be thoroughly cleaned of oxides and any foreign matter as previously specified.

Make tack welds 3⁄16 inch to ¼ inch long where indicated in Fig. 43. Fix the work to be welded in position, and insert plate vertical with principal member uppermost. The welding will be done forward or backward with the latter recommended. Penetration must be to the root of the fillet. The contour should taper gradually to the base metal on either side of the weld.

WELDING PRESSURE VESSELS

The dictionary defines a pressure vessel as *a container (a tank, boiler, or shell) subjected in use to disruptive pressure*. That is to say, pressure vessels are container that must withstand internal or external pressures of a type that could rupture their walls. Consequently, they require special consideration in their design and construction.

This definition for a pressure vessel is admittedly a loose one. Because of recent trends in the design and construction of this type of equipment, it is difficult to make clear-cut categories or classifications for them. For example, certain types of piping will meet all of the *construction* specifications of pressure vessels while failing to meet those of design and usage. Low-pressure vessels, on the other hand, will follow the design of pressure vessels, but not construction requirements or usage.

Essentially, pressure vessels are equipment constructed according to the specifications of the API-ASME Code for *Unfired Pressure Vessels* or the ASME *Power Boiler Code;* designed as cylindrical shapes with conical, spherical or ellipsoidal ends; and used to store substances under pressures generally in excess of 15 pounds per square inch (psi).

Most pressure vessels are classified as *unfired.* That is to say, they are not subjected in service to furnace gases resulting in high rates of heat transfer. However, boilers and certain types of other equipment in the pressure vessel categories are subject to such conditions and are therefore classified as *fired.*

Pressure vessels are subject to many service conditions. Consequently, the metals used in their construction must be capable of withstanding various types of stresses without failure. Examples of these stresses include extremes in pressure or temperature that necessitate the ability to resist metal or structural fatigue and other tendencies toward rupture or deterioration. The principal metals used in pressure vessel construction include low-alloy steels, clad steels, straight chromium steels, as well as several types of copper and copper alloys and other nonferrous metals (e.g. aluminum-manganese). Naturally, the correct identification of the base metal is important for it determines the selection and use of a suitable welding procedure.

Many of the welding processes described in the previous chapters can be used to construct or repair pressure vessels. Even brazing can be used to join thin-walled pressure vessels designed for relatively low pressure service conditions. Forge welding, seamless forging, and riveting, were the three basic methods of pressure vessel construction prior to the advent of arc welding. However, shielded metal-arc welding has gained in popularity since the 1930's until today it represents the predominant method used for constructing and repairing pressure vessels.

SHIELDED METAL-ARC WELDING OF PRESSURE VESSELS

Through extensive research remarkable strides have been made in the pressure vessel field by manufacturing electrodes capable of producing sound, clean highly ductile welds which easily conform to the established standards of the *ASME* and *API* codes.

The fundamentals of shielded metal-arc welding were fully described in Chapter 6 and should be reviewed before attempting to weld pressure vessels. It is obvious, however, that the design and construction specifications (particularly the type and thickness of metal) will determine the selection of electrodes specifically suited for this type of work.

ELECTRODES USED IN PRESSURE VESSEL WELDING

The electrodes used in the arc welding of pressure vessels should have the following characteristics for economical and efficient use:

1. *No porosity.* Electrodes must have sufficient fluxing and de-oxidizing elements in the coating to compensate for the variations in material and the different welding procedures being used.
2. *Good physical properties.* Dense sound deposits having high tensile strength and ductility are prime requisites of good electrodes for welding requiring Xray inspection.
3. *No undercutting.* The weld metal, as it is being deposited, should have a tendency to wash up or lap up the side walls of the groove. This point is extremely important as a large percentage of defects is due to slag inclusions or poor fusion along the side walls. Establish a good welding procedure and adhere to it. Use the proper size rod for the job with a nominal amount of weaving (but keep this down to a minimum).
4. *Good slag.* It is an absolute necessity in good welding practice to use electrodes that contain the exact type of slag for the protection of the weld metal while cooling, and of the correct consistency so that it will not run ahead of the arc.

5. *No objectionable fumes or smoke.* Practically all reliable welding electrode manufacturers have analyzed their electrodes to see if any harmful gases or fumes are given off during welding. Nothing beyond minute percentages has been found,
6. *High deposit efficiency.* Pressure vessel welding requires the use of electrodes capable of depositing metal at a high rate of speed with a minimum of spatter and ease of operation.
7. *Uniformity.* Uniform quality welds are assured under all normal conditions if electrodes are used that have proven their merit over a period of years.

With these electrode characteristics in mind and an understanding of similar variables in the metal or metal alloy being welded, the operator should be able to produce sound clean welds of acceptable strength.

SURFACE PREPARATIONS

Pressure vessels consist basically of shell plates and heads. The flat plates are shaped to their specified contours under pressure by cold or hot forming. This may be done on presses or bending rolls designed for this purpose. Nozzles and other connections are located at positions determined by the pressure vessel design.

All grooves for welding must be straight and as near perfectly uniform as possible. One means of achieving this is by machining the grooves on a planer. The operator should be cautioned that the cutting tools for planing frequently have to be reground. In order to keep the design of the groove as originally selected, gauges (corresponding to the type of groove being used) should be made for checking the tools after grinding. If this is not done, variations in the amount of metal deposited will result which, in turn, will cause the welding costs to vary.

Two types of grooves used in welding pressure vessels are illustrated in Figs. 44 and 45. Fig. 44 shows a type of groove which requires less metal in filling. This groove results in lower electrode and labor costs but requires a backing plate. The type of groove shown in Fig. 45 permits the elimination of a backing strip and

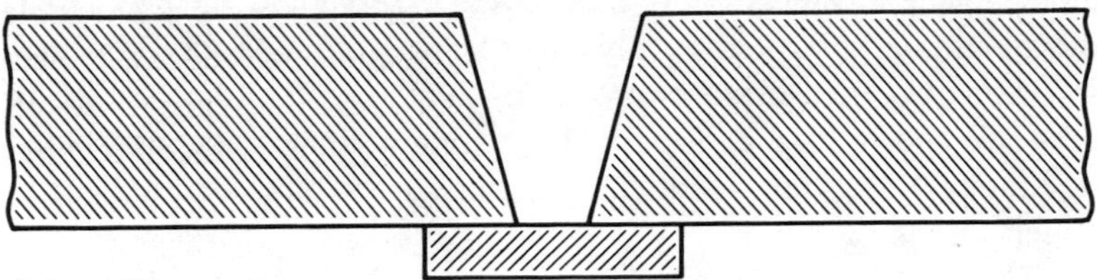

Fig. 44. Pressure vessel groove with backing plate.

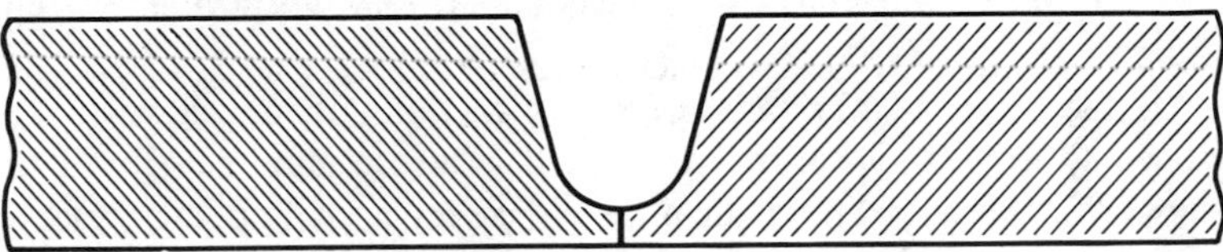

Fig. 45. Pressure vessel groove without backing plate.

provides adequate room for cleaning the spatter of weld metal in the bottom of the groove, thereby insuring better workmanship.

Some designs for vessels specify, in order to eliminate distortion, that the longitudinal seams be machined to a double bevel and the head seams to a single bevel. In any event, the total number of seams in a design should be kept to a minimum. After the machining operation is finished, the shell should be crimped and rolled to size preparatory to tacking and welding.

The metal surfaces should be thoroughly cleaned so that all contaminants (e.g. grease, dirt, oil, etc.) have been removed prior to welding. All scale, slivers, laps, and other surface irregularities, should also be removed. This is particularly important for the interior walls. The cleaning operation may have to be performed several times before welding begins.

WELDING PROCEDURE

All standard shielded metal-arc welding procedures described in Chapter 6 will pertain to the welding of pressure vessels. The pressure vessel seams will have to be tack welded together prior to actual welding. From a practical standpoint (and as a consistent factor of safety), it is recommended that more tack welds be used than is generally necessary. Moreover, care should be taken that the welds are deposited evenly in order to eliminate any unnecessary chipping. If test plates are used, they may be either tacked to the shell and welded as a part of the longitudinal seam, or welded separately using the identical welding procedure as applied to boiler drums.

The use of automatic turning fixtures, which are usually motor driven, is a distinct necessity for the production of uniform weld deposits. The speed or movements of these fixtures must be adjustable so that the point to be welded is constantly before the operator.

If the vessel is fabricated of heavy sections, it may have to be preheated to 300° F before welding. It is also a good safeguard to stress relieve every 1 inch in plate thickness as it is being welded in order to eliminate any possibility of cracking.

The first step in welding pressure vessels should be to deposit full width layers, using sufficient current to keep the slag back of the crater and to be sure of complete fusion along the side walls. The arc length, which is measured by the voltage across the arc while welding, should be held to the correct length so as to keep the coating free of the crater.

Careful cleaning of the layers as they are deposited is important, paying strict attention to the side walls to see that no slag remains along the edges of the layers. This procedure should be repeated until the seam is completely filled and a small reinforcement added.

The next step is to chip out the bottom layer paying careful attention to ensure chipping into good clean metal, and as an added safeguard, the use of depth gauges for checking is advisable. The inside of the shell is then welded and cleaned thoroughly of all spatter as the small particles will show on the radiographs. Machine or grind off the finishing bead so that the films can be properly interpreted. The seams should now be ready for X raying.

WELD INSPECTION

Pressure vessel welds can be tested in a number of ways including:

1. Visual,
2. The magnetic particle method,
3. Ultrasonic testing,
4. Radiography.

By way of example, radiography consists of the X raying of welds for uniformity and good workmanship (Fig. 46). The longitudinal seams are X rayed first according to the required formula. The radiograph is then inspected, any defects repaired, and the heads of the vessel fitted and welded in accordance with the procedure used for welding the longitudinal seams. The head

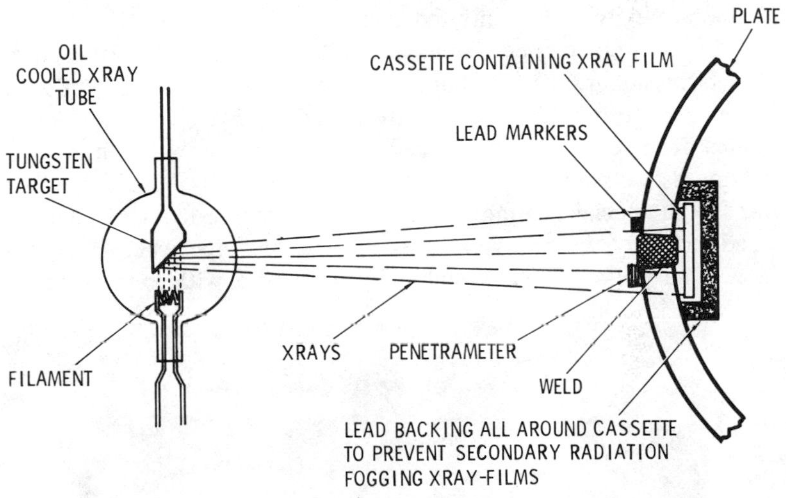

Fig. 46. Exposure arrangement for weld inspection.

seams are then X rayed, any defects repaired, and the final radiographs approved after which the manhole saddle and nozzles are fitted and welded to the vessel.

The vessel should be stress-relieved in the usual manner, i.e. at 1200° F, and held there for one hour for each inch of thickness and slowly cooled in a furnace. The next step is the prescribed hydrostatic test after which the vessel is finished as far as fabrication is concerned.

PLASTIC PIPE WELDING

Thermoplastic materials are gaining increasing importance in the manufacture of pipes and tubing. Their popularity is due to several factors including: (1) a high chemical resistance (particularly to corrosion), and (2) relative ease of assembly and pipe laying.

Thermoplastics are welded with a technique similar to the one used in gas welding (see OXYACETYLENE WELDING, OXYHYDROGEN WELDING, etc.). The plastic filler rod should match the material of the base plastic. Thus, if a polyethylene plastic pipe is being welded, a polyethylene plastic filler rod should be used in the joining procedure.

Plastic pipe and tubing are usually brought to the fusion temperature by a concentrated stream of hot gas or air.

Special units can be purchased for butt welding of plastic pipes 1 inch through 4 inches in diameter. These units can be used in both installation and field work and are of modular design. Heat is obtained from either an air-heated element in combination with a warm gas welding system or a directly electrically heated element used with special heating transformers. A spring-loaded system is used to apply the necessary pressure for the pipe welding operation. The pressure is adjustable by a spindle with vernier scale.

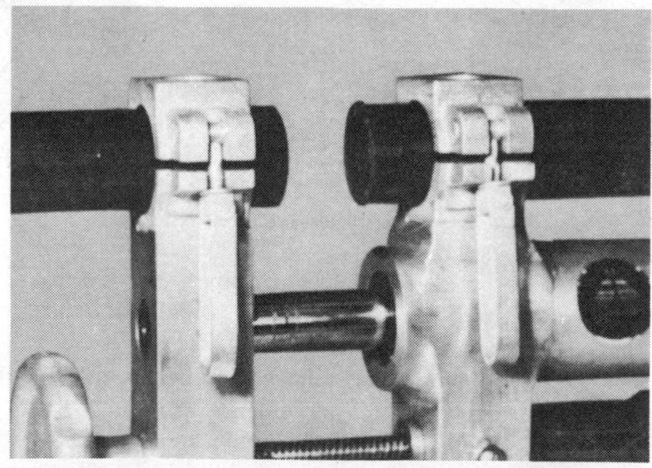

Courtesy MG Welding Products

Fig. 47. Holding the pipe ends in position with clamps.

Courtesy MG Welding Products

Fig. 48. Planing the pipe ends.

Courtesy MG Welding Products

Fig. 49. Heating the pipe ends with an air-heated element.

The four steps in butt welding plastic pipe of this diameter are illustrated in Figs. 47 to 50. Basically they consist of:

1. Clamping in position,
2. Planing the edges,

3. Applying the heat,
4. Applying the pressure.

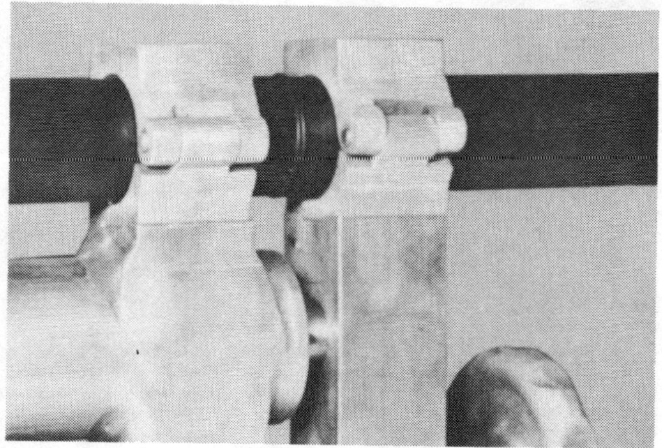

Courtesy MG Welding Products

Fig. 50. The heated pipe edges are fused together under pressure.

HOT-AIR WELDING

Fixed position pipe welds can be made with hot-air welding torches (Fig. 51). These torches may be used for tack welding, hand welding, or high-speed hand welding, depending on the particular requirements of the job (see Chapter 24, WELDING AND SOLDERING PLASTICS for details on these plastic welding techniques).

A typical hot-air welding system (Fig. 52) will consist of the following basic components and supplies:

1. Hot-air welding torch,
2. Control unit,
3. Air blower,
4. Regulating unit,
5. Pressure reducer,
6. Compressed air supply,
7. Welding gas supply,
8. Plastic filler material,
9. Plastic hose, hose clips and distributor piece.

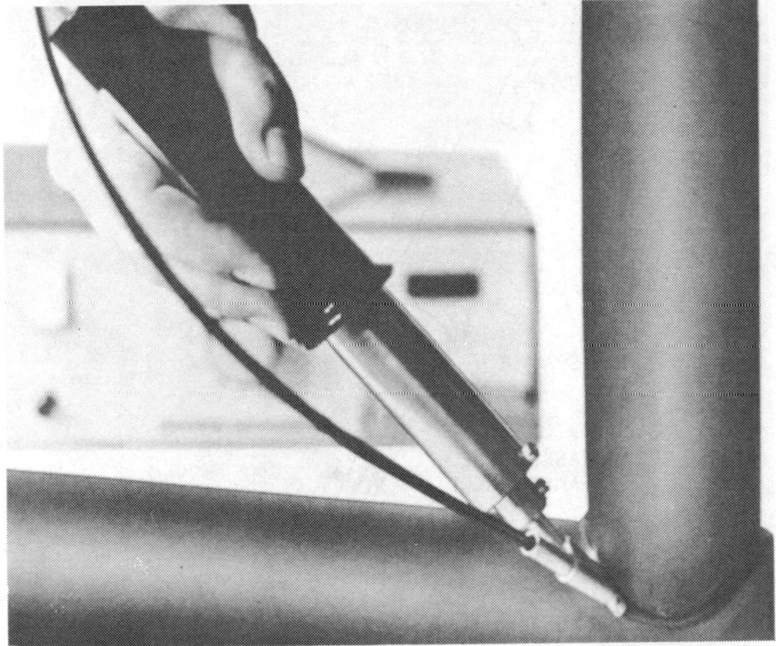

Courtesy MG Welding Products

Fig. 51. Hot-air welding torch being used to make a fixed position pipe weld.

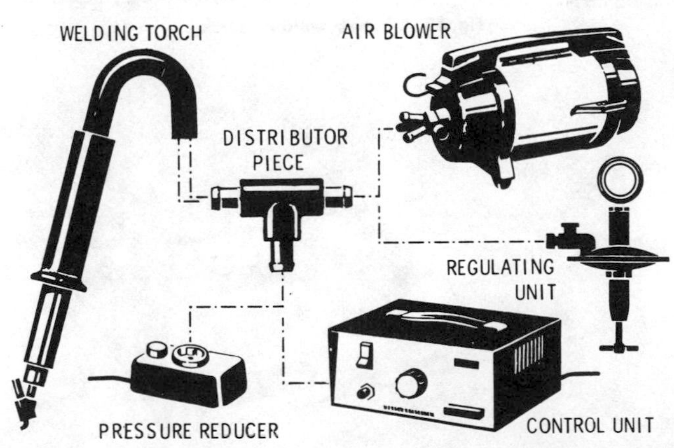

Courtesy MG Welding Products

Fig. 52. Hot-air welding diagram of major components.

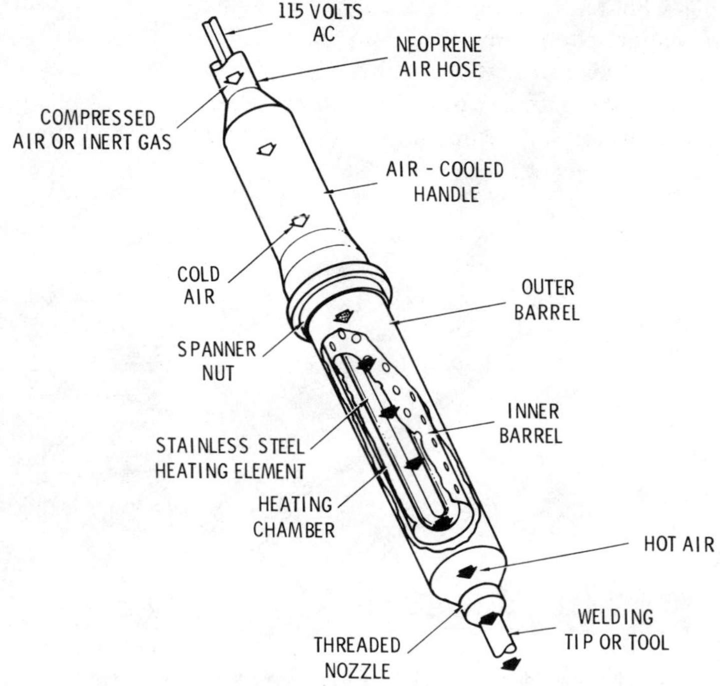

Courtesy Kamweld Products Co., Inc.

Fig. 53. A hot-air welding torch.

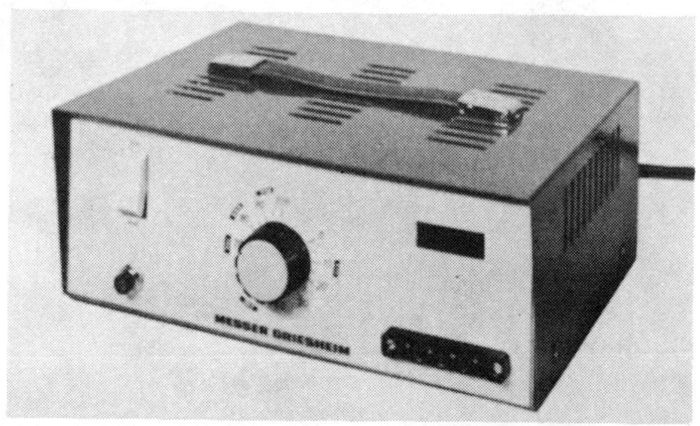

Courtesy MG Welding Products

Fig. 54. Control unit for a hot-air welding system.

The hot-air welding torch (Fig. 53) obtains its heat from a combination of compressed air (or inert gas) and electricity. The compressed air (or inert gas) enters the torch through a hose connected to the handle and immediately passes into a heating chamber where its temperature is considerably raised by an electrically heated stainless-steel heating element. The hot air subsequently passes out of the heating chamber and into the welding tip (or tool) where it heats simultaneously both the base plastic and the filler plastic. Both become fused together in an homogeneous weld.

The air supply can be obtained from either a compressed air line or a compressed air cylinder. An air blower is used to deliver air simultaneously to several torches (usually up to a maximum of three or four torches). Nitrogen (an inert gas) is sometimes used instead of compressed air for welding thermoplastics that are susceptible to oxidation.

A control unit (Fig. 54) compensates for variations in the air pressure and electrical voltage and holds the temperature setting for the hot gas stream constant within a tolerance of ±3° C.

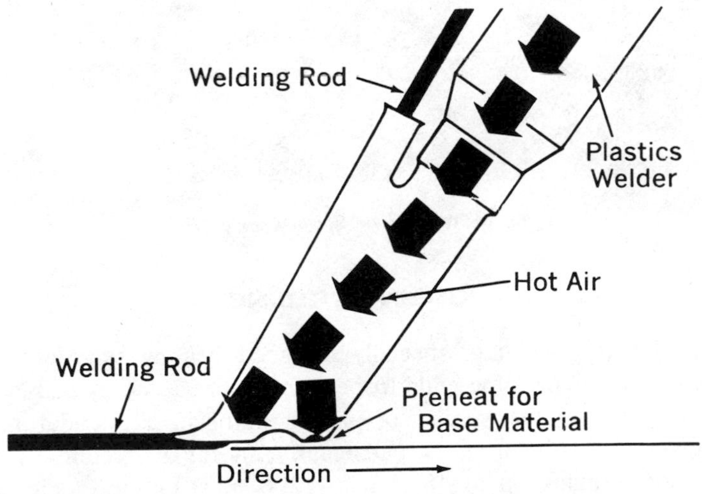

Courtesy Kamweld Products Co., Inc.

Fig. 55. Hot-air welding torch equipped with a device for feeding the welding rod to the surface.

The plastic filler material is available in several different forms (round rod, triangular rod, or flat strip) and must match the composition of the base material. Filler plastic can be added separately by hand (tack welding and hand welding) or by means of a special device attached to the torch (high-speed hand welding) (Fig. 55).

Courtesy MG Welding Products

Fig. 56. Hot-ring welding sewage pipes.

HOT-RING WELDING

In hot-ring welding, specially designed heating rings are inserted between the pipe ends to be joined. These are available in various sizes for the welding of plastic piping in all standardized diameters. This method is particularly useful in welding pipe diameters ranging up to 48 inches (1200 mm) or more. Fig. 56 illustrates the on-site hot-ring welding of sewage pipes of 25-inch diameter.

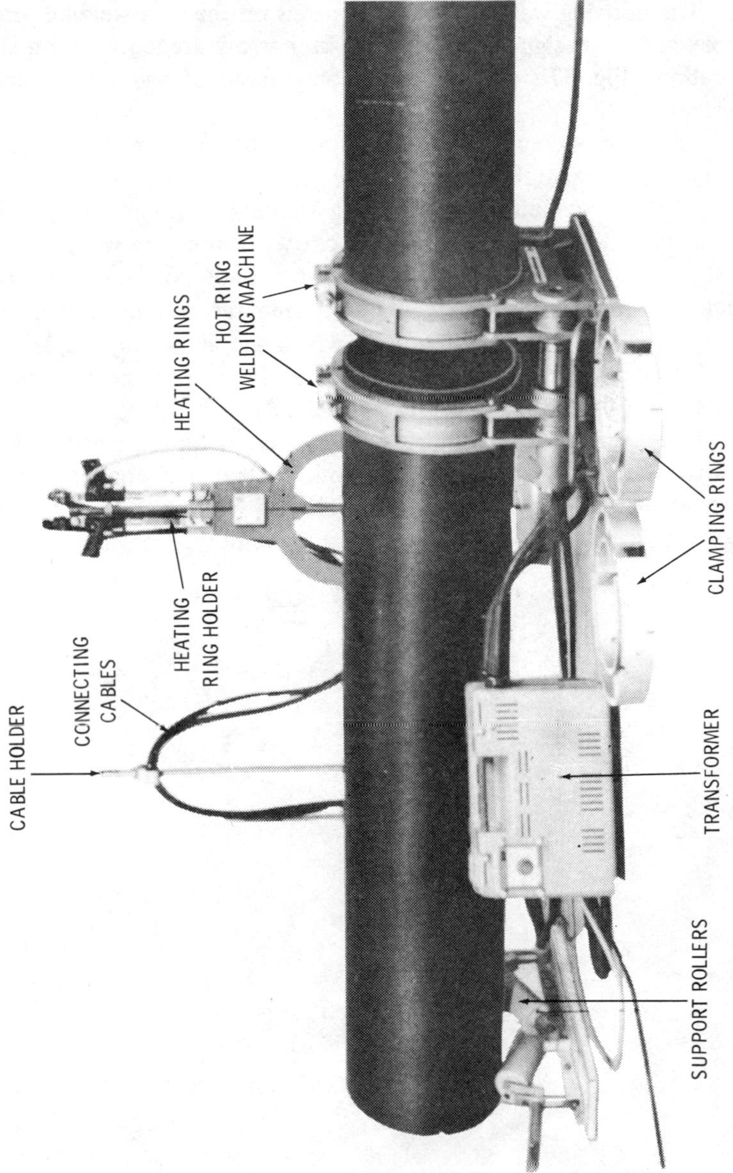

Courtesy MG Welding Products

Fig. 57. Hot-ring welding system showing major components.

The hot-ring welding system operates on the butt-welding principle and is designed to be used in narrow trenches at on-site locations. Fig. 57 shows the basic components of this pipe welding system.

Each components plays an important role in the overall system. For example, the clamping rings are designed to ensure that the pipe remains round during welding. Without these rings some distortion of the pipe might occur when pressure was applied.

The support rollers are used to facilitate positioning (alignment) and movement of the pipes during the welding operation. Two pairs of rollers are required which are hand adjustable.

The heating ring is probably one of the more important components in the hot-ring welding system. It is designed to give an even temperature to the entire surface area of the pipe ends being joined. The rings are made of stainless steel and coated with a special substance to prevent the plastic from adhering to the surface during the heating cycle of the welding operation.

Courtesy MG Welding Products

Fig. 58. Clamping the pipes in position for hot-ring welding.

The required welding temperature must be set on the transformer before the welding operation begins. The temperature of the heating ring should then be checked before each weld is made.

Courtesy MG Welding Products

Fig. 59. Inserting the heating ring between pipe ends.

The procedure for the hot-ring welding of plastic pipe is essentially as follows:

1. Place the pipe sections on the rollers,
2. Adjust the rollers until the pipe ends are level and aligned,
3. Clamp the pipe ends in positions in the hot-ring welding machine (Fig. 58),
4. Use the hydraulic unit to press the pipe ends against the planing tool,
5. Plane the pipe ends smooth and parallel,
6. Remove the planing tool,
7. Insert the heating ring between the pipe ends (Fig. 59),

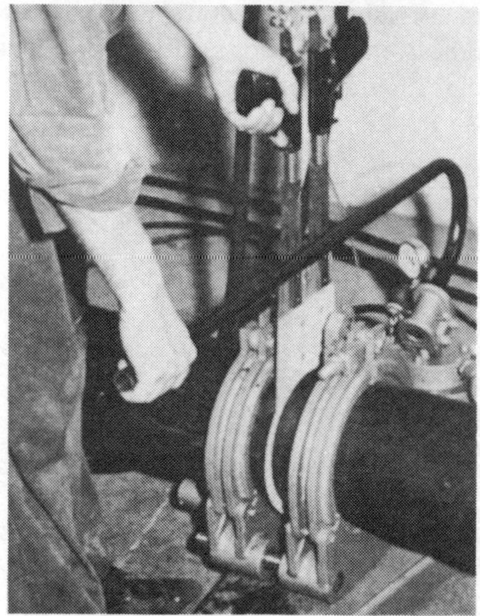

Courtesy MG Welding Products

Fig. 60. Pressing the pipe ends against the heating element.

Courtesy MG Welding Products

Fig. 61. Pressing the heated ends together.

8. Use the hydraulic unit to press the pipe ends against the heating ring with the required pressure (Fig. 60),
9. As soon as the fusion point has been reached, release the pressure and remove the heating ring,

Courtesy MG Welding Products

Fig. 62. The finished weld seam.

10. Use the hydraulic unit to press the heated pipe ends together at a suitable pressure until fused (Fig. 61),
11. As soon as the weld has cooled, release the hydraulic pressure and remove the clamps from the pipe. The completed weld seam should be strong, tight and protrude with only minimum upset (Fig. 62). No finishing work should be necessary.

CHAPTER 18

Welding Tool and Die Steels

Tool and die steels are generally medium- or high-carbon steels containing 0.40% or more of carbon to which varying percentages of alloying elements have been added in order to create specific characteristics. These steels are noted for their hardness and their ability to maintain that hardness under conditions of high temperature. This resistance to the effect of high temperatures is referred to as *red hardness*. Among the tool and die steels, the high-speed tool steels are characterized by high red hardness, whereas water-hardening tool steels have low red hardness.

In addition to hardness, the tool and die steels are required to have such additional characteristics as toughness, heat resistance, corrosion resistance, resistance to wear, and other properties in a variety of combinations that will enable the part to withstand severe service conditions. The properties of a particular tool and die steel will be determined in part by the percentage of carbon and alloying elements (as well as the *types* of alloying elements) in its composition. Perhaps even more important is the means by which it was hardened. It is extremely important that tool and die steels be correctly quenched during the hardening treatment or the correct degree of hardness will not be obtained. For example, extreme brittleness will result from quenching too rapidly; insufficient hardness from quenching too slowly. Quenching, then, is an important factor in the degree of hardness that is imparted to a tool and die steel.

Quenching may be defined as the rapid cooling of a piece of steel from a high temperature (in this case a bright red heat) by immersing it in water or oil, or by subjecting it to some other cooling method.

Not every tool and die steel requires the same rate of cooling. Some require faster rates of cooling while others seem to require slower ones in order to obtain the desired degree of hardness in the steel. Therefore, several cooling media (water, oil, air, etc.) have been devised to meet these varying conditions.

After quenching, the steel will have reached a fully hardened state. Unfortunately, it will also be brittle, lacking in strength and ductility, and subject to cracking and fracture. It is therefore necessary to reduce the hardening stresses and increase those properties that will provide some resilience to severe service conditions. The procedure for doing this is referred to as *tempering*.

It should be pointed out that during quenching the steel is not brought down to room temperature. The quenching procedure will be stopped as soon as the steel has cooled to the 100° F to 200° F range. Air-hardening tool steels are cooled to lower temperatures (100°F to 150° F) than the oil- or water hardening tool steels (150° F to 200° F).

Tempering should follow immediately after the quenching operation has stopped. Essentially, tempering consists of reheating the steel to a temperature recommended for specific characteristics (hardness, toughness, wear resistance, etc.) required by the purpose for which the steel is intended. The steel is then allowed to cool to room temperature.

The preceding paragraphs have presented a very brief description of the complex process by which the hardness and related properties of tool and die steels are obtained. Obviously the heat treatment forms a very important part of this process. Therefore, it is essential to realize that the welding heat and technique can adversely affect the base metal by reversing or otherwise weakening the heat treatment used to produce the tool and die steel.

WELDING TOOL AND DIE STEELS

Since welding tool and die steels is a highly specialized form of welding requiring extensive knowledge of these various steels, it is recommended that an apprenticeship period first be served with a welder who possesses the necessary experience and skills. Beyond the initial period of instruction, it is simply a matter of gaining experience. In any event, it might be wise at this point to provide a number of recommendations that will aid the welder when working with these tool and die steels. These recommendations are as follows:

1. Be certain to correctly identify the tool and die steel being welded. Sometimes the steel will be identified by trade name or special symbols. The manufacturer can then provide the necessary data. For example, he should be able to forward the information that his particular AISI-SAE 01 steel is an oil-hardening tool steel having a quench range of 1450° F to 1500° F (in oil) and a tempering range of 325° F to 500° F. Without identifying information, the welder will have to experiment on a piece of scrap similar in composition or make an educated guess.
2. Never heat tool and die steels too quickly. Fast heating will tend to make the base metal crack.
3. The surface must be thoroughly cleaned before welding. In doing so, it is particularly important not to scratch or nick the surface. Otherwise, cracks or fractures may form at these points.
4. Do not allow the finished weld to cool too rapidly. Rapid cooling usually results in weld cracking. A correct preheating procedure and immediate stress relieving will reduce this tendency sharply.
5. Do not overheat the surface of the tool and die steel. Overheating can result in decarburization (i.e. the loss of carbon).

As the welder gains experience in working with the tool and die steels, he will be able to add considerably to this list of welding recommendations. At this point, they are offered as broad general recommendations and should be considered as only the first of many steps toward acquiring satisfactory skill in this welding specialty.

TYPES OF TOOL AND DIE STEELS

There are actually hundreds of trade names for tool and die steels in use today. Often a single type of steel will be represented by two or more of these trade names. At first this may seem to be terribly confusing, but identifying the name or the composition of the base metal is not of primary importance in welding tools and dies. In any event, it would be unrealistic to expect to find an electrode or welding rod that would match the analysis of each and every base metal. The sheer number of different tool and die steels makes this impractical for manufacturers. Fortunately, matching the analysis of the electrode or welding rod to that of the base metal is not absolutely essential in welding tools and dies. However, it is important to match the electrode or welding rod as closely as possible to the special treatment used in hardening the particular steel.

The *Society of Automotive Engineers* (SAE) and the *American Iron and Steel Institute* (AISI) have classified the tool and die steels into a number of different categories based on such factors as the method of quenching, specific characteristics (e.g. shock resistance) or its intended application. Each category is denoted by a letter symbol followed by a number indicating the specific chemical composition of the steel. These various categories of tool and die steels with their particular letter symbol are as follows:

1. Water-hardening tool steels (W),
2. Oil-hardening cold-work tool steels (O),
3. Air-hardening medium-alloy cold-work tool steels (A),
4. High-carbon, high-chromium cold-work tool steels (D),
5. Hot-working tool steels (H),
6. Tungsten-based high-speed tool steels (T),
7. Molybdenum-based high-speed tool steels (M),
8. Shock-resistant tool steels (S),
9. Carbon-tungsten special-purpose tool steels (F),
10. Low-alloy special-purpose tool steels (L),
11. Low-carbon mold steels (P).

In most cases, the letter symbol apparently represents the first letter in the name of the category. The numeral suffix denotes sub-categories and not the percentages in the chemical composition of the steel. For example, Ol denotes an oil-hardening cold-work tool steel having a chemical composition of 0.90% carbon, 1.00% manganese, 0.50% chromium, and 2.50% tungsten. O2 will be the designation for an oil-hardening steel of slightly different composition (in this case 0.90% carbon, and 1.60% manganese).

This chapter will offer a brief description of the various categories of tool and die steels. More detailed information can be obtained from manufacturers of electrodes and welding rods used in the welding of tool and die steels.

Water-hardening tool steels (W) are divided into three types depending on the principal alloying element. These three types are:

1. Straight carbon steels,
2. Carbon-chromium steels,
3. Carbon-vanadium steels.

All three types also contain approximately 0.25% each of manganese and silicon. The chromium and vanadium are added to increase hardness, wear resistance, and toughness.

Water-hardening tool steels have a greater tendency to crack or change shape than other types of tool steels. They also have an equally high tendency to warp. Moreover, their low red hardness restricts them to cold-work applications. On the other hand, the water-hardening tool steels possess relatively better weldability and machinability than other types of tool and die steels.

Among other applications, water-hardening tool steels are used in the manufacture of woodworking tools, razors, shears, and other tools that require a fine cutting edge.

Oil-hardening tool steels (O) are divided into six basic categories. These six categories of oil-hardening tool steels are:

1. Tungsten steels,
2. Manganese steels,
3. Low-manganese steels,
4. Chromium steels,
5. Chromium-vanadium steels,
6. Silicon steels.

The edge keeness of the oil-hardening tool steels is somewhat less than that of the water-hardening type. However, there is less of a tendency to crack and less change of shape. These steels do not have the shock resistant core of the water-hardening tool steels and therefore do not have a high resistance to impact.

Using oil as a quenching medium prevents the type of build up of excessive stresses one expects to find in the water-hardening tool steels. Consequently, oil-hardening tool steels are most often used in the production of tools and dies having sharp or thin edges that must withstand severe service conditions without cracking or fracturing.

Hot-working tool steels (H) are divided into three basic categories. These are:

1. Chromium-type steels,
2. Tungsten-type steels,
3. Molybdenum-type steels.

Each category is distinguished by the principal alloying element. Some authorities add a fourth category which consists of approximately equal percentages of tungsten and chromium.

These tool and die steels are very tough and possess a strong resistance to abrasive and erosive action. They possess a high impact strength with a concomitant low tendency toward cracking.

The major application of hot-working tool steels is in the manufacture of mandrels, hot shears, hot punches, and hot forging dies. As a result of the last application, these steels are frequently referred to as hot-working *die* steels.

The air-hardening medium-alloy cold-work tool steels (A) possess approximately 1.0% carbon, 5.0% chromium and varying combinations of vanadium, molybdenum, and manganese. Distortion remains at a minimum during heat treatment. Their wear resistance is higher than that of the oil-hardening tool steels. Consequently, these tool steels are frequently used in the manufacture of complex intricate dies.

The high-carbon, high-chromium cold-work tool steels (D) can also be regarded as air-hardening types and are frequently combined with the previously mentioned medium-alloy variety to form a single category. Their composition is similar in all respects except for chromium content which commonly reaches 11.0% to 15.0%. The carbon content is also slightly higher at 1.25%.

These steels have a greater tendency to crack and are not as tough as the medium-alloy type. They are very difficult to machine and great care must be taken when grinding their surfaces lest cracking occur. They are used in the manufacture of parts that must possess a strong resistance to abrasive action.

The *high-speed tool steels* can be divided into tungsten-based types (T) and molybdenum based types (M), and find their principal application in parts produced for high-speed cutting operations. Most high-speed tool steels, today, are of the molybdenum based type. They are characterized by a high red hardness, strong resistance to wear, shock, and vibration, and a long service life.

Shock-resisting tool steels (S) were developed, as the name suggests, for applications that must resist strong impact. They are therefore used in the manufacture of chisels, hammers, riveting tools, crushers, and similar types of tools and equipment. These steels possess good machinability.

The various categories of *special-purpose tool steels* (including the *low-carbon mold steels)* represent a miscellaneous category of tool steels specially developed for specific purposes.

WELDING TOOLS AND DIES

The reason for welding tool and die steels is basically an economic one. It is much less expensive to restore broken or cracked parts by welding than to replace them. This is particularly true of the tool and die steels which must function under severe service conditions. By reducing the downtime of a piece of equipment, the subsequent loss of production is also minimized.

The welding of tools and dies is a specialty that finds application in the following areas:

1. Correcting mistakes caused by design or production error,
2. Altering contours of a piece to conform to design changes,
3. Composite fabrication,
4. Repairing damaged surfaces,
5. Building up worn surfaces.

Some of these applications are illustrated in Fig. 1. There would also appear to be some overlapping with the functions of surfacing

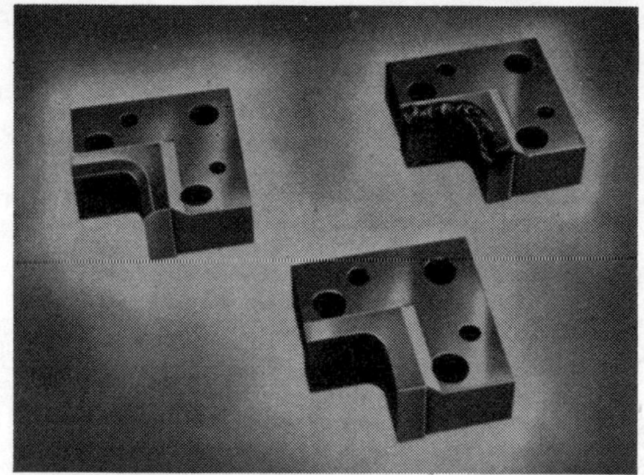

(A) Changing the design.

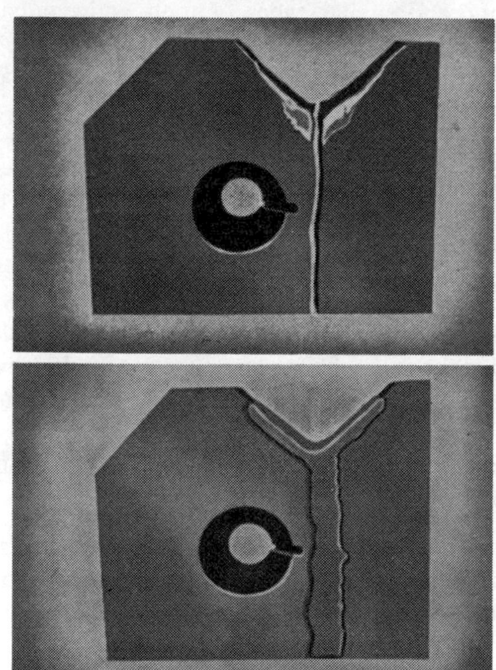

(B) Repairing a fracture.

Courtesy Welding Equipment and Supply Co.

Fig. 1. Changing or repairing mold designs.

and metallizing. This is true to a certain extent in the building up of worn surfaces.

TOOL AND DIE WELDING PROCESSES

A number of different welding and joining processes can be used when working with the tool and die steels. Those that have been used with varying degrees of success include:

1. Shielded metal-arc welding,
2. Oxyacetylene welding,
3. Submerged-arc welding,
4. Atomic-hydrogen welding,
5. Inert-gas-shielded arc welding,
6. Silver brazing.

This chapter will concentrate in the sections that follow, on a description of the shielded metal-arc welding process as it is used in welding tool and die steels. Some additional comments will also be made concerning silver brazing.

The oxyacetylene welding process can be used to weld tool and die steels with an appropriate welding rod and a torch adjusted for a carburizing flame (one with a slight excess of acetylene). Medium-carbon and high-carbon welding rods are recommended. The standard post-welding heat treatment should be employed to minimize the effects of the welding heat.

Submerged-arc welding has been used for building up worn surfaces. A suitable flux and welding rod are necessary and the deposits are applied with standard submerged-arc welding procedures. Neither the preheating nor the post-welding heat treatment is necessary with this welding process.

Atomic-hydrogen welding may be used to repair damaged surfaces of forging dies. The standard welding procedure is used with a suitable filler metal.

Argon is recommended as the shielding gas for the inert-gas-shielded arc welding of tool and die steels. Welding equipment manufacturers will generally recommend a suitable filler metal wire and the appropriate machine settings. The standard welding procedure is used to make the deposits.

TOOL AND DIE ARC WELDING PROCEDURE

The procedure for welding tool and die steels with the shielded metal-arc welding process is basically as follows:

1. Identify the problem,
2. Identify the base metal,
3. Select a suitable electrode,
4. Prepare the surface for welding,
5. Adjust the power source for the proper amperage,
6. Start the arc and begin welding,
7. Peen and clean each bead before depositing the next layer,
8. Stress-relieve the weld deposit,
9. Clean the surface.

Identification of the welding problem is essentially the first step in welding tool and die steels. For example, the operator may be confronted with such problems as fabricating a composite die, changing the contours (edges, corners, etc.) of a tool or die, correcting structural or design errors, prolonging service life, or repairing cracked, chipped, or broken surfaces. These various problems will determine the type of metal alloy and welding technique to use.

Proper base metal identification is as important for welding tool and die steels as it is in other types of welding. The student welder should review the pertinent sections in the introductory chapters of this book.

The selection of the electrode will largely be determined by the type of deposit desired and the type of base metal involved. The welder should use the smallest electrode diameter for the job.

The surface should be initially prepared by grinding away enough metal to *completely* remove the defect and to permit a deposit of 1/8 to 3/16 inch of weld metal. Thoroughly clean the surface so that all contaminants (dirt, grease, oil, etc.) are removed. Once the cleaning operation has been completed, the surface is ready for preheating. Heat the surface to its draw-range temperature. If a small piece is involved, the minimum draw-range temperature is used. Maximum draw-range temperatures are reserved for larger pieces. Table 1 illustrates the draw-range temperatures for five types of tool and die steels. Note that high-

Table 1. Preheat or Draw-Range Temperatures For Five Categories of Tool and Die Steels

Tool and Die Steel Class	Preheat or Draw Temperatures	Recommended Currents (Amps)			
		1/16″	3/32″	1/8″	5/32″
Water-hardening	270-450°F	—	60-90	75-125	100-150
Oil-hardening	250-400°F	30-50	60-90	75-125	100-150
Air-hardening	300-1000°F	—	60-90	75-125	100-150
Hot-working	900-1200°F	—	60-90	75-125	100-150
High-speed Steel	1000°F	—	35-50	55-70	85-100

Courtesy Chemetron Corporation

speed steels have no minimum or maximum preheat temperatures. That is to say, the draw temperature is represented by a single value.

Preheating can be done by means of a temperature-controlled furnace, an oxyacetylene torch or a hot plate. Whatever the heating method used, it is important to maintain an even temperature throughout the piece being welded.

The oxyacetylene torch is most commonly used when a furnace is not available or very large pieces must be preheated. During the preheating operation, the temperature of the metal should be carefully watched. Exceeding the maximum draw-range temperature can seriously impair the effectiveness of the weld deposit. Temperatures can be measured during the preheating operation by means of a pyrometer or by various chemical means such as temperature-indicating crayons, liquids, or pellets. Temperature-indicating crayons leave marks on the surface that will melt as soon as a specified temperature is reached. The liquid and pellets operate on the same principle. Table 2 indicates the great variety of different temperature-indicating crayons that can be purchased. They are sold according to their individual melting temperatures.

The melting of the temperature-indicating crayon will indicate that the desired surface temperature has been reached. A suitable time allowance should be provided to permit the heat to extend throughout the piece of metal. While this is taking place, adjust

Table 2. A List of the Many Temperatures (In Degrees Fahrenheit) in Which Tempilstik Temperature Indicating Crayons Are Available

100	169	250	331	475	900	1400	1950
103	175	256	338	488	932	1425	2000
106	182	263	344	500	950	1450	2050
109	188	269	350	525	977	1480	2100
113	194	275	363	550	1000	1500	2150
119	200	282	375	575	1022	1550	2200
125	206	288	388	600	1050	1600	2250
131	213	294	400	625	1100	1650	2300
138	219	300	413	650	1150	1700	2350
144	225	306	425	700	1200	1750	2400
150	231	313	438	750	1250	1800	2450
156	238	319	450	800	1300	1850	2500
163	244	325	463	850	1350	1900	

Courtesy Welding Equipment and Supply Company

the power source for the proper amperage. As is indicated in Table 1, the recommended current will vary according to the diameter of the electrode and, in the case of high-speed tool steels, the type of tool and die steel.

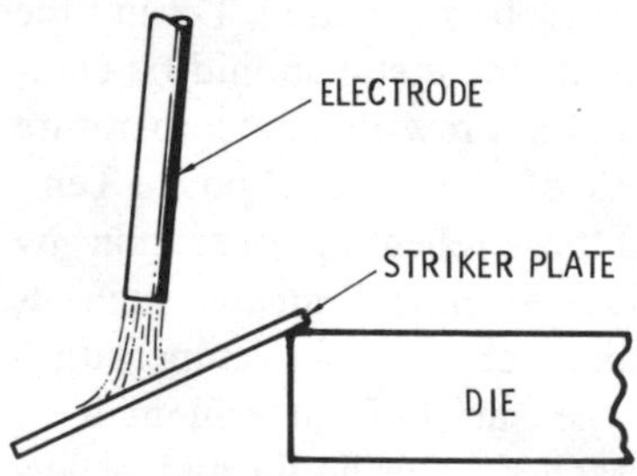

Fig. 2. Striking the arc.

Start the arc on a piece of scrap steel (Fig. 2) and transfer it to the surface of the metal being welded (the procedure for striking the arc is described in Chapter 6). This helps to avoid the possibility of marring the surface of the tool or die.

Maintain a very short arc (as short as possible without extinguishing the arc). Direct the arc against the deposited weld metal

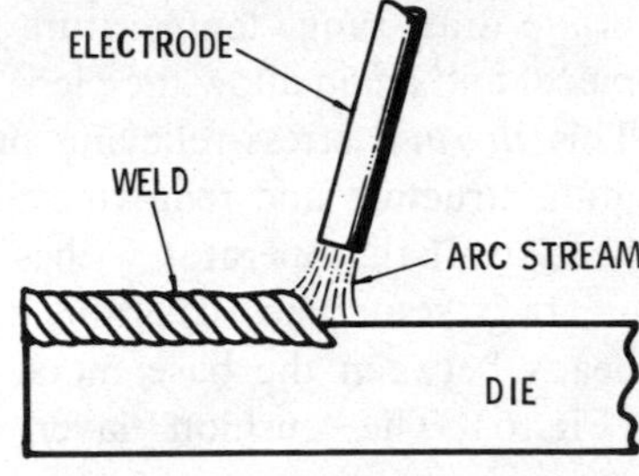

Fig. 3. Directing the arc against the deposited weld metal.

Courtesy Chemetron Corp.

(Fig. 3) and move the electrode in a direction away from edges or corners. This helps prevent the formation of craters at these points. If the arc needs to be restarted, it should also be directed against previously deposited weld metal.

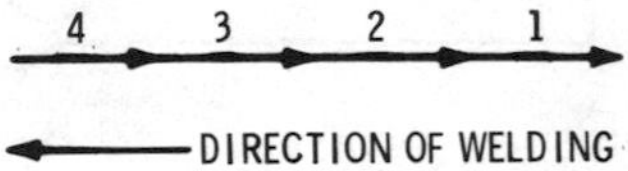

Fig. 4. The backstep sequence.

Either a backstep (Fig. 4) or skip weld sequence (Fig. 5) is recommended for depositing the weld metal.

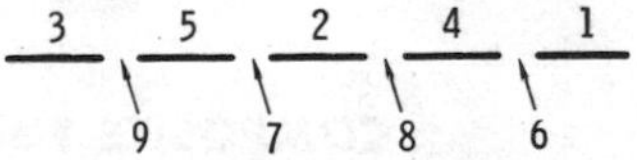

Fig. 5. The skip weld sequence.

Each bead should be stress-relieved by peening immediately after it has been deposited. This is a *mechanical* stress-relieving procedure which involves hitting the bead with a hammer. Care should be taken not to hit the bead too hard, or cracking may occur. It is essential that peening be done while the weld metal is still hot and ductile, otherwise weld cracking may again be the result. Cold peening is definitely *not* recommended.

After each weld has been peened, it is important that the surface be cleaned before making the next pass. Brush the surface, making certain to remove any traces of slag that might combine with (and thereby weaken) subsequent deposits.

When all deposits have been completed, allow the tool or die to cool slowly at room temperature. As soon as the metal has cooled to the recommended temperature, reheat the piece to the appropriate draw-range temperature (taking into account the size of the piece) and again allow the piece to cool slowly to room temperature. This *thermal* stress-relieving procedure will result in refining the grain structure and removing the internal stresses produced during heating. If the operator wishes, the procedure may be repeated.

It is sometimes desirable to deposit a cushion layer of weld beads between the base metal and the final layer of weld metal (Fig. 6). The "cushion" layer(s) will not be of the same composi-

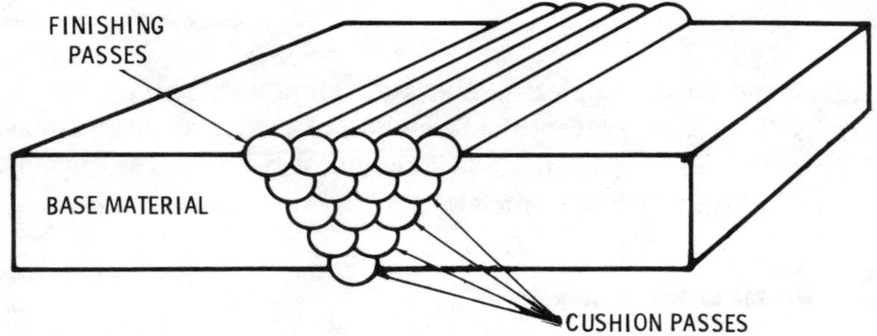

Fig. 6. The cushion layer.

tion as the layer comprising the finishing beads and must therefore be completely covered by the latter.

COMPOSITE TOOL AND DIE FABRICATION

It is sometimes more economical and practical to completely fabricate a tool or die in the shop. If this is the case, a "softer" steel (low-alloy steel or a carbon steel) will be used as the core with the desired tool steel forming the outermost layer. The welding procedure is similar to the depositing of cushion layers described in the section on TOOL AND DIE WELDING PROCESSES in this chapter.

The advantage of composite fabrication is that it creates tools and dies with resilient cores. This factor reduces the tendency to crack or break due to the extreme hardness of tool and die steels. Sometimes a tool or die will be required to perform a variety of functions simultaneously. This will necessitate different types of steel at different points of contact. Composite fabrication is an excellent means of obtaining this kind of varied surface design.

Courtesy Welding Equipment and Supply Co.
Fig. 7. A compositely fabricated female trimmer.

Fig. 7 shows a compositely fabricated female trimmer. The core consists of 1035 steel which serves as a base for a layer of air-hardening tool steel.

SILVER BRAZING APPLICATIONS

Silver brazing can be used to repair certain types of steel that are prone to shattering, cracking, or breaking. These generally include such tools as drills, arbors, punches, saw blades, broaches, and similar types.

A high-strength silver brazing alloy with a low melting temperature is suggested for this operation. Manufacturers of welding equipment and supplies will generally recommend an appropriate silver brazing alloy.

The procedure for silver brazing tools is basically as follows:

1. Prepare the surface by grinding away the defect,
2. Thoroughly clean the surface,
3. Apply a suitable flux to the surface,
4. Pretin the surface with the silver brazing alloy,
5. Reapply the flux,

6. Heat the surface to an appropriate temperature and apply the silver brazing alloy,
7. Allow the part to cool.

The first two steps to the silver brazing procedure have been repeatedly described in other sections of this book. For example, the grinding operation was dealt with in the section on TOOL AND DIE ARC WELDING in this chapter.

A suitable flux will generally be recommended by the manufacturer of the silver brazing alloy. An oxyacetylene torch is used to pretin the surface. Adjust the torch for a carburizing flame (an excess of acetylene). Heat the surface until the temperature is sufficient to "melt" the flux. This will occur at approximately 1150° F, and will be indicated by the flux changing to a liquid state.

When this point is reached, flow the silver brazing alloy onto the surface. The torch should be kept in constant motion during this operation to insure even heating. The initial tinning layer should be thin and evenly spread over the entire joint surfaces.

Apply a second coating of flux and adjust the parts so that they are correctly aligned and fit tightly together. In silver brazing, it is highly desirable that the fit between parts be as close as possible.

Reheat the surface with the oxyacetylene torch. The same torch adjustment (carburizing) as used in the pretinning operation will be used here. Moreover, the torch should be kept in motion to insure even heating. The correct temperature for silver brazing will be indicated by the flux changing to a liquid state and the metal turning to a dull red. When this occurs, flow on more silver brazing metal. If all conditions have been met satisfactorily, the silver brazing alloy will flow into the joint by capillary attraction.

The part may be cooled by burying it in powdered asbestos or some other suitable medium (blown mica, lime, etc.).

CARBIDE TIPPING

Tool tips that are made from one of the refractory metals (e.g. tungsten-carbide tool tips) are very difficult to braze to shanks. The major problem here is the difficulty in wetting a refractory metal. Then, too, refractory metals of the tungsten-carbide type are so brittle they tend to crack as a result of the stresses caused by brazing. It is therefore essential when brazing tungsten-carbide tool tips to use a filler metal designed to increase or promote the wetting action. Nickel is often added to filler metal to improve the wetting action, and silver brazing alloys with at least a 2% to 5% content are recommended for the difficult-to-braze refractory metals.

The following procedure is recommended for silver brazing tungsten-carbide tool tips to shanks (Fig. 8):

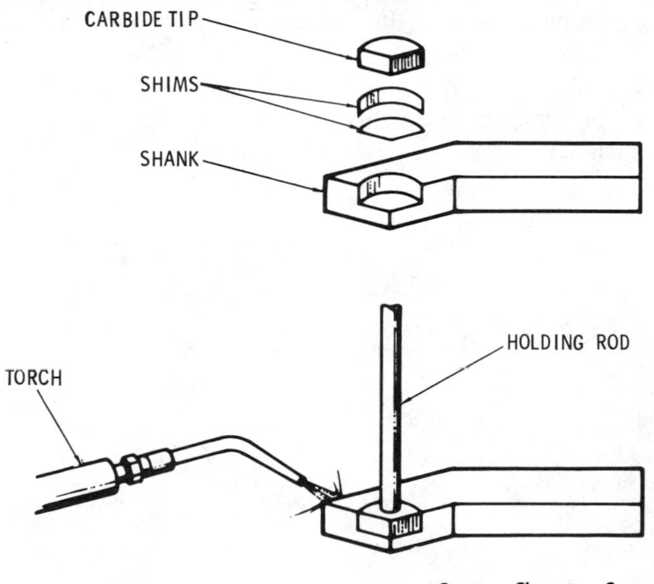

Courtesy Chemetron Corp.

Fig. 8. Silver brazing a carbide tip to a shank.

1. Thoroughly clean the surfaces of the joint,
2. Apply a suitable flux to the carbide tip and the joint surface on the shank,
3. Cut a piece of .005 inch shim to the dimensions of the carbide tip,
4. Coat the shim with flux,
5. Insert the shim in the shank recess,

6. Place the carbide tip on top of the shim,
7. Adjust the oxyacetylene torch for a carburizing flame,
8. Apply the flame to the bottom and sides of the shank (do not direct the flame directly against the carbide tip),
9. Keep the torch constantly in motion,
10. Place a holding rod against the carbide tip. As soon as the metal of the shank has turned a cherry red and the flux has changed to a liquid state, it will be possible to move the carbide tip slightly with the holding rod when the filler metal alloy has reached the proper molten state.
11. Remove the flame and exert a slight pressure against the carbide tip with the holding rod as soon as the molten state is reached,
12. Allow the part to cool by burying it in powdered asbestos or some other suitable cooling medium (blown mica or lime).

CHAPTER 19

Welding Cast And Wrought Iron

Cast iron is a rather complex mixture containing 91% to 94% metallic iron and varying proportions of other elements, the more important of which are carbon, silicon, manganese, sulphur, and phosphorous. The welding operator, in order to handle cast iron jobs successfully, must have a clear understanding of the fundamental factors which determine the physical properties of an iron casting.

Iron and its alloys, including the various types of steel, are all made from iron ores. The differences among these metals are determined largely by the alloy content and the process by which each was produced. For example, one of the basic distinctions between cast iron and steel is the carbon content. Cast iron will contain 2.5% to 4.0% carbon; whereas most of the various steels will have a carbon content of 1.2% or less. Steels that contain carbon in percentages approaching that of cast iron were once referred to as *semi-steels*. This term is now obsolete. The different types of steel and steel alloys are described in Chapter 22. This chapter is concerned exclusively with cast iron and wrought iron.

Iron ore is a combination of iron, oxygen, and other elements which must be regarded as impurities that have to be separated from the iron content. The blast furnace was developed for just this purpose. The iron ore is smelted in the blast furnace to remove the earthen matter and the oxygen. The former combines with a flux (in this case limestone) and forms a liquid slag which separates from the iron. The iron ore is subjected to tremendously high temperatures in the blast furnace (over 3000°F) which causes the oxygen to be released in the form of a gas. The molten iron sinks to the bottom of the blast furnace and is drawn off to cool in special molds. This is the so-called pig iron from which the various cast irons and steels are produced through different types of refining processes (e.g. the puddling furnace, the open hearth furnace, the Bessemer process, etc.). Pig iron contains the highest carbon content (approximately 4%) of all the ferrous metals. Refining consists of reducing the carbon content and adding alloying elements to obtain the desired characteristics. These refining processes are briefly described in the sections of this chapter concerned with each specific type of cast iron.

Not only should the welding operator have a clear understanding of the composition factors which determine the physical properties of cast iron, he should also be aware of the effects produced by the welding process itself. This will be true not only for the ferrous metals (both the irons and steels), but for the nonferrous ones (copper, aluminum, etc.) as well.

While most metals and their alloys can be welded with more or less satisfactory results, the economy and degree of satisfaction of welding various metals may be affected by any one of the following factors:

1. Oxidation,
2. Vaporization,
3. Nonmetallic inclusions,
4. Change of structure,
5. Gas solubility,
6. High thermal expansion,
7. High thermal conductivity,
8. Hot shortness,
9. Thermal conductivity,
10. Electrical conductivity.

Oxidation produces a gaseous oxide of some of the elements, causing gas holes in the weld metal. Oxidation producing solid oxides, which have a melting temperature higher than the metal, cause slag inclusions.

Vaporization will cause the loss of some element in the metal which vaporizes at a temperature lower than the melting point of the metal.

Some metals may contain finely divided *nonmetallic inclusions* which have a melting point higher than that of the metal and, therefore, did not coalescence under the high temperature of the welding flame or arc; they then form visible slag inclusions.

A *change of structure* in some metals may take place during welding, causing change of physical properties or change of resistance to corrosion, etc.

Different elements may affect the solubility of various gases at different temperatures and a decrease in solubility of a gas with a decrease in temperature at the freezing point may cause porosity in weld metal. The burning out or elimination of an element during welding may cause the capacity of the metal for a given gas to decrease and thus cause the gas to be given up producing porosity in the weld metal. Absorption of gases during welding, which forms stable compounds with elements in the metal, alter the composition and physical properties of the weld metal.

High thermal expansion causes high contraction of the weld metal during cooling. This will generally result in thin cracks forming on the weld surface, particularly when the contraction rates of the filler metal and base metal are unequal.

High *thermal conductivity* governs the rate of transfer of heat from the fusion zone. The higher the thermal conductivity of a metal, the faster the heat is carried away from the welding zone. A higher welding temperature is required to counteract the rapid dissipation of heat.

Hot shortness refers to the loss of strength of a metal at high temperatures. The tendency of a metal toward hot shortness may result in hot cracking in the weld area. This can be avoided by increasing the welding speed, laying narrow beads, using a multilayer welding technique, and exercising caution when clamping or jigging.

TYPES OF IRON

There are a number of different types of irons, some more suitable than others for welding. In their approximate order of weldability, the most frequently used types can be listed as follows:

1. Gray cast iron,
2. Malleable cast iron,
3. Nodular cast iron,
4. White cast iron.

The welding processes recommended for each of these four principal types of cast iron are summarized in Table 1. Other types of iron that will be considered in this chapter include the following:

1. Ingot iron,
2. Alloy cast iron,
3. Galvanized iron,
4. Wrought iron.

Gray Cast Iron

Gray cast iron is produced from pig iron by allowing the casting to cool very slowly. The slow rate in cooling allows some of the carbon to separate and form flakes. These graphite flakes give gray cast iron its distinctive color. Unfortunately, they also contribute to making gray cast iron somewhat more brittle than is desired, a weakness that is corrected in nodular cast-iron.

Gray cast iron is an alloy composed of iron, carbon, and silicon. Traces of phosphorous are also usually present. The chemical anyalysis of a gray cast iron will vary in accordance with its composition. For example, the silicon content may vary from 1.0% to 3.0% and the carbon content from 2.5% to 4.0% (in the free or graphitic state). The amounts will vary according to the use for which the gray cast iron is produced.

The tensile strength of gray cast iron ranges from 10,000 to 60,000 psi depending on the particular class. The *American Society for Testing Materials (ASTM)* has devised a classification system based on the tensile strengths of different gray cast irons. A gray cast iron having a minimum tensile strength of 10,000 psi will

Table 1. Summary of Welding Procedures

Cast Iron Type	Procedure	Treatment	Properties
Gray iron	Weld with cast iron	Preheat and cool slowly	Same as original
Gray iron	Braze weld	Preheat and cool slowly	Weld better; heat affected zone as good as original
Gray iron	Braze weld	No preheat	Weld better; parent metal hardened
Gray iron	Weld with steel	Preheat if at all possbile	Weld better; parent metal may be too hard to machine; if not preheated, needs to be welded intermittently to avoid cracking
Gray iron	Weld with steel around studs in joint	No preheat	Joint as strong as original
Gray iron	Weld with nickel	Preheat preferred	Joint as strong as original; thin hardened zone; machinable
Malleable iron	Weld with cast iron	Preheat, and postheat to repeat malleableizing treatment	Good weld, but slow and costly
Malleable iron	Weld with bronze	Preheat	As strong, but heat-affected zone not as ductile as original
White cast iron	Welding not recommended		
Nodular iron	Weld with nickel	Preheat preferred; postheat preferred	Joint strong and ductile, but some loss of original properties; machinable; all qualities lower in absence of preheat and/or postheat

Courtesy The James F. Lincoln Arc Welding Foundation

belong to class 10; one with a minimum tensile strength of 20,000 psi to class 20; and so forth. There are a total of seven classes.

The melting point of gray cast iron is approximately 2150°F. Its ductility is rather low which is indicated by limited distortion at breaks. The impact and shock resistance is almost nonexistent. Gray cast irons generally exhibit ease of machining and weldability. They are also easily cast in a wide variety of forms.

Malleable Cast Iron

Malleable cast iron is obtained by annealing or heating white cast iron over a prolonged period of time and then allowing it to cool slowly. This heat treatment results in a greater resistance to impact and shock, higher strength, and greater ductility. Consequently, it is better able to withstand these types of strain than gray cast iron.

The tensile strength of malleable cast irons range from 30,000 to 100,000 psi depending upon whether the normal base structure is ferrite or pearlite. The lower tensile strengths (53,000 psi or below) belong to the ferrite malleable cast irons; the higher ones (60,000-100,000 psi) to the pearlite group. Most malleable cast iron is produced with a tensile strength of 40,000-50,000 psi. It should be noted that as the tensile strength of a malleable cast iron increases, the ductility decreases.

The carbon content of malleable cast iron is approximately 2% to 3% and must be in the combined form (free or graphitic carbon cannot be malleableized). Malleable cast iron is stronger and tougher than gray cast iron. It possesses good machining characteristics, but because of malleableizing it has limited weldability.

Recommended joining processes for working with malleable cast iron include oxyacetylene welding and braze welding, with the latter preferred because of its lower temperatures. High welding temperatures will cause the malleable characteristic to break down and a reversion to the characteristics of white cast iron. For this reason, welding processes that generate a high heat (e.g. shielded metal-arc, carbon-arc, etc.) are not recommended for working with this metal *unless* temperatures of approximately 1400° F to 1450° F or below are used.

Nodular (Ductile) Cast Iron

Nodular (or ductile) cast iron is so named because the graphite present in its composition takes the form of nodules rather than flakes. This results in giving nodular cast iron greater shock resistance than gray cast rion. The formation of graphite nodules is produced by adding magnesium to the molten metal during the production process.

The melting point of nodular cast iron is slightly below that of gray cast iron. Special heat treatment and the addition of alloying elements results in a metal that combines the more desirable characteristics of both gray cast iron and of steel. The principal alloy added to obtain these characteristics is magnesium.

The tensile strengths of nodular cast iron will range from 60,000 to 120,000 psi, depending on the composition of the base metal, the process used to obtain the module structure, and other factors.

White Cast Iron

White cast iron is produced from pig iron by causing the casting to cool very rapidly. The rate of cooling is too rapid to allow the carbon to separate from the iron carbide compound. Consequently, the carbon found in white cast iron exists in the combined form.

White cast iron is very hard and brittle, and does not lend itself well to machining. As a result, this type of cast iron is not often used in its original state for castings. Because of its hardness, it is frequently used as a wear-resistant outer surface for inner cores of gray cast iron. These white cast iron surfaces are produced by causing the molten metal to flow against heavy iron chills placed in the mold. This causes rapid cooling which, in turn, results in a very hard surface. Castings produced in this manner are referred to as *chilled-iron castings*. A second major use of white cast iron is the production of malleable iron castings.

Generally, white cast iron will have a tensile strength of 40,000-50,000 psi or more. The melting point of white cast iron is 2300° F, or slightly higher than that of gray cast iron.

Oxyacetylene welding is the recommended welding process for working with white cast iron. Arc welding is not recommended.

Ingot Iron

Ingot iron (low-carbon iron) is an almost pure open-hearth iron containing only small amounts of carbon, manganese, and other elements. Because of its low alloy content, ingot iron is soft and malleable. It is easy to weld and exhibits many of the welding characteristics of low-carbon steel. Preheating or annealing is generally not required.

The melting point of ingot iron is approximately 2786° F, which is higher than any of the other irons. Its tensile strength is normally about 45,000 psi.

Ingot iron has excellent welding characteristics, and can be joined by any of the welding processes used today. A low-carbon or medium low-carbon rod should be used when arc welding. When using the oxyacetylene welding process, a welding rod matching the analysis of the base metal is recommended.

Alloy Cast Iron

Nickel, molybdenum, chromium, copper, aluminum, and other elements can be added to gray cast iron to produce an alloy possessing certain specifically desired characteristics (e.g. greater corrosion resistance, higher strength, etc.). For example, a nickel cast iron will have greater corrosion and wear resistance than gray cast iron and a molybdenum cast iron will possess higher tensile strengths.

When welding alloy cast irons, the operator should take into consideration the characteristics of the alloying element and proceed accordingly. Aluminum cast irons, for example, present the problem of aluminum oxides forming on the surface during welding.

Galvanized Iron

Sometimes a zinc coating is applied to the surface of an iron. The product (galvanized iron) is frequently encountered in the form of galvanized-iron pipe.

Braze welding or any of the fusion welding processes can be used to join galvanizing iron. The operator should be cautioned against inhaling the toxic zinc-oxide fumes that are given off during the welding. Welding should be done only where adequate ventilation is provided.

Fast welding speeds should be used; otherwise, the protective zinc coating will be removed by the heat of the welding operation.

Wrought Iron

Wrought iron is produced by melting pig iron ingots in a puddling furnace. This process not only removes almost all of the carbon, but also other impurities such as silicon and manganese. The highly refined iron combines in a mechanical mixture with the iron oxide-silicate slag to produce the fibrous structure of wrought iron. Its good tensile strength, high ductility, high corrosion resistance, and fatigue resistance, are all attributed to its structure.

The tensile strength of wrought iron is about 45,000 psi. Its melting point is approximately 2800°F, or several hundred degrees above the melting point of its own slag. Because the slag melts at a temperature considerably lower than the wrought iron base metal, excellent protection against oxidation is given to the surface during welding. Wrought iron is characterized by an extremely low carbon content. In fact, it has the lowest carbon content among the various commercial irons.

Wrought iron is easy to work and very easy to weld. Oxy-acetylene welding with a neutral flame and a low-carbon steel welding rod will produce high quality welds. It should be noted, however, that the welds do not have a wrought-iron structure. Attempts to use high-carbon steel welding rods to increase weld strength should be avoided.

Both shielded metal-arc and carbon-arc welding have been used to join wrought iron. The operator should avoid excessive penetration of the base metal.

WELDING AND JOINING METHODS

Cast iron and wrought iron can be joined by a number of different methods including:

1. Gas welding (oxyacetylene),
2. Braze welding,
3. Brazing,
4. Shielded metal-arc welding,
5. Carbon-arc welding,

6. Resistance welding,
7. TIG welding,
8. Thermit welding.

Gray cast iron may be either fusion welded with a cast-iron welding rod, or braze welded. For the vast majority of cast-iron welding jobs, braze welding is to be preferred as its lower temperature of application greatly simplifies the work.

The technique for fusion welding gray cast iron can be learned by making practice welds with two pieces of cast-iron about ⅜-inch thick, each piece having a fairly straight edge 5 or 6 inches long. The procedure is as follows:

1. Bevel one edge of each piece to 45° and clean the beveled surfaces with a stiff wire brush.
2. Support the two pieces of cast iron on fire bricks placed on the welding table. A piece of carbon black placed under the vee will help the beginner in controlling the molten metal.
3. Use a welding tip one size larger than for steel of the same thickness.
4. Adjust the welding torch for a neutral flame.
5. Play the flame along the sides of the vee until the entire joint has been thoroughly preheated.
6. Starting at one end, direct the flame at the bottom of the vee until the metal there has melted. The flame should be directed so that the tip of the inner cone is about ⅛ to ¼ inch from the metal surface.
7. When the bottom of the vee is thoroughly fused, the flame should be moved slightly from side to side, melting down the sides gradually so that the liquid metal runs down and combines with the molten puddle at the bottom of the vee.
8. If the metal gets too hot and tends to run away, raise the flame slightly. Move the torch from one side of the vee to the other, keeping both sides melted as well as the bottom.

The proper use of the welding rod and the flux (if the situation requires) is very important. The end of a length of ¼-inch cast-iron welding rod should be introduced into the outer cone of the flame, heated, dipped into the flux, and then placed with the end in the molten puddle. The heat of the molten metal on which the welding flame continues to play will melt the rod gradually, and the surface of the puddle will gradually rise with the addition of this metal.

The rod should never be held above the weld and melted drop by drop into the puddle. Also, be careful to fuse the sides of the vee ahead of the advancing puddle so the molten metal is never forced onto colder metal. If the latter occurs, it will cause an adhesion with little or no strength at that spot.

When gas bubbles or white spots appear in the puddle or at the edges, flux should be added and the flame played around the spot until the impurities float to the top. These are skimmed from the weld with the welding rod. Such impurities adhering to the hot rod are removed by tapping it against the table, thus eliminating them entirely. This removal of dirt must be done carefully and systematically, for impurities left in the weld will constitute defects and result in a joint of little strength.

The rod should be added to the molten metal until that section of the vee is built slightly above the level of the rest of the piece. When one section an inch or so long is built up, the bottom of the vee adjacent to it is next melted and the operation repeated. Of course, care must be taken to keep the end of the built-up section as well as the sides of the vee in complete fusion with the puddle.

Cast iron welding should be carried on as fast as possible. When finished, cover the completed weld with a piece of asbestos paper or bury it in the annealing bin so it will cool slowly.

If preheating is unnecessary or inadvisable, care should be taken not to heat the casting too long or too much at the point of welding at one time. The recommended procedure is to apply the welding heat as briefly as possible, and then to allow the casting to cool for a somewhat longer period. Repeat this procedure until the weld is completed.

Weld in cast iron, if of sufficient thickness, may be strengthened by the mechanical method of studding. Studs of steel and approximately ¼ to ⅜ inch in diameter should be used. The cast iron should be beveled to form a vee, drilled, and tapped along the vee, so that the studs may be screwed into the casting. The studs should project about 3⁄16 to 1⁄4 inch above the cast iron surface. The studs should be long enough to be screwed into the casting to a depth of at least the diameter of the studs (Fig. 1).

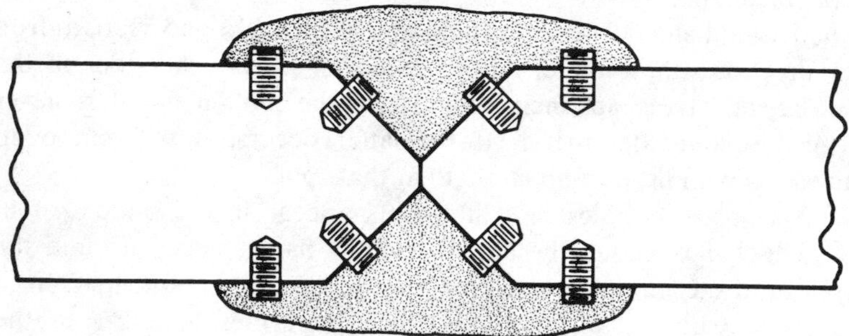

Fig. 1. Usual procedure of studding for cast iron welding.

When the shielded metal-arc welding process is used to weld gray cast iron, the operator should select covered electrodes designed to keep the amount of heat required to a minimum (thus reducing thermal disturbance and resultant hardness).

Usually 1⁄8-inch electrodes are used for keeping the heat down as mentioned. The electrode is made positive, the work negative, and the current is approximately 80 amperes. This apparently too low current is employed to satisfy the heat conditions. The electrode itself will carry considerably more current but the requirements of cast iron welding make the use of higher heat inadvisable.

The electrode to use in cast iron welding is a coated electrode with a steel base. It provides a solid dense weld on cast iron of greater tensile strength than the cast iron itself. It affords an excellent bond or union with the cast iron. Because of the low current used (80 amperes on an 1⁄8 inch electrode), the hardening effect usually present along the line of fusion is materially reduced, the resultant weld being therefore much more machineable than is usually the case where other electrodes are employed.

The welding of gray cast iron should be done intermittently. In some cases "skip" welding is used with a weld not over 8 inches made at one time. Immediately after each bead is deposited, it should be lightly peened, thoroughly cleaned, and allowed to cool before the next bead is applied. Care should be taken to keep the work clean and not allow it to become too hot. A good rule in reference to cast iron is to keep the work clean and cold.

Iron castings may be welded with a carbon-arc and a cast iron filler rod. A proper manipulation of the arc and filler rod when welding in the flat position on heavy castings will produce a fairly machinable weld. By playing the carbon arc about the work, rapid cooling is prevented. By this method, it is possible to float the oxide out of the molten metal. If this is done, hard spots which would cause trouble in the machining operation are eliminated.

A dehydrated borax flux is sometimes used, as it enables the operator to float out some of the undesirable impurities. The cast-iron filler rod used with the carbon electrode is usually of far higher grade material than the base metal.

RESISTANCE WELDING

Resistance welding is recommended for joining wrought iron and low-carbon iron. This is primarily a production line process not generally used in the small workshop. The welds produced by this process have approximately the same physical characteristics as the base metal itself.

These irons are characterized by relatively high thermal conductivities. Consequently, the resistance welding process will have to produce a heat sufficiently high to counteract the effects of the thermal conductivity and, at the same time, produce a satisfactory weld.

THERMIT WELDING

Thermit welding is commonly used for joining wrought iron, low-carbon iron, and alloy cast irons. The key factor in the weldability of an iron by the thermit welding process is the carbon content. It should not exceed 0.06%. Thermit welding is not recommended for joining gray cast iron or malleable cast iron if another method (e.g. oxyacetylene or braze welding) is available. Both of these cast irons can be welded by the thermit welding process, but only if conditions are exactly right.

Welds produced by the thermit welding process are machinable and generally give a good color match. However, due to uneven contraction rates between the filler metal and base metal, thin cracks develop on the surface of the weld. This shrinkage of the metal during cooling is not entirely predictable even when every precaution has been taken.

Thermit welding is frequently used to repair iron castings and to fabricate heavy parts in ship building. Other applications of this welding process (as they relate to the various types of iron) are described in Chapter 10 (THERMIT WELDING).

OTHER JOINING PROCESSES

The following may be included among the processes used to join the various types of iron:

1. Submerged-arc welding,
2. Forge welding,
3. Braze welding,
4. Brazing.

Submerged-arc welding can be used to join wrought iron and low-carbon iron. It is not recommended for gray cast iron, malleable cast iron, or any of the alloy cast irons. A suitable flux (granulated) and filler metal rod must be used.

Both forge welding and braze welding are recommended for joining wrought iron and low-carbon iron. No significant difficulties are encountered with either welding process.

On the other hand, some difficulty is encountered when attempting to braze cast irons. For example, graphitic carbon on the surfaces to be joined will prevent adequate bonding. This problem can be solved by following special cleaning procedures before brazing. Because this joining process does not result in melting the base metal and causing a fusion between it and the filler metal, brazing should not be used where joints of high strength are required.

CHAPTER 20

Welding Copper and Copper Alloys

Copper is a brownish red metal characterized by being soft, tough, ductile, and very malleable. It is one of the best conductors of heat and electricity. It is a useful metal in itself and also in its various alloys such as brass and bronze. The commercial alloys in which copper is a major constituent number in the several hundreds. The applications of copper and its alloys are extensive and include such varied forms as electrical wire and equipment, tubing, tanks, chemical equipment, pipes, sheets, ingots, and castings.

Pure copper wire or sheet is much harder and stronger than copper ingots or castings. The work required to produce the former (often done cold) hardens and strengthens the metal in direct proportion to the amount of rolling, forging or drawing, but copper is always softer than steel. It can be hammered out cold into a thin edge.

Copper has a melting point of 1981° F. It possesses a high coefficient of linear expansion and a tensile strength of approximately 32,000 psi (in the annealed state).

COPPER ALLOYS

Copper can be combined with a number of other alloying elements to produce a wide range of alloys. Some of these elements (e.g. aluminum, beryllium, nickel, silicon, zinc, and tin) form major classifications of copper alloys. Other elements (e.g. cadmium, lead, and arsenic) are added in small amounts to provide special characteristics to a particular alloy. There are approximately nine major classifications of copper and its alloys. These may be listed as follows:

1. Oxygen-bearing coppers,
2. Oxygen-free coppers,
3. Copper-zinc alloys,
4. Copper-silicon alloys,
5. Copper-tin alloys,
6. Copper-nickel alloys,
7. Copper-nickel-zinc alloys,
8. Copper-aluminum alloys,
9. Copper-beryllium alloys.

Oxygen-Bearing Coppers

Oxygen-bearing coppers contain traces of oxygen (in the form of copper-cuprous oxide eutectic) and other impurities that can cause weld porosity during welding unless precautions are taken. These oxygen-bearing coppers can be divided into two basic types: (1) fire-refined copper, and (2) electrolytic tough pitch copper. The latter is the purer of the two and is far more widely used.

The melting point of the oxygen-bearing coppers is approximately 1981° F. There is no difference between the *liquidus* and *solidus* temperatures. The tensile strength ranges up to 38,000 psi (in the annealed state).

Due to the presence of copper oxides, oxyacetylene welding is not recommended. Silver brazing with a slightly oxidizing flame (and a suitable flux) is preferred.

Oxygen-Free Coppers

The oxygen-free coppers are those coppers from which the oxygen has been removed. These coppers can be divided into two basic types: (1) oxygen-free high conductivity copper, and (2) deoxidized copper. In the case of the latter, the oxygen is removed by adding small traces of a deoxidizing agent such as phosphorous, silicon, manganese, or boron to the copper. When the copper is deoxidized with phosphorous, it is commonly referred to as *phosphorized copper* or *phosphor deoxidized copper*. When a deoxidant is used (e.g. phosphorous or silicon) a slight excess is used, so a little remains alloyed in the copper and furnishes protection during welding.

Hydrogen is used as the deoxidizing agent for oxygen-free high conductivity copper. However, instead of being added to the molten copper it functions as a shielding atmosphere during the melting and casting of the metal. This type of copper is also referred to as *OFHC copper*.

The oxygen-free coppers possess a high degree of thermal and electrical conductivity. These coppers are 99.90% pure. In the case of phosphor deoxidized copper, the remainder of the composition consists of traces of phosphorous. If the amount of phosphorous is above 0.01%, it is a high phosphorous-type deoxidized copper and has a melting range of 1977° F to 1981° F. The low-phosphorous type deoxidized copper (those having less than 0.01% phosphorous) and oxygen-free copper have melting ranges of 1981° F (that is, no difference between the solidus and liquidus temperatures).

Oxygen-free high conductivity copper and deoxidized copper are available in a number of forms including bar, pipe, sheet, rod, wire, and tube. These coppers are easier to weld than the oxygen-bearing types, and are generally preferred over the latter whenever copper must be welded. Deoxidized copper is preferred to oxygen-free high conductivity copper because the welds are consistently and uniformly cleaner. These coppers have a tensile strength of approximately 30,000 to 35,000 psi and a yield strength of 7000 to 10,000 psi.

Oxyacetylene welding can be used for joining both oxygen-free high conductivity copper and deoxidized copper. Because of their high thermal conductivity, a large welding tip size is recommended. The flame adjustment should be slightly oxidizing or neutral (when a flux is used). A deoxidized copper rod is used for the filler metal. Larger sections (heavier gauge plate and castings) will require preheating.

Oxygen-free coppers can also be braze welded. A slightly oxidizing flame and a bronze welding rod are recommended for this joining method.

Other methods for welding oxygen-free coppers are:

1. Shielded metal-arc welding,
2. TIG welding,
3. Submerged-arc welding.

It should be pointed out, however, that these last named welding methods are usually restricted to the lighter gauges and require special electrodes, welding rods, and fluxes.

Copper-Zinc Alloys

The copper-zinc alloys (or brasses) provide the largest category of copper alloys. Most of these alloys are composed of copper and zinc in various proportions, although tin or lead is added in small amounts in some grades.

The copper-zinc alloys can be divided roughly into three major categories depending on the proportion of copper to zinc. One of these categories, commonly referred to as the *low brasses,* contains 80% or more copper with zinc in proportional amounts. Examples include the red brasses (85% copper, 15% zinc), zinc bronze, commercial bronze, and ounce metal.

A second major category contains less than 80% copper with zinc in proportional amounts. These are commonly referred to as the *high brasses,* and are represented by such alloys as yellow brass, brazing brass, commercial brass, alpha brass, and admiralty. The brasses in this category are more commonly used than the others. They are available in a variety of shapes including castings (yellow brass), sheets, rods, extruded shapes, and wire.

The third category of copper-zinc alloys is based on an approximate composition of 60% copper and 40% zinc. The alloys in this category include Tobin bronze, Muntz metal, naval brass, and manganese bronze. They are available in sheets, rods, forgings, castings, extruded shapes, and tubes.

Copper-Silicon Alloys

The copper-silicon alloys are also known as the *silicon bronzes* and are represented by a number of different trade names, including *Everdur, Olympic Metal, PMG, Duronze, Herculoy,* and *Tombasil.*

Those copper-silicon alloys containing 94.0% to 96.0% copper, 1.50% to 3.25% silicon and 1.25% or less of zinc, tin, manganese, or iron are the most widely used in industry. They are noted for such qualities as good corrosion resistance and high strength and especially for their weldability. The copper-silicon alloys are more easily welded than any other alloy.

Because of the silicon content, these alloys are characterized by a low thermal and electrical conductivity. As the silicon content increases (up to a maximum of 3.50%), the thermal and electrical conductivity decreases.

The carbon-arc welding process is recommended for joining the copper-silicon alloys. Graphite electrodes and phosphor-bronze welding rods give the best results. The normal precleaning procedures and the use of a suitable flux are also necessary.

Copper-Tin Alloys

The copper-tin alloys are also referred to as *true bronzes, tin bronzes,* or *phosphor bronzes.* These are alloys of copper and tin with occasionally zinc and lead added. Copper varies from 80% to 90% and the tin from 10% to 20%. The greater proportion of tin makes a harder metal but decreases the tensile strength.

When copper-tin alloys high in tin and lead are heated, certain of the constituents melt and exude from the metal at a temperature considerably below the melting point of the alloy.

Copper-tin alloys are characterized by a bronze color and are noted for being high in strength and corrosion resistance. They are available in a variety of forms, including castings, rods, wires, and sheets.

The copper-tin alloys are subject to hot shortness, a slow spreading heat during welding, and a comparatively slow cooling rate. The shielded-arc welding process is recommended for joining these metals. A fast welding speed should be used.

Copper-Nickel Alloys

The copper-nickel alloys are available in three principal grades, each distinguished by the approximate nickel content. The chemical compositions of these three grades are: (1) 90.0% copper, 10.0% nickel, (2) 80.0% copper, 20.0% nickel, and (3) 70.0%

copper, 30.0% nickel. These alloys also contain small amounts of zinc, tin, and manganese though never in amounts exceeding 1.0%. Smaller amounts of iron and lead are also present.

The copper-nickel alloys are available in rods, bars, sheets, wire, and plate. Their thermal and electrical conductivities are low. On the other hand, they are tough, ductile, and possess high tensile strength.

The addition of nickel to these alloys increases their corrosion resistance and directly influences their tensile strength. The nickel content necessitates the use of special fluxes when oxyacetylene welding is employed as a joining process. With the possible exception of carbon-arc and atomic-hydrogen welding, most of the welding processes can be used to join the copper-nickel alloys.

Copper-Nickel-Zinc Alloys

The copper-nickel-zinc alloys are also referred to as *nickel silver*. The latter term derives from the light silvery color obtained by adding 15% to 20% nickel to a copper-zinc alloy. Sometimes traces of lead are also added to increase the machinability.

The melting point of the copper-nickel-zinc alloys is 1930°F. They possess a tensile strength of approximately 60,000 psi (in the annealed state), high corrosion resistance, and low electrical and thermal conductivities (caused by the addition of nickel).

Oxyacetylene and resistance welding are recommended for joining nickel silver. Brazing is another successful method but should not be used where strength is a requirement of the joint.

Copper-Aluminum Alloys

The copper-aluminum alloys are frequently referred to as *aluminum bronzes*. They range in color from yellow to bright yellow, are generally harder than the brasses, and are strongly corrosion resistant.

The composition of the copper-aluminum alloys is approximately 80% to 95% copper and 5% to 20% aluminum. Other alloying elements (e.g. iron, manganese, and nickel) are also added from time to time in an attempt to obtain certain desired characteristics. These additional elements will normally not exceed 1% to 3% of the total composition.

The greater the aluminum content of these alloys the more difficult it becomes to weld them, particularly with the oxyacetylene welding process. Consequently, shielded metal-arc welding has become a commonly used method of joining these metals. The carbon-arc welding process is also extensively used. Submerged-arc welding and resistance welding have been used to a lesser extent.

Copper-Beryllium Alloys

The addition of beryllium under suitable conditions (proper heating, quenching, and reheating) greatly adds to the tensile strength of these alloys. In fact, beryllium copper possesses greater strength than any of the other copper alloys. In addition to its high strength characteristics, this copper alloy is also noted for its resilience and high resistance to wear.

Beryllium-copper consists of approximately 97.0% copper, 2.0% to 2.3% beryllium, 0.5% nickel, and 0.25% iron. It melts at a temperature of 1750°F. The tensile strength is approximately 72,000 psi (in the annealed state) but may be considerably increased by various methods of hardening.

Copper-beryllium alloys are available in a variety of forms including castings, tubes, sheets, rods, forgings, and wires. Their applications include the manufacture of nonsparking tools, plastic molds, and dies.

Shielded metal-arc, carbon-arc and the inert-gas metal-arc welding processes (MIG and TIG) are recommended for joining the copper-beryllium alloys.

WELDING METHODS

Copper can be tested for weldability by heating a piece of it to a bright red, just below the melting point, and then hammering it vigorously on an anvil. If it breaks, it is unsuitable for fusion welding. This identifies it as an electrolytic tough pitch copper that contains traces of copper oxides. Some of these copper oxides tend to gravitate toward the edge of the weld during welding. The result is a weld that is usually weaker than the base metal. Consequently, most welding is done on oxygen-free coppers.

Welding Copper and Copper Alloys

Welding methods commonly used to join copper and the copper-base alloys include the following:

1. Carbon-arc welding,
2. Oxyacetylene welding,
3. Shielded metal-arc welding,
4. Atomic-hydrogen arc welding,
5. Submerged-arc welding,
6. TIG welding,
7. MIG welding,
8. Resistance welding,
9. Flow welding.

Copper and copper alloys can also be joined by braze welding, brazing, and soldering. These methods are discussed in the appropriate chapters.

Carbon-Arc Welding

Many welders recommend the automatic carbon-arc welding process because it ensures smooth dense welds. However, the manual carbon-arc produces very satisfactory results when a high capacity, highly efficient arc welder capable of delivering a uniform current is used. This type of arc welder is necessary to maintain the required stable 40-volt arc.

The compostion of the filler metal will vary according to the physical characteristics required in the welding structure. If the weld must have low electrical resistance, the filler metal may be of pure copper or cadmium copper.

A number of years ago, *Hobart Brothers Company* developed a method for welding ordinary electrolytic or deoxidized copper that used an arc length considerably longer than when welding steel. The name *long arc* was given to this method of welding copper. *High-voltage welding* was also used sometimes because a high voltage is required to produce the long arc.

The length of the arc generally ranges from about ¼ to 1 inch or slightly more, depending on the thickness of the metal. On copper ⅛ inch thick or thicker, the distance from the tip of the carbon electrode to the center of the molten crater in the work metal should be not less than ½ inch. In fact, the greatest strength and all around results in practical work have been attained with an arc length of 1 inch to 1⅛ inch especially when heavier copper plates have been involved.

On copper less than ⅛ inch thick, due to the lower heat used, the arc length should be about ¼ inch which is still a long arc for such light gauge metal. Table 1 gives recommended machine ratings for welding various thicknesses of copper.

A water-cooled electrode holder is recommended because the carbon becomes very hot. The carbon electrode should be ground, or otherwise sharpened to a tapered point with approximately a 1/16 inch diameter at the tip, tapering back for a distance of at least ¾ inch (Fig. 1). The electrode should be placed in the holder so that not more than 2½ or 3 inches of the carbon projects from the end.

Fig. 1. Tapered point of carbon electrode.

Table 1. Recommended Machine Ratings For Welding Various Thicknesses of Copper

Machine Rating	Thickness of Copper
8 kW (200 amperes at 40 volts)	Less than 5/64 in.
12 kW (300 amperes at 40 volts)	Less than 9/32 in.
16 kW (400 amperes at 40 volts)	½ in. or less
24 kW (600 amperes at 40 volts)	Over ½ in.

Copper is nonscaling and noncorrosive so it does not require as much cleaning as steel. However, oil, grease, dirt, and other foreign matter must be removed.

When copper is welded to steel (and this may be successfully accomplished by the long-arc method) the steel should be more carefully cleaned than when welding steel to steel. Best results will be obtained if the steel is ground, wire-brushed, or sand-blasted at the point where copper is to be joined.

It is recommended that all work be so positioned that the welding is in the flat or downhand position. Best results will be obtained when the tip of the welding carbon can be held in the downward position and the filler rod falls into the molten crater as it is melted down. However, expert operators have been able to weld copper in the vertical or semi-vertical positions.

The weld should be backed up with steel or heavy copper (the latter is preferable when practical). The proper work positions backing placements and carbon positions for welding copper with the long-arc method are illustrated in Fig. 2.

Copper plates under 3/16 inch need not be beveled. Plates 3/16 inch and thicker should be beveled and spaced about 1/8 inch apart at the root. Plates 1/4 inch and over should be beveled when possible on both sides.

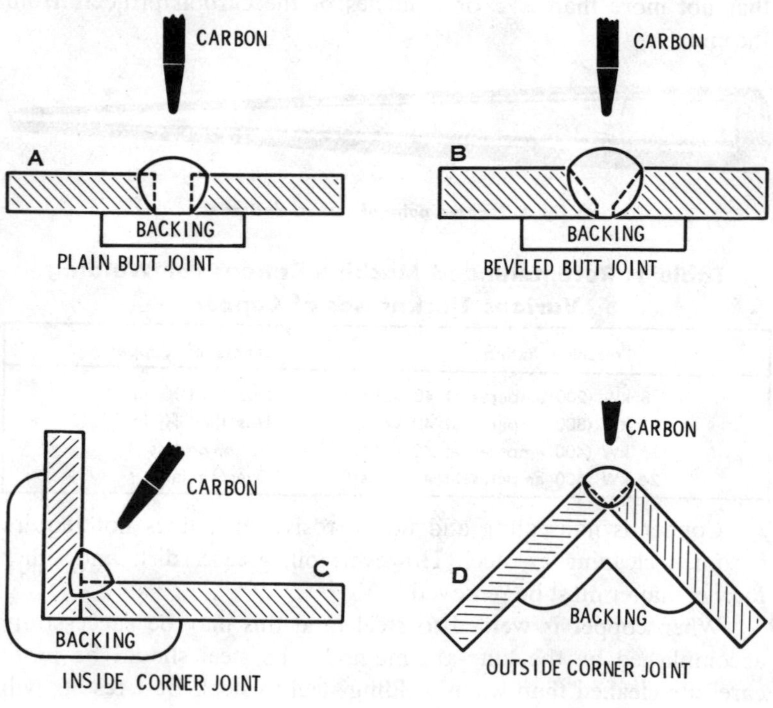

Fig. 2. Position of work,

The filler rod should be held so that it is almost parallel with the work, with the end being melted resting lightly on the work being welded.

Preheating is not necessary on plates up to 3/8 inch thick. Heavier plates should be preheated to a dull red. The preheating is accomplished by applying a gas torch on the work just ahead of the welding arc. A flux is not used.

The quality of the weld depends on the ability of the operator to move rapidly along the joint, fusing the filler rod and copper in the weld without holding the intense heat of the arc in one place long enough to burn the copper. Welding speeds of from 12 to 24 inches per minute are possible.

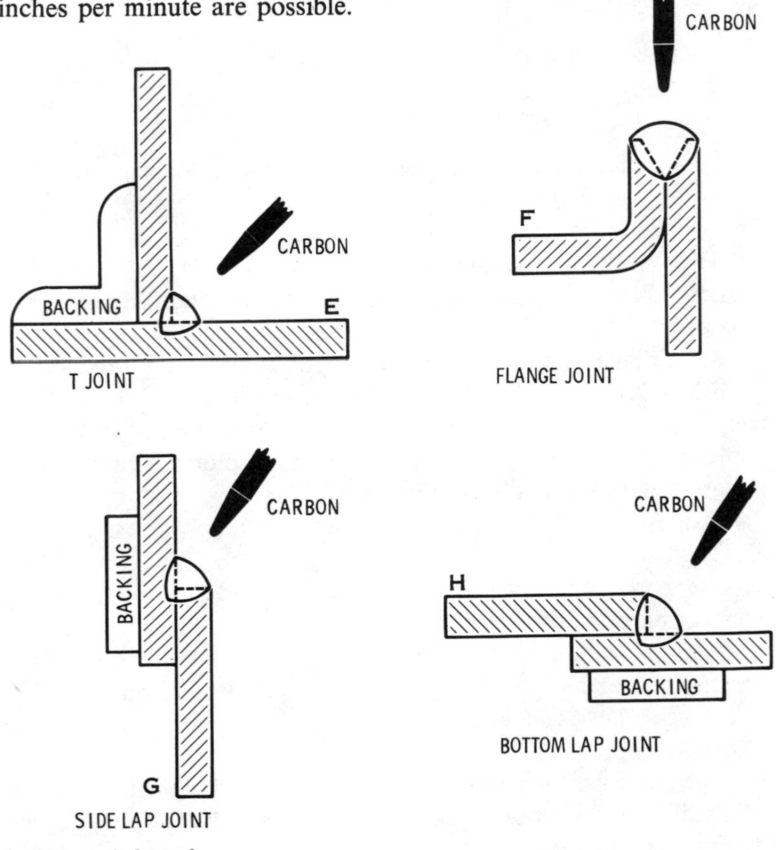

backing, and electrode.

Oxyacetylene Welding

The technique used for the fusion welding of deoxidized copper with deoxidized copper welding rod is very similar to that employed for steel. The same joint designs are also used. Acetylene is the recommended fuel gas for welding copper and its alloys. The other fuel gases do not deliver as much heat as required when welding these metals. A flux is usually necessary.

The work should be preheated. This is particularly true of the larger sections. A neutral flame with a deoxidized copper rod is used.

Oxyacetylene welding is recommended for the following copper alloys:

1. Copper-zinc alloys,
2. Copper-tin alloys,
3. Copper-nickel alloys.

It is not recommended for oxygen-bearing copper, nor for such alloys as copper-tin, copper-aluminum, or copper-beryllium. Braze welding is used quite extensively for joining nickel silver.

The welding tip should be one or two sizes larger than for a similar thickness of steel. This is due to the high heat conductivity of copper.

A welding rod containing traces of phosphorous (approximately 1.0%) is recommended for welding electrolytic and deoxidized copper. The numerous copper alloys require welding rods that are appropriate for their varying compositions. Color matching (weld to base metal) is virtually impossible.

Oxides form during the oxyacetylene welding process and are absorbed into the weld changing its structure. Oxidation can be kept to a minimum by using a flux and a welding rod containing traces of phosphorous.

A suitable flux is almost always necessary when welding the various copper alloys. The type of flux will depend on the alloy being welded. A flux is also recommended for the pure coppers (99.90% copper).

The flux is added prior to welding (usually in the form of a paste brushed onto the surface) and also during welding by dipping the welding rod into the dry flux.

A neutral flame (or one that is slightly oxidizing) is recommended when welding copper and copper alloys. The flame should be positioned vertically over the work with the inner cone held approximately ⅛ to ¼ inch above the surface. Never allow the inner cone to come into contact with the metal. Oscillate the torch tip so that the path of the flame is a continuous movement of small circles. Play the flame against both the welding rod and the work surface in order to equalize the temperature. Otherwise, the welding rod will stick to the base metal.

Shielded Metal-Arc Welding

Copper can be arc welded manually by using a suitable shielded metal-arc electrode. This metal can also be welded manually with a carbon-arc and filler rod, or it can be welded automatically by the carbon-arc process. For shielded metal-arc welding, a phosphor bronze electrode is frequently used.

Preheating is usually not necessary on light copper. On heavier sections, some preheating is advisable due to the high thermal conductivity of copper. The preheating is accomplished by using a carbon electrode with negative polarity and rapidly moving the arc over the area to be welded. As heat builds up in the part being welded, it may be necessary (in some cases) to reduce the current.

The shielded metal-arc welding process is especially recommended for the copper-tin and copper-aluminum alloys. It is also successfully used to join copper-zinc alloys, copper-nickel alloys, copper-beryllium alloys, and copper-silicon alloys. Because of the presence of copper oxides which can cause embrittlement in the area around the weld, the shielded metal-arc welding process is not recommended for the oxygen-bearing electrolytic tough pitch coppers.

Atomic-Hydrogen Welding

Atomic-hydrogen welding has been used to join copper and its alloys but it never gained the extensive use of some of the more popular methods. This is principally due to the fact that the quality of the welds (generally subject to porosity) do not justify the expense. A flux is always required when using the atomic-hydrogen welding process.

Submerged-Arc Welding

Submerged-arc welding has been used to join both oxygen-bearing coppers and oxygen-free coppers, and some of the copper alloys. Those copper alloys that have been successfully joined by submerged-arc welding include the copper-silicon alloys, the copper-aluminum alloys, the copper-nickel alloys, and some of the copper-tin alloys. Steel has also been joined to copper using filler metals matching the composition of the copper base metal.

Submerged-arc welding of copper and copper alloys is restricted primarily to the use of fillet and butt welds on sections of plate (generally ⅛ to ¾ inch thick). The major difficulties encountered when using the submerged-arc welding process are the comparatively low melting points and the high degree of thermal conductivity of copper and the copper alloys. These difficulties have been partially or completely eliminated by the development of granular fluxes suitable for welding this type of metal and its alloys.

TIG and MIG Welding

The necessary heat for welding with the TIG welding process is provided by a very intense electric arc which is struck between a virtually nonconsumable tungsten electrode and the metal surface. The electrodes range in diameter size from 1/16 to ¼ inch depending on the thickness of the metal. TIG welding is given in Table 2.

Because a nonconsumable electrode is used in TIG welding, filler metal must be obtained from another source. On joints where

Table 2. Data for TIG Welding Copper and Copper Alloys

Metal Thickness In.	Tungsten Electrode Size In.	Helium		Argon	
		Current (DC Straight Polarity) Amp	Gas Cu. Ft./Hr.	Current (DC Straight Polarity) Amp	Gas Cu. Ft./Hr.
1/16	½	50-125	10-15	60-150	8-12
⅛	3/32	125-225	14-20	140-280	10-15
3/16	⅛	200-300	16-22	250-375	12-18
¼	3/16	250-350	20-30	300-475	16-25
½	¼	300-550	25-35	400-600	20-30

Courtesy The American Welding Society

filler metal is required, a welding rod is fed into the weld zone and melted with the base metal in the same manner used with oxyacetylene welding. Copper-silicon rods are used with copper and most of the copper alloys. Either a copper-silicon or a copper-nickel rod can be used when welding the copper-nickel alloys. On the other hand, only a copper-aluminum welding rod can be used when welding the copper-aluminum alloys and only a copper-beryllium welding rod when welding the copper-beryllium alloys.

Fluxes are not necessary when using the TIG welding process to join copper and its alloys. The welds are strong, relatively clean, and possess a high degree of ductility.

The MIG welding process is one that has been more recently developed. It uses a consumable electrode and has a much faster rate of deposition than the TIG welding process. The arc is struck between the work and the electrode the latter also functioning as a filler metal. Although TIG welding has enjoyed a much longer use as a method of joining coppers and copper alloys, the MIG welding process is increasing in popularity.

Resistance Welding

Resistance welding is not generally recommended for joining copper and its alloys. When it is used, butt welding is the preferred method and specially developed electrodes are required. Welding time must be very short or the electrodes will have a tendency to stick to the base metal when welding certain types of copper alloys. Those copper alloys that can be easily welded with the resistance welding process are:

1. Copper-silicon alloys,
2. Copper-tin alloys,
3. Copper-nickel alloys.

Flow Welding

Flow welding is a joining process in which the molten filler metal is allowed to flow onto the base metal and into the joint. The temperature of the molten filler metal is sufficient to cause the base metal to soften and fuse together. Flow welding has been used to repair copper and copper-alloy castings.

CHAPTER 21

Welding Aluminum and Aluminum Alloys

Aluminum is a silvery white metal which possesses a high resistance to atmospheric corrosive action, a high thermal and electrical conductivity, a weight which is only one third as much as other commonly used metals, and a high degree of weldability. Due to the many desirable properties of aluminum, it has numerous commercial applications. Aluminum and the aluminum alloys are frequently used in automobiles and aircraft. Additional uses include tanks, cooking utensils, barrels, and a wide variety of other equipment.

Commercially pure aluminum in the annealed state has relatively low mechanical properties. Its tensile strength (13,000 psi) is approximately one fourth to one fifth that of structural steel. It melts at a temperature of 1218° F, and is characterized by a high coefficient of linear expansion.

Its strength may be more than doubled by working the metal cold, that is, by strain hardening, after the cast structure of the ingot has been broken down by hot working. This gain in strength is accompanied by a loss in ductility; the ease of forming is decreased as the amount of cold working is increased.

WELDING PROBLEMS

Oxidation is the greatest problem encountered when welding aluminum. A semitransparent oxide film forms on the surface on exposure to air. This film must be removed to ensure a satisfactory weld. A flux is generally used for this purpose (see USING A FLUX), although mechanical means can be used.

Another welding problem is caused by the high coefficient of linear expansion. Unless certain precautions are taken, the welding heat will cause distortion and buckling. Proper joint design, edge preparation, and preheating (particularly in the case of aluminum castings) will counteract this tendency.

Due to the high thermal conductivity of aluminum, the heat source must be highly concentrated and provide deep penetration. In addition to that, welding aluminum requires a fast weld cycle and a particularly narrow weld zone. All of these welding procedure adjustments will reduce or eliminate the possibility of excessive shrinkage (and the accompanying weld cracking) that generally results from the high thermal conductivity of aluminum.

Aluminum does not show the red heat color that other metals exhibit when reaching their melting-point temperature. It is, therefore, difficult to determine when the proper welding temperature has been reached. Two indicators that the welder should watch for are:

1. The melting of the dry flux,
2. A blistering of the metal surface.

Both of these conditions indicate that a proper welding temperature has been attained.

Finally, it is essential that the surface, welding rods, and electrodes be absolutely clean when welding aluminum.

ALUMINUM ALLOYS

Commercial aluminum consists of 99.0% to 99.6% pure aluminum. The small percentage of other elements (iron and silicon) is regarded as an impurity subject to control.

The tensile strength of the aluminum alloys in the cast or the annealed condition varies, depending on their composition, up to values about double that of commercial aluminum. Aluminum alloys melt at lower temperatures (about 895° F to 1215° F) than does commercially pure aluminum. This must be taken into consideration when welding.

So many aluminum alloys have been developed that it has been necessary to devise special numerical systems for their identification. Alloying elements are added to pure aluminum to change properties (specifically to increase its strength). The major alloying elements are:

1. Silicon,
2. Copper,
3. Magnesium,
4. Zinc,
5. Manganese,
6. Nickel.

Other alloying elements which are used to a lesser degree than the six just mentioned include tin, lead, titanium, berylium, chromium, and cadmium.

Aluminum alloys can be divided into two basic categories:

1. Wrought aluminum alloys,
2. Cast aluminum alloys. The wrought aluminum alloys are the more numerous.

Because it is very difficult to make castings from pure aluminum, it is necessary to add alloying elements to the pure form of the metal to give it the characteristics (a graduated range of freezing stages) necessary for casting. Silicon (ranging as high as 20%) is commonly used for this purpose, although both copper and zinc have also been used as well as combinations (e.g. silicon-magnesium-copper). The tensile strength of heat-treated cast aluminum alloys ranges as high as 45,000 psi. For non-heat-treated cast aluminum alloys it is considerably lower.

The wrought aluminum alloys are formed by adding one or more of the six major alloying elements which results in a tensile strength much higher than pure aluminum. While cast aluminum alloys are limited to the production of aluminum castings (using metal or sand molds), wrought aluminum alloys can be forged, drawn, extruded, or rolled.

Aluminum alloys can also be classified according to whether or not they are treatable. The hardness of nonheat-treatable aluminum alloys cannot be increased by any form of heat treatment. The heat-treatable alloys, on the other hand, do respond to some form of heat treatment. The result is increased hardness and strength.

Among the wrought aluminum alloys, the 1XXX, 3XXX, and 5XXX series (see ALUMINUM IDENTIFICATION SYSTEM) are non-heat-treatable. The 2XXX, 4XXX, 6XXX, and 7XXX series denote heat-treatable types of alloys.

WELDING METHODS

Aluminum and the aluminum alloys can be welded by a number of commercial methods (Table 1), such as:

1. Gas welding (both oxyacetylene and oxyhydrogen flame),
2. Metal-arc welding,
3. Resistance welding (both spot and seam),
4. Carbon-arc welding,
5. Atomic hydrogen welding,
6. TIG welding,
7. MIG welding,

Table 1. Process Chart For Welding Typical Aluminum Alloys

METAL OR ALLOY	Gas	Arc	Resistance	
			Spot	Seam
High-purity aluminum	A	A	A	A
Commercially pure aluminum (1100)	A	A	A	A
Aluminum-manganese alloy (3003)	A	A	A	A
Aluminum-magnesium-chromium alloy (5052)	A	A	B	B
Aluminum-silicon-magnesium alloys (6053, 6061, 6063)	A	A	A	A
Aluminum-copper-magnesium-manganese alloys (3014, 2017, 2024)	No	B	B	B
Alclad 2014 and Alclad 2024	No	B	A	A

A —Welds can be applied generally on a commercial basis.
B —Commercial welding depends on design or special technique.
No—Commercial welding is not feasible.

Courtesy The James F. Lincoln Arc Welding Foundation

8. Aluminum brazing (see Chapter 15. BRAZING AND BRONZE WELDING).

ALUMINUM IDENTIFICATION SYSTEM

In 1954 the Aluminum Association devised a numerical identification system for aluminum alloys which replaced several older commercial designation systems. The rationale for introducing this newer system was to create a broader and more exact means of identifying each alloy and to establish uniformity in their identification throughout the industry.

Where possible, elements of the older system were incorporated in the newer one. A popular identification system is the one devised by the *Aluminum Company of America.* It consists of one- and two-digit numbers in combination with a letter suffix (e.g., S denotes a wrought aluminum alloy). In the new system, a 14S alloy is redesigned 2014. The last two digits (XX14) are numbers identifying an individual alloy. The 2XXX category indicates that copper is the major alloy (up to 5%). In the older identification system, 10S-29S indicated that copper was the major alloy.

The new identification system devised by the Aluminum Association assigns four-digit numbers to wrought aluminum alloys (1XXX-7XXX) and three-digit numbers to cast aluminum alloys (0XX-7XX). Table 2 lists the identification numbers for each of the eight categories of wrought aluminum alloys.

The temper designation of each aluminum alloy is indicated by a letter suffix that appears immediately after the alloy identification number. It is separated from this number by a dash (e.g. 2014-O).

There are four temper categories:

1. Annealed (-O),
2. As fabricated (-F),
3. Strain hardened (-H),
4. Heat treated (-T).

These suffixes indicate the condition of the alloy, specifically the strength and how it was obtained (Table 3).

Table 2. Numerical Designation of Wrought Aluminum Alloys

NUMBER SERIES	MAJOR ALLOYING ELEMENT
1xxx	None (99.0% minimum aluminum)
2xxx	Copper
3xxx	Manganese
4xxx	Silicon
5xxx	Magnesium
6xxx	Magnesium-silicon
7xxx	Zinc
8xxx	Other major alloying elements

Notes: (a) In the 1xxx series, the last two digits indicate the specific minimum content of aluminum in hundredths of 1%. Example: 1075 would have 99.75% minimum aluminum. The second digit indicates the number of individual impurities subject to rigid control.
(b) In all other series, the last two digits are assigned to specific alloys and the second digit merely indicates modifications of the alloy.
(c) The letter x before the number indicates it is an experimental alloy not yet accepted as a permanent addition to the aluminum family.

Courtesy The James F. Lincoln Arc Welding Foundation

GAS WELDING OF ALUMINUM

Fusion welding with an oxyacetylene or oxyhydrogen torch are among the earlier methods employed for joining aluminum or its alloys. In the hands of an experienced welder, the process is simple and rapid. The same types of joints made in any other metal can also be made in aluminum.

Some training will be necessary before a welder can turn out consistently reliable results with aluminum. The metal has distinct characteristics of its own, which involve somewhat different techniques from those required with steel, cast iron, and other metals. This technique, however, is by no means difficult to acquire, since aluminum is one of the most readily weldable of all metals.

SELECTION OF TIP SIZE

For any particular welding job, the selection of correct torch tip size is largely a matter of experience. The size of the tip

Table 3. Interpretation of Letter Suffix to Aluminum Alloy Number Designation

SUFFIX	SIGNIFICANCE
—F	As-fabricated, or as-cast (seldom used except for nonheat-treatable cast alloys)
—O	Annealed, recrystallized (wrought products only)
—H	Strain-hardened (only wrought products not heat-treatable).
	—H1 Strain-hardened only
	—H2 Strain hardened and then partially annealed
	—H3 Strain-hardened and then stabilized (applies only to alloys containing magnesium; these would otherwise age-soften at room temperature)
	The second digit following the H indicates the temper or degree of hardening: Recognizing zero as full annealed and 8 as full hard, 2 is quarter-hard, 4 is half-hard, etc.; 9 means extra hard.
—W	Solution heat-treated, unstable temper. Alloy has been heated and quenched to obtain a solid solution, and no artificial aging treatment has been given. Material is in a condition of unstable temper, for it is slowly aging and becoming stronger at room temperature.
—T	Treated to produce stable temper (other than —F, —O, or —H). The T is usually followed by a number to indicate how the temper was achieved:
	—T2 Annealed (cast products only)
	—T3 Solution heat-treated, then cold-worked (wrought products only)
	—T4 Solution heat-treated, then naturally aged to stabilize it
	—T5 Artificially aged only
	—T6 Solution heat-treated, then artificially aged
	—T7 Solution heat-treated, then stabilized
	—T8 Solution heat-treated, cold-worked, then artificially aged (wrought products only)
	—T9 Solution heat-treated, artificially aged, then cold-worked (wrought products only)
	—T10 Artificially aged, then cold-worked (wrought products only)

Courtesy The James F. Lincoln Arc Welding Foundation

depends largely on the shape and size of the object to be welded, as well as on its thickness, since the larger articles have a greater capacity for heat and a larger radiating surface (Table 4).

The skill of the worker is another factor, since a quick worker will be able to use a larger tip than a slower and less experienced man.

Table 4. Approximate Size of Tips and Relative Gas Pressures Used in Welding Aluminum of Different Thicknesses

OXYHYDROGEN				OXYACETYLENE		
Metal Thickness B & S Gauge	Diam. of Orifice in Tip Inch	Oxygen Pressure Lb./sq. in.	Hydrogen Pressure Lb./sq. in.	Diam. of Orifice in Tip Inch	Oxygen Pressure Lb./sq. in.	Acetylene Pressure Lb./sq. in.
24-22	0.035	1	1	0.025	1	1
20-18	0.045	1	1	0.035	1	1
16-14	0.065	2	1	0.055	2	2
12-10	0.075	2	1	0.065	3	3
1/8-3/16	0.095	3	2	0.075	4	4
1/4	0.105	4	2	0.085	5	5
5/16	0.115	4	2	0.085	5	5
3/8	0.125	5	3	0.095	6	6
5/8	0.150	8	6	0.105	7	7

Courtesy The Aluminum Company of America

FLAME ADJUSTMENT

In welding aluminum and its alloys, an oxyhydrogen flame is often used when it will supply sufficient heat for the job at hand. The oxyhydrogen flame produces a clean and satisfactory joint and usually supplies sufficient heat for welding metal up to 1/4 inch in thickness. A larger tip is used for hydrogen than for acetylene on any given gauge of sheet.

A neutral flame is used with either hydrogen or acetylene. The neutral flame is not only the hottest, but it also prevents excessive oxidation of the molten metals by providing a reducing gas envelope. A slightly carburizing flame may also be used. It, too, has the tendency to reduce oxides to a minimum. Fig. 1 illustrates the various characteristics of both oxyacetylene and oxyhydrogen flames.

USING A FLUX

The oxide film that covers the surface of aluminum has a melting point of approximately 3500°F. Consequently, it will not break down during the welding process. Since this thin film of aluminum oxide will prevent the formation of a suitable weld, it must be removed from the welding zone by mechanical means (rubbing with steel wool, grinding, machining, etc.) or through chemical reduction by means of a flux solution prior to welding.

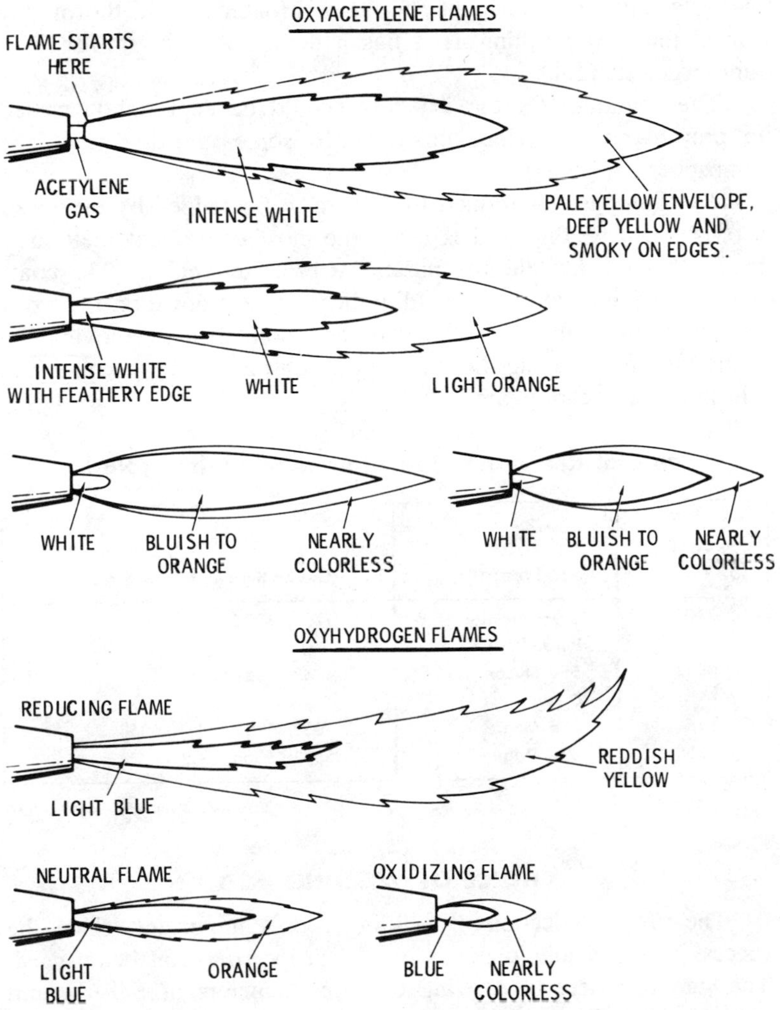

Courtesy Airco Welding Products

Fig. 1. Characteristics of oxyacetylene and oxyhydrogen flames.

Fluxes designed specifically for use with aluminum and aluminum alloys must be used.

The flux is prepared by mixing it with water to the consistency of a thin paste (about two parts of flux to one part of water). A day's supply of the flux paste should be made up each morning. Flux left overnight should be broken up and thoroughly stirred the next morning as it has a tendency to crystallize in a lump when standing.

The container for the dry flux should be kept tightly closed to prevent spoilage since flux absorbs some moisture from the atmosphere.

The flux paste is applied to a seam to be welded by means of a brush. If a welding rod is used, the most convenient method is to dip the rod into the flux paste just prior to welding. The coating of flux is melted by the welding heat. It runs down on the work and flows along the joint ahead of the torch flame, removing the oxide film, and leaving the metal in an absolutely clean condition. Thus the parts readily join.

Table 5. Recommended Aluminum Welding Rods

AWS Class No.	Type	Use on Base Metal, AWS No.
1100	Commercially Pure. Drawn	1100, 3003
4043	5% Silicon, Drawn	4043, 6061, 6063
5154	3½ Mag.	5052, 5154, 6061, 6063
5356	5% Mag.	5052, 5356

Courtesy Airco Welding Products

CHOICE OF WELDING ROD

The proper selection of welding rod is important, since the success of many jobs depends on the proper material being used. The standard sizes of welding rod are diameters of 1/16, ⅛, and ¼ inch. The diameter of the rod should approximate the thickness of the material to be welded. Ordinarily, ⅛-inch diameter rod is suitable for welding any thickness of metal up to ⅛-inch; and 3/16-inch diameter rod for the heavier gauges. However, this welding rod may be obtained in any gauge to suit individual preference. Table 5 lists the recommended aluminum welding rods.

An ER4043 welding rod is often used because of its tendency to reduce the excessive shrinkage stresses common to aluminum. As a general rule, a filler metal with a melting point below that of the base metal will reduce these shrinkage stresses.

PREHEATING

Aluminum sheet ⅜ inch or more in thickness and the larger aluminum castings, should be preheated to 700°F or 800°F in order to avoid heat strains and to reduce the amount of oxygen and acetylene required for the actual melting of the seam. If the base metal, for some distance on either side of the seam, is maintained at a temperature slightly below its melting point, then when the torch is applied, the additional expansion at any one point will be small and unlikely to cause distortion.

During preheating the temperature should not be allowed to exceed 800°F. If the temperature rises above this point, there is the danger of some of the ingredients of the alloy melting and producing "burned" material. In addition, the high temperature might cause large castings to collapse in the preheating furnace.

In many welding shops, pyrometers are used to determine the proper preheating temperature. For shops not equipped with pyrometers, the preheating temperature may be approximated by testing with: (1) pine stick, (2) chalk, or (3) sound. At the proper temperature for welding, a pine stick rubbed on a casting will leave a charmack on it. Chalk marks, made with carpenter's blue chalk, will turn white at the proper temperature for welding. When struck, cold aluminum gives a metallic sound, which becomes duller as the temperature is raised. At the temperature required for welding, there is no longer a metallic ring.

WELDING CASTINGS

In general, the welding of aluminum alloy castings requires technique similar to that used on aluminum sheet and other wrought sections. However, the susceptibility of many castings to thermal strains and cracks, because of their intricate design and varying section thickness, should be carefully considered. In addition, many castings in highly stressed structures depend on heat treatment for their strength, and welding tends to destroy the effect of such initial heat treatments. Unless satisfactory facilities for reheat treatment are available, the welding of such heat-treated castings is not recommended.

Before welding a broken casting, it should be cleaned carefully with a wire brush and gasoline to remove every trace of oil, grease, and dirt. Because of the action of the torch and puddling iron, it is not necessary to tool the crack in the casting or cut out a V-groove (unless, of course, the casting has a very heavy cross section). It is necessary, however, that the stock surrounding the defect be completely melted or cut away before proceeding with the weld. If a piece is broken out, it should be held in correct position by light iron bars and appropriate clamps. The clamps should be so attached that the casting will not be stressed during heating. The aluminum casting is at this point ready for preheating.

If the casting is a large one or one with intricate sections, it should be preheated slowly and uniformly in a suitable furnace prior to welding. If the casting is small, or if the weld is near the edge and in a thin walled section, the casting may be preheated in the region of the weld by means of a torch flame. Cast aluminum should be heated slowly, to avoid cracking in the section of the casting nearest the flame.

Broken pieces should be tack welded into position as soon as the casting has been preheated. The welding of a broken piece should begin at the middle of the break, and should proceed toward the ends. The welding rod must be melted by the torch, as the heat of the molten metal of the weld is not sufficient to melt it. When the weld is finished, the excess molten metal should be scraped off with a puddling iron, and the casting allowed to cool slowly.

Holes in castings are welded in much the same manner as are cracked and broken castings. It is necessary to melt away or cut away the sides of the hole in order to remove all pockets and to permit proper manipulation of the torch.

Aluminum-silicon or aluminum-copper-silicon welding rods are used for welding ordinary castings. For heat-treated alloy castings, it is preferable to use welding rods of the same alloy as the casting.

Flux is used when welding aluminum castings. Puddling alone will merely break up the oxide film and leave it incorporated in the weld, while fluxing will cause the oxide particles to rise to the surface, resulting in a clean sound weld. It is important that the added metal be completely melted and the molten metal thoroughly worked with the end of the welding rod or with a puddling iron. Thus the flux and oxide are worked to the surface of the molten metal and there is very little danger of the finished weld becoming contaminated with particles of flux or other foreign inclusions.

GAS BEAD WELDS

Bead welds made with the gas welding process should be practiced on 1⁄16- and 1⁄8-inch aluminum sheets (1100 or 3003 analysis). A forehand movement of the torch (with very little oscillation) is recommended. Fig. 2 illustrates the basic contours for a correctly formed weld bead. Note the absence of overlapping or undercutting.

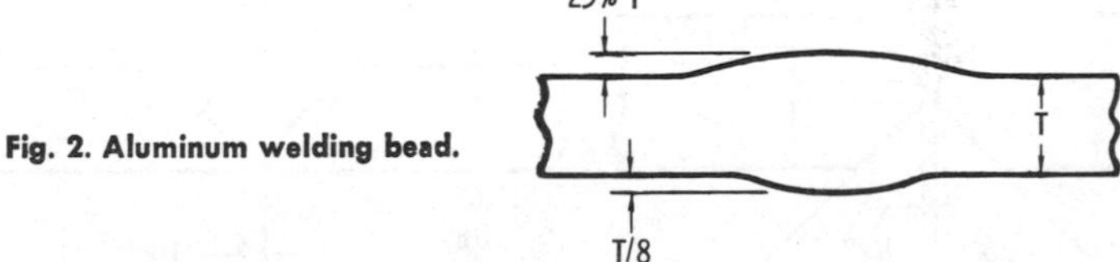

Fig. 2. Aluminum welding bead.

WROUGHT ALUMINUM GAS BUTT WELDS

Butt welds should also be practiced on 1⁄16- and 1⁄8-inch thick aluminum sheets (1100 or 3003 analysis). Fig. 3 indicates two types of edge preparations (flanged and straight) that can be used.

Before welding, clean the edges of the aluminum with sandpaper and immediately apply the flux to the edge, top, and bottom surface of the area to be welded. On thicker pieces of aluminum, the flux should be applied at least 1 inch back from the edges on

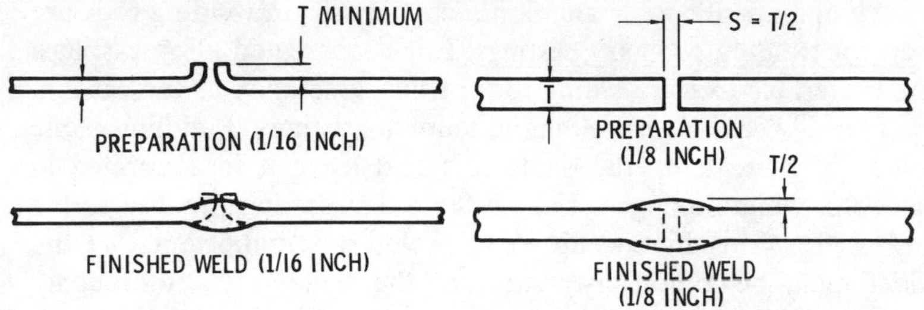

Courtesy Airco Welding Products

Fig. 3. Butt welds on 1/16 and 1/8 inch thick aluminum sheets.

both the top and bottom. A forehand movement of the welding torch is recommended.

Once the beginner has succeeded in making butt welds on aluminum sheet, he should practice making the same weld on 3/16- and 1/4-inch aluminum plate (1100 and 3003 analysis). Because of the thickness of the metal, the edges will have to be beveled. Fig. 4 illustrates suggestions for beveling.

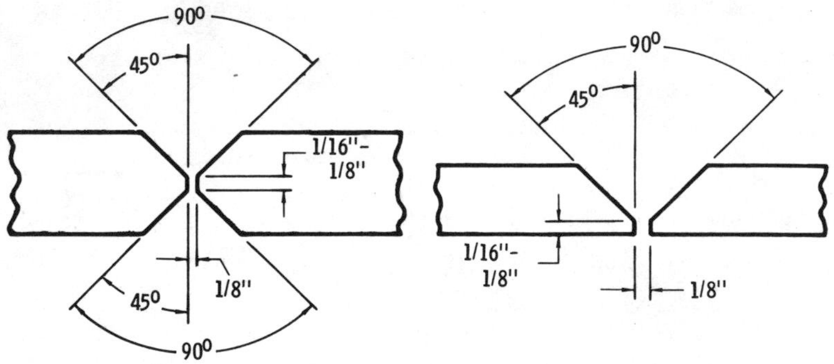

Courtesy Airco Welding Products

Fig. 4. Suggestions for beveling aluminum plate.

The forehand method is also recommended for welding aluminum plate. The cleaning preparation and flux application described for butt welding aluminum sheet apply here, too.

Fig. 5 illustrates several recommended joints for welding aluminum.

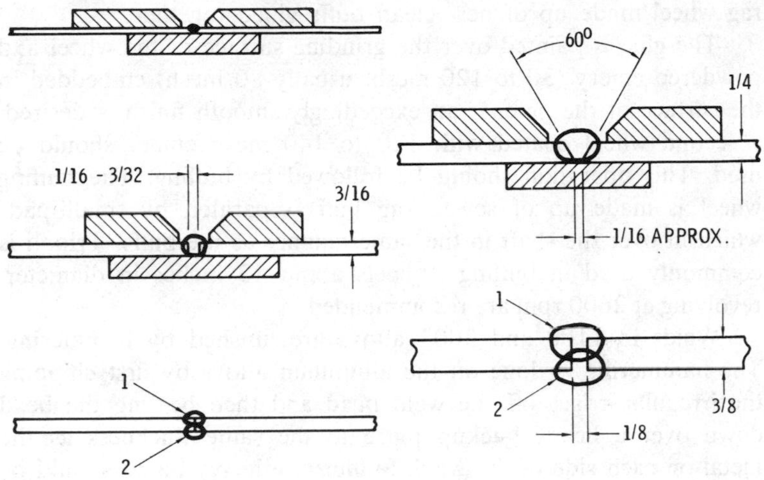

Fig. 5. Various joints for welding aluminum.

CLEANING AND FINISHING WELDS

As soon as the weld is completed and the work has had time to cool, it should be washed thoroughly to remove all traces of flux. This is particularly true of parts that are to be painted subsequent to welding, as the presence of even minute quantities of flux under a paint coating will lift the coating in a short time. On other parts, residual welding flux on the joint may be corrosive to the metal when exposed to atmospheric moisture.

The most satisfactory way of removing the flux is by providing a medium that will dissolve the flux ingredients. Accessible parts may be cleaned with a brush and boiling water applied to the surface of the weld. Inaccessible welds or weld parts are cleaned by immersing the part in a cold solution of 10% sulphuric acid held at 150°F. for 10 minutes. The acid should contact both the inside and outside surfaces.

Welds are improved in appearance by finishing the surface. This may be done by chipping off the greater portion of the excess metal with a pneumatic chisel. The cutting edges of the chisel should frequently be dipped in light oil to prevent aluminum chips sticking to the chisel. After chipping, the joint should be ground on a rag wheel made up of new clean buffs glued together.

The glue is painted over the grinding surface of the wheel and powdered emery (30 to 120 mesh, usually 80 mesh) embedded in the surface of the glue. If an exceedingly smooth finish is desired, a second wheel coated with 120 to 140 mesh emery should be used. This operation should be followed by buffing. The buffing wheel is made up of sewed rag buffs separated by small pads which fit over the shaft in the same manner as the buffs. Tripoli is commonly used in buffing. Wheels about 12 inches in diameter, revolving at 2600 rpm are recommended.

Welds in 1100 and 3003 alloys are finished by hammering. The hammering is done on the aluminum alloys by first chipping the irregular edges off the weld bead and then beating the bead down over a heavy backup piece to the same thickness as the metal on each side of the weld. Sufficiently heavy blows should be used to work the weld throughout the cross section and not merely to peen the surface. This process tends to relieve the contraction strains set up during cooling of the weld and the adjacent metal, as well as to close up any surface porosity that may be present.

METAL-ARC WELDING OF ALUMINUM

Pure aluminum and various aluminum alloys in sheet, forged, extruded, and cast forms can be welded with the metal-arc welding process. A DCRP current is recommended with flux covered electrodes.

The metal-arc welding process is particularly applicable to certain classes of work. There is less bucking and warping of the work with metal-arc welding than with gas welding.

A heavily coated electrode of 5% silicon aluminum alloy is the kind most frequently used for metal-arc welding. The electrode coating should be such that it will dissolve any aluminum oxide that may be formed during the welding process. The coating should also form a very fusible slag to cover the molten weld metal and protect it from oxidation while cooling.

The high melting rate on most aluminum electrodes necessitates rapid welding and sometimes makes it difficult to get sufficient heat into the work. There may be a tendency toward porosity along the line of fusion. This can be eliminated by supplying sufficient heat (which is accomplished by slightly preheating). Metal-arc welding operates most successfully on aluminum thicknesses of ⅛ to 1 inch.

FLUX FOR METAL-ARC WELDING

A proper flux is just as essential for successful metal-arc welding as it is for gas welding. It serves the same purpose, namely, removal of the natural oxide film. In addition, it permits the molten metal to flow smoothly from the end of the electrode, and aids in stabilizing the arc. The flux is prepared and applied in the same way as described for gas welding.

If a dipping trough for the flux is used, it should be made from aluminum or brass. Iron should be avoided, as the chemical action of the flux on that metal will cause contamination. The slag produced by the flux and welding heat should be removed immediately after each pass.

METAL-ARC WELDING PROCEDURE

The exceedingly high temperature of the arc, together with the short flame, is an important factor in developing the improved physical properties of arc welds. The high temperature produces rapid fusion in the work and in the electrode, while the short arc localizes the heat. These properties greatly reduce the width of the heat affected area in the sheet.

The length of the metal arc used is a short one, because it is found to aid in stabilizing. As the currents used in metal-arc welding aluminum sometimes fall in the lower range of the smallest available welding generators, it may be found that the machine does not hold a stable arc. To correct this difficulty, a stabilizer consisting of about ½ ohm can be connected in series with the arc.

In some cases, results may be improved by changing the polarity of the electrode. No general rule can be given, but a suitable polarity for the particular kind or class of work can be easily found by trial (Table 6).

Welding Aluminum and Aluminum Alloys

As was mentioned previously, a DCRP current with flux-covered electrodes is recommended for aluminum metal-arc welds. The electrode should be held approximately perpendicular to the work at all times, with the electrode coating almost touching the molten pool of metal. The arc should be so directed that both edges of the joint to be welded are properly and uniformly heated.

Welding should advance at such a rate to make a uniform bead. Before starting a new electrode the slag should be removed mechanically from the crater of the weld and from approximately one inch of the weld back of the crater.

To start a new electrode, the arc should be struck in the crater of the bead, then quickly moved back along the completed weld for one-half inch, then the welding should proceed forward after the crater is completely remelted. Before starting to weld, the metal should, of course, be properly cleaned.

Table 6. Electrode Size and Machine Setting For Aluminum Metallic-Arc Welds

Thickness Inch	Electrode Diameter Inch	Amperes	Rods Per Pound
0.064	1/8	45-55	32
0.081	1/8	55-65	32
0.102	1/8	65-75	32
0.125	1/8	75-85	32
5/32	1/8 or 5/32	85-100	32-23
3/16	5/32	100-125	23
1/4	5/32 or 3/16	125-175	23-17
5/16	3/16	175-225	17
3/8	1/4	225-300	10.5

Welding is performed in a vertical plane by proceeding either in a downward or upward direction. Either a straight line forward motion or a weaving motion may be used in advancing the arc. Overhead welding is done with a number of straight beads.

In tack welding, the current can be increased approximately 50% above that for continuous welding. The electrode should be given a rotary motion.

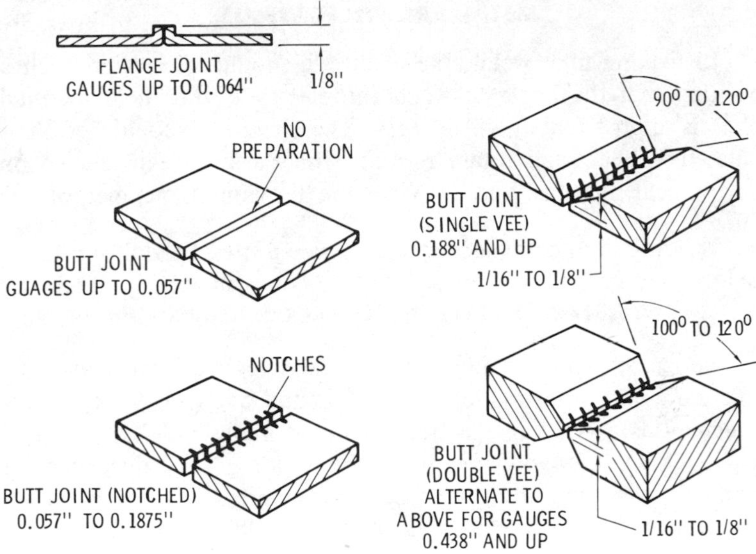

Fig. 6. Preparation of butt joints for gas welding.

METAL-ARC BUTT WELDS

When making a butt weld, the work should be held in position by jigs and backed up by copper. For plates ⅛ inch and thicker, the copper backing should be slightly grooved beneath the joint to be welded. Welding of butt joints in 3/16-inch plate and heavier should be done with two beads. No backing or clamping is required for welding joints in this manner (Fig. 6). Square-groove butt welds in thicknesses of ⅛ to 1 inch are easily welded by the metal-arc welding process.

METAL-ARC LAP WELDS

In this type of weld, the electrode should be given a small rotary motion, playing the arc first on the upper member of the joint and then on the lower member. The electrode should be held in such a position that the angle between the electrode and the horizontal plate is approximately 45 degrees.

METAL-ARC FILLET WELDS

In making fillet welds, the electrode should be held in such a position that the angle between the electrode and the horizontal plate is approximately 45 degrees. The electrode should be manipulated with a small rotary motion with the arc being played on the vertical member and then on the horizontal member of the joint.

RESISTANCE SPOT WELDING OF ALUMINUM

Resistance spot welding is best suited for the non heat-treatable aluminum alloys (1XXX, 3XXX, and 5XXX). Less success is achieved in welding the heat-treatable alloys (2XXX, 6XXX, and 7XXX), but adjustments in the welding procedure can be made to overcome any difficulties encountered. Aluminum alloys in the annealed or soft state should not be welded with this process.

During the resistance spot-welding process, a molten zone is formed between the two aluminum sheets, over a small area directly between the electrodes. In order to keep pressure on this area at a minimum, one of the electrodes should establish only a point contact on its sheet. To do this, the electrode may be finished with either a conical or spherical point. The conical design is more commonly employed, the cone forming a 158° to 166° included angle (machined at an 11° to 7° angle).

Most spot-welding machines designed for other materials are not suited for use with aluminum alloys, since they do not provide a sufficiently high welding current for the work. Machines designed specifically for aluminum alloys supply any desired secondary current up to a value representing the heaviest work which the machine is expected to handle. Table 7 gives figures which may be used as a basis for specifying the maximum current which a machine should supply for any given gauge of work.

RESISTANCE SEAM WELDING OF ALUMINUM

Resistance seam welding of aluminum is similar in almost every respect to resistance spot welding (e.g. the need for higher

Table 7. Machine Settings for Spot Welding Aluminum Alloys

Gauge		Time Cycles	Current Amperes	Electrode Pressure	
B & S No.	Inch			Min. Lb.	Max. Lb.
26	0.016	4	14,000	200	400
24	0.020	6	16,000	300	500
22	0.025	6	17,000	300	500
20	0.032	8	18,000	400	600
18	0.040	8	20,000	400	600
16	0.051	10	22,000	500	700
14	0.064	10	24,000	500	700
12	0.081	12	28,000	600	800
10	0.102	12	32,000	800	1000
8	0.128	15	35,000	800	1200

welding currents, specially designed machines, etc.). A major difference between the two is the type of electrode used. Resistance seam welding uses electrodes formed in the shape of wheels. These produce a series of overlapping welds which result in a pressure-tight seam. Although resistance seam welding may employ either synchronous or nonsynchronous timing, the former is absolutely essential to resistance seam welding, not only to obtain uniformity of successive overlapping welds, but also to meet electrical requirements.

CARBON-ARC WELDING OF ALUMINUM

The carbon arc can also be successfully adapted to welding certain types of joints in aluminum. This method, however, is not as flexible in its application as metal-arc welding.

The use of the carbon arc for welding aluminum at this time is confined to butt joints (either straight or corner) and the simpler

Table 8. Current Values

3/32 in. electrode .... 75 to 100 amperes
1/8 in. electrode .... 125 to 150 amperes
5/32 in. electrode .... 150 to 175 amperes
3/16 in. electrode .... 175 to 200 amperes

Table 9. Recommended Filler Metals For MIG and TIG Welding of Aluminum

BASE METAL	319 355	43 356	214	6061 6063 6151	5456	5454	5154 5254	5086	5083	5052 5652	5005 5050	3004	1100 3003	1060
1060	4145 4043 4047	4043 4047 4145	4043 5183 4047	4043 4047	5356 4043	4043 5183 4047	4043 5183 4047	5356 4043	5356 4043	4043 4047	1100 4043	4043	1100 4043	1260 4043 1100
1100 3003	4145 4043 4047	4043 4047 4145	4043 5183 4047	4043 4047	5356 4043	4043 5183 4047	4043 5183 4047	5356 4043	5356 4043	4043 5183 4047	4043 5183 5356	4043 5183 5356	1100 4043	
3004	4043 4047	4043 4047	5654 5183 5356	4043 5183 5356	5356 5183 5556	5654 5183 5356	5654 5183 5356	5356 5183 5556	5356 5183 5556	4043 5183 4047	4043 5183 5356	4043 5183 5356		
5005 5050	4043 4047	4043 4047	5654 5183 5356	4043 5183 5356	5356 5183 5556	5654 5183 5356	5654 5183 5356	5356 5183 5556	5356 5183 5556	4043 5183 4047	4043 5183 5356			
5052 5652	4043 4047	4043 5183 4047	5654 5183 5356	5356 5183 4043	5356 5183 5556	5654 5183 5356	5654 5183 5356	5356 5183 5556	5356 5183 5556	5654 5183 4043				
5083	NR	5356 4043 5183	5356 5183 5556	5356 5183 5556	5183 5356 5556	5356 5183 5556	5356 5183 5556	5356 5183 5556	5183 5356 5556					
5086	NR	5356 4043 5183	5356 5183 5556	5356 5183 5556	5356 5183 5556	5356 5183 5554	5356 5183 5554	5356 5183 5556						

5154 **5254**	NR	4043 5183 4047	5654 5183 5356	5356 5183 4043	5356 5183 5554	5654 5183 5356	5654 5183 5356
5454	4043 4047	4043 5183 4047	5654 5183 5356	5356 5183 4043	5356 5183 5554	5554 4043 5183	
5456	NR	5356 4043 5183	5356 5183 5556	5356 5183 5556	5556 5183 5356		
6061 **6063** **6151**	4145 4043 4047	4043 5183 4047	5356 5183 4043	4043 5183 5654			
214	NR	4043 5183 4047	5654 5183 5356				
43 **356**	4043 4047	4145 4043 4047					
319 **355**	4145 4043 4047						

NOTE: First filler alloy listed in each group is the all-purpose choice. NR means that these combinations of base metals are not recommended for welding.

Courtesy Chemetron

types of lap joints. In this type of welding, a flux-coated filler rod is inserted between the joint to be welded and the carbon electrode. The welding heat is then transmitted through the rod to the edges of the joint. As the flux becomes molten, it will remove the oxide film with which it comes in contact, allowing the base metal and filler material to fuse together.

This method of arc welding is particularly well adapted to butt joints in the lighter gauge metals from 20 to 14 B. & S. gauge).

CURRENT VALUES

Use only sufficient current to obtain proper fusion as excess heat causes splattering as well as porosity and undercutting. The exact machine setting for each job depends upon the set up and other variables and should be established by practice welds on scrap pieces. Table 8 gives suggested current values for a number of electrode sizes.

ATOMIC HYDROGEN WELDING OF ALUMINUM

This method of welding aluminum is not very common today, although at one time it enjoyed more frequent use. Its principal advantages are: (1) a hydrogen shield that protects the metal surface from the surrounding atmosphere, and (2) the relatively quick attainment of welding temperatures. Despite the protection afforded by the hydrogen shield, it is necessary to use a flux with atomic hydrogen welding. The high cost of this welding process is a contributing factor in its limited use today.

TIG WELDING OF ALUMINUM

Tungsten inert gas (TIG) welding is the most commonly used method of welding aluminum today. Originally helium (heliarc welding) was used as the inert shielding gas, but argon in combination with AC has come to be preferred. The latter provides a much more effective shield than hydrogen or other gases. As a result, a flux is not necessary on aluminum surfaces that have been adequately cleaned. Welding is possible in all positions.

Exceptional arc control and a high degree of visibility are

achieved through the use of the nonconsumable tungsten electrode in TIG welding. As a result, this method of welding is highly suitable for joining the thinner gauges of aluminum without the need for a filler metal. Because TIG welding of aluminum in thicknesses greater than ¼ inch required preheating, a more feasible method of welding larger weldments of this metal was sought. This led to the development of metal inert gas (MIG) welding.

MIG WELDING OF ALUMINUM

Metal inert gas (MIG) welding uses an inert shielding gas and a consumable electrode automatically fed through the nozzle of the welding gun in the form of a metal wire. The high metal deposition rate characteristic of this process makes it exceptionally suitable for medium and heavy weldments. Extensive joint preparation and design is not required due to the arc penetration characteristics of MIG welding. Weld clean-up is kept to a minimum. For the best results, it is recommended that an argon shielding gas be used with DCRP current. Table 9 gives recommended filler metals for MIG and TIG welding of aluminum.

CHAPTER 22

Welding Steel And Steel Alloys

It is not always easy to make a clear-cut distinction between iron and steel. This is particularly true if the carbon content level of either one is close to the 2% level, for the carbon content is frequently used as a factor in classifying them. If the carbon content exceeds 2%, the metal is classified as one of the several irons (e.g. cast iron, wrought iron, etc.). On the other hand, a carbon content of less than 2% indicates that the metal should be classified as a steel.

Steels and steel alloys are produced in a seemingly endless variety of types, each differing according to its specific composition. Steels differ in composition both as a result of the process used to produce them and the alloys (e.g. tungsten, silicon, manganese, etc.) added during the production process in an effort to obtain certain desired physical properties. There is quite literally a steel or steel alloy for almost every industrial purpose.

The fact that there are so many types of steels and steel alloys has resulted in some problems of classification and terminology. These are dealt with briefly in this chapter. More detailed descriptions of the classifications of these metals can be found in any of the studies produced by the *Society of American Engineers* or the *American Iron and Steel Institute* and incorporated in technical reference works (see, for example, *Baumeister and Marks: Standard Handbook for Mechanical Engineers,* Seventh Edition, Chapter 6; or the *American Welding Society: Welding Handbook,* Third Edition, Chapters 27-31B).

STEEL PRODUCTION

Steel is produced from pig iron by removing a number of different impurities. This is accomplished by subjecting the pig iron to extremely high temperatures in one of the following four types of furnaces:

1. An open-hearth furnace,
2. A basic oxygen furnace,
3. A Bessemer converter,
4. An electric furnace.

The impurities combine with other elements to form a slag which is removed from the furnace. As soon as the impurities have been reduced to the desired level, the molten steel is drawn off into a ladle by tapping the furnace. Alloys are added before tapping or after the steel has been poured into the ladle.

The next step is to pour the molten steel from the ladle into iron ingot molds. These steel ingots are then directly hot rolled into blooms, billets, slabs, or skelps.

A *bloom* is an approximately square section of hot rolled ingot steel with dimensions of 6 inches by 6 inches or larger. A bloom (or a steel ingot) can be further reduced in size to a *billet* which is 1½ inches to slightly under 6 inches square. Both blooms and billets are rolled into bars (which can then be rolled into sheets), structural shapes, and a number of other forms.

Slabs are sections of hot-rolled ingot steel in which the width exceeds the thickness by a considerable extent. These slabs are further worked into steel plates or sheets.

Skelp is produced in the form of steel strips, which are used for the manufacture of pipe.

During the production process, ingot steel is subject to the formation of a number of defects that can be classified either as internal (composition) defects, or as surface ones. Examples of the former are: (1) inclusion, (2) piping, and (3) segregation. *Inclusion* refers to the presence of silicates, sulphides, and other impurities in the steel. *Piping* describes the formation of pipes (i.e. cavities) in steel (or iron) ingots during cooling, and are caused by the unequal rates of contraction. Impurities will concentrate in the steel during cooling. This activity is called *segregation.* Unfortunately, there is very little that can be done to correct the formation of these internal defects; they can only be minimized to some extent.

Surface defects, on the other hand, are subject to correction. Examples of surface defects are cracks, laps, seams, and scabs. Cracks are formed generally by uneven surface contraction during the cooling of the steel ingot. *Scabs* are irregular protuberances on the surface of the steel casting caused by splashing when pouring from the ladle or by cracks in the mold. Cracks that are extended by pressure in the direction that the steel is being rolled are referred to as *seams. Laps,* on the other hand, are formed by the folding over of a thin extrusion of steel as it passes through two successive rollers.

Cracks can be filled, and laps, seams, and scabs removed by a number of different methods including machining, grinding, scarfing (deseaming), and flame scaling. These are more fully described in Chapter 13.

MECHANICAL TREATMENTS OF STEEL

Steel can be mechanically treated by hot-working or cold-working the metal. Hot-working refers to the forging (by pressure or hammering), rolling, or shaping of the metal while it is still hot (that is, above the critical or transformation temperature range). Cold-working, on the other hand, occurs at temperatures below the critical range and approaching those of atmospheric conditions.

Both hot-working and cold-working a metal will increase its tensile and yield strengths. However, these two forms of mechanical treatment will have an opposite effect on its ductility. Cold-working will decrease ductility, whereas hot-working will increase it.

HEAT TREATMENTS OF STEEL

Heat treatments are also used to obtain certain desired properties. Essentially, heat-treating the metal consists of heating and cooling it after it has reached the solid state. There are a number of different methods of heat-treating various steels in today's industry.

The more commonly used ones are:

1. Quenching,
2. Hardening,
3. Annealing,
4. Normalizing,
5. Stress relieving.

Quenching is the rapid cooling of a metal usually by immersing it in oil or water. This will cause certain changes in the structure of the metal. For example, carbon steel that is quenched will form a martensite structure.

Hardening (or *quench hardening*) involves heating the steel to about 100°F above the critical temperature range and then quenching it. *Case hardening* refers to the hardening of the outside layer (case) while leaving the core relatively soft.

Annealing is a heat treatment procedure designed to soften steel. This involves heating the steel to approximately 100°F above the critical temperature range, holding it at that temperature for a specific period of time, and allowing it to cool slowly to room temperature.

Normalizing is very similar to annealing except that the steel is held above the critical temperature very briefly and the cooling takes place in air at normal temperatures. Normalizing will result in refining the grain structure of a metal. It is sometimes used after quenching.

One form of heat treatment frequently encountered by the welder is *stress-relieving*. Unlike other heat-treatment procedures described here, the temperature of the metal is not raised above the critical (transformation) range. The metal is uniformly heated to a temperature sufficient to relieve any stresses caused by welding and then it is allowed to cool slowly.

ALLOYING ELEMENTS

In steels exclusively of the alloy type, the carbon content is the common denominator by which different grades of steel are chosen for various commercial uses. The carbon content of the various grades of steel range from approximately 0.04% to 1.65%. High-carbon steels require special technique. The amount of carbon, however, is not the sole determining factor in the weldability of steel. Many additions, impurities, and alloying elements used in steel affect the welding characteristics of the metal.

Some of the elements found in steel which may affect its welding characteristics are aluminum, silicon, carbon, manganese, titanium, nickel, vanadium, chromium, and molybdenum. Steel having a silicon content of about 0.01% or greater may make welding unsatisfactory unless other alloying or deoxidizing elements are present. Also, a content of more than 0.01% of aluminum may cause a poor weld unless the steel also contains a certain proportion of other deoxidizing elements. It should be noted that these other elements can, and do, affect these limits for aluminum and silicon. However, steel having a high aluminum content can be welded satisfactorily if a corrective autogenizer flux or filler is applied during the welding process. In general, the presence of nickel or vanadium may improve the weldability of the steel to a small degree.

STEEL IDENTIFICATION SYSTEMS

The *Society of American Engineers (SAE)* and the *American Iron and Steel Institute (AISI)* have both created numbering identification systems for many of the alloy steels. For the most part, these identification numbers duplicate one another. For example, a 3.50% nickel steel in either system is represented by the numerals 23XX. However, the *AISI* identification system is regarded as a refinement of the other since it indicates the process by which the steel was produced.

Table 1 illustrates the *SAE* steel numbering system. Most of these identification numerals consist of at least four digits. The first digit refers to the general type of steel in terms of its major alloy-

Table 1. The SAE Steel Numbering System

Type of Steel	Numerals (And Digits)
Carbon Steels	1xxx
Plain Carbon	10xx
Free Cutting (Screw Stock)	11xx
Free Cutting, Manganese	X13xx
High-Manganese Steels	T13xx
Nickel Steels	2xxx
0.50% Nickel	20xx
1.50% Nickel	21xx
3.50% Nickel	23xx
5.00% Nickel	25xx
Nickel-Chromium Steels	3xxx
1.25% Nickel, 0.60% Chromium	31xx
1.75% Nickel, 1.00% Chromium	32xx
3.50% Nickel, 1.50% Chromium	33xx
3.00% Nickel, 0.80% Chromium	34xx
Corrosion and Heat Resisting Steels	30xxx
Molybdenum Steels	4xxx
Chromium	41xx
Chromium-Nickel	43xx
Nickel	46xx and 48xx
Chromium Steels	5xxx
Low-Chromium	51xx
Medium-Chromium	52xxx
Corrosion and Heat Resisting	51xxx
Chromium-Vanadium Steels	6xxx
Tungsten Steels	7xxx and 7xxxx
Silicon-Manganese Steels	9xxx

Courtesy The James F. Lincoln Arc Welding Foundation

ing element. For example, 20XX indicates that it is a steel having nickel as its basic alloy (i.e. nickel steel); 30XX indicates that it is a steel having nickel chromium as its basic alloy (i.e. a nickel-chromium steel); and so forth.

The second digit indicates the approximate percentage of the basic alloy contained in that particular steel. Thus, a 23XX indicates that the nickel content is 3.50%. Do not make the mistake of assuming that the identification numeral equates with a content percentage on a one-to-one basis. In the 23XX identification numerals, the second digit does *not* indicate a 3.0% content. The numeral 3 is an *indicator* of a 3.5% nickel steel.

The last two (or three) digits refers to the carbon content of the steel. This is expressed in hundreths of 1%. For example, a 2315 nickel steel indicates a carbon content in the 0.15-0.20% range.

Letter prefixes are used in the *AISI* steel numbering identification system primarily to indicate the process by which the steel was produced, although they are not strictly limited to this purpose in at least two examples (the prefix X and suffix H). The letter prefixes and their meanings are as follows:

1. *A* indicates an open hearth alloy steel,
2. *B* indicates a *Bessemer* carbon steel,
3. *C* indicates a carbon steel (standard open-hearth type),
4. *D* indicates a carbon steel (acid open-hearth type),
5. *E* indicates an electric-furnace steel,
6. *H* (a suffix letter added to the four digit identification number) indicates steels of known hardenability,
7. *X* indicates a variation from the standard analysis.

Table 2. Stainless Steel Numbering Identification System

Series*	Major Alloying Elements	Characteristics
2xx	Chromium-Nickel-Manganese	Non-hardenable Austenitic
3xx	Chromium-Nickel	Non-hardenable Austenitic
4xx	Chromium	(1) Hardenable Martensitic (2) Non-hardenable Ferritic
5xx	Chromium (4 to 6%)	Air-hardenable Martensitic

*Second and third digits in number identify specific type.
Letter suffix identifies modfication of original grade.

Courtesy The James F. Lincoln Arc Welding Foundation

By way of example, an A4620 identification number indicates that this is an alloy steel produced by the open-hearth process (AXXXX); that it is a molybdenum steel (A46XX); and that it has a carbon range of 0.17-0.22% (A4620). If the same identification number were prefixed by an X, this would indicate that the carbon and manganese (not the molybdenum) analysis was slightly different. For example, the carbon range for an X4620 molybdenum steel is 0.18-0.23%.

The American Iron and Steel Institute has devised a three-digit numbering system for identifying the various stainless steels (Table 2). All true stainless steels are classified either in the 3XX or 4XX series. The chromium content of the 5XX series is considerably lower than the others, and there is some question as to whether or not this chromium metal should be classified as a stainless steel.

There does not seem to be any logical determination for the last two digits of this identification system. Consequently, they serve only to indicate individual stainless steel types within each of the four broad classifications.

CARBON STEELS

Carbon steels are steels in which carbon represents the principal alloying elements. Other elements such as manganese, phosphorous, sulphur, aluminum, copper, etc. may be present in unspecified amounts. Carbon steels are also referred to as *plain carbon steels* or *straight carbon steels*. This group of steels can be divided into four subgroups based on the carbon content. These subgroups are as follows:

1. Low-carbon steels,
2. Medium-carbon steels,
3. High-carbon steels,
4. Very high-carbon steels.

All carbon steels contain less than 2% carbon. A carbon content in excess of 2% indicates the carbon analysis for cast iron.

The uses of carbon steel listed in Table 3 represent a wide range of applications extending from the ubiquitous nail to the many automotive component parts.

Table 3. The Uses of Steels on the Basis of Their Carbon Content

Carbon Class	Carbon Range %	Typical Uses
Low	0.05 - 0.15	Chain, nails, pipe, rivets, screws, sheets for pressing and stamping, wire.
	0.15 - 0.30	Bars, plates, structural shapes.
Medium	0.30 - 0.45	Axles, connecting rods, shafting.
High	0.45 - 0.60	Crankshafts, scraper blades.
	0.60 - 0.75	Automobile springs, anvils, bandsaws, drop hammer dies.
Very High	0.75 - 0.90	Chisels, punches, sand tools.
	0.90 - 1.00	Knives, shear blades, springs.
	1.00 - 1.10	Milling cutters, dies, taps.
	1.10 - 1.20	Lathe tools, woodworking tools.
	1.20 - 1.30	Files, reamers.
	1.30 - 1.40	Dies for wire drawing.
	1.40 - 1.50	Metal cutting saws.

Courtesy The James F. Lincoln Arc Welding Foundation

Low-carbon steels (also referred to as *mild steels*) are characterized by a carbon content not exceeding 0.30%. The minimum amount of carbon found in these steels ranges from 0.05 to 0.08%. A carbon content below these minimum levels (i.e. below 0.05%) indicates a type of pure iron referred to as *ingot iron*.

The low-carbon steels have a melting point of approximately 2600°F to 2700°F. The slag contained in these steels will melt at a lower temperature than the metal itself, making the use of a flux unnecessary. The low-carbon steels possess a number of excellent qualities that contribute to their being the most widely used steel in industry today. First, and probably foremost, they are relatively inexpensive. They are also tough and ductile. Finally, the low-carbon steels are generally easy to weld and machine.

Shielded metal-arc welding is commonly used to weld the low-carbon steels. Its greatest application is with mild steel in the low-carbon ranges. As the carbon content increases, it becomes necessary to apply preheat and postheat treatment procedures in order to obtain a satisfactory weld.

Both mild-steel electrodes (E60XX) and low-hydrogen electrodes (e.g. the E6015 or E6016) are recommended for use with the low-carbon steels. These have been classified by the *American Welding Society* and it is a simple matter to select the most suitable electrode for the job at hand.

The oxyacetylene welding process is also frequently used to join the low-carbon steels. Because of the significant difference between the melting temperature of the mild steel and its slag, a flux is generally not used. The selection of a filler metal, however, is important and will depend on the analysis of the base metal. This, in turn, will determine the type of flame used.

There are three ferrous welding rods recommended as filler metals for the low-carbon (mild) steels.

These are:

1. An alloy-steel rod (AWS Classification GA 60),
2. A high-tensile steel rod (AWS Classification GA 60),
3. A low-carbon (mild) steel rod (AWS Classification GA 50).

The alloy-steel rod will produce a weld with a tensile strength in excess of 60,000 psi. The weld produced by the low-carbon steel rod will be somewhat lower with a minimum tensile strength of 52,000 psi.

A neutral flame or one with a slight excess of acetylene is recommended when using the alloy-steel rod. A slightly carburizing flame is also recommended for welding with either a high-tensile or low-carbon steel rod (with the carburizing flame a somewhat higher welding speed will be achieved). An oxidizing flame must be strictly avoided, because it will form oxides on the surface that may be included in the weld and thereby weaken it. Excessive sparking is one indicator of an oxidizing flame.

A backhand welding technique is preferred to a forehand one. In backhand welding, the flame is directed back against the weld where it functions as a shield against contamination and thus reduces the possibility of oxide formation. Using this technique also results in less torch and rod manipulation, less agitation of the metal, and faster welding speeds.

The weld preparation procedures (cleaning, beveling, etc.) described in Chapter 4 definitely apply when welding these steels.

Copper-bearing steels are low-carbon steels that contain approximately 0.20% copper. A copper content exceeding this amount generally causes the development of surface cracking in the base metal surrounding the weld. Copper-bearing steels are welded with the same procedures used for other low-carbon steels.

Copper-bearing steels are low-carbon steels that contain approximately 0.20% copper. A copper content exceeding this amount generally causes the development of surface cracking in the base metal surrounding the weld. Copper-bearing steels are welded with the same procedures used for other low-carbon steels.

The carbon content of the *medium-carbon steels* ranges from 0.30% to 0.45%. These steels are not as easily welded as the low-carbon steels, and this is directly due to the increased carbon content. The medium-carbon steels are harder and stronger than the mild steels, but tend toward weld brittleness in the higher-carbon range.

Weld cracking can be avoided when working with medium-carbon steels in the higher carbon ranges by employing special welding techniques and electrodes suitable for this purpose. The low-hydrogen electrodes are recommended when arc welding medium-carbon steels. The E-6015, E-6016 and E-6018 electrodes require preheating the base metal to reduce weld cracking. These electrodes are used where greater amounts of weld deposit are required. A postheat treatment is also recommended when welding these steels.

The oxyacetylene welding process can also be used to join the medium-carbon steels. A slightly carburizing flame is recommended. A flux is not used. An alloy-steel or high-tensile steel welding rod gives the best results.

Other joining processes that have been used to join this category of steels include: resistance welding, and braze welding.

A *high-carbon steel* contains 0.45% to 0.75% carbon and finds common application in the manufacture of such items as knives, saws, bits, and tools. In other words, these manufactured items require hard tough surfaces exhibiting a high resistance to wear. Some classifications of the carbon steels add a fourth subgroup having an even higher carbon content range (0.75% to 1.50%). These are referred to as *very high-carbon steels*. Due to their extreme hardness and other characteristics that limit weld quality, these carbon steels are seldom welded except for repair purposes.

The major difficulty with the high-carbon (and very high-carbon) steels is that they require a certain amount of heat treatment, and the heat of the gas flame or electric arc can adversely affect this treated surface. These adverse effects of the welding heat can be somewhat rectified by maintaining a fast welding speed and by post heating the surface.

Arc welding is generally limited to repair applications insofar as the high-carbon and very high-carbon steels are concerned. The low-hydrogen electrodes are recommended for use as a filler metal.

Oxyacetylene welding with a carburizing flame and high-carbon welding rods is also frequently used to join these metals.

Other welding processes used to a limited extent for joining high-carbon steels are:

1. Submerged arc welding (very limited use),
2. Forge welding (also very limited use),
3. Thermit welding (widely used to join rails and reinforcing bars),
4. Resistance welding.

LOW-ALLOY STEELS

Low-alloy steels are characterized by toughness and a high degree of strength, and generally exhibit a strong tendency to resist corrosion. On the whole, their physical properties are higher than those of the carbon steels. No single alloying element is present in amounts greater than 5.0%. Even the total of all alloys in a low-alloy steel seldom exceeds this percentage.

Some of the more common low-alloy steels that are being used today are:

1. Nickel low-alloy steels,
2. Nickel-chromium low-alloy steels,
3. Nickel-copper low-alloy steels,
4. Molybdenum low-alloy steels,
5. Manganese low-alloy steels,

6. Chromium low-alloy steels,
7. Chromium-vanadium low-alloy steels,
8. Silicon-manganese low-alloy steels,
9. Tungsten low-alloy steels.

Most of the welding processes can be used to join the low-alloy steels with varying degrees of success. In the case of oxy-acetylene welding, it is primarily a question of selecting a suitable filler metal. In most cases, the composition of the welding rod will match that of the base metal. Sometimes, however, a welding rod different in composition will be selected in order to offset changes resulting from welding. The rest of the welding procedure is similar to the oxyacetylene welding of low-carbon steel.

The major problem encountered when welding low-alloy steels with the shielded metal-arc welding process is the selection of a suitable electrode. This, of course, is determined by many factors including the composition of the base metal, the degree of penetration desired, the desired tensile strength, and the type of weld to be made. The *American Welding Society* has classified the low-alloy steel electrodes which makes it much easier for the welder to select the most suitable electrode.

The successful submerged-arc welding of low-alloy steels requires the selection of both a suitable flux and filler metal. Most manufacturers will recommend the best combination (flux and filler metal) to use for a particular welding job.

Whereas shielded metal-arc welding, oxyacetylene welding, and submerged-arc welding can be used to join any thickness of low-alloy steel, some of the other welding processes are generally restricted to the lighter gauges. These welding processes include MIG welding, TIG welding, and atomic-hydrogen welding.

Resistance spot welding and flash welding have been successfully used on most types of low-alloy steels. The machine control settings are, of course, determined by the base metal composition. Projection and seam welding are somewhat restricted in use. Other joining processes used to a limited extent when working with low alloy steels are brazing and thermit welding.

Nickel low-alloy steels contain approximately 3.00% to 3.50% nickel. The carbon content range is commonly 0.12% to 0.20%;

the manganese, 0.40% to 0.60%; and the silicon, 0.20% to 0.35%. Traces of sulphur also occur (though not exceeding 0.40%).

The nickel is added to the steel in an effort to increase its strength and hardness. As a result, a nickel low-alloy steel with the same tensile strength as a plain carbon steel will commonly possess half its carbon content.

Shielded metal-arc welding with a suitable electrode (e.g. a low-hydrogen electrode, should cracking appear, an E-6012 electrode for thin sheets when reduced penetration is desired, etc.) is usually recommended for working with the nickel low-alloy steels.

The procedure for using the oxyacetylene welding process when welding nickel low-alloy steels is similar to the one used for welding low-carbon steel.

Nickel-chromium low-alloy steels are characterized by a nickel content in the 1.00% to 3.50% range; a chromium content in the 0.40% to 1.75% range; and a carbon content in the 0.20% to 0.55% range. The chromium is added to increase weldability by keeping the carbon content at a low level. It also functions as a hardening agent. Most nickel-chromium low-alloy steels require preheating.

Shielded metal-arc welding with an electrode of the E-7010 or E-7015 type is used with good success. Nickel-chromium low-alloy steels possessing higher-carbon ranges require special electrodes.

The oxyacetylene welding process can also be used to weld these metals. The procedure is identical to that used for joining low-carbon steel.

The nickel content of *nickel-copper low alloy* steels ranges from 1.00% to 2.00%. The copper content is slightly lower at 1.00% to 1.60%—the carbon content generally does not exceed 0.25%. If the carbon exceeds 0.25%, a preheat of 250° F to 550° F is recommended to develop the desired physical properties.

There are approximately 40 *molybdenum low-alloy steels*. These steels can be roughly divided into three principal subgroups, each based on the extent to which they contain carbon, chromium, or nickel. These three subgroups, each based on the extent to which they contain carbon, chromium, or nickel. These three subgroups are:

1. Carbon-molybdenum low-alloy steels (0.20% to 0.70% carbon),
2. Chrome-molybdenum low-alloy steels (0.40% to 1.10% chromium),
3. Nickel-molybdenum low-alloy steels (1.65% to 3.75% nickel),
4. Manganese-molybdenum low-alloy steels (0.60% to 1.60% manganese).

Molybdenum is added to increase the strength of the steel, particularly at high temperatures. It also increases hardenability and reduces some forms of brittleness. In the molybdenum low-alloy steels, the molybdenum content ranges from 0.15% to 0.70% with most types falling within the 0.20% to 0.30% range. Manganese is also present in all molybdenum low-alloy steels. Its content ranges from 0.25% to 1.00%. Traces of phosphorous, sulphur, and silicon also occur in each of these steels.

As is the case with the nickel-chromium low-alloy steels, chromium is added to increase weldability by keeping the carbon content low. The strength of welds in chrome-molybdenum low-alloy steels depends upon the methods of welding and the heat treament after welding. Either the carbon-arc or the shielded metal-arc welding processes can be used—the procedure being the same as for low-carbon steel. The choice of method and electrode used, however, should be based on the analysis of the metal and the desired weld characteristics.

Some welders prefer to use the oxyacetylene welding process for joining these metals. As with arc welding, the procedure is the same as for low-carbon steel.

Manganese low-alloy steels are characterized by a manganese content in the 1.60% to 1.90% range; a phosphorous and sulphur content of not more than 0.040% for each; a silicon content in the 0.20% to 0.35% range; and a carbon content ranging from 0.18% to 0.48%. The manganese is added primarily to increase tensile strength. A secondary effect is an increased resistance to abrasion.

These steels are somewhat more difficult to weld than the other low-alloy steels. A contributing factor is the increased hardness resulting from the interreaction of carbon with manganese.

The lower the carbon content can be kept, the easier the metal will be to weld. Manganese low-alloy steels with high-carbon content require a number of special procedures including preheating and postheating stress relief.

Shielded metal-arc welding with a suitable electrode can be used with above average success to weld manganese low-alloy steels.

A neutral flame with a suitable filler metal can be used in the oxyacetylene welding of these steels. The procedure is the same as that used for welding low-carbon steels.

Chromium low-alloy steels contain approximately 0.20% to to 1.60% chromium. Other alloys include manganese (0.25% to 1.00%), phosphorous (0.040% or under), sulphur (0.040% or under), silicon (0.20% to 0.35%), and carbon (0.12% to 1.10%). The addition of chromium increases the hardenability, tensile strength, and corrosion resistance of low-alloy steels. The higher carbon content produces the usual tendency toward brittleness and cracking. Preheating and postheating reduce this tendency.

Shielded metal-arc welding can be used for welding the chromium low-alloy steels if the carbon content is in the lower ranges (generally 0.12% to 0.20%). Electrode choice depends on such factors as carbon content, desired weld strength, and whether a heat treatment is required. When the carbon content is high (approximately 0.20% to 1.10%), the shielded metal-arc welding process should only be used to make repair welds.

The *chromium-vanadium low-alloy steels* contain at least 0.10% to 0.15% vanadium. The chromium range is 0.70% to 1.10%; the manganese range, 0.70% to 0.90%; the silicon range, 0.20% to 0.35%; and the carbon range, 0.15% to 0.53%. Neither the phosphorous nor the sulphur contents exceed a maximum of 0.040%.

Chromium and vanadium are added to low-alloy steel to increase its hardenability and to impart a grain structure that is finer than that of the standard chromium low-alloy steels.

Shielded metal-arc welding with special electrodes is used to weld chromium-vanadium low-alloy steels. Preheating is recommended before welding.

Silicon-manganese low-alloy steels are tool steels containing 1.8% to 2.2% silicon, 0.70% to 1.00% manganese and 0.50% to 0.65% carbon. Phosphorous and sulphur are present but neither exceed 0.40%. Chromium is also added as an alloy to some of these steels.

The basic objective in adding silicon to a low-alloy steel is to increase its strength. It also increases toughness, hardenability, and resistance to wear.

Tungsten low-alloy steels are tool steels containing approximately 2.0% tungsten, 1.70% chromium, and 0.50% carbon. This is a hard tough tool steel that is commonly used for making cutting tools. These and other tool steels are described in greater detail in Chapter 18.

LOW-ALLOY, HIGH-TENSILE STEELS

Steel manufacturers have produced a great number of low-alloy high-tensile steels under a variety of trade names. These steels do not fall within the classification of the *AISI* low-alloy steels, although the analyses of some of them may be quite similar.

The low-alloy, high-tensile steels contain carbon in varying amounts (0.10% to 0.40%) with a number of other alloying elements such as chromium, manganese, silicon, phosphorous, copper, nickel, and molybdenum. Each of these alloys is present to some degree in the trade name steel. Most of the other steels of this type include some but not all of them.

These steels possess high tensile strength extending from 60,000 to 135,000 psi with the average being in the 70,000 to 80,000 psi range. Examples of common low-alloy high-tensile steel trade names include *T-1, Corten, Silten, Jalten, Hi-Steel, NAX,* and *Otiscoloy.*

The shielded metal-arc welding process is recommended for working with these steels. The type of electrode to be used depends, of course, on the analysis of the steel. For example, mild-steel electrodes will work very well with steels having a low-carbon content (generally under 0.14%). As the alloy content increases (particularly the carbon content), the tendency to develop cracks in the weld increases. Therefore, the operator must select an electrode that will reduce or eliminate the possibility of this happening.

Electrode selection is also determined by the purpose for which the joint is designed. If the operator requires a weld of higher tensile strength, he will have to select an electrode that will impart this quality to the weld (most commonly a low-alloy electrode). In the final analysis, the successful welding of low-alloy high-tensile steels depends on the operator's correct identification of the steel, the selection of the most suitable electrode, and the application of an effective welding technique.

HIGH-ALLOY STEELS

The principal alloying element in high-alloy steels will exceed 0.50%. This is a primary factor in distinguishing them from the low-alloy types. This section will be concerned with three important categories of high-alloy steels, namely:

1. Manganese high-alloy steels,
2. Some tool steels,
3. Stainless steels.

Manganese high-alloy steels contain approximately 10% to 14% manganese, 0.30% to 1.00% silicon, and 1.00% to 1.40% carbon. Traces of sulphur and phosphorous also occur, but do not exceed 0.60% each. Steels of this type are also referred to as *austenitic manganese steel, Hadfield manganese steel, high-manganese steel,* or simply *manganese steel.*

This steel possesses a high tensile strength ranging from 131,000 to 142,000 psi. Its melting point is approximately 2500°F. Because of its high impact resistance, manganese high-alloy steels are used in the production of equipment designed for the crushing and removal of earth, rocks, ores, etc. Manganese steel is also widely used to build up worn parts made from this metal or to hard-surface parts made of other steels.

For building up worn parts of high-manganese steel, a shielded arc should be used of such type that the physical characteristics of the deposited metal will be approximately the same as the base metal. A direct-current reversed polarity (DCRP) should be used.

The exact heat to be used depends on the mass of the part being built up; more heat being required for heavy sections than for thin sections. Suggested amperages for various sizes of electrodes are listed in Table 4. In general, the lower currents should be used as much as possible.

Table 4. Suggested Amperages For Various Sizes of Electrodes

Wire	Amperage
1/8 in.	50-90
5/32 in.	90-130
3/16 in.	130-170
1/4 in.	170-225

Tool steels contain at least 0.40% to 0.50% carbon and exhibit increased hardness through quenching. This last characteristic distinguishes them from the low-alloy steels whose hardness can only be slightly increased. A full description of tool steels can be found in Chapter 18.

The high-alloy steels are generally very difficult to weld, a characteristic due largely to the presence of carbide-forming alloys in their composition. These alloys (e.g. tungsten or molybdenum) are added to increase the hardness of the metal. The hardness of the average high-alloy steel is generally three to four times that of a straight carbon steel.

Stainless steel is a high-alloy steel with a chromium content ranging from 11.5% to 27.0%. In some cases, nickel (3.5% to 22.0%) is also added. These are the so-called chromium-nickel steels and can be divided into three basic categories according to their chromium and nickel content. These three categories are:

1. 18- 8 chromium-nickel steels,
2. 25-12 chromium-nickel steels,
3. 25-20 chromium-nickel steels.

When arc welding stainless steel, an electrode of the same analysis as the base metal should be used. The 25-12 chrome-nickel steel electrodes may also be used satisfactorily. The electrode must be of the coated type, producing a shielded arc and the coating must be free from materials of a carbonaceous nature.

Reverse polarity (electrode positive, ground negative) gives best results in most applications, although in some cases better fusion and penetration have been obtained by using straight polarity in connection with the welding of heavy sections.

Because 18-8 alloys have lower heat conductivity and higher electrical resistance, welding currents used are 10% to 20% lower than those used in welding mild steel. The 18-8 stainless steel has a high coefficient of expansion, which accentuates its tendency to warp. This difficulty can be controlled by jigs or other fixtures to hold the sheets in place during welding and at the same time limit the heat absorbed by the base metal. Careful preheating of large surfaces in special jigs is sometimes used to prevent distortion during welding.

Before starting to weld, the underside of the seam should be painted with flux. In welding light gauge sheet, it is helpful to put flux on the welding rod as well. Welding should be done with a neutral flame. Puddling of the weld should be avoided as it has a tendency to cause oxidation.

The preferable welding technique consists in holding the rod ahead of the torch so that the molten metal melts in place or is melted simultaneously with the base metal.

Under normal welding conditions, the columbium in the columbium-treated welding rod is retained in the weld metal with only a slight loss. Overheating of the molten metal should be avoided as this would result in a loss of columbium.

Wherever possible, joints should be welded toward an edge. If it is necessary to weld away from an edge, begin the weld an inch or two in from the edge and return to this point later to complete the weld to the edge. For thicknesses ⅛ inch and over, it is often advisable to weld by the backhand method since it minimizes distortion.

CAST STEEL

Steel is frequently taken from the steel-making furnace and cast directly into the form in which it is to be used. Typical examples of steel castings include the frames for generators, steam valves, and ship rudder frames.

Steel castings differ from their iron-casting counterparts in having greater ductility and toughness and a higher tensile strength. They can be divided into several categories according to their compositions. These categories are as follows:

1. Carbon-steel castings,
2. Low-alloy steel castings,
3. Manganese-steel castings,
4. Stainless-steel castings.

The variety in composition makes it essential for the welding operator to correctly identify the casting metal before beginning to weld.

The same principles that govern the welding of ordinary structural steel apply when welding cast steel. Special consideration must of course be given to the shape and size of the castings and the effect of contraction and expansion. Steel castings may be welded to rolled steel in a very satisfactory manner.

Small steel castings require no preparation other than beveling the edges to be welded and making certain the surface is clean. Beveling can be done with the cutting torch or chipping hammer; in some instances by grinding. If the cutting torch is used, be certain that the edges are thoroughly cleaned of oxide and scale by a grinding wheel, or with a wire brush before starting to weld.

Larger castings with thick walls (or of such design that localized heat of welding might set up strains in the steel) should be preheated to a bright red after beveling the edges to be welded. Whether preheating should be general or local depends on the design of the casting.

CLAD STEELS

A clad steel (Fig. 1) consists of a thin metal coating applied over a low-carbon steel core. As such, it represents an effort to obtain, in one combined form, physical properties that exist in two different metals. For example, a chromium or nickel cladding provides a greater resistance to corrosion than the mild steel core but the latter contributes ductility and strength to the composite form.

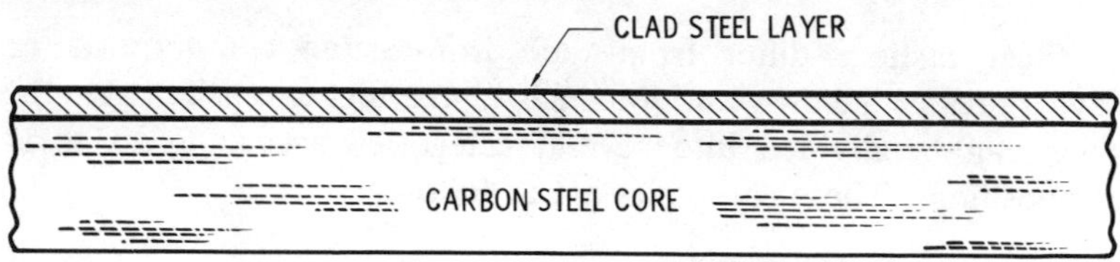

Fig. 1. Composition of clad steel plate.

The thickness of the cladding metal varies and depends on the manufacturer and the use to which the metal is to be put. Usually the thickness of the cladding will range from 5% to 20% of the total thickness of the clad steel plate.

Clad steel plate is produced by bonding a cladding metal (generally one with a high resistance to corrosion) to a backing of low-carbon steel. The usual procedure is to position the cladding and backing metals together by welding and then rolling to the required thickness. The various clad steels include the following types:

1. Chromium-clad steels,
2. Chromium-nickel clad steels,
3. Nickel-clad steels,
4. *Inconel*-clad steels,
5. Monel-clad steels,
6. Cupronickel-clad steels.

The type of cladding determines the procedure for welding a clad steel. In other words, it will be the same procedure as one would use to weld an entire thickness of the cladding metal.

Submerged-arc welding and shielded metal-arc welding are recommended for welding clad steels, particularly the thicker sections. TIG welding, atomic-hydrogen welding and brazing can be used to join the light gauge clad steels. Oxyacetylene welding is not recommended by many operators but it can be used.

Welding stainless-clad steel best typifies the problems involved in working with these composite metals. The operator must exercise caution not to deeply penetrate the core metal when welding the cladding and vice versa (some minor penetration is unavoidable). Most operators prefer to begin with the backing or core side of the clad steel first. The cladding is welded last.

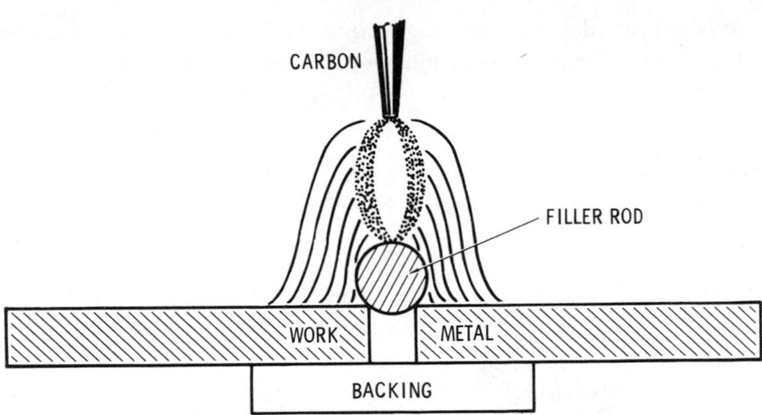

Fig. 2. Carbon-arc butt weld on galvanized metal.

Galvanized steel is produced by *coating* rather than cladding the core metal, but it is nevertheless included among descriptions of the clad steel types. The next section is concerned with welding galvanized steel.

GALVANIZED STEEL

Galvanized steel is steel that has been coated with zinc. In welding, the heat of the arc or flame liberates zinc fumes. Provision should be made for protection of the operator from these fumes, especially through proper ventilation. This precaution should be taken at all times and for every job involving galvanized steel. If the operator should experience any nausea while or after welding galvanized material, he should immediately seek medical help. A recommended measure is to drink several large glasses of milk right after the first feelings of nausea.

The carbon arc can be used successfully for welding galvanized stock. A bronze filler rod is used and the arc played directly on the filler rod so that the galvanizing material is not burned off from the steel alongside the weld, and, the weld being of bronze, the joint itself is also corrosion resistant.

Grade D phosphor bronze makes an ideal filler rod for this purpose because of its high tin content and resulting low melting point which makes it easy to weld very rapidly. The heat of the arc is not maintained very long in one spot. Other types of bronze, such as *Everdur* or *Herculoy* have also been used because of their lower cost.

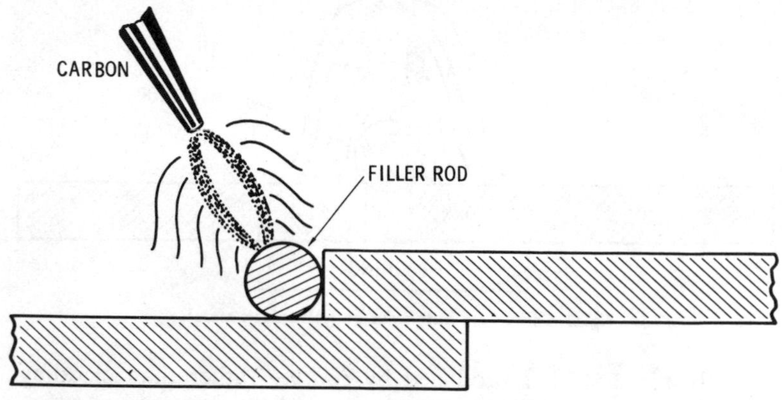

Fig. 3. Carbon-arc lap weld on galvanized metal.

Fig. 2 shows the proper method of setting up for a butt weld on galvanized stock, the backing strip preventing the bronze from running through the joint. The sheets should be spaced apart approximately equal to their thickness. The filler rod diameter should be two or three times the width of the space. The rod, which is generally 86 inches long, should be laid flat in the joint for its entire length. Very low current values should be used (35 to 50 amperes) with ⅛- to ¼-inch carbon electrodes.

The arc should not be started on the galvanized sheets. Start either on the backing bar or on the filler rod and it is very important that the arc be played always on the filler rod—not on the galvanized sheets alongside. Hold a very short arc and proceed rapidly along the seam as the bronze filler rod melts down and fuses with the galvanized steel. Where it is impossible to back up the weld, it should be set up for a lap weld, as shown in Fig. 3.

The procedure for welding galvanized steel, using a mild-steel shielded-arc electrode, is the same as is usually used for ordinary mild steel.

CHAPTER 23

Welding Other Metals

This chapter concerns itself with a discussion of those metals and related alloying elements which have not been treated in previous chapters. It concentrates primarily on their composition, use, weldability, and the methods recommended for welding or joining them.

Many of the metals found in this chapter can be grouped into categories because of characteristics they share with one another. For example, the refractory metals (chromium, columbium, molybdenum, etc.) are particularly noted for their high melting points (3500°F or more). A second major category, the so-called "exotic" metals (molybdenum, titanium, zirconium, etc.), is characterized by a high resistance to corrosion and high strength under high-temperature conditions. Both groups overlap to a certain extent in the case of such metals as molybdenum or titanium.

BERYLLIUM AND BERYLLIUM ALLOYS

Beryllium (formerly glucinum) is a lightweight metallic element characterized by extreme hardness (it can scratch a glass surface) and a color similar to that of magnesium. It has a melting point of 2332°F (1285°C), a density comparable to that of magnesium, and a high electrical and thermal conductivity. It is frequently used as an alloying element with other metals (e.g. copper, nickel, and magnesium) for increased strength, elasticity, and other characteristics. Its light weight (it is lighter than aluminum), tensile strength (about 55,000 psi), and relatively high melting temperature have suited it to applications in the aerospace industries. Other applications include the use of beryllium wire in the production of electrical circuits.

Beryllium is derived primarily from beryllium aluminum silicates. The beryllium metal is prepared by electrolysis. Beryllium (1.0% to 2.5%) is combined with nickel (up to 1.0%) and copper to form the beryllium-copper alloy. This alloy has a relatively high tensile strength up to 180,000 psi. It can be welded with a beryllium-copper filler rod and the carbon-arc welding process. A more detailed discussion of this alloy is found in Chapter 20, WELDING COPPER AND COPPER ALLOYS.

Beryllium is also used in small amounts (0.2%-0.25%) as an alloy in the production of a cast beryllium bronze. Iron, silicon, and cobalt are sometimes added to the beryllium-copper alloy for various desired characteristics.

Beryllium dust and fumes are very toxic. Their extreme danger rests on the fact that there is no known cure (only remission agents) for their effects. *Every safety precaution should be taken when working with beryllium and beryllium alloys.*

LEAD AND LEAD ALLOYS

Lead is a grayish-white metal noted for its softness, ductility, and resistance to corrosion. It is also characterized by low strength (it has a particularly low fatigue strength) and a tendency for fatigue cracking. Commercial lead can be as high as 99.99% pure. The melting point of pure lead is 621°F (327.5°C).

Lead has many applications, including such diverse forms as:

1. Pipes and other fittings used in plumbing,
2. Shielding protection against x-ray and gamma radiation in the dental and medical occupations,
3. Corrosion-resistant linings,
4. Cable sheathings,
5. As an alloy with tin in some solders (low-melting types).

The recommended processes for welding or joining lead are as follows:

1. Gas welding,
2. Carbon-arc welding,
3. Soldering.

Gas welding is the most suitable process for welding lead. It can be performed in all positions, and the welds are normally stronger than the base metal itself. The gas welding process may take the form of oxyhydrogen welding, oxyacetylene welding, or air-acetylene welding. Natural gas in combination with oxygen is also used.

The oxyhydrogen welding process is best suited for the thinner pieces of lead (generally less than ¼ inch thick). Thicker pieces of lead are usually welded with an oxyacetylene torch because of its hotter flame.

A neutral flame with a gas pressure under 5 psi is recommended for best results. The gas pressure will, of course, depend on the size and type of weld. A welding rod matching the analysis of the base metal is used.

MAGNESIUM AND MAGNESIUM ALLOYS

Magnesium is a silvery-white metal having a melting point of 1202°F (1110°C). Pure, sand-cast magnesium has a tensile strength of 12-13,000 psi. Rolled magnesium has a tensile strength approximately double that of the cast type (i.e. about 25,000 psi). By adding alloying elements, the tensile strength can be considerably increased. For example, the addition of 8% to 10% aluminum produces an alloy with a tensile strength of 53,000 psi. A comparison of the properties of magnesium with those of other metals is found in Table 1.

The recommended processes for welding and joining magnesium and magnesium alloys are:

1. TIG welding,
2. MIG welding,
3. Resistance welding,
4. Oxyacetylene welding,
5. Soldering.

Table 1. Comparative Properties of Magnesium and Other Metals

	Approx. Melt. Pt.	Weight (Lb./ Cu. In.)	Approximate Ratios, Magnesium = 1. Weight (Lb./ Cu. Ft.)	Weight Ratio	Thermal Conductivity Ratio	Expansion Ratio
Magnesium	1204	.063	109	1.0	1.0	1.0
Aluminum	1215	.098	170	1.55	1.4	0.9
Copper	1980	.323	560	5.1	2.5	0.62
Steel	2700	.284	490	4.5	0.5	0.45
18-8 Stainless	2600	.286	495	4.56	0.17	0.63
Nickel	2646	.322	560	5.1	0.38	0.52

Courtesy The James F. Lincoln Arc Welding Foundation

Oxyacetylene welding should only be attempted where butt welds are to be made. The flux required by this process has a tendency to run between the sections in a lap weld and cause corrosion at a later date. Many welders will use gas welding only in situations requiring immediate, emergency repairs.

NICKEL AND HIGH-NICKEL ALLOYS

Nickel is a grayish-white metal exhibiting a strong resistance to corrosion and oxidation. It is an important and commonly used alloy in the production of other metals, particularly steels. In toughness and strength, nickel bears a strong resemblance to iron. Because of this characteristic, it is frequently alloyed with the latter in the production of nickel steels. Nickel also functions as an alloy in other types of steels and metals. The melting point is approximately 2647° F (1455° C).

The high-nickel alloys can be divided into a number of different groups depending on the type of principal alloying element. In the case of the almost pure high-nickel alloys, the percentage of alloy is extremely small. Table 2 illustrates the composition of typical nickel alloys.

Duranickel and Nickel 201 are two examples of high-nickel alloys that are very close in purity to commercially pure nickel

(99.5%). This class of nearly pure nickel alloys includes the following percentages:

1. Duranickel 301 (formerly Z Nickel) (94.0 nickel, 4.5 aluminum, 1.5 others),
2. Nickel 201 (formerly L Nickel) (99.5 nickel, 0.20 manganese, 0.30 others),

Table 2. A Comparison of Typical Nickel Alloys

Name	Composition
Nickel	99+% nickel
Monel	67% nickel, 30% copper, 1.5% iron, 1% manganese
"K" Monel	66% nickel, 30% copper, 3% aluminum
"Z" Nickel	94% nickel, 4.5% aluminum
Inconel	80% nickel, 15% chromium, 5% iron
Hastelloy A	56% nickel, 22% molybdenum, 6% iron

Courtesy The James F. Lincoln Arc Welding Foundation

3. Nickel 211 (formerly D Nickel) (95.0 nickel, 4.75 manganese, 0.5 others),
4. Nickel 233 (formerly 330 Nickel) (99.5 nickel, 0.18 manganese, 0.32 others), and
5. Nickel 200 (formerly A Nickel) (99.5 nickel, 0.25 manganese, 0.25 others).

The Monel high-nickel alloys are approximately two-thirds nickel and one-third copper. Small traces of other metals are also found (e.g. iron, manganese, aluminum, etc.) but these seldom exceed two to four percent. Monel is a silvery-white metal possessing greater strength than pure nickel, but a lower impact resistance. Among the various types of Monel metals, the following percentages can be included:

1. Monel (67.0 nickel, 30.0 copper, 1.4 iron, 1.0 manganese, 0.6 others),
2. R Monel (66.0 nickel, 31.5 copper, 1.35 iron, 0.90 manganese, 0.25 others),
3. K Monel (66.0 nickel, 29.5 copper, 2.8 aluminum, 1.0 iron, 0.7 others),
4. S Monel (68.0 nickel, 30.0 copper, 2.0 iron, 4.0 silicon).

Monel flows much more quickly in the welding operation than nickel or even Inconel. A flux is necessary during gas welding and it is generally best to follow the manufacturer's recommendations for the type of flux to use. The Monel class of high-nickel alloys exhibits high tensile strengths which exceed 80,000 psi in most cases (e.g. Monel, 80,000 psi; K Monel, 100,000 psi; S Monel, 140,000 psi, etc.).

The Inconel class of high-nickel alloys combines nickel with chromium as the principal alloying element. Other metals are present in significantly smaller amounts. Among the various types of Monel metals, the following percentages can be included:

1. Inconel (76.0 nickel, 15.8 chromium, 7.2 iron, 1.0 others),
2. Inconel X (73.0 nickel, 15.0 chromium, 6.8 iron, 5.2 others), and
3. Inconel 718 (52.5 nickel, 19.0 chromium, 18.0 iron, 10.5 others).

Incoloy (32.0 nickel, 20.5 chromium, and 46.0 iron) is characteristic of the Inconel class of high nickel alloys, except for its relatively low concentration of nickel. Ni-O-Nel (41.8 nickel, 21.5 chromium, 30.0 iron) shares these characteristics.

Another important class of high-nickel alloys are represented by the trade name *Hastelloy (Haynes Stellite Company).* These alloys are combinations of nickel, molybdenum, and other alloying elements. They are especially resistant to the corrosive attack of various acids.

Examples of *Hastelloy* alloys include the following percentages:

1. *Hastelloy* A (57.0 nickel, 20.0 molybdenum, 20.0 iron, 3.0 others),
2. *Hastelloy* B (65.0 nickel, 28.0 molybdenum, 5.0 iron, 2.0 others),
3. *Hastelloy* C (58.0 nickel, 17.0 molybdenum, 14.0 chromium, 5.0 iron, 5.0 tungsten, 1.0 others).

Illium (62.5 nickel, 21.0 chromium, 7.0 copper, 5.0 molybdenum, 4.5 others) is a nickel-chromium alloy exhibiting a strong resistance to the corrosive attack of certain acids (e.g. nitric acid, sulphuric acid, etc.). It has a melting point of 2350° F, and a tensile strength of approximately 60,000 psi. Table 3 illustrates the properties of some of the nickel alloys.

A description of the copper-nickel alloys (those having less than 50% nickel is included in Chapter 20, WELDING COPPER AND COPPER ALLOYS. Nickel steels are found in Chapter 22, WELDING STEEL AND STEEL ALLOYS.

Nickel and the high-nickel alloys can be welded or joined by a number of different processes with varying degrees of success. Table 4 lists the different welding processes used to weld or join lead and its alloys.

Table 3. Properties of Some of the Nickel Alloys

Material	Tensile Strength psi	Magnetic at Room Temperature	Age Hardening	Remarks
Nickel	60,000 to 165,000	Strong		
Monel	60,000 to 175,000	Mild		
"K" Monel	90,000 to 200,000	No	Yes	
"Z" Nickel	30,000 to 150,000	Slightly when age hardened	Yes	
Inconel	80,000 to 185,000	No		Resists scaling at elevated temperatures
Hastelloy A, B & C	70,000 to 140,000	Slightly	Yes	Resistant to hydrochloric and sulphuric acids

Courtesy The James F. Lincoln Arc Welding Foundation

TIN AND TIN ALLOYS

Tin has a silvery white color, a melting point of approximately 449° F (232° C), and is noted for its softness and ductility.

The use of tin as an alloying element has been known for centuries. Bronze (copper and tin) has been dated to thirty five centuries B.C. with implements found in the Near East. The use of solder (tin and lead) dates from the Roman period. The use of tin as a plating or coating also dates from the Roman period where it was used to coat copper. Its use on iron originated in Germany in the late middle ages.

Tin is combined with many other metals to form commonly used metal alloys. For example, tin is mixed with copper to form bronze and with lead to form solder. Tin foil is produced from a rolled lead-tin alloy. The tin cans used in the food industry consist of thin iron sheets covered on both sides by a coating of tin. Tin is used because it is a nontoxic metal. Pewter is an alloy consisting of at least 91.0% tin, and small amounts of antimony and copper. This is only a small portion of the many metals in which tin functions as an alloying element. Some of the others include (in percentages):

Table 4. Recommended Welding and Joining Processes for Aluminum and Aluminum Alloys

Widely Used Processes	Less Widely Used Processes
Inert-gas-arc (MIG and TIG)	Oxyacetylene gas
Shielded metal-arc	Carbon-arc
Submerged-arc	Furnace brazing
Resistance	Brass brazing
Silver brazing (silver soldering)	
Soft soldering	

Courtesy The James F. Lincoln Arc Welding Foundation

1. Aluminum-tin alloy (79.0 aluminum, 20.0 tin, 1.0 copper) used in the automotive industry for the production of such items as camshaft bearings and connecting rods,
2. Phosphor bronze (95.0 copper, 5.0 tin),
3. Gun metal (88.0 copper, 10.0 tin, 2.0 zinc) used for the production of gears and bearings.

Alloys in which copper constitutes the major element are discussed in detail in Chapter 20, WELDING COPPER AND COPPER ALLOYS. Tin is welded and joined by the same methods that apply to lead (see LEAD AND LEAD ALLOYS).

TITANIUM AND TITANIUM ALLOYS

Titanium is characterized by a silvery color, a high resistance to corrosion and the highest affinity for carbon of all known metals. As a result of this carbon affinity, titanium functions as an alloying element (less than 1%) to stabilize the carbon in steels and thereby prevent cracking.

Another characteristic of titanium is its strong tendency to form carbides. For this reason, it is used in some chromium steels to counteract chromium depletion tendencies.

Titanium has a melting point of 3035° F (1800° C). The titanium alloys are as strong as steel, but are 50% to 60% lighter in weight. Commercially pure titanium has a tensile strength of 45,000 psi, but the hardened titanium alloys may have tensile strengths extending to 200,000 psi. The ductility of this metal is relatively low.

Because of its strong attraction as an absorbent for most solids and gases (except argon and helium), it must be welded under specially shielded conditions.

The welding processes recommended for use when welding titanium and its alloys are:

1. TIG welding,
2. MIG welding,
3. Resistance welding (both spot and seam),
4. Electron-beam welding,
5. Brazing.

TIG welding using argon or helium as a shielding gas and a DCSP current is probably the most suitable method for welding titanium and its alloys. Perfect fit-up as well as a thorough cleaning of the metal surface is required prior to welding.

Cleaning the surface may be done by degreasing, followed with a steel brush and sandblasting. Stainless-steel brushes are recommended, since other types may leave rust or metal deposits on the surface. Sandblasting is used to remove scale that is too difficult for the metal brush. All cleaning must be done immediately prior to welding. The surface must be *thoroughly* cleaned of all contaminants.

In addition to the use of argon or helium as the shielding gas in TIG welding, a trailing shield should also be used until the surface has cooled to approximately 750° F.

Filler metals similar in composition to that of the base metal are used with certain thicknesses. Proper joint fit-up makes this unnecessary for thinner gauges of titanium and titanium alloys.

ZINC AND ZINC ALLOYS

Pure zinc has a melting point of 787° F. Its tensile strength is approximately 18,000 psi. The tensile strength of cast zinc is considerably higher at 40,300 psi.

Zinc is obtained from a sulphide ore by the electrolytic process, or through a complicated procedure of separating the zinc from the ore by reducing, boiling, and condensing.

Zinc is used in its pure form for galvanizing. It is also combined with other metals (aluminum, magnesium, iron, and tin) to form zinc castings. Finally, zinc serves as an alloying element in copper-zinc alloys (e.g. zinc bronze).

Zinc fumes have a nauseating effect, and can be potentially dangerous to the operator. These fumes are released in the form of a white vapor at certain temperatures. Several methods can be used for reducing zinc fumes, including the use of an oxidizing flame, special fluxes, and welding at lower temperatures.

ZIRCONIUM AND ZIRCONIUM ALLOYS

Zirconium is a silvery-white metal with a melting point of 3330°F (1700° C) or almost three times that of aluminum. It has metallurgical characteristics very similar to those of titanium. Pure zirconium has a relatively low tensile strength of approximately 32,000 psi. It exhibits a strong corrosion resistance and excellent thermal stability.

A principal application of zirconium is as an absorber of gases in electronic tubes. It is also used as an alloying element in steel to prevent brittleness and age hardening.

Other applications include:

1. Zirconium sheets, foil, wire, and rod,

2. Surgical tools and components,
3. Chemical equipment.

Its thermal stability and resistance to corrosion make it highly suited for construction in the nuclear field.

Zircaloy-2 and Zircaloy-3 are two of the better known trade names for zirconium alloys. *Zircaloy-2,* a product of *Westinghouse Electric Corporation,* contains 98.28% zirconium. The other elements (tin, iron, chromium, nickel) are added to increase its strength and to reduce the tendency of zirconium to absorb hydrogen. Zirconium reacts to oxygen and nitrogen as well; a factor that makes shielding almost mandatory during the welding process. Argon or helium are used for this purpose in the TIG welding process. For particularly critical work, a dry box should be constructed around the area to be welded and inert gas pumped into the space. The dry box permits closer control of the atmospheric conditions around the weld.

When working outside a dry box, a trailing shield of inert gas that meets the contours of the weldment must be used to ensure proper shielding. One indication that shielding has not been adequate is the change in color of the electrode tip from bright to discolored metal.

CHROMIUM AND CHROMIUM ALLOYS

Chromium has a steel-gray color, a melting point of 3434° F (1890° C), and is generally noted for its strength and high resistance to corrosion. As a result of its strength and corrosion-resistant characteristics, chromium is frequently used as a thin coating over other metals. It is generally applied through an electroplating process.

Chromium is used primarily as an alloy in the production of certain types of steels. These alloy steels include the following types:

1. Chromium steels (3.0% to 30.0%),
2. Chromium low-alloy steels (up to 3.0% chromium),
3. Chromium die steels,
4. Stainless steels.

WELDING OTHER METALS

The methods for welding chromium-alloy and stainless steels are discussed in Chapter 20, WELDING STEEL AND STEEL ALLOYS.

COLUMBIUM (NIOBIUM) AND COLUMBIUM ALLOYS

Columbium (or Niobium) resembles steel in appearance, and is closely associated with tantalum in properties. Columbium has a yellowish-white color, and a melting point of 4474° F (2468° C). Its density is slightly greater than that of iron. The tensile strength of columbium is 48,000 to 59,000 psi in the annealed condition, and up to 130,000 psi as drawn wire.

Columbium exhibits excellent strength, ductility, and corrosion-resistance characteristics. However, its resistance to corrosion is somewhat limited by its strong reaction to oxides. Oxidation becomes a serious problem at temperatures in excess of 400° C, necessitating the use of an inert shield when working the metal at higher temperatures.

Columbium is a useful alloying element in other metals. It is particularly useful in imparting stability to stainless steel. Because of its low resistance to thermal neutrons, columbium is frequently used in nuclear equipment. Columbium is a refractory metal, a characteristic that encourages frequent use of columbium alloys in the production of high temperature-resisting components for missiles, turbines, and jet engines.

Columbium is added to steels in the form of ferrocolumbium (50.0% to 60.0% columbium, 33.0% to 43.0% iron, 7.0% silicon). Columbium (in amounts exceeding 75.0%) is alloyed with such metals as tungsten, zirconium, titanium, and molybdenum to produce columbium alloys that will retain high tensile strength at high temperatures (in the 2000° F to 5000° F range).

Columbium is welded with the same methods employed when welding tantalum (see TANTALUM), and includes such diverse forms as:

1. TIG welding,
2. MIG welding,
3. Resistance welding,
4. Electron-beam welding.

MOLYBDENUM AND MOLYBDENUM ALLOYS

Molybdenum resembles steel in color, and tungsten in most of its physical properties. It has a melting point of 4730° F (2610° C). As such, it comprises one of the refractory metals (those having a melting point above 3600° F) and can be grouped with columbium (niobium), tungsten, and similar metals. Molybdenum is a very important alloying element in the production of iron and steel.

As an alloying element, molybdenum is not generally used in quantities exceeding 4%. Research in the development of molybdenum steels began in the 1890's, but it was not until after World War I that it was found possible to produce such a steel on an economical basis. It received widespread use in the automotive industry, and rapidly expanded the extent of its application to other areas. Molybdenum is now widely used in the production of tool steels and high-speed steels where it contributes to wear resistance, hardness, and strength. Molybdenum is also added to gray iron to increase its tensile strength and hardenability. This occurs in amounts of less than 1.25%.

Molybdenum is added to steel to increase hardness, endurance, corrosion resistance (stainless steel), and the tendency toward deep hardening. Because of the high melting point of molybdenum and its alloys, they are employed in the manufacture of rocket and gas turbine engines. Its greatest use, however, is in the electronics and nuclear industries.

The following welding processes are recommended for welding molybdenum and its alloys:

1. TIG welding,
2. MIG welding,
3. Atomic-hydrogen welding,
4. Electron-beam welding.

Molybdenum is subject to oxidation at temperatures above 1350° F. Certain precautions must be taken to prevent this action from taking place. Since a shielding procedure is a necessity, TIG welding is the most widely used and suitable method for welding molybdenum and its alloys. Either argon or helium (never carbon dioxide) is used as the shielding gas. Electron-beam welding uses a vacuum that is actually more efficient than the shielding gases in protecting the weld from contamination. Unfortunately, it is limited in size to smaller assemblies and work pieces.

TANTALUM AND TANTALUM ALLOYS

Tantalum has a color ranging from whitish to silvery gray, a tensile strength of 50,000 psi, and a melting point of 5425° F. It is characterized by an unusually strong resistance to acids and corrosion. For this reason, it finds useful application in the manufacture of chemical and surgical equipment.

Tantalum and its alloys have many other interesting applications, including its use as filaments in electric light bulbs, and components in jet engines and rocket motors. Because tantalum is a very ductile metal, it can be drawn or rolled without annealing.

Tantalum has an extremely strong reaction to the oxygen in the surrounding atmosphere or to that of the oxygen cutting stream (oxyacetylene cutting torch). For this reason, tantalum loses much of its effectiveness as a high-temperature structural metal.

A thin oxide surface layer similar to the one found on aluminum covers the surface of tantalum. It is this oxide surface layer that contributes to the resistance of tantalum to acids.

The following welding processes are recommended for welding tantalum and its alloys:

1. TIG welding,
2. MIG welding,
3. Resistance welding,
4. Electron-beam welding.

The TIG welding process is the most commonly used and suitable process for welding tantalum and its alloys and the present stage of development. Either argon or helium is used as a shielding gas in the TIG welding process. The size of the vacuum used in electron-beam welding limits this process to small assemblies and sections.

TUNGSTEN AND TUNGSTEN ALLOYS

The color of tungsten ranges from steel gray to a silvery white. It has a melting point of 6039° F (3337° C), the highest of all metals. As a result of this characteristic, tungsten has found applications in the aerospace industry of rocket components that must withstand extreme temperatures. Tungsten is also used in the electronics industry, in welding, in the electrical industry, and in the production of high-speed cutting tools.

Tungsten is an important alloy in the production of steels. Important among these steels are the tungsten tool steels (up to 20.0% tungsten), and particularly the high-speed tool steels. As an alloying element, tungsten contributes to steel such characteristics as hardness, strength, and wear resistance.

The methods recommended for welding tungsten and tungsten alloys are those used for welding tool and die steels (see Chapter 18, WELDING TOOLS AND DIES), and include the following:

1. Arc welding,
2. Electron-beam welding,

VANADIUM AND VANADIUM ALLOYS

Vanadium is a grayish-white colored metal used primarily in steel production and as an alloying element for other metals. Vanadium tends to increase the hardenability of a metal and to reduce or eliminate the harmful effects of overheating.

Vanadium melts at a temperature of 3236° F. Because it is not subject to oxidation (it is highly resistant to corrosion), it serves as a strong deoxidizer in steels. It increases the tensile strength, has little or no effect on the ductility of the metal, and it greatly increases grain growth. In the annealed condition, vanadium has a tensile strength of 66,000 psi.

Some steels in which vanadium is present as a major alloying element are:

1. Python steel (0.25% vanadium),
2. Vasco vanadium steel (0.20% vanadium),
3. SAE 6145 steel (0.18% vanadium).

These and others (e.g. Colonial No. 7, Elvandi, etc.) are referred to in general as *vanadium steels*. This general grouping is frequently divided into subgroups (e.g. *chromium-vanadium steels* or *carbon-vanadium steels)* depending on the alloys present.

Vanadium is derived from several common ores (vanadinite, patronite, etc.) and is available commercially in four forms:

1. Cast vanadium ingots,
2. Pure vanadium (99.5%),
3. Vanadium buttons,
4. Machined vanadium ingots.

OTHER ALLOYING METALS AND ELEMENTS

There are a number of other metals and alloying elements used in the production of metal alloys with which the welder should become familiar. Each of these contribute a particularly desired characteristic to the metal alloy.

These various alloying metals and elements generally include the following:

1. Antimony,
2. Bismuth,
3. Boron,
4. Carbon,
5. Cadmium,
6. Calcium,
7. Cobalt,
8. Manganese,
9. Phosphorous,
10. Silicon,
11. Sulphur (and selenium).

Antimony is a metal used as an alloy in combination with tin and lead for the manufacture of battery grids and bearings. It has a bluish-white color, lacks ductility, and is extremely brittle. The addition of tin and antimony will increase the hardness of lead. The babbit metals are combinations of antimony, tin, copper, and lead. Antimony has a melting point of 824° F (630° C), and a very low thermal and electrical conductivity.

Antimony never occurs as a commercially pure metal. Besides the babbit metals, antimony also functions as an alloying element in type and stereotype metals.

Bismuth is used to harden tin, lead, or copper. It has a grayish-white color, a melting point of 507° F (271° C), and is extremely low in electrical and thermal conductivity. As an alloy, bismuth is sometimes referred to as *expansion metal*. In addition to increasing the hardness of tin, lead, or copper, bismuth forms alloys that are noted for their low melting point.

Bismuth (0.4% to 0.5%) is added to some stainless steels to gain machinability without affecting the resistance to corrosion. In this way, bismuth functions as a replacement for selenium. Bismuth is also added to steel in small amounts to reduce electrical conductivity.

Boron is an alloying element used to increase tool-steel hardening depth. It closely resembles silicon, and is found in combination with other elements. Traces of boron are added to steel to increase strength or promote case hardening. Boron has a melting point of approximately 2400° C.

Cadmium is a bluish-white, ductile metal having a melting point of 608° F (320.9° C). It has several functions in steel production. In small amounts, it is used as an alloy with nickel, tin, copper, or lead, to increase hardness. Many soft solders include cadmium as an alloying element. It is also used as a rust-resistant plating on irons and steels.

Cadmium copper (or cadmium bronze) contains up to 1.20% cadmium. The cadmium increases the strength, but reduces the electrical conductivity. Cadmium copper is used in the production of electrical wire.

Calcium is a soft metal used as an alloy with magnesium, copper, aluminum, or lead to increase hardness. It is a strong deoxidizer and decarburizer for certain metals (both ferrous and nonferrous). Calcium metal has a yellowish-white color, and a melting point of 810°C.

Carbon is a nonmetallic element that functions as a hardening agent in the production of steels. The carbon will chemically combine with the iron during the steel-making process. A metal having less than 0.15% carbon content is referred to as iron.

Steels contain more than 0.15% carbon; tool steels more than 0.60%.

Cobalt has a melting point of 2723° F (1480° C). This temperature and other aspects (high strength, hardness, etc.) give to cobalt the characteristics generally associated with the refractory metals. Cobalt has a white, bluish-tinged color. It can add red-hardness to cutting-metal alloys, a characteristic it shares with tungsten. Pure cast cobalt has a tensile strength of 34,500 psi. Cobalt is used as an alloy in the production of magnet, tool, and cutting steels.

Manganese is the most effective deoxidizer presently in use. As such, it is an important element in steel production. In concentrations from 1% to 15.0%, manganese increases toughness and hardenability.

Manganese has a grayish-white color tinged with reddish shadings, and a melting point of 1245°C. It is used in steel production in the form of ferromanganese and in the production of nonferrous metals in the form of electrolytic manganese metal (99.97% manganese and 0.03% sulfur and iron).

Manganese is used as an alloying element in the production of steel to obtain the following characteristics:

1. Increased toughness,
2. Increased hardenability,
3. Counteraction of embrittlement and hot shortness.

Manganese (1.5%) is added to aluminum to stiffen and strengthen it. High-expansion manganese alloys (combining manganese, copper, and nickel) are extensively used in the electrical and electronics industries. For example, *Chace* alloy 772 (approximately 72.0% manganese, 18.0 copper, 10.0 nickel), a product of the *W. M. Chace Company,* is designed for use in rheostat resistors and other electrical components. Manganese is also used in certain nonferrous alloys (generally 60.0% copper, 20.0 manganese, 20.0 nickel) for similar applications. *Chace* alloy 720 and *Wybdaloy* 720 *(Wyndale Manufacturing Company)* are trade name examples of this type of alloy.

Phosphorous is a nonmetallic element found in iron and steel. The amount must be carefully controlled. Too much phosphorous will cause the iron to become brittle. The correct amount will significantly increase the strength of all carbon steels (usually about 0.05% phosphorous), and the corrosion resistance of low-carbon steels. Phosphorous occuring in concentrations above 0.035% will seriously affect the weldability of a metal unless certain precautions are taken. In arc welding, undesirable effects of high phosphorous concentrations can be counteracted by using a fast travel speed with a low welding current.

Phosphor bronze is a term used to describe any bronze which has been deoxidized by adding small amounts of phosphorous to the molten metal. *Phosphor copper* is a deoxidizing alloy consisting of copper and phosphorous, and is used to deoxidize the brass and bronze alloys.

Silicon is a nonmetallic element having a melting point of 2615°F (1420°C). It increases strength (particularly in the case of the low-alloy steels), deoxidizes, and functions as a cleansing agent. In addition, it increases the hardenability of certain types of steels.

Silicon, in the form of its silica byproduct, forms slags by combining with manganese and iron oxides. The slags are easily removed from the weld area. Silicon is the second most important deoxidizing agent (manganese is first). It also functions as a desulphurizer, keeping sulphurs below undesirable levels. Silicon is introduced into steel during the production cycle in the form of a ferrosilicon. If too much silicon is added, the metal will become brittle. The welding heat can cause surface cavities if silicon is present in amounts exceeding 0.35%.

Silicon is usually added to other metals in amounts up to approximately 25.0%. Examples of these types of alloys are (1) the silicon-manganese alloy (about 65.0% to 70.0% manganese, 12.0% to 25.0% silicon) and (2) silicon-copper (generally 5.0% to 25.0% silicon). The manganese-silicon alloy (73.0% to 78.0% silicon, 20.0% to 25.0% manganese, 1.5% iron, 0.25% carbon) is one of the alloys in which silicon is the major element. Traces of silicon (3.0% to 5.0%) are added to steel to increase wear and acid resistance.

Sulphur is a nonmetallic element that is regarded as an impurity. As such, it is desirable to keep the sulphur content below the

0.06% level. A high level of sulphur will cause porosity and cracking. If sulphur inclusions occur near the surface, the welding heat may cause the sulphur to explode, resulting in a pocket or cavity on the surface. In arc welding, the tendency for this to occur can be reduced by using a low-hydrogen electrode (e.g. the E-6018). Machining qualities of steel can be improved by a sulphur content of 0.1% or slightly more. Selenium will also improve the machinability of a steel. If either sulphur or selenium occur in amounts above 0.35%, the weldability of a metal declines significantly.

CHAPTER 24

Welding Plastics

All plastics can be divided into two principal categories: (1) thermoplastics, and (2) thermosetting plastics. These two groupings are distinguished from one another by their reaction to reheating and reforming.

Thermosetting plastics cannot be reheated and then bent or joined together to form new shapes. Once a thermosetting plastic has cooled after the heat application used in its initial forming, it cannot be reheated to make changes in its form. Examples of thermosetting plastics are epoxy, polyester, silicone, and casein.

Thermoplastics can be repeatedly reheated and reformed. In many respects, thermoplastics resemble metals in their weldability. The types of welds, joint designs, fit-up and the welding procedures (use of welding rod, fusion of the base material with the filler material) all resemble metal welding.

Those plastics in the thermosetting category that are most commonly encountered in welding operations include:

1. Polyvinyl Chloride (PVC) Type I,
2. Polyvinyl Chloride (PVC) Type II,
3. Polypropylene (PP),
4. Acrylonitrile-Butadiene-Styrene (ABS),
5. Polyethylene (PE).

The most commonly used thermoplastics are the two types of polyvinyl chloride (PVC). They possess excellent physical properties, but are somewhat more difficult to weld than thermoplastics.

Table 1. Properties and Applications of Thermoplastics

	Type I PVC	Type II PVC	Modified High-Impact PVC	Type I ABS	Type II ABS	Branched P/E	Medium-Density P/E	Linear P/E	Poly propylene	Flexible Tank Lining	Acrylic
Service Potential											
Impact Strength	F-G	E	G	E	E	E	E	E	G	NA	F
Chemical Resistance	E	G-E	G-E	G	G	G	G-E	G-E	E	G	F
Hardness	E	G	G	G	E	U-F	F	G	G	NA	G
Heat Service Temp.	F-G	F	F-G	G	E	U	U-F	G	E	F-G	G
Working Strength	E	G	E	E	E	F	G	G-E	G-E	F	G
Uses & Availability											
Cement & Fabricate	E	E	E	E	E	U	U	U	U	E	E
Demineralized Water	G	G	G	G	G	G	E	E	E	F-G	G
Duct Work	E	E	E	F	G	NA	U	G	E	NA	G
Etch Stations	E	G	F	F-U	F-U	NA	U	G	E	NA	NA
Fittings	E	E	NA	E	E	NA	NA	G	E	NA	NA
Formed Parts, Thermo.	F-G	G-E	G	E	E	E	E	E	E	F	E
Formed Parts, Injection	G	G	NA	E	E	E	E	E	E	G	E
Gaskets	F	F	NA	NA	NA	G	G	G	G	E	NA
Hoods, Exhaust	E	E	G	NA	NA	NA	U	G	E	NA	F-G
Laboratory Hoods	E	E	G	NA	NA	NA	U	G	E	NA	NA
Laboratory Utensils	E	G	G	NA	NA	G	G	E	E	NA	F

Machined Parts	E	G	NA	G	G	NA	G	G	G	NA	G
Nuts & Bolts	E	G	NA	NA	NA	NA	NA	F	G	NA	NA
Pipe & Tubing	E	E	NA	E	E	F	E	E	E	F-G	F-G
Scrubbers	E	E	G	F	F	NA	U	F	E	NA	NA
Shapes, Extruded	E	E	NA	G	G	G	G	G	G	E	G
Sheet	E	E	E	E	E	E	E	E	E	E	E
Sinks	E	E	G	G	G	NA	G	G	E	NA	G
Stress—Crack	E	G	G	G	G	U-F	F	F-G	E	NA	F-G
Tanks, Lined	F-G	G	F	F	F	F	U	F-G	E	E	NA
Tanks, Self-Supporting	E	E	F	U	U	NA	F-U	G	G-E	NA	G
Table Top	G	E	E	E	E	F	F	F	G	NA	NA
Tumbling Basket	F	F	U	U	U	NA	U	F-G	G-E	NA	F
Valves	E	F	NA	F	F	NA	NA	F-G	G-E	NA	NA
Weld & Fabricate	E	E	G	F	F	G	E	E	E	E	NA

E—Excellent G—Good F—Fair U—Unsatisfactory NA—Not Applicable

A tabulation of usage is always subject to interpretation and special requirements. Ratings are based on performance, availability, or cost. Where lower ratings are assigned in some cases, outstanding performance capability of several other materials were considered, although the lower rated materials would be satisfactoy. An attempt was made to differentiate between non-applicable and unsatisfactory. In the former case, the materials are never or almost never considered for such services due to obvious limitations. In the second case, it designates misapplication of the material in the opinion of the writers and use should be avoided.

Courtesy Kamweld Products Co., Inc.

A listing of the various applications and properties of the major thermoplastics is shown in Table 1.

Thermoplastics have a number of advantages over metal. The principal advantages include reduced cost, strong corrosion resistance (to both water and chemicals), and longer service life under conditions for which they are designed. The stronger resistance to corrosion in combination with the reduced cost and longer service life have made thermoplastics ideal materials for the fabrication of pipes. On the other hand, thermoplastic fabrications and parts lack the structural strength of those made from metal. They also exhibit a lower heat resistance and ductility, and show a greater tendency toward deformation under load.

PLASTIC WELDING PROCESSES

The six basic welding processes used in joining thermoplastics are:

1. Hot-gas welding,
2. Ultrasonic welding,
3. Heated-tool welding,
4. Induction welding,
5. Friction welding.

This chapter is primarily concerned with describing the hot-gas welding and ultrasonic welding processes. These will be dealt with in more detail in the appropriate sections of this chapter.

Heated-tool welding is a thermoplastic joining process in which the heat necessary for fusion temperatures is derived from some sort of heated tool or device. These tools or devices include such diverse forms as electrical strip, heating bars, or hot-plate resistance coil-heaters. Aluminum is the most commonly used surface on these heating tools, because it does not react unfavorably with the hot thermoplastic. For example, the use of a metal such as copper can result in the decomposition of the plastic due to the reaction that is set up between the two.

The heat necessary for heated-tool welding is relatively high, ranging from 400° F to 650° F. When the plastic has reached a temperature sufficient for fusion, the parts are removed from the heating tools and placed together under slight pressure until the plastic cools and a strong joint is formed.

Simple bar-type heating tools can be used to bend thermoplastic sheets to the proper angle by applying sufficient and uniform heat to the areas to be bent. The *Kamweld* heating device shown in Fig. 1 consists of a control box (for temperature control), two

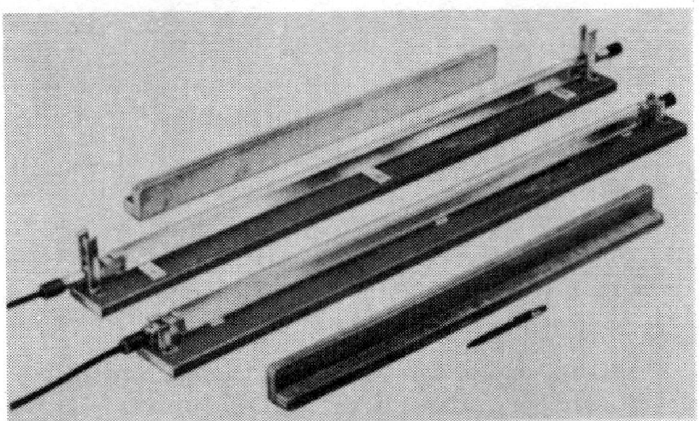

Courtesy Kamweld Products Co., Inc.

Fig. 1. Kamweld heating device.

aluminum heating bars, two base units, and two sheet supports. The heating element extends the length of the bar.

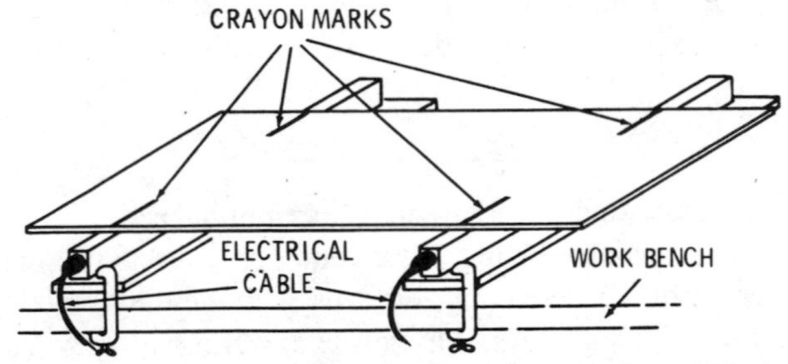

Courtesy Kamweld Products Co., Inc.

Fig. 2. Marking and heating the thermoplastic sheet.

Figs. 2 and 3 illustrate various applications of the heating tool in bending and forming plastic sheet.

Induction welding employs electrically induced heat to produce the fusion temperatures. The electrical current generally flows

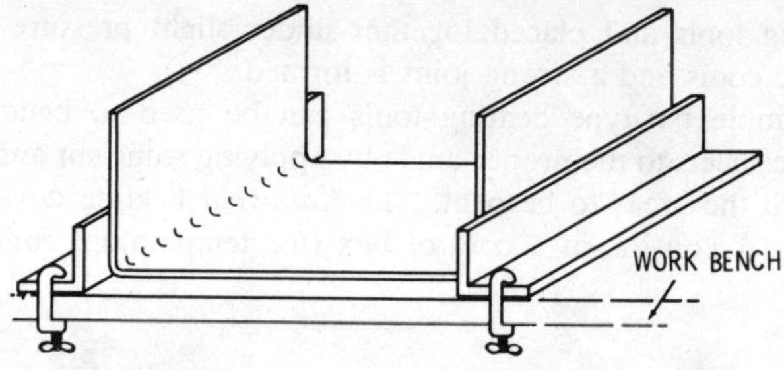

Courtesy Kamweld Products Co., Inc.

Fig. 3. Bending the heated sheet.

through a metallic insert that remains imbedded in the plastic after the welding operation has been completed. Welds produced by this welding process are not as strong as other types. The inclusion of the metallic insert is a decided disadvantage for it tends to weaken the weld. Induction welding is also very expensive, a factor that tends to limit its use.

Friction welding uses friction and pressure applied simultaneously to join surfaces. The friction is generally obtained by spinning the two parts to be welded against one another until the proper fusion temperature is reached. For this reason, friction welding is sometimes referred to as *spin welding*. The spinning may be done on a lathe or a similar device. One of the parts is held stationary while the other spins against its surface. Friction welding is fast, results in strong joints, and produces welds with a good appearance.

TYPES OF WELDS

The types of welds used to join thermoplastics are similar to those found in metal welding. The same edge preparation of the joint (fit-up, root gap, beveling, etc.) used in welding metal is also required for thermoplastics. However, beveling is very essential when attempting to obtain quality welds. Beveling may be even more essential when welding thermoplastics than in metal welding.

Additional plastic is added to the weld with a round rod, an oval rod, a triangular rod, or a flat strip. Examples of typical welds made by using either a round or a triangular rod are shown in Fig. 4. Flexible flat plastic strip is used to weld or repair tank

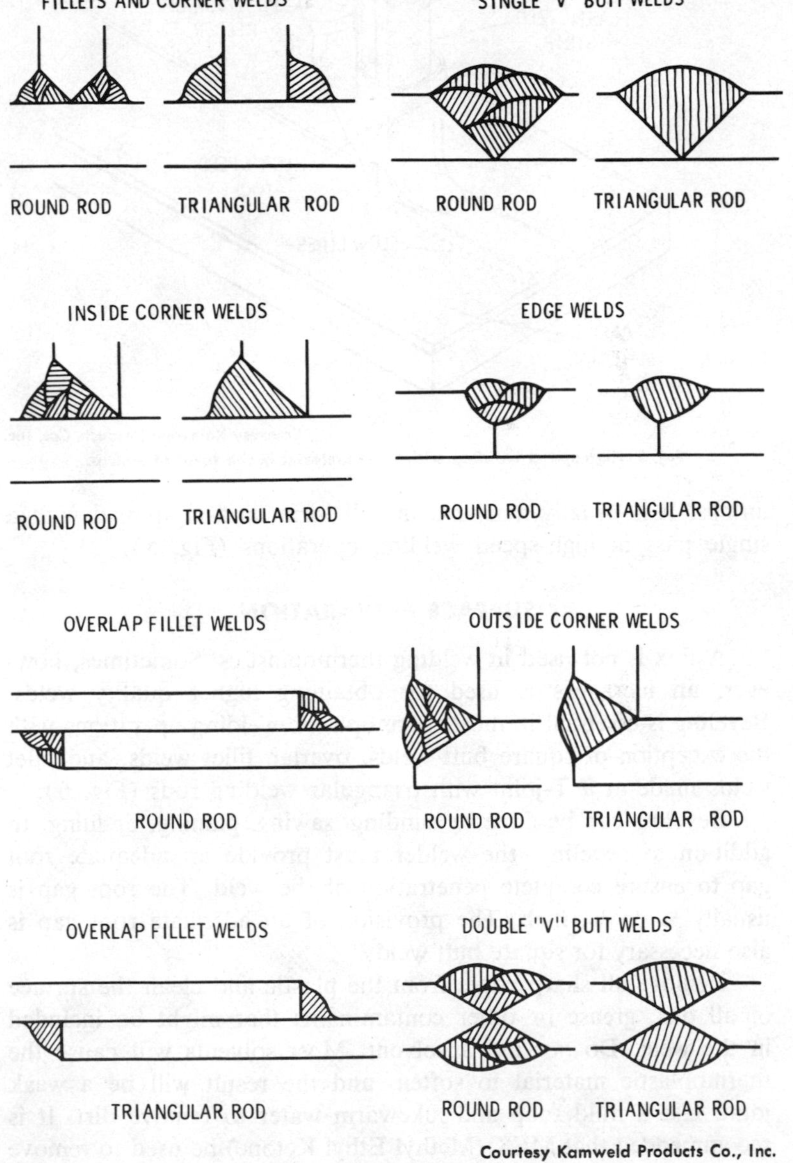

Courtesy Kamweld Products Co., Inc.

Fig. 4. Examples of typical welds.

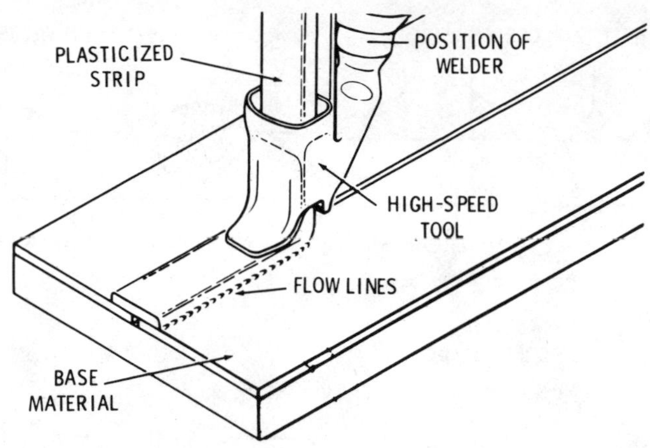

Courtesy Kamweld Products Co., Inc.

Fig. 5. High-speed welding with filler material in the form of a strip.

linings. It is usually available in roll form, and is applied with a single pass in high-speed welding operations (Fig. 5).

SURFACE PREPARATION

A flux is not used in welding thermoplastics. Sometimes, however, an inert gas is used for obtaining higher quality welds. Beveling is essential in most thermoplastic welding operations with the exception of square butt welds, overlap fillet welds, and fillet welds made at a T-joint with triangular welding rods (Fig. 6).

Beveling can be done by sanding, sawing, planing, or filing. In addition to beveling, the welder must provide an adequate root gap to ensure complete penetration of the weld. The root gap is usually 1/32 to 1/64 inch. The provision of an adequate root gap is also necessary for square butt welds.

Remove all sharp edges from the plastic and clean the surface of all dirt, grease or other contaminants that might be included in the weld. Do *not* use a solvent. Most solvents will cause the thermoplastic material to soften, and the result will be a weak joint. Use a mild soap and lukewarm water to remove dirt. It is recommended that MEK (Methyl Ethyl Ketone) be used to remove oil and grease.

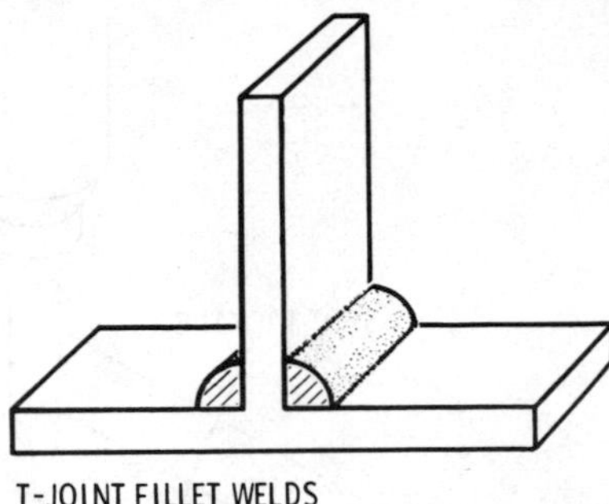

Fig. 6. Weld joints that do not require beveling.

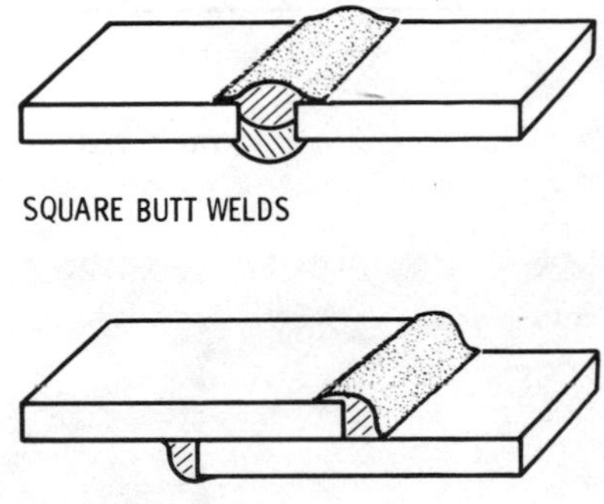

Courtesy Kamweld Products Co., Inc.

HOT-GAS WELDING

Hot-gas welding is one of the most widely used thermoplastic joining processes with remarkably few limitations insofar as materials are concerned.

In hot-gas welding, a heated stream of compressed air or inert gas is directed against the thermoplastic surface. The stream obtains its heat by passing over flame heated or electrically heated coils in the body of the hand welder. No flame or electrical arc touches the surface. The flow-rate of the stream controls its temperature at the point of contact with the surface (the faster the rate of flow, the lower the temperature).

Three basic welding procedures are used in hot-gas welding. These are as follows:

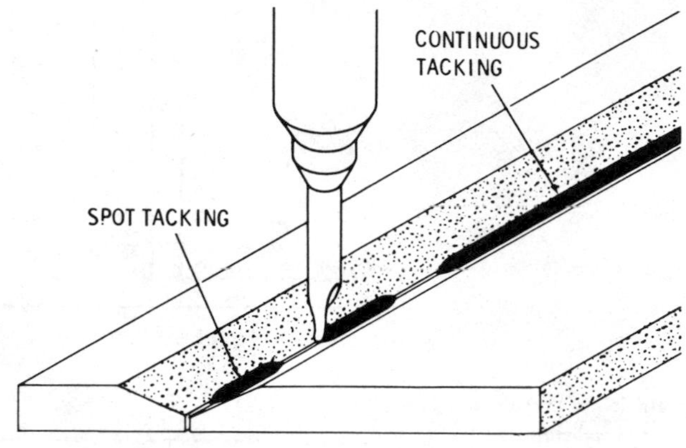

Courtesy Kamweld Products Co., Inc.

Fig. 7. Tack welding procedure.

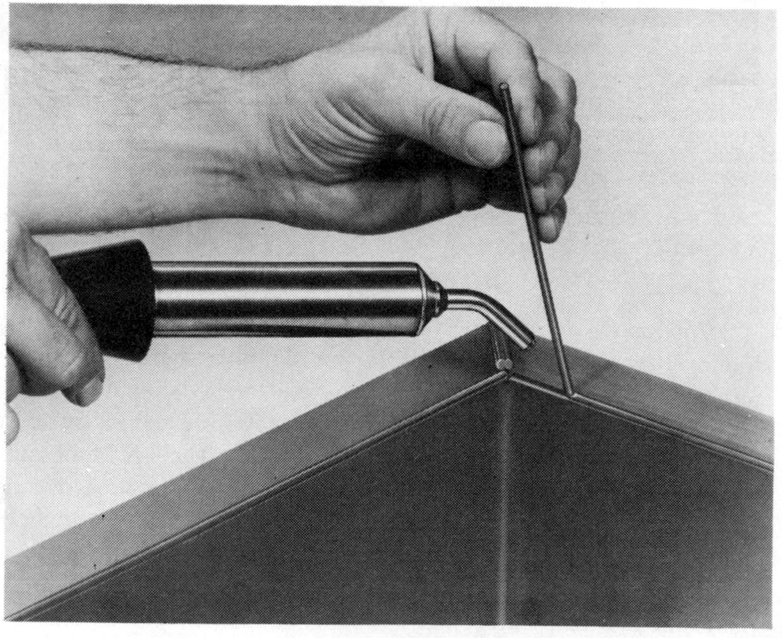

Courtesy Seelye Plastics Inc.

Fig. 8. Hot-gas hand welding.

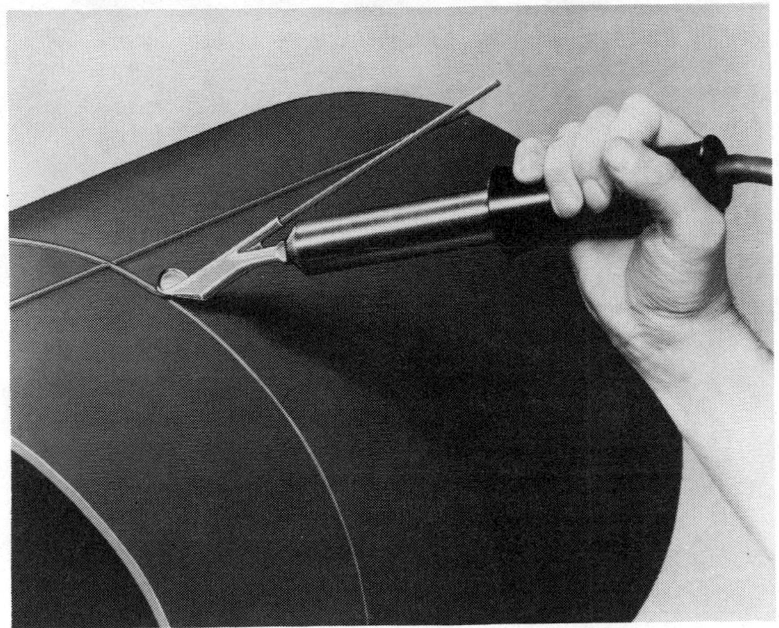

Courtesy Seelye Plastics Inc.

Fig. 9. Hot-gas high-speed welding.

1. Tack welding,
2. Hand welding,
3. Hand high-speed welding.

Tack welding (Fig. 7) is a temporary welding operation used to hold two parts in position until a more permanent weld is produced. Its function therefore resembles that of tack welding metals. No welding rod is used when tack welding thermoplastics. Continuous or spot tacking is possible depending upon the desired strength. A special tack welding tip is screwed into the barrel of the hand welder.

The hand welding procedure (Fig. 8) is used to make permanent welds in difficult areas (corners, short runs, small radii, etc.). Heat and pressure are used to fuse the two parts together. Round, oval, and triangular welding rods are used to provide additional thermoplastic to the weld.

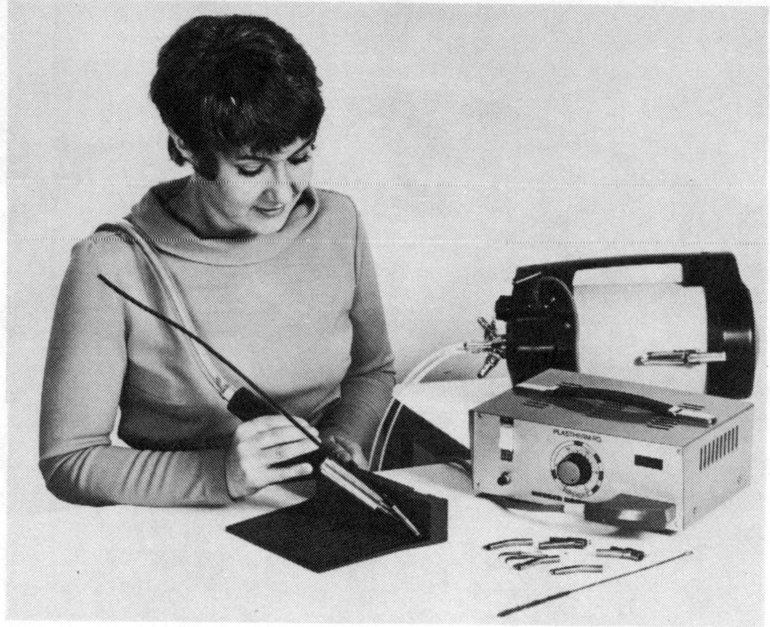

Courtesy MG Welding Products

Fig. 10. Principal components of a hot-gas welding operation.

High-speed hand welding (Fig. 9) uses a flat strip or round welding rod which is fed automatically into the hand welder. The flat strips are used for welding or repairing tank lining and similar types of equipment. All other applications of the high-speed hand welding procedure use the round welding rod.

HOT-GAS WELDING EQUIPMENT AND SUPPLIES

The equipment used in a hot-gas welding operation includes the following basic components (Fig. 10):

1. Hot-gas hand welder,
2. Welding tips,
3. Supply of compressed air or inert gas,
4. Air regulator and pressure gauge,
5. Neoprene air hose with enclosed electric cord.

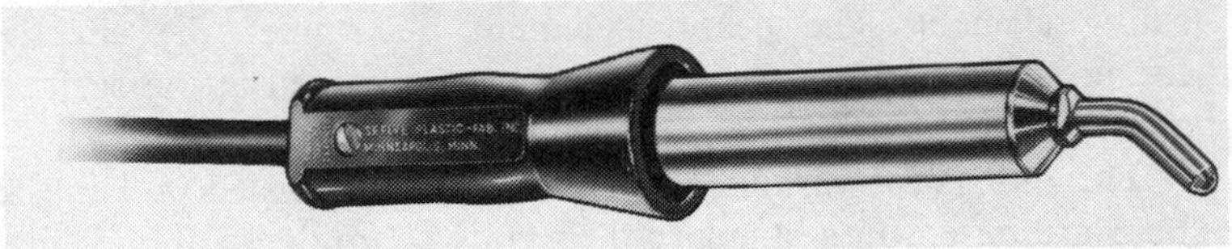

Courtesy Seelye Plastics Inc.

Fig. 11. Hot-gas hand welder.

A typical hot-gas hand welder (Fig. 11) consists of a heating element, an air-cooled handle, a stainless steel barrel, and a heating element connector located in the handle. The heating element is attached by plugging it into the connector.

Heating elements are available in several wattages, generally ranging from 350 to 650 watts. A 350-watt heating element will produce heat in a temperature range of 420° F to 580° F. Temperatures of 700° F to 1000° F will be produced by a 650-watt

Table 2. Differently Rated Heating Elements

320 watts	2.7 amperes
350 watts	3.1 amperes
450 watts	4.0 amperes
550 watts	4.8 amperes
650 watts	5.6 amperes
750 watts	6.5 amperes
800 watts	6.9 amperes

Courtesy Kamweld Products Co., Inc.

Table 3. Air Pressures and Temperatures

Element Watts	Air Pressure lbs. (approx.)	Temp. F. approx. 3/16" from end of round tip
320	2-3	400
340	2-3	410
350	2½-3½	430
450	3-4	540
460	3-4	600
550	4-5	700
650	4½-5½	800
750	5-6	860
800	5-6	900

Courtesy Kamweld Products Co., Inc.

heating element. The heating element determines the effective heating range, and not the degree of heat obtained at the surfaces being joined.

The ratings for various heating elements are listed in Table 2. These are based on a standard 115-volt AC outlet. Table 3 gives the approximate air pressure and welding temperature for each of the wattages in Table 2.

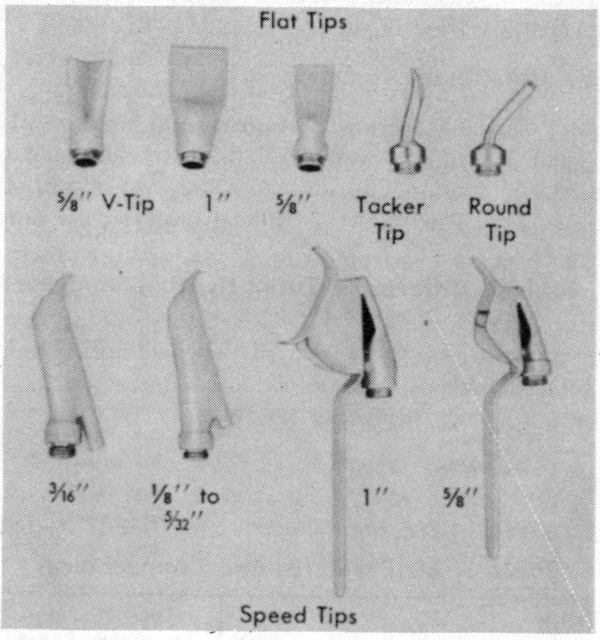

Courtesy Kamweld Products Co., Inc.

Fig. 12. Various types of welding tips.

The welding tips (Fig. 12) are available in a variety of different sizes and shapes depending on the nature of the welding job.

The amount (or degree) of surface heat obtained in welding thermoplastics depends not upon the rating of the heating element, but upon the flow of compressed air or inert gas. Moreover, the faster the rate of flow, the lower the temperature will be. By slowing the rate of flow (and thereby allowing a greater volume of air or inert gas to be heated by the heating element), the temperatures obtained at the surface can be decreased. Exact temperature control is permitted with the air regulator and pressure gauge. A typical gauge is shown with the equipment in Fig. 13.

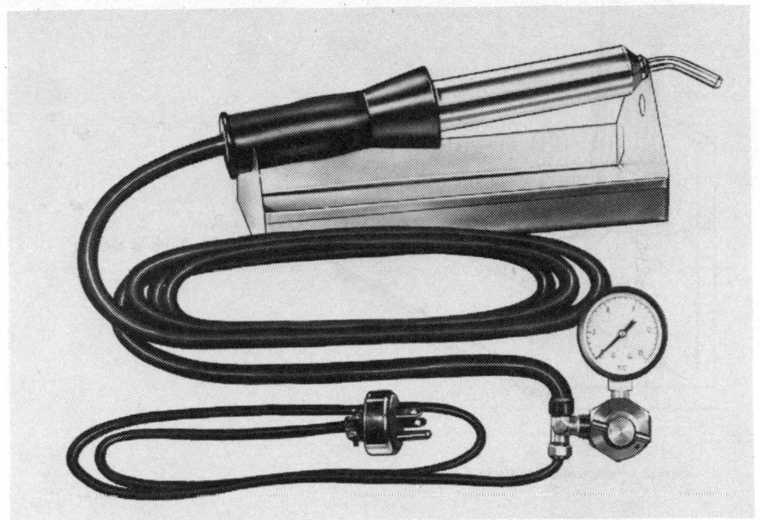

Courtesy Seelye Plastics Inc.

Fig. 13. Typical gauge used with hot-gas welding operation.

The compressed air flow for the welding operation may be obtained from a small air compressor (with a minimum of 2 cfm at 25 psi. A bottled inert gas (e.g. nitrogen) may be substituted for the compressed air.

HOT-GAS WELDING PROCEDURE

The welder should begin by making enough tack welds to hold the parts together in the correct position thereby making special jigs or clamps unnecessary. The following procedure is recommended for tacking:

1. Attach a tacking tip to the barrel of the hand welder.
2. Allow the tacking tip to reach a suitable temperature.
3. Hold the hand welder so that the handle is inclined at an

angle of approximately 80° with the point of the tip just touching the surface.

4. Draw the tip along the seam making spaced tacks of ½ to 1 inch in length (Fig. 14).

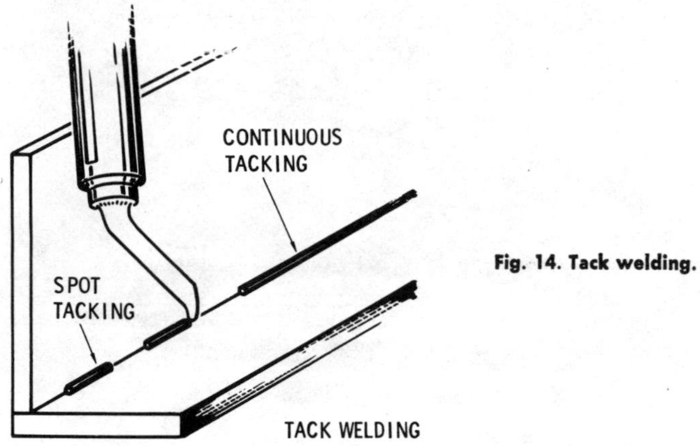

Fig. 14. Tack welding.

Courtesy Kamweld Products Co., Inc.

5. Connect the spaced tacks to form one continuous seam.
6. Allow the tacks to cool. The joint is now ready for regular welding.

After having tack welded the surfaces, the welder is now ready to make a permanent weld. For this purpose, it is necessary to select a suitable welding rod to provide the additional filler plastic. This was not required in tack welding since it was only necessary to temporarily fuse the surfaces together. The composition of the filler plastic should approximate that of the base material.

The procedure for *hand welding* two thermoplastic surfaces is as follows:

1. Prepare the edges and clean the surfaces of the pieces to be joined.
2. Select a suitable welding rod.
3. Select an appropriate heating element (one capable of providing a temperature of 500° F or slightly less) and install it in the hand welder.

4. Attach a tacking tip to the hand welder and perform the necessary tacking operation (refer to the previous paragraphs).
5. After having completed the tacking operation, remove the tip and attach a round tip to the hand welder.
6. Cut the welding rod so that it forms approximately a 60° angle (Fig. 15).

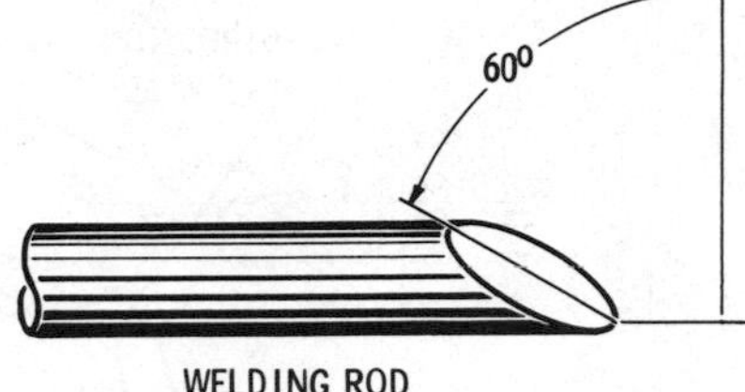

Fig. 15. Cutting the welding rod at a 60° angle.

Courtesy Kamweld Products Co., Inc.

7. Using a weaving or fanning motion, direct the stream of hot air at the surface approximately ½ inch away from the intended starting place of the weld (Fig. 16).
8. The welding rod should be held vertically. Touch the surface of the base material with the welding rod and then pull it away. Continue to do this until the welding rod becomes tacky.

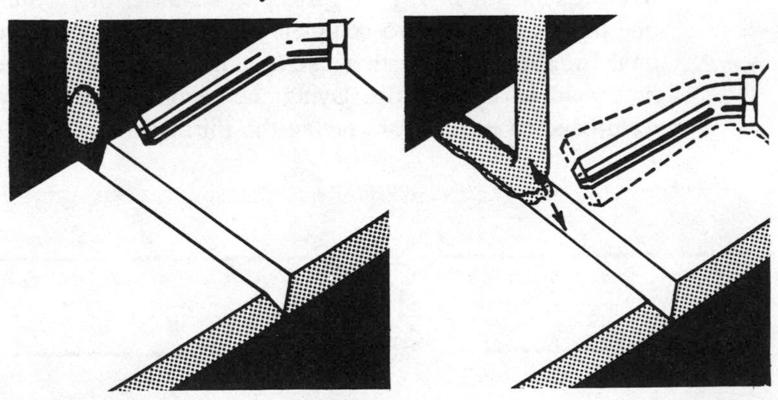

Courtesy Seelye Plastics Inc.

Fig. 16. Starting the weld.

9. As soon as the welding rod becomes tacky, press it vertically against the surface. Direct approximately ⅓ of the heat against the welding rod and the remainder against the base material.
10. Slant the rod in the direction of the weld, as shown in Fig. 17. Avoid stretching the rod excessively as the weld progresses.

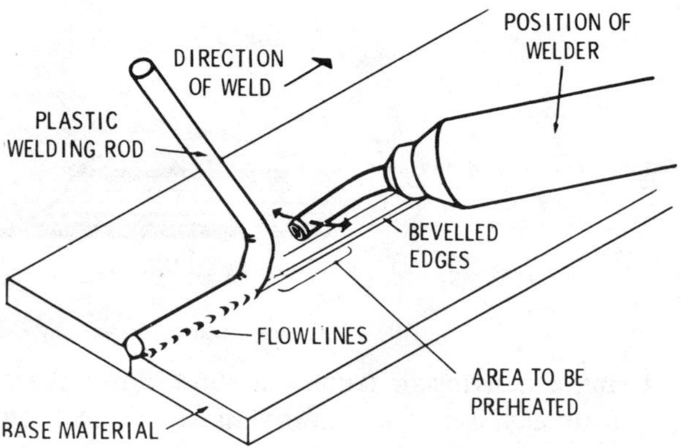

Courtesy Kamweld Products Co., Inc.

Fig. 17. Hand welding procedure.

11. The weld bead should overlap the edges of the joint (or bevel) as shown in Fig. 18. In hand welding, more than one pass is necessary to complete the weld. It will be the final (finishing) passes that overlap the edges. Allow the first weld to cool before laying the second; the second weld bead to cool before laying the third; and so on.

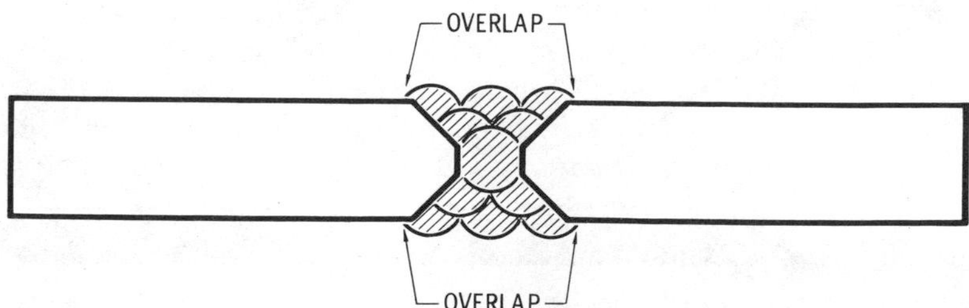

Fig. 18. Cross-section of a weld showing the overlap of the finishing passes.

High-speed hand welding can be accomplished with either a welding rod or strip. Welding strip is used for repairing or welding tank linings and similar equipment. High-speed hand welding with strip is characterized by requiring only a single pass to complete the weld.

The high-speed hand welding procedure (with rod) is essentially as follows:

1. Attach a high-speed welding tip (Fig. 19) to the barrel of the hand welder.

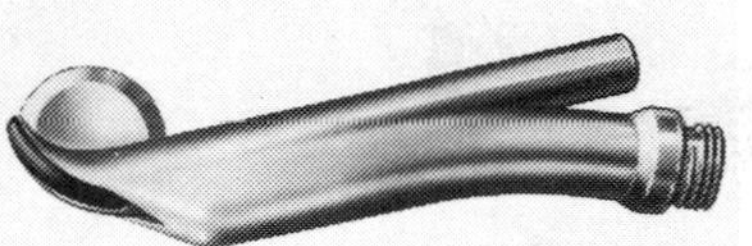

Fig. 19. High-speed hand welding tip.

Courtesy Seelye Plastics Inc.

2. Select a suitable welding rod and cut it at a 60° angle.
3. Adjust the air pressure to the flow rate desired for the job.
4. Turn on the electricity so that thc tip can heat to the welding temperature.
5. Hold the point of the welding tip approximately ½ inch from the surface of the base material.
6. As soon as the surface softens, insert the welding rod through the tip and into the softened plastic (Fig. 20).
7. When the welding rod becomes soft enough, bend it backward with the hand welder so that the latter inclines slightly in the direction of the weld (Fig. 20).
8. The weld progresses at a speed that allows good fusion between the base material and the welding rod. The progress must be continuous and without hesitation. The weld may be ended by bringing the hand welder to a vertical position, cutting off the welding rod with the tip as shown in Fig. 20. An alternative method is to allow the rod to slip out of the welding tip and cut it off with a knife.

All welding with round, triangular, or oval welding rods, is generally accomplished with several passes. This is true whether hand welding or high-speed hand welding is involved. On the other

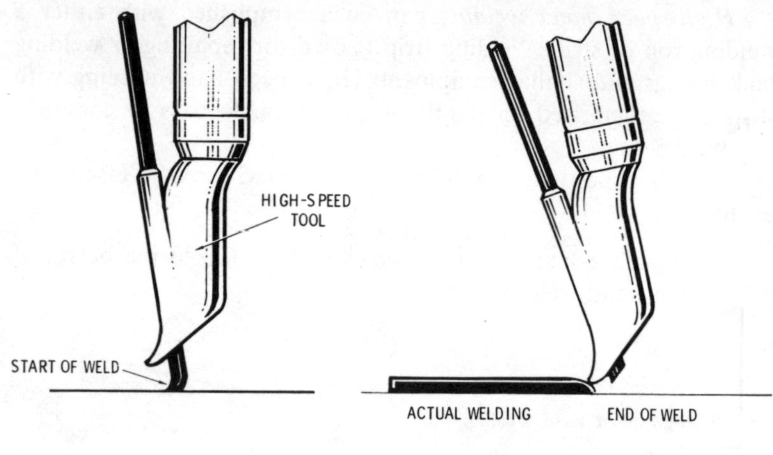

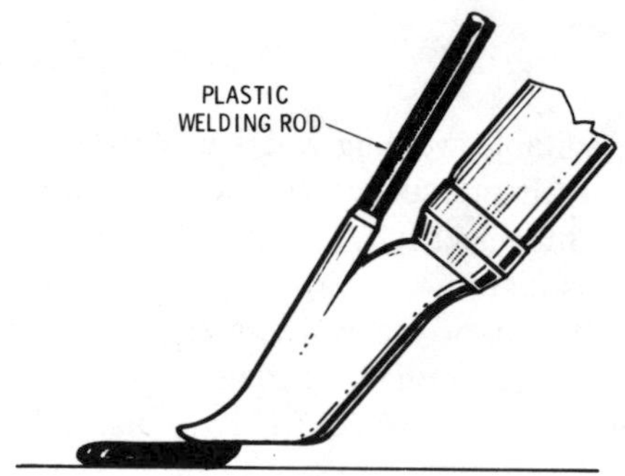

Courtesy Kamweld Products Co., Inc.

Fig. 20. Various angles when using high-speed hand welder.

hand, high-speed hand welding accomplished with flat, flexible welding strip is a *single-pass* operation. When welding with strip as a filler material, follow the same recommendations listed for the other welding procedures except for the following changes:

1. Cut the strip to the desired length (length of weld plus trim allowance) before beginning to weld.
2. Use a slight pressure to hold the strip down as the weld progresses.
3. Stop the weld by permitting the strip to pull through the tip and then cut off the excess.

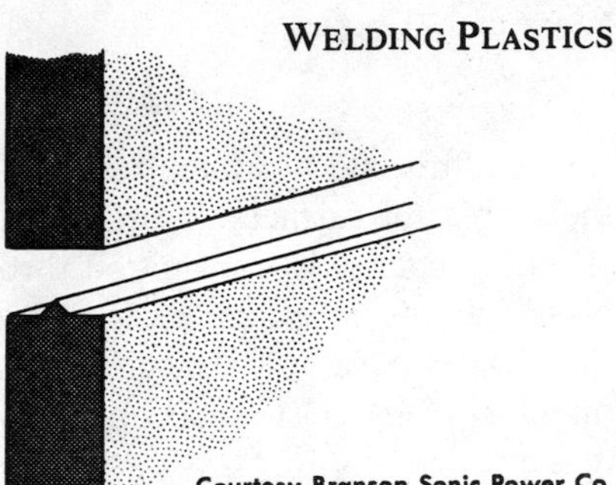

Fig. 21. An ultrasonic welding joint.

Courtesy Branson Sonic Power Co.

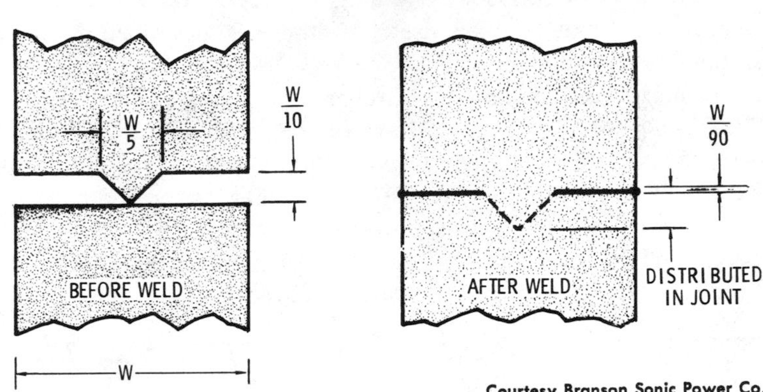

Courtesy Branson Sonic Power Co.

Fig. 22. Simple ultrasonic weld butt joint.

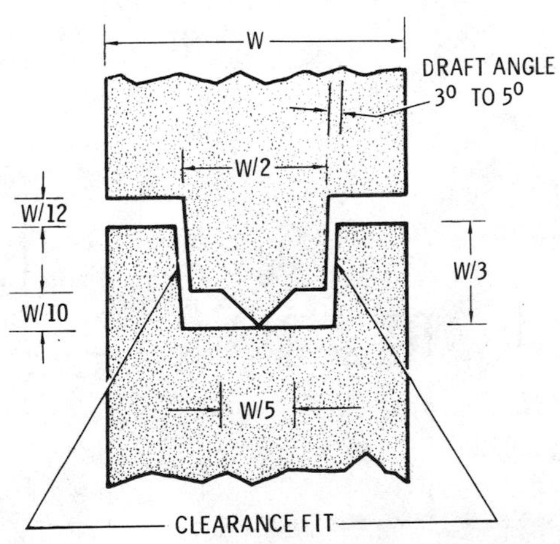

Courtesy Branson Sonic Power Co.

Fig. 23. Tongue-and-groove joint used in ultrasonic welding.

ULTRASONIC WELDING

In ultrasonic welding, the heat required for the fusion of two thermoplastic surfaces is produced by ultrasonic vibrations. This vibratory energy is transmitted through the upper layer of thermoplastic to the interface of the two surfaces where localized frictional heat develops. The heat causes a molded ridge (or energy director) (Fig. 21) to melt and fuse with the surface against which it is pressed.

Variations of ultrasonic welding include ultrasonic inserting, staking, and spot welding. Each of these is described in latter sections of this chapter.

Common joint designs for ultrasonic welding are shown in Figs. 22, 23 and 24. Note the position of the energy director in each case. These are not used in the other variations of ultrasonic welding (staking, inserting and spot welding).

A typical error in ultrasonic welding joint design is excessive beveling (Fig. 25). The result is either incomplete bonding or the expulsion of large amounts of material at the joint.

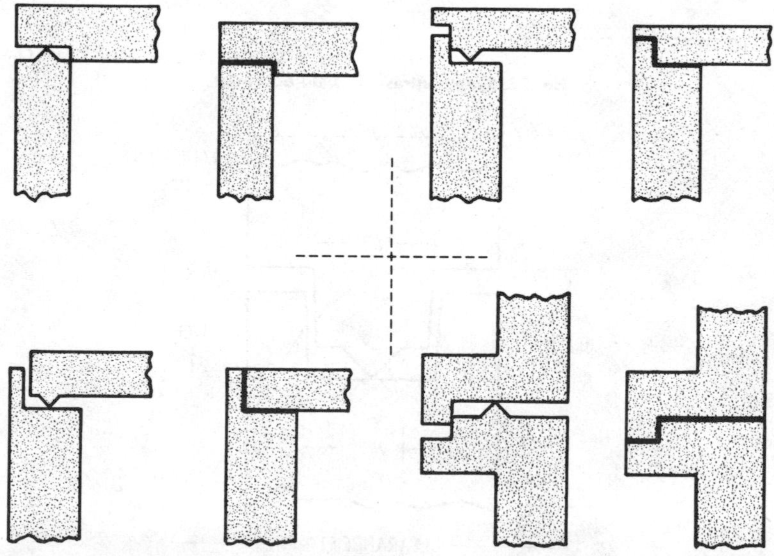

Courtesy Branson Sonic Power Co.

Fig. 24. Examples of other joints used in the ultrasonic welding.

ULTRASONIC WELDING EQUIPMENT AND SUPPLIES

The ultrasonic plastic welder illustrated in Fig. 26 is an example of the type of equipment used for welding thermoplastics. This particular welder converts 115 or 220 volts AC line voltage into 20-kHz electrical energy. The electrical energy is changed into mechanical vibratory energy by means of a converter in the welder. The vibratory energy is then transmitted to a horn which amplifies these vibrations before transmitting them to the surfaces being welded. Fig. 27 represents a schematic of the same ultrasonic welder shown in Fig. 26 illustrating the relationships among the various components.

Horns are designed to transmit mechanical vibratory energy to the thermoplastic being welded. A number of different horn designs are available depending upon the specific application.

The three basic types of horns are:

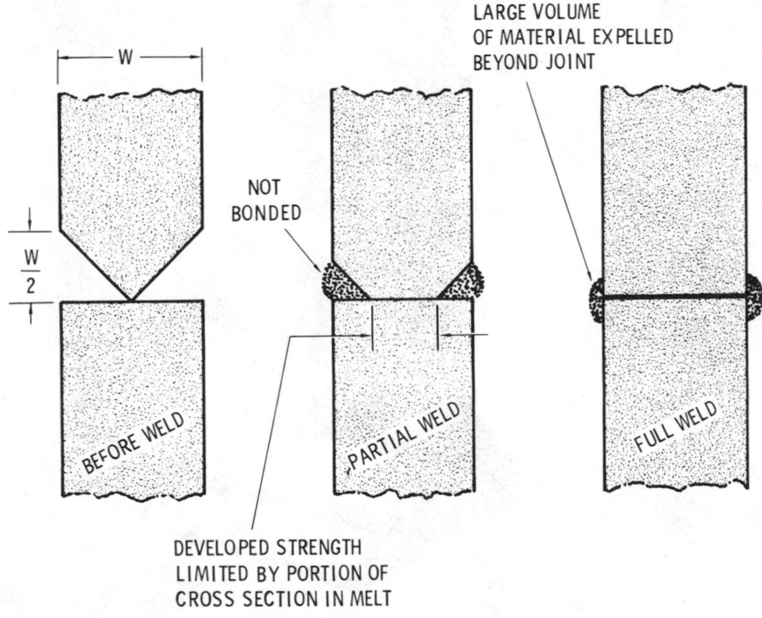

Courtesy Branson Sonic Power Co.

Fig. 25. Results of incorrect joint preparation.

1. Standard horns,
2. Non-standard horns,
3. Booster horns (Fig. 28).

Booster horns are used to increase or decrease the amplitude of the mechanical vibratory energy. Standard horns receive wide application in most ultrasonic welding operations. An ultrasonic welder with a non-standard horn attached is shown in Fig. 29.

Ultrasonic welders can be operated individually or in combination, depending on the requirements of the job. In a typical ultra-

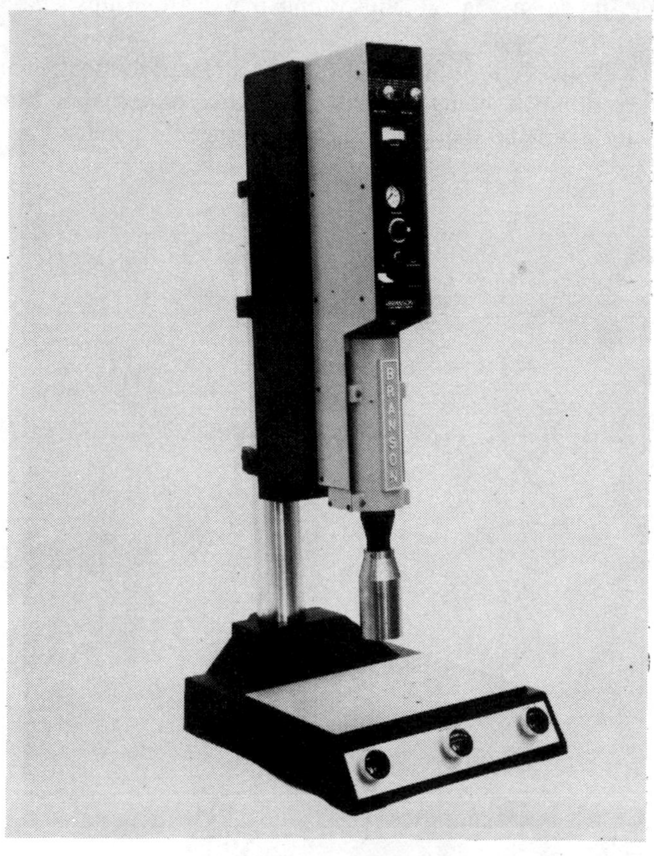

Courtesy Branson Sonic Power Co.

Fig. 26. Ultrasonic plastic welder.

sonic gang welding system, up to as many as eight welding horns can be operated simultaneously with a single-power source.

ULTRASONIC INSERTING

Ultrasonic inserting (Figs. 30 and 31) is used for inserting and encapsulating metal in plastic. A premolded hole, slightly smaller than the metal insert, is formed in the plastic. The metal insert is then driven into the hole. As soon as the surface of the metal insert meets that of the plastic, the ultrasonic vibrations release

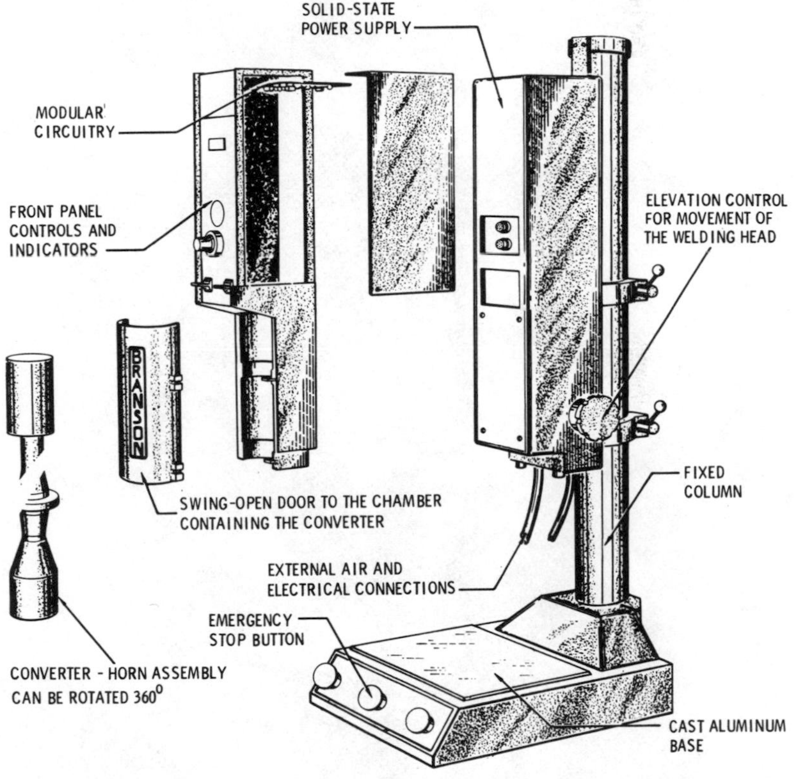

Courtesy Branson Sonic Power Co.

Fig. 27. Schematic of the plastic welder shown in Fig. 26.

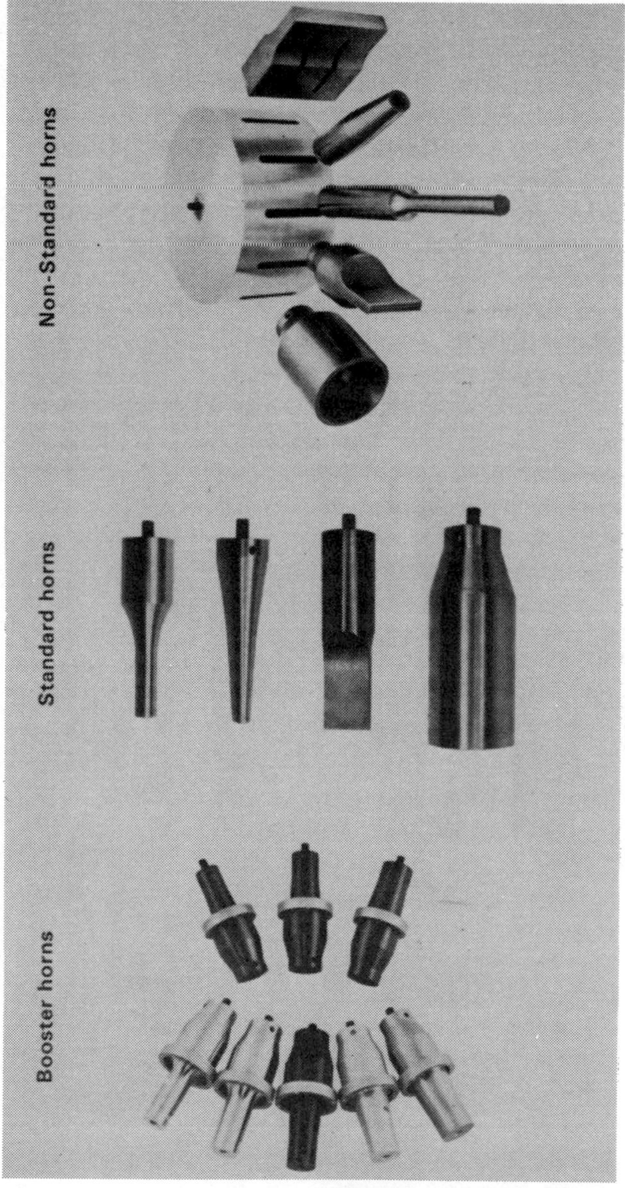

Courtesy Branson Sonic Power Co.

Fig. 28. Examples of three basic types of ultrasonic welding horns.

their energy in the form of heat and melt the plastic which reforms around the insert.

ULTRASONIC STAKING

Ultrasonic staking (Fig. 32) is used primarily in the assembly of plastic and metal. For example, plastic studs are driven into holes prepared in the metal, melted and reformed so that the stud forms a locking head over the metal surface. Figs. 33 and 34 illustrate

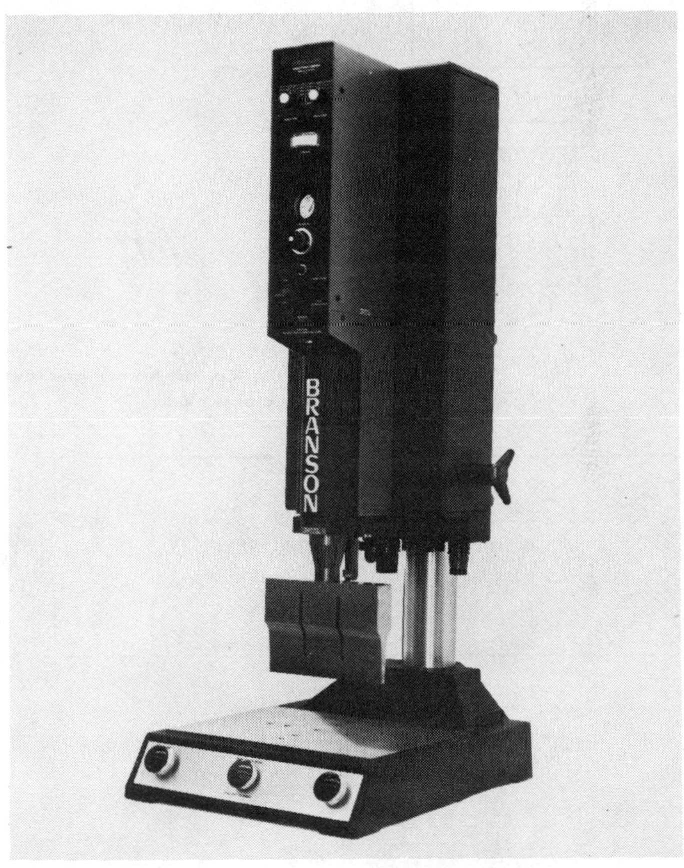

Courtesy Branson Sonic Power Co.

Fig. 29. Ultrasonic plastic welder with a scan welding head.

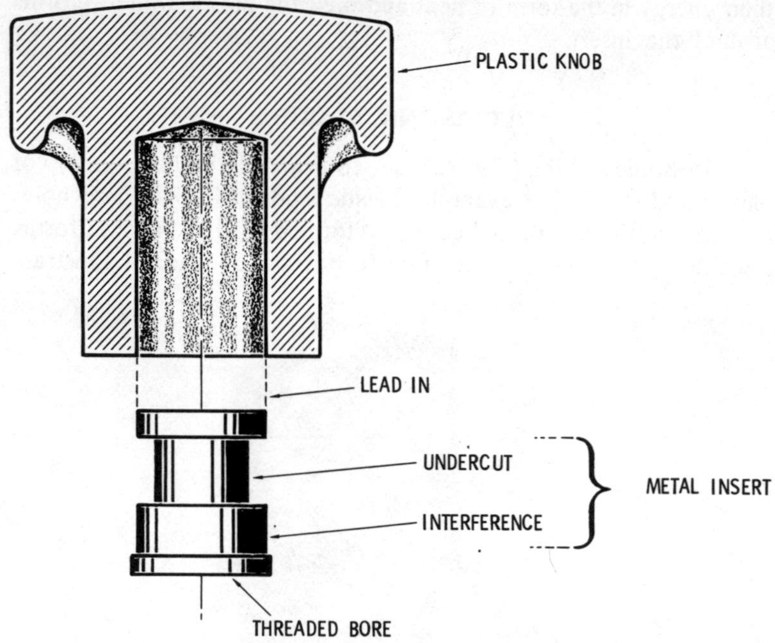

Courtesy Branson Sonic Power Co.

Fig. 30. Metal part inserted in a plastic knob.

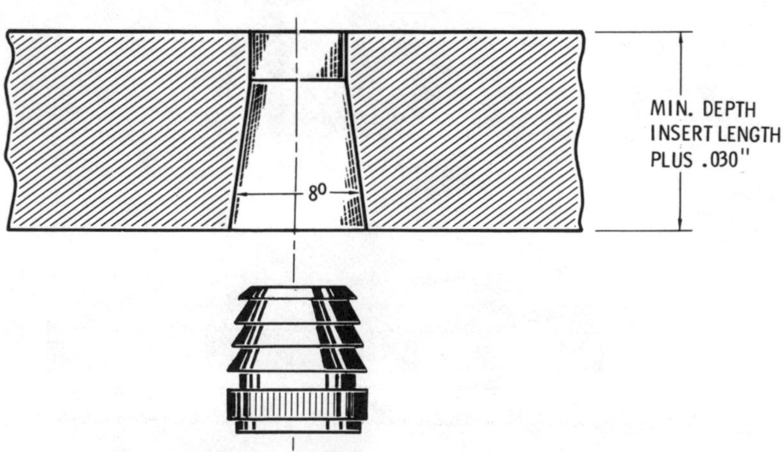

Courtesy Branson Sonic Power Co.

Fig. 31. Tapered inserts.

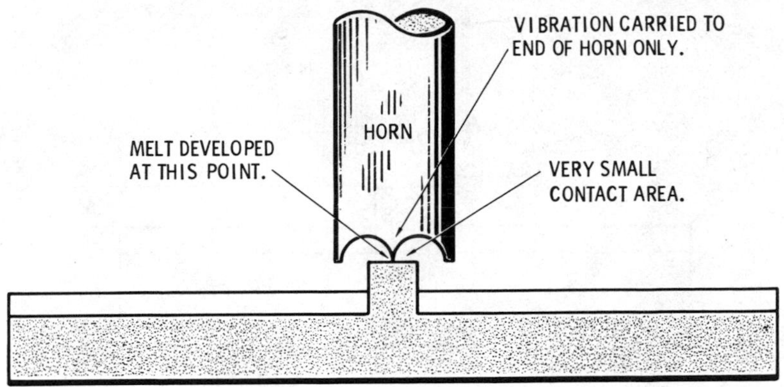

Courtesy Branson Sonic Power Co.

Fig. 32. Ultrasonic staking.

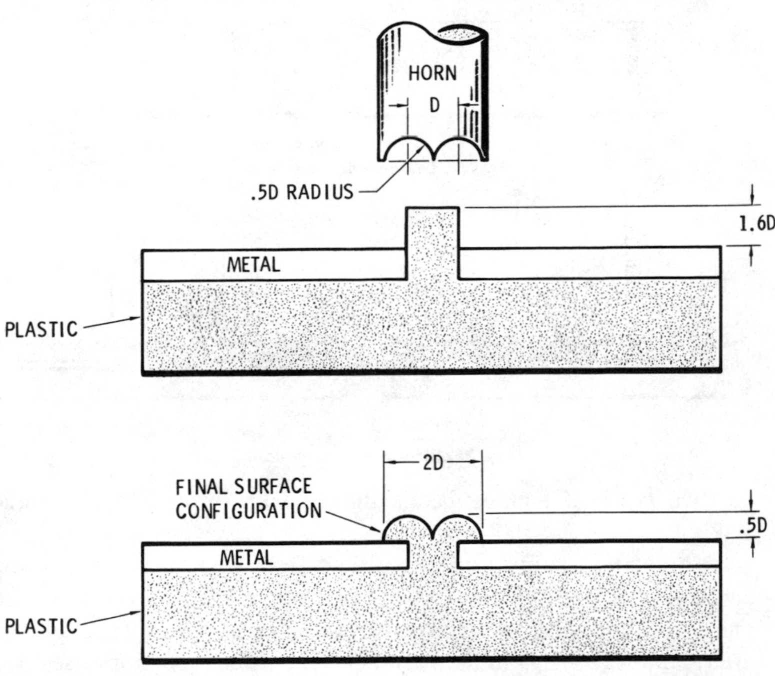

Courtesy Branson Sonic Power Co.

Fig. 33. Standard head form.

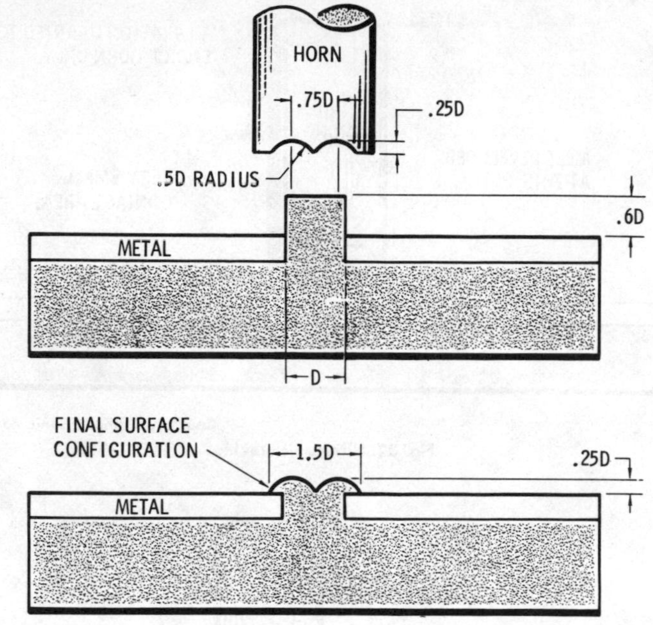

Courtesy Branson Sonic Power Co.

Fig. 34. Low profile head form.

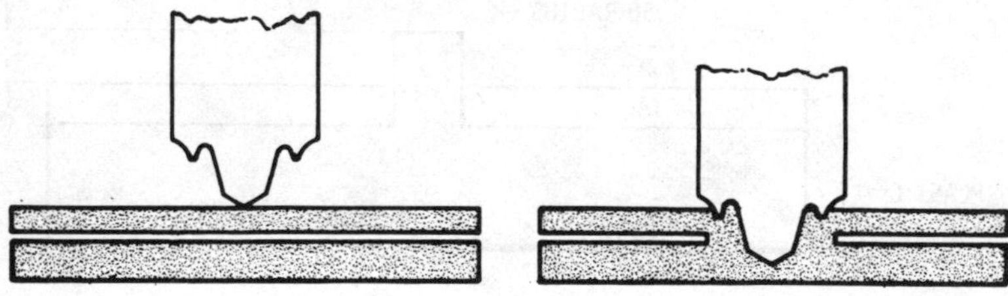

Courtesy Branson Sonic Power Co.

Fig. 35. Ultrasonic spot welding.

the two types of locking heads that can be formed in ultrasonic staking.

Ultrasonic staking is distinguished from ultrasonic welding by employing (1) out-of-phase vibrations generated between the horn and the thermoplastic surface; (2) as small as possible a contact area; and (3) light initial contact. Ultrasonic staking uses the same principle of creating localized heat through the utilization of high-frequency vibrations.

ULTRASONIC SPOT WELDING

In ultrasonic spot welding (Fig. 35), the tip is driven by vibratory energy through the upper sheet of thermoplastic and into the lower sheet to a depth of approximately ½ the thickness of the latter. A special ultrasonic spot welding gun (Fig. 36) is used for

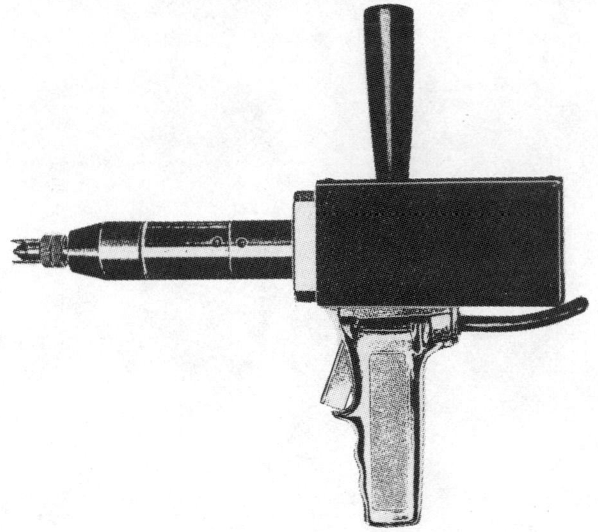

Courtesy Branson Sonic Power Co.

Fig. 36. Ultrasonic spot welding gun.

the application of ultrasonic vibrations to the thermoplastic parts. At the same time as the cavity is being formed, a frictional heat is developed at the interface of the two thermoplastic surfaces. Molten plastic from the upper sheet flows into the cavity formed in the lower sheet bonding the two surfaces together.

CHAPTER 25

Safety and Health Measures

This chapter concentrates primarily upon recommendations concerning safety and health measures for operators working with any of the various joining or cutting processes. Sections on safety equipment and clothing are also included.

The *American Welding Society,* the *American Standards Association,* as well as various governmental and private organizations, have compiled safety standards for the operation of gas and arc welding and cutting equipment. Among the publications produced by these organizations are the following:

1. HANDBOOK ON HEALTH AND SAFETY IN WELDING AND ALLIED PROCESSES. (*International Institute of Welding.* Document IIS/IIW/58-60).
2. SAFETY IN WELDING AND CUTTING (*American Standards Association.* ASA Z49.1-58).
3. SAFE PRACTICES IN THE INSTALLATION AND OPERATION OF OXYACETYLENE WELDING AND CUTTING EQUIPMENT. (*International Acetylene Association*).
4. RECOMMENDED SAFE PRACTICES FOR INERT-GAS METAL-ARC WELDING. (*American Welding Society.* AWS A6.1-58).

The beginner is urged to read these publications and become familiar with their recommendations. The different manufacturers of welding equipment and supplies also publish safety information.

SAFETY AND HEALTH MEASURES

Various aspects of welding and cutting safety will be examined in this chapter. Many of these will appear to be negative in nature. In other words, they stress what should *not* be done when welding or cutting. This should be regarded as the reverse side of the coin of accident prevention. The other, or "positive" side, is the operator's experience and training in the particular welding or cutting procedure being used. Accidents do not need to happen. Understanding their cause is one method of prevention; understanding and practicing correct procedure is another.

EYE PROTECTION

Radiant energy in types and quantities (harmful to the unprotected eye) is involved in many operations such as steel production, heat treatment of metals, and in the many types of joining processes. In order to understand the effect of radiant energy on the eye, it is first necessary to describe its various forms.

Radiant energy is transmitted in the form of waves, and will vary in form by the wavelength or distance between wave crests. The shortest waves are the so-called *cosmic waves,* and are probably not harmful to the eye. The *gamma waves* represent another waveform and are encountered in connection with radium. They are slightly longer than X rays in wavelength.

Other rays that exist in the energy spectrum encountered in industrial operations include: (1) ultraviolet rays, (2) visible rays, and (3) infrared rays. The wavelength of ultraviolet rays varies between 200 and 400 millimiconrs (a millimicron is one-millionth of a millimeter). Between 400 and 750 millimicrons are found in the visible rays, or in other words, those radiations which are distinguished as light.

The color gradations for ultraviolet and visible rays range from violet to red, violet being at the lower end of the spectrum and red at the upper end. Immediately beyond the last visible red of the light rays lies the infrared zone. This region contains the heating rays which the eye cannot see, but which the nerves of the skin feel and appreciate. The short infrared zone extends from 750 to 1250 millimicrons. The long infrared zone is from 1250 to 100,000 millimicrons.

The ultraviolet rays, the visible rays, and the infrared rays are all encountered by the welder during his work. Each of these three wave bands affects the eye in a different way.

Ultraviolet rays affect the cornea, causing an irritation of the outer membrane commonly known as "sand in the eye". If the exposure is of sufficient intensity and duration, the result is extremely painful. It usually takes about six hours after exposure. Normally, however, the inflammation will respond to treatment and will subside after several hours.

Because of the fact that the ultraviolet rays do not produce the sensation of sight, the eye has no way of detecting them. Hence, it is desirable to take utmost precaution by using lens that has been certified and proven to be of the highest standards.

Inasmuch as these rays affect the outer portions of the eye and do not necessarily have to be seen, they can take effect even though the eye is not looking directly at their source. This explains the desirability of providing supervisors and other workmen close to exposed welding operations with goggles that afford adequate absorption and which protect the eyes from rays which might strike the eye from any direction. Since the intensity of these rays varies inversely with the square of the distance, these goggles need not be of exceptionally strong absorptive qualities.

Visible rays emanating from the electric welding arc are of sufficient strength that it is necessary to reduce their intensity in order to prevent glare and fatigue of the retina and optical nerves which are naturally highly responsive to these wavelengths. This problem of reducing the intensity of the visible rays is not so diffcult because this naturally results from the use of any lens which is designed to absorb the ultraviolet and infrared rays which are harmful to the eye.

The problem in lens design is to provide complete absorption of the harmful rays, and at the same time to give a sufficient amount of visible rays to make it possible for the welder to see his work without eye strain.

Infrared rays, found immediately beyond the extreme visible red of the spectrum, are the most dangerous to the eye. Continued exposure to such radiation in sufficient quantity results in permanent injury to the iris, the lens and the retina. One of the most serious injuries is the cataract, which is a gradually increasing opacity of the lens. This danger is of particular significance in connection with the short infrared rays coming from white hot bodies because these rays, between 750 and 1250 millimicrons, can readily penetrate all of the media of the eye and thereby affect those inner parts which are concerned with vision.

Very little infrared radiation of wavelength greater than 1250 millimicrons penetrates beyond the cornea. The majority of infrared radiation from high temperature operations is in this long infrared or red-hot wavelength zone. Hence the greater portion of

Table 1. Suggested Lens Shades for Various Welding and Cutting Operations

Operation	Shade Number
Torch Brazing	3 or 4
Oxygen Cutting	
up to 1 inch	3 or 4
1 to 6 inches	4 or 5
6 inches and up	5 or 6
Gas Welding	
up to 1/8 inch	4 or 5
1/8 to 1/2 inch	5 or 6
1/2 inch and up	6 to 8
Oxyacetylene Flame Spraying	5
Gas Tungsten-Arc Welding (nonferrous)	11
Gas Metal-Arc Welding (nonferrous)	11
Gas Tungsten-Arc Welding (ferrous)	12
Gas Metal-Arc Welding (ferrous)	12
Gas Shielded-Arc Welding (Mig and Tig)	
5 to 75 amp range	5 to 9
75 to 200 amp range	10
200 to 400 amp range	12
above 400 amps	14
Plasma Flame Spraying	12
Shielded Metal-Arc Welding	
up to 5/32-inch electrodes	10
3/16 to 1/4-inch electrodes	12
5/16-inch electrodes and up	14
Atomic Hydrogen Welding	10 to 14
Carbon Arc Welding and Cutting	14

Courtesy Jackson Products

infrared radiation is absorbed in the outer portions of the eye. Therefore, the chief concern in the design of a protective lens is to prevent passage of the infrared rays which are just outside the visible range.

The welder and others associated with welding operations should be provided with glasses designed to provide maximum protection and at the same time to afford adequate vision for proper welding technique. Until recently it has been customary to assume that if the visible rays were cut down to a comfortable intensity, the ultraviolet and infrared rays were also reduced proportionately. The usual practice was to use alternate layers of red and blue glasses to reduce the light intensity to a value consistent with the work being done. Such a procedure may or may not offer complete protection since some glasses absorb strongly the visible rays and yet transmit the harmful infrared or ultraviolet rays quite freely.

Scientific methods for the testing of protective lenses have been developed. These tests have resulted in the setting up of standards for lense of various grades and have led to the development of glass formulas which have materially increased the protective qualities of lenses. Complete and positive protection to the eyes is now available. Suggested shades for various types of welding and joining operations are given in Table 1.

Federal specifications for welding lens not only specify the percentage of rays transmitted by the various shade numbers, but also specify the thickness of the glass and its optical properties.

Proper lens selection is important. There is a vast difference in welding lenses, both as to their value and their effect on welding production. It is impossible to distinguish one lens from another by casual inspection. Scientific tests are required to determine their qualities. It is vitally important when selecting a welding lens to take into account the reputation of the manufacturer and their experience in the welding field.

Federal specifications for welding lenses require that all the lenses shall conform to certain tolerances. For instance, it is specified that there shall be a minimum and maximum percentage of transmitted rays in the visible range, and that there shall be certain limitations on the infrared and ultraviolet rays.

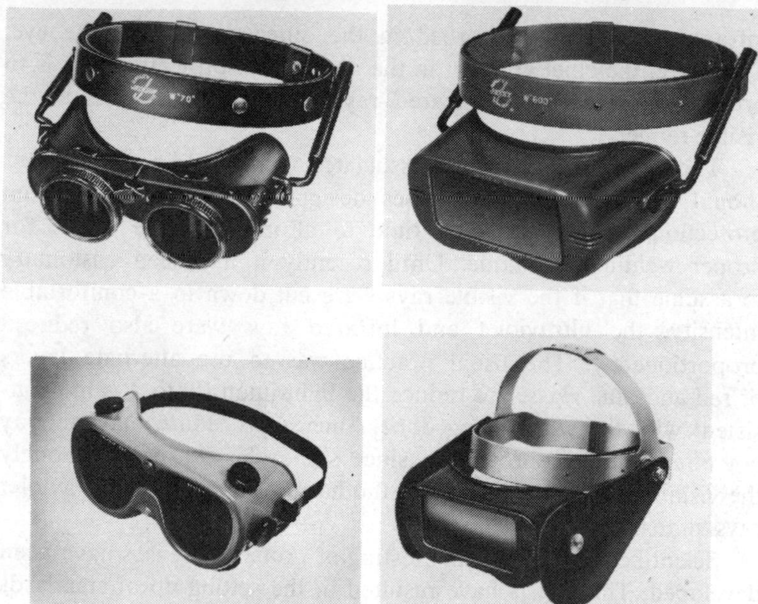

Courtesy Jackson Products

Fig. 1. Examples of various type goggles.

Non-spatter cover glass is a chemically treated glass which protects the lens from spatter yet allows maximum visibility. Spatter does not adhere to this special type of glass, prolonging its life five or ten times that of ordinary glass and maintaining clear visibility. A soft cloth should be used for cleaning.

Goggles are available in a great number of different designs and types. Either glass or plastic lenses may be purchased, and the lenses themselves can be either clear or tinted. Most goggles are either of the round lens or rectangular lens types. Examples of some of these are shown in Fig. 1.

Face shields (Fig. 2) are also useful for protecting the face of the operator against flying sparks and other dangerous matter. They find widespread use in arc welding and cutting. Face shields are available in clear plastic or in different shades of green and can be purchased in several thicknesses (the *ANSI* requires a minimum thickness of .041 inch or more). Face shields provide limited impact and splash protection.

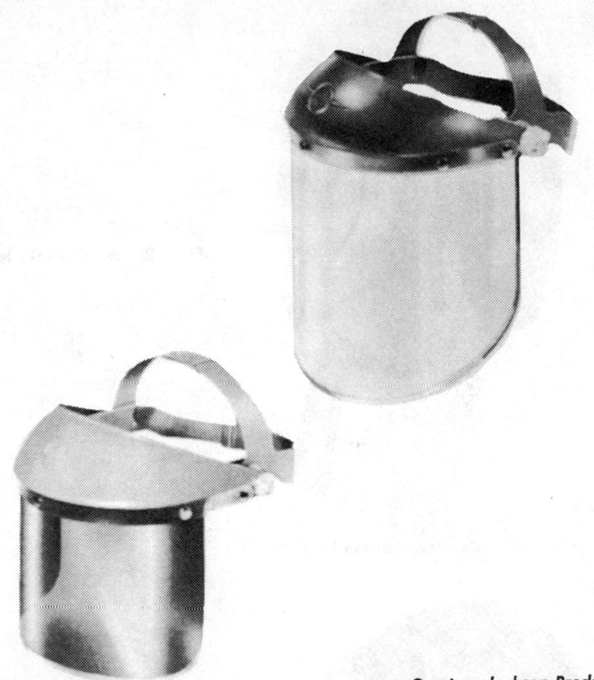

Courtesy Jackson Products

Fig. 2. Face shields used in welding and cutting operations.

Hand shields (Fig. 3) are simple protective devices that are held by the operator in front of his face to prevent being struck by flying sparks or particles of molten metal. Their major disadvantage is that they restrict the use of the welder's hands.

Helmets, safety caps, and other headgear (Fig. 4) have been designed for operators of welding and cutting equipment to protect the head against serious blows. These are available in many different designs, including a great number of sizes. The helmets are made either from molded fiberglas or from metal plate.

GAS WELDING AND CUTTING

The safety recommendations for gas welding and cutting include not only those for the oxyacetylene process, but the other oxy-fuel gas processes as well.

Fig. 3. A typical hand-held shield.

Courtesy Jackson Products

Courtesy Jackson Products

Fig. 4. Typical helmet and safety cap.

Briefly, the *general* safety recommendations for gas welding and cutting are as follows:

1. Never weld in the vicinity of inflammable or combustionable materials.

2. Never weld on containers which have held combustionable or inflammable materials without first exercising the proper precautions recommended by the *American Welding Society.*
   (RECOMMENDED PROCEDURE TO BE FOLLOWED IN WELDING OR CUTTING CONTAINERS WHICH HAVE HELD COMBUSTIBLES. AWS A6.0-52).
3. Never weld in confined spaces without adequate ventilation or individual respiratory equipment.
4. Never pick up hot objects.
5. Never do any chipping or grinding without suitable goggles.
6. Never move individual cylinders unless the valve protection cap, where provided, is in place, hand tight.
7. Never drop or abuse cylinders in any way.
8. Make certain that cylinders are well fastened in their station so that they will not fall.
9. Never use a hammer or wrench to open any valve on a cylinder.
10. Never force connections that do not fit.
11. Never tamper with cylinder safety devices.
12. Always protect hose from being trampled on or run over. Avoid tangle and kinks. Never leave the hose so that it can be tripped over.
13. Protect the hose and cylinders from flying sparks, hot slag, hot objects and open flame.
14. Never allow the hose to come into contact with oil or grease; these deteriorate the rubber and constitute a hazard with oxygen.
15. Be sure that the connections between the regulators, adapters, and cylinder valves are gas tight. Escaping acetylene can generally be detected by the odor. Test with soapy water, never with an open flame.
16. Never use matches for lighting torches (hand burns may result). Use spark lighters, stationary pilot flames, or some other suitable source of ignition; do not light torches from hot work in a pocket or small confined space. Never attempt to relight a torch that has blown out without first closing both valves and relighting in the proper manner.

17. Never hang a torch with its hose on regulators or cylinder valves.
18. Never cut material in such a position that will permit sparks, hot metal or the severed section to fall on the cylinder, hose, legs or feet.
19. When welding or cutting is to be stopped temporarily, release the pressure adjusting screws of the regulators by turning them to the left.
20. When the welding or cutting is to be stopped for a long time (during lunch hour or overnight) or taken down, close the cylinder valves and then release all gas pressures from the regulators and hose by opening the torch valves momentarily. Close the torch valves and release the pressure adjusting screws. If the equipment is to be taken down, make certain that all gas pressures are released from the regulators and hose and that the pressure adjusting screws are turned to the left until free.
21. Never use oil on any regulator gauge. As a combustible substance, oil has a very low flash point.

The safety recommendations for working with the oxygen equipment (cylinders, regulator, hose, etc.) and supply are as follows:

1. Always refer to oxygen by its full name "oxygen" and not by the word "air". This will avoid the possibility of confusing oxygen with compressed air.
2. Never use oxygen near flammable materials, especially grease, oil or any substance likely to cause or accelerate fire. Oxygen itself is not flammable but does support combustion.
3. Do not store oxygen and acetylene cylinders together. They should be separately grouped.
4. Never permit oil or grease to come in contact with oxygen cylinders, valves, regulators, hose or fittings. Do not handle oxygen cylinders with oily hands or oily gloves.
5. Never use oxygen pressure reducing regulators, hose or other pieces of apparatus with any other gasses.
6. Open oxygen cylinder valve slowly.

7. Never attempt to mix any other gases in an oxygen cylinder.
8. Oxygen must never be used for ventilation or as a substitute for "compressed air."
9. Never use oxygen from cylinders without first connecting a suitable pressure reducing regulator to the cylinder valve.
10. Never tamper with nor attempt to repair oxygen cylinder valves unless qualified to do so.

The safety recommendations for working with the acetylene equipment and supply are as follows:

1. Call acetylene by its full name "acetylene" and not by the word "gas." Acetylene is far different from city or furnace gas.
2. Acetylene cylinders should be used and stored valve end up.
3. Never use acetylene from cylinders without a suitable pressure.
4. Turn the cylinder so that the valve outlet will point away from the oxygen cylinder.
5. When opening an acetylene cylinder valve, turn the key or spindle not more than one and one-half turns.
6. Acetylene cylinder key for opening valve must be kept on valve stem while cylinder is in use so that the acetylene cylinder may be quickly turned off in an emergency.
7. Never use acetylene pressure reducing regulators, hose, or other pieces of apparatus with any other gases.
8. Never attempt to transfer acetylene from one cylinder to another nor to refill an acetylene cylinder, nor to mix any other gas or gases in an acetylene cylinder.
9. Should a leak occur in an acetylene cylinder, take the cylinder out in the open air, keeping well away from fires or open lights. Notify the manufacturer immediately if any leaks occur.
10. Never use acetylene at pressures in excess of 15 psi. The use of higher pressures is prohibited by all insurance authorities and by law in many localities.

ARC WELDING AND CUTTING

The hazards involved in arc welding and cutting include the following:

1. Electric shock,
2. Toxic fumes,
3. Radiation from the arc,
4. Flying sparks,
5. Splattering metal,
6. Hot metal.

Electric shock can be avoided by proper handling of the arc welding equipment. The arc welding machine must be properly grounded at all times. The work area should be dry. All insulation (wiring, electrodes, etc.) should be checked, and replaced if found inadequate.

Always work in a well ventilated area. This is the best protection against toxic fumes and dust. If the ventilation is poor, adequate respiratory equipment is necessary.

The electric arc gives off harmful radiation. Goggles with suitable lenses and protective clothing are recommended as protection against these rays, as well as against flying sparks, splattering metal, and hot metal.

A brief outline of the safety recommendations for the arc welding and cutting processes is as follows:

1. Keep the working area and floor clean and clear of electrode stubs, scraps of metal, and carelessly disposed tools.
2. See that cable connections are tight and that cables do not become hot.
3. Never look at an electric arc with the naked eye.
4. Never weld while wearing wet gloves or wet shoes.
5. Never use electrode holders with defective jaws.
6. Never leave the electrode holder on the table or in contact with a grounded metallic surface (place it on the support provided for that purpose).
7. Never weld on closed containers or on containers that have held combustible materials.
8. Never allow an arc welding machine to rest on a dirt floor.

9. Operate arc welding machines and equipment only in clean, dry locations.
10. Insofar as possible, protect arc welding machines in the field from weather conditions.
11. Always install arc welding machines in compliance with the requirements of the *National Electrical Code* and the codes of the local government (county, city, etc.).
12. Make certain that the arc welding machine is properly grounded.
13. Use the proper terminals on the arc welding machine for the power line voltage connection.
14. Never work on the wiring of an arc welding machine unless qualified to do so.
15. Cutting should be done away from the operator to avoid the possibility of molten metal spraying into the face or onto clothing.

UNDERWATER WELDING AND CUTTING

Most underwater work involves cutting operations with either a gas torch or an electric arc. However, many of the recommended safety precautions listed in this section will apply equally to both cutting and welding operations.

In addition to the standard practice followed in the use of all gas welding and cutting equipment, the following are recommended in the interest of safety and best performance when using a gas torch:

1. Only a qualified diver assisted by an experienced tender and a trained torch man should use underwater equipment.
2. The diver should be a qualified operator on the surface before taking training under water.
3. The training should be given under the supervision of a qualified instructor, with standard equipment, and under the most favorable diving conditions.
4. Detailed operating instructions should be obtained from the manufacturer and followed without deviation.

5. Before starting any cutting, be sure that there are no highly combustible or explosive materials, whether gases, liquids, or solids, adjacent to the point of cutting or within a radius of at least thirty feet.
6. The torch should be lit and the adjustments tested on the surface before it is used under water.
7. At the end of each day's work, the torch should be cleaned thoroughly and dried inside and outside, including the oxygen high-pressure valve, seat and spring.
8. Owing to the difficult footing and poor visibility usually prevailing under water, the diver should handle the torch with care, stay completely clear of the torch hose and avoid excessive slack in the lines.
9. The orifices in the cutting tips should be kept clean but care should be used in cleaning them so as not to distort or enlarge them.
10. The oxygen regulators used should be adequate for the delivery of the needed volume without freezing.
11. The diving gear used should be in good condition and should be equipped with a loudspeaker telephone designed for uninterrupted two-way communication. A trained torch attendant should be on duty continuously on the surface at the oxygen and the gas regulators.
12. No work of any kind should ever be permitted on the surface over the space in which the diver is working within a radius less than the depth of operations.

The safety precautions governing the use of an electric arc in underwater welding and cutting operations include all of the appropriate recommendations made in the previous section with the following amendments and additions:

1. The cable and torch insulation and all joints in the circuit should be checked for current leakage at frequent intervals.
2. The diver should not permit any part of his body or gear to become a part of the electric circuit.
3. While alternating current may be used, if the diver's body or gear inadvertently enters the circuit, the resulting shock is more pronounced than with direct current.

4. The diver should inspect his helmet and other metallic parts of his gear regularly for deterioration resulting from electrolysis and place his ground connection at points with reference to his position while cutting that will reduce this to a minimum.
5. Always wear diver's rubber gloves or mittens, telephone the surface to shut off current before changing electrodes, and keep it shut off except when actually cutting. Be sure you close the hinged frame holding the welding lens on the outside of the helmet before striking the arc, and remove the electrode before taking the torch under water or returning it to the surface.
6. When using carbon or ceramic electrodes, special care must be taken to prevent breakage in handling and while cutting as both electrodes are brittle.
7. Qualified divers who are acceptable welders on the surface may learn to operate the arc welding or cutting torches under water with fair results after practicing for a day or two.

OTHER WELDING AND CUTTING PROCESSES

In resistance welding, submerged-arc welding, and similar welding and cutting processes, it is necessary to operate complicated welding machines. This involves working around moving parts, and there is always the danger of injury to the fingers and hand. It is essential then that caution be exercised when operating this type of equipment.

Electrical shock is always a danger when working with welding machines. The operator is cautioned against touching the wiring of the welding machine unless qualified to do so. Moreover, it is absolutely necessary that the welding machine be properly grounded at all times.

SAFE PRACTICES IN SOLDERING

Do not allow the soldering fluxes to come in contact with the skin or clothing. Acid fluxes will ruin clothing or injure the skin.

The eyes must be protected against solder splatter with suitable goggles. This type of eye injury is a frequent cause of production slowdown in the electronics industry when these precautions are not taken.

Never solder on containers (tanks, pressure vessels, etc.) that may have contained flammable or volatile substances. There should be no flammable or volatile fumes in the area when soldering. Good ventilation will minimize the danger.

Good ventilation in the soldering area is also important to reduce the danger of toxic fumes given off during the soldering operation.

VENTILATION

It is absolutely necessary that proper ventilation be provided for each welding and cutting operation. The fumes given off during welding or cutting can be injurious to the operator's health. Some fumes, such as those produced when working with zinc, lead or cadmium, can be toxic. These conditions also hold true for braze welding, brazing, and soldering.

In the case of permanent welding stations, the size of the working area is an important consideration. Overcrowding will reduce the effectiveness of any ventilation system. The ventilating system itself (exhaust fans, etc.) should be designed to maintain the level

Courtesy Dockson Corp.

Fig. 5. Examples of respirators used in welding and cutting.

of toxic fumes and other contaminants at or below the maximum permitted level. Individual respiratory equipment is sometimes necessary when room ventilation is inadequate (Fig. 5).

PROTECTIVE CLOTHING

Trousers, shirts, and other clothing should be made of a flame resistant material. It is recommended that the trousers be cuffless. If the operator is arc welding, then the clothing must be thick enough to minimize or prevent penetration by the dangerous radiation given off by the arc. This radiation cannot be seen, but is nevertheless present. It results in skin burns similar to those suffered in too much exposure to the sun.

Sleeve cuffs should be tight against the wrist to prevent entrapment of flying sparks or molten particles. For this purpose, elastic bands or gauntlet cuffs are recommended.

The types of protective clothing available from manufacturers is quite extensive. For example, asbestos sleeves with snap fasteners at the wrist and adjustable leather straps at the top of the arm can be purchased for minimum protection. More extensive types of protective clothing include leather jackets; cape sleeves with detachable bib; waist, bib, or split-leg aprons; and leather spats for protection of the ankles.

Black, flame-resistant cotton twill is often recommended for use with inert gas arc welding. Protective clothing made from this type cloth is cheaper and lighter than leather.

CHAPTER 26

Welding Symbols

The *American Welding Society* has developed a standardized set of symbols for welding and brazing. These welding symbols provide the operator with a number of different important points of information including:

1. The type of joint to use,
2. The type of weld,
3. The size of the weld,
4. The amount of deposited metal to use,
5. The location of the weld.

The fact that welding symbols have been standardized means that a common "language" has been provided for everyone concerned. It is a very *specific* means of communications. As such, it greatly reduces the possibility of error or confusion on the part of the operator.

The welding symbols in this chapter, and the instructions for their use, are included through the courtesy of the *American Welding Society*. This does not constitute a complete listing of welding symbols; only the more common ones were used. A complete listing can be found in the WELDING HANDBOOK, a publication of the *American Welding Society*.

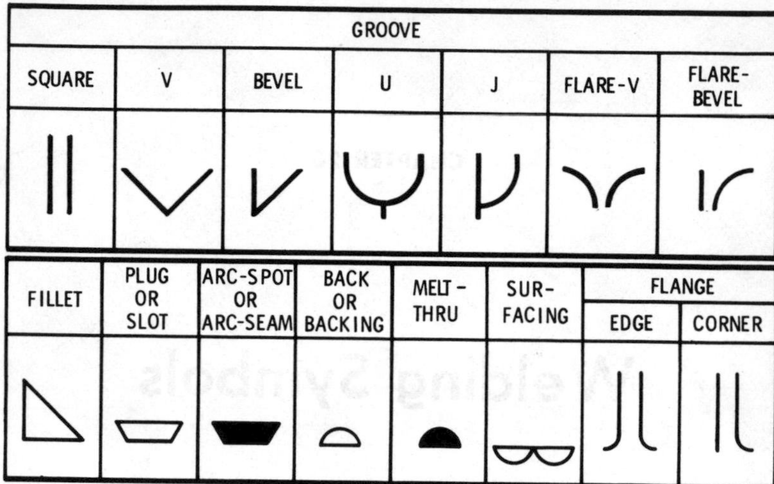

Basic arc and gas weld symbols.

TYPE OF WELD			
RESISTANCE-SPOT	PROJECTION	RESISTANCE-SEAM	FLASH OR UPSET

Basic resistance weld symbols.

WELD ALL AROUND	FIELD WELD	CONTOUR	
		FLUSH	CONVEX

Supplementary symbols.

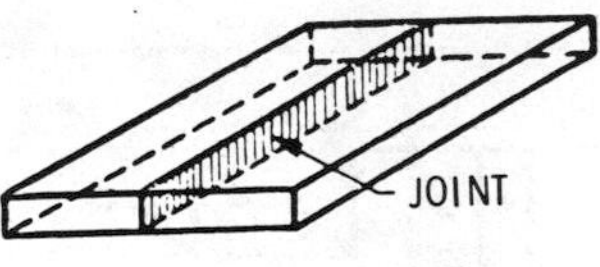

(A) BUTT JOINT

APPLICABLE WELDS

- SQUARE-GROOVE
- V-GROOVE
- BEVEL-GROOVE
- U-GROOVE
- J-GROOVE
- FLASH
- UPSET

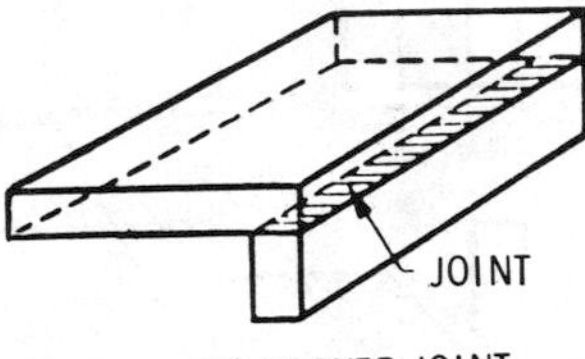

(B) CORNER JOINT

APPLICABLE WELDS

- SQUARE-GROOVE
- V-GROOVE
- BEVEL-GROOVE
- U-GROOVE
- J-GROOVE
- FILLET
- FLASH

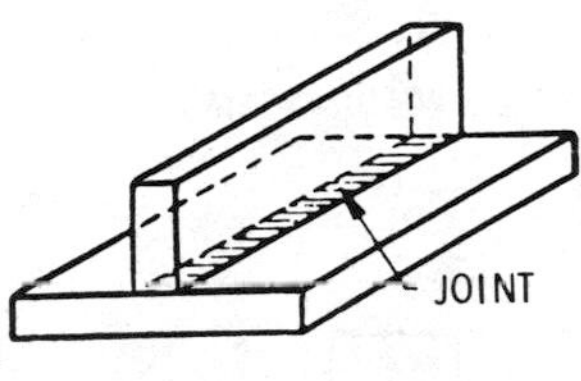

(C) TEE JOINT

APPLICABLE WELDS

- BEVEL-GROOVE
- J-GROOVE
- FILLET
- SLOT
- PLUG

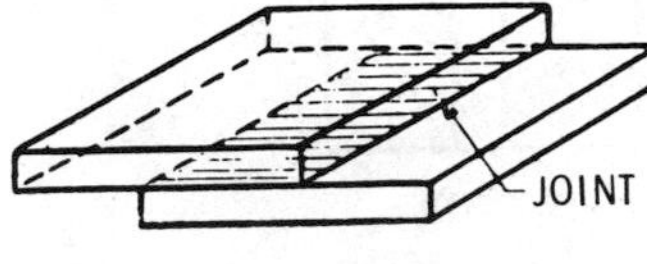

(D) LAP JOINT

APPLICABLE WELDS

- FILLET
- BEVEL-GROOVE
- J-GROOVE
- SLOT
- PLUG
- SPOT
- PROJECTION
- SEAM

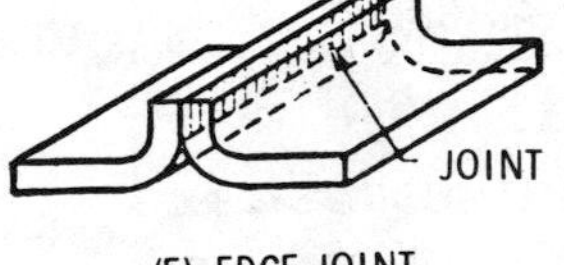

(E) EDGE JOINT

APPLICABLE WELDS

- SQUARE-GROOVE
- BEVEL-GROOVE
- V-GROOVE
- U-GROOVE
- J-GROOVE
- SPOT
- SEAM
- PROJECTION

Basic types of joints.

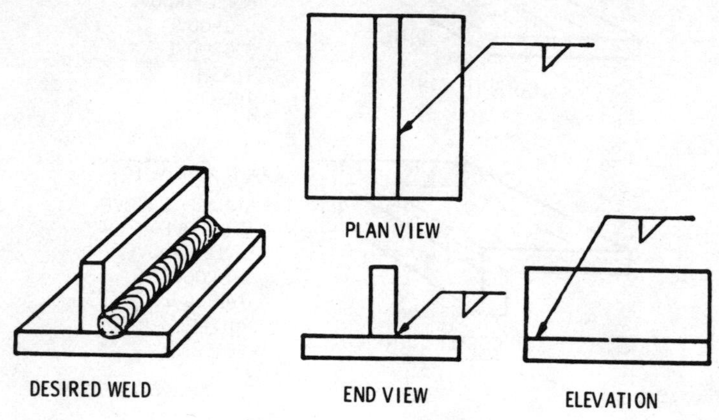

(A) ARROW-SIDE FILLET WELDING SYMBOL

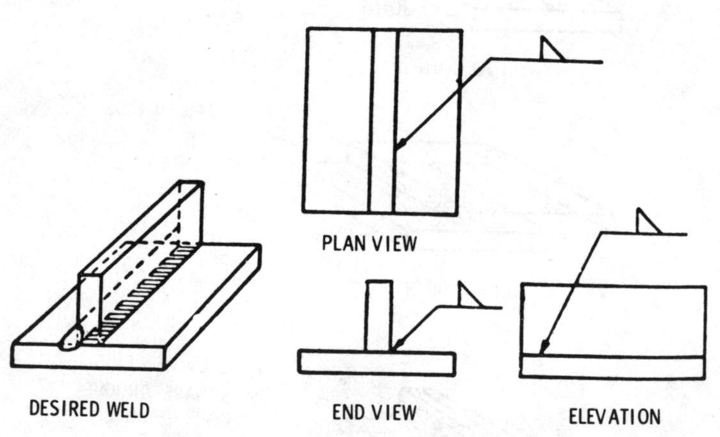

(B) OTHER-SIDE FILLET WELDING SYMBOL

Application of fillet welding symbols.

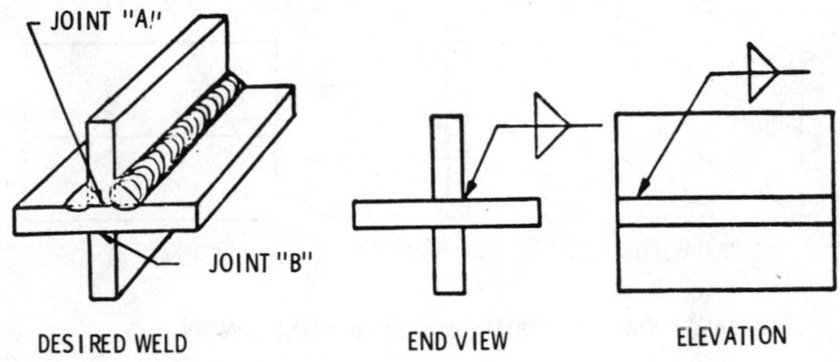

(C) BOTH SIDES FILLET WELDING SYMBOL FOR ONE JOINT

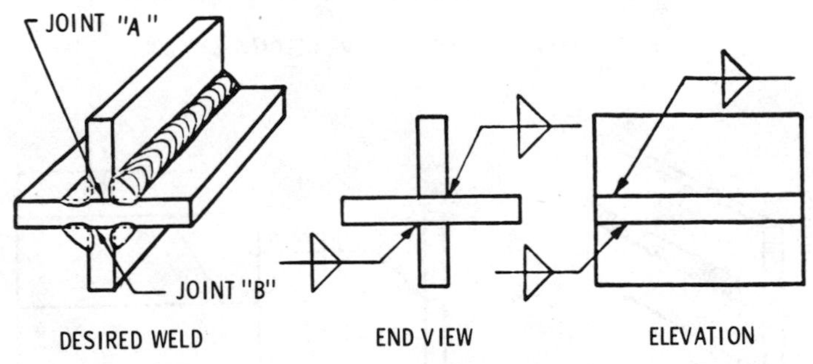

(D) BOTH SIDES FILLET WELDING SYMBOL FOR TWO JOINTS

Application of fillet welding symbols.

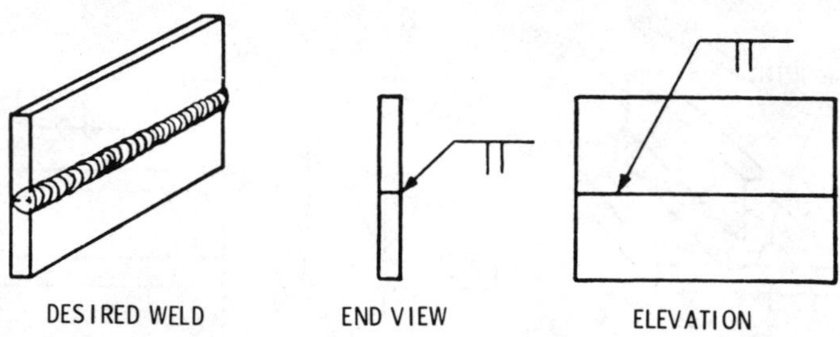

(A) ARROW-SIDE SQUARE-GROOVE WELDING SYMBOL

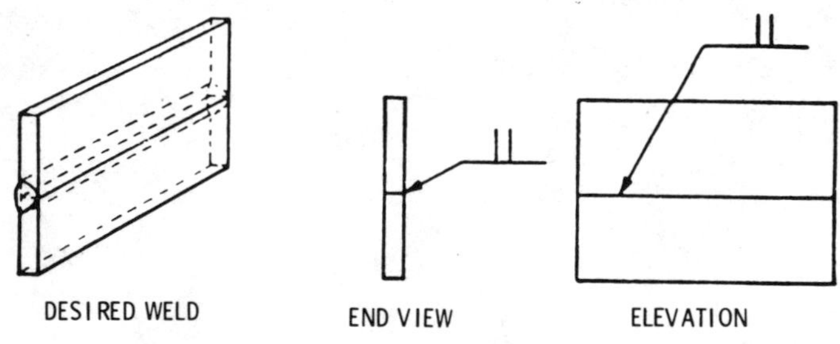

(B) OTHER-SIDE SQUARE-GROOVE WELDING SYMBOL

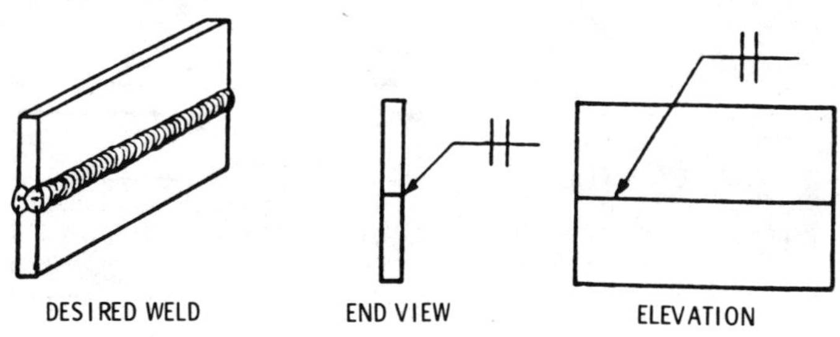

(C) BOTH-SIDES SQUARE-GROOVE WELDING SYMBOL

Application of square-groove welding symbols.

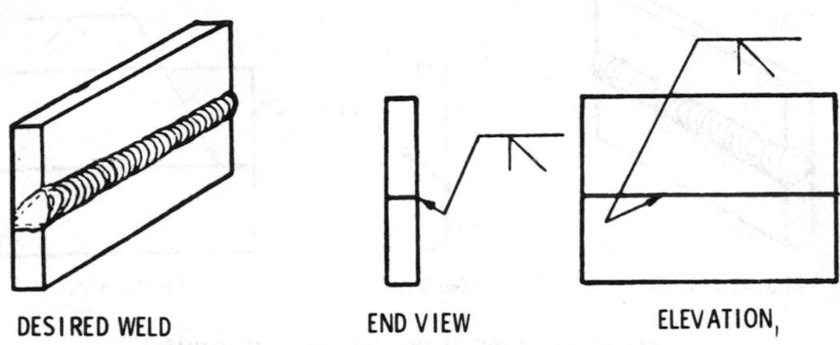

(A) ARROW-SIDE BEVEL-GROOVE WELDING SYMBOL

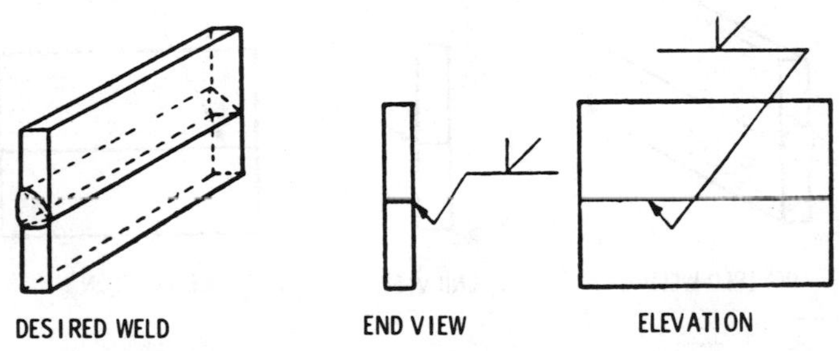

(B) OTHER-SIDE BEVEL-GROOVE WELDING SYMBOL

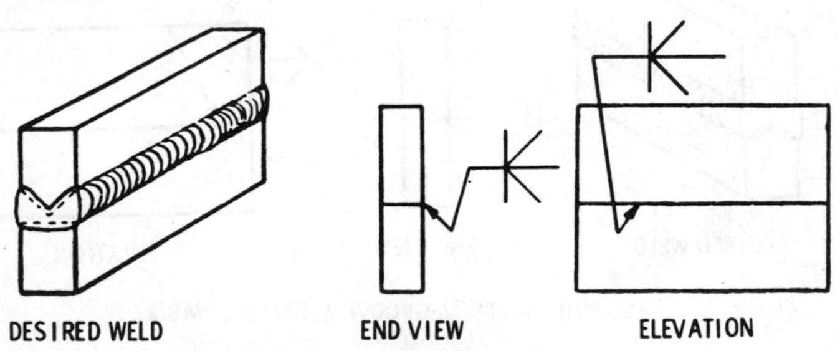

(C) BOTH-SIDES BEVEL-GROOVE WELDING SYMBOL

Application of bevel-groove welding symbols.

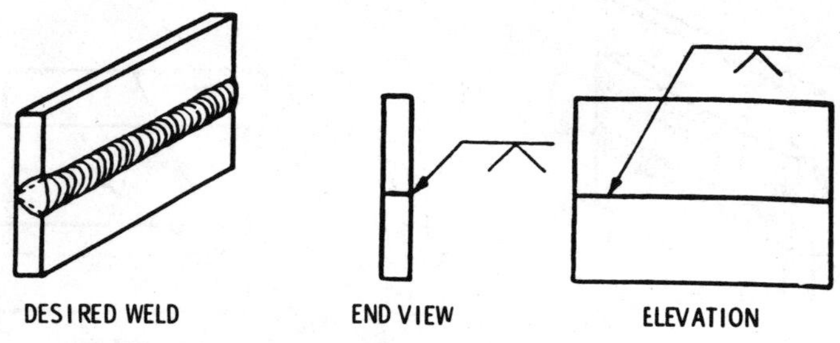

(A) ARROW-SIDE V-GROOVE WELDING SYMBOL

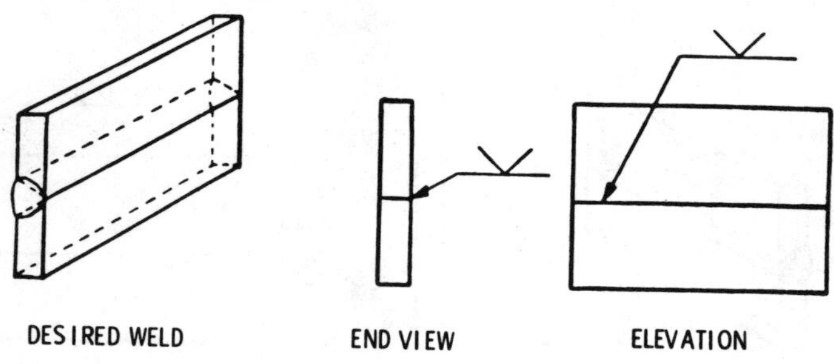

(B) OTHER-SIDE V-GROOVE WELDING SYMBOL

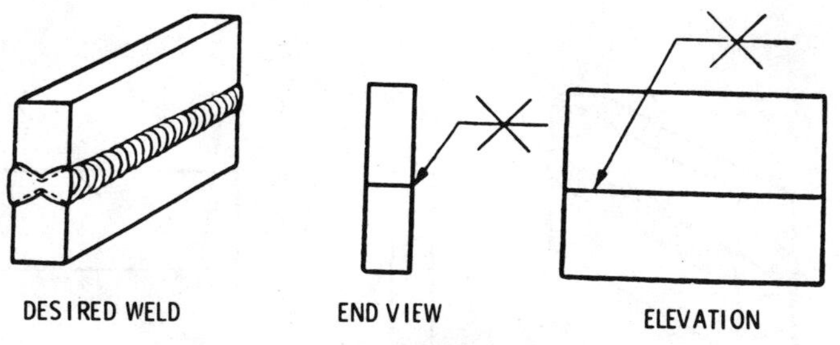

(C) BOTH-SIDES V-GROOVE WELDING SYMBOL

Application of V-groove welding symbols.

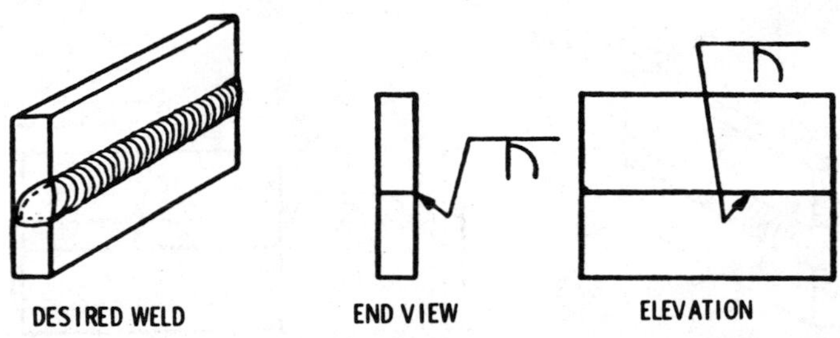

(A) ARROW-SIDE J-GROOVE WELDING SYMBOL

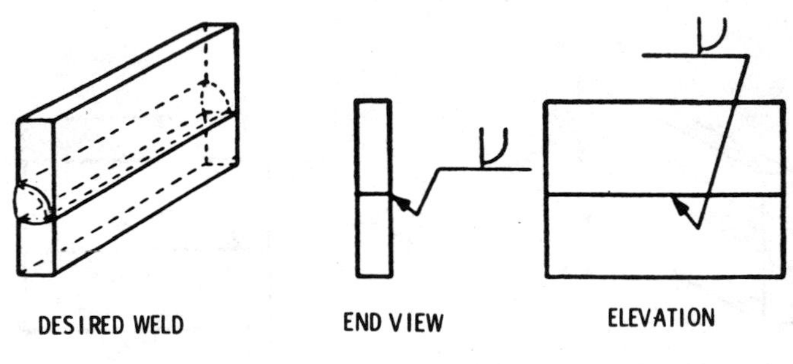

(B) OTHER-SIDE J-GROOVE WELDING SYMBOL

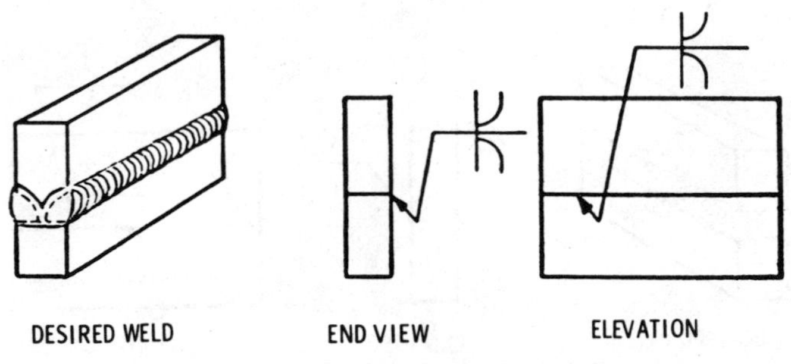

(C) BOTH-SIDES J-GROOVE WELDING SYMBOL

Application of J-groove welding symbols.

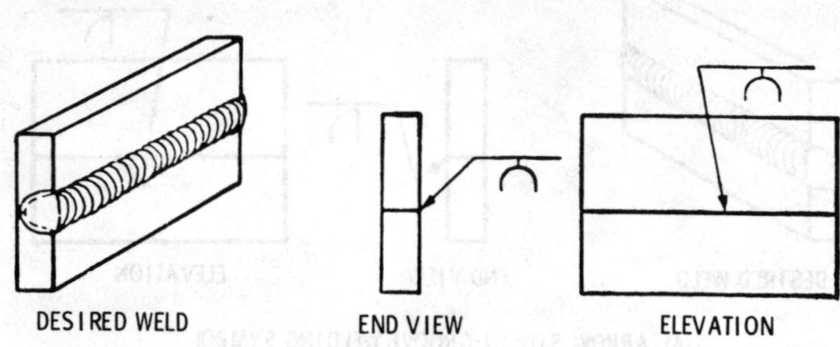

(A) ARROW-SIDE U-GROOVE WELDING SYMBOL

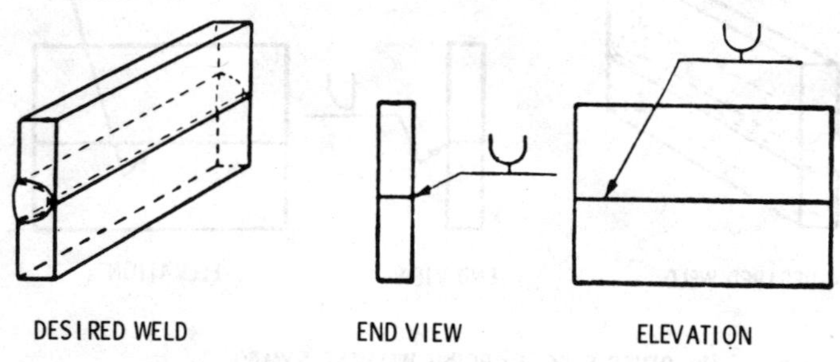

(B) OTHER-SIDE U-GROOVE WELDING SYMBOL

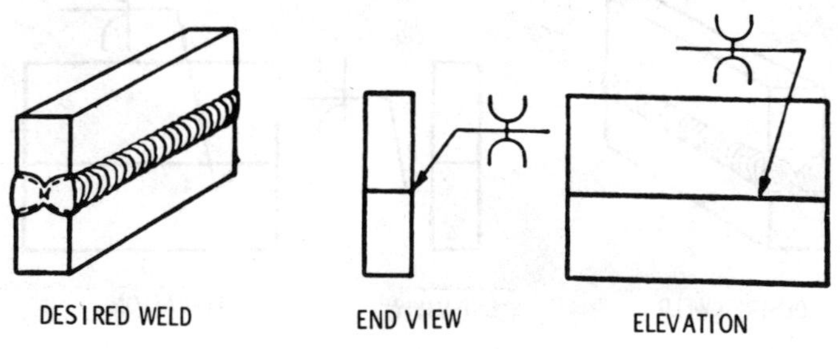

(C) BOTH-SIDES U-GROOVE WELDING SYMBOL

Application of U-groove welding symbols.

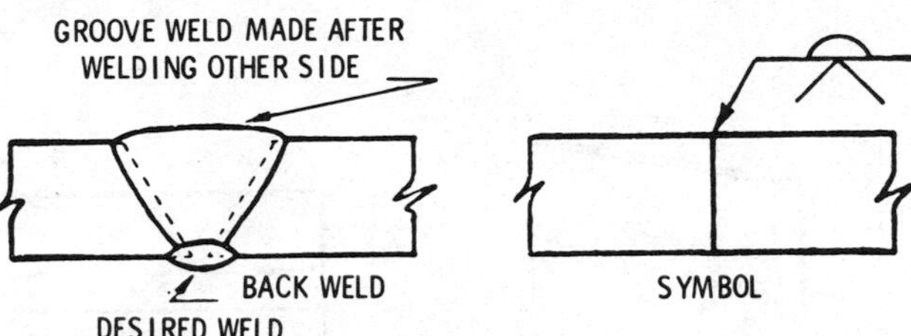

(A) USE OF BEAD WELD SYMBOL TO INDICATE SINGLE-PASS BACK WELD

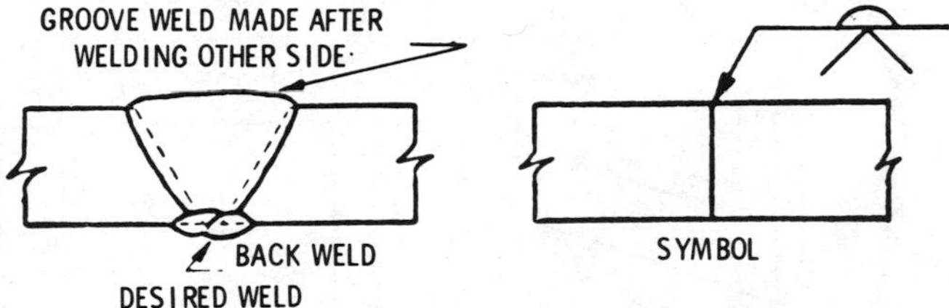

(B) USE OF BEAD WELD SYMBOL TO INDICATE MULTIPLE-PASS BACK WELD

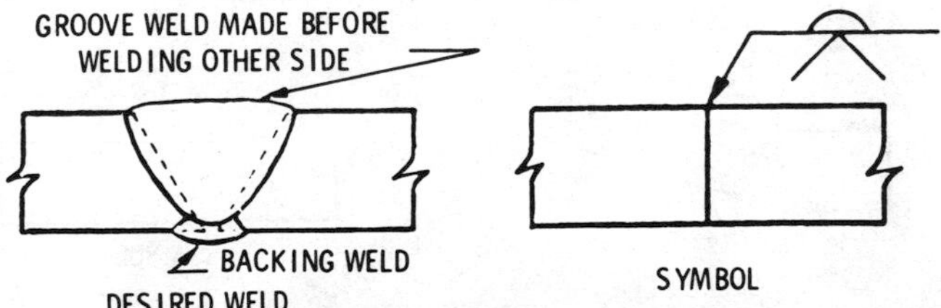

(C) USE OF BEAD WELD SYMBOL TO INDICATE SINGLE-PASS BACKING WELD

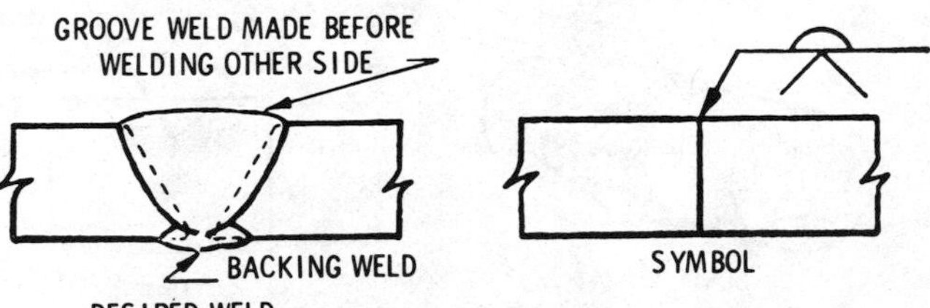

(D) USE OF BEAD WELD SYMBOL TO INDICATE MULTIPLE-PASS BACKING WELD

Bead weld symbols to indicate bead-type back and backing welds.

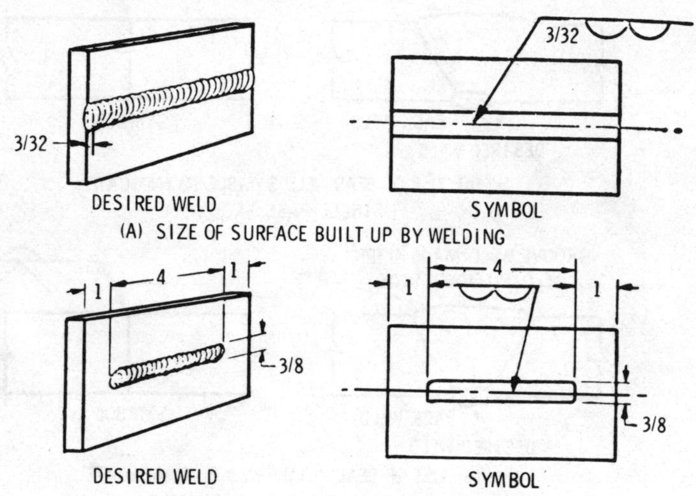

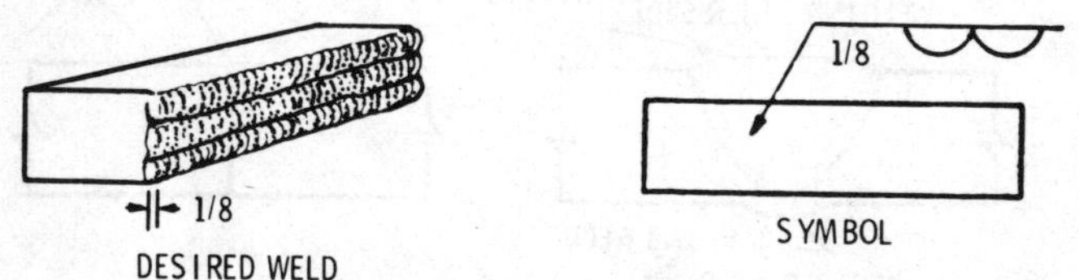

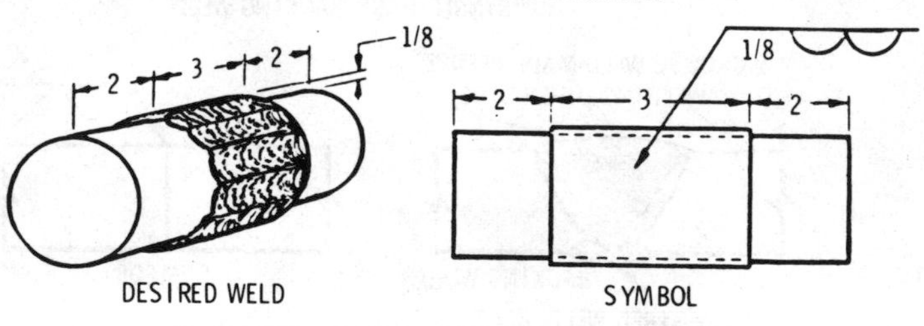

Dual bead weld symbols to indicate surface built up by welding.

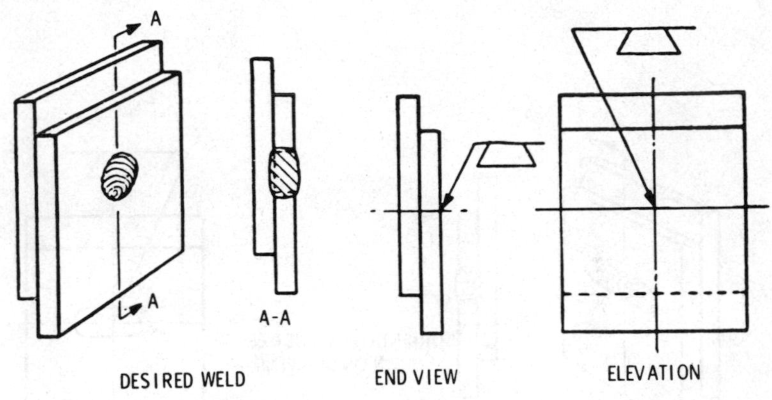

(A) ARROW-SIDE PLUG WELDING SYMBOL

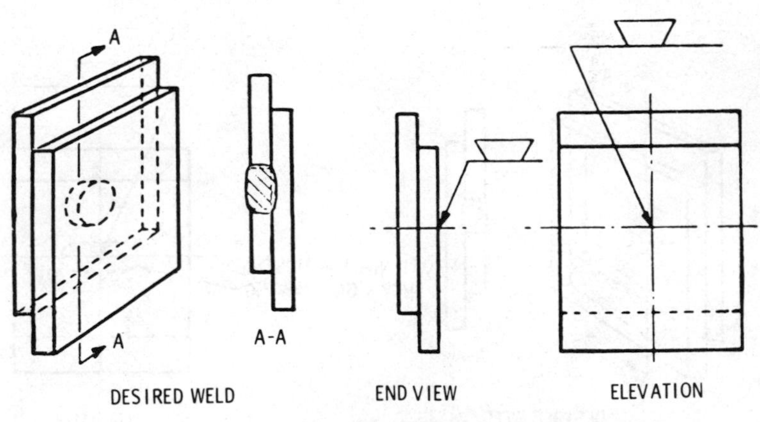

(B) OTHER-SIDE PLUG WELDING SYMBOL

Plug welding symbols.

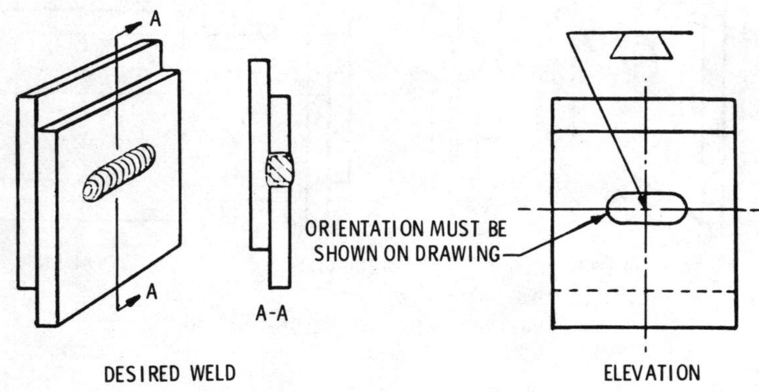

(A) ARROW-SIDE SLOT WELDING SYMBOL

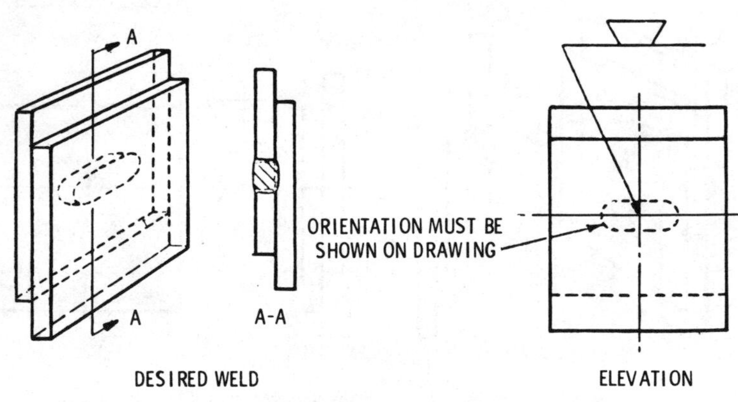

(B) OTHER-SIDE SLOT WELDING SYMBOL

Slot welding symbols.

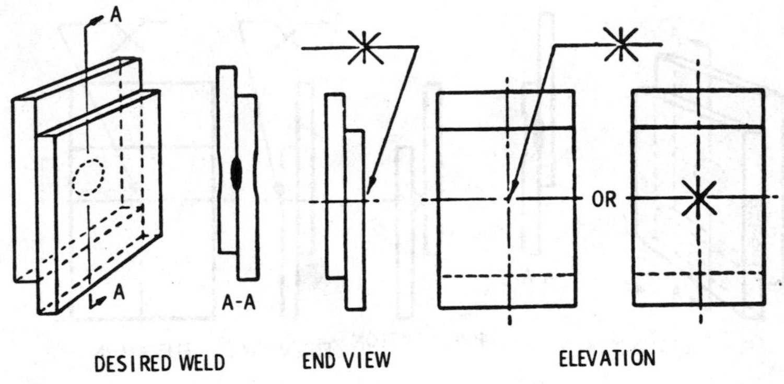

(A) SPOT WELDING SYMBOL

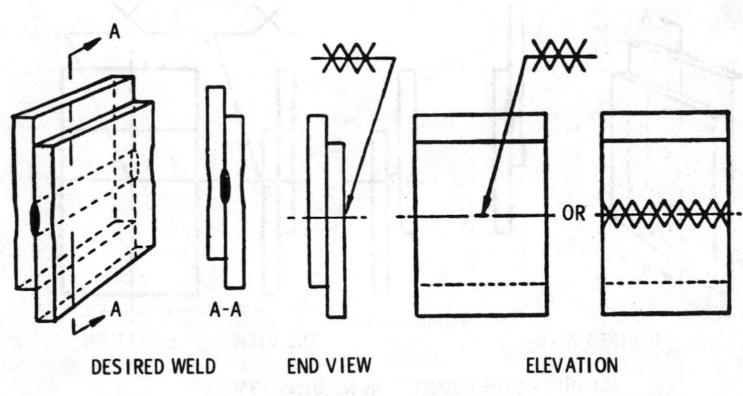

(B) SEAM WELDING SYMBOL

Spot and seam welding symbols.

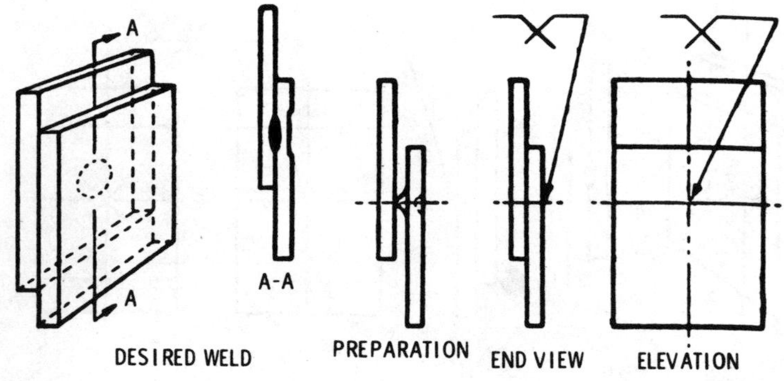

(A) ARROW SIDE PROJECTION WELDING SYMBOL

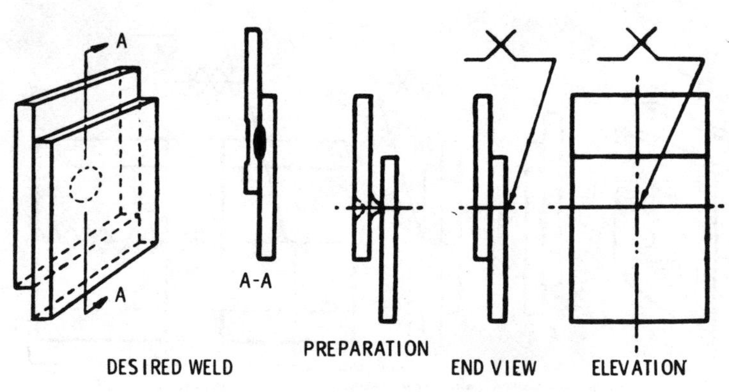

(B) OTHER SIDE PROJECTION WELDING SYMBOL

Projection welding symbols.

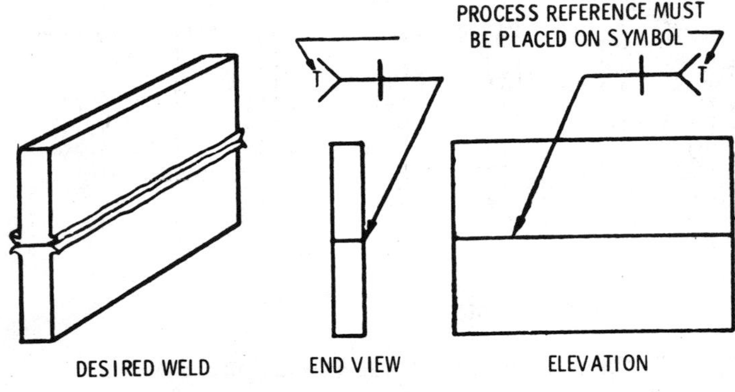

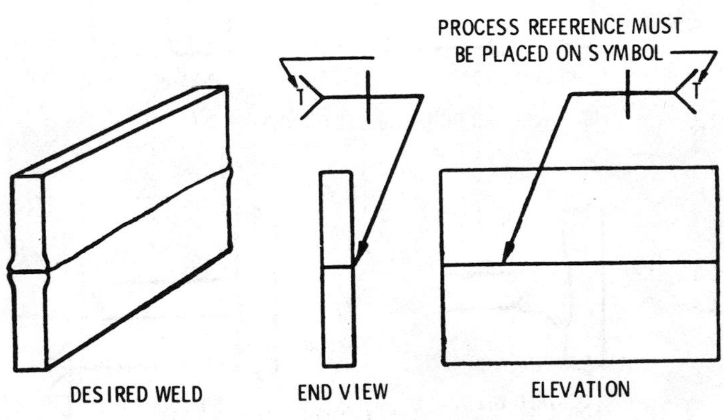

Flash and upset welding symbols.

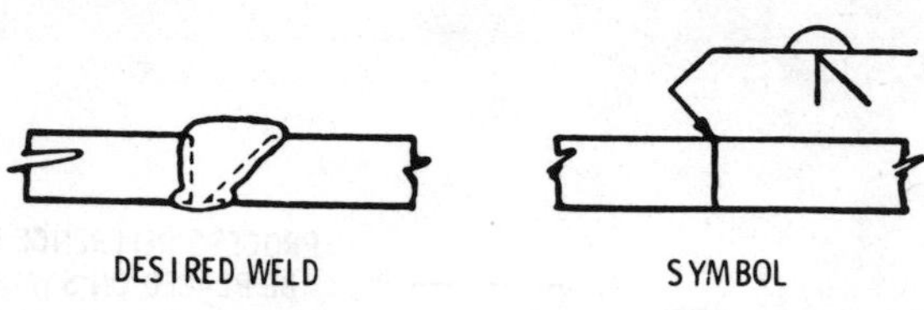

(A) SINGLE BEVEL GROOVE AND BEAD WELD SYMBOLS

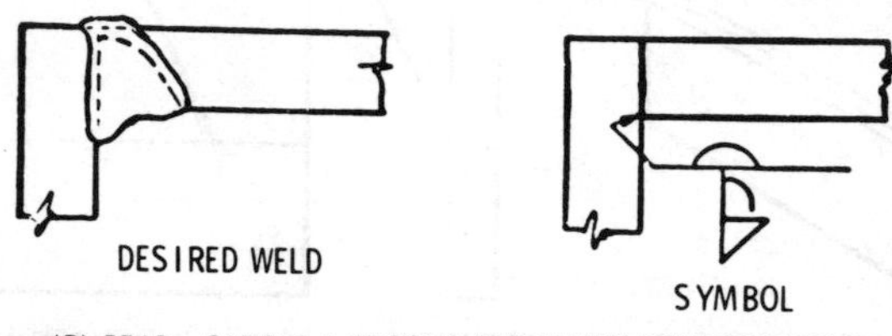

(B) BEAD SINGLE J GROOVE AND FILLET WELD SYMBOLS

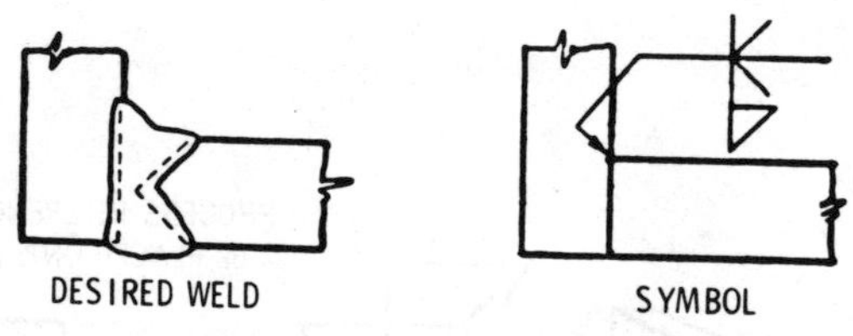

(C) FILLET AND DOUBLE BEVEL GROOVE WELD SYMBOLS

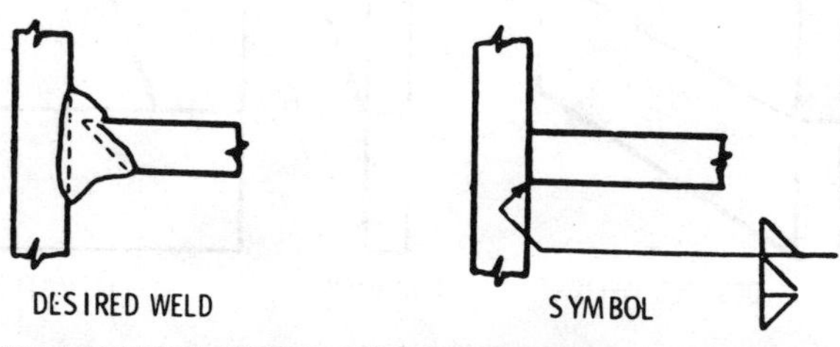

(D) SINGLE BEVEL GROOVE AND DOUBLE FILLET WELD SYMBOLS

Combination of weld symbols.

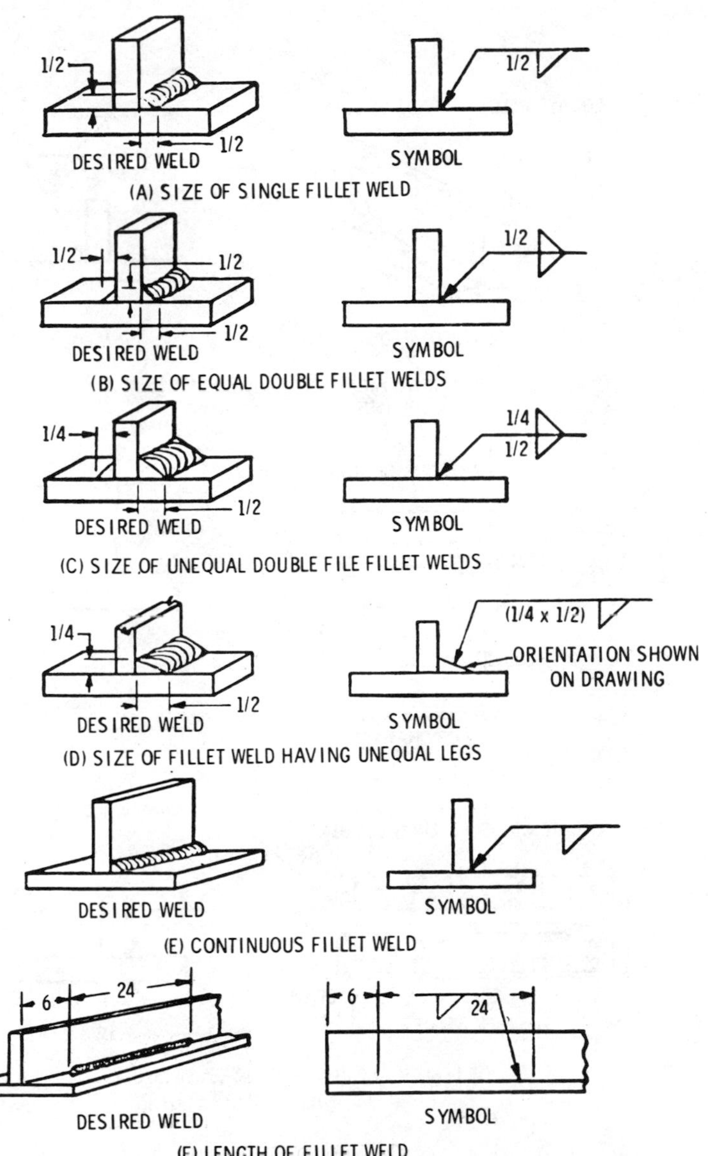

Application of dimensions to fillet welding symbols.

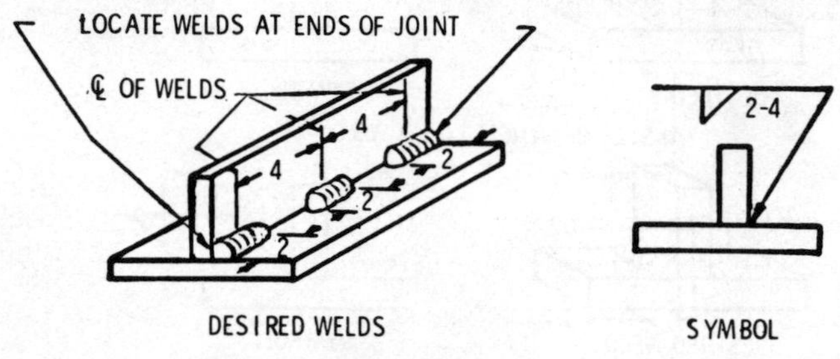

(A) LENGTH AND PITCH OF INCREMENTS OF INTERMITTENT WELDING

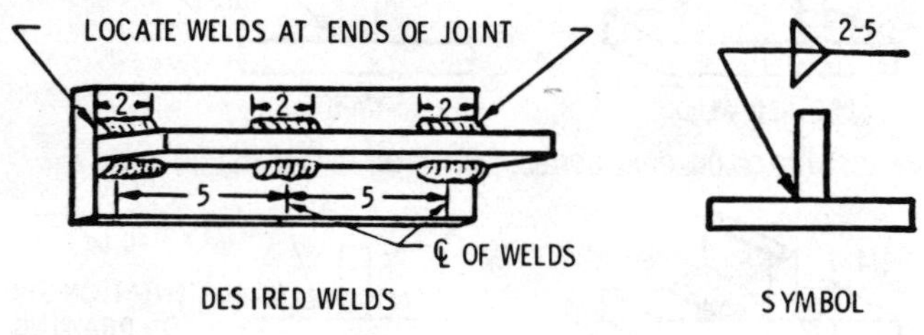

(B) LENGTH AND PITCH OF INCREMENTS OF CHAIN INTERMITTENT WELDING

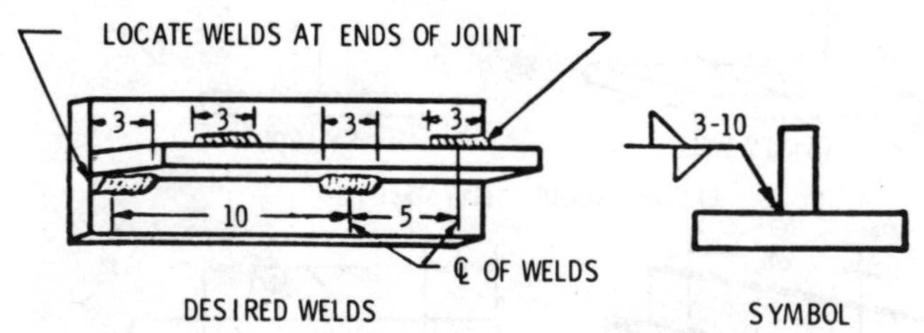

(C) LENGTH AND PITCH OF INCREMENTS OF STAGGERED INTERMITTENT WELDING

Dimensions to intermittent fillet welding symbols.

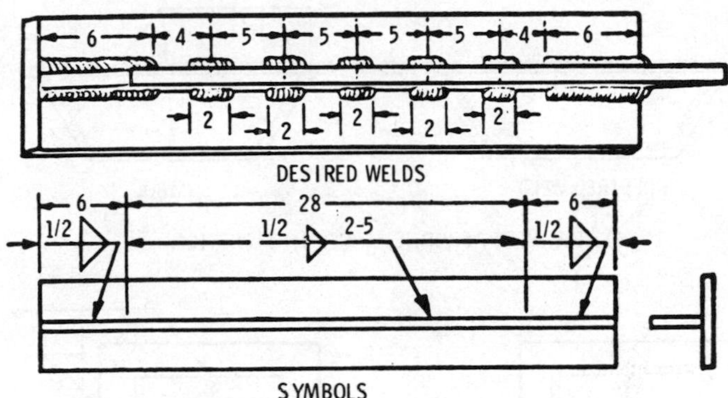

(A) COMBINED INTERMITTENT AND CONTINUOUS WELDING

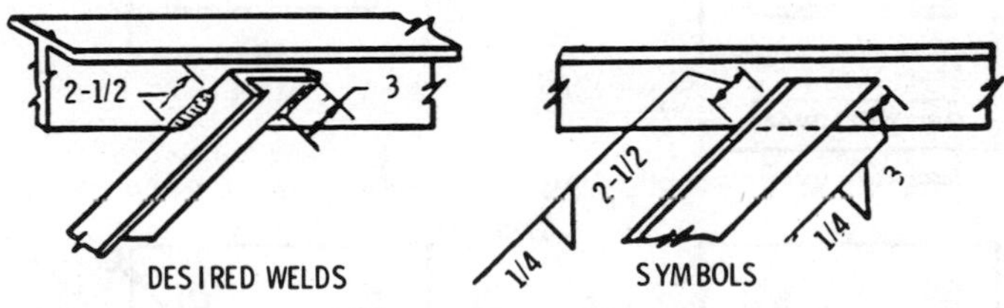

(B) WELDS DEFINITELY LOCATED

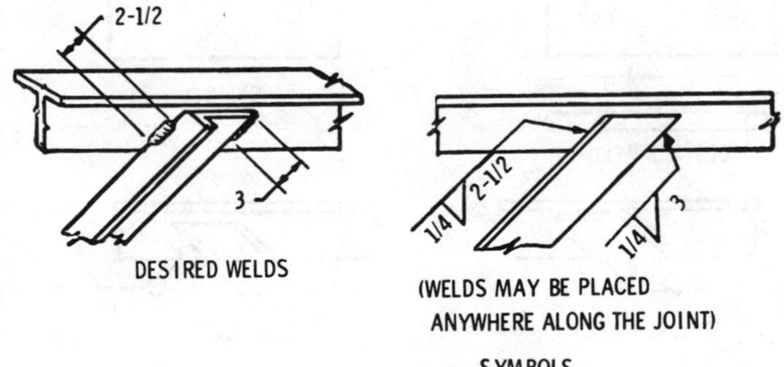

(C) WELDS APPROXIMATELY LOCATED

Location and extent of fillet welds.

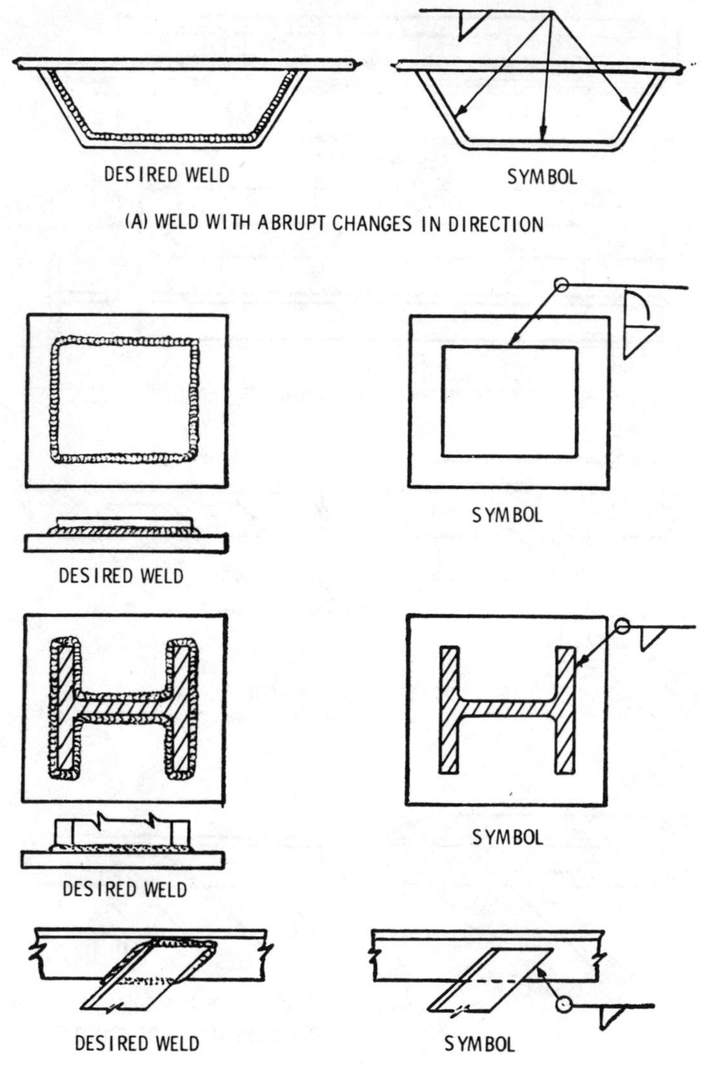

Designation of extent of welding.

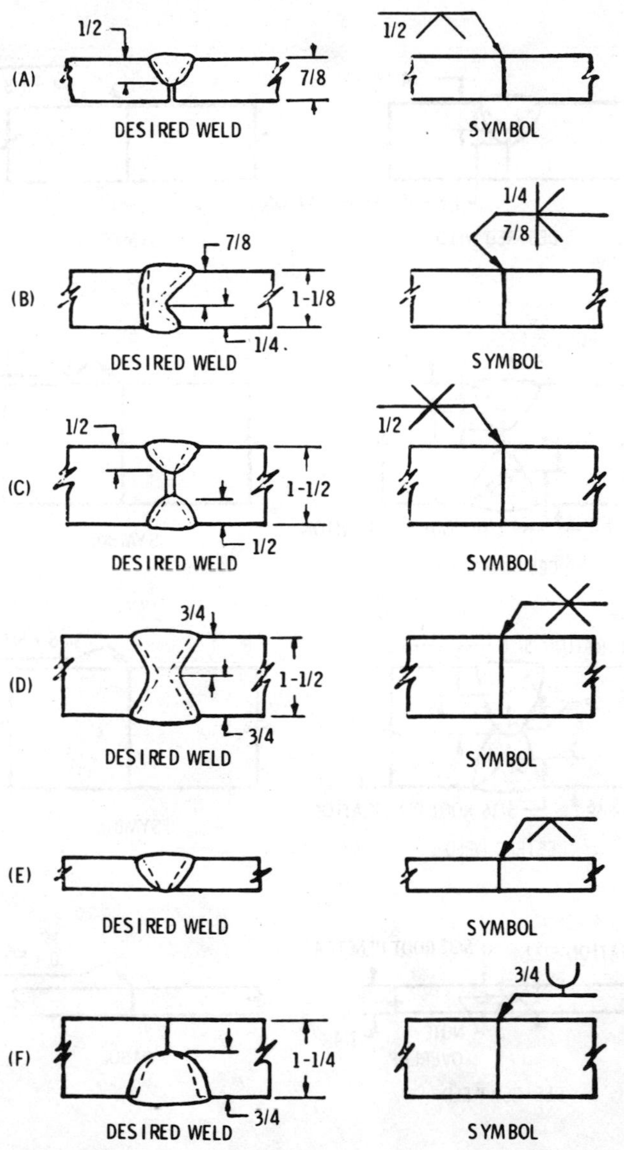

Designation of size of groove welds with no specified root penetration.

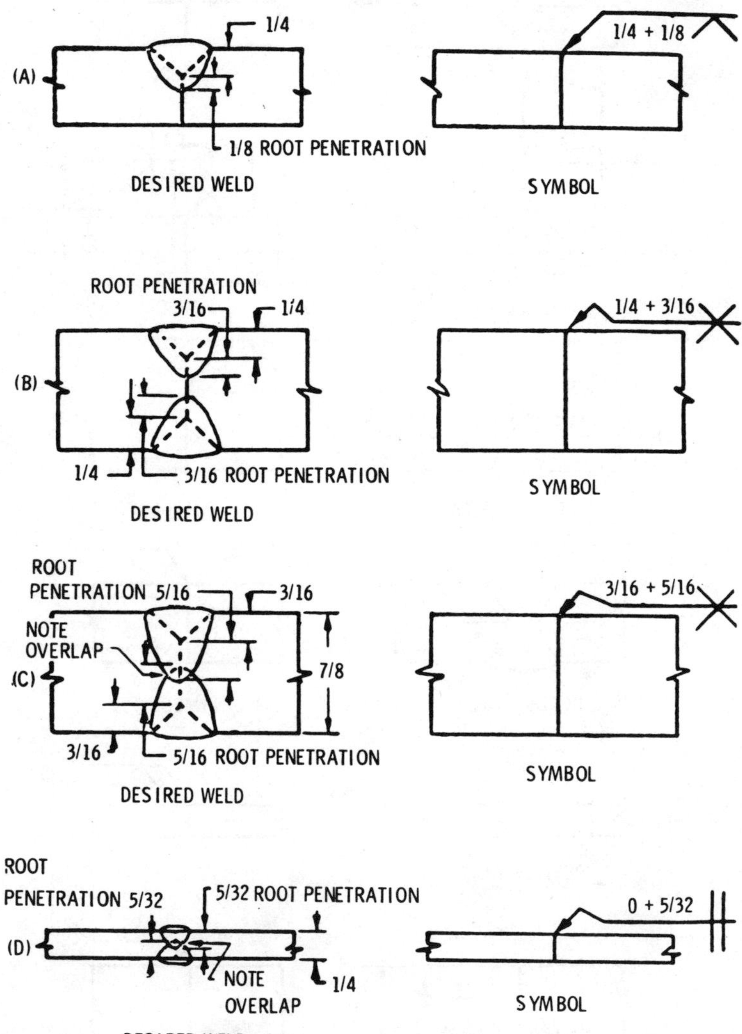

Designation of size of groove welds with specified root penetration.

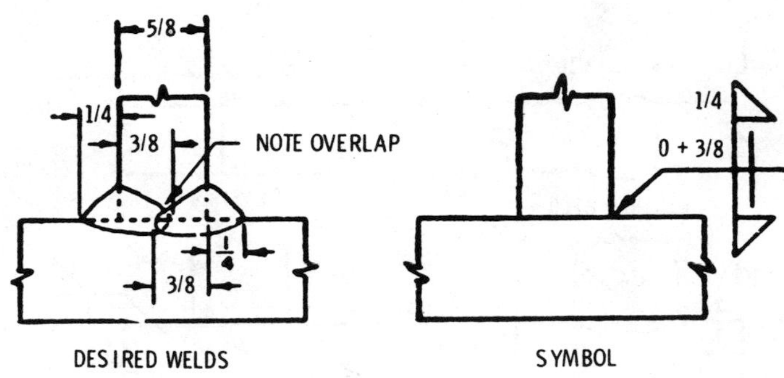

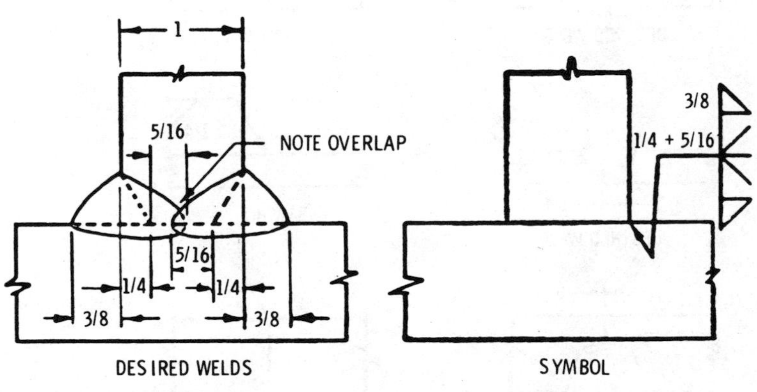

Size of combined welds with specified root penetration.

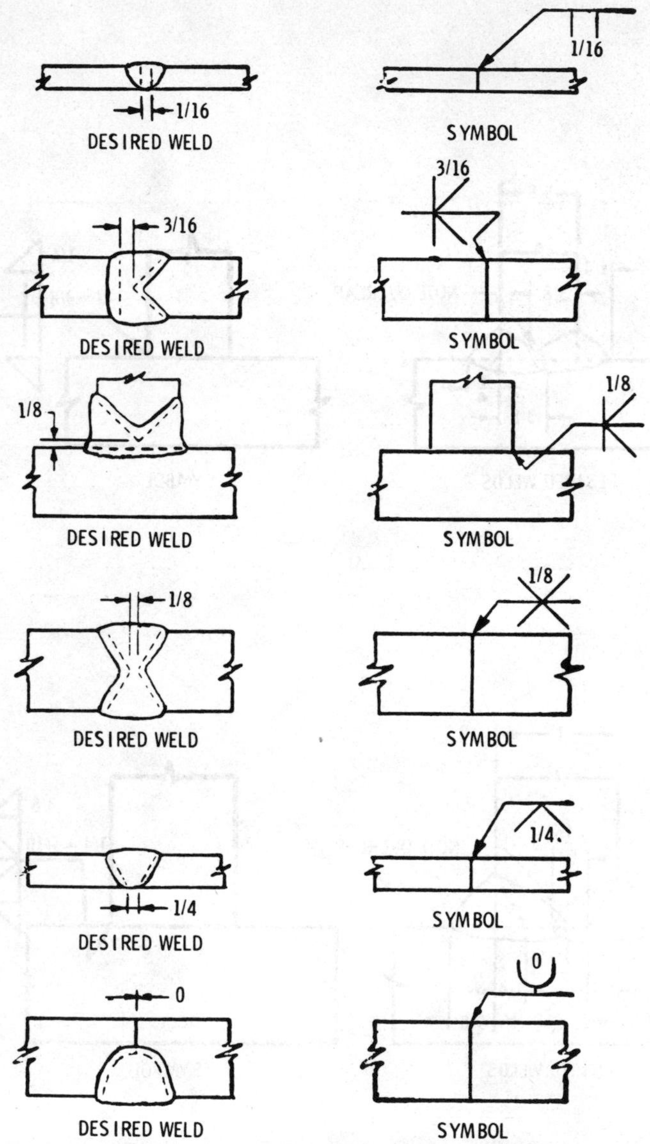

Root opening of groove welds.

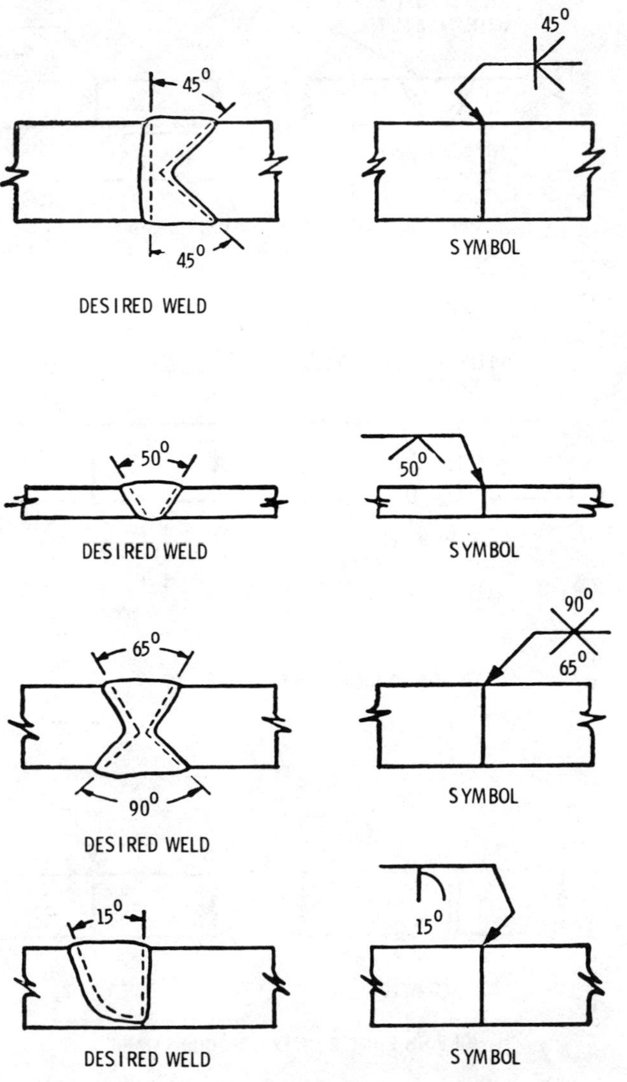

Groove angle of groove welds.

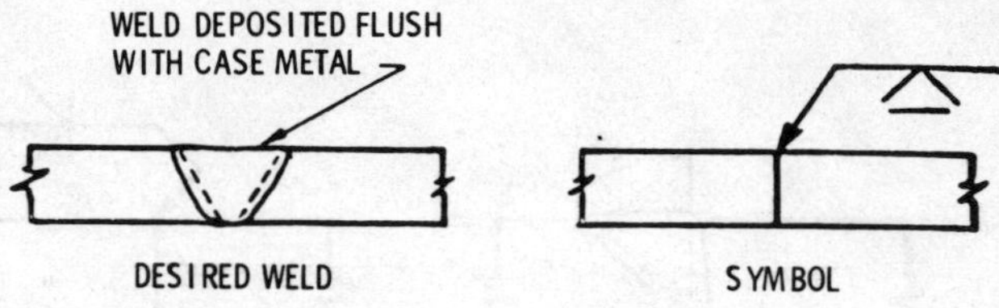

(A) ARROW SIDE FLUSH CONTOUR SYMBOL

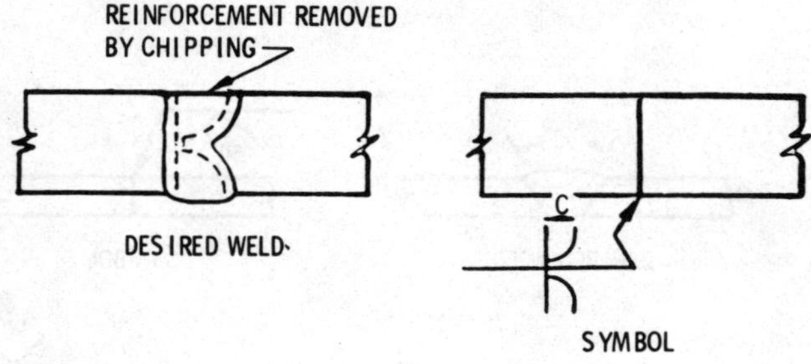

(B) OTHER SIDE FLUSH CONTOUR SYMBOL

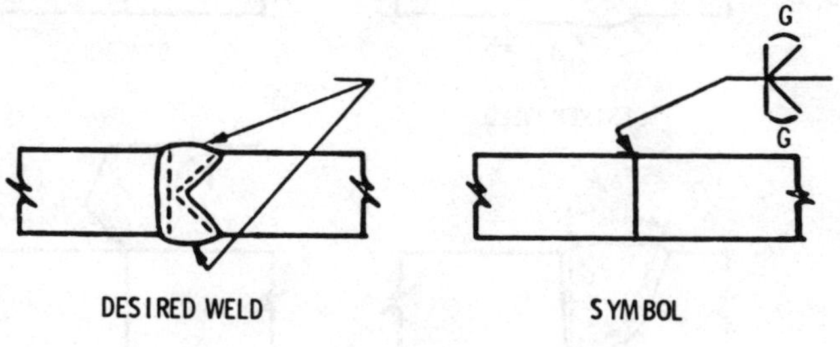

(C) BOTH SIDES CONVEX CONTOUR SYMBOL

Application of flush-and-convex contour symbols to groove welding symbols.

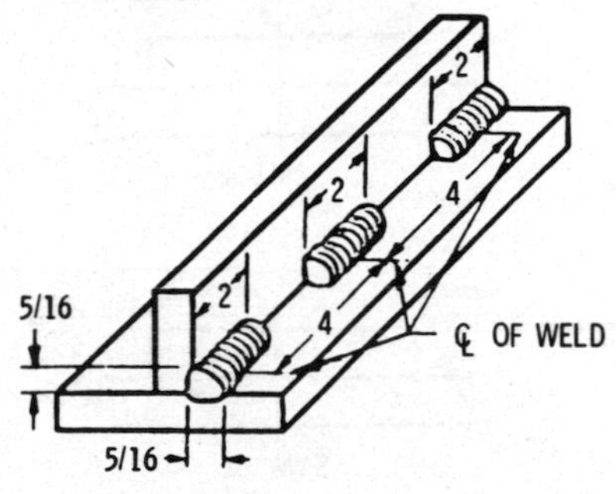

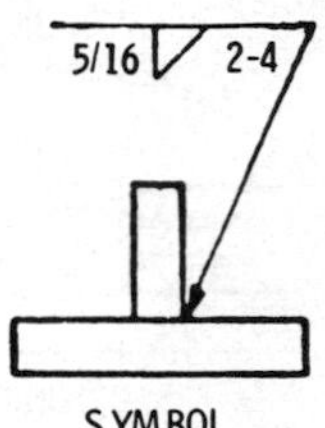

(A) FILLET WELDING SYMBOL SHOWING USE OF COMBINED DIMENSIONS

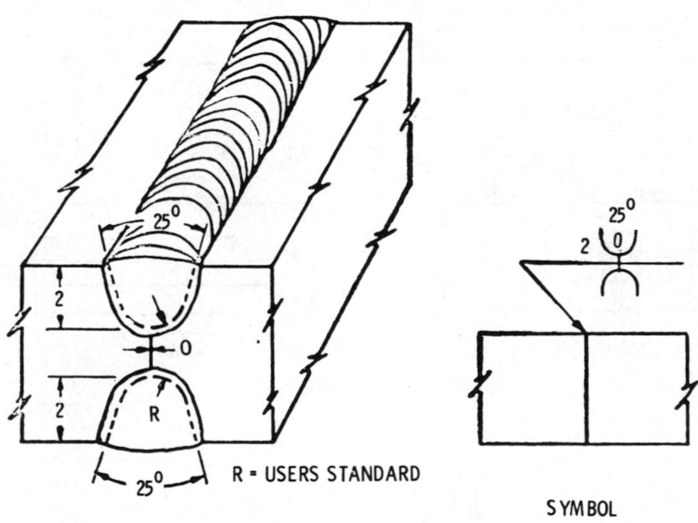

(B) GROOVE WELDING SYMBOL SHOWING USE OF COMBINED DIMENSIONS

Dimensions of fillet and groove welding symbols.

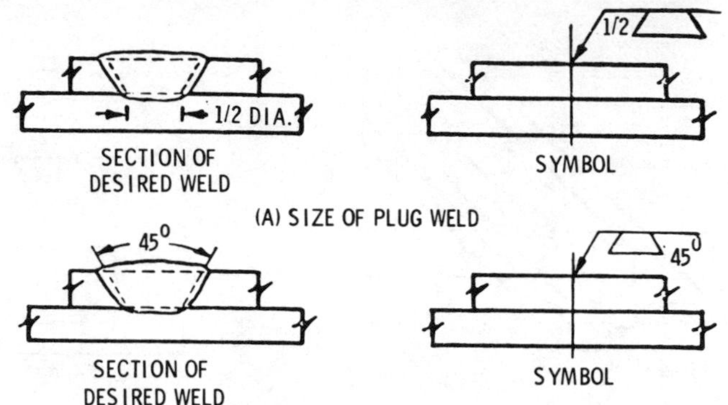

(B) INCLUDED ANGLE OF COUNTERSINK OF PLUG WELDS

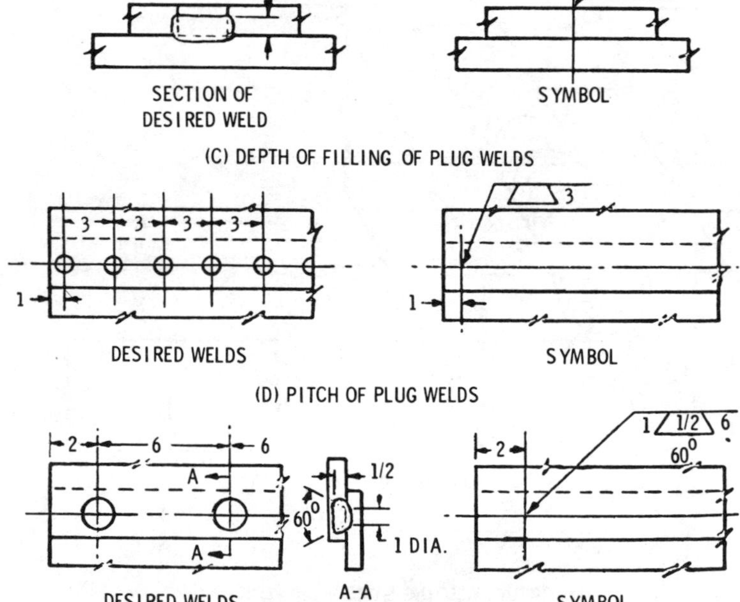

Dimensions to plug welding symbols.

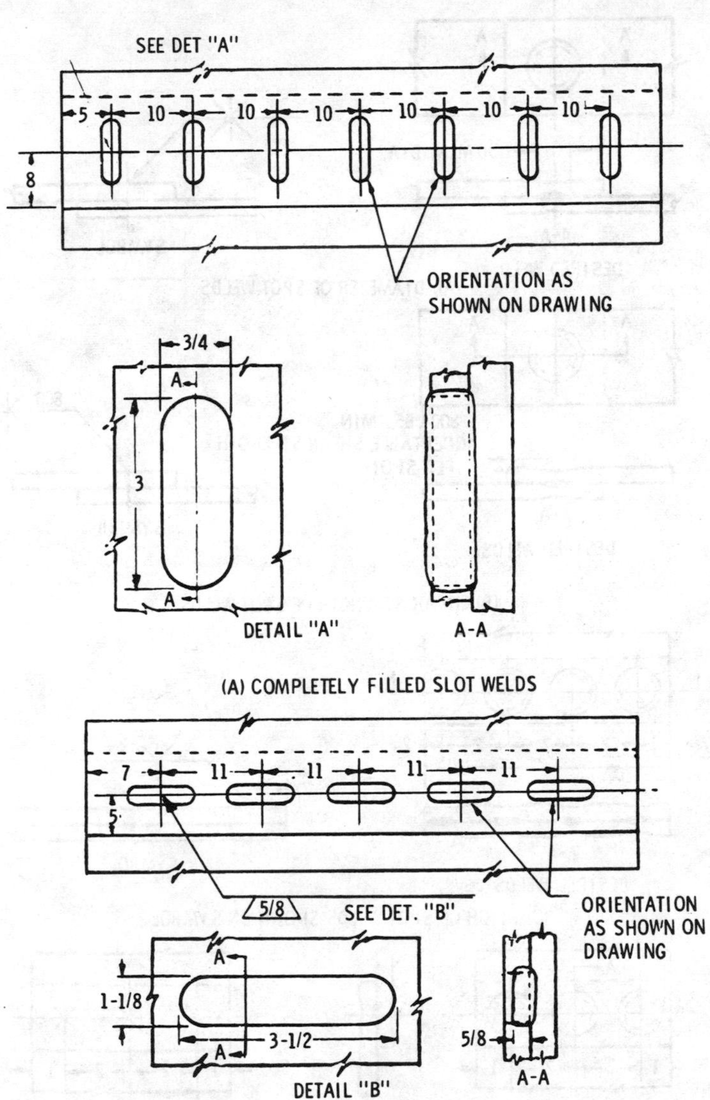

Dimension to slot welding symbols.

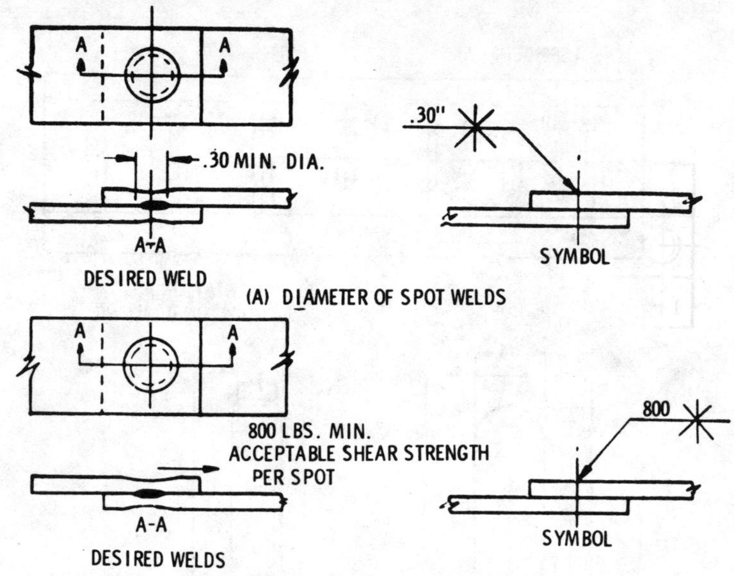

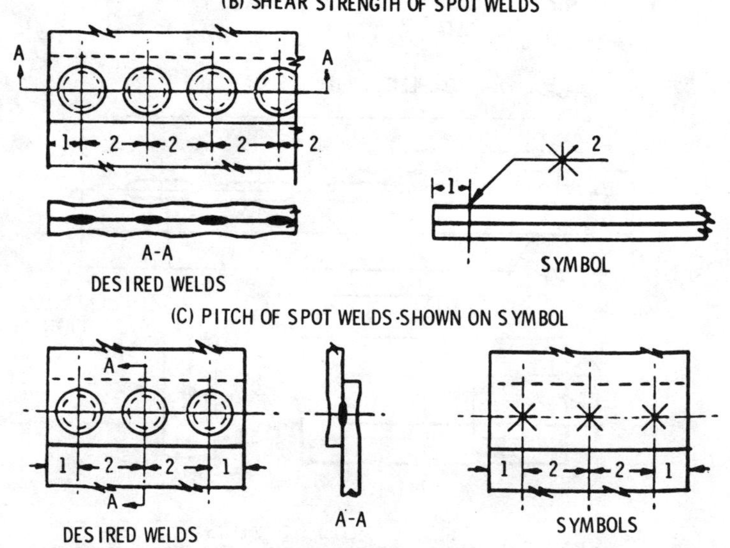

Dimensions to spot welding symbols.

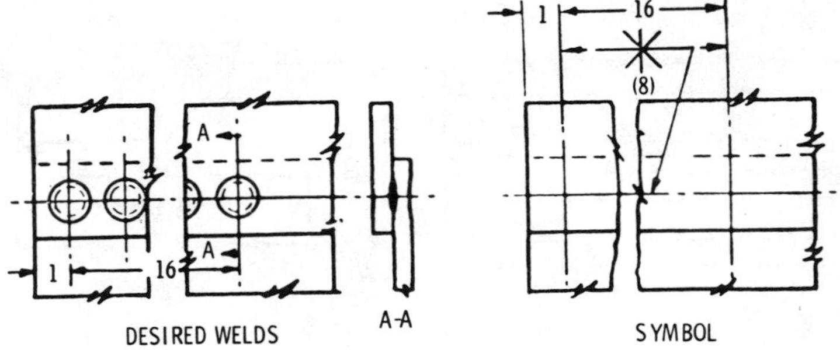

(E) EXTENT OF SPOT WELDING

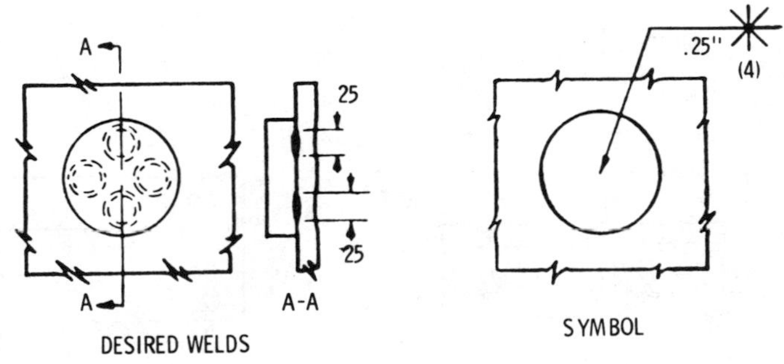

(F) SPECIFIED NUMBER OF SPOT WELDS LOCATED AT RANDOM

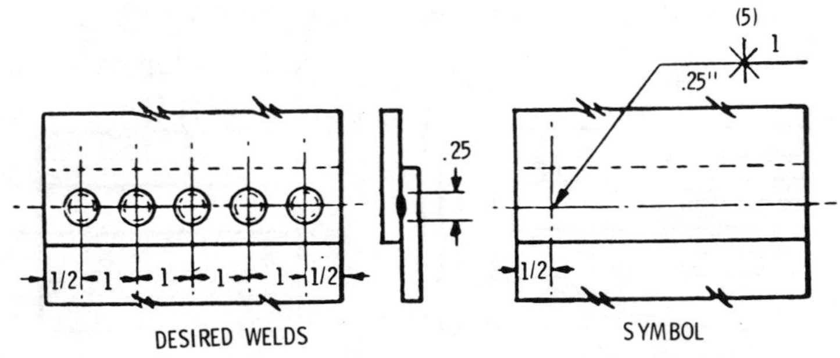

(G) SPOT WELDING SYMBOL SHOWING USE OF COMBINED DIMENSIONS

Dimensions to spot welding symbols.

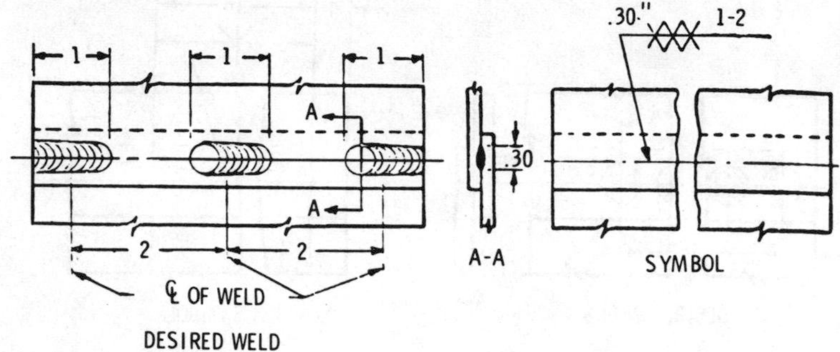

(A) LENGTH AND PITCH OF INTERMITTENT SEAM WELDS

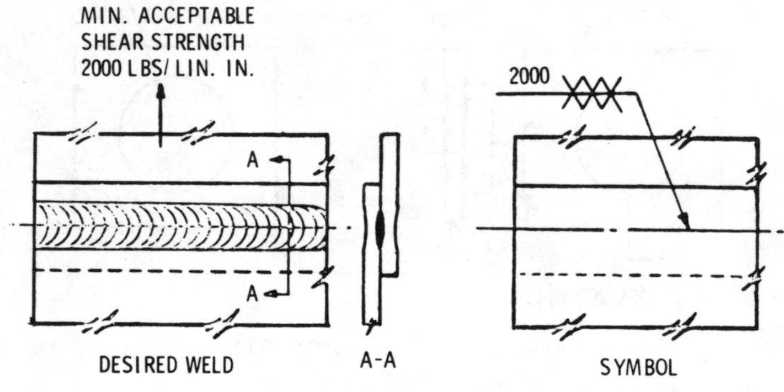

(B) STRENGTH OF SEAM WELDS

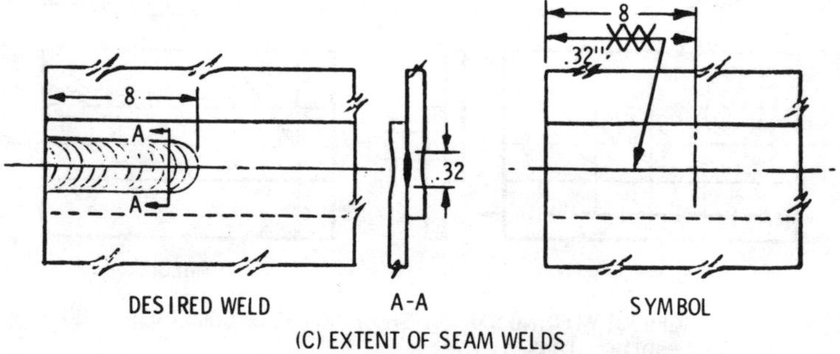

(C) EXTENT OF SEAM WELDS

Dimensions to seam welding symbols.

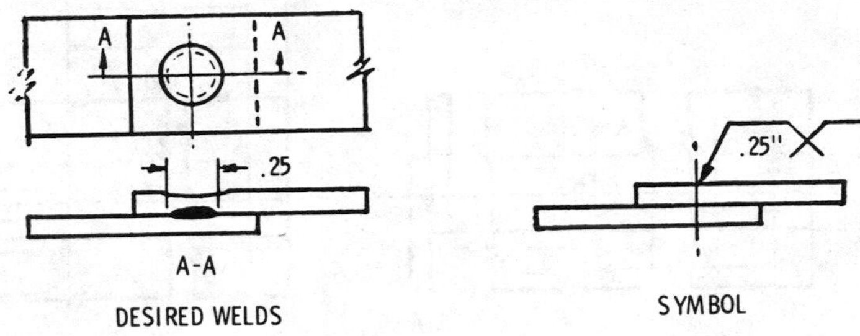

(A) DIAMETER OF PROJECTION WELDS

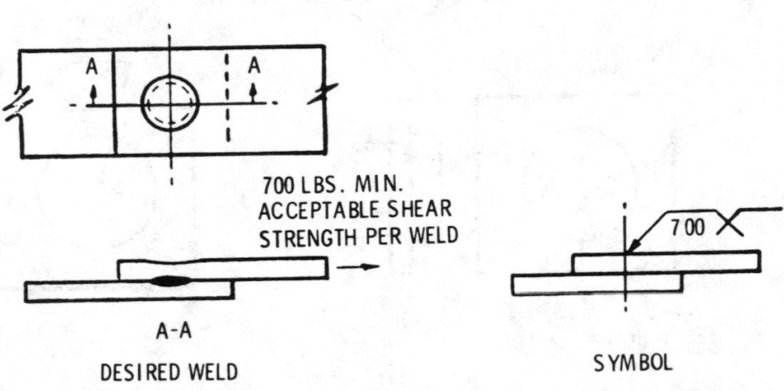

(B) SHEAR STRENGTH OF PROJECTION WELDS

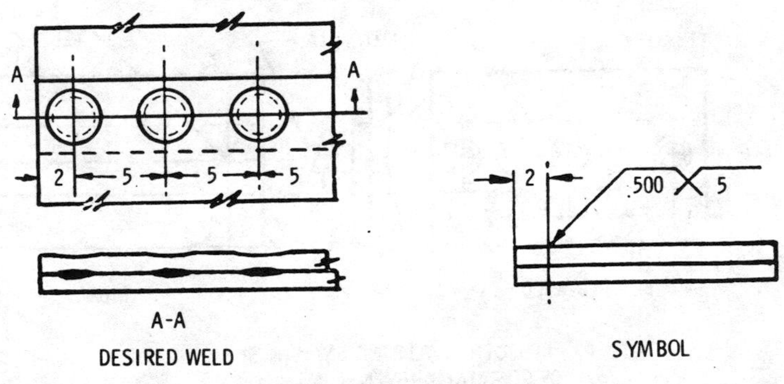

(C) PITCH OF PROJECTION WELDS

Projection welding symbols.

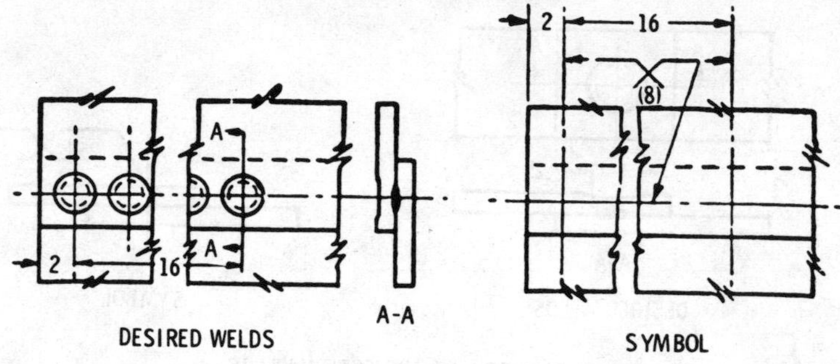

(D) EXTENT OF PROJECTION WELDING

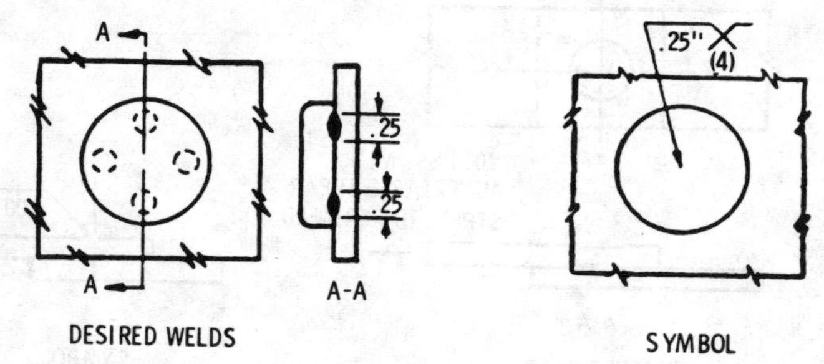

(E) SPECIFIED NUMBER OF PROJECTION WELDS LOCATED AT RANDOM

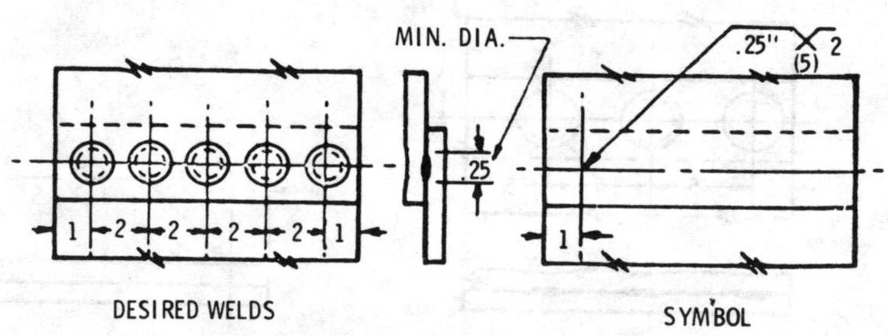

(F) PROJECTION WELDING SYMBOL SHOWING USE OF COMBINED DIMENSIONS

Projection welding symbols.

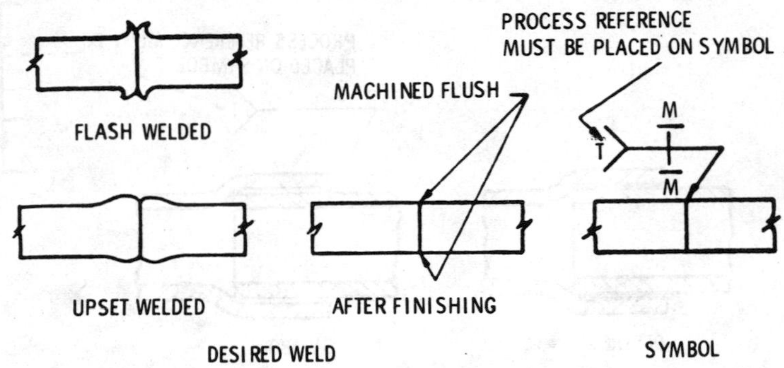

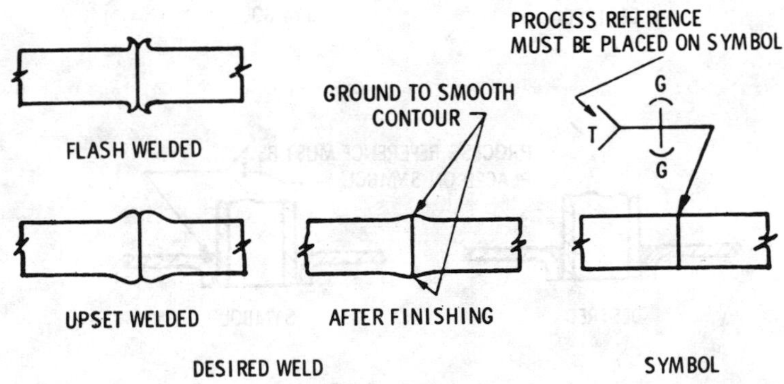

Application of flush-and-convex contour symbols to flash and upset welding symbols.

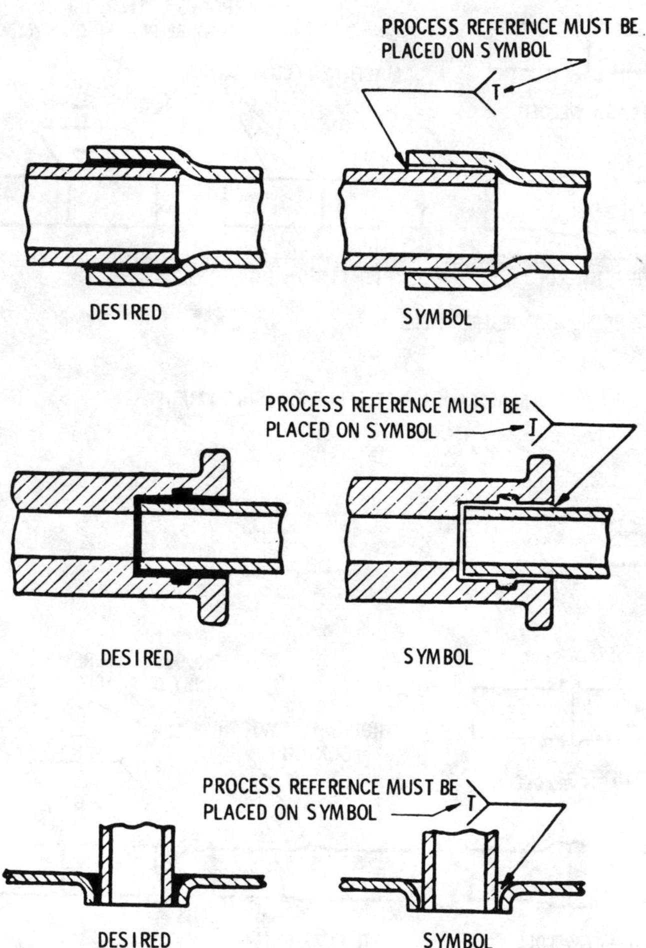

Application of brazing symbols.

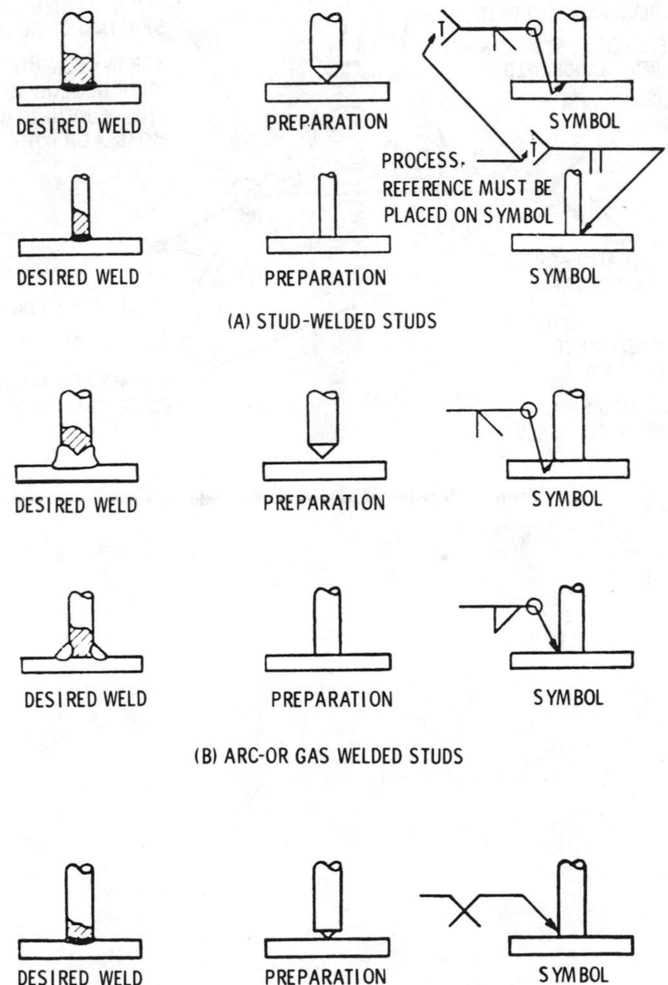

Use of welding symbols to indicate the welding of studs.

WELDING SYMBOLS

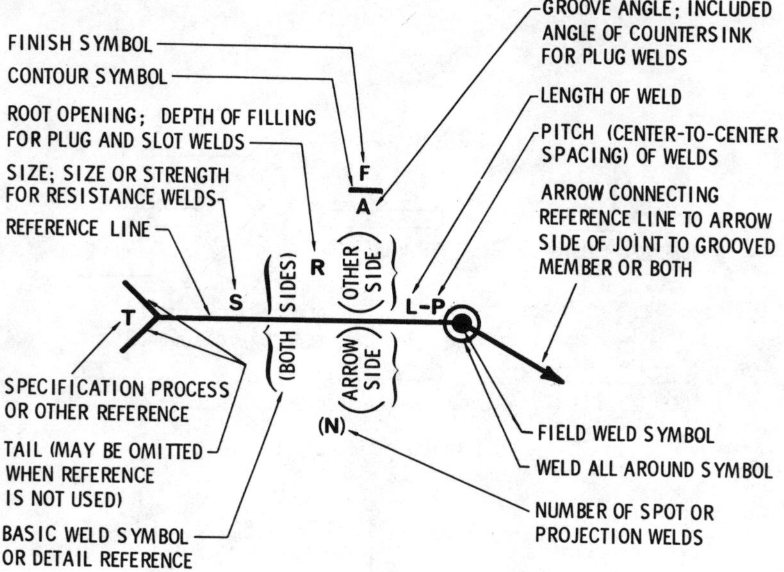

Standard location of elements of a welding symbol.

CHAPTER 27

Testing and Inspecting Welds

There are many different kinds of tests used to determine the strength, hardness, and other properties of welds. Most of these tests require special equipment and trained qualified personnel for their effective use. A few such as bend or hammer tests can be performed by the beginner.

It is not the intention of this chapter to describe each and every method of testing and inspecting welds. Only some of the more common types will be examined. For those who wish to pursue this subject further and at greater depth, it is recommended that they consult the various publications of the *American Welding Society* (STANDARD METHODS FOR MECHANICAL TESTING OF WELDS, INSPECTION HANDBOOK FOR MANUAL METAL ARC WELDING, etc.).

All forms of testing and inspecting welds can be grouped into two basic categories: (1) nondestructive tests, and (2) destructive tests. The nondestructive tests are methods provided to determine the suitability of a weld for the service conditions to which it will be subjected. These tests are forms of inspection that do not affect or alter the structure or appearance of the weld. Consequently, they are not as thorough in determining the properties of a weld as are the destructive tests. On the other hand, destructive tests are designed to destroy the weld in the quest for information about weld properties. The beginner will be primarily concerned with such simple destructive tests as the bend or hammer tests.

NONDESTRUCTIVE TESTS

Almost every type of nondestructive test requires trained qualified personnel and special equipment for its effective use. Visual inspection is probably the only exception to this rule.

The nondestructive tests that will be described in this chapter include the following:

1. Visual inspection,
2. Magnetic particle,
3. Penetrant,
4. Radiographic,
5. Ultrasonic,
6. Eddy current.

Visual Inspection

Some authorities do not include visual inspection in the nondestructive category of testing but prefer to regard it as a third category distinct from the other two. The reason for making this distinction is that visual inspection involves no direct and active testing of the weld, whereas nondestructive and destructive testing do make such an examination.

Visual inspection involves the close examination of the weld surface and joint with or without a magnifying glass. This form of inspection is limited to an *external* examination of the surface. There is no way of examining the internal structure of the weld. Among other things, a visual inspection would be expected to include the following information:

1. Weld bead size,
2. Weld bead contour,
3. Crater deficiencies,
4. Degree of fusion (surface),
5. The absence or presence of cracks, undercutting and shrinkage cavities.

Magnetic Particle Testing

Magnetic particle testing is a form of weld inspection in which surface or near-surface flaws are located by means of an induced magnetic field and its effect upon finely divided magnetic particles distributed across the surface. Magnetic particle testing will only work on materials that can be magnetized.

The magnetic particles may be applied either wet or dry. The former condition makes possible the detection of very fine cracks. Dry particles are preferred for subsurface inspection. It is recommended that the magnetic particles be applied to the surface while the magnetizing current is still on. After the magnetic particle inspection has been completed, the piece of metal should be demagnetized.

Penetrant Inspection Method

Penetrant inspection is a method of locating defects open to the surface. Examples of the types of defects detected by the penetrant inspection method are:

1. Fatigue cracks,
2. Shrinkage cracks,
3. Crater cracks,
4. Pits,
5. Pores (porosity),
7. Incomplete fusion.

The penetrant inspection method is limited to defects *open* to the surface. It has no subsurface inspection capabilities. The penetrant inspection method can be divided into two basic types: (1) dye penetrant inspection, and (2) fluorescent penetrant inspection.

Fluorescent penetrant inspection involves the application of a fluorescent liquid to the metal surface. A short time is allowed for the penetrant to sink into any openings, and then the excess is removed and the surface dried. A black light source directed against the surface will indicate those areas (defects) penetrated by the fluorescent liquid.

Dye penetrant inspection is similar in application to the fluorescent penetration method. A liquid dye is sprayed onto the surface and allowed time to penetrate the openings. The excess is then removed, the surface dried, and a developer is sprayed over the area. The developer is stained by any dye residue, clearly marking defects on the surface.

Radiographic Inspection

Radiographic inspection is a nondestructive method of detecting internal weld defects, and is based essentially on the ability of short-wave radiations (X rays or gamma rays) to penetrate objects opaque to ordinary light.

Not all the shortwave radiation will penetrate the weld. Some of the radiation will be absorbed by the weld itself, and the amount absorbed depends on the thickness and the density of the weld. Internal weld defects will appear as variations in the amount of radiation absorbed by the weld. These variations will appear on a radiograph (an image picture of the absorption activity of the weld).

Ultrasonic Weld Inspection

The purpose of ultrasonic weld inspection equipment is to generate a sound wave (with a pitch or frequency so high that it is inaudible to the human ear) for inspection purposes.

The principle of ultrasonic weld inspection is based on the echo principle. Sound waves are produced by using a radio-frequency current to vibrate (or expand and contract) a quartz crystal 5,000,000 or more times per second. Sound waves of this type are capable of penetrating solid materials like X rays. When their frequencies conform with the resonance values of different inspection materials, they are echoed or reflected back to their source by non-resonant media (such as oxide deposits or gas pockets within a weld). The reflections are amplified like radio signals and used to produce visual images on a cathode tube (similar to the picture tube on a television set). The same unit that sends the ultrasonic sound waves also acts as their receiver. A trained and qualified operator is necessary to interpret the oscilloscope wave pattern, and determine the location and nature of the weld defect. Ultrasonic weld inspection can be used to inspect both the internal and surface structure of a weld.

Eddy Current

The eddy current nondestructive method of testing uses electromagnetic energy for detecting and locating weld defects.

One or more AC coils are brought into contact with the metal surface to be inspected. As a result of the proximity of the AC coils, eddy currents (those that move contrary or opposite the main current) are induced in the metal. The magnetic field of the induced eddy currents opposes the magnetic field of the AC coil (or coils). The result is an increase in the impedance (resistance) of the coil. Coil impedance can be measured. Weld flaws passed under the AC coil will change the coil impedance. Special recording equipment immediately signals the operator of the change.

The eddy current method can be used to inspect both ferrous and nonferrous metals. It is particularly useful in detecting internal cracking, weld porosity, and poor fusion.

DESTRUCTIVE TESTS

Destructive tests are methods of determining the properties of a weld in such a way that the weld itself is usually destroyed. In most cases, the destructive test requires special equipment and personnel with the proper qualifications and training. A few tests (e.g. bend tests) may be done satisfactorily by the beginner after a brief period of instruction.

The destructive tests that will be described in this chapter are as follows:

1. The bend test,
2. The impact test,
3. The tensile test,
4. The nick-break test,
5. The hardness test,
6. The etch test.

The Bend Test

The destructive bend test may be used to determine a number of weld properties including ductility, weld penetration, tensile strength, and fusion. It is an easy and inexpensive test to apply

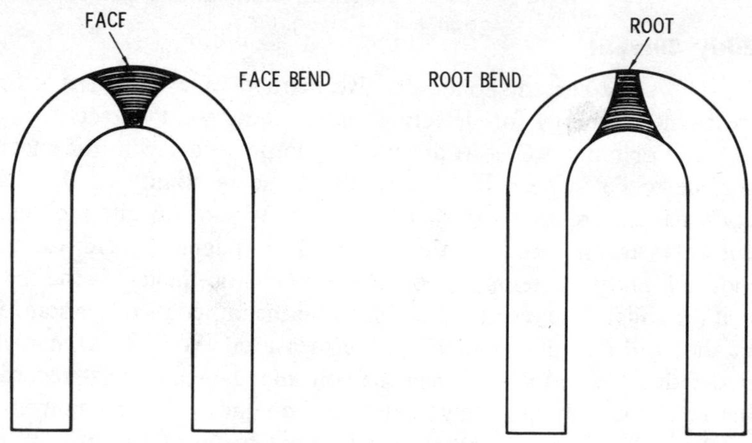

Courtesy Airco Welding Products

Fig. 1. The guided bend specimens.

and can be learned by the beginner with a minimum of instruction.

The two basic types of bend tests are: (1) the guided bend test, and (2) the free bend test. The guided bend test can also be divided into two types depending in which direction the specimen is bent. These two types are: (1) the guided *face-bend* test, and (2) the guided *root-bend* test (Fig. 1).

All guided bend tests are made in specially designed jigs (Fig. 2). The surface of the specimen to be tested is first ground smooth so that all weld reinforcement is removed (Fig. 3). The specimen is then placed across the die supports and bent by depressing the plunger until it forms the shape of a U. If a guided face-bend test is being conducted, the specimen is placed across the die supports so that the face is down. In a guided root-bend test, the root faces down.

Root-bend tests are used primarily to determine the degree of weld penetration. Face-bend tests are used to inspect a wider range of weld properties including the amount of weld porosity, the degree of fusion, and the absence or presence of inclusions. Any weld defects greater than ⅛ inch in any dimension will be considered as weld failure.

The free bend test (Fig. 4) is used to measure the ductility of the weld metal. The bending is performed in any type of machine or vise capable of exerting a sufficiently large compressive force.

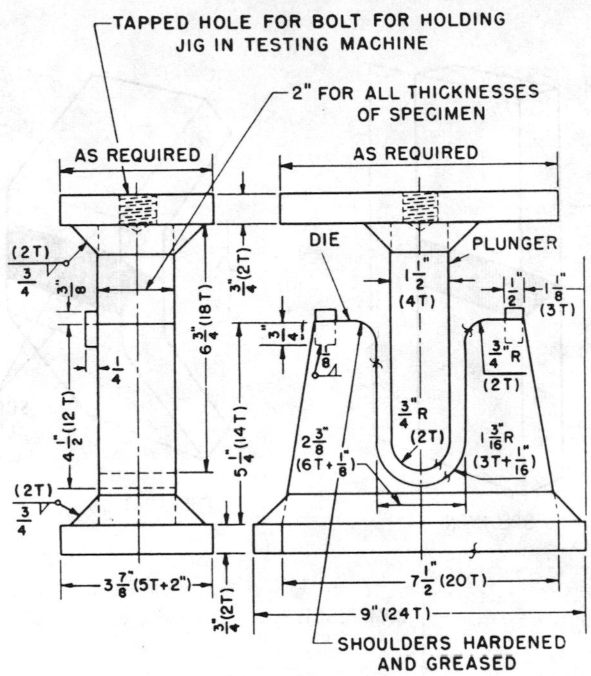

Courtesy Airco Welding Products

Fig. 2. A guided bend test jig.

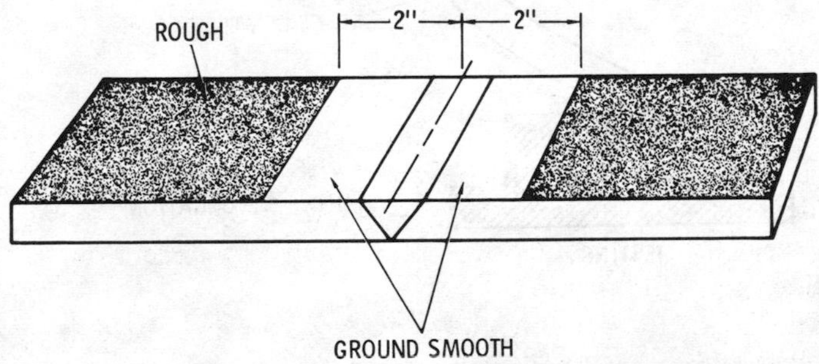

Courtesy Airco Welding Products

Fig. 3. A properly machined specimen.

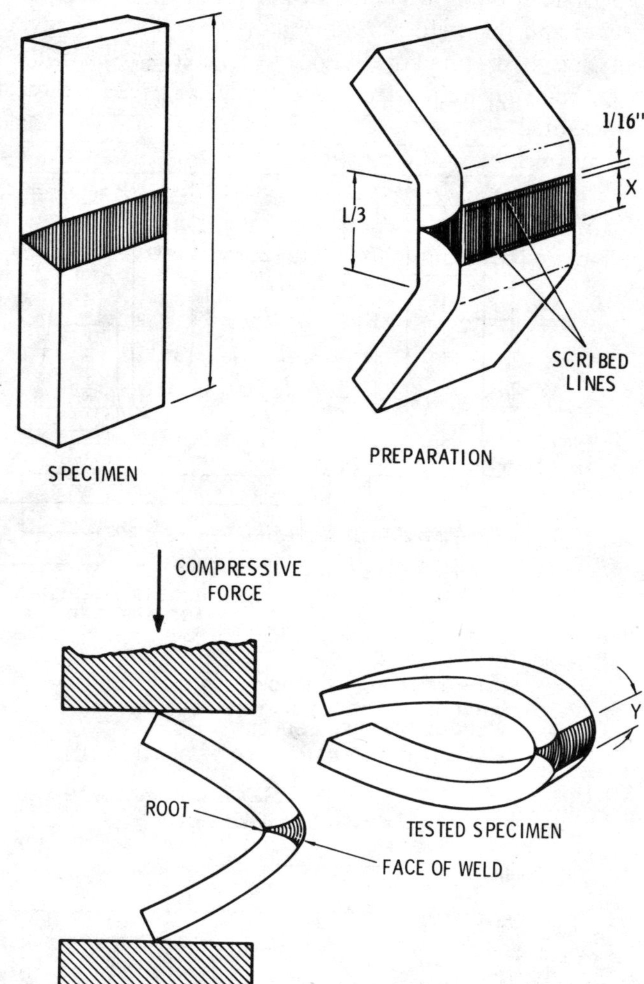

Courtesy Airco Welding Products

Fig. 4. The free bend test.

The specimen being tested must be machined so that the surface is clean and the weld reinforcements have been removed. As shown in Fig. 4, the machined surface is first marked with lines 1/16 inch in from the weld edges. The distance between these lines is then measured (distance X). After taking the initial measurement, the operator bends the specimen to an angle of approximately 30°. The bends should occur at the third points. The specimen is then bent so that the bend occurs along the center of the weld. The specimen is bent until flat or until cracks exceeding 1/16 inch occur.

The elongation percentage is determined in the following manner:

1. Measure the distance between the scribed lines *after* the specimen has been bent flat. This is designated "distance Y" to distinguish it from the initial measurement (i.e. distance X).
2. Subtract distance X from distance Y.
3. Divide the results of step 2 by distance X.
4. Multiply the results of step 3 by 100.

The elongation percentage must be a 15% minimum or the weld has failed the free bend test. In addition, no cracks greater than 1/16 inch should occur.

Impact Test

Impact testing involves the rapid application of a predetermined load against a metal surface. The purpose of impact testing is to determine the amount of impact a specimen will absorb before fracturing.

The two basic types of impact testing are: (1) the Izod method, and (2) the Charpy method. The Izod method involves the striking of a notched specimen supported vertically in a vise. The notch is made 1/3 the distance from the end of the specimen. The specimen is placed in a vise so that the notch faces the direction from which the impact blow is delivered and is positioned so that the notch is roughly on the same level as the top of the vise.

In the Charpy method, the notch is made in the middle of the specimen. The specimen is then placed with each end resting

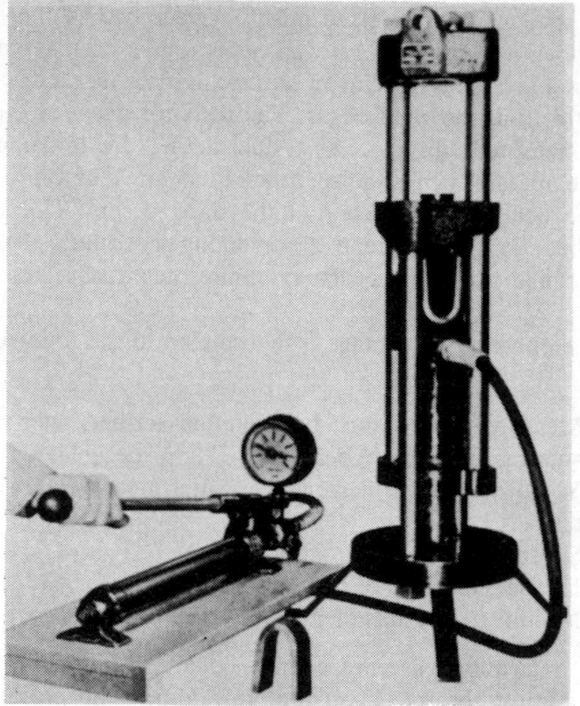

Courtesy Airco Welding Products

Fig. 5. A tensile and guided bend testing machine.

on supports so that the notch faces down over an opening and away from the direction of the blow.

Tensile Test

The tensile test is used to determine the tensile strength and ductility of a weld. The specimen to be tested is placed in a machine similar to the one shown in Fig. 5 and stretched until it is pulled apart. The tensile strength is expressed as the pounds per square inch required before the breaking point is reached.

Fig. 6 indicates the manner in which a typical tensile test is conducted. Before the specimen is subjected to the test, the operator must first measure its width and thickness and then determine the cross sectional area (calculated by multiplying the width

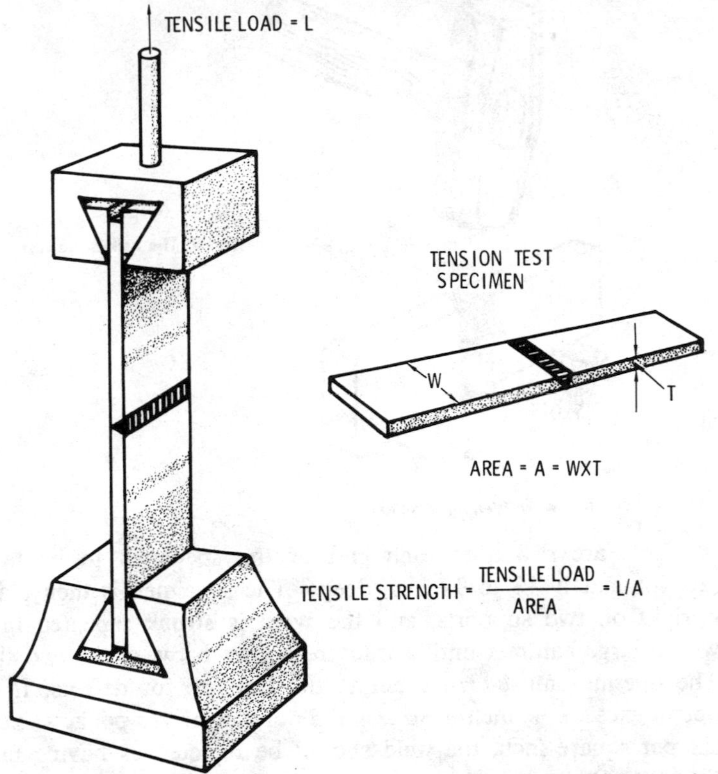

Courtesy Airco Welding Products

Fig. 6. The tensile strength test.

times the thickness and expressed in square inches). The point at which the specimen breaks is recorded in pounds on the testing machine. Tensile strength is calculated by dividing the tensile load (the number of pounds of tension required to break the specimen) by the cross sectional area. The specimen is usually required to withstand 90-100% of the minimum specified tensile strength of the base metal.

Nick-Break Test

The nick-break test (Fig. 7) is used to determine weld ductility, the degree of fusion, the presence of slag inclusions, and weld porosity.

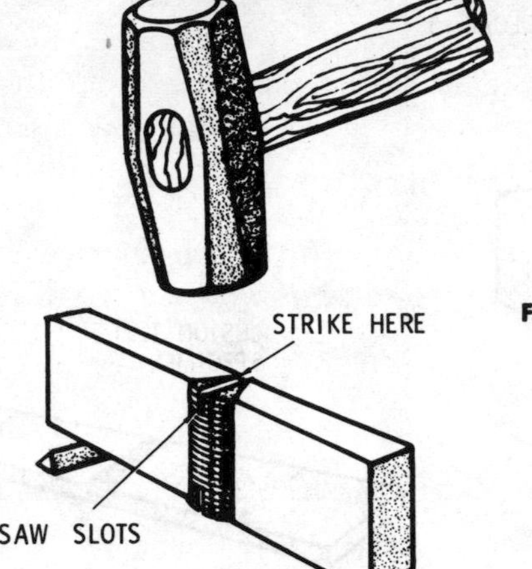

Fig. 7. The nick-bend test.

Courtesy Airco Welding Products

Slots are sawed at each end of the specimen to be tested (approximately ¼ to ⅜ inch deep). The specimen is then placed upright on two supports, and the weld is struck repeated blows with a large hammer until a fracture occurs between the two slots. The operator must then examine the fracture for defects. If any defect exceeds 1/16 inch in size or the number of gas pockets exceed six per square inch, the weld should be rejected as having failed the test.

Hardness Test

Hardness indicates the ability of a metal to resist denting, scratching, abrasion, or penetration. If the welder can determine the hardness of a metal, the tensile strength can also be determined through a conversion table. The four methods for testing the hardness of a metal are: (1) the Brinell test, (2) the Rockwell test, (3) the Vickers test, and (4) the Shore Scleroscope. Each of these testing methods was described in detail in Chapter 2.

Etch Test

Etching is a form of inspection used to determine the depths of penetration and fusion and the grain structure of the weld and the adjacent metal. This is a destructive testing method, because it requires the obtaining of a cross section of the weld for the test. This is accomplished by cutting the weld across its diameter, filing and grinding until smooth, and finishing (polishing) with sand or emery cloth (begin with No. 1 and finish with No. 00 or No. 000). The surface is then cleaned and a 10% to 20% solution of aluminum persulphate applied. After allowing the solution a few minutes to penetrate, wash the surface with water and allow it to dry.

Appendix

Decimal and Millimeter Equivalents of Fractional Parts of an Inch

Inches		Inches	mm	Inches		Inches	mm
	1-64	.01563	.397		**33-64**	.51563	13.097
1-32		.03125	.794	**17-32**		.53125	13.097
	3-64	.04688	1.191		**35-64**	.54688	13.890
1-16		.0625	1.587	**9-16**		.5625	14.287
	5-64	.07813	1.984		**37-64**	.57813	14.684
3-32		.09375	2.381	**19-32**		.59375	15.081
	7-64	.10938	2.778		**39-64**	.60938	15.478
1-8		.125	3.175	**5-8**		.625	15.875
	9-64	.14063	3.572		**41-64**	.64063	16.272
5-32		.15625	3.969	**21-32**		.65625	16.669
	11-64	.17188	4.366		**43-64**	.67188	17.065
3-16		.1875	4.762	**11-16**		.6875	17.462
	13-64	.20313	5.159		**45-64**	.70313	17.859
7-32		.21875	5.556	**23-32**		.71875	18.256
	15-64	.23438	5.953		**47-64**	.73438	18.653
1-4		.25	6.350	**3-4**		.75	19.050
	17-64	.26563	6.747		**49-64**	.76563	19.447
9-32		.28125	7.144	**25-32**		.78125	19.844
	19-64	.29688	7.541		**51-64**	.79688	20.240
5-16		.3125	7.937	**13-16**		.8125	20.637
	21-64	.32813	8.334		**53-64**	.82813	21.034
11-32		.34375	8.731	**27-32**		.84375	21.431
	23-64	.35938	9.128		**55-64**	.85938	21.828
3-8		.375	9.525	**7-8**		.875	22.225
	25-64	.39063	9.922		**57-64**	.89063	22.622
13-32		.40625	10.319	**29-32**		.90625	23.019
	27-64	.42188	10.716		**59-64**	.92188	23.415
7-16		.4375	11.113	**15-16**		.9375	23.812
	29-64	.45313	11.509		**61-64**	.95313	24.209
15-32		.46875	11.906	**31-32**		.96875	24.606
	31-64	.48438	12.303		**63-64**	.98438	25.003
1-2		.5	12.700	**1**		1.00000	25.400

Decimal Inch Equivalents of Millimeters and Fractional Parts of Millimeters

mm	Inches	mm	Inches	mm	Inches	mm	Inches
1-100 =	.00039	33-100 =	.01299	64-100 =	.02520	95-100 =	.03740
2-100 =	.00079	34-100 =	.01339	65-100 =	.02559	96-100 =	.03780
3-100 =	.00118	35-100 =	.01378	66-100 =	.02598	97-100 =	.03819
4-100 =	.00157	36-100 =	.01417	67-100 =	.02638	98-100 =	.03858
5-100 =	.00197	37-100 =	.01457	68-100 =	.02677	99-100 =	.03898
6-100 =	.00236	38-100 =	.01496	69-100 =	.02717	1 =	.03937
7-100 =	.00276	39-100 =	.01535	70-100 =	.02756	2 =	.07874
8-100 =	.00315	40-100 =	.01575	71-100 =	.02795	3 =	.11811
9-100 =	.00354	41-100 =	.01614	72-100 =	.02835	4 =	.15748
10-100 =	.00394	42-100 =	.01654	73-100 =	.02874	5 =	.19685
11-100 =	.00433	43-100 =	.01693	74-100 =	.02913	6 =	.23622
12-100 =	.00472	44-100 =	.01732	75-100 =	.02953	7 =	.27559
13-100 =	.00512	45-100 =	.01772	76-100 =	.02992	8 =	.31496
14-100 =	.00551	46-100 =	.01811	77-100 =	.03032	9 =	.35433
15-100 =	.00591	47-100 =	.01850	78-100 =	.03071	10 =	.39370
16-100 =	.00630	48-100 =	.01890	79-100 =	.03110	11 =	.43307
17-100 =	.00669	49-100 =	.01929	80-100 =	.03150	12 =	.47244
18-100 =	.00709	50-100 =	.01969	81-100 =	.03189	13 =	.51181
19-100 =	.00748	51-100 =	.02008	82-100 =	.03228	14 =	.55118
20-100 =	.00787	52-100 =	.02047	83-100 =	.03268	15 =	.59055
21-100 =	.00827	53-100 =	.02087	84-100 =	.03307	16 =	.62992
22-100 =	.00866	54-100 =	.02126	85-100 =	.03346	17 =	.66929
23-100 =	.00906	55-100 =	.02165	86-100 =	.03386	18 =	.70866
24-100 =	.00945	56-100 =	.02205	87-100 =	.03425	19 =	.74803
25-100 =	.00984	57-100 =	.02244	88-100 =	.03465	20 =	.78740
26-100 =	.01024	58-100 =	.02283	89-100 =	.03504	21 =	.82677
27-100 =	.01063	59-100 =	.02323	90-100 =	.03543	22 =	.86614
28-100 =	.01102	60-100 =	.02362	91-100 =	.03583	23 =	.90551
29-100 =	.01142	61-100 =	.02402	92-100 =	.03622	24 =	.94488
30-100 =	.01181	62-100 =	.02441	93-100 =	.03661	25 =	.98425
31-100 =	.01220	63-100 =	.02480	94-100 =	.03701	26 =	1.02362
32-100 =	.01260	....		....		....	

Wire Gauge Standards

Wire gauge no.	Decimal parts of an inch						
	American or Brown & Sharpe	Birmingham or Stubs wire	Washburn & Moen on steel wire gauge	American S. & W. Co.'s music wire	Imperial wire gauge	Stubs steel wire	U.S. standard for plate
0000000	0.651354	....	0.4000	....	0.500	....	0.500
000000	0.580049	....	0.4615	0.004	0.464	....	0.46875

Wire Gauge Standards (Cont'd)

Wire gauge no.	American or Brown & Sharpe	Birmingham or Stubs wire	Washburn & Moen on steel wire gauge	American S. & W. Co.'s music wire	Imperial wire gauge	Stubs steel wire	U.S. standard for plate
	Decimal parts of an inch						
00000	0.516549	0.500	0.4305	0.005	4.432	....	.0.43775
0000	0.460	0.454	0.3938	0.006	0.400	....	0.40625
000	0.40964	0.425	0.3625	0.007	0.372	....	0.375
00	0.3648	0.380	0.3310	0.008	0.348	....	0.34375
0	0.32486	0.340	0.3065	0.009	0.324	....	0.3125
1	0.2893	0.300	0.2830	0.010	0.300	0.227	0.28125
2	0.25763	0.284	0.2625	0.011	0.276	0.219	0.265625
3	0.22942	0.259	0.2437	0.012	0.252	0.212	0.250
4	0.20431	0.238	0.2253	0.013	0.232	0.207	0.234375
5	0.18194	0.220	0.2070	0.014	0.212	0.204	0.21875
6	0.16202	0.203	0.1920	0.016	0.192	0.201	0.203125
7	0.14428	0.180	0.1770	0.018	0.176	0.199	0.1875
8	0.12849	0.165	0.1620	0.020	0.160	0.197	0.171875
9	0.11443	0.148	0.1483	0.022	0.144	0.194	0.15625
10	0.10189	0.134	0.1350	0.024	0.128	0.191	0.140625
11	0.090742	0.120	0.1205	0.026	0.116	0.188	0.125
12	0.080808	0.109	0.1055	0.029	0.104	0.185	0.109375
13	0.071961	0.095	0.0915	0.031	0.092	0.182	0.09375
14	0.064084	0.083	0.0800	0.033	0.080	0.180	0.078125
15	0.057068	0.072	0.0720	0.035	0.072	0.178	0.0703125
16	0.05082	0.065	0.0625	0.037	0.064	0.175	0.0625
17	0.045257	0.058	0.0540	0.039	0.056	0.172	0.05625
18	0.040303	0.049	0.0475	0.041	0.048	0.168	0.050
19	0.03589	0.042	0.0410	0.043	0.040	0.164	0.04375
20	0.031961	0.035	0.0348	0.045	0.036	0.161	0.0375
21	0.028462	0.032	0.0317	0.047	0.032	0.157	0.034375
22	0.025347	0.028	0.0286	0.049	0.028	0.155	0.03125
23	0.022571	0.025	0.0258	0.051	0.024	0.153	0.028125
24	0.0201	0.022	0.0230	0.055	0.022	0.151	0.025
25	0.0179	0.020	0.0204	0.059	0.020	0.148	0.021875
26	0.01594	0.018	0.0181	0.063	0.018	0.146	0.01875
27	0.014195	0.016	0.0173	0.067	0.0164	0.143	0.0171875
28	0.012641	0.014	0.0162	0.071	0.0149	0.139	0.015625
29	0.011257	0.013	0.0150	0.075	0.0136	0.134	0.0140625
30	0.010025	0.012	0.0140	0.080	0.0124	0.127	0.0125

Wire Gauge Standards (Cont'd)

	Decimal parts of an inch						
Wire gauge no.	American or Brown & Sharpe	Birmingham or Stubs wire	Washburn & Moen on steel wire gauge	American S. & W. Co.'s music wire	Imperial wire gauge	Stubs steel wire	U.S. standard for plate
31	0.008928	0.010	0.0132	0.085	0.0116	0.120	0.0109375
32	0.00795	0.009	0.0128	0.090	0.0108	0.115	0.01015625
33	0.00708	0.008	0.0118	0.095	0.0100	0.112	0.009375
34	0.006304	0.007	0.0104	....	0.0092	0.110	0.00859375
35	0.005614	0.005	0.0095	....	0.0084	0.108	0.0078125
36	0.005	0.004	0.0090	....	0.0076	0.106	0.00703125
37	0.004453	....	0.0085	....	0.0068	0.103	0.006640625
38	0.003965	....	0.0080	....	0.0060	0.101	0.00625
39	0.003531	....	0.0075	....	0.0052	0.099	
40	0.003144	....	0.0070	....	0.0048	0.097	

Metal Weights

Material	Chemical Symbol	Weight, in Pounds Per Cubic Inch	Weight, in Pounds Per Cubic Foot
Aluminum	Al	.093	160
Antimony	Sb	.2422	418
Brass	—	.303	524
Bronze	—	.320	552
Chromium	Cr	.2348	406
Copper	Cu	.323	450
Gold	Au	.6975	1205
Iron (cast)	Fe	.260	450
Iron (wrought)	Fe	.2834	490
Lead	Pb	.4105	710
Manganese	Mn	.2679	463
Mercury	Hg	.491	849
Molybdenum	Mo	.309	534
Monel	—	.318	550
Platinum	Pt	.818	1413
Steel (mild)	Fe	.2816	490
Steel (stainless)	—	.277	484
Tin	Sn	.265	459
Titanium	Ti	.1278	221
Zinc	Zn	.258	446

Colors and Approximate Temperature for Carbon Steel

Color	Temperature
Black Red	990°F
Dark Blood Red	1050
Dark Cherry Red	1175
Medium Cherry Red	1250
Full Cherry Red	1375
Light Cherry, Scaling	1550
Salmon, Free Scaling	1650
Light Salmon	1725
Yellow	1825
Light Yellow	1975
White	2220

Metric Measures

The metric unit of length is the meter = 39.37 inches.

The metric unit of weight is the gram = 15.432 grains.

The following prefixes are used for sub-divisions and multiples: Milli = 1/1000, Centi = 1/100, Deci = 1/10, Deca = 10, Hecto = 100, Kilo = 1000, Myria = 10,000.

Metric and English Equivalent Measures

MEASURES OF LENGTH

Metric		English
1 meter	=	39.37 inches, or 3.28083 feet, or 1.09361 yards
.3048 meter	=	1 foot
1 centimeter	=	.3937 inch
2.54 centimeters	=	1 inch
1 millimeter	=	.03937 inch, or nearly 1-25 inch
25.4 millimeters	=	1 inch
1 kilometer	=	1093.61 yards, or 0.62137 mile

MEASURES OF WEIGHT

Metric		English
1 gram	=	15.432 grains
.0648 gram	=	1 grain
28.35 grams	=	1 ounce avoirdupois
1 kilogram	=	2.2046 pounds
.4536 kilogram	=	1 pound

1 metric ton 1000 kilograms	=	.9842 ton of 2240 pounds 19.68 cwt. 2204.6 pounds
1.016 metric tons 1016 kilograms	=	1 ton of 2240 pounds

MEASURES OF CAPACITY

Metric		English
1 liter (= 1 cubic decimeter)	=	61.023 cubic inches .03531 cubic foot .2642 gal. (American) 2.202 lbs. of water at 62° F.
28.317 liters	=	1 cubic foot
3.785 liters	=	1 gallon (American)
4.543 liters	=	1 gallon (Imperial)

English Conversion Table

Length				
Inches	×	.0833	=	feet
Inches	×	.02778	=	yards
Inches	×	.00001578	=	miles
Feet	×	.3333	=	yards
Feet	×	.0001894	=	miles
Yards	×	36.00	=	inches
Yards	×	3.00	=	feet
Yards	×	.0005681	=	miles
Miles	×	63360.00	=	inches
Miles	×	5280.00	=	feet
Miles	×	1760.00	=	yards
Circumference of circle	×	.3188	=	diameter
Diameter of circle	×	3.1416	=	circumference
Area				
Square inches	×	.00694	=	square feet
Square inches	×	.0007716	=	square yards
Square feet	×	144.00	=	square inches
Square feet	×	.11111	=	square yards
Square yards	×	1296.00	=	square inches
Square yards	×	9.00	=	square feet
Dia. of circle squared	×	.7854	=	area
Dia. of sphere squared	×	3.1416	=	surface
Volume				
Cubic inches	×	.0005787	=	cubic feet
Cubic inches	×	.00002143	=	cubic yards
Cubic inches	×	.004329	=	U. S. gallons
Cubic feet	×	1728.00	=	cubic inches

Cubic feet	× .03704	=	cubic yards
Cubic feet	× 7.4805	=	U. S. gallons
Cubic yards	× 46656.00	=	cubic inches
Cubic yards	× 27.00	=	cubic feet
Dia. of sphere cubed	× .5236	=	volume

Weight

Grains (avoirdupois)	× .002286	=	ounces
Ounces (avoirdupois)	× .0625	=	pounds
Ounces (avoirdupois)	× .00003125	=	tons
Pounds (avoirdupois)	× 16.00	=	ounces
Pounds (avoirdupois)	× .01	=	hundredweight
Pounds (avoirdupois)	× .0005	=	tons
Tons (avoirdupois)	× 32000.00	=	ounces
Tons (avoirdupois)	× 2000.00	=	pounds

English Conversion Table

Energy

Horsepower	× 33000.	=	ft.-lbs. per min.
B. t. u.	× 778.26	=	ft.-lbs.
Ton of refrigeration	× 200.	=	B. t. u. per min.

Pressure

Lbs. per sq. in.	× 2.31	=	ft. of water (60°F.)
Ft. of water (60°F.)	× .433	=	lbs. per sq. in.
Ins. of water (60°F.)	× .0361	=	lbs. per sq. in.
Lbs. per sq. in.	× 27.70	=	ins. of water (60°F.)
Lbs. per sq. in.	× 2.041	=	ins. of Hg. (60°F.)
Ins. of Hg (60°F.)	× .490	=	lbs. per sq. in.

Power

Horsepower	× 746.	=	watts
Watts	× .001341	=	horsepower
Horsepower	× 42.4	=	B. t. u. per min.

Water Factors (at point of greatest density—39.2°F.)

Miners inch (of water)	× 8.976	=	U. S. gals. per min.
Cubic inches (of water)	× .57798	=	ounces
Cubic inches (of water)	× .036124	=	pounds
Cubic inches (of water)	× .004329	=	U. S. gallons
Cubic inches (of water)	× .003607	=	English gallons
Cubic feet (of water)	× 62.425	=	pounds
Cubic feet (of water)	× .03121	=	tons
Cubic feet (of water)	× 7.4805	=	U. S. gallons

Cubic inches (of water)	×	6.232	=	English gallons
Cubic foot of ice	×	57.2	=	pounds
Ounces (of water)	×	1.73	=	cubic inches
Pounds (of water)	×	26.68	=	cubic inches
Pounds (of water)	×	.01602	=	cubic feet
Pounds (of water)	×	.1198	=	U. S. gallons
Pounds (of water)	×	.0998	=	English gallons
Tons (of water)	×	32.04	=	cubic feet
Tons (of water)	×	239.6	=	U. S. gallons
Tons (of water)	×	199.6	=	English gallons
U. S. gallons	×	231.00	=	cubic inches
U. S. gallons	×	.13368	=	cubic feet
U. S. gallons	×	8.345	=	pounds
U. S. gallons	×	.8327	=	English gallons
U. S. gallons	×	3.785	=	liters
English gallons (Imperial)	×	277.41	=	cubic inches
English gallons (Imperial)	×	.1605	=	cubic feet
English gallons (Imperial)	×	10.02	=	pounds
English gallons (Imperial)	×	1.201	=	U. S. gallons
English gallons (Imperial)	×	4.546	=	liters

Metric Conversion Table

Length

Millimeters	×	.03937	=	inches
Millimeters	÷	25.4	=	inches
Centimeters	×	.3937	=	inches
Centimeters	÷	2.54	=	inches
Meters	×	39.37	=	inches (Act. Cong.)
Meters	×	3.281	=	feet
Meters	×	1.0936	=	yards
Kilometers	×	.6214	=	miles
Kilometers	÷	1.6093	=	miles
Kilometers	×	3280.8	=	feet

Area

Sq. Millimeters	×	.00155	=	sq. in.
Sq. Millimeters	÷	645.2	=	sq. in.
Sq. Centimeters	×	.155	=	sq. in.
Sq. Centimeters	÷	6.452	=	sq. in.
Sq. Meters	×	10.764	=	sq. ft.
Sq. Kilometers	×	247.1	=	acres
Hectares	×	2.471	=	acres

Volume

Cu. Centimeters	÷	16.387	=	cu. in.
Cu. Centimeters	÷	3.69	=	fl. drs. (U.S.P.)

Cu. Centimeters	÷	29.57	= fl. oz. (U.S.P.)
Cu. Meters	×	35.314	= cu. ft.
Cu. Meters	×	1.308	= cu. yards
Cu. Meters	×	264.2	= gals. (231 cu. in.)
Litres	×	61.023	= cu. in. (Act. Cong.)
Litres	×	33.82	= fl. oz. (U.S.J.)
Litres	×	.2642	= gals. (231 cu. in.)
Litres	÷	3.785	= gals. (231 cu. in.)
Litres	÷	28.317	= cu. ft.
Hectolitres	×	3.531	= cu. ft.
Hectolitres	×	2.838	= bu. (2150.42 cu. in.)
Hectolitres	×	.1308	= cu. yds.
Hectolitres	×	26.42	= gals. (231 cu. in.)
Weight			
Grams	×	15.432	= grains (Act. Cong.)
Grams	÷	981.	= dynes
Grams (water)	÷	29.57	= fl. oz.
Grams	÷	28.35	= oz. avoirdupois
Kilo-grams	×	2.2046	= lbs.
Kilo-grams	×	35.27	= oz. avoirdupois
Kilo-grams	×	.0011023	= tons (2000 lbs.)
Tonneau (Metric ton)	×	1.1023	= tons (2000 lbs.)
Tonneau (Metric ton)	×	2204.6	= lbs.
Unit Weight			
Grams per cu. cent.	÷	27.68	= lbs. per cu. in.
Kilo per meter	×	.672	= lbs. per ft.
Kilo per cu. meter	×	.06243	= lbs. per cu. ft.
Kilo per Cheval	×	2.235	= lbs. per h. p.
Grams per liter	×	.06243	= lbs. per cu. ft.
Pressure			
Kilo-grams per sq. cm.	×	14.223	= lbs. per sq. in.
Kilo-grams per sq. cm.	×	32.843	= ft. of water (60°F.)
Atmospheres (international)	×	14.696	= lbs. per sq. in.
Energy			
Joule	×	.7376	= ft. lbs.
Kilo-gram meters	×	7.233	= ft. lbs.
Power			
Cheval vapeur	×	.9863	= h. p.
Kilo-watts	×	1.341	= h. p.
Watts	÷	746.	= h. p.
Watts	×	.7373	= ft. lbs. per sec

Equivalent Temperature Readings for Fahrenheit and Centigrade Scales

Fahrenheit Degs.	Centigrade Degs.	Fahrenheit Degs.	Centigrade Degs.	Fahrenheit Degs.	Centigrade Degs.	Fahrenheit Degs.	Centigrade Degs.
—459.4	—273.	—21.	—29.4	17.6	— 8.	56.	13.3
—436.	—270.	—20.2	—29.	18.	— 7.8	57.	13.9
—418.	—260.	—20.	—28.9	19.	— 7.2	57.2	14.
—400.	—240.	—19.	—28.3	19.4	— 7.	58.	14.4
—382.	—230.	—18.4	—28.	20.	— 6.7	59.	15.
—364.	—220.	—18.	—27.8	21.	— 6.1	60.	15.6
—346.	—210.	—17.	—27.2	21.2	— 6.	60.8	16.
—328.	—200.	—16.6	—27.	22.	— 5.6	61.	16.1
—310.	—190.	—16.	—26.7	23.	— 5.	62.	16.7
—292.	—180.	—15.	—26.1	24.	— 4.4	62.6	17.
—274.	—170.	—14.8	—26.	24.8	— 4.	63.	17.2
—256.	—160.	—14.	—25.6	25.	— 3.9	64.	17.8
—238.	—150.	—13.	—25.	26.	— 3.3	64.4	18.
—220.	—140.	—12.	—24.4	26.6	— 3.	65.	18.3
—202.	—130.	—11.2	—24.	27.	— 2.8	66.	18.9
—184.	—120.	—11.	—23.9	28.	— 2.2	66.2	19.
—166.	—110.	—10.	—23.3	28.4	— 2.	67.	19.4
—148.	—100.	— 9.4	—23.	29.	— 1.7	68.	20.
—139.	— 95.	— 9.	—22.8	30.	— 1.1	69.	20.6
—130.	— 90.	— 8.	—22.2	30.2	— 1.	69.8	21.
—121.	— 85.	— 7.6	—22.	31.	— 0.6	70.	21.1
—112.	— 80.	— 7.	—21.7	32.	0.	71.	21.7
—103.	— 75.	— 6.	—21.1	33.	+ 0.6	71.6	22.
— 94.	— 70.	— 5.8	—21.	33.8	1.	72.	22.2
— 85.	— 65.	— 5.	—20.6	34.	1.1	73.	22.8
— 76.	— 60.	— 4.	—20.	35.	1.7	73.4	23.
— 67.	— 55.	— 3.	—19.4	35.6	2.	74.	23.3
— 58.	— 50.	— 2.2	—19.	36.	2.2	75.	23.9
— 49.	— 45.	— 2.	—18.9	37.	2.8	75.2	24.
— 40.	— 40.	— 1.	—18.3	37.4	3.	76.	24.4
— 39.	— 39.4	— 0.4	—18.	38.	3.3	77.	25.
— 38.2	— 39.	0.	—17.8	39.	3.9	78.	25.6
— 38.	— 38.9	+ 1.	—17.2	39.2	4.	78.8	26.
— 37.	— 38.3	1.4	—17.	40.	4.4	79.	26.1
— 36.4	— 38.	2.	—16.7	41.	5.	80.	26.7
— 36.	— 37.8	3.	—16.1	42.	5.6	80.6	27.
— 35.	— 37.2	3.2	—16.	42.8	6.	81.	27.2
— 34.6	— 37.	4.	—15.6	43.	6.1	82.	27.8
— 34.	— 36.7	5.	—15.	44.	6.7	82.4	28.
— 33.	— 36.1	6.	—14.4	44.6	7.	83.	28.3
— 32.8	— 36.	6.8	—14.	45.	7.2	84.	28.9
— 32.	— 35.6	7.	—13.9	46.	7.8	84.2	29.
— 31.	— 35.	8.	—13.3	46.4	8.	85.	29.4
— 30.	— 34.4	8.6	—13.	47.	8.3	86.	30.
— 29.2	— 34.	9.	—12.8	48.	8.9	87.	30.6
— 29.	— 33.9	10.	—12.2	48.2	9.	87.8	31.
— 28.	— 33.3	10.4	—12.	49.	9.4	88.	31.1
— 27.4	— 33.	11.	—11.7	50.	10.	89.	31.7
— 27.	— 32.8	12.	—11.1	51.	10.6	89.6	32.
— 26.	— 32.2	12.2	—11.	51.8	11.	90.	32.2
— 25.6	— 32.	13.	—10.6	52.	11.1	91.	32.8
— 25.	— 31.7	14.	—10.	53.	11.7	91.4	33.
— 24.	— 31.1	15.	— 9.4	53.6	12.	92.	33.3
— 23.8	— 31.	15.8	— 9.	54.	12.2	93.	33.9
— 23.	— 30.6	16.	— 8.9	55.	12.8	93.2	34.
— 22.	— 30.	17.	— 8.3	55.4	13.	94.	34.4

Equivalent Temperature Readings for Fahrenheit and Centigrade Scales (Cont.)

Fahrenheit Degs.	Centigrade Degs.	Fahrenheit Degs.	Centigrade Degs.	Fahrenheit Degs.	Centigrade Degs.	Fahrenheit Degs.	Centigrade Degs.
95.	35.	134.	56.7	172.4	78.	211.	99.4
96.	35.6	134.6	57.	173.	78.3	212.	100.
96.8	36.	135.	57.2	174.	78.9	213.	100.6
97.	36.1	136.	57.8	174.2	79.	213.8	101.
98.	36.7	136.4	58.	175.	79.4	214.	101.1
98.6	37.	137.	58.3	176.	80.	215.	101.7
99.	37.2	138.	58.9	177.	80.6	215.6	102.
100.	37.8	138.2	59.	177.8	81.	216.	102.2
100.4	38.	139.	59.4	178.	81.1	217.	102.8
101.	38.3	140.	60.	179.	81.7	217.4	103.
102.	38.9	141.	60.6	179.6	82.	218.	103.3
102.2	39.	141.8	61.	180.	82.2	219.	103.9
103.	39.4	142.	61.1	181.	82.8	219.2	104.
104.	40.	143.	61.7	181.4	83.	220.	104.4
105.	40.6	143.6	62.	182.	83.3	221.	105.
105.8	41.	144.	62.2	183.	83.9	222.	105.6
106.	41.1	145.	62.8	183.2	84.	222.8	106.
107.	41.7	145.4	63.	184.	84.4	223.	106.1
107.6	42.	146.	63.3	185.	85.	224.	106.7
108.	42.2	147.	63.9	186.	85.6	224.6	107.
109.	42.8	147.2	64.	186.8	86.	225.	107.2
109.4	43.	148.	64.4	187.	86.1	226.	107.8
110.	43.3	149.	65.	188.	86.7	226.4	108.
111.	43.9	150.	65.6	188.6	87.	227.	108.3
111.2	44.	150.8	66.	189.	87.2	228.	108.9
112.	44.4	151.	66.1	190.	87.8	228.2	109.
113.	45.	152.	66.7	190.4	88.	229.	109.4
114.	45.6	152.6	67.	191.	88.3	230.	110.
114.8	46.	153.	67.2	192.	88.9	231.	110.6
115.	46.1	154.	67.8	192.2	89.	231.8	111.
116.	46.7	154.4	68.	193.	89.4	232.	111.1
116.6	47.	155.	68.3	194.	90.	233.	111.7
117.	47.2	156.	68.9	195.	90.6	233.6	112.
118.	47.8	156.2	69.	195.8	91.	234.	112.3
118.4	48.	157.	69.4	196.	91.1	235.	112.8
119.	48.3	158.	70.	197.	91.7	235.4	113.
120.	48.9	159.	70.6	197.6	92.	236.	113.3
120.2	49.	159.8	71.	198.	92.2	237.	113.9
121.	49.4	160.	71.1	199.	92.8	237.2	114.
122.	50.	161.	71.7	199.4	93.	238.	114.4
123.	50.6	161.6	72.	200.	93.3	239.	115.
123.8	51.	162.	72.2	201.	93.9	240.	115.6
124.	51.1	163.	72.8	201.2	94.	240.8	116.
125.	51.7	163.4	73.	202.	94.4	241.	116.1
125.6	52.	164.	73.3	203.	95.	242.	116.7
126.	52.2	165.	73.9	204.	95.6	242.6	117.
127.	52.8	165.2	74.	204.8	96.	243.	117.2
127.4	53.	166.	74.4	205.	96.1	244.	117.8
128.	53.3	167.	75.	206.	96.7	244.4	118.
129.	53.9	168.	75.6	206.6	97.	245.	118.3
129.2	54.	168.8	76.	207.	97.2	246.	118.9
130.	54.4	169.	76.1	208.	97.8	246.2	119.
131.	55.	170.	76.7	208.4	98.	247.	119.4
132.	55.6	170.6	77.	209.	98.3	248.	120.
132.8	56.	171.	77.2	210.	98.9	249.	120.6
133.	56.1	172.	77.8	210.2	99.	249.8	121.

Index

N

O

The Audel® Mail Order Bookstore

Here's an opportunity to order the valuable books you may have missed before and to build your own personal, comprehensive library of Audel books. You can choose from an extensive selection of technical guides and reference books. They will provide access to the same sources the experts use, put all the answers at your fingertips, and give you the know-how to complete even the most complicated building or repairing job, in the same professional way.

Each volume:

- **Fully illustrated**
- **Packed with up-to-date facts and figures**
- **Completely indexed for easy reference**

APPLIANCES

REFRIGERATION: HOME AND COMMERCIAL

Covers the whole realm of refrigeration equipment from fractional-horsepower water coolers, through domestic refrigerators to multi-ton commercial installations. 656 pages; 5½ x 8¼; hardbound. **Cat. No. 23286 Price: $9.50.**

AIR CONDITIONING: HOME AND COMMERCIAL

A concise collection of basic information, tables, and charts for those interested in understanding, troubleshooting, and repairing home air conditioners and commercial installations. 464 pages; 5½ x 8¼; hardbound. **Cat. No. 23288 Price: $7.50.**

HOME APPLIANCE SERVICING, 3rd Edition

A practical book for electric & gas servicemen, mechanics & dealers. Covers the principles, servicing, and repairing of home appliances. 592 pages; 5¼ x 8¼; hardbound. **Cat. No. 23214 Price: $8.95.**

REFRIGERATION AND AIR CONDITIONING LIBRARY—2 Vols.

Cat. No. 23305 Price: $15.95

OIL BURNERS, 3rd Edition

Provides complete information on all types of oil burners and associated equipment. Discusses burners—blowers—ignition transformers—electrodes—nozzles—fuel pumps—filters—Controls. Installation and maintenance are stressed. 320 pages; 5½ x 8¼; hardbound. **Cat. No. 23277 Price: $7.50.**

Use the order coupon on the back page of this book.

AUTOMOTIVE

AUTO BODY REPAIR FOR THE DO-IT-YOURSELFER

Shows how to use touch-up paint; repair chips, scratches, and dents; remove and prevent rust; care for glass, doors, locks, lids, and vinyl tops; and clean and repair upholstery. 96 pages; 8½ x 11; softcover. **Cat. No. 23238 Price: $5.95.**

AUTOMOBILE REPAIR GUIDE, 4th Edition

A practical reference for auto mechanics, servicemen, trainees, and owners Explains theory, construction, and servicing of modern domestic motorcars. 800 pages; 5½ x 8¼; hardbound. **Cat. No. 23291 Price: $12.95.**

CAN-DO TUNE-UP™ SERIES

Each book in this series comes with an audio tape cassette. Together they provide an organized set of instructions that will show you and talk you through the maintenance and tune-up procedures designed for your particular car. All books are softcover.

AMERICAN MOTORS CORPORATION CARS

(The 1964 thru 1974 cars covered include: Matador. Rambler. Gremlin. and AMC Jeep (Willys).). 112 pages; 5½ x 8½; softcover. **Cat. No. 23843 Price: $7.95.**

Cat. No. 23851 Without Cassette **Price: $4.95**

CHRYSLER CORPORATION CARS

(The 1964 thru 1974 cars covered include: Chrysler, Dodge, and Plymouth.) 112 pages; 5½ x 8½; softcover. **Cat. No. 23825 Price $7.95.**

Cat. No. 23846 Without Cassette **Price: $4.95**

FORD MOTOR COMPANY CARS

(The 1954 thru 1974 cars covered include: Ford, Lincoln, and Mercury.) 112 pages; 5½ x 8½; softcover. **Cat. No. 23827 Price: $7.95.**

Cat. No. 23848 Without Cassette **Price: $4.95**

GENERAL MOTORS CORPORATION CARS

(The 1964 thru 1974 cars covered include: Buick, Cadillac, Chevrolet, Oldsmobile, and Pontiac.) 112 pages; 5½ x 8½; softcover. **Cat. No. 23824 Price: $7.95.**

Cat. No. 23845 Without Cassette **Price: $4.95**

PINTO AND VEGA CARS,

1971 thru 1974. 112 pages· 5½ x 8½; softcover. **Cat. No. 23831 Price: $7.95.**

Cat. No. 23849 Without Cassette **Price: $4.95**

TOYOTA AND DATSUN CARS,

1964 thru 1974. 112 pages; 5½ x 8½; softcover. **Cat. No. 23835 Price: $7.95.**

Cat. No. 23850 Without Cassette **Price: $4.95**

VOLKSWAGEN CARS

(The 1964 thru 1974 cars covered include: Beetle. Super Beetle. and Karmann Ghia.) 96 pages; 5½ x 8½; softcover. **Cat. No. 23826 Price: $7.95.**

Cat. No. 23847 Without Cassette **Price: $4.95**

AUTOMOTIVE AIR CONDITIONING

You can easily perform most all service procedures you've been paying for in the past. This book covers the systems built by the major manufacturers, even after-market installations. Contents: introduction—refrigerant—tools—air conditioning circuit—general service procedures—electrical systems—the cooling system—system diagnosis—electrical diagnosis—troubleshooting. 232 pages; 5½ x 8½; softcover. **Cat. No. 23318 Price: $5.95.**

Use the order coupon on the back page of this book.

DIESEL ENGINE MANUAL, 3rd Edition

A practical guide covering the theory, operation, and maintenance of modern diesel engines. Explains diesel principles—valves—timing—fuel pumps—pistons and rings—cylinders—lubrication—cooling system—fuel oil and more. 480 pages; 5½ x 8¼; hardbound. **Cat. No. 23199 Price: $8.95.**

GAS ENGINE MANUAL, 2nd Edition

A completely practical book covering the construction, operation, and repair of all types of modern gas engines. 400 pages; 5½ x 8¼; hardbound. **Cat. No. 23245 Price: $7.95.**

OUTBOARD MOTORS & BOATING, 3rd Edition

Provides the information you need to maintain, troubleshoot, repair, and adjust all types of outboard motors. Explains the basic principles of outboard motors and the functions of the various engine parts. 464 pages; 5½ x 8¼; softcover. **Cat. No. 23279 Price: $6.95.**

BUILDING AND MAINTENANCE

ANSWERS ON BLUEPRINT READING, 3rd Edition

Covers all types of blueprint reading for mechanics and builders. This book reveals the secret language of blueprints, step-by-step in easy stages. 312 pages; 5½ x 8¼; hardbound. **Cat. No. 23283 Price: $6.95.**

BUILDING A VACATION HOME

From selecting a building site to driving in the last nail, this book explains the entire process, with fully illustrated step-by-step details. Includes a complete set of drawings for a two-story vacation and/or retirement home. Softcover. 192 pages; 8½ x 11; softcover. **Cat. No. 23222 Price: $7.95.**

BUILDING MAINTENANCE, 2nd Edition

Covers all the practical aspects of building maintenance. Painting and decorating; plumbing and pipe fitting; carpentry; heating maintenance; custodial practices and more. (A book for building owners, managers, and maintenance personnel.) 384 pages; 5½ x 8¼; hardbound. **Cat. No. 23278 Price: $7.50.**

COMPLETE BUILDING CONSTRUCTION

At last—a *one-volume* instruction manual to show you how to construct a frame or brick building from the footings to the ridge. Build your own garage, tool shed, other outbuilding—even your own house or place of business. Building construction tells you how to lay out the building and excavation lines on the lot; how to make concrete forms and pour the footings and foundation; how to make concrete slabs, walks, and driveways; how to lay concrete block, brick and tile; how to build your own fireplace and chimney: It's one of the newest Audel books, clearly written by experts in each field and ready to help you every step of the way. 800 pages; 5½ x 8¼; hardbound. **Cat. No. 23323 Price: $19.95.**

GARDENING & LANDSCAPING

A comprehensive guide for homeowners and for industrial, municipal, and estate groundskeepers. Gives information on proper care of annual and perennial flowers; various house plants; greenhouse design and construction; insect and rodent controls; and more. 384 pages; 5½ x 8¼; hardbound. **Cat. No. 23229 Price: $7.95.**

CARPENTERS & BUILDERS LIBRARY, 4th Edition (4 Vols.)

A practical, illustrated trade assistant on modern construction for carpenters, builders, and all woodworkers. Explains in practical, concise language and illustrations all the principles, advances, and shortcuts based on modern practice. How to calculate various jobs. **Cat. No. 23244 Price: $24.50**

Vol. 1—Tools, steel square, saw filing, joinery cabinets. 384 pages; 5½ x 8¼; hardbound. **Cat. No. 23240 Price: $6.50.**

Vol. 2—Mathematics, plans, specifications, estimates 304 pages; 5½ x 8¼; hardbound. **Cat. No. 23241 Price: $6.50.**

Vol. 3—House and roof framing, laying out foundations. 304 pages; 5½ x 8¼; hardbound. **Cat. No. 23242 Price: $6.50.**

Vol. 4—Doors, windows, stairs, millwork, painting. 368 pages; 5½ x 8¼; hardbound. **Cat. No. 23243 Price: $6.50.**

Use the order coupon on the back page of this book.

PLUMBERS AND PIPE FITTERS LIBRARY—3 Vols.

A practical, illustrated trade assistant and reference for master plumbers, journeymen and apprentice pipe fitters, gas fitters and helpers, builders, contractors, and engineers. Explains in simple language, illustrations, diagrams, charts, graphs, and pictures, the principles of modern plumbing and pipe-fitting practices. **Cat. No. 23255 Price $19.95**

Vol. 1—Materials, tools, roughing-in. 320 pages; 5½ x 8¼; hardbound. **Cat. No. 23256 Price: $6.95.**

Vol. 2—Welding, heating, air-conditioning. 384 pages; 5½ x 8¼; hardbound. **Cat. No. 23257 Price: $6.95.**

Vol. 3—Water supply, drainage, calculations. 272 pages; 5½ x 8¼; hardbound. **Cat. No. 23258 Price: $6.95.**

PLUMBERS HANDBOOK

A pocket manual providing reference material for plumbers and/or pipe fitters. General information sections contain data on cast-iron fittings, copper drainage fittings, plastic pipe, and repair of fixtures. 288 pages; 4 x 6; softcover. **Cat. No. 23339 Price: $5.95.**

QUESTIONS AND ANSWERS FOR PLUMBERS EXAMINATIONS, 2nd Edition

Answers plumbers' questions about types of fixtures to use, size of pipe to install, design of systems, size and location of septic tank systems, and procedures used in installing material. 256 pages; 5½ x 8¼; softcover. **Cat. No. 23285 Price: $5.50.**

TREE CARE MANUAL

The conscientious gardener's guide to healthy, beautiful trees. Covers planting, grafting, fertilizing, pruning, and spraying. Tells how to cope with insects, plant diseases, and environmental damage. 224 pages; 8½ x 11; softcover. **Cat. No. 23280 Price: $8.95.**

UPHOLSTERING

Upholstering is explained for the average householder and apprentice upholsterer. From repairing and reglueing of the bare frame, to the final sewing or tacking, for antiques and most modern pieces, this book covers it all. 400 pages; 5½ x 8¼; hardbound. **Cat. No. 23189 Price: $7.95.**

WOOD FURNITURE: Finishing, Refinishing, Repairing

Presents the fundamentals of furniture repair for both veneer and solid wood. Gives complete instructions on refinishing procedures, which includes stripping the old finish, sanding, selecting the finish and using wood fillers. 352 pages; 5½ x 8¼; hardbound. **Cat. No. 23216 Price: $8.50.**

ELECTRICITY/ELECTRONICS

ELECTRICAL LIBRARY

If you are a student of electricity or a practicing electrician, here is a very important and helpful library you should consider owning. You can learn the basics of electricity, study electric motors and wiring diagrams, learn how to interpret the NEC, and prepare for the electrician's examination by using these books. **Cat. No. 23324 Price: $43.95.**

Electric Motors, 3rd Edition. 528 pages; 5½ x 8¼; hardbound. **Cat. No. 23264 Price: $8.95.**

Guide to the 1978 National Electrical Code. 672 pages; 5½ x 8¼; hardbound. **Cat. No. 23308 Price: $9.95.**

House Wiring, 4th Edition. 256 pages; 5½ x 8¼; hardbound. **Cat. No. 23315 Price: $6.95.**

Practical Electricity, 3rd Edition. 496 pages; 5½ x 8¼; hardbound. **Cat. No. 23218 Price: $7.95**

Questions and Answers for Electricians Examinations, 6th Edition. 288 pages; 5½ x 8¼; hardbound. **Cat. No. 23307 Price: $6.95.**

Wiring Diagrams for Light and Power, 3rd Edition. 400 pages; 5½ x 8¼; hardbound. **Cat. No. 23232 Price: $6.95.**

ELECTRICAL COURSE FOR APPRENTICES AND JOURNEYMEN

A study course for apprentice or journeymen electricians. Covers electrical theory and its applications. 448 pages; 5½ x 8¼; hardbound. **Cat. No. 23209 Price: $7.95**

Use the order coupon on the back page of this book.

RADIOMANS GUIDE, 4th Edition

Contains the latest information on radio and electronics from the basics through transistors. 480 pages; 5½ x 8¼; hardbound. **Cat. No. 23259 Price: $7.50.**

TELEVISION SERVICE MANUAL, 4th Edition

Provides the practical information necessary for accurate diagnosis and repair of both black-and-white and color television receivers. 512 pages; 5½ x 8¼; hardbound. **Cat. No. 23247 Price: $8.95.**

ENGINEERS/MECHANICS/ MACHINISTS

MACHINISTS LIBRARY, 2nd Edition

Covers modern machine-shop practice. Tells how to set up and operate lathes, screw and milling machines, shapers, drill presses, and all other machine tools. A complete reference library. **Cat. No. 23300 Price: $23.00**

Vol. 1—Basic Machine Shop. 352 pages; 5½ x 8¼; hardbound. **Cat. No. 23301 Price: $7.95.**

Vol. 2—Machine Shop. 480 pages; 5½ x 8¼; hardbound. **Cat. No. 23302 Price: $7.95.**

Vol. 3—Toolmakers Handy Book. 400 pages; 5½ x 8¼; hardbound. **Cat. No. 23303 Price: $7.95.**

MECHANICAL TRADES POCKET MANUAL

Provides practical reference material for mechanical tradesmen. This handbook covers methods, tools, equipment, procedures, and much more. 256 pages; 4 x 6; softcover. **Cat. No. 23215 Price: $4.50.**

MILLWRIGHTS AND MECHANICS GUIDE, 2nd Edition

Practical information on plant installation, operation, and maintenance for millwrights, mechanics, maintenance men, erectors, riggers, foremen, inspectors, and superintendents. 960 pages; 5½ x 8¼; hardbound. **Cat. No. 23201 Price: $11.95.**

POWER PLANT ENGINEERS GUIDE, 2nd Edition

The complete steam or diesel power-plant engineer's library. 816 pages; 5½ x 8¼; hardbound. **Cat. No. 23220 Price: $12.95.**

WELDERS GUIDE, 2nd Edition

This new edition is a practical and concise manual on the theory, practical operation, and maintenance of all welding machines. Fully covers both electric and oxy-gas welding. 928 pages; 5½ x 8¼; hardbound. **Cat. No. 23202 Price: $11.95.**

WELDER/FITTERS GUIDE

Provides basic training and instruction for those wishing to become welder/fitters. Step-by-step learning sequences are presented from learning about basic tools and aids used in weldment assembly, through simple work practices, to actual fabrication of weldments. 160 pages· 8½ x 11; softcover; **Cat. No. 23325 Price: $7.95.**

Use the order coupon on the back page of this book.

FLUID POWER

PNEUMATICS AND HYDRAULICS, 3rd Edition

Fully discusses installation, operation, and maintenance of both HYDRAULIC AND PNEUMATIC (air) devices. 496 pages; 5½ x 8¼; hardbound. **Cat. No. 23237 Price: $8.50.**

PUMPS, 3rd Edition

A detailed book on all types of pumps from the old-fashioned kitchen variety to the most modern types. Covers construction, application, installation, and troubleshooting. 480 pages; 5½ x 8¼; hardbound. **Cat. No. 23292 Price: $8.95.**

HYDRAULICS FOR OFF-THE-ROAD EQUIPMENT

Everything you need to know from basic hydraulics to troubleshooting hydraulic systems on off-the-road equipment. Heavy-equipment operators, farmers, fork-lift owners and operators, mechanics—all need this practical, fully illustrated manual. 272 pages; 5½ x 8¼; hardbound. **Cat. No. 23306 Price: $6.95.**

HOBBY

COMPLETE COURSE IN STAINED GLASS

Written by an outstanding artist in the field of stained glass, this book is dedicated to all who love the beauty of the art. Ten complete lessons describe the required materials, how to obtain them, and explicit directions for making several stained glass projects. 80 pages; 8½ x 11; softbound. **Cat. No. 23287 Price: $4.95**

BUILD YOUR OWN AUDEL DO-IT-YOURSELF LIBRARY AT HOME!

Use the handy order coupon today to gain the valuable information you need in all the areas that once required a repairman. Save money and have fun while you learn to service your own air conditioner, automobile, and plumbing. Do your own professional carpentry, masonry, and wood furniture refinishing and repair. Build your own security systems. Find out how to repair your TV or Hi-Fi. Learn landscaping, upholstery, electronics and much, much more.

HERE'S HOW TO ORDER

1. Enter the correct catalog number(s) of the book(s) you want in the space(s) provided.

2. Print your name, address, city, state and zip code, clearly.

3. Detach the order coupon below and mail today to:

Theodore Audel & Company
4300 West 62nd Street
Indianapolis, Indiana 46206
ATTENTION: ORDER DEPT.

All prices are subject to change without notice.

ORDER COUPON

Please rush the following book(s).

Write book catalog numbers at left. (Numbers are listed with titles.)

NAME ____________

ADDRESS ____________

CITY ____________ STATE ____________ ZIP ____________

☐ Payment Enclosed ____________ Total
(No Shipping and Handling Charge)

237

☐ Bill Me (Shipping and Handling Charge will be added)

Add local sales tax where applicable.

Litho in U.S.A.

HERE'S HOW TO ORDER

Select the Audel book(s) you want, fill in the order card below, detach and mail today. Send no money now. You'll have 15 days to examine the books in the comfort of your own home. If not completely satisfied, simply return your order and owe nothing.

If you decide to keep the books, we will bill you for the total amount, plus a small charge for shipping and handling.

1. Enter the correct catalog number(s) of the book(s) you want in the space(s) provided.
2. Print your name, address, city, state and zip code, clearly.
3. Detach the order card below and mail today. No postage is required.

Detach postage-free order card on perforated line

FREE TRIAL ORDER CARD

☐ Please rush the following book(s) for my free trial. I understand if I'm not completely satisfied, I may return my order within 15 days and owe nothing. Otherwise, you will bill me for the total amount plus a small postage & handling charge.

Write book catalog numbers at right.

(Numbers are listed with titles)

NAME______________________________

ADDRESS______________________________

CITY____________________STATE________ZIP________

☐ Save postage & handling costs. Full payment enclosed (Plus sales tax, if any.)

Cash must accompany orders under $5.00.
Money-Back guarantee still applies.

333

DETACH POSTAGE-PAID REPLY CARD BELOW AND MAIL TODAY!

Just select your books, enter the code numbers on the order card, fill out your name and address, and mail. There's no need to send money.

15-Day Free Trial On All Books . . .

First Class
Mail
Permit No.
1076
Indianapolis, Ind.

BUSINESS REPLY MAIL no postage necessary if mailed in the United States

Postage will be paid by:

Theodore Audel & Company

4300 West 62nd Street
Indianapolis, Indiana 46206

ATTENTION: ORDER DEPT.